Pears
Encyclopædia of Gardening
FLOWERS, TREES & SHRUBS

Pears Encyclopædia of Gardening

FLOWERS, TREES & SHRUBS

PELHAM BOOKS

First published in Great Britain by
PELHAM BOOKS LTD
52 Bedford Square
London, W.C.1
1972

7207 0249 6

Set and printed in Great Britain by
Tonbridge Printers Ltd
Peach Hall Works, Tonbridge, Kent
in Times ten on twelve point
and bound by James Burn
at Esher, Surrey

Contents

List of Illustrations

Between pages 384 and 385

LINE DRAWINGS IN TEXT

General Entries

The entries in this encyclopaedia are arranged alphabetically by the botanical names of the plants, with common names included in the form of cross-references. In addition, there are a number of general entries on larger groups of plants, gardening techniques, etc. These entries, which appear in the overall alphabetical listing, are as follows:

Annuals
Artificial Lighting
Bamboos
Bonsai
Bulbs
Cactus
Climbing Plants
Conifers
Curcurbita
Diseases
Everlasting Flowers
Ferns
Floating Plants
Foliar Feeding
Glossary of Botanical Terms
Grass
Ground Cover Plants
Hanging Baskets
Hardening
Heath
Hedging Plants
Mist Propagation
Oxygenating Plants
Pests
Plant Pots
Shrubs
Silver Foliage
Soil, acid/calcareous
Topiary
Trees
Trough Gardens
Tubs
Vases
Wall Plants
Water Garden
Window Boxes

ABELIA

Hardy in all but the most exposed gardens, it is deciduous, with *A. floribunda* partially evergreen. *A. chinensis* bears small white flowers tinted with rose and with the scent of honeysuckle whilst *A. triflora* grows up to 10 ft. (2.7 m) tall and has flowers of similar colouring and scent. The flowers appear in June and are borne on the new season's wood. Plant in March in a well-drained loam and after flowering, cut out the older wood. Propagate by layering or by cuttings of the new wood, rooted in a sandy compost under glass.

ABIES

The Silver Firs: evergreen trees with flat, needle-like leaves arranged in two rows and with erect, cylindrical cones. Possessing extreme hardiness, they require a deep, moist soil though several species are lime-tolerant. They are planted as single specimens rather than in groups. Plant October-March and no pruning is necessary. They do not take well to town garden conditions though several species, i.e. *A. homolepsis* and *A. veitchii* will grow well on the outskirts of industrial towns.

SPECIES AND VARIETIES

Abies balsamea. The Balsam Fir: native of North America and which makes a small slender tree with dark green leaves, silver on the underside and bears erect violet cones. The buds and leaves have a balsamic scent. From incisions made in the bark, a substitute for Balm of Gilead is obtained from which Friar's Balsam is made.

A. cephalonica. Native of mountainous regions of north Greece, it is the best Silver Fir for a chalk soil. It forms long slender branches clothed with dark green awl-shaped leaves, silver on the underside. The 6 in. (15 cm) long cones change from green to red.

A. homolepis. Native of Japan where it reaches a height of 120 ft. (36 m), its branches are borne regularly up the trunk whilst its shoots are clothed in short leaves spirally arranged and which are striped with silver on the underside.

A. lowiana. Low's White Fir: it makes a broad conical tree with its horizontal branches wider at the base and in its native Oregon will attain a height of 300 ft. (90 m). The young shoots are olive-green with the blunt-ended leaves borne in two ranks. The mahogany-brown cones are borne in terminal whorls.

A. veitchii. Intolerant of a lime-laden soil, it grows fast in a deep moist soil and does well in or near towns. The leaves are glossy and densely arranged in two rows, glaucous above, silvery beneath whilst the purple-brown cones have heart-shaped scales.

ABUTILON

Native of Chile, *A. vitifolium* requires a warm garden or to be planted against a sunny wall, in a well-drained soil containing peat or leaf mould. Plant late in April and if the shoots are damaged by frost, cut back to a healthy bud. *A. vitifolium* will grow 20 ft. (6 m) tall with grey vine-like leaves and throughout summer bears pale lavender flowers in trusses. There is a white flowered form, 'album'. Propagate by cuttings removed with a 'heel' and root under glass.

ACACIA (Mimosa)

Native of Australia, it is best grown in sheltered gardens of south-west England and Ireland, where it will grow up to 30 ft. (9 m) or more in height, bearing from January to March spikes of fluffy globular flowers of golden-yellow which are in great demand by the cut-flower trade. It requires a loamy soil containing peat and is propagated by cuttings of the new wood, removed with a 'heel' in July and rooted under a cloche or frame.

Species of beauty are *Acacia baileyana,* which has fern-like glaucous leaves and bears multitudes of golden-yellow flowers; and *A. dealbata* which also has fern-like leaves and bears racemes of yellow flowers.

ACANTHOPANAX

Small deciduous trees, native of China and Japan, usually with spiny stems. They have deeply-lobed leaves and greenish-white flowers followed by handsome black fruits. They require a light loamy soil and are propagated by root cuttings in spring. They will root more quickly if treated with hormone powder and placed under glass.

A. ricinifolius is outstanding in its beauty and once planted will require little or no attention. Sometimes listed as *Kalopanax pictus.*

ACANTHUS

An interesting race of plants which originate from south-eastern Europe, and the complicated pattern of their long, crinkled leaves, was adapted by the builders of ancient Greece for their stone carvings. The plants which bloom in July and August, are hardy and will thrive in any well-drained loam and in a position of partial shade. If the soil is light, plant in November; if heavy, plant in spring, allowing 2 ft. (60 cm) between the plants.

SPECIES

Acanthus mollis latifolius. Above its huge leaves it sends up its sturdy, upright flower spikes of pinky-blue to a height of 3 ft. (90 cm) or more.
A. spinosus. It has rosettes of spiny leaves and bears erect flower stems 3 ft. (90 cm) tall. The blooms are white and purple, and are at their best in August.

ACER (Maple)

Tall-growing trees which are perhaps the most useful of all trees for the small garden for they are of neat habit, yet with their large leaves their rich colourings are revealed to advantage. Besides the beauty of the foliage throughout summer, the leaves take on rich new tints in autumn. They are amongst the first trees to open their leaves and amongst the last to lose them. It is the Japanese maples, *Acer palmatum*, which possess the most attractive foliage and one of the best is 'Purpureum', its shining crimson-purple leaves turning scarlet in late autumn. In contrast, plant with it the variety 'Dissectum', which has sharply-toothed, brilliant green foliage. Another lovely form is 'Osakazuki', its large bronzy-green leaves turning brilliant scarlet in autumn. More slowly growing is *Linearilobum* 'Major Halswell', which has narrow claw-like foliage and retains its attractive coral-red colour throughout summer.

The Norway maples, *A. platanoides*, are hardier, and for an exposed garden 'Goldsworth Purple' and 'Drummondii', with its creamy-yellow variegated foliage, are oustanding. Similar is *Acer colcium* 'Rubrum', which forms a neat upright tree. As the leaves unfold, they are of a brilliant crimson colour, turning bright green in summer and rich golden-orange in autumn. This is one of the most interesting trees of the garden and should be more widely planted.

Excellent for a small garden is *A. negundo* 'Argentea', which makes a small dense tree, its foliage being coloured pearly-grey and silvery-white. There is also another form 'Aurea', which has golden-green leaves striped with sulphur. Yet another attractive form is *A. ginnala,* a native of the banks of the Amur River, the small, dark green leaves having striking red ribs and turning brilliant scarlet in autumn.

ACHILLEA

A. eupatorium. Requiring a light, sandy soil and a position of full sun, this plant should be included in every border, for the large flat golden flower heads are borne from early July until October. The deeply-serrated silvery-green foliage provides additional interest. The bloom is as long lasting in water as it is when dried, whilst in the border the plants remain colourful for twelve weeks or more. Plant in November and propagate by division of the roots at the same time of the year. Plant 3 ft. (90 cm) apart.

VARIETIES

CANARY BIRD. An excellent variety for planting towards the front of the border, for the small flat bright yellow heads are held on 20 in. (50 cm) stems. In bloom June to August.

CORONATION GOLD. Like a smaller edition of 'Gold Plate'. It grows only 3 ft. (90 cm) tall and is in bloom from June until late in September. Its compact habit, rich colour and long flowering season make it one of the best of all hardy border plants.

FLOWERS OF SULPHUR. A German hybrid, the flowers are of a softer shade of yellow and are borne on 3 ft. (90 cm) stems. It blooms earlier than the other *A. eupatorium* hybrids; from mid-June until August.

GOLD PLATE. Well-named, for its huge flat heads are very much like golden plates.

PARKER'S VARIETY. Similar to 'Gold Plate' but the blooms are more globular. It is long lasting when dried.

A. millifolium. The native Yarrow, known to our gardens since the middle ages, when it was used to stem the flow of blood from wounds and as an antidote to aching teeth. It has pretty serrated foliage and grows to a height of 2 ft. (60 cm). It is in bloom from mid-June until August, planting being done during November in ordinary soil.

CERISE QUEEN. One of the few true cerise coloured flowers of the border and very free flowering.

FIRE KING. A recent introduction, the blooms are larger than those of 'Cerise Queen' and of true crimson colouring.

A. ptarmica. The Sneezewort, and though William Robinson rightly describes it as being one of the best of white flowers, care must be taken in its introduction to the border on account of its creeping rootstock. The roots run freely beneath the soil making it difficult to work amongst the plants, and they will interfere with neighbouring plants. The plant appreciates a cool, moist soil, very different from that favoured by *A. eupatorium*. It requires a soil containing humus rather than manure, and will appreciate a 1 oz. per sq. yd. dressing of sulphate of potash in April. The mint-like roots are best planted in March and should they spread too rapidly, they should be exposed and severed with a sharp knife.

PERRY'S WHITE. Very similar to 'The Pearl' though coming into bloom several weeks earlier.

THE PEARL. Produces its myriads of pure white button-like blooms from June until August and is a valuable plant for providing cut bloom and as a contrast in the border to the more brilliantly coloured summer flowers.

ACHIMENE

The tubular flowers with their flat open ends are borne on graceful stems up to 18 in. (45 cm) in length whilst the glossy leaves which are small and narrow are borne at regular intervals along the stem. The best way to grow the plant is over an inverted pot, supporting the shoots with small twigs and allowing them to drape over the side in cascade fashion. The plants may also be used for indoor hanging baskets.

SOWING THE SEED

The seed is small and must be germinated in a temperature of not less than 70°F. (21°C.). Where growing cool house plants in the same greenhouse, rather than raise the temperature to the requirements of these warm house plants, it will be advisable, and at the same time more economical, to use a propagator which will be heated either by oil or electricity. If seed is sown towards the end of February and a temperature of 68°F. (20°C.) maintained, it will germinate by the end of March. As the transplanted seedlings require a similar temperature for at least another month a second propagator, possibly less elaborate, may be used until they are ready for moving to small pots that in which the seeds were germinated also being used for the transplanted seedlings.

The John Innes Sowing Compost is suitable for sowing the seeds, which should be only just covered with a little of the compost or silver sand. Small seed pans which have been washed clean in a disinfectant should be used, having first been crocked so that there will be the necessary drainage. Sow as thinly as possible having first made the compost moist, then give a light sprinkling and place in the propagator. The seed must not be allowed to dry out but in the closed progagator there will be considerable humidity and the need for watering will be reduced to a minimum.

As soon as the seedlings are large enough to handle they must be moved to pans or boxes containing the John Innes No. 1 potting compost. The young plants should be kept comfortably moist and grown on in a similar temperature, but not exceeding 70°F.

Again, when the young plants have made sufficient growth and before overcrowding takes place, the temperature of the propagator should be gradually reduced to around 62°F. (17°C.). The plants are now ready for their size 48 pots containing a compost made up of 3 parts fibrous loam, 1 part each decayed manure and leaf mould and ½ part coarse sand. They should be planted five to a pot. This will provide a more pleasing effort than if planted individually, for they are of tall, graceful habit rather than dwarf and bushy as are the other members of this group of exotic plants.

BRINGING ON THE PLANTS

It will now be the end of April and as the plants will be out of the propagator it will be necessary to do everything possible to harbour sufficient heat to provide a night-time temperature of 58°F. (14°C.). If the greenhouse has been lined with polythene this will prevent the loss of considerable heat harboured during daytime. Until the end of May, day-time ventilation should be given on only the warmest days and the ventilators should be closed by late afternoon to enable the greenhouse to retain the maximum amount of heat by nightfall. Should a cold period be experienced it will be necessary to employ artificial heat at night until the end of May.

Until the end of May, keep the plants on the dry side and give water no later than mid-afternoon to enable excess moisture to dry off the plants before temperatures begin to drop at nightfall. As the plants come into bud and bloom towards the end of June they will need copious amounts of moisture, and the greenhouse should be damped down twice daily to create the right amount of humidity these plants require. When coming into bloom the plants may be placed in a window indoors until early October when flowering is coming to an end and the foliage begins to turn yellow. Watering should be gradually reduced until the foliage has died down at the end of November when the pots are placed on their side in a frost-free room. An attic room or cupboard in the home will be suitable. They should be started into growth again at the beginning of April or the grub-like tubers may be placed in dry sand stored in a warm room during winter.

Of the named varieties, *A. longiflora* bears large blooms of deepest purple, whilst *A. coccinea* bears rich scarlet flowers which are a striking contrast to the waxy white blooms of *A. gloxiniaeflora*. The new 'Michelssen' hybrids which grow only 6 in. tall are ideal for small pots.

'Tarantella' bears scarlet flowers and 'Joanna Michelssen' flowers of deep salmon.

ACIDANTHERA

In the British Isles it is autumn-flowering, and at its best in October. Being native of Abyssinia, the acidanthera needs sunlight and a position sheltered from the wind; also a soil which is well-drained. Peat and leaf mould are not necessary, but they appreciate, like the gladiolus, some decayed manure worked into the soil in March and some coarse sand or grit. Not fully hardy, the corms should be lifted in November as soon as the foliage has died down. Early April is the correct time for planting, not before, as the soil will not be warm enough for the corm to commence growing. Plant the corms 3 in. (7.5 cm) deep and 6 in. (15 cm) apart, a border or specially prepared bed being suitable. Where protection from winds can be given there should be no need to stake the stems which reach a height of about 3 ft. (1 m).

SPECIES

Acidanthera murielae. It bears large, pure, glistening white flowers which dangle like snowdrops and which have an attractive crimson blotch at the base of the petals. They carry a sweet perfume and are excellent as cut flowers. Inexpensive to purchase, the corms should be planted in quantity for the cut-flower market, for the blooms last well in water if cut just before they are fully open. As the blooms may not open until late autumn, their planting should be confined to a southerly garden or to a sunny corner.

ACONITUM

It is called Monk's Hood on account of the hooded flowers and during early Tudor times it was known as Wolf's Bane, for its poisonous roots were used as a bait to exterminate wolves. Though all parts of the plant are poisonous, it is cultivated for the beauty of its flowers, as well as commercially for the drug which is extracted from its roots.

The aconitum forms crowns which are readily divided, and should be planted in November. The plant likes a semi-shade; it may be planted in groups at the edge of a woodland garden, or in a border shaded from full sun by nearby property. It requires a damp, heavy soil enriched with peat or leaf mould.

SPECIES AND VARIETIES

Aconitum fischeri. It is an attractive plant for the front of a border, bearing its spikes of large deep purple-blue hooded flowers from mid-July until the end of August. Plant near dwarf golden-rods, or *Helenium* 'Wyndley', which grow to a similar height and bloom at the same time.

A. lycoctonum. It is of ancient origin and should be planted near *A. wilsonii*. Its pale Jersey cream 'hoods' make a delightful contrast. Plant 2 ft. (60 cm) apart.

A. wilsonii. (Barker's Variety). It bears its branched spikes of amethyst blue on 5 ft. (1.5 m) stems and remains in bloom from early August until October.

ACROLINIUM

A hardy annual which deserves to be better known. The seed is sown early in September in favourable districts, or under frames or cloches in March. It is an everlasting flower growing to a height of 15 in. (38 cm) and bearing delicate rose-pink flowers which are long lasting when cut and dried in autumn. There is also a double form, *flore plena*, and a white counterpart, *album*. When fully opened the bloom is cut with as long a stem as possible, almost at ground level and is made into bunches which are hung suspended for a fortnight in an airy room (not a greenhouse) before being used for home decoration.

ADENOCARPUS

Native of Spain, *A. decorticans* is closely related to the broom and should be given the protection of a sunny wall in the British Isles. It is evergreen and grows 6–7 ft. (about 2 m) tall with pale green trifoliate leaves, and in May and June bears racemes of golden-yellow. Plant early March in a well-drained sandy loam and as with the broom, propagate by seeds sown in small pots for the plants do not transplant readily.

ADENOPHORA

A. farreri bears its spikes of large deep blue campanula-like flowers during June and July. The Gland Bellflower as it is called, is a colourful plant to use behind the front row border plants. It likes a rich loamy soil to which has been incorporated some decayed manure or peat. Because of its parsnip-like rootstock it does not like disturbance. Plant in spring and propagate by root cuttings. It grows 3 ft. (90 cm) fall.

ADONIS

With its anemone-like flowers and serrated foliage, this is a delightful plant. There is both a summer and an autumn flowering species, but as the former, *A. aestivalis,* is slow to germinate, seed should be sown late in August the previous year and this may mean that it is best confined to a warm garden. Seed of *A. autumnalis,* is sown early in April. Thin out the plants to 10 in. 25 cm) apart. They reach a height of 12 in. (30 cm).

AECHMEA

These are tender perennials with leaves of arresting beauty, being long and strap-like, terminating into a sharp point and with silver horizontal stripes. *A. chantini* bears spikes of scarlet and gold tubular flowers which arise to a height of 18 in. (45 cm) from the centre of a rosette of handsome leaves whilst *A. fasciata* has leaves alternately banded with grey and green and from clear pink spiky bracts are borne bright blue flowers. The beauty of flowers and bracts are retained for six months or more.

Seed is sown in the John Innes compost in a temperature of 72°F. (22°C.) and when large enough, the seedlings are transplanted to small pots containing the John Innes potting compost which should be kept moist whilst the plants are grown on in a temperature of 60°F. (16°C.). After six months, they are moved to larger pots, containing the John Innes compost enriched with decayed manure and leaf mould when the plants are grown on in the sunny window of a warm room. They are amongst the most striking of all house plants.

AEONIUM

They are amongst the most interesting of the succulents, for they form flat rosettes which possess their own particular beauty. They are similar in appearance and in their cultural requirements to the echeverias, requiring a small pot, for they like to get their roots around the side. The compost should be composed of fibrous loam and peat in equal parts, to which is incorporated a small quantity of sharp sand and some mortar. The compost should be kept just moist throughout the year.

Aeonium canariense forms a large handsome rosette of pointed, heart-shaped leaves of a lovely shade of grey, whilst *A. tabulaeforme* grows into a rosette as large as a dinner plate and almost as flat. Of more upright habit is *A. arboreum* 'Atropurpureum' which makes a plant of upright, bushy habit, the leaves being of a lovely shade of deep purple. It bears pale yellow flowers in spring.

Propagation is by offsets which form around the plants and which should be removed whilst quite small.

AESCULUS

The Horse Chestnut, native of Greece and with its large divided fan-like leaves and candle-like spires of blossom borne in early summer, is one of the best of ornamental trees. The flowers are followed by polished mahogany 'conkers'. It requires a heavy rich loam and is propagated from the fruits ('conkers') sown in pots in spring. It does well under town conditions and is attractive at all seasons for in autumn the foliage takes on rich yellow and bronze colourings.

A. californica makes a low, spreading tree and bears erect spikes of white or pink flowers which are sweetly scented whilst *A. hippocastanum* 'Flore Plena', is the double flowered form of the Common Horse Chestnut but does not set seed ('conkers'). It is obtainable as white or red flowered.

AETHIONEMA

This is one of the loveliest of all chalk-loving plants. It likes a sunny, dry soil where it will thrive even where chalk is lacking. Where other plants may prove difficult this is one which will flourish. The outstanding variety is the hybrid 'Warley Rose', a dwarf shrublet bearing rich carmine-pink flowers throughout summer. Earlier in bloom is *A. iberideum* which bears white flowers and silvery-grey foliage. Another dainty species is *A. kotschyanum* which covers itself in a mat of silvery-pink flowers. Where the soil is lacking in humus, the aethionemas will do well.

AFRICAN CORN LILY, *see* Ixia

AFRICAN DAISY, *see* Arctotis and Dimorphotheca

AFRICAN LILY, *see* Agapanthus

AFRICAN VIOLET, *see* Saintpaulia

AGAPANTHUS

The Blue African Lily, though not quite hardy in exposed gardens, is a delightful subject for the midsummer border and does well in tubs. Those species producing their blooms in different shades of blue may be planted with the pure white form; whilst the same combination of colours looks most attractive when in tubs. The agapanthus likes a soil containing plenty of sand and some humus, in the form of peat or leaf mould. The bed must be well-drained and the bulbs planted in a sunny position. During winter a heavy peat mulch should be used for covering the plants and in addition, fronds of bracken placed over them. Plant 4 in. (10 cm) deep during April and they will come into bloom early in July. When planting in large pots or tubs, broken, but not crushed, charcoal should be used in the soil to keep it sweet. In winter the tubs should be removed to a dry cellar or outhouse until the following April to protect the plants from frost.

SPECIES

Agapanthus inapertus. It is a beautiful species, producing deep blue flowers with black anthers, and is at its best during August. In heavy clay soil, the plants should be lifted during winter and stored in peat and loam.

A. pendulinus. This is a tall-growing variety, the almost black blooms being carried on 6 ft. (2 m) stems. It is particularly attractive planted in the shelter of a shrubbery with a southerly aspect.

A. umbellatus. The hardiest of the agapanthus species and a flower most suitable for cutting. There is no need to lift in winter if planted in beds and given the protection of a wall or hedge. It bears bright blue flowers. The species *albus* of purest white is also lovely when cut.

AGAVE

The plants require a large pot and plenty of room and are not suitable for a small room. As they are less fleshy than most succulents, they must be given more water during winter, though the compost should not be kept in a saturated condition. They will also tolerate more shade than the aloes, to which the plants are closely related, requiring a similar compost and a minimum winter temperature of 40°F. (4°C.). They are propagated by detaching the plantlets which form around the plant.

The best form is *A. americana marginata* which makes a striking plant with its long spiked leaves which are glossy and of rich olive-green striped with cream. Also interesting and unusual is *A. filifera*, which forms a beautifully rounded rosette of as many as fifty or more long, thin, pointed fibrous leaves at the end of which is a drooping short fibre which produces a most graceful effect. Another of beauty is *A. victoriae-reginae*, which is slower growing and remains more compact. The leaves, which are almost square at the ends, are attractively variegated.

AGERATUM

An annual, it is one of the most useful of bedding plants for summer display. Sow in gentle heat early in the year, prick off when the seedlings are large enough to handle and grow on in a temperature of 50°F. (10°C.). Harden before planting out in June, 6 in. (15 cm) apart. Plant in sun or semi-shade.

The new tetraploid Blue Mist grows 6 in. (15 cm) tall and makes a rounded plant with pale green almost heart-shaped leaves and bears its fluffy misty-blue flowers, like tiny powder puffs, in compact trusses from late June until September. 'Blue Chip' is more vigorous and bears large trusses of lavender-blue flowers. 'North Sea' is of different colouring, the flowers being of bright rosy-mauve. For contrast, 'White Cushion' makes a fluffy mound of tiny white balls. For a trough the most compact of all is 'Little Blue Star' which bears its purple-blue flowers at a height of only 3 in. (7.5 cm).

AGROSTEMMA

A hardy annual, it blooms best from a sowing made in autumn or early spring where it is to bloom. The novelty 'Milas', bearing large flowers of a pleasing shade of soft lilac, on 3 ft. (90 cm) stems is the one to grow for cutting. Sow the seed thinly and space to 4in. (10 cm) apart. The bloom should not be picked until fully open when it will remain fresh in water for several days.

AILANTHUS

The Tree of Heaven, it is native of China and is deciduous with large pinnate leaves. It will reach a height of 60 ft. (18 m). The inconspicuous flowers are followed by bright red winged fruits in autumn. It is a valuable tree for a town garden and requires no attention apart from thinning out overcrowded and dead wood. Propagate by suckers or from cuttings rooted in sandy compost under glass.

A. glandulosa has large pinnate leaves with glands at the base of the leaflets.

AKEBIA

Akebia quinata, like the honeysuckle, is of twisting habit, and will attain a height of 20 ft. (6 m). It may be said to be partly evergreen but in a severe winter the plant will lose its leaves after Christmas. The leaves are of a vivid green colour, whilst the flowers which are vanilla-scented and of a rich chocolate colour are produced during May and June in long racemes, resembling those of the wistaria. Long purple fruits follow the flowers early in autumn, making it a most colourful plant to cover a trellis.

ALANGIUM

A. chinensis, native of China is a deciduous tree growing 6–8 ft. (2–3 m) tall with maple-like leaves and in June and July bears fragrant lily-like flowers of pure white with gracefully recurving petals. Plant in spring in a loamy soil. Do no pruning but cut back to a healthy bud, any shoots damaged by frost. Propagate by layering in autumn.

ALDER, *see* Alnus

ALECOST

Sometimes called Costmary, this is a herb which few seem to realise exists at all. It is perennial and will be too strong growing for the small herb 'patch', but its leaves are so useful to put in stews and soups and to flavour ale. The leaves are long with serrated edges and they smell like mint. Frances Bardswell says 'like weak mint sauce' which may be more accurate. It is a useful herb in that its flowers may also be used for their fragrance when dried.

It is increased by division and should be planted in spring 2 ft. (60 cm) apart.

ALE-HOOF, *see* Glechoma

ALKANET, *see* Anchusa

ALLIUM

Several of the ornamental onions of which there are upwards of one hundred species, are most colourful in the border early in summer. They like a sunny position and ordinary soil is suitable. They are useful plants for a dry, sandy soil and in such a soil they are at their best. Early November is the time to plant, when the bulbs are set 3 in. (7.5 cm) deep and sprinkle some sand around each bulb. The taller species should be planted towards the back of a border or in the shrubbery; the dwarf forms being suitable for the rockery. They are ideal flowers for the labour-saving garden in that they may be left completely untouched for many years before they need to be lifted and divided. They are also valuable for naturalising.

SPECIES

Allium albopilosum. It produces its heads of star-like flowers of deep pink during June, on 2 ft. (60 cm) stems.

A. anceps. A dwarf species bearing rich pink flowers on 6 in. (15 cm) stems during August and is a valuable late-summer rock garden plant.

A. azureum. It has rush-like leaves and bears large ball-shaped heads of sky-blue flowers on long stems during June. A hardy species, native of Siberia and useful for cutting.

A. karataviense. From Afghanistan, it is a suitable rock-garden plant. It produces its lilac flowers throughout May, but its chief claim to beauty lies in its grey, crimson-tinted foliage.

A. moly. The best species for naturalizing in grass or the shrubbery for it increases rapidly and so should not be planted too near other choice bulbs. Called the Golden Garlic, its compact bright yellow heads are produced during June on 20 in. (50 cm) stems. (see plate 1).

A. neapolitanum. A good cut flower for it bears pure white sweetly scented flowers on 2 ft. (60 cm) stems early in summer. Not as hardy as most species.

A. ostrowskianum. Probably the best of the alliums for it produces its compact heads of deep lilac blooms on 6 in. (15 cm) stems, rather like *Primula denticulata.*

ALMOND, *see* Prunus

ALNUS (Alder)

A native tree which grows well in wet ground, the Common Alder bears catkins in March and as it withstands clipping, may be used as a hedge or specimen tree. It has toothed oval leaves. It may be planted in swampy ground where few other plants will survive. Propagation is usually by layering.

Alnus glutinosa 'Aurea', the golden-leaf alder is most attractive and to plant with it is *A. incana* 'Acuminata', its leaves being like grey velvet on the underside.

ALOE

They make large plants and are grown for their attractive foliage. The plants are happy where placed in a shaded position during winter so long as they are given some sunlight in summer. Probably the best is *A. variegata*, the handsome Partridge-breasted Aloe, which forms a rosette of triangular spiked leaves which is dark green, flecked right across with fawn, similar to the markings on the breast of a partridge. *A. arborescens* is tree-like in habit, growing 2 ft. (60 cm) tall and the same distance in width, the long spiky leaves being produced from a woody stem. *A. mitriformis* is similar in form, its toothed leaves being triangular in shape, whilst *A. humilis* makes a dwarf group of rosettes in compact form. *A. eru* is also attractive, having long, thin pointed leaves and bearing yellow flowers in spring. *A. striata* is most interesting, its grey leaves being edged with pink and which blend attractively with its coral-pink flowers which are produced in summer.

The aloes, like the haworthias and gasterias, like a compost made up of 3 parts fibrous loam, 3 parts leaf mould or peat, 3 parts sharp sand and 1 part each mortar rubble and crushed brick. During summer they should be well watered, and though they will require little moisture in winter, the compost must never be allowed to become quite dry.

Propagation is simple and is by means of the tiny plantlets which appear round the parent plant like a hen with its chickens. They may easily be lifted and detached with their roots, to be planted into a small pot in a similar compost to that described.

ALONSOA

An annual, it may be raised in gentle heat or sown late in April where it is to bloom. The dainty flower spikes resemble miniature sidalceas. The best form is *A. warscewiczii* 'Compacta', which grows to a height of 15 in. (38 cm) and should be allowed the same distance to mature. Pinch back the growing point to encourage bushy growth.

ALOYSIA

The lemon-scented verbena, *A. citriodora* should be confined to sheltered gardens in the south-west of the British Isles. It is a charming plant for growing against a low wall because it rarely exceeds a height of 7 ft. (2 m). Its purple flowers appear in August. It is hardier and more fragrant if grown in a poor, sandy soil, so give no manure and no humus. This is a plant that will withstand the hottest conditions and is thus valuable for a sun-baked wall. It is the leaves which exhale a refreshing perfume, especially when brushing against each other in a slight breeze.

Plant from pots early in April and where conditions are not completely favourable the lower parts of the plant should be covered with sacking or straw from November until March. Propagation is by means of cuttings, whilst pruning should be confined to the removal of dead wood in March and the shortening back of long shoots in autumn.

ALPINE VIOLET, *see* Viola

ALSTROEMERIA

The tuberous-rooted Peruvian Lily is one of the finest of all border plants and where it is happy, it grows like a weed. The roots show similar hardiness to those of the eremurus. In a cold, badly-drained soil they are apt to decay but in a deep loam, thoroughly enriched with humus in the form of decayed manure and peat, the roots may be deeply planted so that they will be quite untroubled by frost. The Peruvian Lily delights in an annual top dressing of humus and the best bed the author ever grew was situated alongside a greenhouse, and over which boxes of seed compost were continually emptied. After five years the roots must have been 3–4 ft. (about 1 m) deep and each summer the bed was brilliant with a mass of lily-like blooms of deep orange-red, borne on erect 3 ft. (1 m) stems. Of all cut flowers, alstroemeria always maintains its popularity with florists.

Before planting, work in as much grit as possible if the soil is in any way heavy, and as when planting *Eremurus,* it is advisable to spread out the spider-like roots over a small mound of compost containing sand and peat. Plant the roots 15 in. (38 cm) apart and 10 in. (25 cm) deep. Plant only half that depth if the soil is heavy. Where planting in beds make the trenches 12 in. (30 cm) deep and 6 in. (15 cm) wide and into the bottom incorporate plenty of humus. Plant in spring so that root action can take place at once. A 1 oz. per sq. yd. dressing of sulphate of potash given in April each year will greatly benefit the plants and ensure depth of colour in the blooms. Propagation is by division of the tuberous roots, taking care not to cause them damage. The bloom should be cut when the buds are showing colour, cutting them with as long a stem as possible and bunching them into fives.

SPECIES AND VARIETIES

Alstroemeria aurantiaca. This is the species grown in quantity for market, of which there are a number of varieties. It bears deep yellow flowers, spotted with brown.

DOVER ORANGE. The best of the *aurantiaca* group, bearing flowers of the richest orange-bronze and is much sought after by florists.

LIGTU HYBRIDS. The flowers appear in shades of orange raspberry-pink, and flame.

A. haemantha. An attractive species bearing vivid blood-red orange-throated blooms but is not as prolific and vigorous as *A. aurantiaca.*

ALTHAEA

From the time when it was introduced to this country by the Crusaders, *A. rosea* has adorned the walls of country cottages everywhere, for it grows to a height of 6–7 ft. (about 2 m) and it is at its best against a wall, or at the back of a sheltered border. Though it will readily seed itself, it is perennial. The plants may also be raised from seed sown in April in drills, the young plants being moved to their permanent quarters in November to bloom the following summer. Plant 2 ft. (60 cm) apart, in groups at the back of the border. The best strain is 'Chater's Double', obtainable in pastel shades of peach, primrose, pink, salmon, etc. Unless the garden is sheltered the plants will require support or they may be blown down, the stems being severed at the base.

ALYSSUM

ANNUAL

The Sweet Alyssum, so called from the scent of its flowers which resembles new mown hay. It may be used as an edging or to carpet a bed of scarlet geraniums, begonias or salvias for it forms an almost prostrate mat 8 in. (20 cm) across. Raise plants under glass by sowing seed early in the year in gentle heat. Harden and plant out in June 9 in. (22 cm) apart. Alternate with blue lobelia if required.

'Snow Cloth' makes a neat mound 2 in. (5 cm) high and 8 in. (20 cm) across. Plant alternately 'Royal Carpet', studded in flowers of brightest purple, or 'Violet Cloud' which makes a dome-like plant with reddish-purple flowers. 'Pink Heather', well named is also neat and attractive, whilst the taller growing 'Bouquet', bears flowers of brightest blue on 18 in. (45 cm) stems and is a valuable cut flower.

PERENNIAL

Gold Dust, as the perennial *Alyssum saxatile* is called, is well-known for it provides a brilliant golden contrast to the aubretias during early summer. There are other varieties even lovelier; one is called 'Dudley Neville' of compact habit with the flowers of a rich shade of cream, whilst 'Flore Pleno' is later flowering and fully double, producing a very rich effect. These alyssums will flourish in as dry and as baked a position it is possible to find for them.

AMARANTHUS

This hardy annual, Milton's 'Immortal Amaranth', *A. caudatus* now rarely to be seen, was at one time the most popular of all plants and was named the Purple Flower-gentle. Growing 3 ft. (1 m) tall, Gerard said that the plume-shaped flowers were 'of a shining light purple, with a gloss like velvet, but far passing it . . . and being tathered, do keep their beauty a long time, in-so-much that being set in water, it will revive again as at the time of gathering', hence it was regarded by the ancients as the symbol of immortality. It is Spenser's 'Sad amaranthus in whose purple gore meseems I see Aminta's wretched fate'. It takes its name from a Greek word meaning 'incorruptible'. *A. caudatus* was known to Gerard who called it the Great Purple Flower-gentle. It was also known as Love-lies-bleeding or Princes' Feather on account of its drooping crimson plumes.

Another form, *A. tricolor* was grown in Tudor gardens. It is the three-coloured amaranthe, native of the East Indies and named Floramor in Tudor times. Gerard said that 'every leaf resembleth in colour the most fair and beautiful feather of a parot . . . a stripe of red, a line of yellow, a dash of white, a rib of green . . .' The plants are raised under glass, sowing in gentle heat in February and planting out, after hardening, in June when they reach their full beauty during the early weeks of autumn.

T. tricolor 'Molten Fire' has striking foliage, being blotched with shades of crimson-red, bronze and amaranthe purple. It makes a compact plant 2 ft. (60 cm) tall.

The true Princes' Feather is *A. hypochondriacus* which reached this country from Virginia in 1684. he plants require ample supplies of moisture during dry weather.

AMARYLLIS

A genus of a single species, more commonly called the Belladonna Lily. It is one of those valuable bulbous plants which will bloom out doors in a sheltered position early in autumn. In the more exposed parts, the bulbs may not be entirely hardy and should be covered with peat or bracken during early November; or be lifted, cleaned and stored until replanted early in May. In the west, the bulbs may be planted 6 in. (15 cm) deep and left undisturbed for years, when they will produce their lovely pink lily-like flowers on leafless stems from late August until October. They like shelter rather than shade, such as provided by wattle hurdles or a sparsely planted shrubbery. Like so many bulbous plants they like a cool, moist root system, but some sunshine above to ripen the bulbs. Work plenty of humus into the soil and dig deeply as the bulbs need a 6 in. (15 cm) covering of soil over them. They should be planted early in May, in groups, spacing the bulbs 8 in. (20 cm) apart. When growing, they require large amounts of moisture. The flowers reach a height of 2 ft. (60 cm) and where shelter cannot be given, some staking will be necessary for the stems are easily broken. Where the plants grow well, the blooms may be marketed with local florists where they find a ready sale. The blooms do not transport well.

This is also a valuable plant for a cool greenhouse, the bulbs being planted three to a large pot, or one to a small pot in November, so that they may bloom late in summer. They enjoy a rich compost and a little decayed manure, mixing it well into the loamy soil. When planting the bulbs, either in the open ground or in pots, they should be surrounded with sand or peat before filling in the soil round them. Firm planting is essential and when in pots, the nose of the bulbs should be just above the level of the compost.

Amaryllis belladonna is native of South Africa. It produces in summer, green strap-like leaves from which emerge in autumn nodding trumpets borne in clusters. An improved form *A. purpurea major*, produces very large blooms of a rich rose-pink colour whilst *A. parkeri* produces up to a dozen blooms on each stem of an exquisite shade of rose, shaded buff at the base.

AMELANCHIER

An interesting hardy deciduous tree which does well in a calcareous soil. The species *A. canadensis* forms a thick but narrow bush bearing clusters of snow-white flowers early in May, followed by crimson berries in August when the foliage takes on vivid orange colourings. It should be planted in March, 2–3 ft. (60–90 cm) apart and when established should be cut back to 12 in. (30 cm) of ground level to make it more bushy. This is done in November or mid-March. The plants should then be allowed to grow away with little more attention than the occasional cutting back of the loose shoots.

Another species, *A. vulgaris* bears racemes of large, creamy-white flowers, followed by blue-black berries and the same exquisite foliage colourings in autumn.

AMERICAN ALOE, *see* Agave

AMERICAN SWEET FERN, *see* Comptonia

AMORPHA

A tender deciduous shrub, *A. fruticosa* is related to the Broom and grows 3 ft. (90 cm) tall. Known as the False Indigo, it has fern-like foliage and from July until September bears slender racemes of indigo-blue flowers with golden anthers. Plant in March, in a well-drained sandy soil and in an open, sunny position. The young shoots may be cut back by frost but others will appear from the base as suckers which may be used for propagation.

AMPELOPSIS

The Virginian Creeper of which there are a number of forms, including one which is fully evergreen, are self-clinging and are at their best on a stone wall where the brilliant colouring of the leaves is accentuated. Those who have seen this plant growing on the outer walls of the New Court of St John's College, Cambridge, will appreciate its brilliance when against a stone wall. If the plant has a fault, it is that it loses its foliage too soon in autumn, often by early October.

The ampelopsis is a member of the vine family, but is rarely listed as such. Unlike the ornamental vines, the ampelopsis is self-clinging, attaching itself to a wall by means of discs which act in the same way as a suction dart. *A. veitchii* bears smaller and neater leaves than the true Virginian Creeper, *A. quinquefolia*, and these take on richer colourings.

It is sometimes known as the Boston Ivy. Bearing even neater leaves is the new 'Beverley Brook' variety; whilst *A. henryana* bears silver and rose-coloured leaves, brilliantly coloured crimson on the underside. The latter variety is best grown on a north wall in the southern part of England, for it may not be quite hardy elsewhere. *A. striata* is an attractive self-clinging evergreen form and quite hardy.

Plant in November, in a soil enriched with humus and not lacking for lime. Plant pot-grown plants, for the ampelopsis hates root disturbance. As the foliage, except for *A. striata*, is deciduous no clipping is necessary, apart from the removal of excessive growth about a window or where interfering with spouting. The ampelopsis (and the ivy) is best grown by itself, for it will prove rather too rampant where grown with other wall plants, and the attractive tracery of their stems and the unusually rich colouring of the foliage will be more appreciated without competition. The Virginian Creepers are essentially wall plants rather than for growing up trellis or fences.

AMUR CORK TREE, *see* Phellodendron

ANAPHALIS

For a large rockery or for the front of a shrubbery or a small border, this is a charming plant of everlasting type yet one which no one seems to grow and only in a few gardens is it to be found. It grows well in all soils, especially those of a sandy nature and is quite hardy, whilst retaining its foliage over a long period. Plant in spring 18 in. (45 cm) apart and propagate by cuttings, rooted in a frame in May.

SPECIES

A. nubigena. It grows to a height of 15 in. (38 cm) and forms a compact plant which could be used to edge a large border. From July until early November it covers itself in clusters of fluffy white flowers which like all the everlastings, are papery to the touch. The foliage too is of a bright silver-grey which is retained until almost the year end.

A. triplinervis. It grows to a height of only 12 in. (30 cm) and is slower growing than *A. nubigena,* but bearing the broader silvery-grey foliage, and clusters of fluffy white flowers during the same period.

A. yedoensis. Growing to a height of nearly 2 ft. (60 cm), it is a plant of upright habit, the white everlasting flowers being borne on long stems during August and September.

ANCHUSA

ANNUAL

It is a striking bedding plant, growing 15 in. (38 cm) tall. It is best treated as biennial, like the antirrhinum, sowing seed in a frame in August and pricking out the plants in October. Plant out in April 12 in. (30 cm) apart in a sunny position. *Anchusa capensis* 'Blue Bird' comes into bloom early July and continues until September, bearing masses of brilliant blue forget-me-not-type flowers amongst grey-green foliage.

PERENNIAL

Whilst *A. italica* is a plant for the back row of the border, there are two forms which are amongst the loveliest of all border plants and which grow only 15 in. (38 cm) high. Both should be freely planted about the border in groups and they prefer a well-drained sandy soil. They come into bloom with the doronicums and so should be planted together. Plant early in spring.

SPECIES AND VARIETIES OF PERENNIAL ANCHUSA

Anchusa caespitosa. In bloom from early May until July, it makes a compact, rounded plant and bears its gentian-blue flowers with their striking white eye on branching 15 in. (38 cm) stems.

A. italica. A member of the borage family, *A. italica* is a native of south Europe and reached this country at the beginning of the nineteenth century. In its native haunts the roots were, in ancient times, used for the extraction of a crimson dye, from which the plant takes its name. Propagation is by sowing seed or by division of the roots. Spring is the best time to plant, allowing 2 ft. (60 cm) between the plants. The anchusa prefers a light, sandy soil containing some humus. In a heavy, badly-drained soil the plants may tend to die back during a winter of excessive rain. The plants remain in bloom from mid-June until August. It grows 4 ft. (1.2 m) tall.

MORNING GLORY. It bears its very dark blue flowers on 5 ft. (1.5 m) stems over a period of ten to twelve weeks.

OPAL. The flowers are of a lovely pale blue colour.

PRIDE OF DOVER. Brilliant sky blue of compact habit.

ROYAL BLUE. The rich gentian-blue flowers are freely produced in compact pyramidal heads. It grows 3 ft. (90 cm) tall.

A. myosotidiflora. So named because of the likeness of the brilliant blue flowers to those of the forget-me-not and which are in bloom during the same month of April. The plants grow only 12 in. (30 cm) tall and remain compact through summer, bearing a second crop of flowers during August.

ANDROSACE

The several species of this plant will thrive in a chalky soil; they prefer some limestone chippings placed round their crowns after planting and a small quantity of lime worked into the soil. *A. carnea*, with its neat habit and fleshy-pink flowers blooms late in spring, whilst *A. carnea haggeri* bears deeper coloured flowers. *A. chamaejasme*, the dwarf jasmine, bears white flowers with yellow centres during mid-summer. Flowering well into autumn is *A. lanuginosa*, which bears hairy green leaves and bloom of pale lavender with an attractive red eye. Another is *A. villosa* which produces its clusters of white flowers during spring. The androsaces require a gritty, well-drained soil, and an open sunny situation.

ANEMONE

To would-be anemone growers the most important advice is to refuse all large corms. These have not only become acclimatized to the soil and surroundings and to the climate in which for several years they have been growing, but they have become 'woody' and tough, whilst their best flowering days have long since passed. This same advice may also be given to growers of gladioli, freesias and begonias; it does not follow that the largest corm will produce the best bloom. This may be true of bulbs, but not of corms. The anemone growers of the west country plant a corm of 1–2 cm or 2–3 cm size, no larger than a garden pea, so that anything four times as large and of a woody, knobby appearance will mean that it has seen its best days and should be avoided. Anemones grow readily from seed and the 2 cm planting-size corm is attained within twelve months of sowing. The 3–4 cm corm may also be grown. They will cost almost twice as much as the smaller sizes and little will be gained by planting them in spring, though this should be the size used for frame or pot culture. Anything above the 5 cm should be left alone for no good will come of its planting.

Another cause of failure may be due to planting too thickly. With the smallest corms, which are also the most vigorous, one is tempted to plant, like peas, much too closely. The anemone must be given room to develop its vigorous rooting system so instead of taking out a shallow trench and sowing the corms, make a drill 3 in. (7.6 cm) deep and carefully place the corms 2 in. (5 cm) deep and 4 in. (10 cm) apart, allowing 12 in.

(30 cm) between the rows. Some prefer to plant two corms together at distances of 10 in. (25 cm) placing them 2 in. (5 cm) apart.

When to plant is important, especially where the flowers are being grown for profit. Those who do not cover with glass in winter should aim to have the corms in bloom early in spring and planting is done during July. In a heavy soil it often happens that the 3–4 cm corm is used for this late planting. The grower with cloches or frames should set out the corms during June so that they will commence to bloom late in autumn, when they may then be covered during November. About fourteen weeks will elapse from sowing to flowering time. The amateur may prefer to obtain a late summer crop by sowing in early May, when the corms will bloom from mid-August until severely cold weather prevents the formation of more buds. If the rows are covered with decayed leaves or peat this will provide some winter protection and will later act as a mulch. The plants will continue to bloom throughout summer and may well produce an amount of bloom on strong stems throughout the following year. Like strawberries, the question of just when to dig them up and replant must be left with the size and quantity of bloom. Generally two years will see them past their best though much will depend upon the strain of corm and the preparation of the land.

PREPARATION OF THE SOIL

The anemone produces a larger, more deeply coloured bloom and a greater quantity of buds when grown in a heavy loam than in a sandy soil. This is by no means to say that it enjoys a heavy, wet soil, for drainage it must have, and any soil not well-drained will be useless for anemones. Should the soil be unduly heavy, it may be better to plant on raised drills or in raised beds so that no excess moisture will hang about the roots during winter. In any case, the soil will need some preparation. A heavy soil will need lightening by the addition of some coarse sand, some peat or leaf mould, or even some spent hops which are obtainable from a brewery. Some lime rubble or hydrated lime is also necessary, for contrary to popular belief, the anemone does not favour an acid soil. A soil which is of an acid nature will give only a light crop. When growing anemones in a town garden where deposits of soot and sulphur constantly make the soil acid, the rows should be given a yearly application of lime applied before a mulch is given in early autumn. Anemones will grow well in town gardens provided these details are attended to.

Those growing in light soils will have no drainage problems, but a light soil will usually produce a small bloom on a thin stem. Humus therefore is necessary and this should be by way of well-rotted manure such as old mushroom-bed compost, or a compost made up by mixing pig or poultry manure in the dried state with peat. This is often applied at the rate of twenty tons per acre with professional growers and augmented by a five cwt. per acre dressing with hoof or fish meal when the plants are coming into bloom. The amateur should add what is available to the soil before planting. Lime is essential, and the following spring give a dressing of a 1 oz. per yd. row of sulphate of potash which is raked into the ground between the rows. The professional grower will apply as much as four cwt. of potash to the acre. And whether growing in light or heavy land, or in town or country, do not forget to provide a light mulch when the corms have been flowering for six months. Peat seems to be the best and good advice is to sprinkle the drills with peat just prior to planting the corms.

Where the plants are to be covered with lights, closer spacing of the rows is permissible, with the corms spaced slightly further apart in the rows. The width of the bed should correspond to the width of the lights used, which will be either 4 ft. (1 m) or 6 ft. (2 m). One method is to have the plants coming into full bloom by late autumn when boards are placed around the beds and held into place by pegs. These will take the lights towards the end of October. Should the weather be unduly cold during mid-winter, soil should be banked up round the outside of the boards and straw should also be placed both round and over the lights. As anemones resent any stuffy conditions there must be ample ventilation wherever glass is used. The glass should be removed entirely on all sunny days. Under cloches anemones flower to perfection for here they receive the much-appreciated protection from winds. Though their name is derived from the Greek, anemos, wind-flower, the anemone does in fact, detest strong winds and the growers in the west country spend considerable time and expense in providing them with protection.

RAISING PLANTS FROM SEED

Anemones may readily be raised by sowing the seed in pans of sterilised loam to which is added some peat and some sand. The pans should be placed in a cold frame early in April or the seed may be sown direct to a seed bed made up in the frame. Or, they may be sown under cloches at the same time of the year. It is better to sow the seed in drills rather than broadcast and as the seeds are covered with fur, they are best mixed with some sand to ensure more even distribution. The seed should be covered with only the smallest amount of sand after sowing. Professional growers who raise their own seed usually mix a quantity of radish seed when sowing on a large scale in the open early in summer. The early germination of the radish seed enables the rows to be noticed during hoeing operations, until the anemone seed germinates. Those planting the small corms in drills will also find a sowing of radish seed useful in determining the rows until the first growth is observed.

When large enough to handle, the seedlings may be thinned and those remaining may be left to flower the following summer or they may be lifted at the year end and the small pea-size corms dried and replanted in March. In very exposed areas, this lifting and drying in October and replanting in spring is often done with established corms.

Seed in pans in a frame should be kept moist and shaded from strong sunshine during summer. Seeds of several of the species used for naturalizing may be sown under trees or in any suitable place where they will not be disturbed either when sown or when in bloom. *Anemone apennina*, *A. nemorosa* and *A. blanda*, may all be sown in their natural surroundings.

When sowing in the open, remember to provide some protection from winds, otherwise the fern-like foliage may, in a drying wind, become shrivelled. Those growing anemones for cutting should remove all blooms whether used for market or not, otherwise they will form seed and soon cease flowering altogether.

SPECIES AND VARIETIES

Anemone apennina. Producing their blooms of clearest blue during March and April, it is hardy and will bloom to perfection in shade. It should be massed under trees or in the shrubbery where it may be left undisturbed. The pale green fern-like foliage adds to its attraction. It is also a lime lover.

A. blanda. In sheltered positions it will bloom from February until the end of April and though the blooms are similar to *A. apennina* in colour and size, this species requires full sun. It is at its best planted round young standard apple and ornamental trees. *A. blanda* has also a rooting system similar to that of *A. apennina*, being tuberous with a creeping habit. With *A. apennina* it should be planted in September to flower early in spring. There is a bright rose-pink variety which looks lovely planted with the blue form.

A. coronaria. Of the single or De Caen anemones, 'His Excellency' is crimson-red with a white contrasting centre; 'Hollandia' is brilliant scarlet; 'The Bride', white; and 'Mr Fokker', deep blue. Of the St Brigid or double anemone, 'Lord Lieutenant' is purple-blue; and 'The Governor', crimson-red.

A. fulgens. Increasing rapidly if left undisturbed, this is an ideal plant for a grassy bank or rockery, or for planting beneath trees where it may receive as much of the early summer sunlight as possible. The blooms are brilliant scarlet, the anthers jet black.

A. japonica. Native of the interior of China, it did not reach Britain until the nineteenth century, the early form producing a bloom of white flushed with pink. The first pure white variety, 'Honorine Jobert' originated in a French garden a century ago. With its handsome hairy grey-green leaves and masses of anemone-like flowers held on branched stems above the foliage, it is a charming plant for a shrubbery, for the border, or for planting in semi-shaded beds. The bloom is produced from early August until the end of October, and so long as the soil is retentive of summer moisture, the plants will flourish anywhere. In a dry, sandy soil, lacking moisture, the plants will tend to be stunted and the full beauty of the blooms is not revealed. Plant in November, 2 ft. (60 cm) apart and propagate by root cuttings or by division.

HERZBLUT. A German introduction, bearing over a long period blooms of deep rose-red on 2 ft. (60 cm) stems.

LOUISE UHINK. The large pure white blooms are held on 3 ft. (90 cm) stems and remain long in bloom.

MARGARETTA. The deep rose-red flowers are almost fully-double and are early to bloom.

MAR VOGEL. The semi-double blooms are of a lovely clear pink shade, produced on 2 ft. (60 cm) stems.

PROFUSION. Possibly the finest variety; of compact habit growing to a height of only 20 in. (50 cm) it is ideal for the front of the border. The lovely rose-pink flowers are freely produced from August until November.

SEPTEMBER CHARM. Of neat, compact habit, its large shell-pink flowers are borne on 20 in. (50 cm) stems.

A. nemorosa. The true Wood Anemone enjoys a soil containing plenty of leaf mould and a position of almost complete shade. The colour is a delicate pale blue. It should be given plenty of lime rubble when planted in autumn.

A. ranunculoides. This is a delightful little species, bearing dwarf, bright yellow flowers set amidst pale green foliage. A variety called 'Superba' has attractive bronzy leaves. Plant early autumn to flower in March and April. A delightful rock garden plant.

ANGELICA

A. officinalis may live for several years but is usually treated as a biennial the seed being sown in August as soon as collected in the position where the plant is to mature. It will grow up to 6 ft. (2m) tall and may be grown at the back of the border, where it will appreciate shade given to its roots. It will seed itself and may even become a nuisance. The stems are cut from June onwards, when young, to candy and use in confectionery, later to flavour apple tarts and tomato chutney. Its stems may also be used for stewing with those of rhubarb and will greatly add to the flavour. It should be said that the wild form is almost flavourless.

Bees visit its white flowers whilst its leaves and roots are used in medicine. Parkinson tells us that the dried roots powdered and taken in water 'will abate the rage of lust in young persons', whilst it will also help 'tremblings and passions of the heart'. Taken hot, the leaves infused in boiling water to which lemon juice and honey are added, it is an excellent drink for a cold.

ANGELICA TREE, *see* Aralia

ANNUALS

FOR A BORDER

Amaranthus caudatus
Anchusa capensis 'Blue Bird'
Antirrhinum 'Giant Ruffled'
Arctotis grandis
Aster 'Duchess'
Aster 'Super Princess'
Calendula 'Geisha Girl'
Cleome pungens 'Rose Queen'
Cosmea 'Sunset'
Helenium 'Sunny Boy'
Hibiscus trionum
Kochia childsii
Larkspur, Stock-flowered
Lavatera 'Loveliness'
Love-lies-bleeding
Malope 'Aphrodite'
Marigold 'Donbloon'
Marigold 'Hawaii'
Molucella laevis
Nicotiana 'Lime Green'
Poppy 'Dannebrog'
Poppy 'Pink Chiffon'
Scabious 'Monarch Cockade'
Sunflower 'Dwarf Sungold'
Tithonia 'Torch'
Venidium fastuosum

FOR A DRY SOIL

Cladanthus arabicus
Dimorphotheca
Eschscholtzia
Lupin 'Dwarf Pixie'
Mesembryanthemum
Nasturtium
Phacelia campanularia
Portulaca
Sanvitalia procumbens
Ursinia

FOR CUTTING

Ageratum 'Bouquet'
Agrostemma 'Milas'
Antirrhinum 'Double Formula'
Arctotis grandis
Aster 'Duchess'
Aster 'Pink Magic'
Calendula 'Geisha Girl'
Calendula 'Radio'

FOR CUTTING (*continued*)

Chrysanthemum 'Cecilia'
Cornflower 'Blue Diadem'
Cosmea 'Goldcrest'
Cosmea 'Sunset'
Gaillardia picta 'Lorenziana'
Gypsophila elegans 'Rosea'
Jacobaea 'Sutton's Rose'
Larkspur, Stock-flowered
Layia elegans 'Cutting Gold'
Leptosyne stillmanii
Marigold 'Climax'
Nigella 'Miss Jekyll'
Sweet Sultan 'Giant'
Tithonia 'Torch'
Zinnia 'Dahlia-flowered'

FOR EDGING

Ageratum 'Blue Chip'
Ageratum 'North Sea'
Alyssum 'Little Dorrit'
Alyssum 'Pink Heather'
Anagallis 'Grandiflora'
Aster 'Daisy Mae'
Brachycome
Chrysanthemum multicaule
Eschscholtzia, Min: Primrose
Leptosiphon
Linaria maroccana 'Fairy Bouquet'
Lobelia Mrs Clibran
Marigold, French 'Golden Frills'
Nemophila insignis
Phlox drummondii 'Twinkle'
Sweet William 'Wee Willie'
Tagetes signata 'Carina'
Virginia Stock

FOR INDOOR POT CULTURE

Alonsoa warscewiczii
Amaranthus tricolor
Balsam 'Tom Thumb/Elfin'
Cacalia coccinea
Calceolaria gracilis
Calendula 'Orange Coronet'
Celosia 'Dwarf Golden Plume'
Celosia 'Dwarf Red Plume'
Collinsia bicolor
Diascia barberae
Exacum affine
Mignonette
Nolana 'Lavender Gown'
Petunia
Salpiglossis 'Splash'
Schizanthus
Stocks
Torenia fornnieri

FOR A TROUGH AND ALPINE GARDEN

Ageratum 'Little Blue Star'
Alyssum minimum
Aster 'Thousand Wonders'
Dianthus sinensis
Marigold 'Dainty Marietta'
Nolana 'Lavender Gown'
Sanvitalia procumbens
Sweet William 'Wee Willie'
Tagetes signata 'Golden Gem'
Verbena 'Delight'
Zinnia 'Thumbelina'

QUICK TO BLOOM

Bartonia aurea
Candytuft, Red Flash
Eschscholtzia
Limnanthes
Linaria
Nemesia
Nemophila
Omphalodes linifolia
Phacelia campanularia
Ursinia
Virginia Stock

WHICH GROW WELL SOWN WHERE THEY ARE TO BLOOM

Alonsoa warscewiczii
Anchusa capensis
Bartonia aurea
Brachycome iberidifolia
Calendula
Candytuft
Clarkia
Collinsia bicolor
Convolvulus major
Cornflower
Cosmea
Echium
Eschscholtzia
Eucharidium
Felicia bergeriana
Gamolepsis
Godetia
Jacobaea
Larkspur
Layia elegans
Leptosiphon
Leptosyne
Limnanthes
Linaria
Linum
Lupin *hartwegii*
Matthiola bicornis
Mignonette
Nasturtium
Papaver rhoeas
Phacelia campanuloides
Ursinia
Virginia Stock
Viscaria oculata

ANOMATHECA

A. cruenta is a delightful plant for the late summer rock-garden, like a miniature montbretia and with grass-like foliage. It is in bloom from mid-July until September. Though a native of South Africa, it is hardy given the protection of rockery stone, even in the north, or it may be planted on a sunny slope. A single bulb will rapidly increase, the vivid salmon-coloured blooms with crimson blotches at the base of each petal looking attractive when planted in groups. It enjoys a sunny position wherever it is planted and a soil containing liberal quantities of sand or grit.

Plants may be raised from seed sown in a cold frame late in March, or sow in a deep box in a frame, allow the foliage to die down naturally at the end of summer then plant six of the small corms to a size 60 pot the following March. Stand in a cold frame until the buds form, then transfer to a light position in the home. The dainty blooms are held on 10 in. (25 cm) stems and should be supported by small thin twigs inserted into the soil amongst the foliage.

ANTHEMIS

A large genus of shrubby plants, native of the north temperate regions and with handsome leaves twice or thrice pinnately divided and releasing a pleasantly pungent smell when handled. The flowers of white or yellow are borne at the ends of the branched stems. They are easily grown plants which grow well in ordinary well drained soil and are readily raised from seed.

SPECIES AND VARIETIES

Anthemis cupaniana. It grows 15 in. (38 cm) tall and makes a bushy plant with handsomely cut silvery foliage which tends to become green in autumn. It requires a dry, sandy soil and a sunny position. Its white flowers are pleasantly scented.

A. nobilis. The Chamomile and a plant of almost prostrate form which in olden times was planted to make a lawn with its dense matted growth whilst when trodden or sat upon, it released a refreshing smell of ripe apples. It takes its name from the Greek meaning 'fresh apples'. From the flowers, a drink is made to sooth tired nerves.

Plants are readily raised from seed sown in a frame or outdoors. The plants should be set out 6 in. (15 cm) apart and may be used to make a lawn or to cover a bank. They may be rolled in and will withstand clipping to keep them low.

A. tinctoria. It is a border plant and will also make a colourful display planted in beds for cutting. The best form is *A. sancti johannis,* which comes true from seed and bears its deep orange daisy-like flowers on 2 ft. (60 cm) stems throughout summer. The plants are not easily divided, and as they are readily raised from seed, this is the usual way of propagation. Seed is sown in the open in May for planting in the border the following spring for they do not like autumn transplanting. They grow well in dry, sandy soil and should be planted 2 ft. (60 cm) apart.

If named varieties are required, they are increased from cuttings taken in early summer and rooted in a closed frame where they remain over winter. Excellent is 'Beauty of Grallagh', with its bushy habit and which bears its deep golden-yellow blooms from June until September; and 'Grallagh Gold' which bears lemon-yellow flowers and is of more dwarf habit.

ANTHERICUM

A. liliastrum is St Bruno's Lily, named after the founder of the Carthusian Order. It is an excellent plant for the border, bearing its pure white lily-like flowers on 2 ft. (60 cm) stems during May and June. A native of the high meadowlands of the Alps, the bulbs require a light, well-drained soil, preferably containing some sand and peat to help with the retention of summer moisture.

Plant in November which is the most suitable time to increase the stock by lifting and dividing the bublets. Plant 3 in. (7.5 cm) deep and 8 in. (20 cm) apart in groups of four or five bulbs. The bulbs should be planted near those early flowering plants such as *Anchusa myosotidiflora* or *Centaurea montana* which provide protection with their early growth and which come into bloom at the same time.

ANTIRRHINUM

One of the most popular of all bedding plants, it is perennial but is best treated as biennial, the seed being sown in a cold frame late in August, the seedlings pricked into boxes and wintered under a frame. In this way the plants may be planted out early in May and will come into bloom towards the end of June. The growing point should be removed to encourage bushy growth. Another method is to sow seed in heat in January, or over a hot-bed in March, but the plants will be a month later to bloom and will not be so robust.

For bedding, the Intermediate varieties are popular as they grow to a height of 15 in. (38 cm). Outstanding is the silvery-pink 'Malmaison' and 'Guardsman', is orange-scarlet with yellow lips, 'Eclipse', is deep crimson, and 'Golden Queen' is deep yellow. There is also a strain of more dwarf habit for bedding, the plants growing 9 in. (22 cm) high. Amongst the best are 'Pink Exquisite', 'Yellow Prince' and 'Cherry', but the colour range is considerable.

For those plants troubled by Rust Disease, there is a new range of rust-resistant varieties, which grow to a height of 12 in. (30 cm). 'Pink Freedom', with its almond blossom flowers; 'Golden Fleece', yellow and apricot; and 'Victory', buff and orange are outstanding.

For edging or carpeting, the 'Tom Thumb' and 'Magic Carpet' hybrids are delightful, having a large colour range and an almost prostrate habit. The variety 'Delice', introduced by Unwins grows 9 in. (22 cm) tall and the same across and completely covers itself in tiny spikes of creamy-apricot. Similar in habit are the 'Floral Cluster Fl' hybrids from Japan, obtainable in scarlet, yellow and pink. The most compact of all are the 'Little Gem' hybrids, growing 4 in. (10 cm) tall and which are ideal for edging small beds or for trough gardens and window boxes.

For cutting, plant the Fl hybrid 'Mme Butterfly' strain, the 1970 All-America Award Winner, which grows 3 ft. (90 cm) tall, bearing spikes of large double blooms which remain fresh in water for at least a week.

APHELANDRA

A. squarrosa is one of the most striking of all indoor plants. The leaves are almost 12 in. (30 cm) in length and are borne from a single main stem which attains a height of 2 ft. (60 cm). The leaves are of most interesting shape, being 3 in. (7.5 cm) broad at the centre and terminating to almost a point at the extreme ends. The brilliant green leaves have pure white veins running from the mid-rib to the margin, whilst the plants bear terminal flowers of brush-like appearance and yellow in colour.

Native of South America, the plants like a winter temperature of not less than 50°F. (10°C.), so should be grown in a centrally-heated house away from draughts and in a position of partial shade.

The plants require an 'open' compost, made up of 2 parts fibrous loam, 2 parts peat and 1 part grit and make sure that the pot is well crocked. Though the plants enjoy plenty of moisture in summer, they should be given limited amounts in winter.

Propagation is by means of small side shoots growing from the main stem. They will require a humid atmosphere in which to root and are best rooted in a propagator.

APRICOT, *see* Prunus

AQUILEGIA

Though named varieties of the Columbine, a plant mentioned by both Chaucer and Shakespeare, and native to Britain, are readily increased by root division, the new mixed strains produce so delightful a colour range that growing from seed is the recognized method of propagation. Seed is sown in spring, in pans or boxes under a frame, or in the open where it is covered with a sheet of glass. Like most hardy plants which readily grow from seed, the blooms will tend to lose quality if kept more than two or three years in the border, so regular sowings should be made. Sow in a sandy compost and cover the seed very lightly. Prick out the seedlings into deep boxes, or to a partially shaded position in the open, for aquilegias enjoy similar treatment as do pansies and polyanthus.

There are now a number of new strains, with the long-spurred bell-shaped blooms borne on 3 ft. (90 cm) stems. Outstanding are the 'McKana Giant Hybrids' from America, the flowers being 4 in. (10 cm) across with the spurs 4 in. (10 cm) in length and embrace every imaginable colour. Where a self colour is required, 'Crimson 'Star', 'Copper Queen' and 'Blue King' may be expected to come reasonably true from seed. This plant will be quite happy in a partially shaded corner or planted beneath young trees.

For the alpine garden is *A. alpina* 'Hensol Harebell' with its rich purple bells, a plant which will come true from seed, whilst a rare, but easily cultivated species is *A. bertoloni* which bears its blue and white flowers on 3 in. (7.5 cm) stems. The scarlet and golden blooms of '*A. canadensis* are also outstanding on the rockery and if possible plant them with the dark crimson-flowered *A. rubicunda*. A magnificent species is *A. scopularum* which produces its large spurred flowers of sky blue on 3 in. (7.5 cm) stems from June to September.

ARABIS

So common is *A. albida* with its grey foliage and masses of pure white flowers, that the arabis has recently been looked upon to be more useful for covering an unsightly bank than to be planted alongside more refined plants on the rockery. There are, however, a number of charming species with a neat and equally free-flowering habit which should not be neglected. One is *A. coccinea* which bears masses of red flowers throughout summer, the plant rarely reaching a height of more than 3 in. (7.5 cm). Plant with it *A. carducharum* of similar habit but bearing white flowers. *A. sundermannii* bears rich rose-coloured flowers on equally dwarf stems.

ARALIA

Native of China and known as the Angelica Tree, it makes a spreading tree 30 ft. (9 m) tall with large compound leaves 2 ft. (60 cm) long and the same across. The tree has spiny stems and bears large panicles of white flowers in August and September. It does best in a moist soil and is propagated from suckers.

A. chinensis has stout spiny stems and is valuable for its freedom of flowering at an unusual time.

ARALIA, *see also* Fatsia

ARAUCARIA

A genus of tall trees, native of South America, *A. araucana* being known as the Monkey Puzzle or Chile Pine. Of extreme hardiness, its branches radiate from the main trunk whilst the scale-like leaves, borne along the branches are prickly pointed and spirally arranged so that monkeys are unable to climb the tree. The large globular cones are of about 6 in. (15 cm) diameter and are usually borne separately (males and females on different trees). Though often planted in town gardens early in the century, they are rarely seen to advantage, preferring a clean, moist atmosphere. They grow better in western Britain. Plant in April in a deep, well drained loam. Propagation is by seed sown when ripe, under cloches. The young plants are moved to their permanent quarters when two years old.

ARBOR-VITAE, *see* Thuya

ARBUTUS

The Strawberry Tree, one of the most handsome of all trees, bears orange-red edible fruits like those of the strawberry. It belongs to the Heather family and requires an acid soil. It is highly tolerant of exposed coastal areas, especially of southern Ireland. It has dark glossy leaves and is evergreen. The flowers are white and pitcher-shaped whilst the fruit is borne on the previous season's wood and on the flowers. Propagate by seed or cuttings which should be rooted in small pots as the arbutus does not like root disturbance.

A. andrachne has attractive crimson stems and bears white flowers in spring; whilst *A. unedo*, native of Killarney bears its flowers and fruits at the same time, in early autumn.

ARCTOTIS

The African Daisy like all South African annuals requires an open, sunny situation and a dry, poor soil. Sow the seed in gentle heat early in the year and plant out 8 in. (20 cm) apart early in June, after hardening. The best strain is the 'Harlequin', the plants growing 12 in. (30 cm) tall and bearing their large daisy-like flowers in orange, red, apricot, yellow and intermediate shades and zoned with contrasting colours.

ARISTOLOCHIA

A. sipho is known as the Dutchman's Pipe on account of its peculiar curved flowers which are bent like a siphon. This vigorous climber will twine up a lath frame to a height of 18 ft. (5.4 cm), the yellowish-brown flowers appearing during June from the axils of the large oval leaves.

Planting should be done in March, in a sandy soil containing peat or leaf mould, and the plants appreciate a midsummer mulch. They require plenty of room and pruning should consist of the removal of any long, straggling shoots after flowering.

ARROWROOT PLANT, *see* Maranta

ARMERIA

A. cephalotes is the perennial evergreen thrift with its dense grass-like foliage and is a valuable plant for the front of a border for not only is its foliage evergreen, but it bears masses of tiny rounded flowers on 18 in. (45 cm) leafless stems, during April and May. The variety 'Bee's Ruby' bears a much deeper pink bloom than the type. It will flourish in ordinary loam and is quite happy in a sandy soil. Plant 12 in. (30 cm) apart in November and propagation is by division of the roots at this time.

ARTEMESIA

A genus of two hundred species of hairy herbs or sub-shrubs with alternate pinnately dissected leaves and bearing their flowers in drooping racemes.

SPECIES

Artemesia abrotanum. The Lad's Love or Southernwood, it has attractive fern-like leaves which retain their rich, pungent aroma long after they are dried. Mixed with leaves of the Cotton Lavender, it may be placed amongst clothes to keep away moths, whilst it is used to form the base of most pot-pourris.

In a shady situation, it will make a plant 3–4 ft. (about 1 m) tall and will grow to a similar distance in width. Both the stems and leaves are covered with silvery hairs which give it a most attractive appearance in the garden, whilst the yellow flowers, the size of buttons, are produced in close panicles. The form 'Lambrook Silver' is more heavily silvered,

the foliage being enhanced by cascading flowers of the colour and appearance of mimosa.
A. absinthium. The bitter Wormwood is a handsome plant of the roadsides of the British Isles with grey foliage and bearing heads of button-like yellow flowers. The stems and foliage are covered in silky down and the whole plant is extremely aromatic. It is used in the manufacture of absinthe and as a tonic beer. In earlier times it was used as a flea repellent and strewn over cottage floors for no insect would approach it.
A. argentea. Native of Madeira, it grows 12 in. (30 cm) tall, its finely divided leaves which are pleasantly aromatic, being densely covered in white hairs to give the appearance of filigree lace dusted with silver. The flowers are borne in July.
A. chamaemelifolia. A hardy perennial, its country name is Lady's Maid. It has attractive feathery foliage which is refreshingly aromatic, whilst the plants may be clipped to form a dwarf hedge to surround a bed of herbs.
A. dracunculus. It is known as Tarragon or French Tarragon though it is a native of Asia. Like Southernwood it loses its leaves during winter but is fully perennial and is happiest in a dry, poor, sandy soil. Seed is both difficult to obtain and to germinate and propagation is usually by cuttings which will readily root in sandy soil. The leaves, 'of the colour of cypress' possess a hot, rather bitter flavour and no more than two or three should be shredded and mixed with a salad. Or they may be used after being pickled in vinegar. Steeped in vinegar they provide a pleasing condiment to use with fried fish, whilst the best tartar sauce is made from tarragon leaves.
A. glacialis. A plant of charm, it is well-named for it appears through the melting snow in all its silvery beauty and is delightfully fragrant. It grows 6 in. (15 cm) tall.
A. lanata. A plant of spreading habit with silvered filigree foliage which gives it a moss-like appearance. It grows but a few inches high and keeping its foliage through winter is valuable for protecting early bulbs.
A. ludoviciana. Growing 30 in. (76 cm) tall, it is a border plant of beauty forming tall graceful spires of silvery-green, its leaves being indented at the end.
A. maritima. The Sea Wormwood is to be found on cliffs and coastal strips mostly in south-east England. It is a perennial plant, resembling *A. absinthinum* but is smaller in all respects with twice-pinnatifid leaves with numerous blunt segments and covered in down. The bright orange flowers are borne in heads and may be erect or drooping. The flowers are borne from July until September. The plant has a sweet aromatic scent.
A. odoratissimum. Native of south Europe, it grows 2 ft. (60 cm) tall

and has beautiful fern-like leaves of grey-green, covered in silvery hairs to give the plant a whitish hue from a distance. It blooms during the latter weeks of summer, bearing its milk-white flowers in crowded heads and they diffuse a delicious fruity scent for some distance around.

A. onthifolia. It forms a graceful pyramid 4 ft. (1.2 m) high, its grey-green leaves being divided into thread-like segments. It bears panicles of large white flowers late in summer.

A. pedemontana. A tiny sub-shrub which grows 4 in. (10 cm) tall and retains its silky silvery foliage throughout winter thus making it useful as an edging plant to provide contrast to brilliantly coloured salvias and geraniums, in summer, and dwarf tulips in winter.

A. pontica. This is the Roman Wormwood, used for flavouring brandy. An infusion of the flowers in summer provides a most health-giving drink. With its beautiful silvered filigree foliage, it is an attractive garden plant growing 18 in. (45 cm) tall.

A. schmidtii nana. A hybrid, it is a valuable rock garden plant for it is of creeping habit, its beautiful silver, crimson in autumn, foliage being strongly aromatic. In summer, the plant has the appearance of a frosted web. (see plate 2).

A. 'Silver Queen'. A hybrid of outstanding beauty and a valuable border plant growing 2 ft. (60 cm) tall. It has deeply cut silvery-grey foliage and is strongly pungent.

A. splendens. With its dense growth, silvered foliage and trailing habit, it has the appearance of a laced cushion lying on the ground. The leaves are finely divided whilst they grow so closely together as to completely hide the soil beneath them.

A. stelleriana. Native of Siberia, it is an artemesia of unusual form for it can be made to trail over a wall or against a stake, like a cascade chrysanthemum and having a silvery-grey lobe-shaped leaf of similar shape. It bears pale yellow flowers in summer whilst its foliage has a fruity pungency.

A. tridentata. Known as the Sage Brush, it is a handsome evergreen plant with fragrant foliage like that of the sweet briar. The leaves are thick and velvet-like and covered with grey down, like those of sage, whilst it bears small daisy-like flowers in autumn. It grows 6–7 ft. (about 2 m) tall and is native of the west states of America.

A. villarsi. It makes a graceful compact bush 18 in. (45 cm) tall with the appearance of a tiny grey cypress tree. Its foliage is most pungent.

A. vulgaris. Mugwort, an aromatic plant of the English countryside. It grows 3–4 ft. (about 1 m) tall and has purple stems with dark green leaves, downy on the underside. The yellow flowers are borne in dense spikes during September. It was a favourite plant with wayfarers during olden times for it was supposed to prevent fatigue. There is an old saying that mugwort placed in the shoes will keep a man from weariness, 'though he may walk forty miles a day'. It was also considered a valuable food for poultry, horses and goats.

ARTIFICIAL LIGHTING

Flowering plants will respond to the lighting of a greenhouse during the winter months in the same way that hens do when their houses are lit in winter, by laying more eggs. Chrysanthemums and pelargoniums especially, respond to artificial lighting. Where a temperature of 50°F. (10°C.) can be maintained and the plants given an occasional watering with liquid manure to keep them growing, the lighting is switched on for six hours each day from 6.00 p.m. beginning in November. By the month's end, pelargoniums given this treatment will begin to form new flower buds and will shortly afterwards, come into bloom. Though the quality of bloom and depth of colour is not the equal of summer flowering plants, the display will provide adequate returns for the small sum spent on the lighting and the plants will remain in bloom throughout winter. They should be given a rest period during February and March and be cut back, before being brought into new growth again. Do not dry off but reduce watering to a minimum whilst resting.

By the use of artificial lighting, chrysanthemums may be made to produce bloom the whole year. Certain varieties respond to different degrees of treatment so that it is possible to have the plants in bloom at a given date by varying the hours of lighting. In general, the mid-season and late varieties make most growth during the time of long daylight hours and as the days begin to shorten, they form their flower buds. With most varieties (there are a few exceptions), the buds form when daylight falls below thirteen to fourteen hours, with the result that they form their buds early in autumn and most have finished flowering before Christmas. Thus, by providing the plants with lighting, the 'daylight' hours may be extended and the blooming delayed until the requisite time. 100-watt tungsten filament lamps are spaced 6 ft. (1.8 m) above the plants and to have chrysanthemums in bloom throughout the year, they will require two additional hours of lighting in April and May, and also in August and September three hours in October and March; four hours in November and February; five hours in December and January. As with tomatoes, the periods of darkness should be seven to eight hours daily, the method being to use a time switch to give alternating periods of light and darkness throughout.

To bring the plants early into bloom, during the period April-August, black polythene sheeting is used to induce them to form flowering buds.

The sheeting is fixed inside the roof of the house and is drawn across at 6.00 p.m. each day and drawn back at 7.30 a.m. Or the sheeting may be kept in place longer if the buds are late in forming.

The African Violet or Saintpaulia will also bloom well under artificial lighting. Indeed, it may be grown entirely without seeing daylight. The plants are grown in tiers or on shelves, like mushrooms in a cellar, the only light being fluorescent lighting. The plants are placed in their pots on trays covered with pebbles which are always kept moist and a temperature of 60°F. (16°C.) is maintained. Each tray is lighted by an 80-watt tube fitted to a 6 ft. (1.8 m) long reflector which is suspended 12 in. (30 cm) above the plants, one hundred of which are accommodated to each tray. The lights are switched on at 8.00 a.m. and turned off at 11.00 p.m. upon retiring for the night. With fifteen hours of artificial light, the plants will remain healthy and be continually in bloom. In this way, a cellar can become a valuable source of income.

Tomatoes may be made to bear heavier crops by artificial lighting and where this is used with soil heating, plants may be grown on to bear twenty or more trusses of fruit. By the use of mercury vapour lamps of 400-watts which are fixed at a height of 3–4 ft. (about 1 m) above the benches and 4 ft. (1 m) apart, seed may be sown early in December, a time when the shorter hours of daylight normally prevent the satisfactory raising of plants. With artificial lighting, the plants receive seventeen hours of what amounts to bright sunlight, out of twenty-four, the use of a time switch ensuring the necessary seven hours of total darkness. It must be remembered too, that vapour lamps also generate heat and where used with soil heaters or where the air temperature of the greenhouse is artificially raised, this will be adjusted accordingly by thermostat control. The result will be a saving in heating costs.

Even earlier crops may be obtained by the 'double lighting' technique. Seed is sown under vapour lamps during the last days of October. At the end of December they are planted in soil-heated troughs and a month later, a further four to five weeks of lighting, seventeen hours a day is given during which time the plants form their first truss of fruit which will ripen by early April. The plants are grown on until they have produced twenty or more trusses, until the end of October when the house is cleared and cleaned and the procedure begins again.

Additional cropping may be obtained from cucumbers in the same way. By sowing late in November under vapour lamps, the first fruits will be ready before the end of February. The plants will be short-jointed with a good rooting system and with a rapid break of laterals. Cucumbers do not need a daily rest period, and may be given twenty-four hours of lighting.

ARUM (LILY), *see* Zantedeschia

ARUNDO

A. conspicua is the Giant New Zealand Reed, and bears cream, pink-tinted plumes. It is similar to the Pampas Grass though it is larger and more free flowering. It likes a moist, rich soil, where at the back of a large border it will give an added splendour to that part of the garden. Particularly attractive is it when planted against a background of dark trees. Give the roots a heavy mulch in November and disturb as little as possible. The Arundo bears its graceful plumes from early August until November and is most attractive growing near tall purple michaelmas daisies.

ASCLEPIAS

In a sunny position, and in a light, sandy soil enriched with some peat and decayed manure, the tuberous-rooted Butterfly Weed from North America, *A. tuberosa,* is one of the most showy of plants. From the end of July until mid-September it bears corymbs of tiny vivid orange starry flowers. The tubers should be planted 6 in. (15 cm) deep and it is advisable in the less favourable districts to cover the crowns with ashes during winter. Plant in spring, allowing 18 in. (45 cm) between the plants. Propagate by division of the tubers.

ASH, *see* Fraxinus

ASPARAGUS, DECORATIVE

There are two well-known species used for decorative purposes and readily raised from seed. *A. sprengeri* bears shoots up to 4 ft. (1 m) in length and which are covered in small, bright green spine-like leaves, whilst *A. plumosus* is an excellent greenhouse plant of semi-climbing habit sending out its shoots up to 8 ft. (2.5 m) and bearing flat, frond-like leaves or sprays which are widely used by florists for mixing with carnations and sweet peas. The form *A. plumosus nanus* is a plant of compact habit and makes a most attractive pot plant with its pale green feathery foliage. Those plants of trailing habit are best grown in hanging baskets or draped over wires fixed to the roof and extending the full length of the greenhouse. The plants are evergreen and continue to produce their shoots for many years if grown well. They provide a pleasing background for other greenhouse plants whilst they will provide 'fern' for mixing with cut flowers for the home.

As with the Grevillea, the seed must be fresh and this is imported each year in large quantities from Italy. Also, like the Grevillea, it requires a temperature of between 60°–65°F. (16°–18°C.) in which to germinate. March is the best time to sow. Where there is not sufficient heat, sowing should be delayed until early June, covering the boxes with glass, natural warmth being sufficient to cause germination.

All members of the asparagus family love loam of a fibrous quality and they prefer leaf mould to peat, so where possible make up one's own compost, using 2 parts loam, 1 part leaf mould and 1 part coarse sand and make sure that the compost is sterilized before sowing. The seed is large and presents no difficulty in its handling. It should be lightly, but completely, covered, whilst if the compost is kept comfortably moist, germination will begin in about three weeks from sowing.

The seedlings should be transplanted when large enough to handle, moving them to 3 in. (7.5 cm) pots containing the John Innes Potting Compost in which leaf mould has been substituted for the peat.

During summer, the plants will require ample supplies of moisture whilst they must be shaded by whitening the glass. The young plants will make rapid headway and will prove to be gross feeders, so as soon as the pots are filled with roots, they should be transferred to size 48 in which they will remain. Here, the John Innes No. 2 Compost should be used, against using leaf mould of good quality in place of the peat.

The plants will require a winter temperature of between 48°–50°F. (9°–10°C.) during which time they will require very little moisture though at no time must they be allowed to become dry at the roots or the foliage will turn yellow and the 'spines' will drop.

During summer, water copiously and feed once each week with dilute liquid manure. On all suitable occasions admit as much fresh air as possible and damp down the greenhouse whenever a hot sun is shining. The plants should also be syringed daily. In fact, everything should be done to prevent a dry atmosphere. The plants will be kept vigorous and the foliage fresh by careful feeding and by judicious cutting of the fronds.

ASPHODELUS

An uncommon plant, it is in bloom from early June until September bearing its stately eremurus-like spikes of bright golden-yellow flowers above tufts of grass-like foliage. The plants thrive in ordinary well-drained soil and are ideal for the back of a small border. Plant in November, 2 ft. (60 cm) apart and propagate by root division.

ASPIDISTRA

A genus of six species, native of south-east Asia and taking their name from the Greek, *aspidiseon* a little shield, an allusion to the flat shield-like style which covers the aperture formed by the six perianth-leaves. They are herbaceous plants with thick creeping rhizomatous roots from which arise smooth glossy dark green leaves about 15 in. (38 cm) long and which have long stalks. The flowers appear just above the surface of the soil and are bell-shaped, purple or yellow in colour. The plants are not frost hardy but where grown indoors they will remain healthy for many years in the dry atmosphere of a sitting room and provided the temperature does not fall to freezing point. For a dull room or for a window facing north, they are without equal in their tolerance of such conditions.

The plants like to have their roots confined in a restricted area and should be given a size 48 pot. This will cause the roots to send up a succession of new leaves. The plants require a compost made up of 2 parts fibrous loam and 1 part each leaf mould and sand. Plant the roots 2 in. (5 cm) deep and make the compost firm about them. During summer they are copious drinkers but in winter, give only sufficient moisture to keep them alive. If given too much water at this time the foliage will turn yellow. Each year top dress with a mixture of fresh loam and decayed manure which will do much to maintain the constant formation of fresh leaves of rich colouring. They will benefit from regularly wiping clean with lukewarm water to remove deposits of soot and dust. This will also prevent the appearance of red spider.

SPECIES AND VARIETIES

Aspidistra lurida. Native of China and Japan, it is the species grown in most Victorian homes and made famous in the song by Miss Gracie Fields. It has thick rhizomatous roots from which arise lance-shaped leaves of dark glossy green 18 in. (45 cm) long. The variety *Variegata* has silver streaks running from base to apex and a well grown plant has an appearance of great beauty. An occasional watering with dilute liquid manure will greatly enhance the colouring and health of the leaves.

ASTER

A. AMELLUS

This is one of the oldest of plants and requires different cultural treatment than other perennial asters. The plants grow to a height of about 20 in. (50 cm) and bloom from mid-August until October. Besides their value for planting towards the front of the border where their grey-green foliage is as attractive as their flowers, they are one of the most valuable plants to provide cut bloom, having quite a different habit from other asters. Being natives of the Mediterranean area, the plants are not quite so hardy as the others. They require a soil retentive of summer moisture, but one well drained in winter. A light soil well enriched with humus is most suitable. They like nitrogenous manure, and not being robust growers there is little need to worry about them making too much foliage even with an excess of nitrogen. Wool shoddy, hops, and decayed farmyard manure are all suitable. Nor must the plants be allowed to lack lime. Soil which may tend to be of an acid nature must be given a liberal dressing during winter. An open situation suits these asters best and being of compact habit they do not require staking.

This is one plant, scabious is another, which will not tolerate autumn planting. April, when established plants may be divided, is the only time to plant, when new growth is beginning. Plants set out in autumn will almost certainly die off, especially during a severe winter and this fact has greatly reduced the popularity of this fine cut flower.

When growing in beds for cut bloom, plant 16 in. (40 cm) apart in the rows and allow a similar distance when planting in the border.

PROPOGATION OF A. AMELLUS

Plants of the *amellus* section are not as easily propagated by division as are other members of the aster family (see plate 3), for frequently they are to be seen bearing their bloom from a single main trunk. A better method is to take cuttings in early May, when the new shoots are about 4 in. (10 cm) long and to root them in frames as for michaelmas daisies. They should remain undisturbed until the following April when they are planted out.

VARIETIES OF A. AMELLUS

BESSIE CHAPMAN. Strong growing, the bloom is large and of deepest violet-blue.

BLUE STAR. A recent introduction of compact habit and bearing flowers of a clear mid-blue colour.

JACQUELINE GENELRIER. The large blooms are of a lovely shade of soft orchid-pink. Mid-season.

KING GEORGE V. An old favourite, but owing to its freedom of flowering and its large deep blue flowers, is still one of the best of the blues.

LADY HINDLIP. A lovely variety, the bloom being of deep rosy-pink, most pleasing under artificial light.

MAUVE BEAUTY. The very large blooms of deep lilac-mauve are produced over a long season.

MOERHEIM GEM. One of the best varieties, the bloom being of a deep purple-blue colour.

NOCTURNE. A new variety and a new colour in this section. The blooms are large and of a bright rosy-lavender colour. Early.

RED FIRE. A continental variety and the nearest to a crimson-red. Late.

SONIA. Very free flowering, the bloom being of a delightful shade of deep pink.

ULTRAMARINE. The flowers are small, but produced in great profusion and are of a striking shade of mid-blue.

VANITY. The large flowers are of deepest violet-blue. Mid-season.

ANNUALS

The value of this plant for bedding and for cutting is enormous, there being something to suit every taste. Seed may be sown either in gentle heat or in a cold frame in March. Germination and plant growth is rapid and the plants will be ready to go out at the end of May, after the usual transplanting and hardening. A sowing may also be made under cloches early in April or in the open ground at the month end. The aster is a late summer flowering plant and so has plenty of time to make growth through the early summer before coming into bloom towards the end of July. As the plants are of bushy habit allow 9 in. (22 cm) spacing and they like a soil containing plenty of humus.

One of the best asters for bedding is the new 'Dwarf Queen' strain, wilt-resistant, and making a short, bushy plant. It is obtainable in rose, crimson, white, blue and mauve. It is the first aster to come into bloom.

Equally early and for edging is 'Thousand Wonders' which grows 6 in. (15 cm) tall and makes a rounded bush 12 in. (30 cm) across, covered in fully double blooms 3 in. (7.6 cm) across and of deep rose-pink. Compact and dainty too, is 'Daisy Mae' which bears single flowers with a double row of petals, of soft rosy-purple and blooms from July until September.

With their tiny, button-like flowers obtainable in the usual aster colour range, the 'Lilliput' asters, growing to a height of 15 in. (38 cm) and with a branching habit, are excellent for bedding and valuable as pot plants. Of similar form are the new 'Remo' asters with their very rich colours, whilst the dwarf 'Waldersee' strain flowering over a long period at a height of 9 in. (22.8 cm) is most attractive. All these button asters look most delightful when edged with the silvery-grey leaf *Cineraria maritima*.

Also suitable for bedding are the 'Ostrich Plume' asters with their long, twisted petals providing a shaggy appearance. The blooms are large and produced with freedom.

For cutting, outstanding is 'Pink Magic' with its fully double flowers and long quilled petals. It is of a warm shade of salmon-pink. Excellent too, are the 'Duchess' asters in yellow and crimson. Growing 2 ft. (60 cm) tall, the large incurving flowers are like Japanese chrysanthemums.

A. NOVAE-BELGII (Michaelmas Daisy)

There has been seen no greater advancement amongst hardy border plants during the past quarter of a century than that concerning the perennial aster, or michaelmas daisy as it is more affectionately known. Until the early 1930s the only plants to be found in gardens grew to a height of 5 ft. (1.5 m) and covered themselves in tiny pale mauve blooms. The plant had its value because it would grow anywhere and there were few autumn flowering plants at that time. Then came the late Mr Ernest Ballard's lovely introductions which brought about a new popularity in this flower, and for the first time it was realized that it could be capable of some exciting things. Amongst Mr Ballard's introductions was the tall purple-blue 'Princess Marie Louise'; the deep rose-pink 'Miss Muffet'; and 'Dazzler' with its blooms of brilliant rosy-red, a new colour break in this flower. Then followed the first of the 'Gayborder' varieties, the semi-double pink 'Gayborder Supreme', with its neat habit being one of the finest border and cut flower plants ever introduced, and still a favourite. In the immediate post-war years came Mr Ballard's famous trio 'Peace', 'Prosperity' and 'Plenty' with their enormous but refined blooms and with a depth of colour never previously seen. They were to bring the same popularity to this plant as that enjoyed by the delphinium and lupin.

The culture of *A. novae-belgii* presents few difficulties. The michaelmas daisy is a moisture loving plant and like the phlox is always at its best in a heavy, humus-laden soil, or where growing in a light soil blooms at its best only during a wet season. A light soil should have some humus incorporated and this may take the form of material from the compost heap or used hops, peat or leaf mould. Where the soil is not retentive of summer moisture, growth will be stunted and the blooms small.

Planting may be done at any time from the end of October until the end of March, but as the plants must not lack moisture during their first weeks in the border, where the soil is light, November planting is to be preferred. A dry spring could prevent the plants from getting away to a good start which would mean that they would provide little colour their first year. Set the plants 2 ft. (60 cm) apart where planting at the back of the border; 18 in. (45 cm) apart where planting towards the front, spreading out the roots. Plant deeply so as to prevent the roots drying out by strong winds, or from being lifted from the ground by early frosts before they have taken a firm hold. Where the border is exposed, it is preferable to plant those varieties of more compact habit, using those which grow only to a height of 3–4 (about 1 m) at the back. The mid-border plants will require only the minimum of staking, just sufficient to prevent the plants from 'spreading' when in full bloom when their rich colourings would not show to the best advantage. A strong stake should be placed at the back of the plant so that it remains unseen. Around the plant, green twine is loosely tied. The stems should in no way be bunched together. The more compact front of the border varieties should not require supporting.

PROPAGATION OF A. NOVAE-BELGII

The plants should be divided every three to four years. It is possible to divide the plants without lifting them entirely. Merely remove pieces with a trowel. When planting in groups of three, one plant may be lifted and divided in November after two years, another after three years and the other, the following year, so that at no time will there be any shortage of bloom in the border, nor will there be any plants which have lost vigour. When dividing older plants it is the outside parts of the plant which are the most vigorous, the woody centre part should not be retained.

Older plants are divided by the usual method of using two forks, but this is necessary only where there is a large amount of old wood at the centre. Plants which are lifted and divided frequently, may readily be split up into small offsets. These may be grown on in a bed for twelve months before re-planting in the border. When required for immediate re-planting the roots should be divided into only three or four pieces, each containing several new shoots.

Where it is considered necessary to increase a new variety quickly, cuttings 3 in. (7.5 cm) in length may be taken early in May and inserted into a compost of peat and sand in a frame.

BACK ROW VARIETIES (3½–5 ft.: 1.0–1.5 m) OF A. NOVAE-BELGII

BLANDIE. With its semi-double flowers of ivory-white it has become the most popular white for the back of the border. Mid-season. 3–4 ft. (about 1 m).

BLUE GOWN. A clear mid-blue and valuable in that with 'Hilda Ballard' and 'Stella Lewis' it is the last to come into bloom. Late. 4–5 ft. (about 1.5 m).

GAYBORDER BLUE. The best mid-blue for the back of the border bearing very large flowers. Mid-season. 4–5 ft. (about 1.5 m).

GAYBORDER SPIRE. The beautiful mauve-pink flowers are produced in large pyramids. Late. 4–5 ft. (about 1.5 m).

GAYBORDER VIOLET. The deep violet-purple blooms are produced in profusion. Mid-season. 3–4 ft. (about 1 m).

HILDA BALLARD. The blooms are large and of a beautiful petunia-purple colour. Late. 3–4 ft. (about 1 m).

MAID OF ATHENS. An old variety, but still a valuable back row plant on account of the pure shell-pink colouring of the blooms. Early. 4–5 ft. (about 1.5 m).

MELBOURNE MAGNET. The fully double blooms are of a soft shade of lavender-blue. Late to bloom. 3–4 ft. (about 1 m).

PICTURE. A new variety, the large semi-double blooms are of a beautiful shade of carmine-pink. Late. 3–4 ft. (about 1 m).

PRINCESS MARIE LOUISE. An old favourite, bearing flowers of pure sky-blue. Early. 4–5 ft. (about 1.5 m).

STELLA LEWIS. The blooms are of a deep pink flushed with cerise. Late. 4–5 ft. (about 1.5 m).

TAPLOW SPIRE. The deep rose-pink blooms have an attractive golden centre. Late. 4–5 ft. (1.5 m).

MID-BORDER VARIETIES (3½–4 ft.: 1 m). OF A. NOVAE-BELGII

DORIS'S DELIGHT. The semi-double blooms are of a beautiful shade of petunia-crimson. Mid-season. 3 ft. (90 cm).

ERNEST BALLARD. The semi-double flowers are of a deep crimson-pink of enormous size. Mid-season. 3¼ ft. (1 m).

GAYBORDER SUPREME. Excellent for cutting, the semi-double blooms are of a bright rose-pink. Mid-season. 3½ ft. (1 m).

HARRISON'S BLUE. A fine variety, the blooms being of a deep amethyst-blue colour. Late. 3½ ft. (1 m).

JANET MCMULLEN. The deep rose-pink flowers are of exceptional size and quality. Mid-season. 3¼ ft. (1 m).

JUST SO. A most attractive variety for flower arrangement, bearing large semi-double blooms of a unique shade of sky-blue flushed with rose. It makes a short plant and is of bushy habit.

LADY PAGET. The blooms are not large but of an interesting shade of orchid-pink. Late. 3½ ft. (1 m).

MARIE BALLARD. Of bushy habit with plenty of healthy foliage, the large fully double blooms are of a lovely shade of soft powder blue 3½ ft. (1 m).

MODERATOR. A fine new variety, its large fully double blooms being of a brilliant violet-colour. Late. 3½ ft. (1 m).

ORLANDO. A new variety bearing semi-double blooms of clear rose-pink. Early 3½ ft. (1 m).

PEACE. The rose-mauve flowers are huge, with shaggy petals. Mid-season. 3½ ft. (1 m).

PERSUASION. Very free flowering, the large double blooms are of soft sky blue with a purple cast at the centre. Lovely under artificial light.

PETUNIA. An older variety unsurpassed in its colour. The semi-double blooms are of petunia-red. Mid-season. 3½ ft. (1 m).

PLENTY. The enormous flowers are of a clear mid-blue. Mid-season. 3½ ft. (1 m).

ROMANCE. The small semi-double blooms are borne in great profusion at a height of 3½ ft. (1 m) and are of a soft salmon-pink flushed with gold at the centre.

THE ARCHBISHOP. One of the famous 'clerical' varieties, all of which grow to a height of just under 4 ft. (1m) and are ideal back row plants for a small border or for an exposed garden. They are amongst the loveliest of all michaelmas daisies. This variety bears a huge shaggy bloom of deep violet, very like an annual aster. Early.

THE BISHOP. The large semi-double blooms are of a dusky purple-plum colour with a yellow eye. Mid-season. 3¾ ft. (1 m).

THE CARDINAL. Very free flowering, the large blooms are of a deep rosy-carmine shade. Mid-season. 3¾ ft. (1 m).

THE DEAN. The huge warm cerise-pink blooms are produced in abundance and have shaggy petals. Late. 3¾ ft. (1m).

THE RECTOR. The very large blooms are of a rich claret-red colour, making it one of the best of all varieties. Mid-season. 3½ ft. (1m).

THE SEXTON. The large blooms are of a beautiful luminous blue colour. Early. 3¾ ft. (1 m).

TWINKLE. The large refined blooms are of a lovely shade of cyclamen-purple. Early 3¼ ft. (1 m).

FRONT OF THE BORDER VARIETIES (2–3 ft. : 0.75 m), OF A. NOVAE-BELGII

The modern trend being for small gardens and plants which do not require staking, some of the best of modern asters are of dwarf, compact habit and have dark glossy green foliage which is highly resistant to mildew.

BEECHWOOD CHALLENGER. The crimson-red blooms are small, but with attractively quilled petals. Mid-season. 2 ft. (60 cm).

BONINGALE WHITE. The best white variety, the large double blooms being of glistening white and enhanced by the healthy dark green foliage.

CHARMWOOD. One of a new trio of a similar habit to 'Winston Churchill', being excellent for cutting. The blooms are of a vivid rose-pink colour. Late. 3 ft. (1 m).

CHARTWELL. The richly coloured blooms are of deep violet-purple, flushed with crimson. Late. 3 ft. (90 cm).

CHEQUERS. Of similar habit to 'Winston Churchill', the flowers are of richest violet-purple. Early. 2 ft. (60 cm).

CRIMSON BROCADE. It makes a bushy plant 3 ft. (90 cm) tall and covers itself in masses of crimson-red flowers which are borne over a long period.

EVENTIDE. An outstanding variety, bearing large flowers of a deep purple-blue colour. Early. 3 ft. (90 cm).

FONTAINE. Similar in habit to 'Tapestry' and just as lovely. The large shaggy blooms being of a lovely shade of tawny-pink with an orange eye. Very early. 2¾ ft. (80 cm).

GAYBORDER ROYAL. Best described as doge-purple, the blooms are of exquisite form. Mid-season. 2½ ft. (75 cm).

JOHN SHEARER. A new variety bearing a bloom of deep wine-blue. Early. 3 ft. (90 cm).

LASSIE. The semi-double blooms with their conspicuous yellow centre are of a pure shade of apple-blossom pink. Early. 3 ft. (90 cm).

LITTLE JOHN. A compact new variety bearing blooms of a pleasing shade of mauve-pink. Early. 2 ft. (60 cm).

MELBOURNE EARLY RED. A link with 'Little John' and the shell-pink 'Angela Peel', between the dwarf hybrids and those of taller habit. The garnet-red flowers are produced with freedom. Early. 2 ft. (60 cm).

PRUNELLA. Raised at Portsmouth by Mr M. J. Todd, it makes a bushy plant only 2 ft. (60 cm) tall and bears large double blooms of claret-purple.

RED SUNSET. Of more compact habit than the earlier 'Beechwood' varieties, the petunia-crimson blooms are borne in great profusion. Early. 2½ ft. (75 cm).

ROYAL VELVET. Very new and with its bright violet blooms, freedom of flowering and resistance to mildew is certain to become a favourite. Late. 3 ft. (1 m).

TAPESTRY. Ernest Ballard's favourite. The habit is ideal, requiring no staking; the huge shaggy blooms being of a glorious shade of pastel-pink. Very early. 2½ ft. (74 cm).

VIOLET LADY. Of bushy habit, the blooms are borne in profusion and are of a deep violet-blue colour. Early. 2 ft. 60 cm).

WINSTON CHURCHILL. A truly magnificent variety, the blooms are not large, but borne in profusion and are of a vivid shade of beetroot-red. Mid-season. 2¾ ft. (0.8 m).

DWARF HYBRIDS (9–18 in.: 22–45 cm). OF A. NOVAE-BELGII

These are ideal plants for the very front of a border or for autumn bedding. The more dwarf growing are ideal plants for a rockery.

AUDREY. The large blooms are of a delicate shade of pastel-blue which completely smother the plant. Mid-season. 15 in. (38 cm).

AUTUMN PRINCESS a new variety, the semi-double blooms being of a brilliant lavender-blue. Early. 15 in. (38 cm).

CHATTERBOX. Throughout autumn it bears its semi-double blooms of soft lilac-pink in a beautifully rounded bush. (30 cm).

HEBE. Not nearly so well-known as it should be. The bloom is of a rich shade of strawberry-pink. Mid-season. 10 in. (25 cm).

JENNY. The best 'red' dwarf, growing 15 in. (38 cm) tall and bearing masses of semi-double blooms of brilliant crimson-red.

LADY IN BLUE. A new variety of compact habit and bearing masses of semi-double mid-blue flowers. Late. 9 in. 22 cm).

LITTLE BLUE BABY. A perfect miniature, the tiny rounded plants being smothered in sky-blue flowers. Early. 6 in. (15 cm).

LITTLE RED BOY. Until there is a real crimson variety this, with its rose-pink blooms is a good substitute. Mid-season. 12 in. (30 cm)

NIOBE. The single pure white blooms are borne in profusion. Mid-season. 9 in. (22 cm).

PINK LACE. A new variety likely to become a favourite, the double blooms being of a lovely shade of clear pink. Mid-season. 15 in. (38 cm).

ROSE BONNET. Also new, the rounded plants are smothered in misty pink blooms. Late. 10 in. (25 cm).

FLOWERING TIMES OF ASTER NOVAE BELGII

Early	*Mid-season*	*Late*
'Chequers'	'Blandie'	'Blue Gown'
'Eventide'	'Doris's Delight'	'Charmwood'
'John Shearer'	'Ernest Ballard'	'Chartwell'
'Lassie'	'Gayborder Blue'	'Gayborder Spire'
'Little John'	'Gayborder Supreme'	'Harrison's Blue'
'Maid of Athens'	'Gayborder Violet'	'Hilda Ballard'
'Melbourne Early Red'	'Janet McMullen'	'Lady Paget'
'Orlando'	'Peace'	'Melbourne Magnet'
'Princess Marie Louise'	'Petunia'	'Moderator'
'Red Sunset'	'Plenty'	'Picture'
'Tapestry'	'The Bishop'	'Royal Velvet'
'The Archbishop'	'The Cardinal'	'Stella Lewis'
'The Sexton'	'The Rector'	'Taplow Spire'
'Twinkle'	'Winston Churchill'	'The Dean'
'Violet Lady'		

OTHER SPECIES AND VARIETIES

A. acris. In bloom from early August, it is of compact habit and is very free flowering, known to amateur gardeners as the Blue Starwort but sadly neglected by flower growers. William Robinson said that it was 'more precious than a gentian . . . the prettiest flower of its colour'. Growing to a height of about 20 in. (50 cm) the plant forms dense clouds of blue, 'a very poem of flowers' it has been called. It remains in bloom until well into autumn and is a delightful plant for mixing with Korean chrysanthemum and gladioli. There is a white counterpart of the blue *A. acris*, 'Mrs Berkeley', which grows to a height of 15 in. (38 cm) and a dwarf form for the front of the border which bears rosy-mauve blooms on 12 in. (30 cm) stems.

A. alpellus. The result of a cross between *A. alpinus* and *A. amellus*, the variety 'Triumph' bears its violet-blue flowers on 12 in. (30 cm) stems from mid-June until early October.

A. cordifolius. Growing to a height of 3–4 ft. (about 1 m) it should be planted where its long arching sprays do not suffer from autumnal winds. It comes into bloom mid-September, and stays colourful until early November. The variety 'Silver Spray', is outstanding with the denseness of its long sprays of silvery-lilac, whilst 'Edwin Becket' bears sprays of lilac-blue.

A. ericoides. It is known as the Heather-leaf Starwort, on account of its small pointed leaves. The bloom is borne in long sprays throughout autumn and on 2½ ft. (75 cm) stems. Of several varieties, 'Brimstone' makes a sturdy bush smothered in sprays of tiny yellow flowers; 'Delight', bears pure white sprays; and 'Ringlove', sprays of rosy-lavender. A new variety, 'Golden Spray' is delightful used towards the back of the border where it reaches a height of 3–4 ft. (about 1 m).

A. hybridus luteus. A hybrid between *Aster acris* and the *Solidago* (Golden Rod) and possessing the best qualities of both plants. It is now known as *Solidaster luteus*, the densely packed sprays of golden-yellow appearing in early August and remains in bloom until late autumn. The sprays are superb when massed with late summer flowers.

A. laterifolius. Similar to *A. diffusus horizontalis* in that its bloom is borne on short stems which are horizontal to the main stem. Both species are most attractive for mixing and will find a ready sale with the florist.

A. laterifolius. 'Finale' bears 3–4 ft. (about 1 m) sprays covered in rich mauve flowers, whilst of the *A. diffusus* varieties, that with the peculiar name of 'Coombe Fishacre Hybrid', bearing clouds of rosy-lavender flowers is outstanding.

A. lynosyris. Often known as Aster Goldilocks, for it produces its golden flowers in great freedom, like *A. luteus*, on 2 ft. (60 cm) stems and during the same period.

A. novae-angliae. Owing to it being a difficult plant to hybridise this form has never attained the popularity of *A. novi-belgii*. It is a plant of looser habit and bears its blooms in terminal clusters so that the bloom is borne only on the very top of the plant. Growing to a height of 4–5 ft. (about 1.5 m) however, and being happy in a clay soil, the plant has a value for back row planting where the soil is not too kind. It requires the same treatment as *A. novi-belgii*.

BARR'S PINK. An old variety bearing bloom of a deep rose-pink over a prolonged period. Early. 3–4 ft. (about 1 m).

HARRINGTON'S PINK. Its blooms are of a purer shade of pink than any variety. Late. 3–4 ft. (about 1 m).

SURVIVOR. A new American variety bearing very bright rose-pink flowers. Late. 4–5 ft. (about 1.5 m).

TREASURE. The most compact variety and bearing a bloom of delicate lilac-mauve. Mid-season. 3 ft. (90 cm).

A. yunnanensis

NAPSBURY. Similar in habit and constitution to those of the *amellus* section. Demanding spring planting, it should be treated in the same way, the plants being set out 10–12 in. (25–30 cm) apart as it is of compact habit. The bloom is borne in profusion from late July and is of a deep purple-blue with a conspicuous yellow eye, making up a most striking bunch. Like the *amellus* asters it does not like unnecessary disturbance. A bed should be left down for four years and top dressed each June with well decayed manure.

ASTILBE

A. arendsii. It is the species from which the garden hybrids have been raised, the plants being noted for their compact plumes and freedom of flowering. The first varieties come into bloom towards the middle of June, the last flower in September, so that by judicious selection it is possible to have plants in bloom throughout summer. The plants seem to thrive in a moist soil, particularly one of an acid nature. This condition may be partly provided by packing peat around the roots at planting time. Though appreciating plenty of moisture during summer, for which reason this plant always does best in the damp climate of the north-west, it does not like a badly drained soil during winter. Plant late in autumn, allowing 2 ft. (60 cm) between the plants and propagate by root division.

ATTLEYA. It grows to a height of 3–4 ft. (about 1 m) and bears plumes of a delightful shade of orchid-pink.

DUSSELDORF. The bright salmon-coloured plumes are borne on 2½ ft. (75 cm) stems during July and August.

ERICA. A lovely variety, its compact spikes of clear pink are produced from early June until late July.

FEDERSEE. One of the few varieties which will flourish in a sandy soil. The rosy-red plumes are produced in profusion during July and August.

FIRE. The colour of the plumes is a bright salmon-scarlet freely produced through July and August.

GERTRUDE BRIX. Early to bloom, its plumes are of an intense crimson-red, a striking companion to 'White Gloria'.

IRENE ROTSEIFER. Growing to a height of 3–4 ft. (about 1 m) its long plumes of pale pink are produced during August.

JO ORPHURST. A very vigorous and late flowering variety bearing long plumes of ruby-red into September.

RED SENTINEL. A new variety bearing flowers of brick-red throughout July.

WHITE GLORIA. Its thick white plumes are borne during July and August and provide a striking contrast to the red varieties.

A. simplicifolia. With its long flowering season, forms of this lovely plant could be used to edge a border where the soil is moist or of an acid nature. The dwarf, but attractive, plumes are produced from mid-June until early September, the dainty, finely-cut foliage of the plant adding to its appearance.

ATROTSEA. Coming into bloom at the end of June, its glowing pink plumes are held on 12 in. (30 cm) stems.

DUNKELLACHS. A magnificent plant having dark green foliage and bearing a plume of lilac-pink.

A. praecose alba. In bloom early in June and bearing broad white heads, attractively tinged with green.

ASTRALAGUS

A. angustifolius is the Goat's Thorn, a dwarf, slow-growing shrub, native of the Mediterranean. It is covered with spines and has pinnate leaves and white pea-shaped flowers tinted with violet. They are borne in May and June. Growing only 15 in. (38 cm) tall, it requires a sheltered sunny position and a well-drained soil containing some lime rubble. Plant in November and propagate from seed or by cuttings rooted under glass.

ASTRANTIA

A. carnolica prefers a moist soil and partial shade where during July and August it will bear its nodding starry white flowers tinged with pink on 3–4 ft. (about 1 m) stems. The flowers are surrounded with green bracts which turn purple with age. Plant 2 ft. (60 cm) apart in November.

ATHROTAXIS.

The Cedar of Tasmania is a half-hardy conifer with scale-like leaves and which is slow growing. *A. cupressoides* is the best species for British gardens and should be confined to sheltered gardens in the west. It is a small erect tree with closely-imbricated leaves and bears handsome small cones.

Plant in April, into deeply dug ground containing peat or leaf mould. Propagate from seeds sown when ripe or by cuttings removed from the tips of the branches and inserted in sandy compost under glass.

AUBRIETIA

The most popular of all rock plants for the dry wall, paving and for the rockery. The aubrietias will grow almost anywhere but love best a sunny position and a soil containing plenty of humus. Wall or Rock Cress, as it is called, will bloom from April until mid-summer and will quickly form a large clump with its semi-trailing habit which should be kept tidy by trimming after it has flowered. Aubrietia is easily raised from seed but the named varieties should be propagated either by layering or from cuttings. Varieties are numerous, there must be at least a hundred, but some of the largest flowered and of the most outstanding colours are:

CARNIVAL. Rich deep purple.
GLORIOSA. Large silver-pink.
KELMSCOT GEM. Double brilliant scarlet.
LILAC TIME. Lovely lilac-mauve.
MRS J. BAKER. Lavender-blue with distinct white eye.
RED CARNIVAL. Bright crimson-red.
RUSSELL'S CRIMSON. Rich wine-red.
STUDLAND. Lavender-blue, a beauty.

AUCUBA

A. crotonifolia, with its large, yellow and green mottled leaves, is often called the Variegated Laurel although it is of no relation. It is valuable for planting with *Laurus rotundifolia,* for it provides a pleasant contrast to its vivid green foliage and is of similar habit. It grows well in all soils and under town garden conditions.

AURICULA

The auricula may be divided into three types: the Alpine, the Show and the Border. All are hardy perennials and in the open ground will eventually form large plants with broad ear-shaped leaves, either glossy grass-green in colour or silver-green where covered in farina. The main root from which the fibrous rootlets grow, is thick and woody and is known as a 'carrot' for it resembles the vegetable. At soil level, the carrot forms a thick neck around which the offsets form. To propagate, the offsets are removed, each with a few fibrous roots and these are grown on in small pots or are planted out in open ground beds.

The flower stems arise to a height of 6–9 in. (15–22 cm) at the end of which is borne, on short footstalks, the individual blooms known as 'pips'. Usually these have six petals but may be as many as nine, overlapping and symmetrical so that the bloom is flat and circular and about 1 in. (2.5 cm) in diameter.

ALPINE

The Alpine auriculas are to be distinguished by their freedom from meal which is usually present on the flowers and foliage of the show and border varieties. The alpines or 'Pures' as they were called, to distinguish them from those with meal are usually of more vigorous habit than the Shows and are not so demanding in their culture. They are best grown in the cold greenhouse or frame to protect the beauty of the blooms from adverse weather but being free from meal (which will 'run' if made wet) they may be grown, like the borders, as an edging to a path or to beds of roses. If grown in pots, they may be taken indoors just before the blooms begin to open.

The Alpines may be divided into two groups: those bearing a bloom with a white or light centre and those with a gold centre. The body colour, darkest round the eye, shades out to the petal edges. The gold-centred varieties are especially showy for the brightness of the gold centre is enhanced by the deep body colouring.

Early July is the best time to re-pot old plants or offsets. Use a small pot, scrubbed clean and a compost made up of 8 parts fibrous loam; 2 parts leaf mould or old mushroom bed compost; and 2 parts decayed cow manure. Add a small amount of powdered charcoal to act as a sweetener. Charcoal is able to absorb ammonia gases in the compost and convert it into nitrogen. Plants should be re-potted yearly so that the compost may be re-fortified and the roots inspected for pest or disease. The compost should be in a friable condition and the plant inserted up to the lower leaves. Keep the compost comfortably moist but not wet and during warm weather, spray the plants and protect from strong sunlight. This is important until the plants become re-established in the pots. If growing in a greenhouse or frame, ventilate freely at all times, except during damp, foggy weather. As they form their flower stems in spring, Alpines and Shows should be confined to a single flower truss, the main one, all others being removed. Only where there is over-crowding should any buds be pinched out but the Alpines produce a larger number of 'pips' than the Shows and to maintain the balance, it may be necessary to remove one or two.

As the flowers begin to open, water with care, so as not to damage the flowers. At this time it is important to shield them from the direct rays of the sun.

All types of auricula are readily raised from seed if freshly harvested seed is sown in April. Use a box or earthenware pan and the John Innes sowing compost, which should be freshly prepared. Sow the seed with finger and thumb, covering it only lightly and water the compost when necessary, from the base. The compost should not be allowed to dry out at any stage.

When the seedlings are large enough to handle, transplant to boxes, spacing them 1 in. (2.5 cm) apart and grow on outdoors or in a frame until ready for small pots in August. They will come into bloom the following April and May when all auriculas are in bloom.

VARIETIES OF THE ALPINE AURICULA

BOOKHAM FIREFLY. One of the most popular of the alpines since its introduction in 1913, the ground colour being glowing crimson shading to maroon and with a gold centre.

DOWNTON. An exhibitor's favourite, the velvety crimson ground colour shading to buff and with a gold centre.

GOLDEN GLORY. Raised by the late Mr R. H. Briggs, the rich golden-brown colour shades out to old gold at the edges and it has a small gold centre.

GORDON DOUGLAS. One of the easiest of the alpines, it forms a large truss, the flowers of deep violet-blue having a creamy-white centre.

LADY DARESBURY. Raised by Mr C. Faulkner, it is the finest light-centred alpine ever raised, the colour of rich wine-red, shading to cerise, the large blooms having eight petals.

MIDAS. The blooms are of the colour of brown ale, shading out to golden-brown and with a gold centre.

MRS HEARN. A beautiful alpine with a pale cream centre and ground colour of grey-blue, shading out to Cambridge blue.

SHOW

As a florist's flower, i.e. a flower for the specialist to bring to the highest Exhibition standard, the Show auricula is supreme, though is more difficult to manage than either the Alpine or Border type. To protect its paste-like centre which will 'run' like white water colour paint, if wet, the plants must be grown under glass and this has tended to weaken their constitution, though a number of varieties are a century old and are still in cultivation.

The Shows are divided into three sections: edged, selfs and fancies.

The degree of mealiness (paste) determines the colour of the edging. All edged auriculas are really green but the density of the minute hairs changes the green edging to grey or white. At the end of each hair is a small globe from which white wax or resin is secreted and the denser the concentration of hairs, the whiter their appearance as at the centre of the bloom. Between the centre and the edging is what is known as ground or body colour which accentuates the paste. The edged auriculas consist of green, grey, and white edged varieties.

The selfs are devoid of any meal at the edging though they have the paste centre which distinguishes them from the light-centred alpines. The paste is also found at the base of the leaves forming the calyx. The selfs have uniformity of colouring from the edge of the paste to the petal edge, the colour ranging from pale yellow to scarlet. The ground colour of the edged auriculas is black, brown, crimson or violet. The blooms diminish in size in ratio to the degree of intensity of edging. The green edged blooms are the largest, the white edged being the smallest. The green edged also make the largest flower truss.

The fancy auricula is rarely seen for it is not recognised on the show bench as are the other shows and alpines. In some fancies, the green edging continues to the paste centre; in others, there is a bright yellow zone between the centre and edging, whilst some may have a scarlet ground and a small green edge. All show auriculas have the paste-like centre and must be protected from rain and careless watering when in bloom. Otherwise their culture is the same as for the Alpine auriculas grown in pots.

Varieties of the show auricula:

GREEN-EDGED SHOW AURICULAS

BROCKENHURST. It forms a truss of six or seven pips which open together. The six petals are nicely rounded; the tube perfectly round; the paste smooth whilst the edging is brilliant green.

LINKMAN. One of the easiest greens to manage, distinguished by the wire-edge of black around the tube, whilst the black ground colour extends from the paste to almost the petal edges which are beautifully rounded.

LONGDOWN. One of the late Mr C. Haysom's introductions, forming a well-proportioned truss. The paste is dense, the tube round, the edging dark green with black body colour.

GREY-EDGED SHOW AURICULAS

LOVEBIRD. The blooms are small but the paste, ground colour and edging are well proportioned and clearly defined.

SEAMEW. The medium-sized blooms are beautifully rounded; the tube is free of notches, the paste smooth, with a small amount of clearly defined body colour.

SHERFIELD. The premier grey but difficult to manage. The tube is round, the paste smooth with the edging so densely covered in farina to give it almost a white edged classification.

WHITE-EDGED SHOW AURICULAS

ACME. Introduced a century ago, its bloom is the acme of perfection, the paste being dense and smooth with well defined body colour and heavy mealing at the edge.

DOROTHY MIDGELEY. The pip is large, the seven petals over-lapping with paste and body colour of clear outline. The well-formed truss numbers five to eight pips.

SILVERSLEY. The premier white and easily managed. The tube is perfectly round, the paste dense, the body colour black with the heavily mealed petals nicely rounded.

SELF-COLOURED SHOW AURICULAS

ALICE HAYSOM. One of the best of the modern selfs with a nicely rounded tube, smooth paste and bright red ground colour.

BLOXHAM BLUE. Raised by Rev. Oscar Moreton when vicar of Bloxham, Norfolk, and the best blue self with dense paste and body colour of deep violet-blue.

INNOCENCE. A pale yellow self of exquisite form with smooth paste whilst the pips make up a well-balanced truss.

MARY WINN. The best deep yellow self with a well-defined tube, smooth paste and making a beautifully rounded bloom.

OWER. It makes a large rounded truss with dense and clearly defined paste whilst the ground colour is clear golden-yellow. The foliage is heavily mealed.

ROSEBUD. It makes a handsome circular bloom whilst its rose-pink colouring is unique.

SCARLET PRINCE. An outstanding crimson, being bright scarlet-crimson, free from tawny colouring with paste and tube clearly defined.

BORDER

In bloom during the early weeks of summer the garden or border auricula makes a delightful edging to a border though will have finished blooming before the border reaches its best. For this reason the taller growing varieties should be used in groups at the front of the border rather than for edging. These old-fashioned plants, known to late Tudor gardeners, may be seen in this way in the borders of Hardwick Hall, Derbyshire, where they provide charming May colouring.

Though the plants are tolerant of partial shade, they do repay for liberal quantities of humus in the soil, in the form of peat and decayed manure. The auricula is an extremely hardy plant and will thrive under town conditions, but it is a mistake to allow the plants to remain undisturbed for years and without the addition of any humus in the soil. Under these conditions, the plants will form a woody rootstock and will bear a bloom of only poor quality. November is the most suitable planting time, setting out the plants 8 in. (20 cm) apart. Propagation is by division of the roots every three or four years, but each year the plants should be given a mulch of peat, leaf mould or decayed manure.

VARIETIES OF BORDER AURICULAS

Blue Velvet. This is an attractive border auricula, the large heads being made up of several blooms of a deep velvety-blue, which have a striking white centre.

Broadwell Gold. The frilled petals are of a lovely lemon-yellow with a white centre and remain long in bloom.

Old Dusty Miller. This is the name given to a group of very old garden auriculas, the foliage, blooms and stems being covered in an attractive silvery meal. The blooms, obtainable in red, yellow, purple and blue are borne on 10 in. (25 cm) stems.

Southport. The dainty bloom is of a pure terra-cotta with a deep yellow centre and is delightfully fragrant.

AUSTRALIAN BLUEBELL CREEPER, *see* Sollya

AUTUMN CROCUS, *see* Colchicum

AVENS, *see* Geum

AZALEA

INDOOR

The evergreen *A. indica* is grown in large numbers for forcing under glass to supply the Christmas trade, the plants, already in bud, mostly coming from Belgium where they grow well in the peaty, acid soil. They will have usually been grafted on to *Rhododendron luteum* stock. A suitable compost for potting will be 3 parts peat; 1 part each loam and coarse sand. A size 60 pot is suitable. Do not use a large pot or flowering may be delayed. If the roots are long, trim back before potting on or about October 1st. Stand on the greenhouse bench and give a thorough soaking but employ no heat until November 1st, when a temperature of 65°–70°F. (18°–21°C.) should be provided. The plants will start to open their bloom mid-December. At all times, azaleas require copious amounts of water. Also, if young shoots arise from the leaf axils, pinch these out when 1 in. (2.5 cm) long, otherwise they will take from the quality of the flowers. Two good Christmas azaleas are 'Paul Schame' (salmon pink) and 'Mme. van Damme' (pink and white). There are also double flowering forms.

OUTDOOR

They prefer partial shade and a light soil free from lime and well provided with peat. Groups of six or more will provide a brilliant display in the woodland garden, shrubbery or in small beds if the dwarf 'Kurume' hybrids are planted.

Plant the deciduous azaleas in autumn; the evergreens early in spring. They will bloom from mid-April to the end of June.

'Ghent' azaleas are deciduous. They are the latest to bloom, being at their best during June. Growing 4–5 (about 1.5 m) tall with thin, twiggy growth, they bear honeysuckle-like flowers with a powerful scent.

BOUQUET DE FLORE. Pink, shaded deeper at the edges.

GLORIA MUNDI. Dark reddish-orange.

GRANDEUR TRIOMPHANTE. Rosy-lavender.

SANG DE GENTBRUGGE. Blood-red.

There are also double flowering forms:

CORNEILLE. Cream, shaded soft pink.

IL TASSO. Rosy-red, tinted salmon.

PHOEBE. Pale yellow.

RAPHAEL DE SMAT. White, flushed rose.

'Kurume' azaleas bloom from late in April and are evergreen. They grow 18–20 in. (45–50 cm) tall:

ADY WERY. It has flowers of dark vermilion-red.

AZUMA-KAGOMI. The hose-in-hose flowers are deep pink.

BLAEW'S PINK. Deep pink, shaded lighter pink.

HATSUGIRI. Bright crimson-purple.

IRO-HAYAMA. Rosy-lavender, paler at the centre.

SAKATA RED. Fiery scarlet.

Mollis azaleas are deciduous. They bear their scentless flowers early in May in large trusses before the leaves appear. They grow about 4ft. (1.2 m) tall.

Anthony Koster. Pale yellow, flushed orange.

Comte de Gomer. Clear deep pink.

Dr Ooosthoek. Deep crimson.

Frans van der Bom. Orange shaded salmon pink.

Hugo Hardyzer. Scarlet-red.

Salmon Queen. Deep salmon-apricot.

AZARA

Native of Chile, it requires a warm garden in the British Isles or shelter from cold winds, though it is happy in partial shade. It is evergreen and is best planted early in spring in a well drained loam containing peat or leaf mould. *A. microphylla* has small shiny leaves and during the latter weeks of winter and early in spring, bears tiny pale yellow vanilla-scented flowers followed by bright orange berries.

Against a west wall in a favourable district, *Azara gilliesii* is a valuable sweetly-scented wall plant, growing to a height of 8–9 ft. (2 m) in a well-drained soil enriched with leaf mould or peat. The deep green glossy leaves are evergreen, whilst the flowers are unusual, having no petals but numerous stamens bunched together. They carry a delicious vanilla-like perfume. As the flowers are produced during February and March, plant in November. Prune by cutting out any dead wood after flowering.

Propagate by layering in autumn or by cuttings, taken with a 'heel' and rooted under glass.

BABIANA

In flower during June and July, it is an interesting bulb for the sunny sheltered garden. It does best in gardens of the west which enjoy mild conditions throughout winter. In the north they do not achieve best results until transferred to a cold greenhouse where they will produce their violet-blue flowers in profusion throughout spring. But in gardens other than the most exposed, the bulbs should winter if given some protection by way of a mulch or a covering with bracken or ashes. Or the bulbs may be lifted as for gladioli during November and planted again in April after wintering in a dry, frost-proof room. Plant 4 in. (10 cm) deep into a well-drained, loamy soil containing plenty of sand.

Indoors they make excellent pot plants, being potted during August and moved to a protected cold frame during September. Early in December they should be moved to a warm greenhouse or into the home. The blooms are of a rich blue colour very similar to the Tritonias but having broader, hairy foliage. They are sweetly perfumed and freely produced and equally important, they are inexpensive to obtain.

SPECIES

Babiana disticha. From South Africa, it produces its steely blue flowers on 9 in. (22 cm) stems throughout July.
B. plicata. Requires a sunny position preferably on the rockery where it will bear its sturdy purple-blue flowers throughout June.
B. rubro-cyanea. A lovely species bearing tufted blue and crimson flowers in profusion during midsummer.
B. villosa. It produces its 6 in. (15 cm) spikes of rich blue crocus-like flowers during midsummer and is an ideal cool greenhouse plant.

BABY BLUE EYES, *see* Nemophila

BACCHARIS

B. halimifolia is a deciduous shrub growing 9–10 ft. (about 3 m) tall with sage-green leaves and bearing dull white flowers like those of groundsel. It is tolerant of sea spray and makes an efficient hedge for a coastal garden. Of rapid growth, it may be cut hard back if getting thin at the base but normally, prune only straggling shoots.

BALLOON FLOWER, *see* Platycodon

BALLOTA

If cut back in spring, *B. pseudo-dictamnus* will produce an abundance of new shoots. It has cordate leaves which appear as if covered in white wool. It makes a low shrub 2 ft. (60 m) tall and in July, bears pretty pink flowers in whorls. Plant in spring and propagate from cuttings inserted in a sandy compost under glass.

BALSAM, *see* Impatiens

BAMBOOS

For waterside planting, though they do not like to have their roots in perpetually wet land, the evergreen bamboos are amongst the most beautiful plants of the garden. Many species may be grown in the British Isles but although hardier than generally believed, they are not suitable for a wind-swept garden. Plant them where they may be protected from prevailing winds and in the warmer parts, though they will usually grow well on the western side of Britain.

As with all evergreens, spring planting is recommended, early April being the best time. They like a deep loamy soil into which peat, leaf mould or other humus-forming materials have been incorporated. During dry weather give copious waterings and syringe the foliage. Propagate by division of the roots at this time and as they will often die after flowering and which may not take place for many years, it is advisable to lift and divide every three or four years so that there will be continuity of stock plants. At the same time frequent division will ensure healthy plants, free of dead wood at the centre. Bamboos like a yearly mulch of decayed manure or of other humus materials.

The name bamboo covers a number of different genera of plants of similar habit and appearance and which have now been grouped under their different botanical characteristics:

ARUNDINARIA

The chief characteristics are that the plants increase by stolons whilst they have round straight stems.

A. anceps. It is of vigorous habit, reaching a height of 12 ft. (3.6 m) and with thick straight stems which are brown when mature.

A. angustifolia. A neat and slender species growing 12 in. (30 cm) tall, the leaves being brilliant green on both sides.

A. intermedia. From the Himalayas, it sends up its canes to a height of 10 ft. (3 m) and has bright green leaves. A slightly more tender species.

BAMBUSA

B. disticha. A Japanese species growing 2 ft. (60 cm) tall and making a dense tuft. Its leaves have hairs at the margins and are about 2 in. (5 cm) long, arranged in two rows on opposite sides of the stem.

PHYLLOSTOSTACHYS

Their distinctive characteristic is their stems which are alternatively flattened and rounded on both sides whilst they make neat, compact clumps from which arise graceful arching stems.

P. niger boryanus. It forms a dense mass of leafy stems of pale greenish-yellow, blotched with darkest purple whilst they later change to shining purple-black.

PLEIOBLASTUS

Asiatic and characterised by their low, creeping habit.

P. pumilus. Native of Japan with thin arching canes 2 ft. (60 cm) tall and small dark green leaves.

P. variegatus. Similar in habit to *P. pumilus,* its leaves are striped with white.

P. viridi-striatus. It grows 3 ft. (1 m) tall with thick olive-green zigzagged stems and leaves striped with bright yellow and with a circle of hairs at the top of the sheath.

PSEUDOSASA

The canes arch at the top.

P. japonica. A hardy and easily grown bamboo, reaching a height of 10 ft. (3 m) with pale green arching canes.

SASA

Characterised by their round stems which arch to soil level and by their large leaves.

S. tessellata. A Japanese species with leaves larger than any other bamboo, being 18 in. (45 cm) long and 3 in. (7.6 cm) wide. The stems grow 5ft. (2 m) tall and arch to the ground.

SINARUNDINARIA

A hardy genus with round arching canes and small leaves.

S. murieliae. Named after Muriel, daughter of E. H. Wilson, it forms arching canes 10 ft. (3 m) tall and has narrow leaves of palest green.

S. nitida. Similar to *S. murieliae* in habit but with purple canes and smaller leaves.

BAPTISIA

B. australis is a little known border plant which is long-lived in ordinary well-drained soil and is very free flowering. It soon forms a thick bushy plant and during June and July bears masses of rich indigo-blue flower spikes which greatly resemble small lupins. The blooms are enhanced by the deep grey-green colour of the foliage. Plant in spring, 2 ft. (60 cm) apart. It grows 3–4 ft. (about 1 m) tall.

BARBERRY, *see* Berberis

BARTONIA

The large golden-yellow blooms of *B. aurea,* with their numerous fluffy stamens give this flower a most attractive appearance. It is a valuable plant for a dry, light soil and has a long flowering period, from mid-June until late September. Sow early April and thin to 1 ft. (30 cm) apart, for the plants attain a height of 18–20 in. (45–50 cm).

BASIL, *see* Ocymum

BAYBERRY, *see* Myrica

BAY LAUREL, *see* Laurus

BEAUTY BUSH, *see* Kolkwitzia

BEECH, *see* Fagus

BEGONIA

It is one of the most valuable of all plants for indoor and garden decoration, and may be divided into five main groups:

Large flowering tuberous rooted
Multiflora or small flowered tuberous rooted
Rhizomatous, which includes *Begonia rex*
Cane-stemmed
Fibrous rooted or winter flowering

LARGE FLOWERING TUBEROUS ROOTED BEGONIA

Propagation is either by sowing seed or by division of the tubers at planting-time. For exhibition or for greenhouse display, the tubers are potted into size 48 pots in early April, a cold-house being quite suitable. If the tubers have become large and woody, they should first be started into growth in boxes of loam and peat, then when growth can be seen, the tubers should be divided with a sharp knife, taking care to see that each piece of tuber contains a growing shoot. The divided tubers should then be potted into size 60 pots containing a compost made up of fibrous loam, peat and some coarse sand, just pressing the tuber into the compost and in no way covering it.

Tubers being grown for bedding purposes, which are generally sold with the new growth about 1 in. (2.5 cm) high, are almost always started into growth by placing as many as two dozen in a box containing a compost of loam, peat and sand. Early April is the correct time, for the plants should not be bedded out until June 1st or until fear of frost has departed. By then they will have made plenty of roots and some foliage.

When once the plants have made some growth they will prove copious drinkers and should be kept supplied with moisture, especially when grown for glasshouse display. They are not stove plants by any means, in fact, they are happiest in a temperature of from 55°–60°F. (13°–16°C.) and should the midsummer sunshine be unduly severe the greenhouse should be shaded with lime-wash.

To obtain the utmost beauty from the display, begonias indoors should be staged so that where two rows of plants are being grown on the greenhouse staging, those at the back are raised by placing the pots on wooden blocks.

Begonias may be brought into bloom in deep frames, but the plants are never at their best indoors for they do not like the dry, shaded conditions of a living-room and tend to drop their buds. Frame-grown plants may be brought into a sunny window for short periods, so that their brilliance may be enjoyed as much as possible.

As soon as the first buds are noticed, the plants will respond to an application of dilute manure water once a week.

At the end of the flowering period, they should be dried off gradually and the pots may then be stood on their sides during winter without removing the tubers, or the tubers may be shaken out in October and stored in boxes of peat or sand in a frost-proof room, an attic being ideal.

Begonias may quite easily be grown from seed sown during March in heat or early in May in a cool house. Using the John Innes Seed Compost, the seeds are merely pressed into the compost, watered and covered with a sheet of glass. Provided a temperature of 60°F. (16°C.) can be maintained, they will readily germinate, and as soon as large enough to handle, the seedlings should be pricked off into seed-boxes and grown on by later transplanting to small pots.

Begonias which are of exhibition strain will produce large blooms which, if not given support, may flop over, causing the breaking of the stems. Wire supports should be placed into the soil to which the heads are to be supported as soon as the buds are opening.

VARIETIES OF LARGE FLOWERING TUBEROUS ROOTED BEGONIAS

ALDWYTH BERRY. The habit of the plant is compact, the bloom being of a rich yellow shade.

BALLET GIRL. A lovely variety, the white waved petals are edged with rose-pink.

BLITHE SPIRIT. Comes into bloom when most have finished and bears a salmon pink bloom of perfect formation.

DIANA WYNYARD. Produces a huge bloom of purest white.

EL ALAMEIN. Outstandingly brilliant, the colour is rich crimson.

FLAMBEAU. The most vivid scarlet of perfect form.

FLORENCE BIGLAND. Attractive soft apricot.

FRANCES POWELL. A vigorous grower bearing deep pink flowers of great depth.

HERCULES. A new free-flowering variety, bearing bloom of an unusual shade of salmon-red.

SUSAN HOLT. Palest cream of excellent habit.

WAYNE PARKER. The flowers are of great depth, the petals edged rosy pink, like a picotee carnation.

MULTIFLORA OR SMALL FLOWERED TUBEROUS ROOTED BEGONIAS

Though the large-flowered begonias may be used for summer bedding display, it is the *multiflora* type with its less formal habit and continuous flowering in all weather that interests us here. In spite of being widely used for our corporation parks, they remain relatively unknown, yet are amongst the most valuable of all summer bedding plants and are amongst the easiest to cultivate. The small rosette-like blooms are held in sprays of a dozen or more, whilst they possess the rich clear colourings unknown in any other plant but the begonia. The blooms are accentuated by the dark green leaves which are of true begonia form and, being natives of the humid tropical areas of South America, no amount of rain will trouble the plants, as with the geranium and calceolaria which prefer dry conditions.

Making neat, bushy plants 8 in. (20 cm) tall, the *multiflora* begonias are ideal plants for a small garden, for no plant bears more bloom in comparison to its size, nor does so over a longer period.

The tubers should be started into growth about the last day of March; they will grow in a cold house or frame though but slowly, and it will be July before they come into bloom. With gentle heat they will be ready for planting out in early June, and will come into bloom by the middle of the month. The compost for starting the tubers into growth should consist of fibrous turf loam to which has been added some peat and coarse sand, in the ratio of 2 parts turf to 1 part each peat and sand. Begonias like some lime added to the soil, and to each bushel of compost add 1½ oz.

superphosphate and ¾ oz. lime. The tubers may be started in boxes, being merely pressed into the compost, and for the first weeks before they commence to make growth will require little water. When they are making growth they will prove to be copious drinkers.

This same demand for moisture will continue through summer, and if the plants are to be set out in a light, sandy soil, it must be fortified with farmyard manure, peat, spent hops, or any similar form of humus. A heavy soil, which is preferable provided it is well-drained, should be opened up by adding some grit and some peat. But, above all, some lime is essential, for begonias are never happy in the acid soils of our cities. A 2 oz. per sq. yd. dressing with bone meal will provide food over the entire summer months.

The plants should be set out about 7th June, when all fear of frost has gone. Plant 9 in. (22 cm) apart, placing the tubers 3 in. (7.5 cm) below the soil surface and taking care not to break the brittle stem. From the time the plants are set out they must be watered copiously, and should be given a mulch of peat in late July to conserve soil moisture through the summer. All dead blooms should be removed as they form so that they will not set seed, but this is not necessary with the multifloras. At their best in a season of damp, humid days, when other plants such as geraniums and calceolarias look anything but happy, begonias are particularly suited to our English climate in spite of their South American origin.

Lifting and storing begonias calls for care. The plants should be lifted with only their foliage and blooms removed, the stems being intact so that they can be placed in a frost-proof room which is dry and well ventilated, and there be left to die down gradually. Care must be taken in the same way with lifting as in planting, so that the stems are not broken. If the plants are lifted at the beginning of October, the stems will come away from the tubers naturally by the month end. The tubers should then be placed in boxes of dry peat, where they will remain through winter.

As the tubers become older they will increase in size and, if not divided, will tend to die back. The best method is to divide any large tubers (as for indoor varieties) when they have commenced growth in spring, for it is necessary for each piece to contain a shoot. The cut portions should be dusted with sulphur or charcoal to help the cut to heal over. Or the stock may be increased by taking cuttings as the young shoots appear in spring, generally a single shoot from an overcrowded tuber being removed. Side shoots may also be removed when 3 in. (7.5 cm) long as they appear on the main stems. Take care not to damage the 'eye' which will be seen at the base of each shoot, and which must be removed with it.

In the same way as with geraniums, the shoots are placed around the sides of a size 60 pot containing a mixture of sand and peat and, if kept moist, will root in from three to four weeks. The best time to take cuttings is during June and July; if taken later they may not root too readily. The

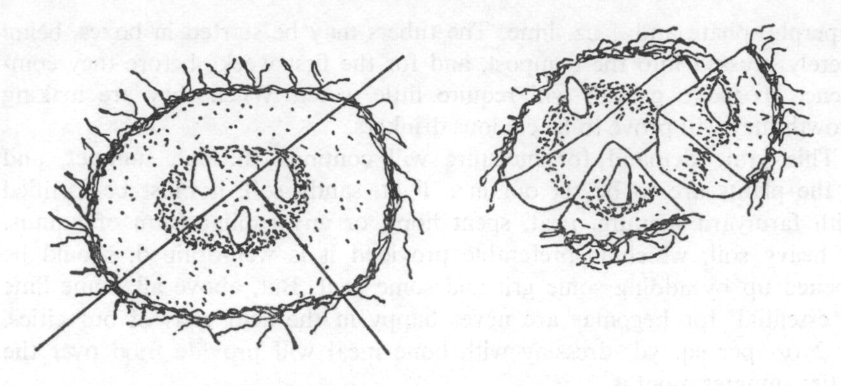

FIG. 1 *Dividing a tuberous begonia*

rooted plants should be potted into small pots containing a compost made up of loam and decayed manure, together with a little peat and some grit.

Of the *multifloras*, 'Red Thousand Beauties', each plant of which covers itself in masses of vivid red blossoms, 'Flamboyant (cherry-red), 'Madame Richard Galle' (orange), and 'Madame Helen Harms' (double yellow), are outstandingly fine bedding varieties.

RHIZOMATOUS BEGONIAS

Begonia rex has been well-named, for though few realize its wonderful qualities, it is one of the most beautiful and easily grown of all indoor plants. The plant was first observed growing in north India, and just a century ago the earliest plants reached this country, arriving at the nursery of John Simmons, who, along with a collection of orchids, took the original plant to London to be auctioned at Steven's Auction Rooms. Attending the sale was one Jean Linden, a nurseryman of Brussels and a famous grower of exotic plants, which were in great demand at that time to beautify the conservatories of the wealthy. Linden purchased the plant for 10,000 francs, and, naming it King of Begonias, introduced it to commerce the following year. His stock soon earned considerably more than the price he gave for the original plant, and within twelve months well over half a million francs had been paid for his plants. To the original plant with its rather dull foliage, Linden crossed *Begonia xanthina pictafolia*, which had beautifully silver-marked leaves, to lay the foundation of the exquisite *Begonia rex* hybrids we know today. But it was not until the turn of the century that *B. decora* was used for crossing with the Rex hybrids to impart on its offspring rich shades of bronze, crimson, purple and pink to bring a new interest to the *Begonia rex*. At that time, however, *B. rex* was unknown as a house plant. It was widely used for summer bedding and was grown in the conservatory, being given hothouse conditions. It is still believed by many that these conditions are necessary for the successful culture of the plant. Yet nothing could be further from the truth.

Greatly in their favour, is the ability of the plants to revive if for some reason they have been left without water for longer than intended. Geraniums will survive but drop their leaves, whilst African Violets will never recover. The *Begonia rex*, however, will completely revive, and will continue to grow as if nothing had happened if the pots are immersed in a bowl of cold water for several hours.

During the summer months the plants should be kept away from the direct rays of the sun, which will tend to bleach the exquisite leaf colourings. The *Begonia rex* is, in fact, at its loveliest only where given a diffused light, and is therefore far more beautiful seen in the home than in the greenhouse where it may be grown beneath the benches, otherwise the glass should be shaded during the midsummer months.

A well-grown plant of a named variety in a size 48 pot will give years of pleasure in addition to providing a constant supply of new stock and normal house conditions suit it to perfection. The plants require the minimum of care, re-potting only in alternate years, and the removal of decayed leaves so that fresh leaves may constantly take their place. They require sufficient water to keep the compost in a comfortably moist condition, and the exclusion of frost. Keeping the plants in diffused light will reduce the need for watering to no more than twice a week during summer and only about once every three weeks in winter, though much depends upon conditions indoors. In the dry atmosphere created by central heating, or by a gas fire, the plants will use up more quickly the moisture in the soil, and besides the need for more frequent waterings, they will greatly benefit from an occasional spraying of the foliage. The best method is to stand the pots in a bowl in the kitchen sink so that the roots become thoroughly moistened and at the same time the foliage should be sprayed. When removed to the living-room excess moisture should be shaken from the foliage.

The *Begonia rex* is most attractive used as a wall plant, suspended in natural wrought-iron or painted brackets. Against a cream-washed wall the wonderful colourings are seen to advantage, especially those varieties whose leaves are marked with crimson or pink. The blooms, if any are borne, will be quite insignificant and it is entirely for their evergreen foliage that the plants are grown. The size of the foliage and vigour of the plant is controlled by the size of the pot in which it is grown, whilst some varieties, i.e. 'Vesuvius' and 'Himalaya', (see plate 5), are more robust than others. For the small room, those of compact habit should be grown, 'Cotswold' and 'Green Velvet' being two suitable varieties, but the colour scheme of one's rooms should be considered when making a selection, and in one room it may be better to grow those varieties where crimson and pink predominate, whilst in another room the various shades of green and silver may appear more pleasing used together.

Begonia rex is also a most suitable plant for a summer display in a window box and especially where the box is away from the direct rays of the midday sun. Like all members of the begonia family it is well able to withstand long periods of wet weather and no amount of buffeting by strong winds will cause it trouble. It is no more a stove-house plant, as it is described in the old garden books, and only frost will cause damage to the plants.

PROPAGATION OF RHIZOMATOUS BEGONIAS

The plant forms a rhizome, or thick fleshy root, which tends to grow out of the soil to form a stem-like root from which the leaves appear on fleshy stems. After two years the plants will tend to become overcrowded in the pots, with the result that the leaves will begin to grow out small so that they do not reveal their full beauty. When this occurs the plants should be shaken from the pots and the rhizomatous roots divided into small sections each containing a leaf. These are replanted into size 60 pots for growing-on.

The plants may be increased more rapidly from leaf cuttings. The best type of leaf is one that is neither too large nor too small and, as when removing dead leaves, those selected for propagation should be removed close to the rhizome by means of a sharp knife. The stem is then removed at the point of contact with the leaf, the leaf being placed on a flat surface with the underside uppermost. Leaf propagation may be done at almost any time of the year but rooting will take place more quickly when the weather is warmer; if the leaves are removed in May or June, the young plants will have made good size by winter. If one or two leaves are carefully removed from each plant this should not harm its appearance and others will soon take their place.

Shallow boxes containing a mixture of moist peat and sand, made quite level, should have been made ready in advance. Short cuts should then be made across the veins of the leaf, spacing them evenly and making about a dozen cuts to each leaf, for it is at the points where the cuts are made that the plantlets are formed. The cuts should be made with a penknife just below the points where the side veins join the main arteries. The leaves are then placed on the compost underside down and they may be held in position either by small pebbles placed on top of the leaf or by thin wire stretched across.

The leaves will benefit from a gentle misty spraying each day, and if placed in a sunny window they should be shaded from sunlight, for it is important that the compost, especially at the surface, is never at any time allowed to dry out. If this should happen the tiny rootlets will shrivel and die. In about a month tiny bulbils will be seen from the places where the cuts were made and shoots will appear. Soon afterwards the remainder of the leaf will have completely decayed, leaving tiny plantlets which should be lifted with care and transferred to small pots for growing-on. By late autumn the plants will have grown large enough to be transferred to size 60 pots in which they will remain for two years.

Yet another method of propagation is to cut right across the leaves and to insert them into the compost along the edges where the cuts have been made, or triangular pieces may be cut and inserted at the point where it has been cut across a vein. Rooting will soon take place. In fact, no plant is easier to propagate and manage.

CARE OF RHIZOMATOUS BEGONIAS

The young plants will flourish in the John Innes Potting Compost, which may be obtained already made up from local nurserymen or a sundries shop, but take care to see that it is fresh. Where it is possible to make up one's own compost for begonias the following is a reliable formula:

3 parts fibrous turf loam
1 part decayed manure
1 part peat
1 part grit or coarse sand

Do avoid soil from a town garden and artificial fertilizers which, if of a nitrogenous nature, will cause the foliage to grow large and coarse whilst the stems will be too soft to support the leaves. The *Begonia rex* prefers an 'open', gritty compost, whilst it loves to have its roots in contact with the side of the pot. It is therefore important not to use a pot which appears too large.

As the fleshy roots will tend to grow in a horizontal position, it is important to plant so that the rhizome is as near to the centre of the pot as possible, which may mean planting the roots slightly to one side. As the rhizome makes growth it will protrude over the side of the pot and if too far, being brittle, it is liable to be broken off. When planting, the top of the rhizome should rest on the top of the compost which should be packed around the roots though not too compressed. Give the plant a thorough soaking and until it has become established stand the pot in a cool, shaded place.

It is not advisable to spray the foliage during winter unless the plants are growing in a warm room, for if the moisture does not dry off quickly it will cause spotting of the foliage, which may also turn brown at the edges.

As the leaves become old they will begin to curl, whilst the markings will begin to fade. They should be removed close to the rhizome before they become too old, so that others may take their place and the healthy appearance of the plant is maintained.

After twelve months the plants should be given a top dressing, using either the John Innes Compost or the formula previously suggested. The rhizomatous roots will have begun to push themselves out of the compost and so should be partially covered again whilst the top dressing will maintain the supplies of nourishment and will prevent the fleshy roots from becoming shrivelled. In another twelve months' time the pots will have become a mass of fleshy roots which should be carefully shaken from the compost, divided and re-potted. Or the plants may be 'knocked' out of the pots and, without disturbing the roots, may be re-potted into a larger pot. If re-potting presents a difficulty and the plants continue to form their rhizomes which extend well beyond the side of the pot, the end of the rhizomes may be removed with a sharp knife. Each piece should contain a leaf, and if inserted in a small pot it will soon form a sturdy plant. It is possible to do just about everything with *Begonia rex* both in its propagation and with its culture. Its powers of recovery from what would appear to be a completely dead plant being truly amazing. It is, in fact, the most fool-proof of all indoor plants and it is surprising that it is not more widely grown, for ordinary house conditions suit the plants admirably.

The number of varieties provide a wide choice, all of which are lovely and worth growing. Here are some of the best, including 'Glory of St Albans', which has survived since the introduction of *Begonia rex* a century ago.

VARIETIES OF RHIZOMATOUS BEGONIAS

CHICAGO. A very striking variety, the silver leaves being margined and blotched with carmine-red.

COTSWOLD. Of compact habit, the silver leaves have deep green markings.

EBFORD. The bottle-green leaves have attractive mauve and silver veins. Of medium vigour.

ETNA. A most striking variety of compact habit, the leaves being crimson-brown, splashed with silver and pink.

EVEREST. Extremely beautiful, the leaves are of brilliant silver with deep green central veins.

FIREFLUSH. Most striking, the leaves being bright green with darker edges over which are interesting red hairs giving the foliage a crimson sheen.

GLADE. A vigorous grower, the rich bright green leaves are shaded with bottle-green.

GLOIRE DES ARDENNES. A vigorous large-leaved variety which is bright crimson-red in colour.

GLORY OF ST ALBANS. Found as a chance seedling at Sanders' Nursery,

St Albans. The leaves are of a vivid rose colour, margined olive-green and with a small olive-green centre.

GREENSLEEVES. The neat sage-green leaves are attractively dotted with dark green.

GREEN VELVET. A lovely plant, the deeply indented leaves are of dark green and black.

HALDON. Most striking in that the leaves alternate between dark and light green in wide bands.

HELEN LEWIS. A striking American variety of robust habit and having large crimson and white leaves.

HIMALAYA. Of vigorous habit, the large green leaves are heavily splashed with silver.

ICELAND. One of the most vigorous varieties, the large silver leaves with their deep green veins are tinted with purple.

LA MARQUISE. The pale green foliage with its deep green veins is spotted with silver and pink.

LA PERLE DE MORTEFONTAIN. A striking variety, having bright reddish-pink leaves, mottled dark crimson at the edges.

LYMPSTONE. An attractive variety of medium habit, the bright silver leaves have deep green veins.

MOUNTAIN STREAM. A lovely variety of compact habit with glittering leaves of silver, pink and green.

MRS HATCHER. Of vigorous habit, the large leaves are crimson-red with darker veins.

PURPLE RAY. A striking variety, it sends up its flower spike 12–15 in. (30–38 cm) above the foliage. The extremely pointed leaves are purple and silver.

REMILLY. Of medium vigour, the silver leaves are beautifully tinted or flushed pale mauve.

ROSY MORN. Of compact habit, the leaves are of a lovely shade of rose-pink.

ROUGEMONT. Outstanding in that the leaves are of a rich cherry-red colour and have an attractive pink and brown border.

SILVER QUEEN. A beautiful variety having large silver leaves strikingly veined deep green.

VESUVIUS. Of vigorous habit, the large crimson-brown leaves are attractively splashed with pink.

VICTORY. An outstanding variety, the bright olive-green leaves being margined with bronze and blotched with silver and pink.

WALLINGBROOK. A vigorous grower and interesting in that its large leaves have three margins of different shades of green with bottle-green centres.

WELCOME. A lovely variety of medium habit, the pink and mauve leaves being bordered with silver.

WOODBURY. Interesting in that the rich purple leaves have black dots along the veins.

There are also a number of other foliage-type begonias of rhizomatous habit, requiring the same cultural conditions except that they do require a warmer temperature, the living-room of a house suiting them admirably, but draughts and low temperatures should be avoided. Most are hybrids which have at some time been crossed with *Begonia rex* and are extremely interesting and beautiful.

SPECIES AND VARIETIES OF RHIZOMATOUS HABIT BEGONIAS

ABEL CARRIERE. The result of a *B. rex-B. evansiana* cross and is a most pleasing hybrid. The silvery leaves are mottled with pink after the manner of 'Frances Kay', whilst it bears bright red flowers. It takes its erect habit from *B. evansiana.*

B. boweri. Discovered as recently as 1948 in Mexico, this is one of the most beautiful of all house plants and difficult to describe. The small gracefully pointed leaves are held in drooping fashion on long stems which intertwine with each other. The edges of the leaves are attractively waved and are almost black, the centre being vivid green.

B. caroliniaefolia. This is one of the first begonias to reach Europe, which it did a century ago from Mexico. It has wide-lobed glossy-green toothed leaves formed from an upright rhizome, and during summer bears dainty pink flowers in long arched sprays. It has been frequently used for crossing.

SILVER STAR. This beautiful plant with its upright rhizome was raised in America by Mrs Frey from a *B. caroliniaefolia,-B. leibmanni* cross and introduced in 1945. In habit it is similar to both its parents and has six deeply serrated, lobed leaves which are glossy and of an olive-green colour overlaid with silver. The leaves are held on stems 10 in. (25 cm) long, whilst in April white star-like flowers are held above the foliage. It requires the same culture as for other rhizome begonias.

B. crestabruchi. An unusual variety having the appearance of curly kale. The edges of the vivid green leaves are hairy and red and are extremely crested or curled.

B. diadema. Crossed with *B. rex* it produced the interesting variety, 'Filigree'. The serrated leaves have a bright centre and striking silver border. It is an interesting hybrid, its dark green leaves being vividly splashed with silver and being of beautiful filigree shape. The plant is of upright habit.

B. foliosa. It is more fibrous rooted and makes an excellent house plant. Its leaves are the smallest of the species and borne on both arched and drooping stems like a Maidenhair fern.

FRANCES KAY. A beautiful hybrid, the result of a *B. rex-B. evansiana* cross, having the same upright habit of *B. evansiana* and bearing handsome long, pointed leaves of deep green, heavily mottled with silver.

B. leibmanni. Introduced from Mexico in 1939, it has a low, prostrate rhizomatous rootstock. The many-lobed leaves are rounded and are of a deep green shade blotched with silver.

B. manicata. A native of Mexico, it is an outstanding plant. The rhizomes grow tall and erect, from which are borne numerous smooth green leaves and olive veins. The tiny drooping shell-pink flowers are borne in numerous sprays, as many as a dozen or more, well above the foliage. A lovely seedling 'Thisbe', raised in California, has broader leaves than the parents which are attractively hairy. The flowers are deep pink.

CANE-STEMMED BEGONIAS

Unlike the Rex begonias, which are tolerant of quite cold conditions and comparative neglect without showing undue signs of distress, both the cane-stemmed and hirsute groups demand a temperature of not less than 50°F. (10°C.) during winter. Under colder conditions the cane-stemmed begonias will drop their exquisite leaves though the plants themselves will not perish. Plants of the hirsute group will not bloom in temperatures below 50°F. (10°C.). Apart from one or two species, of which *B. cathayana* is an outstanding example and which demands a temperature of around 70°F. (21° C) during winter, both the cane-stemmed and hirsute begonias will prove happy in temperatures of between 50° and 60°F. (10°–16°C.) which is quite possible to attain in the house or garden room of the average enthusiast, either by the use of electric or central heating. The cane-stemmed species require a greater degree of moisture than plants of the hirsute group, which are able to tolerate the same dry conditions as the Rex begonias though not so cool a temperature. Happiest in partial shade, though by no means in full shade, plants of both groups will bloom in the home and will not drop their buds or blooms in the same way as will the tuberous begonias growing under sitting-room conditions. The plants enjoy best an even temperature of 55°F. (13° C.) throughout the year, and sufficient moisture to be provided by standing the plants in trays of moist sand, though where the temperature is likely to fall below 55°F. (13°C.) moisture should be reduced accordingly. Where temperatures are likely to fall below 50°F. (10°C.) water should be almost entirely withheld until conditions become warmer.

Where possible, the best way of using the plants for house decoration is to keep them in a warm greenhouse or garden room and to take them into the home for periods of six to eight weeks, transferring them to the warm greenhouse where they may be brought on again by giving them a temperature of 60°F. (16°C.) and additional moisture to increase humidity. The plants may then be taken into the house for another period.

Plants of both the cane-stemmed and hirsute groups are of considerably more robust habit than most of the Rex hybrids. Where growing in their native country, many will attain a height of 12 ft. (3.5 m), whilst in pots, *B. lucerna* and 'President Carnot' will reach a height of 5–6 ft. (about 1.5 m) and may be grown against a wall or lattice screen in much the same way as indoor geraniums of robust habit. The vigour of a plant may, to a great extent, be governed by the size of pot in which it is growing, whilst not all species and varieties reach a height of 5 ft. (1.5 m) in their pots. The cane-stemmed group may be divided into three sections, those of tall upright habit, those of medium height and spreading habit, and those of dwarf, bushy habit which are few in number. Plants of the first group are more suited to a greenhouse or garden room, those of more dwarf habit being ideal plants for the home, though only one or two, such as *B. glaucophylla*, with its semi-trailing habit, will be suitable for a wall bracket. Those of robust but spreading habit, such as *B. lucerna*, make ideal plants for a table or stand in much the same way as the aspidistra was used in Victorian days.

Both the cane-stemmed and hirsute groups are propagated by means of stem cuttings. These are removed from the top of the stems, cutting them immediately below a joint or node. The lower leaves should be carefully trimmed, leaving only two upper leaves which have been reduced in size. Before inserting the cuttings around the sides of a pot to root, dip the base in flowers of sulphur and plant in a compost composed of peat and sand in equal quantities. Though the cuttings may be removed and rooted at almost any time of the year, they will root more quickly in the late spring and early summer months, being ready for individual potting within a month from their removal from the plant.

GROWING-ON CANE-STEMMED BEGONIAS

The cane-stemmed and hirsute begonias should be given a heavier compost than required by most other begonias. This will encourage the plants to grow 'hard' or short-jointed, so preventing too vigorous growth. This will keep them in bounds and where growing in a warm room moisture evaporation of the compost will take place more slowly. The compost should be composed of 5 parts, fibrous loam, 1 part decayed manure, 1 part coarse sand, and 1 part peat or leaf mould to which is added a dusting of steamed bone flour. This is the compost to be used for potting into size 60 and later into size 48 pots in which the plants will continue to grow. If the plants are to be kept within reasonable limits, a larger pot should not be used, though for plants growing in the garden house and where the winter temperature does not fall below 55°F. (13°C.) very large pots may eventually be used, when the plants may attain a height of 5 ft. (1.5 m) and grow 2 ft. (60 cm) thick, sending up their sturdy 'canes' like bamboos, and will be a wonderful sight.

For the first move into individual 2½ in. (65 cm) pots, the John Innes Potting Compost should be used, the plants being grown-on in a position of full light and under moist, warm conditions. They will be ready for the size 60 pots as soon as the soil ball appears to be full of roots and before the plants become pot-bound. By the end of summer the plants should be in their final pots where required to make a limited amount of growth and will by then have formed sturdy plants for use in the home. They will come into bloom the following summer and will continue to bloom intermittently throughout their life.

Though the plants appreciate moisture, the compost which will be slow to dry out should never be over-watered and during the winter months only limited amounts will be needed. Though the cane-stemmed and hirsute begonias are extremely resistant to disease, stem rot may occur where the roots are kept too moist for too long a period.

To maintain the shape of a plant when it has been two years or more in its pot, unduly long shoots may be cut back in spring, when with a little care the plants may be built up into a most pleasing shape. After two years in the final pots, the plants will require re-potting. A small quantity of soil should be removed from the roots, though they should not be unduly disturbed. The fresh compost should be similar to that previously used, and whilst firm planting should be done, the plants should not be made too firm. Loose potting will, however, encourage the plants to be long jointed and too tall.

Watering the plants once each month during summer with dilute liquid manure, obtainable as a proprietary make, will accentuate the rich glossy colouring of the foliage and maintain the plants in a healthy condition. Guard against placing the plants in a draughty position, for unlike the Rex begonias they will tend to drop their leaves if exposed to cold air.

Staking will be necessary for the taller-growing species and those of upright habit, although the stems will be considerably stronger than those of the tuberous begonias. Small canes, which will be difficult to distinguish from the cane-like growth of the plant, should be inserted in the pot in a slightly outwards direction so that the shape of the plant will be of pleasing appearance whilst permitting a free circulation of air to reach the whole of the plant. It will also enable the canes to be ripened by the sun and so prevent 'soft' top growth.

Where the plants are being grown in a greenhouse during summer or in a garden room considerably exposed to the sun's rays, some shading will be necessary during the period from June until early September. The plants should be shaded during midday by stretching lengths of canvas over the roof or whitening the glass. A conservatory should be fitted with lath blinds which may be let down when the sun is at its hottest. Complete shading must not be allowed, for the plants must have some sun during summer to ripen the wood.

Whilst the weather is warm and sunny, the house should be constantly damped down, withholding water during dull periods, whilst at night a cold, clammy atmosphere should be prevented by providing additional ventilation. Regular spraying of the foliage during dry, sunny weather will prevent an attack of red spider or mite.

SPECIES AND VARIETIES OF TALL CANE-STEMMED BEGONIAS

B. aconitifolia. A native of Brazil, it was introduced to Europe only at the beginning of the century. It has large lobed leaves which are dark green splashed with silver. It bears few blooms which are large and of a blush-white colour.

B. coccinea. This was one of the first of all begonia species to be discovered and has been widely used for hybridizing. It makes a plant of tall, erect growth having smooth stems and long bright green leaves, edged red. It blooms profusely, bearing its coral-red flowers in clusters. A lovely form, *B. chasta,* is of the same habit and bears clusters of large pure white flowers. Another, 'Mrs M. Armstrong', bears large pale pink flowers, whilst 'Martha' bears a profusion of small deep crimson flowers. Another very lovely variety is called 'Erna', its leaves being attractively spotted with silver, whilst the small flowers are of a coppery-red colour. It makes a plant of medium habit.

B. maculata. This was the first begonia to reach Europe with silver-spotted leaves. Like *B. coccinea,* it makes a plant of tall, erect habit, its long pointed leaves being attractively spotted with silver. It bears a few flowers which are large and pendulous and of a lovely shade of coral-pink.

Raised at Kew Gardens, *B. maculata* 'Wrightii' has the same cane-like stems and spotted leaves, whilst its beautiful greenish-white flowers are borne in large pendulous clusters. Another variety, 'Annabella', has lovely soft pink flowers.

PRESIDENT CARNOT. Like 'Lucerne', it is derived from *B. corallina,* which so resembles *B. coccinea* as to be almost indistinguishable. It makes a tall plant of almost climbing habit, has toothed dark green leaves tinged with red and bears large trusses of carmine-red flowers.

B. undulata. A native of Brazil, it forms tall, smooth cane-like stems and has bright green pointed leaves with undulating margins. It is free flowering, the blooms being borne in clusters, and are pure white.

SPECIES AND VARIETIES OF BUSHY CANE-STEMMED BEGONIAS

B. albo-picta. Of low, branching habit, its leaves are long and narrow and of bright glossy green spotted with silver. The flowers are greenish white. A variety 'Rosea' bears rose-pink flowers, whilst a seedling 'Clemmence' bears clusters of dark red flowers.

B. argentea-guttata. In spite of its name, it is a lovely plant, of bushy, branching habit. Its long, pointed leaves are extremely toothed and are of an olive-green colour, spotted with silver. The large handsome flowers are of rich creamy white. Derived from *B. albo-picta,* it was raised by Lemoine in 1888 and is a most valuable house plant, but so rarely seen.
B. glaucophylla. Of bushy, almost trailing habit, it was found in Belgium in 1865. The leaves are glossy and almost grey in colour, whilst the flowers, borne in profusion in spring, are of a bright coral-red colour, borne in clusters close to the stems. The leaves have an attractive satiny sheen. Crossed with *B. coccinea,* the result was that fine upright variety 'Marjorie Daw', which has bright pink leaves and bears large clusters of clear pink flowers. It has a similar habit to *B. glaucophylla* and if not grown in a wall container should be trained up a trellis. 'Marjorie Daw 'crossed back with *B. glaucophylla* has produced that fine hybrid *B. glacdaw,* which forms an upright plant of bushy, pyramidal habit. Its beautifully ruffled glossy green leaves are spotted with silver, whilst it bears a few rosy-pink flowers. It is an excellent house plant and is of easy culture. Another plant of the same crossing, *B. glaucoppola,* is of similar habit though the leaves are broader.
B. medora. A seedling believed to have *B. albo-picta* for a parent. Though of upright habit, the laterals tend to droop. The leaves are long and heavily spotted with silver, whilst the few flowers are large and of a lovely shade of shell-pink.

HIRSUTE BEGONIA

Comparatively unknown as house plants, the plants are able to tolerate the dry conditions of a room and will withstand lower temperatures than members of the cane-stemmed group. The stems, foliage and flowers are bearded or hairy, even the ovaries of the flowers being hairy. The plants possess almost the robustness of the Rex begonias yet bear bloom with much greater freedom. They are, in fact, amongst the most beautiful plants in cultivation.
B. alleryi. A *B. metallica-B. gigantea* cross, it has more of the appearance of a pelargonium than of a begonia, the vivid green leaves being deeply veined and extremely toothed at the margins. Raised in France in 1905, it grows 2 ft. (60 cm) tall and bears its clusters of deep pink flowers in profusion.
B. angularis. It grows to a height of 3–4 ft. (about 1 m) and is sparsely foliaged. The large oval leaves tend to droop and are of rich green with grey veins, being toothed at the edges. The small white flowers are borne in short clusters. A form known as *B. acutangula* is of more vigorous, sturdier habit and bears larger leaves.
B. duchartrei. It has *B. scharffiana* for a parent and makes a tall but com-

pact bushy plant with dark green, pointed leaves and bears very large white flowers in clusters. There is also a pink flowering form.

B. gigantea. This is an older species to be found in the Himalayas. From the base it bears a profusion of stems which carry large toothed leaves. The white flowers are borne in clusters at the end of the stems.

B. haageana. It is also known as *B. scharffi,* named in honour of Dr Scharff who, accompanied by Monsieur Haage, found the plant in Brazil. The leaves are large and of a lovely rich olive-green colour and are not so hairy as others in the group. Growing up to 2 ft. (60 cm) tall, it forms a dense bushy plant and bears clusters of large bearded flowers of pale rose-pink.

A number of *B. haageana* seedlings have resulted in some outstanding plants and one of the best is *B. alphonse,* raised in California. The broad, deep green leaves are covered with white hairs, the large shell-pink flowers being held on rich red stems. Another, 'Gertrude', will attain a height of 2 ft. (60 cm), its large bearded white flowers being accentuated by its large bottle-green leaves.

A hybrid of *B. haageana, B. drosti* is a most beautiful plant and under suitable conditions will grow nearly 5 ft. (1.5 m) tall. Of bushy habit, the plant has dark green leaves and bears arching sprays of rose-red flowers. Its main attraction, however, is that every part of the plant is densely covered with white hairs, which gives it a snowy appearance.

B. luxurians. This is a most interesting begonia in that its leaves are deeply-lobed and drooping, having the appearance of a palm tree. Tall growing, it forms a most graceful plant, its small white flowers being borne in clusters. It has passed on many of its characteristics to a seedling, 'Mrs F. Scripps', which attains a similar height and has the same many-lobed olive-green leaves but bears pink flowers.

B. metallica. An excellent house plant, it grows no more than 2 ft. (60 cm) tall and makes a compact bushy plant, covered in hairs. The leaves are toothed and of a bronzy-green colour with metallic purple veins which give the leaves an interesting metal-like appearance. The flowers, borne in large clusters, are large and conspicuous (see plate 6).

A seedling named *B. suprea* makes a bushy plant covered in white hairs and has large olive-green leaves. It bears large pale pink flowers. A seedling from this plant, 'Nelly Bly' is a most outstanding form. It makes a large bushy plant, its toothed leaves being small and pointed and of an attractive bottle-green colour. The bearded flowers are of a rose-pink shade.

B. sanguinea. An early Brazilian discovery, it takes its name from its red stems. The glossy leaves are olive-green and crimson underneath. The pure white flowers are borne on long, arching stems.

B. scharffiana. It is a most attractive plant of low, spreading habit with velvet-like leaves, margined and veined with crimson. The leaf and flower stems are also red, the blooms being palest pink, borne in arching sprays. A seedling, 'Loma Alta', is of vigorous, upright habit, its handsome pointed leaves being more than 12 in. (30 cm) length.

B. vedderi. Its long tapering stems grow to a height of 3 ft. (90 cm) and are furnished from top to bottom with pointed toothed leaves which have deep crimson veins. The leaves are olive-green, the flowers pale pink. A most beautiful plant, it does well in the house.

FIBROUS ROOTED OR WINTER FLOWERING BEGONIAS

Like the *multifloras,* the fibrous rooted begonia *B. semperflorens* may be grown in the greenhouse, in the garden, or in window-boxes. They are best treated as half hardy annuals, the seed being sown in a temperature of 65°F. (18°C.) in January when the plants may be set out in June. Whether growing indoors or outside, they will come into bloom early in July and continue until the end of October. A sowing may also be made in June when the seed will germinate under natural conditions to make plants which will come into bloom at the year end and will continue to give colour indoors until the end of spring. For this purpose the two best forms to grow from seed are the carmine-red and pure white varieties of 'Winter Romance', making bushy little plants 6 in. (15 cm) tall and bearing masses of tiny blooms about the size of a shilling.

The fibrous-rooted begonias are strictly of perennial habit and those which have flowered through summer and autumn should be cut back and wintered as dry as possible, though moisture should not be entirely withheld. They will winter quite happily in a temperature of 45°F. (7°C.), then with the warmer days of spring, they should be given more water and brought into bloom once again.

VARIETIES OF FIBROUS ROOTED OR WINTER FLOWERING BEGONIAS

CRIMSON BEDDER. A most striking variety growing 8 in. (20 cm) tall with crimson-bronze metallic foliage and bearing flowers of brightest crimson.

KARIN. It forms a shrubby little plant 6 in (15 cm) high with chocolate coloured leaves and bears contrasting flowers which are of more pure white than those of any other fibrous begonia.

LUCIFER. Of vigorous habit, it grows 9 in. (22 cm) tall with striking shiny black foliage and bearing deep scarlet-crimson flowers.

PINK PROFUSION. It grows 6 in. (15 cm) tall and bears its salmon-pink flowers in great abundance.

SAGA I. It makes a tiny rounded bush-like plant only 4 in. (10 cm) tall with bronzy green leaves, whilst it bears its tiny crimson flowers in clusters.

STUTTGART. A beautiful new hybrid having dark bronzy-green foliage and bearing flowers of bright carmine-pink.

BELLADONNA LILY, *see* Amaryllis

BELLFLOWER, *see* Campanula

BELLS OF IRELAND, *see* Molucella

BELOPERONE

Native of Mexico, *B. guttata* is known as the Shrimp Plant on account of the shrimp-like form of its flowers. It is a hard-wooded shrubby plant with small oval hairy leaves and it bears pinkish-yellow flowers in long terminal racemes. It will bloom throughout the year in a winter temperature of 50°F. (10°C.) and whilst the plants require ample supplies of moisture in summer, very little must be given in winter. Use the John Innes Potting Compost No. 1, plus additional peat for the plants require a compost which tends to be slightly acid. The plants will appreciate an occasional syringing in summer and a feed with dilute manure water. Also shorten back the stems in early summer if they become too long.

BERBERIDOPSIS

Berberidopsis corallina if given a warm position and winter covering for its roots, will survive a northern winter. In the south it prefers a position away from the direct rays of the sun, the ideal being a west or east wall, where it will grow to a height of 10 ft. (3 m) and bear its drooping coral-red fuchsia-like blooms, with their attractive red stems, from July until September. A native of Chile, it is known as the Coral Plant from the colour of its blooms, whilst it bears prickly, bright evergreen leaves. This plant likes a lime-free peaty soil, so incorporate some peat and leaf mould into the soil at planting time, and give a peat mulch each summer. March is the best time to plant.

BERBERIS

The evergreen and deciduous Barberries are amongst the most useful of all plants for they will grow in an exposed position or in partial shade and in almost any soil, including acid or calcareous. Outstanding in the brilliance of its foliage is *B. thunbergii* which bears large trusses of golden-yellow flowers followed by red berries. It will grow 4 ft. (1 m) tall and makes a valuable low hedge. The form 'Atropurpurea' has reddish-purple foliage and is deciduous as is *B. montana,* which bears the largest flowers of all the barberries, like small narcissus. It grows 6–7 ft. (2 m) tall and blooms in May.

The hybrid *B. x stenophylla,* in bloom at the same time and forming graceful arching branches 8 ft. (2.5 m) tall, is an evergreen which makes an elegant hedge and will withstand hard clipping. It has deep green spiny leaves and has for one parent, the widely planted *B. darwinii* which has glossy evergreen leaves and bears its orange-yellow flowers during winter and spring. They have the scent of lily-of-the-valley and are followed by blue berries. It may be planted on a bank or used as a low hedge.

The evergreen species are planted in March, the deciduous species November to March. When established, they will appreciate a top dressing of decayed manure and the only pruning necessary is to thin out any unduly long shoots. Most species are propagated by layering or by cuttings of the half-ripened wood.

BERGAMOT, *see* Monarda

BERTOLONIA

These trailing plants are natives of Brazil and should be grown only where a minimum winter temperature of 52°F. (11°C.) can be provided. They strike freely in small pots placed in a propagating frame. The fully grown plants should be given a size 60 pot and a compost composed of 2 parts peat, 1 part fibrous loam and 1 part silver sand and they must be kept moist at the roots. They prefer a fairly light position, a table near a window but away from the direct rays of the sun and where they may be allowed to trail their beautiful hairy leaves over the sides.

Where it can be obtained, *B. houtteana* is perhaps the easiest form. Its deeply ribbed olive-green leaves are spotted with rose. *B. pubescens* is also attractive, its olive-green leaves being covered in long white hairs and have a pale brown stripe down the centre.

BETHLEHEM SAGE, *see* Pulmonaria

BETONICA

Happiest in a rich sandy soil, the Betony with its erect spikes of rosy-purple, in bloom from early June until September is a plant beloved by bees. It may be used as an edging, for its foliage does not spread over the grass verge. When used in this way, plant 12 in. (30 cm) apart during November. Propagation is by root division.

SPECIES

B. grandiflora 'Superba'. The thick, erect spikes of rose-purple grow to a height of 10 in. (25 cm) and make this an ideal edging plant.

B. spicata 'Rosea'. Growing to a height of 20 in. (50 cm) and bearing short spikes of lilac-rose flowers, it could well be more widely planted towards the front of the border.

BETONY, *see* Betonica

BETULA

Hardy and elegant and growing well almost anywhere, the Silver Birch is the finest of all forest trees for the small garden, at its loveliest planted in groups of three or four. No tree is more striking in winter with its silvery bark and around it should be planted purple winter and spring flowering crocuses and irises and winter flowering heaths. It grows well in a poor, dry soil and will need no pruning.

Of many attractive species, *B. albo-sinensis*, the Chinese Birch is outstanding in that its bark is brilliant orange, whilst *B. lutea* is known as the Yellow Birch of North America on account of its bright yellow bark. Of the Silver birches, *B. pendula* 'Tristis', the Sad Birch, has slender drooping branches whilst 'Young's Weeping' makes a beautiful flat-topped weeping tree. Where space is limited, plant the form, *fastigiata* which is like a Lombardy poplar with silver bark.

BIGNONIA, *see* Campsis

BILLBERGIA

Though native of the tropical forests of Brazil, it will be happy given a winter temperature of 50°F. (10°C.) and does well where the home is centrally heated. It requires an even temperature rather than one excessively high, whilst it is happy in partial shade.

B. nutans forms rosettes of long parchment-like leaves and on 12 in. (30 cm) stems bears spikes of long green flowers edged with purple and which have striking golden stamens. *B. zebrina* is also a handsome species, its leaves being banded with white, whilst *B. rhodocyanea* has silver leaves and bears delicate pink flowers. *B. vittata* has deep blue flowers with a contrasting red calyx.

The plants may be grown in partial shade but require sunlight to bloom well. Propagation is by means of offsets which are removed in May and planted in small pots containing the John Innes Potting Compost. It blooms at its best during winter and early spring and should be re-potted when the offsets are removed.

BIRCH, *see* Betula

BIRTHWORT, *see* Aristolochia

BITTER WORMWOOD, *see* Artemisia

BLACK-EYED SUSAN, *see* Thunbergia

BLANKET FLOWER, *see* Gaillardia

BLEEDING HEART, *see* Dicentra

BLOOMERIA

B. auria is not a common plant and may be difficult to obtain, but is so lovely that it should be grown wherever possible, at the top of a rockery or on a dry wall where its vivid yellow umbels, borne on stems 9 in. (23 cm) in length will provide a most striking effect during the late spring months. Plant 4 in. (10 cm) deep into a soil which contains a substantial amount of peat and some decayed manure. Native of California, during a severe winter it may not be completely hardy in all parts so provide it with a mulch of strawy manure in December.

BLUEBELL, *see* Endymion

BLUE HIMALAYAN POPPY, *see* Meconopsis

BOCCONIA

B. cordata, the Plume Poppy, is a most attractive plant for the back of a large border and in spite of its exotic appearance will flourish in ordinary soil. Plant 3 ft. (90 cm) apart in November, the huge panicles of coral-pink being borne on 5–6 ft. (about 2 m) stems from July until September. Propagation is by root division in autumn.

VARIETIES

BEE'S FLAME. Of recent introduction, it bears its large plumes of salmon-orange above bronze scalloped leaves, and is at its best during August.

CORAL PLUME. This is outstanding, the blooms being enhanced by its silvery fig leaf-like foliage.

BONSAI

This, the growing of trees in a confined space, is a hobby which has its origin in earliest times. In Japanese homes, bonsai trees are placed on shelves in drawing-rooms in the same way that western homes display china figures, often on glass shelves and lit by concealed lighting. In their shapes, the trees are modelled on those growing naturally, about rocks and in exposed positions where they become gnarled and stunted. The shaping of the trunk has various methods of classification.

CHOKKAN

This is the solitary upright tree usually trained in pyramidal form. The tree should be slow growing with small leaves. The following are suitable:

Chamaecyparis obtusa: Hinoki Cypress
Cryptomeria japonica: Japanese Cedar
Ginkgo biloba: Maidenhair Tree
Juniperus chinensis: Chinese Juniper
Pinus contorta: Beach Pine
Pinus densiflora: Japanese Red Pine
Pinus parviflora: Japanese White Pine

Pseudolarix amabilis: Golden Larch

Dwarf conifers as planted in rock gardens are not suitable for bonsai, having lost the characteristics to be found in the type.

In Chokkan, the tree is persuaded to grow into one of several forms. To grow into pyramidal shape, select a tree of about 2 ft. (60 cm) tall and cut back to the lowest branch. Wire is then coiled round the branch and the lower portion of the trunk, to bring the branch into an upright position. The branch then becomes the new leader and the process is repeated by shortening and pinching the leader, to keep the tree at the required height and to encourage side shoots to form.

The wide conical or umbrella shape is obtained by raising plants from seed and when 9 in. (22 cm) high, pot separately and keep the stem straight by tying to a cane. The first branch should be formed at a height of about 6 in. (15 cm), the lower branches being removed. Throughout summer the branches should repeatedly be pinched back to two or three leaves and they are then trained upwards by wire as described. In a few years, a tree with a thick trunk and a spreading head will be obtained. In Chokkan, the trunk is the primary object of beauty and so that it may be fully appreciated, all branches growing to the front and hiding the trunk, should be removed.

KYOKKUKAN

This is the solitary tree with a twisted trunk, the Japanese Red and White Pines being most suitable to train in this way. Flowering crabs may also be used.

With Kyokkukan, bonsai trees may be formed more quickly than by any other method. In autumn, obtain and pot four year pines, leaving on only those leaves borne during the past summer. Around the base of the trunk wrap a small piece of sacking and around it wind copper wire. Then at a point equi-distant from the base and lower branch, bend the plant and hold into position by wrapping round it the wire. Then bend backwards and forwards, tying with the wire so that eventually the tree is much reduced in height and will have a weather-beaten appearance. The branches are then brought into shape by lightly wrapping thin wire around and bending into the required position. The tips of each branchlet should be made to turn upwards at a slightly different height, to give the plant a more pleasing outline.

SHAKAN

This is the tree with its trunk at an oblique angle for which the Japanese pines, the Yeddo spruce and the maples are suitable.

To grow at an oblique angle, obtain a tree with a low branch growing at right angles and for this purpose, the willow, actinidia and akebia are suitable. Then remove all that part of the plant above the chosen branch which is held in position by the insertion of a cane to which it is tied.

HANKAN

The tree has a gnarled trunk. To obtain this effect, use the Yeddo spruce, wistaria or apricot. Around the trunk wrap copper wire to a point above the lower branches. Then twist round the trunk three or four times and hold in place with the wire.

With the Hanoi method, only the Japanese White Pine is used and this is cleft-grafted on to seedlings of the Red Pine. The procedure is carried out almost entirely by the villagers of Hanoi who have given the name to the method. The graft is made about 6 in. (15 cm) above the base of the stock plant and it is this lower 6 in. (15 cm) that is twisted or coiled around several times. Each coil is held into place by tying to a small but stout wooden peg, inserted into the soil. The plants remain in the open for a further twelve months when the sticks are removed and the plants are transplanted into fresh ground. At this move, the roots are pruned back whilst the grafted head is bent down, each branchlet being held into place by pegs. After another twelve months, the leading shoot is removed and the other branchlets brought into the required positions by coiling with wire. In about five years, the plants will have taken on the appearance of wind-swept pines with weatherbeaten trunks, this being the most familiar of all the Bonsai styles.

KENGAI

The trunk grows down in cascading style in the manner of the Cascade Chrysanthemum. For this, the Japanese pines, *Cydonia japonica* and the maples are suitable.

The method is to plant in a container at an angle of 45° and so that the soil ball is slightly higher than the surrounding compost. The tree is held into position by a stake. The main shoot is then brought over the side of the container and held in position by wire fastened round the container. So that the wire will not cut through the bark, wrap round a piece of cloth before fixing the wire into position. It is also important to hold in place the soil ball by wire otherwise it will lift out of the container.

HANKENGAI

Here the trees are trained to grow horizontally. The method comes somewhere between Shakan and Kengai. The trunk is trained obliquely with the branches cascading downwards are previously described.

GROWING REQUIREMENTS OF BONSAI

When a single tree is grown in a container, this is known in bonsai as Ippon-ue and the tree of whatever style it represents, is known as Tankan. Where two trees are growing in the container, this is Sokan and here the effect will be greater if one tree is taller than the other when the style is known as Yose-ue. Where acacias and robinias are grown, they may be allowed to form suckers, the best of which are retained and grown on into trees. Although the 'trees' appear to be growing separately, they will be connected by their roots, a style which is known as Ne-tsuranari. Another method is to obtain a young tree with low horizontal branches. These are laid on the surface of the soil and pegged into position, the leader shoot (above the branches) being removed. From suitable buds, shoots will arise and roots will be formed on the underside which will grow down into the compost. When the shoots are several inches high, they should be pinched back to two buds to encourage them to grow bushy.

Where a tree has no horizontal branches, it may be stripped of all its branches and have the leader pinched out. It is then laid flat on the surface of the compost with its roots partially exposed. Along the upper part of the trunk, the buds will 'break' and form shoots with roots forming on the underside and they will make individual trees. This method is known as Ikada-buki.

The maples, white pine and other conifers may be induced to grow with their roots clasped to a stone placed in a container, a practice known as Ishi-zuki. Stones should be used which are neither too soft nor too hard and should have several fissures or openings into which the roots may be pressed. The method is to obtain some clay soil which is brought to a paste. This is used to fill in the cracks into which the roots are placed with their ends in the compost around the stone. Cover them with more clay and with sphagnum moss which will help to retain moisture about the roots. Conclude the operation by binding together with thin but strong twine. This may be removed after twelve months during which time the moss should be kept moist when the roots will have become tightly fixed into the stone. During the years ahead, they will grow thick and gnarled to give an appearance of plants growing in a rocky outcrop.

Unglazed containers are the most suitable for bonsai but glazed containers may be made suitable by the use of additional grit in the compost. But whatever type of container is used, drainage holes are essential. These should be covered with crocks to prevent them being filled in by compost. Over the crocks, place a 1 in. (2.5 cm) layer of grit, then a layer of compost, depending upon depth of container. This should be made up of 2 parts fibrous turf loam; 1 part leaf mould and 1 part sand. Then top up with more grit.

Bonsai containers are shallow, so before planting, trim back any long roots to encourage a fibrous rooting system and spread out the roots carefully. Re-potting may take place about every third year which is best done during a dormant period and when any tap roots are trimmed back. It is more realistic to use a few larger pieces of stone on the surface which should not be made too level. It should appear as natural as possible.

Correct drainage will facilitate watering which must be done with care, never leaving the compost for too long in a saturated condition. This means giving more water in summer when the roots are active and can utilise it, than in winter. Never water when not necessary and this may be determined by tapping the side of the container. If there is a dull thud, give no water until the tapping produces a hollow ringing sound. If moisture remains long on the surface, it is an indication that the compost is too firm.

BORDER PLANTS

FLOWERING PERIOD OF HARDY BORDER PLANTS

January
Helleborus niger

February
Helleborus niger

March
Helleborus orientalis
Omphalodes verna
Polyanthus
Primula denticulata
Pulmonaria angustifolia
Pulmonaria saccharata

April
Anchusa myosotidiflora
Armeria cephalotes
Auricula
Caltha polypetala
Caltha palustris
Centaurea montana
Helleborus orientalis
Mertensia paniculata
Mertensia virginica
Omphalodes verna
Polyanthus
Primula aurantiaca
Primula denticulata
Primula rosea
Pulmonaria angustifolia
Pulmonaria saccharata
Veronica gentianoides

May
Anchusa caespitosa
Anchusa myosotidiflora
Anthericum liliastrum
Aquilegia
Armeria cephalotes
Auricula
Caltha polypetala
Caltha palustris
Camassia cusickii
Campanula glomerata
Cardamine pratense
Centaurea montana
Centaurea pulchra major
Convallaria majalis
Coreopsis
Dicentra eximia
Dicentra spectabilis
Dornicum carpetanum
Euphorbia pilosa major
Geranium endressii

Geranium macrorrhizum
Lilium tenuifolium
Mertensia paniculata
Mertensia virginica
Nepeta mussinii
Omphalodes cappadocica
Omphalodes verna
Paeonia lobata
Paeonia mlokosewitchii
Paeonia officinalis
Paeonia peregrine
Paeonia sinensis
Pentstemon deutus
Polyanthus
Polygonum bistorta 'Superbum'
Primula chungensis
Primula florindae
Pyrethrum
Ranunculus gramineus
Thalictrum aquilegifolium
Tiarella cordiflora
Trolius europaeus
Veronica gentianoides

June
Achillea millefolium
Achillea ptarmica
Achillea serrata
Adenophora farreri
Althaea
Anchusa caespitosa
Anchusa italica
Anchusa myosotidiflora
Anthemis
Anthericum liliastrum
Asphodelus lureus
Astilbe arendsii
Astilbe simplicifolia
Baptisia australis
Betonica grandiflora
Betonica spicata
Camassia esculenta
Camassia leichtlinii
Campanula glomerata
Campanula lactiflora
Campanula latifolia
Campanula persicifolia
Cardamine
Centaurea montana
Centaurea pulchra major
Centaurea ragusina
Centaurea ruthenica
Centranthus ruber
Chrysanthemum maximum
Chrysogonum virginianum
Convallaria majalis
Coreopsis grandiflora
Coreopsis verticillata
Cynoglossum nervosum
Delphinium
Delphinium belladonna
Dicentra eximia
Dicentra formosa
Dicentra spectabilis
Doronicum carpetanum
Dracocephalum prattii
Eremurus bungei
Eremurus himalaicus
Eremurus olgae
Eremurus robustus
Erigeron
Eriophyllum caespitosum
Eryngium oliverianum
Eryngium planum
Geranium endressii
Geranium grandiflorum
Geranium macrorrhizum
Geranium pratense
Geranium sylvaticum
Geum
Helenium bigelowi
Hemerocallis
Hesperis matronalis
Heuchera
Incarvillea delavayi
Incarvillea olgae
Inula glandulosa
Inula royleana

June (*continued*)
Iris germanica
Iris kaempferi
Iris sibirica
Lilium candidum
Lilium hansonii
Lilium pomponium
Lilium pyrenaicum
Lilium umbellatum
Lupinus polyphyllus
Lychnis chalcedonica
Lychnis coronaria
Lythrum salicaria
Lythrum virgatum
Malva moschata
Meconopsis cambrica
Mimulus
Nepeta mussinii
Nepeta Souvenir de André Chaudron
Nomocharis mairei
Nomocharis pardanthina
Oenothera speciosa
Oenothera youngii
Ostrowskya magnifica
Paeonia sinensis
Pansy
Papaver orientale
Pentstemon barbatus
Pentstemon campanulatus
Pentstemon deutus
Pentstemon heterophyllus
Pentstemon schonholzeri
Polemonium coeruleum
Polygonatum multiflorum
Polygonum bistorta 'Superbum'
Potentilla
Primula chungensis
Primula florindae
Prunella grandiflora
Pyrethrum
Ranunculus gramineus
Santolina serratifolia
Sidalcea malvaeflora
Spiraea aruncus
Stachys lanata
Thalictrum aquilegifolium
Thalictrum dipterocarpum
Tiarella cordiflora
Tradescantia virgianiana
Trollius europaeus
Verbascum
Veronica amethystina
Veronica incana
Veronica spicata
Viola

July
Acanthus mollis latifolius
Achillea clypeolata
Achillea eupatorium
Achillea millefolium
Achillea ptarmica
Achillea serrata
Achillea taygetea
Aconitum fischeri
Aconitum napellus
Adenophora farreri
Althaea
Anaphalis nubigena
Anaphalis triplinervis
Anchusa italica
Anchusa myosotidiflora
Anthemis
Artemesia lactiflora
Artemesia ludovociana
Artemesia villarsi
Asphodelus lureus
Aster alpellus
Astilbe arendsii
Astilbe simplicifolia
Astrantia carnolica
Baptisia australis
Betonica grandiflora
Betonica spicata
Bocconia cordata
Campanula glomerata
Campanula lactiflora
Campanula latifolia

Campanula macrantha
Campanula persicifolia
Centaurea babylonica
Centaurea dealbata
Centaurea gymnocarpa
Centaurea macrocephala
Centaurea pulchra major
Centaurea ragusina
Centranthus ruber
Chrysanthemum maximum
Chrysogonum virginianum
Coreopsis grandiflora
Coreopsis verticillata
Cynoglossum nervosum
Delphinium belladonna
Delphinium ruysii
Dianthus barbatus
Dicentra eximia
Dicentra formosa
Dicentra spectabilis
Dracocephalum hemsleyanum
Dracócephalum prattii
Echinacea purpurea
Echinops humilis
Echinops nivalis
Echinops ritro
Echinops sphaerocephalus
Eremurus bungei
Eremurus himalaicus
Eremurus olgae
Eremurus robustus
Erigeron
Eriophyllum caespitosum
Eryngium oliverianum
Eryngium planum
Fritillaria imperalis
Funkia fortunei
Funkia japonica aurea
Funkia medio variegata
Gaillardia grandiflora
Galega officinalis
Galtonia candicans
Geranium endressii
Geranium grandiflorum
Geranium pratense
Geranium sylvaticum
Geum
Gypsophila paniculata
Helenium
Heliopsis gigantea
Hemerocallis
Heuchera
Inula afghanica
Inula glandulosa
Inula royleana
Iris germanica
Iris kaempferi
Iris sibirica
Kephalaria tartarica
Kniphofia uvaria
Ligularia clivorum
Lilium amabile
Lilium brownii
Lilium chalcedonicum
Lilium croceum
Lilium hansonii
Lilium martagon
Lilium regale
Lilum superbum
Lilium umbellatum
Lupinus polyphyllus
Lychnis chalcedonica
Lychnis coronaria
Lysimachia clethroides
Lysimachia punctata
Lythrum salicaria
Lythrum virgatum
Malva alcea
Malva moschata
Meconopsis baileyi
Meconopsis cambrica
Mimulus
Monarda didyma
Nepeta mussinii
Oenothera speciosa
Oenothera youngii
Papaver orientale
Pentstemon barbatus

July (*continued*)
Pentstemon campanulatus
Pentstemon heterophyllus
Pentstemon schonholzeri
Phlomis russelliana
Phlox decussata
Polemonium coeruleum
Polygonum bistorta 'Superbum'
Potentilla
Poterium obtusum
Primula florindae
Prunella grandiflora
Salvia glutinosa
Salvia haematodes
Salvia nemorosa
Saponaria officinalis 'Flore Plena'
Scabiosa caucasica
Scutellaris canescens
Sidalcea malvaeflora
Solidago canadensis
Spiraea aruncus
Spiraea palmata
Spiraea venustum
Stachys lanata
Statice latifolia
Thalictrum dipterocarpum
Thalictrum flavum
Thalictrum glaucum
Tradescantia virginiana
Verbascum
Veronica alata
Veronica incana
Veronica teucrium

August
Acanthus mollis latifolius
Acanthus spinosus
Achillea eupatorium
Achillea millefolium
Achillea ptarmica
Achillea serrata
Achillea taygetea
Aconitum fischeri
Aconitum lycoctonum
Aconitum wilsonii
Alstroemeria aurantiaca
Alstroemeria haemantha
Althea
Anaphalis nubigena
Anaphalis triplinervis
Anaphalis yedoensis
Anchusa italica
Anemone japonica
Anthemis
Arundo conspicua
Asclepias tuberosa
Asphodelus lureus
Aster acris
Aster alpellus
Aster amellus
Aster linosyris
Aster luteus
Astilbe arendsii
Astilbe simplicifolia
Astrantia carnolica
Betonica grandiflora
Betonica spicata
Bocconia cordata
Campanula glomerata
Centaurea dealbata
Centaurea gymnocarpa
Centaurea macrocephala
Chrysanthemum Korean
Chrysanthemum maximum
Chrysanthemum rubellum
Chrysogonum virginianum
Cimicifuga racemosa
Coreopsis grandiflora
Coreopsis verticillata
Delphinium belladonna
Delphinium ruysii
Dicentra formosa
Dierama pulcherrimum
Dracocephalum hemsleyanum
Dracocephalum prattii
Echinacea purpurea
Echinops humulis
Echinops nivalis

Echinops ritro

Erigeron

Eriophyllum caespitosum

Eryngium oliverianum

Funkia fortunei

Funkia japonica aurea

Funkia medio variegata

Gaillardia grandiflora

Galega cocinalis

Galtonia candicans

Gaura lindheimeri

Geranium endressii

Geranium pratense

Geum

Gynerium argenteum

Gypsophila paniculata

Helenium

Helianthus

Heliopsis gigantea

Hemerocallis

Heuchera

Inula afghanica

Inula glandulosa

Inula royleana

Kniphofia uvaria

Ligularia clivorum

Lilum auratum

Lilium batemanniae

Lilum henryi

Lilum maxwill

Lilium pardalinum

Lilum speciosum

Lilum superbum

Lilium tigrinum

Lobelia cardinalis

Lychnis chalcedonica

Lysimachia punctata

Lythrum salicaria

Lythrum virgatum

Malva alcea

Meconopsis cambrica

Mimulus

Monarda didyma

Montbretia

Nepeta mussinii

Oenothera speciosa

Oenothera youngii

Pentstemon barbatus

Pentstemon campanulatus

Pentstemon heterophyllus

Pentstemon schonholzeri

Phlomis russelliana

Phlox decussata

Physostegia speciosa

Platycodon grandiflorum

Polemonium coeruleum

Polygonum affine

Polygonum amplexicaule

Potentilla

Poterium obtusum

Salvia azurea

Salvia glutinosa

Salvia haematodes

Salvia nemorosa

Salvia uliginosa

Saponaria officinalis 'Flore Plena'

Scabiosa caucasica

Scutellaria canescens

Sidalcea malvaeflora

Solidago canadensis

Spiraea palmata

Spiraea venustum

Stachys lanata

Statice incana

Statice latifolia

Thalictrum dipterocarpum

Thalictrum flavum

Thalictrum glaucum

Tradescantia virginiana

Verbascum

Veronica alata

Veronica subsessilis

Veronica teucrium

September

Achillea eupatorium

Achillea taygetea

Aconitum lycoctonum

September (*continued*)
Aconitum wilsonii
Alstroemeria aurantiaca
Amaryllis belladonna
Anaphalis nubigena
Anaphalis yedonensis
Anemone japonica
Anthemis
Arundo conspicua
Asclepias tuberosa
Asphodelus lureus
Aster acris
Aster alpellus
Aster amellus
Aster cordifolius
Aster ericoides
Aster laterifolius
Aster luteus
Aster novae-angliae
Aster novi-belgii
Betonica spicata
Bocconia cordata
Centaurea dealbata
Centaurea macrocephala
Chrysanthemum Korean
Chrysanthemum rubellum
Chrysogonum virginianum
Cimicifuga cordifolia
Cimicifuga racemosa
Delphinium belladonna
Delphinium ruysii
Dierama pulcherrimum
Dracocephalum hemsleyanum
Echinacea purpurea
Eriophyllum caespitosum
Eryngium plaum
Gaillardia grandiflora
Geranium endressii
Geranium pratense
Geum
Gynerium argenteum
Helianthus
Heliopsis gigantea
Kniphofia uvaria
Helenium
Liatris callilepis
Liatris pycnostrachya
Liatris spicata
Ligularia clivorum
Lilium speciosum
Lilium tigrinum
Lobelia cardinalis
Lythrum salicaria
Malva alcea
Montbretia
Oenothera speciosa
Oenothera youngii
Pentstemon barbatus
Pentstemon schonholzeri
Phlox decussata
Phygelius capensis
Physalis franchetti
Physostegia speciosa
Physostegia virginiana
Platycodon grandiflorus
Polygonum affine
Polygonum amplexicaule
Potentilla
Poterium obtusum
Salvia azurea
Salvia haematodes
Salvia nemorosa
Salvia uliginosa
Scabiosa caucasica
Solidago canadensis
Statice latifolia
Stokesia cyanea
Tradescantia virginiana
Veronica alata
Veronica crinita
Veronica subsessilis
Veronica teucrium
Yucca filamentosa

October
Achillea eupatorium
Amaryllis belladonna
Anaphalis nubigena

Anaphalis triplinervis
Anemone japonica
Arundo conspicua
Aster acris
Aster amellus
Aster cordifolius
Aster ericoides
Aster laterifolius
Aster novae-angliae
Aster novi-belgii
Centaurea dealbata
Chrysanthemum Korean
Chrysanthemum rubellum
Cimicifuga cordifolia
Eriophyllum caespitosum
Geranium endressii
Gynerium argenteum
Helenium
Helianthus
Liatris callilepis
Liatris spicata
Montbretia
Pentstemon schonholzeri
Phygelius capensis
Physostegia speciosa
Physostegia virginiana
Salvia haematodes
Salvia uliginosa
Scabiosa caucasica
Schizostylis coccinea
Solidago canadensis
Stokesia cyanea
Tradescantia virginiana
Veronica crinita
Yucca filamentosa

November
Aster cordifolius
Aster novi-belgii
Chrysanthemum Korean
Physostegia virginiana
Scabiosa caucasica
Veronica crinita

December
Chrysanthemum Korean
Helleborus niger

BORONIA

B. megastigma is an evergreen shrublet, native of Australia, which grows 18 in. (45 cm) tall and bears during early summer, yellow and crimson flowers which possess a sweet fragrance. It may be raised from seed by using a propagator for it requires a temperature of 65°F. (18°C.) and even so, the seed may take about ten weeks to germinate. However, once germination has taken place, the plants will be quite happy in considerably lower temperatures, for the plant is almost hardy, requiring a winter temperature of 48°F. (9°C.) and from the end of June until early September, the plants may be allowed to stand outdoors in a partially shaded position. It is, however, best to allow them to bloom indoors first, when their rich perfume will scent the greenhouse.

As the plants resent root disturbance, the seed should be sown one to a 3-inch (7.5 cm) pot containing a compost made up of 2 parts peat or leaf mould and 1 part coarse sand. Barely cover the seed with a little peat or leaf mould dust and keep moist, otherwise the seeds will take even longer to germinate.

The plants should be grown on until the following spring, watering sparingly during winter, A close, stuffy atmosphere must be avoided. The plants should be moved to size 48 pots in April, using the John Innes Potting Compost. Pot firmly and pinch out the leading shoot to encourage the plants to 'break' at the base and grow bushy. After flowering in early summer the plants should be repotted and placed outdoors until September, during which time they should be kept well watered.

BOTTLEBRUSH PLANT, *see* Poterium

BOUGAINVILLEA

It may be used to train up the wall of a lean-to greenhouse. It bears its greenish-white flowers with their richly coloured bracts during late summer and autumn, and will grow to a height of 8 ft. (2 m) It requires a winter temperature of between 45° and 50°F. (7°–10°C.) during which time almost all moisture must be withheld. With the increase in the warmth of the sun in spring, watering should be resumed and a top dressing of peat and decayed manure be given. To keep the plant vigorous, prune back early in March those side shoots formed the previous year, to within 2 in. (5 cm) of their base. The new hybrids are more free flowering and they like a compost made up of 3 parts turf loam, 1 part leaf mould or peat and coarse sand.

BOUVARDIA

Native of Mexico, the Scarlet Bouvardia, *B. ternifolia* makes a compact bushy plant which will bloom well in the sunny window of a warm room. It bears its flowers, like red jasmine, in terminal racemes and will bloom during spring and summer after a winter rest. After flowering, keep the compost on the dry side but do not dry off completely. Early in spring, cut back the shoots several inches and commence watering when the plants will come quickly into bloom in a temperature of 50°F. (10°C.). Propagate by cuttings removed in spring and which are rooted in a sandy compost around the side of a pot. When rooted, pinch back the plants to encourage bushy growth. They will come into bloom in twelve months and will remain continuously in bloom only in a temperature of 68°F. (20°C.) when it will make new wood throughout the year.

BOWSTRING HEMP, *see* Sansevieria

BRACHYCOME

An annual known as the Swan River Daisy, is native of Australia and enjoys a light, dry soil and a sunny position. It attains a height of only 9 in. (22 cm) and bears its blue, purple and lavender blooms from mid-June until October. Sow early in May and thin to 6 in. (15 cm). *B. iberidifolia* 'Blue Star' hybrid strain is the best.

BRACHYGLOTTIS

B. repanda is a New Zealand evergreen with large leaves, downy on the underside and in July and August bears greenish-white flowers scented like Mignonette. It will grow 15 ft. (4.5 m) tall and should be given a sunny situation, protected from cold winds, or a sunny wall. Plant in April in a rich loam and do the minimum of pruning. Propagate by cuttings inserted under glass.

BRAMBLE, *see* Rubus

BRODIAEA

Known as the Californian Hyacinths the brodiaeas are exquisite little plants in flower from the end of May until mid-July. The blooms are carried on sturdy stems, up to half a dozen gentian-like flowers appearing on the top of each stem. They are not completely hardy and must be given a mulch in early December and over the clumps should be placed a quantity of weathered ash. There are, however, exceptions to the rule of hardiness. *Brodiaea coccinea* is able to withstand a severe winter in a well-drained soil, and *B. uniflora*, the Tritelia of our gardens, is completely hardy. All may be grown outdoors unprotected in the south and west where they may be grown to perfection in a sandy, well-drained soil, but for the more northerly garden they should be given winter protection. Being natives of California they must be given a sunny position and a soil which will warm up quickly in early summer. To achieve this, plenty of sand, shingle and leaf mould should be well worked into the soil before planting takes place in September. Some thoroughly rotted manure they enjoy too, for they are plants which should not be disturbed when once established. A mixture of dry poultry manure, peat and some strawy farmyard manure, thoroughly mixed and to which some loam and coarse sand has been added, will prove ideal.

Where the bulbs are being grown in grass or in the shrub border, the same compost should be placed into each sq. ft. of ground where the bulbs are to be set. Plant 4 in. (10 cm) deep, placing some sand around each bulb. Plant 4 in. (10 cm) apart and in groups of four. Several varieties are ideal for the rockery and are at their best when planted about stones with a large proportion of shingle around the bulbs and above them for winter protection.

SPECIES

Brodiaea bridgesi. Taller than the other species, growing to a height of 2 ft. (60 cm) and producing its pale mauve flowers throughout June.

B. coccinea. It is one of the outstanding plants of the garden. The brilliant red blooms are attractively tipped with green, making a most striking display throughout June and July. They are carried on 12in. (30 cm) stems.

B. crocea. This is a variety well suited to the rock garden for the dainty yellow flowers are carried on 6 in. (15 cm) stems and borne in profusion during May and early June.

B. grandiflora. Dwarf of habit and of an enchanting sky-blue colour, this species produces its bloom in profusion throughout June.

B. laxa. One of the taller-growing Brodiaeas and a most valuable cut flower. So inexpensive are the bulbs that they could well be planted for commercial cut-flower production with a view to testing the market for this flower. The tubular blooms are of richest purple.

B. multiflora. It bears umbels of deep mauve flowers in profusion. Of dwarf habit, it is at its loveliest throughout July.

B. peduncularis. Rather less hardy than the others but should be planted for its violet and white flowers which are produced on 12 in. (30 cm) stems.

B. tubergeni. Raised from a *B. peduncularis-B. laxa* cross, this new introduction produces flowers of the palest porcelain blue which appear in profusion.

If required for pot culture, several of the species give good results in this way, the best being *B. multiflora, B. grandiflora* and *B. crocea*. They should be potted during November, given cold-frame protection and will come into bloom late in spring if taken into a warm greenhouse or the home during the early new year.

BROOM, *see* Cytisus

BROWALLIA

Natives of South America, they are perennials which may readily be raised from seed and are usually given half-hardy annual treatment. *B.*

speciosa 'Major' is the best-known. It grows about 16 in. (40 cm) tall and makes a bushy plant, covered for weeks in blue tubular flowers which have a white throat. For autumn flowers, sow seed in March in the John Innes Compost and germinate in a temperature of 60°F. (16°C.). For late winter and spring bloom, sow in July. Move to small pots when the seedlings can be handled and to flowering size pots four to five weeks later containing the John Innes Potting Compost. Water sparingly in winter and provide a temperature of 50°F. (10°C.).

BRUNSVIGIA

A genus of thirteen species, native of tropical and South Africa, named in honour of the House of Brunswick and differing from Crinum and *Amaryllis belladona* in the irregular flowers. The bulbs are large, often measuring 6–8 in. (15–20 cm) in diameter and they are more tender, requiring in the British Isles and north Europe, the protection of a warm greenhouse or garden room during winter. The tunicated bulbs have strap-like leaves which appear after the flowers which are funnel-shaped and are borne in large umbels on a 15 in. (38 cm) scape. From twelve to fifty comprise the umbel and they are arranged evenly around the scape on long pedicels.

They bloom in July and are potted in spring, one 4 in. (10 cm) bulb to a size 60 pot and using a mixture of fibrous loam, decayed manure, leaf mould and coarse sand in equal parts. They require firm planting and should have the neck and shoulder just above the level of the compost. They should be given a temperature of 55°F. (13°C.) to start them into growth and as the sun increases in strength they will require increasing supplies of moisture. Make sure that the pots are well 'crocked' before planting and to keep the compost sweet it is advisable to place a few small pieces of charcoal about the crocks. The pots should be sufficiently large as to allow a space of about 1 in. (2.5 cm) between the side and the bulb. Do not plant too deeply as water may enter the neck and cause it to decay.

About six weeks after planting, the buds will be seen pushing up through the wall of the bulb and possibly two stems will be produced in a season depending upon the size of bulb. It will be advisable to turn the pots each day so that the stems grow up straight and though the plants will require ample supplies of moisture the compost must not become stagnant or the roots will decay. They will require a sunny window until showing colour when they should be moved to a position of partial shade. After flowering, the bulbs continue to make lush growth and watering should continue until they begin to die back in winter. They are then given almost no water until they are brought into bloom again in spring.

After flowering, remove the dead flowers and place the pots outdoors, if possible beneath the eaves of a house and in a position of full sun for the bulbs to ripen, without which they will not bloom the following year. For the bulbs to form roots and bear flowers at the same time, they require all the help possible and will benefit from a weekly application of dilute manure water until the leaves begin to wither.

If well-cared for, the bulbs will last twenty years or more but should be re-potted every three or four years taking care to disturb the roots as little as possible.

In the warmer parts, the bulbs may be planted outdoors in a sunny position, preferably at the base of a wall. They are planted 4 in. (10 cm) deep and should be protected during winter by a mulch of decayed manure and leaves which will also provide valuable plant food.

Brunsvigia josephinae and most other species rarely produce offsets and propagation is usually by sowing seed in a sandy compost and in a propagator. They will however, take twelve to fifteen years to bear flowers.

SPECIES

Brunsvigia gigantea. Syn: *B. orientalis.* The bulb may grow to 6 in. (15 cm) across and from it arises a 12 in. (30 cm) scape at the top of which is formed a circular umbel of twenty to thirty funnel-shaped flowers held on long pedicels. The umbel may measure up to 20 in. (50 cm) across whilst the flowers are brilliant red. They bloom in spring in their native Cape Peninsular and in August and September in the British Isles, Europe and North America. The plant has leaves 12 in. (30 cm) long and 4 in. (10 cm) wide.

B. josephinae. Introduced in 1814, it is the most free flowering species. It has grey-green strap-like leaves 2 ft. (60 cm) long and 2 in. (5 cm) wide and it blooms during July and August, bearing on an 18 in. (45 cm) scape, a large circular umbel of about thirty scarlet blooms, sometimes twice that number. The funnel-shaped blooms are about 3 in. (7.5 cm) long. The form minor grows only 9 in. (22 cm) tall and bears twenty to forty smaller pale red blooms in a symmetrical umbel.

BRYOPHYLLUM

Interesting and in some cases quite beautiful when in bloom, they are known as Air Plants for they reproduce themselves by forming numerous tiny plants along the edges of the leaves which fall away at the correct time and will quickly root when in contact with soil. The plants also bear attractive pendent flowers of tubular shape.

B crenatum is one of the best forms for it grows only 12 in. (30 cm) tall. It has round grey leaves and bears its yellowish-orange flowers throughout winter. *B. pinnatum* grows almost twice as tall, having oval leaves and bearing its tubes of green during the winter. *B. tubiflorum* is also tall-growing. It has most attractive leaves and bears its magenta-brown flowers during winter, at which time the plants should be given as sunny a position as possible.

The compost should be composed of 3 parts fibrous loam, 3 parts peat, 1 part silver sand and 1 part lime rubble, for the plants are never happy in an acid soil. A deep seed pan is best for potting so that there will be plenty of room around the plant for the plantlets to take root as they fall from the leaves.

A winter temperature of 45°F. (7°C.) should be provided, whilst the plants will not tolerate draughts.

BUCKTHORN, *see* Rhamnus

BUDDLEIA

B. davidii, the Butterfly Bush, so much frequented by butterflies and bees, is an outstanding late summer and autumn flowering shrub. It makes an abundance of cane-like stems which should be cut back after flowering, and apart from its grey-green leaves bears its long flower spikes in graceful arching form. It is hardy anywhere and grows well in all soils. Effective where planted together are 'Royal Red' (purple-red) and 'White Profusion'. Another excellent combination is 'Empire Blue', with its spikes of powder-blue, and the pink-flowered 'Fascination'.

B. globosa is a native of Chile and except during severe weather is evergreen with attractive grey-green foliage. During May and June it bears its striking orange flowers, which are the size of small golf balls and are sweetly scented. *B. globosa* will require almost no pruning and is one of the most attractive plants of the shrubbery.

Another lovely species is *B. alternifolia* a graceful small-leaved shrub, its arching branches clothed in June with drooping lilac flowers which are sweetly scented.

Propagate by cuttings of the new season's wood, inserted in a frame (or in the open, in a garden with a mild winter climate) in sandy compost.

BULBOCODIUM

It is often mistaken for a crocus and is often taken for a spring-flowering species of the autumn crocus, the colchicum, for it comes into bloom before producing its leaves. It has another similarity in that it is called the 'spring' meadow saffron. It appears to be a colchicum flowering at the wrong time! The plant is most valuable for early March flowering in almost any position in the garden. It is little troubled by soil conditions though it grows best in one that is well-drained, and it should be planted in a position where it is able to receive the early spring sunshine. It is not such a lover of the cool, shady places as the Dog's Tooth Violet. It is at its best on a rockery, or planted in drifts beneath a wall facing the sun. Plant the bulbs 3 in. (7.5 cm) deep in autumn.

B. vernum produces its lavender-pink flowers in March and April. When open the blooms are star-shaped which adds to their charm. The bulbs soon become established and they will increase rapidly.

BULBS, INDOORS

Where growing bulbs indoors one must foresee the numerous pitfalls which may result in disappointment. There are two methods by which the bulbs may be brought into bloom: they may either be planted as bulbs or corms at the appropriate time, and after a period of darkness during which they become well rooted, brought into a warm room; or for those with a garden, they may be lifted from the open ground as soon as plant growth appears above the soil, potted, and brought into bloom in a warm room. The latter method is the simplest, for by the time the plants are ready for lifting, the flower buds will already have formed; and all that is required will be to introduce them to a warm room and to keep them comfortably moist at the roots. They will need no special compost for the plants will occupy the pots for only a short time. Any friable loam will be suitable. The plants should be given a covering of dried leaves in early winter to provide protection against hard frost and to make it easier to lift them as soon as plant growth is seen.

It will be early in February that the spring-flowering plants will be ready for lifting; and this will often be a time of hard frost. The plants may either be lifted individually or in small clumps and in this way the snowdrops, scillas, dog's tooth violets, miniature daffodils and dwarf irises may be brought early into bloom. Remember, when lifting, that most of the bulbs planted outdoors will be at least 4 in. (10 cm) deep, which means that it will be necessary to place the trowel 6 in. (15 cm) below the surface if the roots are not to be harmed. Plants to be lifted in this way and taken indoors should not be subjected to high temperatures. The outdoor spring-flowering plants are extremely hardy and will not tolerate forcing conditions. Until they have become re-established in the pots, the plants should be placed in an outhouse or frame where they will remain for a week; then they may be taken indoors and placed in the window of a room that is not too warm. After flowering, the bulbs should be replanted into the same positions from which they were lifted and the same bulbs should not be taken indoors again for at least two years. This method will not, of course, be suitable for those living in a small town house or flat with little or no space for the growing of bulbs outdoors. In such cases, the bulbs may be planted in pots or earthenware bowls in early autumn and they will never see the open ground, unless maybe after they have bloomed, when they could be handed on to someone with a garden, to flower outdoors in the years to come.

THE PLUNGE BED

Those who have a small garden or courtyard will be able to root their bulbs outdoors in a plunge bed, and this is the most successful method (see plates 8 and 9). No bulb will give a good account of itself unless it first builds up a sturdy rooting system, and this means that it must be prevented from making top growth until it has formed its roots. Light must, therefore, be entirely excluded and this may be done either by placing the bulbs beneath a bed of ashes outdoors or in a cool, dark place indoors. The former method has much to recommend it, for where the pots or pans can be placed on a bed of ashes to encourage drainage, with a 6 in. (15 cm) covering of ashes over the top to exclude light, the bulbs will be cool and may be kept moist but not so wet as to cause them to decay. Where possible, the pots should be placed in ashes made firm, or preferably placed over a hard surface so that worms will not penetrate through to the pots. Sand or grit may be used as an alternative to ashes.

As the plunge bed will be exposed to the elements until the end of November, the bulbs will receive ample supplies of moisture, without which they will be unable to form a vigorous rooting system. This is the reason why there are so many failures with bulbs confined indoors, where a dry cupboard or somewhere beneath the stairs will be the only available place in which to subject them to the dark. Better would it be to place them beneath the kitchen or scullery sink away from warmth, for here the bulbs would be growing in a cooler atmosphere. Where possible, make up a plunge bed outdoors even if no more than half a dozen pots are being grown, for the bulbs will receive ample supplies of moisture and remain cool. In such conditions they will build up vigorous roots which will be seen through the drainage holes when the pots are removed from the plunge bed; and once in a warm room, the bulbs will be stimulated into growth without delay.

The bulbs should be introduced to the light by degrees. First, the ashes are shaken from the pots, which should be cleaned with a damp cloth before going indoors. There they should be placed in half light for several days, otherwise the foliage, which will have appeared already, may turn brown at the edges and so spoil the display. Where the plants are growing in the dry atmosphere of a room heated by a gas fire or by central heating, stand the pots or bowls in earthenware plant pot 'saucers', which are glazed on the inside and so may be kept filled with water.

At all times the compost should be kept comfortably moist; and as the plants make growth and come into bloom this will mean that more frequent waterings will be necessary. There can be no hard-and-fast rule as to watering; this must depend upon whether the pots are placed in a sunny window or one facing north; upon the compost used; and upon the temperature of the room. Also, some bulbous plants will require more moisture than others, hyacinths especially. But do not conclude that the bulbs require moisture whenever the surface becomes dry, for lower down the compost may still be damp. First give the pot a tap with a piece of wood, and if there is a dull 'thud', water should be withheld until the tapping produces a 'ringing' sound; then give a thorough soaking by watering round the side of the pots. To prevent plant growth from being drawn to the light, it will be advisable to turn the pots round a little each day. In this way foliage and flower stems will grow straight and sturdy.

BULBS IN BOWLS

Bulbs rooted in a plunge bed will rarely suffer from lack of moisture, neither should they be troubled by excess moisture remaining about the roots if the pots are well crocked and placed on a bed of ashes. Bulbs grown in bowls without drainage may suffer from both defects, with the result that the flowering display will prove disappointing. Where there is not adequate drainage, it is difficult to give the correct amount of moisture for the bulbs to make vigorous root growth, yet without causing them to decay. Moisture is necessary to start the roots into growth and failure in this respect combined with too warm air atmosphere will stimulate the bulbs into early growth at the expense of root formation. Lack of moisture will cause stunting, the blooms being borne on too short a stem, which will greatly detract from the display. It is all too commonly believed that if the bulbs are placed in warmth as soon as they are planted in the bowls, they will come into bloom more quickly. But where the bulbs are stimulated into growth before the roots have formed, stunting will result, and in some cases it may cause the bulbs to be quite 'blind'.

Never should the bulbs be placed in a room where there is heat until they have had two months in which to form their roots under cool conditions. Place them in a cellar, garage or outhouse; or, where provided with drainage holes, there is nowhere better than in a bed of ashes exposed to the elements. Lack of light and completely cool conditions are essential factors in the first eight weeks of all newly-planted bulbs to be flowered indoors; and with most of the miniature bulbs, such as the daffodils, snowdrops, irises and chionodoxas, there is preference for cool conditions throughout. The secret of success with all bulbs to be brought into bloom indoors is to allow them plenty of time to form their roots. Introduce them to the light as slowly as possible and keep the compost neither too dry nor too moist. Extremes, both of temperature and of moisture must be avoided.

Specially prepared fibre, which is generally mixed with oyster shell and charcoal to maintain sweetness where containers are without drainage holes, is the usual medium for indoor rooting. Bulb fibre is composed mainly of dry peat or coconut fibre which will require a considerable amount of moisture to make it damp amd friable before the bulbs are planted. Never plant into dry bulb fibre, for it is quite impossible to add the necessary moisture afterwards; and this is where so many go wrong. To give that little extra 'body' to the compost, so that moisture requirements may be better controlled, try to mix in a small quantity of pasture loam; a bucketful brought home in the boot of the car is sufficient for a dozen pots or bowls when mixed with the fibre. While forming roots, the bulbs will require little moisture in such a compost, but more should be given as soon as growth commences. And be sure to keep them cool.

After flowering, the bowls should be placed in a shady corner outdoors, perhaps beneath the greenhouse bench or in a cellar or garage, so that the foliage can be dried off and the bulbs may be stored until required again; or they may be divided and replanted in the open ground as soon as they have finished flowering.

FORCING BULBS

There are a number of bulbs which will withstand warm conditions either in a heated greenhouse or in the home. The tiny 'Duc van Thol' tulips, obtainable in scarlet, white and yellow, are grown under specialized forcing conditions by the million by nursery growers to supply the Christmas and New Year market, the plants being transferred when in bud and bloom from the forcing boxes to fancy bowls. It is possibly that because these charming plants are usually presented in this way, that enthusiasts have come to believe that all bulbs will prove more successful where grown throughout in a fancy bowl where better results will be obtained by using an earthenware container provided with ample drainage holes.

The early flowering hyacinths 'Rosalie' and 'Vanguard' will also stand up to hard forcing to flower in the early New Year, while a number of dwarf early single tulips, which grow about nine inches tall, may be brought into bloom during the latter days of March. Amongst the most suitable are 'Proserpine', silky carmine-pink; 'Joffre', deep golden yellow; and the scarlet-flowering 'Brilliant Star'. Both the 'Duc van Thol' tulips and the early hyacinths will often have been 'prepared' so that they will come quickly into bloom when subjected to heat. To achieve this condition, they will have been kept in cold storage, and should be planted almost immediately they are received. But this does not mean they may be taken into heat at once. It is more important that the bulbs should be allowed fully eight weeks in which to form the strong rooting system which is vital if they are to be subjected to a high temp. The 'Duc van Thols' should be planted in boxes or pots about October 1st, together with the other early tulips and hyacinths; and after eight weeks under cool, dark conditions, they may be introduced to a warm greenhouse or warm room, and will be happy in a temperature of up to 60°F. (16°C.) In such a temperature the bulbs will come into bloom by the end of December, but in ordinary warm room temperatures, not until late in January. The hyacinths and the other early tulips, will not bloom until March and should not be taken indoors until mid-December.

For forcing, top-quality bulbs will be required and they will not be of use for forcing again, though they may still be planted in the garden. Deep earthenware bowls or pans are the most suitable containers and five or six 12 cm. bulbs may be planted in each. The bulbs should be handled carefully as tulips bruise easily. When planted, the top of the bulbs should be just covered with the compost. Water thoroughly and place the bulbs beneath ashes either outside or in a cellar or garage where they should require no further watering, until taken into a warm temperature (see plate 11). The cooler the bulbs are kept whilst rooting, the more rapidly will they respond to warmth; but do not subject them to sudden changes of temperature.When ready for forcing, the bulbs should first be given a temperature of about 45°F. (7°C.) and this should be increased gradually until a maximum of 60°F. (16°C.) may be achieved. As the bulbs make growth they will require copious amounts of water under these conditions, and if lacking moisture at any time flowering will be delayed and stunted growth may result.

As the plants come into bloom, they may need some support, for the stems will not be so sturdy as where 'grown cool'. Wire supports, procurable from any sundriesman, should be used, and will be almost invisible. As soon as the plants come into bloom they should be given reduced temperatures so as to prolong the flowering season as much as possible, though fluctuations of temperature must be avoided and draughts eliminated, otherwise the stems may collapse.

Bulbs may also be grown indoors without soil or fibre. Hyacinths are grown in special glass jars, the lower part of which is filled with water and in the upper part is placed the bulb, the base of which rests just above the water level into which the roots will grow. In another type, the bulb sits between three lips or brackets so that it cannot topple over when moved or when in bloom. The containers may also be of transparent plastic.

Smaller bulbs such as crocus, scillas, grape hyacinths and miniature daffodils may be grown in what is known as 'acorn' glasses. They are only slightly larger than a matchbox and the bulbs are grown in the same way. They should be 'planted' in the glasses at the normal time and placed in a dark cool place while the lower portion of the container is filled with roots. This will take about two months but each week check for water evaporation and add more as required. Then gradually introduce to the light when the bulbs will come into bloom.

Prepared bulbs of the 'Paperwhite' narcissus may be planted in bowls and covered with pebbles, only the neck of the bulb being left uncovered. They may be placed directly in the window with no time in darkness and kept moist. If planted in October, they will bloom early in the New Year.

BULBS, IN THE ALPINE GARDEN

It is in the alpine garden that the miniature bulbs are to be seen at their best, for it is here that they find conditions similar to those of their native haunts. Often high above sea level, they are to be found snuggling up against large boulders for protection from the cold winds of spring. They will remain healthy and vigorous and bloom year after year if planted in a well-drained soil though more small bulbs are lost each year through excess moisture remaining about the roots than for any other reason. They grow naturally in pockets of sandy loam overlying rubble, where sufficient moisture will be retained in summer and excess will drain away in winter. In such conditions, the plants will remain healthy almost indefinitely. Dwarf conifer trees may be used to provide a background; the smallest of the alpine bulbs may be used for a trough garden or window box.

Plant liberally for most of the miniatures are admirable in having foliage that is neat and tidy and which dies back without looking untidy. In a town garden perennial plants are often troubled by deposits of soot, so the fact that the foliage dies down after flowering is a great help. In this respect the compact 'Juliae' primroses are similar, and for a few weeks in winter they lose their foliage entirely. They should be planted with the bulbs on a town rockery to provide a pleasing carpet for the spring-flowering bulbs.

To maintain both bulbs and plants in a healthy condition, they should be provided with a mulch of peat or leaf mould to which is added a small amount of decayed cow manure or used hops. This is worked into the soil around the plants and over the bulbs, where they are not covered with carpeting plants. The foliage of the plants may be partially hidden by pressing it beneath nearby plants or by pinning it to the soil as it dies back. It may also be hidden, by using plants of shrubby habit which grow about 6 in. (15 cm) tall. If the spring and early summer-flowering plants are planted in the same group as those that bloom in later summer and early autumn, this will help to hide the dying foliage of the early-flowering bulbs, which must not be entirely removed until it has had time to die back.

It is advisable to remove from the rockery all leaves which may have fallen from nearby trees in late autumn and which become lodged about the stones and plants. They not only look untidy but will become saturated, and may cause nearby plants and bulbs to decay should the winter be unduly wet.

Provided both plants and bulbs are given the care necessary for their continued health, they should remain vigorous and free flowering for many years without the need to lift and divide. Overcrowding should, however, be guarded against and where this occurs, the bulbs may be thinned by lifting a number from each clump so as to allow more room for those left in the ground. Those which have been lifted may be divided and replanted elsewhere.

NAME	FLOWERING TIME	HEIGHT
Allium cyaneum	July-August	4 in. (10 cm)
Anomatheca cruenta	July-August	6 in. (15 cm)
Brodiaea crocea	May-June	6 in. (15 cm)
Brodiaea minor	June	6 in. (15 cm)
Bulbocodium vernum	March	3 in. (7.5 cm)
Chionodoxa luciliae	March-April	4 in. (10 cm)
Chionodoxa sardensis	March	4 in. (10 cm)
Chionodoxa tmoli	April	4 in. (10 cm)
Crocus ancyrensis	February-March	2 in. (5 cm)
Crocus balansae	March	1–2 in. (2.5–5 cm)
Crocus fleischeri	February-April	2 in. (5 cm)
Crocus korolkowi	January-March	1–2 in. (2.5–5 cm)
Crocus laevigatus	December-February	3 in. (7.5 cm)
Crocus medium	October	3 in. (7.5 cm)
Crocus minimus	March-April	1–2 in 2.5–5 cm)
Crocus nudiflorus	October-November	3 in. (7.5 cm)
Crocus ochroleucus	November-December	2 in. (5 cm)
Crocus olivieri	April-May	2 in. (5 cm)
Crocus sativus	October	2 in. (5 cm)
Crocus vernus	March	2 in. (5 cm)
Cyclamen alpinum	December-February	3 in. (7.5 cm)
Cyclamen atkinsii	December-March	4 in. (10 cm)
Cyclamen cilicium	September-November	3 in. (7.5 cm)
Cyclamen graecum	July-September	4 in. (10 cm)
Eranthis cilicica	February-March	2 in. (5 cm)
Eranthis hyemalis	January-February	2 in. (5 cm)
Erythronium dens-canis	March-April	6 in. (15 cm)
Erythronium hendersonii	April	4 in. (10 cm)
Fritillaria citrina	April-May	6 in. (15 cm)

NAME	FLOWERING TIME	HEIGHT
Fritillaria pudica	April-May	4 in. (10 cm)
Galanthus allenii	April	4 in. (10 cm)
Galanthus byzantinus	December-January	4 in. (10 cm)
Galanthus latifolius	March-April	3 in. (7.5 cm)
Galanthus nivalis	February-March	5 in. (12.5 cm)
Hyacinthus azureus	April	5 in. 12.5 cm)
Hyacinthus dalmaticus	April	4 in. (10 cm)
Iris bakeriana	January-February	6 in. (15 cm)
Iris danfordiae	February	4 in. (10 cm)
Iris histrioides	March	6 in. (15 cm)
Iris reticulata	March-April	6 in. (15 cm)
Muscari argaei album	May-June	4 in. (10 cm)
Muscari armeniacum	April-May	6 in. (15 cm)
Muscari polyanthus album	April-May	6 in. (15 cm)
Narcissus bulbocodium	March-April	6 in. (15 cm)
Narcissus canaliculatus	April	6 in. (15 cm)
Narcissus capax plenus	April	6 in. (15 cm)
Narcissus juncifolius	April	4 in. (10 cm)
Narcissus minimus	February-March	4 in. (10 cm)
Narcissus nanus	March-April	6 in. (15 cm)
Narcissus rupicola	April	4 in. (10 cm)
Narcissus triandrus	April-May	6 in. (15 cm)
Narcissus watieri	May	4 in. (10 cm)
Puschkinia scilloides	March-May	4 in. (10 cm)
Scilla bifolia	February-March	6 in. (15 cm)
Scilla tubergeniana	February-March	5 in. (12.5 cm)
Scilla verna	April	4 in. (10 cm)
Tulipa australis	April	6 in. (15 cm)
Tulipa batalinii	April	4 in. (10 cm)
Tulipa biflora	March-April	5 in. 12.5 cm)
Tulipa dasystemon	April-May	6 in. (15 cm)
Tulipa linifolia	May	6 in. (15 cm)
Tulipa maximowicizii	April	6 in. (15 cm)
Tulipa persica	May	3 in. (7.5 cm)
Tulipa pulchella	March-April	4 in. (10 cm)
Tulipa wilsoniana	May	6 in. (15 cm)

BULBS, OUTDOORS

There are three ways of growing bulbs in grass. They may be: (a) planted in grass which is rough and is cut only twice a year; (b) planted in a lawn which must be kept neat and tidy; or (c) planted in grass beneath tall trees, where lack of sunshine means that the sparse grass will require little or no attention. In each case choose vigorous, long-living bulbs and plant them where they will give the best display.

PLANTING IN ROUGH GRASS

Grass must be kept under control if the garden is to look tidy whilst the bulbs must be allowed time to die back naturally before their foliage is removed.

Bulbs in coarse grass should have finished flowering by the end of May. This will allow time for the foliage of the May-flowering bluebell, the summer snowflake and the later-flowering tulip species to die down before the grass is cut and made tidy in summer. Autumn-flowering bulbs may also be planted in coarse grass, but choose those which bloom as early in autumn as possible. This will allow the foliage to die down before the grass is made tidy towards the year's end. Alternatively, the autumn-flowering bulbs may be confined to a corner of the coppice or orchard where the grass is cut in mid-summer only. For the same reason, bulbs for planting in a lawn should be those which finish flowering by the end of April so that their foliage has time to die back before it is necessary to give the grass its first cut.

If bulbs growing in grass are to remain vigorous and free flowering, as well as increasing each year, they must be treated with the utmost respect after flowering. This is the time when the sap from the leaves is returning to the bulb, and to allow it to do so gradually will mean a vigorous bulb, capable of continuing its reproduction and flowering the following year. If the foliage is removed too soon, the bulb will quickly lose vitality and may stop flowering altogether. After flowering, most bulbous plants tend to become untidy, the leaves turning first yellow then brown, so that there is always the temptation to remove them before they have fully died back. For this reason, bulbs are best grown in a border or shrubbery, under mature trees where little else will grow, or around the base of small trees, where there will be no urgency about removing their foliage.

Plant with a view to enjoying as long a period of flowering as possible but many of the miniature bulbs, those needing an open, sunny situation those which bloom on too short a stem, will be unsuited for planting in coarse grass. Select those which come into bloom late in spring, with others to continue the display through the summer and into autumn. Those blooming during midsummer must be omitted where it is necessary to cut the grass after the late spring and early summer bulbs have died back. The orchard or wild garden could be colourful with bulbs in bloom during April and May and again during September and October, with the grass-cutting done round about July 1st, at its point of maximum growth.

At this date the late spring-flowering bulbs will have died back and those which are to bloom in autumn will not have begun to form their leaves. If cutting is delayed for more than a day or two, the autumn-flowering bulbs will be seriously damaged. The flowering season may be prolonged where it is also possible to plant miniature bulbs in short grass, for they bloom during the first days of the new year and provide colour until those growing in rough grass are ready to come into bloom.

First to bloom will be the snowdrops. A 3 cm bulb will produce a bloom during its first season but the best method of planting is to lift and divide the clumps immediately after flowering, replanting into grass with a trowel. Attractive as are the dwarf varieties for naturalising, those of taller habit should also be planted, though considerably more expensive. A form of *Galanthus nivalis* 'Arnott's Seedling', bears its sweetly-perfumed flowers on 12 in. (30 cm) stems; *G. allenii*, with its egg-shaped blooms, is also long in the stem; *G. imperati* 'Atkinsii', which bears a long-petalled bloom on a 10 in. (25 cm) stem.

Both the spring- and summer-flowering snowflakes should be grown in the wild garden. *Leucojum vernum* grows only 8 in. (20 cm) tall, its pretty, drooping, bell-shaped flowers appearing before the end of March. It remains in bloom until early May when *L. aestivum*, Parkinson's 'Bulbous Violet' takes over, its elegant white flowers tipped with green appearing on 12 in. (30 cm) stems.

Nor should the exquisite *Erythronium tuolumnense* be omitted, for it comes into bloom in March, before any other plant excepting the snowdrop; its yellow blooms, like golden butterflies, being borne on 12 in. (30 cm) stems. But it does require protection from cold winds.

The Wood Lilies (Trilliums), natives of North America, are also delightful plants for a wild garden, loving shady places and a moist soil. They bloom during April and May on 12 in. (30 cm) stems. One of the most interesting is *T. erectum*, with glossy, claret-coloured flowers of most disagreeable smell. Plant with it the 'Wake Robin', *T. grandiflorum*, which bears a large three-petalled bloom of purest white above handsome shiny leaves. Many of the bulbs so valuable for naturalising have white flowers, and their purity is accentuated by the green of the grass.

Another wild garden bulb is the Star of Bethelehem, Ornithogalum, of which *O. narbonense* and *O. nutans*, flowering in early summer, are outstanding, bearing their spikes of silvery white on 9 in. (22 cm) stems, the blooms closing up with the approach of evening.

A number of the taller-growing daffodils are admirable subjects for planting in coarse grass; those of more dwarf habit being grown on a rockery, towards the front of a shrubbery, or in the short grass on a sunny bank. The Lent Lily, the English wild daffodil, which has a creamy white perianth and pale yellow trumpet, should be planted in profusion, for the bulbs are inexpensive. The blooms are borne on 8 in. (20 cm) stems.

Another, blooming early, is the hybrid 'W. P. Milner', its almost pure white flowers borne on 10 in. (25 cm) stems. Charming too, is *N. cernuus*, with blue-green foliage and bearing nodding flowers of silvery white on 10 in. (25 cm) stems. The elegant *N. cyclamineus* hybrid, 'February Gold', with bright orange trumpets on 12 in. (30 cm) stems during March, should also be planted; to continue the display until early June, plant *N. gracilis*, like a tiny Jonquil with sweetly-scented flowers. All the miniature daffodils take a year or more to become fully established, and they should be disturbed as little as possible.

The miniature daffodils are best confined to grassy banks. Here *N. minor* makes its appearance towards the end of March and the taller-flowering *N. gracilis* finishes blooming early in June, the last of the daffodils. Plant in groups of six bulbs, not too deep, with the smallest at the top of the bank, the tallest at the bottom. With those daffodils growing in long grass, plant *Muscari latifolium*, a most distinctive Grape Hyacinth, bearing sky-blue and dark-blue bells on 12 in. (30 cm) stems.

The tulip species should find a place in the wild garden, for the species live longer than the large-flowered Dutch hybrids. They will multiply in the same way as the snowdrops, scillas and other naturalised bulbs. *T. praestans* is an excellent choice, and throughout April bears its brilliant orange-scarlet blooms three or four to a 12 in. (30 cm) stem. To extend the season well into May choose *T. saxatilis*, with shining dark green leaves and flowers of a lovely shade of lilac, on 10 in. (25 cm) stems. Another handsome tulip is *T. whittallii* 'Major', which bridges the gap between the appearance of these species. Its urn-shaped flowers are of vivid orange. Also suitable for planting in coarse grass is *T. orphanidea* from Greece. The pointed petals are of mahogany colour shaded with purple. Also for the wild garden is *Allisum karataviense*, a beautiful and interesting member of the garlic family, and although *A. moly* is always at its best in the shrubbery, it need not be excluded from the wild garden. *A. karataviense* has metallic leaves tinted with red and its ball-shaped flowers borne on 12 in. (30 cm) stems, are of a lovely shade of lilac-grey. Also in bloom early in June is the sweetly-scented White Garlic of south Europe, *A. neapolitanum*. It grows 12 in. (30 cm) tall and is excellent for cutting, but as it flowers late it might be better in the shrubbery.

Choicest of all plants for the wild garden are the fritillaries, especially the Snake's-Head Fritillary, its long, drooping flowers mottled to resemble the head of a snake. *F. meleagris* in its numerous lovely forms flowers during May and has strong wiry stems. The blooms are less fragile than they appear to be. *F. pallidiflora*, from Siberia, has yellow flowers beautifully chequered inside with purple. *F. meleagris* was known as the 'Ginny Hen Flower', because its markings resembled those on the feathers of the guinea fowl. It was later named *F. caperonius*, after Noel Caperon, a Frenchman who found it growing in the meadows near Orléans. Parkinson sadly tells us that 'he was, shortly after the finding, taken away in the Massacre'.

There is a wide choice of scillas, in blue, pink and white. The tiny *S. bifolia* and the Siberian Squill are best in border or rockery; the Spanish and English Wood Hyacinths, the familiar bluebells of our woodlands, are best confined to the wild garden. Here they multiply with great rapidity, so much so that they will become a nuisance if planted in the border. *S. hispanica* is a more refined form of our native bluebell. *S. nutans* 'Blue Queen' bears dozens of pyramidal spikes of porcelain-blue bells on 12 in. (30 cm) stems. Plant with it but not too close, the beautiful white form, 'Alba', and the delicate pink, 'Franz Hals', which grows slightly taller.

PLANTING UNDER TREES

In gardens with mature trees there will be an opportunity of planting the hardy cyclamen where they are always happiest, in those pockets formed around the boles of trees. They seem to enjoy the protection from cold winds afforded by the trees and will flourish in a light, loamy soil enriched with leaf mould. But they like some lime rubble about their roots and some should be incorporated before the corms are planted just beneath the surface of the soil.

Where possible, plant two or three corms each of half a dozen species and there will be flowers the whole year round apart from the several weeks of midsummer. But it is during midwinter, when the grass is short, that they will be most appreciated.

The hardy cyclamen is one of the few plants to grow well under conifers. Several of the species have blooms which are fragrant and when established, they seed themselves freely, forming close mats of crimson, pink and white; the leaves, with their silver markings, are also most striking.

The corms must not be too deeply planted; put no more than an inch of compost over them and make sure that you do not plant them upside down. The only exception is *C. europaeum*, which likes to be set deeper. Set the corms 9 in. (22 cm) apart so that established clumps will not be overcrowded. Plant the spring-and summer-flowering species in September and those that bloom in autumn and winter in July.

First to bloom, early in spring, is *C. coum*, with rounded leaves of bottle green and small flowers of vivid carmine-red. There is also a lovely white form, 'Album', which is scarce and hard to find. Then in April and May comes *C. repandum*, an exquisite plant with leaves marbled with silver and bright crimson flowers. For July and August, choose *C. europaeum*, with its sweetly-scented flowers of rosy-red, appearing before the leaves, and to follow *C. cilicium* with flowers of purest pink. *C. neapolitanum*, also for July and August, is worth planting for its foliage alone; it seems well able to survive with the minimum of moisture. For winter flowering there is *C. atkinsii*, which produces its welcome pale carmine-pink flowers in a sheltered corner throughout the winter.

The hardy cyclamen are also attractive in the dell or woodland garden where the ground is almost fully shaded when the trees come into leaf. Here, the Dog's Tooth Violets and the Winter Aconite will flourish. The Wood Anemone, *A. nemorosa*, will multiply rapidly, quickly covering the ground with its single white blooms. Plant with it, as generously as possible, the variety 'Royal Blue', which bears large flowers of a lovely shade of lavender-blue on 6 in. (15 cm) stems. They will succeed the scillas and will be in bloom at the same time as the Wood Lily, *Trillium erectum*, with deep crimson flowers on 12 in. (30 cm) stems. The flowers are large, the foliage luxuriant. One bulb will cost about the price of a packet of cigarettes, but will give years of pleasure. One or two carpeted with the white Wood Anemone will make a most arresting picture in the half shade of the dell.

The miniature daffodils and snowdrops will be perfectly happy in the woodland garden; so will *Tulipa sylvestris*, where the ground is not entirely shaded. Bulbs of *Corydalis cava* and *C. angustifolia*, with their graceful fern-like foliage, are also at home there, as are Solomon's Seal and various ferns. The foliage of the bulbs will die back almost unnoticed, hidden by the ferns and the foliage of the cyclamen and erythronium.

Bulbs planted in a lawn or beneath trees should be set out in generous drifts. Sun-lovers are naturally best in unshaded grass, the shade-tolerant ones under trees. The following will be happy in partial shade, for it must be remembered that when most of them are in bloom, the trees will not yet be fully clothed in their summer green:

Anemone nemorosa	*Hyacinthus amethystinus*
Corydalis cava	*Muscari botryoides*
Cyclamen neapolitanum	*Narcissus*
Eranthis	*Ornithogalum nutans*
Erythronium	*Scilla*
Galanthus	

Though the bulbs will have finished flowering by mid-April, the foliage will not have died back by the time the grass is ready for its first cutting. But if tidiness matters, there is no harm done if you remove that portion of the foliage which has turned brown and will by then have nothing to contribute to the future vigour of the bulb, or the foliage may be kept tidy by tying it together into a loose knot. Until all the foliage has died down, it should never be removed close to the ground. This will mean that if a lawn is making more growth than usual during the early weeks of summer, you must compromise by cutting the grass around and between the bulbs wherever possible. For t is reason it is better to plant in groups or drifts rather than here and there all over the lawn. Later, the blade of the mower may be raised as high as possible for the first cutting made over the bulbs when the foliage has almost died back. Afterwards, the lawn may be cut as closely as required, for by then the foliage will have played its part. If the foliage has to be removed before it has completely died back, removal should be gradual. Bulbs having large, coarse foliage should never be planted in a lawn but should be confined to the shrubbery.

PLANTING IN GRASS

The bulbs should be planted after the grass has been cut reasonably short as late in autumn as possible. Planting will be easier and the grass will still be short when most of the spring bulbs are in bloom. Cutting late will also ensure that the grass will not require cutting again until early the following summer, when the foliage of the bulbs has had ample time to die back.

As selective weed killers of the hormone type, and based on MCPA potassium salt, are now widely used to keep a lawn free from weeds, their possible ill effects on bulbs must not be overlooked.

The weed killer Verdone may be safely used during autumn, when spring-flowering bulbs are dormant and the foliage has died back and has been removed. The preparation should not, however, be used when the leaves of the bulbs are showing above ground. In any case, autumn is the most effective time for using the selective weed killers, when the ground is still warm from the summer sunshine and the autumnal rains cause the weeds to make vigorous growth. One treatment given at this time of the year should be sufficient to keep the lawn free from weeds, and the bulbs will in no way be harmed.

There are two methods of planting. Either a section of turf is lifted by cutting it out with a sharp spade to the required measurements, or the bulbs are inserted singly, with a special bulb trowel or the bulb-planting tool suitable for planting small bulbs. It will save much time and energy-expenditure where the garden is large. A bulb trowel is so designed that a small circular piece of turf, and soil to the correct depth, may be removed with ease. The bulb is then dropped in and the turf replaced and trodden firm.

Where a section of turf is removed with a spade or is rolled back by cutting on three sides, the soil should be stirred and a little peat and sand placed over the surface into which the bulbs will be pressed. Six to eight bulbs may be planted where about 1 sq. ft. of turf has been removed, for we shall be dealing only with small bulbs, some very small indeed. Do not plant in straight lines, and do not remove turves with geometrical exactitude, for you aim at a natural-looking arrangement: where planting singly, set groups of half a dozen, spacing the bulbs several inches apart; and do not plant too closely, for overcrowding will soon occur.

As each opening is made in the ground, drop in a little prepared compost before the bulb is placed into position (the right way up). It is important to ensure that the base of the bulb is actually in contact with the compost, so that root action may begin promptly. If the bulb is in an air pocket it may not form satisfactory roots and will take longer to become established. For this reason a pointed tool should never be used for taking out a hole for bulbs. It is a good plan to mark out the ground by means of plant labels printed with the name of the bulb before any planting is done, so that the same ground will not be planted over more than once. Plan as you would before planting the herbaceous border, so that the purple-blue grape hyacinths may bloom with the yellow-and white-flowered minature daffodils, and let the spring snowflakes mingle with the scillas.

Bulbs are the ideal plants for the labour-saving garden, for even if for some reason the garden is left derelict for any length of time it will come alive again through its bulbs just as soon as a clearance has been made, even though almost all other plant life will have been choked out of existence.

CORRECT PLANTING DEPTHS

Planting depths must not be haphazard, for though certain liberties may be taken with the larger bulbs, the miniatures must be planted with care. Some plants produce a corm rather than a bulb, e.g. anemone and aconite, and corms should be planted no more than 2 in. (5 cm) deep; other depths may vary between 3 and 4 in. (7.5–10 cm) depending upon the size of bulb and soil conditions. All bulbs should be planted one inch deeper in sandy than in heavy-texture soil. The following depths are suggested for any ordinary loamy soil when planting in short or rough grass:

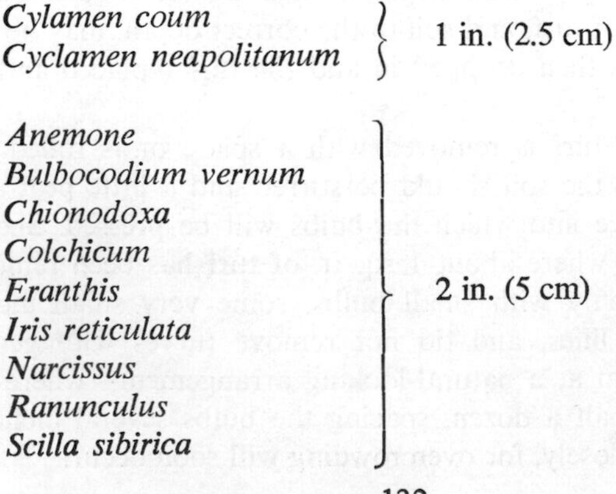

Cylamen coum	1 in. (2.5 cm)
Cyclamen neapolitanum	
Anemone	2 in. (5 cm)
Bulbocodium vernum	
Chionodoxa	
Colchicum	
Eranthis	
Iris reticulata	
Narcissus	
Ranunculus	
Scilla sibirica	

Bulbs, Outdoors

Name	Depth
Allium moly *Corydalis* *Crocus* *Erythronium dens-canis* *Hyacinthus amethystinus* *Leucojum vernum* *Muscari* *Ornithogalum nutans* *Scilla nutans*	3 in. (7.5 cm)
Brodiaea grandiflora *Cyclamen europaeum* *Fritillaria meleagris* *Puschkinia scilloides* *Scilla tubergeniana* *Sternbergia lutea* *Trillium* *Tulip*	4 in. (10 cm)
Galanthus	5 in. (12.5 cm)
Lilium	6 in. (15 cm)

BULBS FOR PLANTING IN A LAWN

NAME	FLOWERING TIME	HEIGHT
Chionodoxa gigantea	March-April	5 in. (12.5 cm)
Chionodoxa luciliae	March	4 in. (10 cm)
Chionodoxa sardensis	March	4 in. (10 cm)
Crocus balansae	March	2 in (5 cm)
Crocus biflorus	March	3 in. (7.5 cm)
Crocus candidus	April	2 in. (5 cm)
Crocus chrysanthus	February-April	4 in. (10 cm)
Crocus corsicus	March	3 in. (7.5 cm)
Crocus fleischeri	February-April	2 in. (5 cm)
Crocus imperati	January-March	3 in. (7.5 cm)
Crocus laevigatus	December-February	3 in. (7.5 cm)
Crocus niveus	December-March	4 in. (10 cm)
Crocus sieberi	February-March	3 in. (7.5 cm)
Crocus vernus	March	2 in. (5 cm)
Eranthis cilicica	February-March	2 in. (5 cm)
Eranthis hyemalis	January-March	2 in. (5 cm)
Galanthus 'Colesbourne'	February-March	4 in. (10 cm)
Galanthus imperati	December-January	6 in. (15 cm)

NAME	FLOWERING TIME	HEIGHT
Galanthus latifolius	March-April	3 in. (7.5 cm)
Galanthus nivalis	February-March	5 in. (12.5 cm)
Hyacinthus amethystinus	March-April	8 in. (20 cm)
Iris bakeriana	January-February	6 in. (15 cm)
Iris danfordiae	February	4 in. (10 cm)
Iris reticulata	March-April	6 in. (15 cm)
Narcissus bulbocodium	March-April	6 in. (15 cm)
Narcissus cyclamineus	March	8 in. (20 cm)
Narcissus lobularis	February-March	8 in. (20 cm)
Narcissus minimus	February-March	4 in. (10 cm)
Narcissus nanus	March-April	6 in. (15 cm)
Narcissus 'Rip van Winkle'	March	6 in. (15 cm)
Puschkinia libanotica	March-April	6 in. (15 cm)
Scilla bifolia	February-March	6 in. (15 cm)
Scilla sibirica	February-March	8 in. (20 cm)
Scilla tubergeniana	February-March	5 in. (12.5 cm)
Tulipa tarda	April	3 in. (7.5 cm)
Tulipa urumiensis	April	6 in. (15 cm)

BULBS FOR A LIME-LADEN SOIL TO PLANT IN A SHRUBBERY

NAME	FLOWERING TIME	HEIGHT
Allium anceps	August	6 in. (15 cm)
Allium karataviense	May	10 in. (25 cm)
Allium moly	June	12 in. (30 cm)
Allium ursinum	April	12 in. (30 cm)
Anemone appennina	March-April	6 in. (15 cm)
Anemone blanda	March-April	4 in. (10 cm)
Chionodoxa gigantea	April	5 in. (12.5 cm)
Chionodoxa luciliae	March-April	4 in. (10 cm)
Cyclamen atkinsii	December-March	4 in. (10 cm)
Cyclamen graecum	July-September	6 in. (15 cm)
Cyclamen libanoticum	March-April	6 in. (15 cm)
Cyclamen neapolitanum	September-November	4 in. (10 cm)
Cyclamen repandum	April-May	6 in. (15 cm)
Galanthus imperati	December-January	6 in. (15 cm)
Galanthus olgae	October-November	6 in. (15 cm)
Galanthus plicatus	March	8 in. (20 cm)
Iris bakeriana	January-February	6 in. (15 cm)
Iris histrioides	March	6 in. (15 cm)
Iris reticulata	March-April	6 in. (15 cm)
Narcissus cernuus	April	9 in. (23 cm)
Narcissus juncifolius	April	4 in. (10 cm)

NAME	FLOWERING TIME	HEIGHT
Narcissus lobularis	February-March	8 in. (20 cm)
Narcissus odorus	May	8 in. (20 cm)
Narcissus triandrus	April	6 in. (15 cm)
Oxalis adenophylla	May-June	3 in. (7.5 cm)
Oxalis bowieana	June-July	9 in. (22 cm)
Puschkinia scilloides	March-May	4 in. (10 cm)

BULBS FOR AN ACID SOIL

NAME	FLOWERING TIME	HEIGHT
Anemone nemorosa	April-June	6 in. (15 cm)
Anemone sylvestris	May-June	12 in. (30 cm)
Corydalis cava	April-May	8 in. (20 cm)
Corydalis decipiens	May	9 in. (23 cm)
Corydalis densiflora	March-April	6 in. (15 cm)
Corydalis wilsonii	April-May	8 in. (20 cm)
Eranthis cilicica	February-March	2 in. (5 cm)
Eranthis hyemalis	January-February	2 in. (5 cm)
Scilla amethystina	May-June	8 in. (20 cm)
Scilla pratensis	May-June	7 in. (17.5 cm)

BUPLEURUM

A semi-evergreen, *B. fruticosum* is not quite hardy north of the Thames except in south-west Scotland but given some protection, it is a valuable plant for a coastal garden and grows well in a chalky soil. Growing 6 ft. (2 m) tall, it has attractive blue-green foliage and bears in August and September bright yellow flowers in terminal umbels. Plant in spring and propagate by cuttings of the half-ripened wood inserted in a sandy compost under glass.

BURNING BUSH, *see* Dictamnus

BUSH ANEMONE, *see* Carpentaria

BUSH HONEYSUCKLE, *see* Diervilla

BUSY LIZZIE, *see* Impatiens

BUTTERFLY BUSH, *see* Buddleia

BUTTERFLY WEED, *see* Asclepias

CACALIA

The Tassel Flower, a native of Africa, is an annual and should be given a position of full sun and a dry soil. *C. coccinea* bears tassels of bright scarlet whilst there is an equally striking golden form, *C. aurea.* Seed is sown late in April in the open or late March in gentle heat as this plant transplants readily. As it grows to a height of 18 in. (45 cm), it should be thinned to 12 in. (30 cm) apart.

C. coccinea, with its vivid tassel-like flowers borne on 12 in. (30 cm) is a showy plant for a greenhouse or for the home. It does best when made to flower in autumn from an early summer sowing.

CACTUS

The culture of cacti is for the specialist, for the plants demand completely different conditions from ordinary greenhouse plants. They like an abundance of sunlight in summer and dry conditions during winter. The greatest success will be obtained if they are grown (with other succulents) in a small greenhouse devoted entirely to themselves and where a minimum winter temperature of 48°F. (9°C.) can be maintained.

There is considerable interest to be enjoyed by raising cacti from seed for it is possible to see the plants in all stages of growth and some very interesting transformations take place. It is also the cheapest method of obtaining a collection. Cacti present little difficulty if fresh seed is sown. Messrs. Thompson & Morgan and most specialist growers such as Neal's of Worthing, offer packets of mixed cactus seed, which may be expected to give high germination.

SOWING THE SEED

Early May is the best time to make a sowing, for the seeds require a temperature of between 65°–75°F. (18°–24°C.) in which to germinate reasonably quickly. Earthenware seed pans should be used which may be covered with glass, or better still, are placed in a propagating frame to encourage germination. The seed may also be germinated in a warm cupboard indoors, the pans being moved to the greenhouse as soon as germination has taken place and gradually getting the seedlings accustomed to the light by providing plenty of shade at first.

Adequate drainage is essential and before filling the pans with compost, a layer of broken crocks or crushed brick should be placed at the bottom so that the compost does not block the drainage holes. The compost for seed sowing should also be as porous as possible and one made up of 4 parts silver sand, 1 part crushed lime rubble and 1 part peat dust, will be ideal. Cacti are great lime lovers and this, in some form, should never be omitted from their diet. The compost should be used quite dry and should be pressed tightly down making the surface completely level so that it is about 1 inch below the rim of the pot. There will be no need to sterilise the compost as each of the ingredients will be almost sterile. For this reason unsterilised loam or leaf mould should never be used.

The compost should be given a thorough soaking after which the seeds may be sown. As they are small, a more even distribution will be obtained if the seed is first mixed with some dry silver sand before scattering over the surface with finger and thumb. The seeds will work their way down amongst the grains of sand and covering will not be necessary nor will it be advisable, unless giving a very light covering of peat dust.

For reliable germination, and with certain cacti species this may take many months, the seed must never be allowed to become dry though watering must be done with care. The surface of the compost may be kept moist by the use of a mist-like syringe and this should be used almost daily, removing the glass and brown paper covering so that the surface may be moistened. Otherwise the compost may be kept in a comfortably moist condition by immersing the base in water whenever necessary, but this should never be done until the compost is showing signs of dryness. To give unnecessary waterings will cause the tiny seedlings to decay.

In a temperature of 70°F. (21°C.), the seed will begin to germinate in about two weeks, but it may take the whole of summer before complete germination has taken place, and during this time those seeds which have germinated first should not be disturbed. It is important to shade the seedlings from strong sunlight, for though cacti are sun lovers, for them to be exposed to the direct rays of the sun beneath glass will cause scorching. The brown paper may be kept in position until almost all the seed has germinated, though it should be removed during sunless days. Afterwards the glass may be whitened, or sheets of brown paper may be tacked on the inside of the greenhouse.

As the first seeds to germinate make headway, it will be necessary to raise the glass and paper covering of the seed pans or they may be dispensed with entirely, relying on the warmth of the midsummer sunshine to bring about germination of the other seeds. To allow the earliest seedlings sufficient room to grow, 1 inch should be allowed from the surface of the compost to the rim of the pan.

The *mammillarias* and *rebutias* appear as tiny, rounded, bead-like plants with no cotyledons. These are the first leaf-like growths characteristic of other cacti seedlings. From the fleshy stem of the *opuntias* come these same seed leaves, between which appear the flat, rounded pads covered with spines. It is important to be able to distinguish between the various forms, for those with cotyledons should be transplanted at the end of summer when all seeds should by then have germinated, whilst those of bead-like form are best left undisturbed until the following spring for they are much slower growing. If those forming seed leaves are moved, the others will have more room to develop as they continue to make slow growth in a winter temperature of around 48°F. (9°C.). During this time they should not be kept quite dry, being given only sufficient moisture to keep them growing slowly.

TRANSPLANTING

Care must be taken in transplanting and growing-on the seedlings. They are moved to individual pots or to seed pans containing a good depth of compost. Special concrete containers 6 in. (15 cm) deep and to any length and width may also be used, after first being treated on the inside with potassium permanganate solution. Probably the best method is to move into small cylindrical pots which should not be glazed.

Each pot should be well crocked and for a compost use 3 parts fibrous loam and 1 part each crushed brick, coarse sand, decayed manure, peat and limestone. The brick and sand will provide porosity, for a well-drained soil is essential, whilst the peat and loam will prevent the too rapid evaporation of moisture. Cacti also require some food if they are not to be too slow growing, and their diet must never lack lime. Such a compost as that suggested will supply all the necessary ingredients. Bone meal, also slow acting, may be used in place of the decayed manure, but artificial fertilisers are best avoided.

It is also important not to use a pot which is too large, for cacti, like auriculas, grow and bloom best where the roots are pot-bound. As cacti in the seedling stage are liable to be attacked by Pithium baryanum or damping off disease, the loam and manure (where used) should be sterilised. Also, to guard against an attack, the seedlings should be watered with Cheshunt Compound or orthocide containing Captan.

The pots should be filled several days before transplanting the seedlings, to allow the compost time to consolidate. The compost should be watered after the pots are filled so that it will be in just the right condition to take the seedlings.

Transplanting should be done with care for it will be noticed that the seedlings form a single carrot-like tap root which may be easily broken with rough handling. The tiny plantlets should be removed from the seed pan with a small smooth-ended stick or cane. This is also used to make the hole into which the seedling is replanted in the pot. Care must also be taken not to cause bruising to the body of the seedling which may happen should there be any undue pressure in handling. Make the seedlings firm in the pots and water in with Cheshunt Compound or orthocide.

To conserve moisture and to reduce to a minimum the need for watering, the pots should be placed on trays of shingle which is kept continually moist during warm weather, or they may be placed in shallow boxes of moist peat.

GROWING ON THE PLANTS

The young plants will need careful shading from hot sunlight by whitening the greenhouse glass, but this should be removed in October. They will also require moisture, but wait until the compost begins to dry out before giving more. During autumn, the plants will require less moisture and little during the winter months, but during their first winter they should not be given a rest period as is necessary with established plants. They should be kept growing on in a temperature of around 48°F. (9°C.), keeping the compost on the dry side, but not completely dry, though moisture should not be allowed to come into direct contact with the plants. During winter, the atmosphere of the greenhouse must be kept as dry as possible, though it will not be necessary to ventilate as in summer when the plants require ample fresh air.

The following April, spraying of the plants may recommence, whilst they should be given more moisture for during summer they will continue to make quite rapid growth and will by now have revealed their true form. Those species such as the *mammillarias*, the *rebutias* and the *astrophytums*, which appear to sit on top of the soil in ball-like fashion, will by the end of summer have attained the size of golf balls. They will then be ready to move into the small size 60 pots in which they will remain and some will first begin to bloom the following year, two years after the seed was sown.

For potting-on, a similar compost should be prepared, the plants being carefully shaken from the small pots so that the roots are disturbed as little as possible. The plants should occupy the same soil level in their new pots. When re-potting, care will be necessary in handling the plants on account of the spines. Leather gloves should be used, or a wad of paper should be wrapped round the plant at soil level, which will enable the plant to be held without disturbing the soil whilst it is turned upside down and shaken from the pot. To facilitate moving the plant from its original pot, the compost should not be too dry, nor should it have recently been watered.

The plants will be grown-on in their new pots, giving plenty of ventilation on all suitable occasions and shading from the direct rays of the sun. Then at the beginning of winter, moisture should be reduced to a minimum in order that the plants can be given a rest period. They should not be completely dried off but only sufficient moisture should be provided to keep the plants alive. It must be remembered that by their succulent nature they are able to survive long periods of drought. During winter a dry atmosphere is essential.

The plants will take on a reddish tint which shows that they are ripening correctly. They bear their bloom during spring and early summer and at this time require copious amounts of moisture. Towards the end of March, the giving of additional moisture should recommence so that the flower buds will develop. If moisture is withheld at this time, the buds will shrivel and fall. The plants will require most moisture when in full bloom. The flowers are replaced by attractive berries which remain on the plant for some considerable time.

CACTI EASILY RAISED FROM SEED

Astrophytum. Known as the Star Cactus, the plants have the appearance of a half orange and are composed of orange-like segments, most of them being almost spineless. Their bright yellow flowers are borne at the top of the plant. They are interesting and grow readily from seed.

Cereus. They grow tall and upright, the rounded stem being composed of four segments. The stems are greyish-blue, but the plants do not bloom until they have attained a large size.

Echinocactus. The almost globular plants with their yellow spines are amongst the best of the cacti to raise from seed, for they bear their yellow or reddish-purple blooms at an early age. The plants like a richer than usual soil and rather more sunshine than most cacti.

Echinopsis. The plants are of similar shape to the *astrophytums,* though wider at the top, whilst the segments are not so pronounced. The plants like a rich soil and plenty of moisture during summer and will bear their large flowers when quite small. *E. eyriesii* bears attractive pale pink scented flowers, whilst *E. fibrigii* bears handsome large white blooms.

Gymnocalycinum. It resembles *echinopsis* and blooms early and well. The plants have short tubercles and strange recurved spines whilst the ribs are clearly defined. The plants are quite fast growing and enjoy an abundance of sunlight.

Mammillaria. They form cylindrical plants covered with nipple-shaped tubercles, borne in rings around the plant. At the end of each nipple is produced a flower. These are small and almost stemless and star-shaped,

being of yellow or white, though those of *M. erythrosperma* and *M. bombycina* are crimson. The fruits or berries of the former are red, those of the latter are white. The plants require very little moisture at all times and rather more lime in their diet than most cacti (see plate 12).
Opuntia. They grow into large plants forming rounded flat pads, borne one on top of another. Some have large spines whilst others are spineless and few of them bear flowers. *O. salmiana* is an exception to this for it bears large yellow flowers, but not until the plants are of a considerable age. *O. microdasys* is an interesting species of which there are three varieties, bearing tufts of small hairy spines in rust, yellow and white and which are known as glochids.
Rebutia. It makes an almost ball-like plant having less pronounced nipples than those of the *mammillaria.* Its beautiful large white or yellow flowers are borne in rows all round the plant.

CAESALPINIA

Closely related to Robinia, *C. japonica* is a deciduous shrub armed with prominent spines. It has pretty acacia-like leaves of soft jade green and in June bears its pale yellow flowers in long racemes. The flowers are enhanced by their crimson anthers. Not hardy north of the Thames unless given the protection of a wall, the plants require a sunny sheltered position and a deep loamy soil. Plant in spring and prune by cutting back unduly long shoots after flowering. It is usually propagated by seed sown in small pots in a frame.

CALADIUM

Though they are not evergreen and the beauty of their foliage may be enjoyed for only eight months of the year, these tuberous-rooted plants are very suitable for growing in a centrally heated room where winter warmth will extend the season until the year end. Their large leaves shaped like arrow-heads are similar in shape and colouring to those of the *Begonia rex*, but are borne on long slender stems often up to 20 in. (50 cm) long. The plants are to be found in their natural state along the banks of the Amazon, where the leaves will measure up to 18 in. (45 cm) across, though only about half that width in our homes.

Their culture is in no way difficult but they must have warmth. The tubers are started into growth in April on a shelf above a radiator or near the warmth of a fire and should be planted ½ in. (1.25 cm) deep into small pots containing a mixture of moist peat and sand, but until they have begun to make growth they should be kept almost dry or they will decay. As soon as the shoots can be seen, the tubers should be removed to size 60 pots containing a compost composed of 1 part loam, 2 parts peat, 1 part sand or grit and 1 part decayed manure. Place in a light position and give more moisture as the plants make growth. The leaves will begin to fade early in December, when water should be gradually withheld until they have completely died down. The tubers should be stored in sand in a warm cupboard, or better still, they should be kept in their pots.

The leaves are as remarkable in their colouring as those of the *Begonia rex*. 'Ace of Hearts' is rose-red with crimson ribs and a wide pale green margin, whilst 'Bleeding Heart' has a grey leaf with a striking scarlet centre. 'Dr Meade' is equally beautiful, the leaf being rich bronzy red, whilst 'Marie Moir' has a white leaf blotched with crimson and a wire edge of green. 'Lord Derby' bears a pink leaf veined with green.

CALAMINTHA

It is *Calamintha acinos* which grows only 4 in. (10 cm) tall and bears its blue and purple flowers in whorls. It is happy in dry conditions and may be planted on a bank or between crazy paving. Both the leaves and the flowers emit a rich minty fragrance when crushed and particularly following a shower. A sprig or two placed in the handbag will remain pleasantly refreshing on a warm summer day.

CALANDRINIA

An annual and because its tiny crimson flowers open only when in the direct rays of the sun, this dainty plant must be grown in a position of full sun. *C. umbellata* is a charming plant on the rockery growing to a height of 6 in. (15 cm) and the same distances across. Sow the seed late in April where it is to bloom and thin to 6 in. (15 cm) apart.

CALCEOLARIA

ANNUAL

The annual form, *C. gracilis* should be raised in gentle heat early in the year and after hardening is planted out in June, 12 in. (30 cm) apart. With its dainty dangling pouches of lemon-yellow and bright green foliage, it is valuable for bedding, being in bloom July-August and growing 10 in. (25 cm) tall. It also makes a pleasing pot plant.

PERENNIAL

The Continental introductions of Weiser and Virnich make plants only 10 in. (25 cm) tall and bear masses of brilliantly coloured blooms for at least four months. On a single plant more than 200 blooms will be open together and each measures more than 1 in. (2.5 cm) in width. The large pouch-shaped blooms are of brilliant yellow, spotted with crimson. The blooms are borne in large trusses on stems which are slightly splayed to reveal their full beauty and are enhanced by the bright yellow-green foliage. The bushy habit is obtained by the careful pinching out of the side shoots at an early stage.

The *multiflora compacta* strain is obtainable in orange, carmine, blood red and yellow spotted, the seed coming true to colour. The Monarch strain, in which the huge flowers are exquisitely spotted is also excellent, the plants growing 15 in. (38 cm) tall.

SOWING THE SEED FOR PERENNIAL CALCEOLARIA

As for primulas, cool treatment throughout is the secret of success with calceolarias. Seed is sown in June in boxes or pans containing a compost made up to 2 parts sterilised fibrous loam and 1 part silver sand to which has been mixed 1 oz. per bushel of superphosphate. Calceolaria seed likes a well-drained, sandy compost.

The seed must not be covered. Sow thinly and water gently in, then cover with a sheet of brown paper to exclude light and hasten germination. This, if the compost has not been allowed to dry out, will take place within a month, the paper being removed as soon as the seedlings appear. It is important to shield them from strong sunlight. Water with care, giving just sufficient to keep them growing, otherwise the seedlings may damp off. When large enough to handle, transplant into boxes containing a similar compost but to which has been added some moist peat.

The seedlings should be grown as cool as possible, shading and damping down the greenhouse whenever the weather is warm. Also, give ample ventilation. Soon the seedlings will have grown into rosette-like plants and will be ready for moving to size 60 pots in which they will bloom.

It is important that the potting compost be porous and well drained, and to the John Innes No. 1 Compost, half as much again by way of sand, should be used if the loam is of a heavy nature. Also, it will be advisable to substitue leaf mould for the peat as calceolarias require a rich compost. A little decayed manure may also be used with advantage, but excessive nitrogen should be avoided or plant growth may become 'soft' and this will encourage mildew. The young plants should not be set too deeply nor too firmly.

BRINGING-ON THE PERENNIAL CALCEOLARIA

The young plants should be given careful attention as to watering and ventilation and until mid-October they will prove to be heavy drinkers. Even so they should be watered only when necessary, some plants requiring more or less water than others: it will be advisable to tap the pot of each plant separately, giving moisture only where required. During winter, water sparingly. At this time, fresh air should be admitted on all suitable occasions, whilst a temperature of between 45°–50°F. (7°–10°C.) should be maintained.

It will be advisable to protect the plants against green fly attack by dusting with 'Lindex' and this should be repeated as the plants begin to form their buds in early spring.

When the plants begin to make new growth in spring, the growing points should be removed to encourage bushy growth and later the new shoots must also be stopped. As the sun gathers strength, the plants will require more moisture and as much ventilation as possible. They should be fed once each week with soot water, whilst those which have grown bushy may be moved to a larger pot containing a compost as previously described, using the John Innes Base at double strength.

Towards the end of April the plants will come into bloom and it will be no longer necessary to employ artificial heat. At this time, dust with 'Lindex' against green fly.

CALENDULA

One of the easiest and most popular of all annuals. It may be sown in mid-September in a cold frame or under cloches, the plants being pricked-off in March into beds 5 ft. (2 m) wide, with the plants 8 in. (20 cm) apart. They will bloom early in June, and like the cornflower will continue well into August. Or seed may be sown in frames or under cloches in early March when the plants will bloom in mid-July. A sowing may also be made in May or early June and the plants transferred to frames or cloches in August, a time when frames are not much in use, when they will come into bloom late in autumn. In a favourable district the plants will continue to bloom until the year's end. Do not give the calendula an excess of nitrogenous manure or the plants will make foliage at the expense of bloom, though it does appreciate a well-nourished soil. The calendula is never happy in a dry soil.

For cutting, 'Radio' with its long quilled petals of deepest orange is outstanding for it will bloom almost the whole year if successional sowings are made. Equally good is 'Geisha Girl' which has incurving petals and bears large double blooms of glowing orange flushed with red. For bedding, 'Orange Coronet' makes a rounded compact plant 10 in. (25 cm) tall and is the earliest to bloom. Its round double blooms with quilled petals are of bright golden-orange.

CALIFORNIAN FUCHSIA, *see* Zauschneria

CALIFORNIAN HYACINTH, *see* Brodiaea

CALIFORNIAN LILAC, *see* Ceanothus

CALIFORNIAN NUTMEG, *see* Torreya

CALIFORNIAN ORANGE, *see* Carpentaria

CALIFORNIAN POPPY, *see* Eschscholtzia and Romneya

CALIFORNIAN REDWOOD, *see* Sequoia

CALLIOPSIS

An annual and for bedding is 'Golden Sovereign', which bears masses of flowers on 9 in. (22 cm) stems, whilst for cutting *C. atrosanguinea* growing to a height of 3 ft. (90 cm) and the maroon and yellow, 'Dazzler', flowering on 18 in. (45 cm) stems, are also useful. Seed is sown late in August or early March under cloches, the plants being thinned to 8 in. (20 cm) apart. They bloom from early July until the frosts.

CALLITRIS

The Cypress Pine of Tasmania, *C. tasmanica* is suitable for growing only in the mildest parts of the British Isles. It makes a slender cypress-like tree and has horizontal drooping branches covered in scale-like leaves. It requires a dry sandy soil and is intolerant of lime. It should be planted in April. Propagation is by seed or from cuttings removed from the tips of the young shoots in August and inserted in a sandy compost under glass.

CALTHA

ALPINE

These are tiny species of the familiar Marsh Marigold which prefer a damp, partially shaded place and a slightly acid peaty soil. *C. palustris* 'Flore Pleno' is the buttercup we all know but in an attractive double form, whilst *C. leptosepala* bears white buttercup-like flowers during May and June. With creeping habit and equally useful about crazy paving as on the rockery is *C. sagitta*, one of the few alpines to bear a green flower.

BORDER

For a moist soil, a clay soil enriched with humus suits them admirably, the Marsh Marigolds are showy plants for the front of a border by waterside and come into bloom early in April, continuing until June. Plant in November, 15 in. (38 cm) apart and propagate by root division.

SPECIES OF BORDER CALTHA

C. palustris. The golden cups are borne on 15in. (38 cm) stems; the double form, 'Flore Pleno', being even more colourful.
C. polypetala. This is the Giant King Cup which bears its large golden-orange blooms on 18 in. (45 cm) stems.

CAMASSIA

Native of North America, they are true dual-purpose plants, delightful in out-of-the-way places in the garden, even lovelier in the home. They are at their best when growing in grass in the wild garden. In bloom in May, when the selection of flowers is limited, camassias should receive more attention than they do. Bearing flowers in various shades of blue or cream, like miniature, loose-belled hyacinths and with hyacinth-like foliage, the various species look most graceful when planted together in a position sheltered from wind. They thrive in all soils provided drainage is good. Plant the bulbs in October, 4 in. (10 cm) deep and 6 in. (15 cm) apart; no closer, for they will increase rapidly. They are hardy and need no winter protection nor are they worried by either slugs or birds. For this reason they are suitable for the town garden. They appreciate some coarse sand worked around the bulbs at planting-time and a small quantity of leaf mould worked into the soil.

SPECIES

Camassia cusickii. It produces purple-blue flowers with their delightful golden anthers on 2 ft. (60 cm) stems during May. There is a lovely creamy-white form which blooms at the same time.

C. esculenta. A free-flowering species, producing its navy blue flowers early in June, and is of compact habit. If the spikes are cut when in the bud stage they gradually open and last a considerable time in water. When established as many as a dozen spikes will appear from each clump.
C. fraseri. A tall-growing variety which produces its blue star-like blooms on long stems throughout May.
C. leichtlinii. It produces tall spikes of rich cream-colour flowers during May which provide a delightful contrast when planted with the blue-flowering species.

CAMELLIA

A genus which includes *C. sinensis* which provides the tea of commerce. Native of the Philippine Islands, China and Japan, the camellia requires an acid soil and protection from cold winds. They should also be planted where the early sunshine does not reach the plants, until frost has had time to leave the buds. Camelias are copious drinkers and during summer require large quantities of moisture whilst they will also appreciate a daily syringing of the foliage. Like most peat-loving plants, e.g. azaleas, they will be happiest in a position of partial shade, such as a clearing amongst trees or shrubs.

Camellias are usually grown in pots and are planted out in early July, generally a month of rain and humidity and so they will get away to a good start. The plants may be grown in their pots indoors when, in a temperature of 50°F. (10°C.) they will bear their symmetrical carnation-like flowers, double or single, through winter and spring in the case of *C. sasanqua*; and spring and early summer for *C. japonica*.

SPECIES AND VARIETIES

Camellia japonica. Native of south-east China and Japan, it grows 10–20 ft. (3–6 m) tall and has large glossy deep green foliage. It bears its wax-like flowers from March until May depending upon site and situation. It is best planted out early in April.

ADOLPHE AUDUSSON. In the semi-shade of a woodland glade, it is a blaze of colour in spring, its symmetrical semi-double blooms being of blood-red with prominent stamens.

AUGUSTO PINTO. A most beautiful variety bearing large double blooms of a lovely shade of deep pink, edged with white.

FUROAN. The flowers are large and single, being of bright peach-pink with conspicuous golden stamens.

NOBILISSIMA. One of the earliest, the blooms are purest white with a high centre and of perfect symmetry.

TRICOLOR. The large double blooms are of blush-white, flaked with carmine and rose.

C. reticulata. It has handsome leathery net-veined leaves 4 in. (10 cm) long and 2 in. (5 cm) wide which act as a pleasing background for the large brilliantly coloured flowers.

CHANG'S TEMPLE. The flowers measure 6 in. (15 cm) or more across and are of paeony-type with waved petals. The colour is deep pink blotched with white.

PAOCHUCHA. A cultivar of compact habit bearing large semi-double blooms of orient red.

PURPLE GOWN. It has large broad leaves and bears handsome double blooms of rich paeony-purple.

C. sasanqua. At its best against a wall, it bears deliciously scented flowers from October until April and which are enhanced by the broad dark green leaves.

FORTUNE'S YELLOW. Introduced by Robert Fortune in 1848, it bears large flowers of rich Jersey cream colour with a central boss of petaloids.

ROSEA PLENA. A camellia of vigorous habit bearing large double blooms of soft China pink.

SHISHIGASHIRA. The blooms are small but are borne in profusion and are of deepest carmine with a central boss of petaloids.

CAMPANULA

ALPINE

In bloom from early May until late in September, the dwarf campanulas are amongst the most dainty and interesting plants of the garden. Those of extremely dwarf habit, e.g. *C. allionii* and *C. saxifraga,* bearing almost stemless blooms are delightful plants for the alpine house or scree garden, whilst those growing slightly taller may be grown in the trough garden or window-box or on the rockery. They are also most attractive used as an edging to a path or drive or to border a rose-bed or shrubbery, but here only those having a long flowering period should be grown, a good example being *C. pulloides* which produces its rich violet bells from June until September and makes a compact plant 6 in. (15 cm) tall.

The extreme hardiness of the campanulas and their tolerance for partial shade, their ease in propagation and ability to quickly make a large plant in ordinary soil should ensure them more popularity than they enjoy, for in every way they are ideal plants for the small, labour-saving garden. Most of them are extremely perennial, requiring little or no attention when once they have been planted. Their ease in propagation is also greatly in their favour, for to increase one's stock all that is necessary is to lift a root during autumn or early in spring and to divide into numerous small pieces each containing roots. It is quite possible to propagate by removing the outer portions of a plant without in any way lifting it, though being so fibrous rooted, the plants may readily be established if not allowed to lack moisture.

PROPAGATION OF ALPINE CAMPANULA

The small pieces may be potted into 3 in. (7.5 cm) pots containing the John Innes Potting Compost, or a suitable compost may be mixed by using 2 parts decayed turf loam, 1 part coarse sand or grit and 1 part leaf mould, to which is incorporated some well-pulverized lime rubble (mortar), for almost all campanulas appreciate some lime in the soil.

The rooted pieces may also be transplanted to open-ground beds prepared in a similar manner, setting out the pieces 4 in. (10 cm) apart. It is important to keep the soil moist. The advantage of growing in small pots is that they may be moved at almost any time of the year except when the ground is frozen. It matters little if the plants are in bloom. With their vigorous fibrous rooting system campanulas may be likened to primulas which of all plants, are amongst the easiest to re-establish. If they have a fault, it is in their limited colour range, the blooms being white or in various shades of blue, purple and mauve, but their diversity of form makes them so interesting. Quite a number of them, too, remain in bloom for only a short period, no more than four to five weeks, so it is advisable to grow as large a number of species and varieties as possible. In this way there will be plants in bloom through summer and autumn. Where growing for garden display it will be advisable to concentrate on those which remain longest in bloom, such as *C. turbinata* and *C. pusilla,* which will bloom from mid-summer until the end of September.

A number of the campanulas prefer a crevice between two stones to protect them from excessive moisture or they should be confined to the alpine house or frame. These are the most dwarf forms which make tiny green mats and bear almost stemless nodding bells. These plants require somewhat drier conditions and where growing on a rockery or in pots or pans, they appreciate scree conditions to prevent soil splashing and to encourage drainage. Pockets of soil should be specially prepared, and after planting cover the surface around the plant with shingle. This will also act as a mulch during summer and will also suppress annual weeds. Campanulas are at their best under such conditions, whilst they love to get their roots around and beneath large stones.

To plant campanulas along the top of a wall, using those of trailing habit and plant into pockets of well-prepared soil or the plants will suffer from lack of moisture in a dry summer. A little cow manure and some leaf mould will prove adequate. Those of trailing habit are *C. caphalonica* and *C. poscharskyanum* and its numerous hybrid forms which also look most attractive trailing over the sides of a trough.

September is a suitable time to propagate, soon after the plants have finished blooming and to allow them sufficient time to become established before the arrival of frosts and winter rains. The plants are diverse in form, and for them to be seen to advantage, plant in groups of three, about 8 in. (20 cm) apart, planting the white-flowered forms between those which bear purple or blue flowers. They will also show up well if those bearing star-like flowers are planted between the large cup-flowered varieties. There are also those which bear dainty bells whilst others have grey foliage. With some forms the blooms are small and are borne several to a stem, whilst others are borne singly. Each one is a beauty in its own particular way.

Being fibrous rooted, the plants may be readily transported, for the soil is not easily shaken from the roots, whilst there will be little fear of the stems being damaged.

These dwarf campanulas will all flourish in a chalk and lime-laden soil, in fact they will only show to best advantage where lime is present in the soil.

SPECIES AND VARIETIES OF ALPINE CAMPANULA

(a) denotes suitable for the alpine house.

A. arvatica (a). A native of Spain, it is a tiny creeper growing no more than 2 in. (5 cm) tall and from June until August covers itself in large luminous violet stars. There is also a lovely white form, 'Alba'. *C. arvatica* in particular likes a limy soil and good drainage.

C. carpatica. One of the best of all campanulas, forming large dense clumps of bright green and flowering throughout summer on 6 in. (12 cm) stems.

BLUE MOONLIGHT. It bears huge cups of a delicate shade of grey-blue on 6 in. (15 cm) stems.

BRESSINGHAM WHITE. A recent introduction, the large pure white blooms are held on erect 8 in. (20 cm) stems and remain long in bloom.

CARPATHIAN CROWN. The blooms are of palest china-blue which open flat, like saucers.

CHEWTON JOY. The cup-like flowers are of pale lavender and have an attractive green centre.

HOLDEN SEEDLING. A variety of dwarf, compact habit, it bears large handsome bells of palest blue.

ISOBEL. Of taller habit than the others, it is very free-flowering and bears large flat blooms of violet-purple.

MOUNT EVEREST. Of dwarf, compact habit, it bears pure white saucer-like blooms on 4 in. (10 cm) stems.

MRS DE FRERE. It bears large upright saucer-shaped blooms of deepest blue. The habit is neat and compact.

RIVERSLEA. Its large saucers of rich true blue borne on 8 in. (20 cm) stems remain in bloom for nearly sixteen weeks.

WHEATLEY VIOLET. It has beauty of form, bearing large upturned cups of rich violet-mauve.

WHITE STAR. Free and long in bloom, it bears its large white cups tinged with blue on 8 in. (20 cm) stems.

C. elatinoides. It makes a tiny compact plant and throughout summer bears its lovely violet bells in dainty sprays on 4 in. (10 cm) stems.

C. garganica. It loves the sun and rather dry conditions and does well on a wall. The best form is 'Hirsuta', which has downy grey foliage and bears pale blue star-like flowers on 4 in. (10 cm) stems. 'W. H. Payne' is a striking variety, the violet flowers having a contrasting white centre.

C. isophylla. It should be given the protection of a crevice, or it makes a beautiful plant for a hanging basket but it does not enjoy an excess of winter moisture about its roots. Its powder-blue flowers have an attractive lilac-grey centre and are borne in profusion throughout summer. (see plate 13).

C. medium, the Canterbury Bell. Flowering at a height of nearly 3 ft. (1 m) and in bloom through the early part of summer, the biennial campanula medium is best planted on the herbaceous border, or against a wall or fence protected from the prevailing winds. To allow the plants a full twelve months to develop, seed is sown early in June either broadcast or in shallow drills and should be just covered with soil. Or sow in boxes covered with or under glass. The plants should be moved to prepared beds for growing on when large enough to handle. There are two types, the single and the double, or cup-and-saucer type, obtainable in blue, mauve, pink and white. The strain, 'Bells of Holland' grows only 18 in. (45 cm) tall and is suitable for bedding.

C. muralis major. A lovely trailing plant which is happy in a sunny position and is ideal for a wall. Throughout summer it covers itself in masses of deep violet flowers.

C. nitida (a). An unusual and interesting species, its stiff spikes of pure blue cups arise from dark green rosettes. The form 'Coerulea Plena' bears double cups of deep blue, whilst 'Alba' bears creamy-white flowers from June until August. It likes a crevice or a position near a large stone and plenty of grit about its roots for drainage.

C. pilosa major. A most attractive plant, having shiny green foliage and bearing large open bells of lovely pale blue during June and July.

C. poscharskyanum. Of trailing habit, it bears masses of grey-blue flowers all summer. 'Lisduggan' is a pretty pink-flowered form, whilst 'E. H. Frost' bears masses of pure white star-like flowers and has large pelargonium-shaped leaves.

C. pulla (a). It is of the Harebell type and a most delightful plant, though it blooms only during June and July. It is of most compact habit, forming tiny green hummocks and bearing its nodding bells on 3 in. (7.5 cm) stems, making it an ideal plant for a trough garden. It bears satiny pink flowers. Lovely, too, is 'Lilacina', which bears its dainty rosy-lilac blooms on only 2 in. (5 cm) stems; whilst 'G. F. Wilson' bears its violet bells on erect stems and in great profusion.

C. pulloides (a). A beauty for the scree, bearing its huge cup-like blooms of tyrian-purple on 4 in. (10 cm) stems from June until August.

C. punctata. An interesting plant, it blooms in May and bears loose spikes of white tubular bells, flushed with pink, at a height of 12 in. (30 cm.) It likes a sandy, well-drained soil and a sunny position.

C. pusilla. This is the lovely Harebell or 'Fairies Thimbles', delightful plants for the side of a path or for planting in shade. They are plants of neat, compact habit, and bear their nodding bells on 4 in. (10 cm) stems from June until the end of September.

MARY MILNE. A pure white form of the cream-flowered *C. pusilla* 'Alba'.

MIRANDA. The bells are large and fat and of a lovely shade of icy-blue.

MISS WILLMOTT. It bears its medium-blue flowers in great profusion.

OAKINGTON BLUE. Outstanding, the large bells being of a deep shade of purest blue.

C. radeana. A lovely campanula, bearing its bells of deepest violet-blue with long golden anthers on 8 in. (20 cm) stems. *C. kemulariae* is similar but of taller habit whilst the blooms are larger.

C. raineri-hirsuta (a). Similar in habit to the *carpaticas*, it is worthy of planting if only for its ash-grey foliage. It should be grown in the alpine house or scree, where throughout summer it bears its large upturned cups of china-blue on 1 in. (2.5 cm) stems.

C. rotarvatica (a). It enjoys conditions of the alpine house or scree, where it will bloom from July until autumn, its dainty deep navy-blue bells being borne on 3 in. (7.5 cm) stems above its bright green foliage.

C. saxifraga (a). A fine variety for the scree or alpine house, bearing large open bells of deepest violet during May.

C. spruneriana. A lovely plant, it comes into bloom in May, bearing, on 9 in. (22 cm) stems, masses of silvery mauve-pink tinted bells.

C. turbinata. A campanula of great beauty, it blooms from July until September bearing large blue cups 2 in. (5 cm) across and held on 6 in. (15 cm) stems. Even lovelier is 'Jewel', which has a small dainty leaf and bears enormous violet cups.

C. warleyensis. Valuable in that it is late flowering. Of prostrate habit, it bears double blooms of lavender-blue. 'Bressingham Double' is a deeper-coloured form, the blooms being carried on arching stems.

HYBRIDS OF ALPINE CAMPANULA

BIRCH HYBRID. A lovely little plant, bearing its deep blue cups on 6 in. (15 cm) stems from May until well into August.

BLUE TIT. It blooms from June until September, its bright blue bells being held on erect 6 in. (15 cm) stems.

BUNYARD'S VARIETY. The habit is compact and free-flowering, covering itself in saucer-like bells of pale blue.

GLANDORE. Of vigorous habit, it bears its clear blue star-like flowers in long racemes from June until August.

HANNAH. A new hybrid which bears a profusion of pure white bells during July and August.

JOHN INNES. Of vigorous trailing habit, its stems are covered in large pale blue star-like flowers throughout summer.

LISDUGGAN. This is a new hybrid of trailing habit, bearing throughout summer, racemes of soft pink bells.

LYNCHMERE. A valuable new hybrid being late-flowering, and which bears masses of soft violet-blue flowers on 4 in. (10 cm) stems.

MOLLY PINSENT. It is late-flowering, its deep violet-blue cups being held on 5 in. (12.7 cm) stems.

NORMAN GROVE. It makes a little hummock of olive-green leaves and during late summer bears its lavender-blue cups on 3 in. (7.6 cm) stems.

R. B. LODER. A lovely Harebell hybrid, bearing erect sprays of double-blue bells during August and September.

STANSFIELDII. This lovely plant has attractive crinkled leaves and bears its dainty trumpet-shaped blooms on 3 in. (7.6 cm) stems throughout summer. The flowers are of a lovely shade of lavender-grey.

SYMONSII. It has pale green foliage and bears masses of pale blue cups during July and August.

WOCKII. Slow growing, it makes a tiny shrublet which is covered in small lavender-blue bells during July and August.

ZET. A dainty variety for the trough garden which covers itself in brilliant blue flowers throughout summer.

BORDER

For the border there are no more beautiful or long-lived plants than the tall Bell-flowers, most of which were cultivated before Tudor times. There are numerous species and varieties suitable for planting in beds, or no staking. They prefer an ordinary soil which is not too heavy and is period. Because of their upright habit, the border species require little or no staking. They prefer an ordinary soil which is not too heavy and is well drained, for they are not too happy where there is an excess of winter moisture. Plant in March, which is also the best time to propagate by root division. Plant 2 ft. (60 cm) apart for the tall species; 20 in. (50 cm) for those of more dwarf habit.

SPECIES AND VARIETIES OF BORDER CAMPANULA

C. lactiflora. Not long-lived except in a light, sandy soil. Its long, graceful flower spikes reach a height of 3–4 ft. (1 m) and are in bloom from June until August. 'Pritchard's' variety with its deep blue flowers is excellent, whilst 'Loddon Anna' bears bells of a delicate shade of flesh pink.

C. latifolia 'Brantwood'. A lovely plant for midsummer, its royal purple bells being borne on sturdy 2 ft. (60 cm) stems.

C. macrantha. During July and August it bears long spikes of amethyst-blue bells on 3–4 ft. (1 m) stems and does well in all soils. There is a lovely white form, 'Alba'.

C. persicifolia. The Peach-leaved Bell-flower, its thin, narrow leaves resembling the foliage of the peach. It is a plant which has been cultivated in England at least since early Tudor times, bearing large open bells of mid-blue, arranged almost the whole way up the stems. It blooms from the end of May until late August and there are many lovely varieties:

DELFT BLUE. Its name describes the soft china-blue colour of its bells.

FLEUR DE NEIZE. Its lovely pure white double blooms are borne on 3 ft. (1 m) stems during June and July.

GARDENIA. It bears flat double bells of pure white on 2–3 ft. (1 m) stems from May until late August.

PRIDE OF EXMOUTH. Growing to the same height as 'Gardenia', it bears fully double bells of a lovely shade of lavender-blue from June until early September.

TELHAM BEAUTY. It grows to a height of just over 3 ft. (1 m) and bears very large bells of pure sky-blue.

WIRRAL BELLE. Its beautiful double flowers of rich bluebell blue are borne on 3 ft. (1 m) stems during July and August.

CAMPION, *see* Lychnis

CAMPSIS

Known as the 'Bignonia', the 'Tecoma', and the 'Trumpet Flower', it is one of the few self-clinging climbing plants, and for a warm position there is none more striking. It may be planted on a south wall, but it cannot be said to revel in a sunbaked situation, though it appreciates warmth and sunshine. It is deciduous and when established makes rapid growth, attaining a height of 25 ft. (7 m) or more and becomes as wide as it grows tall. Plant in March in a soil containing some peat or leaf mould. Throughout summer the plants should not lack moisture, and if growing in full sunshine, they should be given a mulch of lawn mowings during early summer and again during July and August. The plants should be pot grown and for the first years almost no pruning will be necessary. Afterwards, the cutting away of any unwanted shoots should be done.

A stone or white-washed wall clothed in the vivid orange-scarlet trumpets of *Campsis grandiflora* will present a most arresting picture, but for the plants to bloom in such a manner several years may well have elapsed since their planting. It is, however, worth waiting for, the clusters of blossom appearing from early July until autumn. Another form, *C. radicans*, is of less vigorous habit, and bears rust-coloured flowers which are yellow on the inside; their loveliness is accentuated by the grey-green foliage.

CANDYTUFT, *see* Iberis

CANTERBURY BELL, *see* Campanula

CAPE COWSLIP, *see* Lachenalia

CAPE GOOSEBERRY, *see* Physalis

CAPE HYACINTH, *see* Galtonia

CAPE LILY, *see* Crinum

CAPE PRIMROSE, *see* Streptocarpus

CAPSICUM

The Ornamental Peppers make excellent indoor pot plants, growing short and bushy and bearing their pointed fruits which change from yellow to crimson during winter. Sow the large seeds individually in small pots in gentle heat in February and move to larger pots containing the John Innes Potting Compost early in summer. At all times, do not allow the seeds to lack moisture or germination will not take place, whilst the plants must be kept growing by maintaining correct moisture and shielding them from strong sunlight.

To encourage a bushy plant, pinch out the growing point when moving to their final pots and for the flowers to set their fruits well, hand fertilise with a camel hair brush as they open. Afterwards, regular syringing of the plants will help them to set their fruit and as they do so, they should be given ample supplies of moisture during summer to enable the fruits to swell.

A compact and free fruiting variety is 'Frips' which grows 8 in. (20 cm) tall, the fruits changing from pale cream to orange and crimson.

CARAGANA

Small trees or shrubs, native of Siberia and Manchuria, they are known as the Pea trees on account of their pea-shaped flowers. They like a well drained sandy soil and are tolerant of town garden conditions. Propagation is by seed sown in July when ripe or by cuttings, taken in July and rooted in a frame or under cloches.

The best species is *C. arborescens* which grows to 20 ft. (6 m) in height and has its leaves divided into numerous linear leaflets. Early in summer it bears masses of yellow sweet pea-like flowers. Also valuable is *C. decorticans* of similar habit and also bearing similar yellow flowers.

CARAWAY, *see* Carum

CARNATION, *see* Dianthus

CARUM

For it to bear seed, *Carum carvi* must be treated as biennial, the seed being sown in July the previous year. Oil extracted from caraway seeds is used to make the liqueur Kummel, whilst the seeds with their strong aromatic taste are used in Ireland to flavour bread and on the continent to flavour cakes. The plant grows 3–4 ft. (1 m) tall and in Shakespeare's time it was grown for flavouring apples. It was Mr Justice Shallow who said, 'We will eat a pippin of last year's graffing, with a dish of caraways', and from the account book of Sir Edward Dering (1626) there is this conclusion to a meal – 'Apples and caraways'. Caraway seeds were at one time taken with fruit to relieve indigestion and for this reason the seed was served with roast apples in the Hall of Trinity College, Cambridge. An excellent remedy for flatulence is to crush some caraway seed and add to half a pint of boiling water. Allow to get cold and take a teaspoonful whenever necessary.

CARDAMINE

The Cuckoo-flower is useful for planting in a damp soil and in a position of partial shade, where during May and June it will bear a profusion of lilac stock-like flowers. Plant in autumn and allow 18 in. (45 cm) between the plants.

Cardamine latifolia grows 20 in. (50 cm) tall and continues to bear its lilac flowers well into July whilst *C. pratensis* 'Flore Plena' is the double form of the Cuckoo-flower and begins to bloom in April.

CARDINAL FLOWER, *see* Lobelia

CARDIOCRINUM

A genus of three species, native of the Himalayas and east Asia which at one time were included in the genus *Lilium*. They differ in that their bulbs have few scales whilst the seed capsules are toothed. They are plants of the dense woodlands of Assam and Yunnan, where the rainfall is the highest in the world and they grow best in shade and in a moist humus laden soil. The basal leaves are cordate, bright green and glossy; the flowers trumpet-like with reflexed segments. They are borne in umbels of ten to twenty on stems 10–12 ft. (about 3 m) tall. In their native land, they are to be found growing with magnolias and rhododendrons.

The bulbs are dark green and as large as a hockey or cricket ball. They should be planted early in spring away from a frost pocket. Plant with the top part exposed and 2 ft. (60 cm) apart, three bulbs together in a spinney or in a woodland clearing presenting a magnificent sight when in bloom. They require protection from the heat of summer and a cool root run, they are also gross feeders so the soil should be well enriched with decayed manure and should contain a large amount of peat or leaf mould. The bulbs will bloom in July and will begin to grow in the warmth of spring. By early June, the flower stems will have attained a height of 8–9 ft. (about 2.5 m) or more and will be bright green with a few scattered leaves. The basal leaves will measure 10 in. (25 cm) wide, like those of the arum. The flowers last but a few days but in the warmer parts are replaced by attractive large seed pods whilst the handsome basal leaves remain green until the autumn. The stems are hollow.

PROPAGATION

After flowering and the leaves have died back, the bulb also dies. Early in November, it is dug up when it will be noticed that three to five small bulbs are clustered around it. These are replanted 2 ft. (60 cm) apart with the nose exposed and into soil that has been deeply worked and enriched with leaf mould and decayed manure. They will take two years to bear bloom but if several are planted each year there will always be some at the flowering stage.

To protect them from frost, the newly planted bulbs should be given a deep mulch after planting with decayed leaves or peat whilst additional protection may be given by placing over the mulch, fronds of bracken.

Plants may be raised from seed sown in a frame in a sandy compost or in boxes in a greenhouse. It should be sown in September when harvested and it will germinate in April. In autumn, the seedlings will be ready to transplant into a frame or into boxes, spacing them 3 in. (7.5 cm) apart. They need moisture whilst growing but very little during winter when dormant. In eighteen months, in June, they will be ready to move to their eventual flowering quarters such as a clearing in a woodland where the ground has been cleaned of perennial weeds and fortified with humus and plant food. Plant 2 ft. (60 cm) apart and protect the young plants with low boards erected around the ground. They will bloom in about eight years from sowing time.

SPECIES

Cardiocrinum giganteum. Native of Assam and the east Himalayas where it was found in the rain saturated forests by Dr Wallich in 1816. It was first raised from seed and distributed by the Botanic Gardens in Dublin and it was first seen flowering in the British Isles, at Edinburgh in 1852. Under conditions it enjoys, it will send up its hollow green stems (which continue to grow until autumn) to a height of 10–12 ft. (about 3 m), each with as many as ten to twenty or more funnel-shaped blooms 6 in. (15 cm) long. The flowers are white, shaded green on the outside and reddish-purple in the throat. Their scent is such that when the air is calm, the plants may be detected from a distance of one hundred yards. Especially is their fragrance most pronounced at night. The flowers droop downwards and are at their best during July and early August. The large basal leaves which surround the base of the stem are heart-shaped and shortly stalked.

CARMICHAELIA

Hardy, deciduous, erect or prostrate broom-like shrubs, native of New Zealand and requiring a sunny situation and a well drained sandy soil. *C. flagelliformis* grows 4 ft. (1.2 m) tall and makes a spreading bush with flattened pale green branchlets which are covered in small purple-blue pea-like flowers during June and July. *C. petriei* grows 5 ft. (1.5 m) tall with stout branches and it bears its fragrant violet-blue flowers in dense racemes during midsummer.

Plant in spring direct from pots for like the legumes, it resents root disturbance. Do no pruning.

Propagation is by seed or cuttings removed in August and inserted in sandy compost under glass.

CARPENTERIA

The Californian Orange, *C. californica* is closely related to the hydrangea but is evergreen with long narrow leaves, downy on the underside. Against a sunny wall it will grow 8 ft. (2.5 m) tall and in July bears large white flowers in dense clusters, the flowers having striking golden stamens. In the British Isles it requires shelter from cold winds.

Plant in spring, in a well drained loam and do not prune except to remove dead wood. Propagate by cuttings of the new season's wood inserted in a sandy compost under glass.

CARPINUS

The hornbeam similar to beech in that it retains its dead foliage through winter, though it is not of so bright a shade of brown. Whereas the beech is an excellent plant for a dry, shallow or chalky soil, the hornbeam thrives in a moist, heavy clay soil. In this respect hornbeam and beech are the deciduous counterparts of yew and thuya, one requiring dry, the other moist conditions.

The hornbeam, *Carpinus betulus*, is quicker gowing than beech. It will require little attention for several years after planting, other than the removal of loose shoots to keep the height even. When established, the plants withstand quite as hard clipping as the beech, though this is rarely necessary. November is the time to plant, for then the interesting golden catkins, borne by the male, may be enjoyed during March and April. The hornbeam should be planted 15 in. (38 cm) apart. It is a delightful tree for a town garden and does well in a smoke-laden atmosphere.

CARYOPTERIS

A deciduous shrub, requiring protection north of the Thames and which grows best against a warm wall where it will attain a height of 6 ft. (2 m). It likes a soil containing plenty of peat or leaf mould. *C. x clandonensis*, a hybrid, is the hardiest and is valuable in that it bears its clusters of bright blue flowers in August and September.

Plant in spring and trim back the old flowering shoots in March. Propagate by division or by cuttings of the young shoots removed in July and rooted in sandy compost under glass.

CASSIOPE

Evergreen and requiring a peaty soil, several of the species are delightful early summer flowering plants, none being finer than *C. fastigiata* with its nodding white bells, whilst even more dwarf is *C. selaginoides* with its moss-like foliage and tiny nodding bells of finest white.

CASTANEA

This is the Sweet Chestnut which grows well in light soils and will attain a height of 60–70 ft. (about 20 m). The glossy dark green lanceolate leaves with their toothed edges take on brilliant golden shades in autumn. The male flowers which appear in July, are borne in long erect catkins and are scented. The females produce the edible 'chestnuts'. No pruning is necessary. Propagation is by sowing the chestnuts in autumn, preferably where they are to remain.

C. sativa is the Sweet Chestnut and the best form is 'Marron de Lyon', which makes a more compact tree and fruits at an early age.

CATALPA

Valuable late-summer flowering deciduous trees, native of the southern United States, and west China and as they form symmetrical heads, make suitable trees for lawn planting beneath which bulbs may be grown. The slender leaves have acuminate lobes, whilst the flowers resemble those of the foxglove. The trees may be shaped and kept symmetrical by autumn pruning, at the same time removing any dead wood. Propagation is by cuttings, removed in July with a 'heel' and rooted under glass, preferably in a propagating frame.

C. bignonioides is the most widely grown species, bearing late in July, spikes of white flowers spotted with mauve, like Horse Chestnut 'candles'. The variety 'Aurea' has handsome golden leaves. *C. fargesii* has smaller leaves whilst it bears its flowers in corymbs of twelve or more. They are lilac coloured, spotted with crimson, brown and yellow.

CATCHFLY, *see* Silene

CATMINT, *see* Nepeta

CEANOTHUS

There are two forms of the ceanothus, one deciduous, the other evergreen. The deciduous are less tender than the evergreen species and bloom late in summer and in autumn, whilst the evergreen species, with but one or two exceptions, bloom during May and June. Two or three species or hybrids planted together will bear flowers from early summer until the end of autumn, making a glorious sight with their blooms of blue and pink.

Both forms are happy in ordinary garden soil and in a town garden, for they are not troubled by slightly acid conditions. They like some peat about their roots together with a small quantity of decayed manure. Plant the deciduous species in October and the evergreens in spring; both should be pruned during early summer. Prune the deciduous species immediately after flowering in autumn, and the evergreens at the end of June, for they bloom on the new season's wood. Pruning consists of shortening back all shoots, otherwise they will gradually become weaker and bear few flowers. Any dead wood must also be removed as it forms, but the evergreens should not be as heavily pruned as the deciduous species.

SPECIES AND VARIETIES

One of the finest evergreen forms is *Ceanothus veitchianus*. This is the most tender variety, being a hybrid from the almost equally tender *C. dentatus*. Like all the ceanothus, the tiny feathery blooms are clustered together to form compact heads, a dozen or more to a spray. Like *C. dentatus*, the flowers are of deep blue, the evergreen leaves being attractively notched and are of a glossy deep green. Both attain a height of 12 ft. (3.5 m).

Possibly the hardiest of the evergreens is *C. cyaneus*, which may be planted against a north wall in a warm district; or against a south wall in a more exposed garden. It will reach a height of about 15 ft. (4.5 m) and will bear a mass of clear Cambridge-blue flowers, which are the largest of all the species. Also amongst the hardiest is *C. thyrsiflorus*, the tallest and most vigorous ceanothus, which bears attractive pale grey-blue flowers. An exception to the rule that the evergreens bloom early is *C. burkwoodii*, which blooms at the same time as the deciduous species, from late July until October, producing large sprays of vivid mid-blue on wood formed the same season.

Of the deciduous forms, *C. arnouldii* bears large leaves and sprays of powder-blue flowers, but it is the hybrids of *C. spinosus* that make the most vigorous wall plants. Of these, possibly the best known is Gloire de Versailles, with its large sprays of powder-blue. Autumnal Blue bears rich mid-blue flowers, whilst 'Charles Detriche' has an almost navy-blue flower – a delightful contrast to the paler shades. Interesting too is 'Albert Pittet', which has flowers of a lovely dusky-pink. The ceanothus requires only the minimum of support for it is of shrub-like habit.

CEDAR, *see* Cedrus

CEDRUS

A genus of evergreen trees with needle-like leaves, the cedar is a native of North Africa, the near east and the Himalayas. For its hardiness, longevity and ease of culture, it is the finest of all conifers for specimen planting in a lawn and none excels it in beauty. Plant in April, retaining as much soil as possible about the roots. It requires an open, sunny situation and a deeply dug well drained soil. Propagate from seed or by cuttings removed from the tips of the young branches and inserted in sandy compost under glass.

SPECIES AND VARIETIES

Cedrus atlantica. The Mt Atlas Cedar of Algeria, it is a rapid and easy grower with spreading horizontal branches. The variety aurea has golden leaves whilst 'glauca' is the Blue Cedar, one of the most handsome of all coniferous trees.

C. deodara. The Deodar or Indian Cedar of the Himalayas is of pyramidal habit with pendent feathery branches, the leaves being brightest green whilst the form aurea is the Golden Cedar. Like *C. atlantica*, the lower branches are formed almost at ground level and are retained whilst the tree reaches a considerable height. Allow space for it to develop for the lower branches will decay if there is over-crowding nor will the tree be seen in its full beauty.

C. libani. The Cedar of Lebanon with fan-like spreading branches clothed in numerous short dark green leaves which give the tree a sombre appearance. The handsome smooth cones are first purple, then brown. The form 'Pendula Sargentii' makes a smaller tree and has drooping branches. It is suitable for specimen planting in a lawn.

CELASTRUS

It is an extremely hardy plant of climbing habit, which will attain a height of 20 ft. (6 m). The plants are valuable in that their flowers are followed by brilliantly coloured fruits, which are retained through winter, whilst the deciduous foliage turns a rich yellow colour in autumn and remains thus for several weeks before falling. With their twisting vine-like habit, the plants are suitable for growing up a trellis, against an old tree, over a pergola, or on a lath frame against a wall.

Probably the most colourful species is *Celastrus articulatus*, a native of China, which bears greeny-white flowers in early summer followed by large golden-orange berries in autumn. They are a glorious sight amidst the golden foliage, and, as the leaves fall, the fruits open to reveal vivid yellow inner skins and scarlet seeds. Another species is *C. scandens*, two plants of which have to be placed together before they can produce bloom and fruits which are similar to those of *C. articulatus*.

March is the best time to plant and prune, for by then the winter display will have ended.

CELMISIA

These are hardy perennials which like a slightly acid soil, but a dry, sunny position. They are like dwarf cosmeas, having feathery foliage and their flowers are like small marguerites. The best form is *C. sessiliflora* with its yellow-green foliage and large, pure white flowers borne in mid-summer.

CELOSIA

An annual, *C. plumosa* may be raised in a temperature of 60°F. (16°C.) early in the year and makes an ideal pot plant. The Dwarf Red Plume and Golden Plume grow only 10 in. (25 cm) tall and bear their bright feathery plumes over several weeks of summer; or they may be planted out for bedding in early June after hardening. Plant 12 in. (30 cm) apart, alternating the colours for a brilliant display. Celosias enjoy a humid atmosphere where growing under glass.

C. cristata, the Cockscomb is equally handsome in pots or used for bedding. It requires greater care in its culture, it being important that growth is not checked at any stage. Keep growing in a temperature of 60°F. (16°C.) and transfer to the pots before exhausting the sowing compost. The variety 'Prairie Fire' grows 15 in. (38 cm) tall and bears flowers of brilliant scarlet which appear in pyramid form with a larger flower surrounded by smaller 'combs'. Resistant to adverse weather, it is highly suitable for bedding.

CENTAUREA

ANNUAL

Centaurea cyanus, the Cornflower is an annual and though for long a popular flower for cutting, the dwarf forms are ideal for bedding, making bushy plants and growing to a height of only 10 in. (25 cm). Seed may be sown in gentle heat in early February or in a cold frame in late summer to come early into bloom the following year. Or sow in a frame or under cloches in early March. Plant 9 in. (22 cm) apart, setting out the plants in April.

The first of the dwarfs was 'Jubilee Gem', with its rich cornflower blue flowers; then followed the lovely 'Lilac Lady' and 'Rose Gem', all of which look most charming planted together. As a contrast to plant with them or with scarlet salvias, the double white-flowered 'Snowball' is most effective.

The taller growing varieties are always in demand for cutting, especially the pink which is more popular than the blue. The seed should be sown in early October in favourable districts and a month earlier elsewhere, possibly at the end of August in the most exposed gardens. Sow 2 ft. (60 cm) apart and thin out the seedlings to 12 in. (30 cm) apart in March for the cornflower makes a tall, bushy plant. In early gardens the first bloom may be picked about June 1st from an autumn sowing. In less favourable areas, late June will see the first bloom. The best varieties for cut bloom are those of the 'Monarch Ball' strains, long stemmed, extremely double, and very early flowering. 'Pink Ball', shell-pink; 'Red Ball', rosy-red; and 'Blue Ball', devoid of the usual purple colouring are all valuable. Outstanding too, is 'Blue Diadem'.

Centaurea moschata or Sweet Sultan is one of these hardy annuals which dislikes transplanting so should be sown where it is to bloom. Sow thinly in drills 9 in. (23 cm) apart late in August and thin out the seedlings to 6 in. (15 cm). Should the weather be severe, wattle hurdles should be placed over the rows but otherwise the plants require no attention. They come into bloom in July and require no staking. The bloom is cut when just opening. The flowers are sweetly scented.

PERENNIAL

It is a valuable border plant, perhaps more for its hairy silver foliage than for its blooms. The plants are best propagated either by stem or root cuttings, the fleshy roots often proving difficult to eradicate where the plant is no longer required in the border. A greater use could well be made of those species bearing yellow flowers which are very much more attractive than those bearing flowers of purple-blue, as *C. montana* Purpurea. Plant in November, 18 in. (45 cm) apart for those of more dwarf habit; allowing 2 ft. (60 cm) for the taller species. The plants will flourish in ordinary soil.

SPECIES OF PERENNIAL CENTAUREA

C. babylonica. In a moist soil it will grow to a height of nearly 6 ft. (2 m), having large furry silver-white leaves and bearing sprays of brilliant yellow flowers.

C. dealbata. It is one of the longest flowering of all plants. Its leaves appear as if white-washed, its cerise-pink flowers being borne on 2 ft. (60 cm) stems from July until November.

C. gymnocarpa. It has the most silvery foliage of all the cornflowers and bears small flowers of rose-purple during July and August.

C. macrocephala. A lovely plant, having long toothed grey foliage and bearing golden-yellow flowers on 3–4 ft. (about 1 m) stems from early June until September.

C. montana Violetta. This is a much better variety than the type, bearing purple-blue flowers from mid-April until mid-June; hence its value in the border.

C. pulchra major. It blooms from May until July bearing its bright rose-pink flowers on 2 ft. (60 cm) stems above lovely grey foliage.

C. ragusina. It grows to a height of only 1 ft. (30 cm) and has richly silvered leaves and flowers of brilliant gold. A fine midsummer flowering plant for the front of the border.

C. ruthenica. A handsome plant having long, finely cut leaves of mealy-green and bearing pale yellow flowers on 3–4 ft. (about 1 m) stems during June and July.

CENTURY PLANT, *see* Agave

CEPHALOTAXUS

The Cluster or Large-leaved Yew, has shortly stalked leaves, spirally arranged in two ranks. Like the Yew, it grows well in shade and in a calcareous soil provided it has depth. The best form is *C. harringtonia drupacea,* native of Japan which grows 10 ft. (3 m) tall and bears large ovoid fruits like the damson. It grows as a spreading bush rather than as a tree but the form *fastigiata* grows erect. Plant in April. Propagation is from seed sown after being left two years on the tree to ripen; or from cuttings of the young shoots inserted in sandy compost under glass.

CHAENOMELES

The Flowering Quince, or Cydonia, *C. japonica* is one of the most beautiful of flowering shrubs. It may be grown against a wall as a specimen bush, or trained as a hedge. It bears its camellia-like flowers during March and April, and these later form greenish-yellow fruits which make excellent jelly if removed in September. The cydonia is an extremely hardy plant which grows well in almost any soil.

To form a hedge, 'Snow', with its large pure white flowers, and 'Rowallane', pure scarlet, make an arresting picture when planted alternately, with the shoots trained along strong wires. Also excellent are 'Falconet Charlotte', which bears double salmon-pink flowers; 'Elly Mossel', rich crimson flowers, and 'Boule de Feu', semi-double flowers of brilliant orange-scarlet.

To encourage the flowers to appear in profusion on short spurs, the ends of the shoots should be pruned back immediately after flowering each year.

CHAMAECYPARIS

The False Cypress is distinguished from the true Cypressus by its flat branchlets and leaves arranged in two ranks. Included in the genus is *retinospora* and Lawson's Cypress of which there are numerous varieties of garden value. Many are fast-growing trees, valuable for planting as a hedge or to hide an unsightly view, or they may be planted in groups about the garden. Most are hardy but to prevent the foliage from browning, protect from cold north-easterly winds.

Plant in spring, in a deep, moist soil and allow the plants space to develop. Propagate by seed sown when ripe (usually in spring); or by cuttings removed from the tips of the shoots in July. These are inserted in a sandy compost under glass. Plants raised from seed will appear in eight weeks but should be grown on in their boxes until the following spring before being planted out.

SPECIES AND VARIETIES

Chamaecyparis lawsoniana. Lawson's Cypress, it makes a tall tree of pyramidal form and possesses extreme hardiness. The closely imbricated leaves are of dark glossy green whilst the branchlets are drooping and fern-like. The pea-like cones are borne in profusion. The whole plant has a pleasant cedar-like perfume. There are many varieties of all sizes and shapes and with a wide variation of foliage. Allumii has glaucous foliage but tends to grow lanky and sparce. Bowleri has pendent branches and is known as Smith's Weeping Cypress whilst the popular 'Fletcheri' makes a close feathery pyramid of glaucous blue. Similar in its light and feathery habit is 'Hillieri' but which has golden foliage, whilst 'Pottenii' makes a conical tree of soft sea-green. 'Triomphe de Boskoop' bears glaucous-blue foliage and lutea, foliage of golden-yellow.

C. nootkatensis. Native of Vancouver Island, it is distinguished from *C. lawsoniana* by its drooping branchlets which are covered with small dark green leaves whilst the globular cones have a projection at the centre. The variety 'Pendula', is one of the most elegant of weeping evergreens whilst 'Lutea' has foliage of soft yellowish-green. Its hybrid *C. leylandii* is rapid growing and is tolerant of sea winds and a calcareous soil. It will withstand clipping in spring when used as a hedge.

C. obtusa. Native of Japan, it is slow growing with spreading fan-like branches and blunt scaly leaves which are borne in whorls of four and closely pressed to the branches. It is the Hinoki Cypress used in Bonsai as Chokkan (denoting an upright tree) and is used with the Japanese Cedar and the pines. Several varieties are valuable rock garden plants, 'Nana' making a rounded mound of feathery glaucous foliage; and 'Pygmaea', a tree of Bonsai form, with fan-like branchlets formed horizontally.

C. pisifera. The Sarawa Cypress of Japan which makes an open, pyramidal tree of slender form, the branchlets being clothed in four rows of scale-like leaves, marked beneath with two white lines. 'Plumosa' is conical in habit with feathery glaucous foliage whilst 'Plumosa Aurea' has golden foliage.

C. thyoides. The White Cedar, a quick-growing plant, of slender form, the branchlets being covered with closely imbricated pale green leaves and with pea-like cones. It is intolerant of a calcareous soil. The variety 'Glauca' is of silvery hue whilst 'Variegata' has leaves marked with gold and 'Ericoides' has heath-like foliage which turns bronze in winter.

CHAMOMILE, *see* Anthemis

CHEIRANTHUS

Of all biennials the wallflower, *Cheiranthus cheiri*, seems the most difficult to grow well. It is usually sown too late in summer, and not transplanted as soon as it should be. The old country house gardeners would sow in April, to allow the plants twelve months before coming into bloom and would transplant into a bed of fine soil 4 in. (10 cm) deep over a layer of stone or slate to prevent the formation of tap-roots, but early and regular transplanting will overcome this need. Fresh seed should be used for quick germination, the seedlings being thinned out as soon as large enough to handle. Sow thinly in shallow drills 12 in. (30 cm) apart.

An attractive display may be obtained by filling a circular bed with various contrasting colours, with the deep velvety-red, 'Belvoir Castle' in the centre, surrounded by 'Ivory Queen'. Then a circle of 'Fire King', 'Orange Bedder', 'Eastern Queen' and 'Cloth of Gold', the whole making a display of great splendour. Or they may be used for inter-planting a bed of late flowering tulips. About the garden they are essentially at their best when massed, and should be grown as near to the house as possible so that their scent may be enjoyed from indoors.

WALLFLOWER, SIBERIAN

This is the vivid orange Siberian wallflower, *Cheiranthus allionii*, also obtainable in a rich yellow form, 'Golden Bedder'. The plant grows in rosette style, almost like the sempervivum and does not require the same attention to detail as the ordinary wallflower. It comes into bloom when the wallflowers are finishing, and will continue through summer. For this reason they should be planted either in beds to themselves or in groups to the front of the herbaceous or shrub borders. The plants may be allowed to remain in the seedling beds until early spring.

CHERRY, *see* Prunus

CHERRY PIE, *see* Heliotropium

CHILEAN BELLFLOWER, *see* Nolana

CHILEAN CROCUS, *see* Tecophilaea

CHILEAN LANTERN TREE, *see* Crinodendron

CHILEAN YEW, *see* Podocarpus

CHILE PINE, *see* Araucaria

CHIMONANTHUS

C. fragrans is the Wintersweet, a hardy deciduous shrub which bears, during the coldest of winter days, purple-yellow wax-like flowers of delicious scent. Several sprays taken indoors and the ends placed in damp sand will scent a large room. The plants require protection from cold winds which may cause them to die back. No pruning is necessary apart from tipping back the side shoots after flowering; this will ensure the continual formation of new wood which will carry bloom the following year.

CHINCHERINCHEE, *see* Ornithogalum

CHINESE DOVE TREE, *see* Davidia

CHINESE HAWTHORN, *see* Osteomeles

CHINESE LANTERN, *see* Physalis

CHIONANTHUS

Handsome small deciduous trees which are native of west China and the United States and which grow well in a sunny situation and in ordinary soil. *C. virginicus* is the best known, being the Fringe Tree of North America, so called for the fringe-like petals of its white flowers. These are borne in panicles in June and July though flowers are not borne until the trees are well established. Plant November to March.

CHIONODOXA

It increases more rapidly than any other bulb and blooms in March and April. Its native haunt is Greece and it is from the Greek *chioni*, meaning snow, and *doxa*, glory, that it takes its name, Glory of the Snow. It is a valuable flower in that it appears unmindful of the cold winds and sleet showers of early spring. Unlike the crocus it refuses to close up its flowers as if protecting itself from the elements whilst planted in pots and allowed to bloom in a frame or cold house in March, it makes a welcome addition to the spring display.

Ideal for the rockery or for the front of a border or shrubbery where they will receive their fair share of sunshine, the chionodoxas are planted in September in groups, planting them 2 in. (5 cm) deep. The plant is not so tolerant of damp, shady conditions as the snowdrop or winter aconite; it requires a more open situation and enjoys a soil containing plenty of leaf mould or peat and above all, one containing some sharp sand or grit which it enjoys in its native haunts. As they quickly reproduce themselves both from offsets and self-sown seed, a light mulch during early winter will help the bulbs to retain their vigour and depth of colour and will encourage them to come into bloom early. A mixture of decayed manure and peat or leaf mould will be ideal.

For pot culture, use a compost containing leaf mould and plenty of grit and give the bulbs no undue forcing.

A pleasing way with chionodoxas indoors is to plant half a dozen bulbs of all the species and varieties and colours together in a large seed pan. When taken indoors in the early new year they will come into bloom late in January. Starting with *C. sardensis* the various species come into bloom in turn right through the later winter and early spring months. Cover the outside of the pan with a piece of old 'black-out' material and the effect will be astonishing. Or, the same bulbs may be planted in a large but shallow glass or painted pottery bowl. Remember to place some broken crocks along the bottom before filling with compost which will help with drainage and a good idea is to mix into the compost a small handful of crushed charcoal which will keep the soil sweet over a long period.

Of such distinctive colouring are the chionodoxas that it is preferable to plant them in drifts, keeping them entirely to themselves. They are rarely troubled by either mice or birds.

SPECIES

Chionodoxa gigantea (Grandiflora). It bears large flowers of the clearest sky-blue and is at its best throughout April. The colour is accentuated when planted with *C. gigantea* Alba.

C. luciliae. Growing to a height of 4 in. (10 cm), it bears vivid blue flowers which have an attractive white eye. The flowers, borne in March are carried on dainty sprays and are freely produced. The form Alba bears white flowers whilst Rosea, so rarely found in gardens, bears flowers of shell pink. A new variety of great charm is Pink Giant which produces its spikes of rich shell pink flowers on 9 in. (23 cm) stems and is excellent for cutting.

C. sardensis. From Sardinia, it bears flowers of a most striking colour, being similar to the blue of *Gentiana sino-ornata*. The flowers which are carried on loose sprays, are very freely produced and bloom early in March.

C. tmoli. From the Levant, it produces a larger flower and is more prolific. It blooms early in April.

CHLOROPHYTUM

A beautiful and unusual plant and though it should not be classed as of trailing habit, its graceful form and the fact that from its runners dainty little plants are formed in mid-air, makes it an idea plant for a hanging basket or pot either in the home or in a garden room. It bears long narrow glossy leaves which droop in a most graceful fashion and are striped with cream and pale green. It is known as the Spider Plant, for at the end of the runners are formed small spider-like plants which are replicas of the parent, in much the same way as Mother-of-Thousands. These plants should be detached and planted in 3 in. (7.5 cm) pots, or they may be pegged into the moss around the hanging basket, holes being made if polythene is used, where they will form an abundance of root and will quickly clothe the basket with their foliage.

A minimum winter temperature of 42°F. (5°C.) should be provided and never at any time should the plant be over-watered. It will, in fact, exist for long periods entirely without water and during winter will need little. The form best known is *C. capense variegatum*, which will form a large plant covering a large basket when fully grown, so no more than a single plant should be used for a basket. An 'open' compost is necessary, one containing 3 parts fibrous loam and 2 parts silver sand.

CHOISYA

The Mexican Orange Flower, *C. ternata* is hardier than usually believed but the plant does like a warm wall on which to ripen its wood to enable it to withstand winter weather and bloom well. It thrives in a dry, sandy soil when protected from cold winds, and will attain a height of 8 ft.

(2.5 m). An evergreen, it has a long flowering season, bearing its small, orange blossom-like flowers from mid-May until autumn, and will continue to bloom until Christmas in the south-west. With their attractive golden stamens, the blooms carry the rich honey-like fragrance of the hawthorn.

Plant in April which is also the best time for pruning, cutting out any decayed wood and shortening unduly long shoots.

CHRISTMAS BOX, *see* Sarcococca

CHRISTMAS CACTUS, *see* Zygocactus

CHRISTMAS ROSE, *see* Helleborus

CHRISTMAS TREE, *see* Picea

CHRYSANTHEMUM

ANNUAL

Like the calendula, the annual chrysanthemum, comes into bloom in midsummer and flowers until early autumn, and might be more popular than it is, for it has a longer flowering season than most plants, lasts well in water and does not drop its petals. True, the plants do not winter well unless under glass, but this cannot be the reason for their lack of popularity. Probably the modern varieties of *C. tricolour,* will help to bring this flower back into favour.

Seed is sown under glass in September, preferably broadcast in a frame, and there the plants remain through winter for planting out 9 in. (23 cm) apart early in April. No staking is required. The bloom is cut when almost open.

Excellent for cutting is *C. coronarium* 'Golden Crown', of a bright buttercup yellow, the bloom being full, the petals frilled, and held on 2 ft. (60 cm) stems. In the *C. tricolour* section, 'The Sultan', coppery-scarlet; 'John Bright', pure golden-yellow; and 'Northern Star', pure white with a zone of pale yellow, are outstanding.

INDOOR

Queen of autumn flowers, there is no crop more profitable, first from the sale of the bloom and later from the sale of cuttings and plants, nor is there any flower able to produce such a ravishing display of colour during the dull winter months. The plant is ideal to fit in the glasshouse routine where tomatoes are grown in summer, but where heat can be given the period of flowering may be extended from late November until the end of winter. Then in early spring the greenhouse may be utilised profitably to house stools from which cuttings are taken. But even in a cold house a wonderful display of bloom may be obtained between early October and Christmas, a display equally as brilliant as may be seen in a heated house. The amateur will almost certainly concentrate on the huge Japanese varieties, growing plants in pots and by the judicial use of the dwarf and taller growing varieties will be able to arrange them in tiers so as to show off their great beauty to the full. The catalogues will give those varieties most suitable for showing in public and for beautifying the home. There are literally hundreds but a few of the most beautiful and easily grown indoors are:

MID SEASON: INDOORS

ALAN ROWE. The finest white incurved for exhibition and market with tight, solid blooms.

AUTUMN TINTS. A reflexed decorative for early flowering under glass, the large blooms being of rich deep bronze.

COPPER GLOBE. The tightly incurving blooms are of bright coppery bronze, produced in profusion.

FAIR LADY. The incurving blooms have great substance and are of clearest pink.

HERBERT CUERDEN. The reddish-bronze flowers are tightly incurving to form almost a ball of perfect symmetry.

JOY HUGHES. The finest early indoor reflexed pink, bearing blooms of exhibition quality.

LONDON GAZETTE. A reflexed crimson of great substance with a gold reverse.

ROCKWELL. A handsome reflexed cream with large solid blooms.

SUN VALLEY. The finest early indoor yellow, the blooms being tightly incurving and of bright primrose yellow.

LATE FLOWERING: INDOOR

AMERICAN BEAUTY. Very late and the best white incurve of its flowering period.

FRED SHOESMITH. A superb white incurve with small foliage and a tremendous cropper.

LILAC LOVELINESS. A bright lilac-pink sport of 'Loveliness' with slightly incurving blooms.

LOULA. The bright red reflexing blooms have a golden reverse and are borne with great freedom.

RIVALRY. A late yellow incurved of great substance.

RIVAL'S RIVAL. The large solid incurved blooms are of deep old gold, touched with bronze.

ROSE HARRISON. The finest and most prolific late reflexing pink.

ST MORITZ. A pure white incurving for late November.

WORTHING SUCCESS. A late reflexing variety of a lovely shade of shell pink.

SINGLE INDOOR VARIETIES

CHESSWOOD BEAUTY. Bright crimson 'sport' from 'Mason's Bronze' and equally prolific. November.

DESERT SONG. Introduced in 1930 and never surpassed. The shaggy blooms of rich terra cotta, tipped with gold measure 8 in. (20 cm) across. November.

FIREBIRD. Very free flowering, the blooms are of rich orange shaded gold at the centre. November.

MARY LOW. It has superseded 'Molly Godfrey' as the best clear pink single. Late October.

MASON'S BRONZE. Prolific, the evenly-shaped blooms being clear terra-cotta, overlaid bronze. November.

NERINA. Rich purple with an inner zone of white. November.

LARGE EXHIBITION (JAPANESE) INDOOR VARIETIES

ALBERT SHOESMITH. An incurving yellow of substance with broad petals. Secure first crown bud.

BIRMINGHAM. Of excellent habit, the bright crimson blooms with a gold reverse have broad slightly incurving petals. Stop mid-May and take first crown bud.

CONNIE MAYHEW. The huge incurved blooms are of a lovely shade of soft creamy-primrose. Plants break naturally. Take first crown bud.

DORRIDGE APRICOT. Of dwarf habit, it makes a large incurving bloom of apricot-bronze. Stop late April and again in June for second crown bud.

EASTERN QUEEN. The large incurving blooms are of pale shell pink with a dusky sheen. Stop end of April for first crown bud.

GOLDEN WEDDING. The incurving blooms with their interlacing petals are of rich golden-yellow. Stop early May for first crown bud.

GREEN GODDESS. Of unique colouring, the huge incurving blooms being of palest green. Stop early April for first crown bud.

KEITH LUXFORD. The large reflexing blooms of rose-pink have a silvery reverse. Stop early April for first crown bud.

LADY DOCKER. It bears an exhibition bloom with broad incurving petals of chestnut bronze shaded with salmon. Stop mid-April and take first crown bud.

OWEN SUTTON. A semi-incurve with broad wine-red petals with a buff reverse. Stop early May for first crown bud.

SHIRLEY LAVENDER. The large incurved blooms with their curling and interlacing petals are of clear lavender-mauve. Plants break naturally; secure first crown bud.

WOOLMAN'S PERFECTA. A white incurve of outstanding form. Stop mid-March and take second crown bud.

WOOLMAN'S YELLOW. A magnificent yellow of reflexing habit with firmly textured petals. Plants break naturally; secure first crown bud.

AMERICAN SPRAY INDOOR VARIETIES

They bloom late October until Christmas and should not be disbudded:

AMBER LONG ISLAND BEAUTY. An orange-amber 'sport' of 'Long Island Beauty'.

CHRISTMAS GREETING. Bright crimson of reflexed form.

CORSAIR. The white flowers, like golf balls are borne in profusion.

GALAXY. The single reflexed blooms are of deep orange-bronze.

MARIGOLD. It bears a profusion of golden-yellow reflexed blooms.

MEDALLION. The double pompon-shaped blooms are of a lovely shade of antique pink.

VIBRANT. The double ball-shaped blooms are of rich old gold.

MAKING A START WITH CHRYSANTHEMUMS

With chrysanthemums as against pot plants, tomatoes and most other glasshouse plants, a small area of ground outdoors must be at one's disposal for here the plants will be set out either in their pots or in the open ground during the summer months and are taken indoors around October 1st, where they bloom until the year ends. If no ground is available, the plants may be grown in pots and allowed to stand out in a yard or in a place where they may receive full sunlight and protection from winds. The compost used in the potting should consist of 3 parts turf loam and the balance made up of decayed manure, a little peat and a sprinkling of bone meal.

For healthy stock, cuttings should be purchased from a reliable source in November or December, rooted in gentle heat and planted outside in spring (see plates 14–17).

Newly-purchased cuttings should be placed in bowls of cold water for several hours before inserting them in boxes of a specially prepared mixture of sand and granulated peat. Being almost sterile this compost does not encourage damage from the dreaded damping off disease, the spores of which are present in garden soil. Some growers use sterilised soil with a very small amount of peat and sand added. This latter compost has the advantage in that the plants will not need moving as soon as rooted as will be the case when using only peat and sand. If possible have the boxes of much the same size and in each, place rooting compost to a depth of 2 in. (5 cm). Make this as level as possible, and using a dibber, make the hole for each cutting and press firmly in when it has been inserted. Each box will take forty-eight cuttings. Give a soaking of water and where heat is used maintain a temperature of 45°F. (7°C.). Care should be taken to shield the newly-inserted cuttings from the sun by placing sheets of paper over them. Provided the atmosphere is kept moist and the cuttings are watered with care, they will begin to take root after about three weeks. Give the cuttings a dusting of green flowers of sulphur once a week to prevent them 'damping off'. Where heat is used, cuttings may be inserted at intervals throughout the winter from November until March, but should this be done, a cold frame must be available to which the earliest rooted plants are removed to make way for others. Young plants raised in heat should have a gradual hardening-off in a frame before being planted into the open ground. Some growers plant into small pots instead of directly into the frames. From there they are removed into larger pots and are lifted outside into a position where they are to remain until taken into the house for flowering in October.

COMPOST PREPARATION FOR CHRYSANTHEMUMS

Chrysanthemums are heavy feeders and it is essential that the soil which is being used for their cultivation during the summer months has first received a heavy dressing of manure. This should be at least ten tons per acre and on poor land as much as fifty tons would not be too much. Old mushroom bed compost is useful, especially for plants in pots, for it makes an ideal fertilizer. Spent brewers' hops give excellent results, also shoddy, and where either of these two sources of plant food can be obtained, they should be stored under cover for a few months before use. A one oz. per sq. yd. dressing of sulphate of potash will bring out the colour of the blooms to a greater degree. The land should be dug over thoroughly and allowed to be broken down by frost during the winter months; it should be ready for planting in April. The plants are placed in rows 18 in. (45 cm) apart each way, well firmed and given a thorough soaking of water. Pinching out the tops should not be done until the plants have been in the ground for ten days, otherwise they receive too big a shock from being moved and 'stopped' at the same time. If possible, the plants should be put out in damp weather and it is vital to give them some form of protection from cold winds which, in most parts of the country, prevail in late spring. Either wattle hurdles or canvas sheets firmly fixed to stout stakes will answer the purpose or even a row of hedging plants such as privet or beech, would suffice. Frost is also a worry at this time of year and care should be taken to water the plants in time for the foliage to dry off before early evening.

CARE OF CHRYSANTHEMUMS

After ten days or so the plants should have their growing points removed, and again after three weeks for later flowering varieties.

At this time, a 5 ft. (1.5 m) cane should be inserted into the soil about 2 in. (5 cm) from each stem and the plants securely (but not too tightly) tied with raffia. By early July a large amount of side growths will be seen on each flowering stem at the leaf joints. These must be removed with a sharp knife and each week from that time onwards until the plants begin to bloom. On each stem will be observed a number of buds. Only the best one should be retained, the others being removed with finger and thumb. When plants are being grown in pots, care must be taken to see that they do not suffer from lack of water.

Those plants being grown in the open ground for indoor flowering should be removed to the house mid-October. First cut round each plant (with a spade) about 6 in. (15 cm) from the stem and allow them to stand like this for a week before they are lifted from the ground. This allows tiny roots to form where the incisions have been made, which will ensure that each plant is lifted with a good ball of soil to the roots. Take care when bringing the plants into the house for the buds are easily broken off at this stage. Place in rows closely together, pressing a little soil around each and making them firm. Give the roots a thorough soaking with cold water and, where necessary, stake the plants exactly as done when they were growing outside. Care should be taken not to give too much water to the foliage, otherwise mildew may set in. A temperature of about 50°F. (10°C.) is satisfactory.

OUTDOORS

Those flowering outdoors will do so between mid-July and early December depending upon the weather. They may be divided into two groups:

(a). Large flowering types which are disbudded and grown on exactly as for indoor-flowering decoratives, staking and disbudding, only that they bloom where they are planted. Afterwards, they are cut back to within

3 in. (7.6 cm) of soil level and may be grown on without lifting for a second year or more; or the roots are lifted and planted in boxes of compost and placed in a greenhouse or a frame for propagation from cuttings or offsets.

EARLY FLOWERING OUTDOOR VARIETIES

ARGUS. Outstanding in its colour, the reflexed blooms being of rich chestnut-bronze. September.

ASTA LEE. Free flowering and easy, the blooms being of a lovely shade of carnation-pink, pleasing under artificial light. September.

AUTUMN FANTASY. It has small dark green foliage and bears bloom of crimson-red with a gold reverse. August to September.

BAGSHOT YELLOW. The large reflexing blooms are of deep golden-yellow. August to September.

CADLEIGH. Early flowering, the large handsome blooms are of rich deep crimson. August to September.

ERNIE BALDWIN. An exhibition incurved of great substance, the blooms being of rich deep golden-yellow. September.

ESCORT. The slightly reflexed blooms with their stiff petals are of vermilion-red. The best red for cutting. August to September.

EVELYN BUSH. A large incurving white of great purity, unsurpassed for early bloom. August to September.

FAIR DAWN. Of dwarf habit and excellent for pot culture, it bears a bloom of icing-sugar pink. August to September.

FLORINDA. Dwarf growing and weatherproof, the blooms of deep purple-pink are outstanding under artificial light.

FRINGE. Early and of dwarf habit, the blooms are of rich crimson-red, tipped with gold and with gold shading at the centre.

GOLDEN DAYS. The standard yellow for it is refined and withstands all weathers. The bloom is of clear bright yellow with long tapering petals. September.

JULIE ANN. Of dwarf compact habit, it bears well proportioned blooms incurved at the centre and of soft salmon-pink. August to September.

MARGARET ZWAGER. An outstanding pink bearing large slightly incurving blooms of soft shell pink. September.

NICHOLAS ZWAGER. The large reflexing blooms are the richest crimson with a gold centre. September to October.

ORANGE SWEETHEART. Slightly incurving, it makes a large firm-textured bloom of brightest orange. September.

PROMISE. An exhibition incurved of faultless form, its bright pink flowers having a silver reverse. September.

SCARLET EMPEROR. The best early scarlet, the blooms being borne on sturdy stems and in profusion. August to September.

SHANTUNG QUEEN. Semi-incurving, the blooms are of deep pure gold. August to September.

STARTLER. Pillar box red, the blooms are borne over eight weeks in great profusion. Mid-August to mid-October.

THOMPSON'S PINK. Tolerant of adverse weather, it is one of the best of the early pinks, bearing bloom of a lovely shade of soft lavender-pink. August to September.

TIBSHELF HEIRLOOM. It bears a bloom of exhibition quality of deep old rose with an incurving gold centre. September.

TRACY WALLER. The blooms are bright pink with attractively interlacing petals. September.

YELLOW CRICKET. Incurving and ball-shaped, the flowers are of clear daffodil yellow. August to September.

YELLOW EVELYN BUSH. A 'sport' of 'Evelyn Bush' with the same dwarf habit and bearing a large incurving bloom of pale lemon-yellow. September.

(b). Spray or small flowering varieties, including the Pompons and Koreans. They may be treated like any other border plant, being given only sufficient staking to keep the blooms from being spoilt by wind and require no disbudding. They are amongst the easiest of all plants to manage and will give rich colour to the garden at a time when there is little else in bloom.

The plants may be treated as semi-permanent, propagating in exactly the same way as for other border plants, whereas to maintain the quality of bloom of the large-flowering varieties, annual propagation must take place from rooted offsets or cuttings. The spray varieties may be considered labour-saving plants, requiring neither the disbudding nor the continual removal of side shoots. For bedding, too, the new Korean Cushion chrysanthemums make a startling display from mid-August until the end of October.

The spray chrysanthemums suitable for border colours may be divided into three classes: (i) Koreans, (ii) Pompoms, and (iii) varieties of *Chrysanthemum rubellum*, which may be considered intermediate between the first two. All require exactly the same culture, and are extremely hardy, flowering until the end of November. It should be said, however, that the very late-flowering varieties such as 'Crimson Bride', which comes into bloom late in October, should be confined to southern gardens, for adverse weather experienced in the north from that time onwards may mean that the buds do not open at all, and in any case, fog and soot deposits will harm the blooms. Those varieties which bloom after the first few days of October should be avoided where growing in the less favourable districts. The late September-flowering varieties will, in any case, continue to bloom until well into November, when the border is cleared. Nowhere do they look more attractive than when planted near Michaelmas Daisies, their blue and purple colourings providing a pleasing contrast to the reds, bronzes and yellow shades of the Koreans, and in no way do they clash. The bright green serrated foliage of the spray chrysanthemum greatly contributes to its charm, whilst the strong sage-like aroma of the foliage should ensure their inclusion in a border of fragrant plants. The plants require a sunny, open situation. Chrysanthemums grow best in a rich, well-manured soil; of all flowering plants they are possibly the greatest feeders. It should be said, however, that the soil need not be so rich for the sprays as for the large-flowering varieties, for size of bloom is not so important a consideration. In poor soil, however, the blooms will be smaller than where growing in soil of good 'heart', and where this is so the blooms will lack colour, and especially that lustre which makes them so colourful when cut.

The young plants should be obtained towards the end of April and should be planted 20 in. (50 cm) apart, either in beds or in the border. Where growing in the border, plant them near those which bloom early and which will give early spring protection. These plants will have finished blooming and will have begun to die back before the chrysanthemums make their bushy growth. Firm planting is essential and, as the chrysanthemums will be small when set out, mark their position with a stick. It is essential that the plants do not suffer from lack of moisture during their early days in the open ground, and during July and August weekly applications of manure water will enhance the size and colour of the blooms.

After flowering, the plants should be given a mulch of peat or decayed manure, which is placed around the roots, but not over them; by November, next year's flowering shoots will have begun to appear, a second batch following in May when the soil begins to warm.

For two or three years the plants may remain undisturbed, after which time they should be lifted and divided in the usual way by using two forks. The old woody centre should be discarded. The plants may be lifted early in November each year, or in alternate years, when propagation may be carried out in several ways. Those who do not possess a greenhouse, and when it is not considered necessary to propagate on a commercial scale, may remove the green shoots with a portion of root attached. These are planted 4 in. (10 cm) apart in a frame, over which is placed a glass light or a sheet of plastic material. There the young plants remain throughout winter, and will be ready for planting out again in April. The plants should be given almost no water during winter and, to prevent mildew, they should be treated once a month with flowers of sulphur. Just before the plants are set out they should have the growing point removed to encourage them to build up a bushy plant. This is better done a week before they are moved, so as not to cause them any undue check. Plants obtained through the post and which have not already been 'stopped', should not have the growing point removed until they have become firmly settled in the ground.

Where it is not possible to use a frame, the plants should be left undisturbed throughout winter. In April, when the new shoots have appeared, the plants may be lifted and the rooted offsets pulled away from the old root and replanted at once.

KOREANS

Those varieties growing about 2 ft. (60 cm) tall should be planted in the small garden, those of taller habit being confined to the larger garden, or to the market garden. The following will prove suitable to provide autumn colour in the small garden:

CALIPH. A superb variety, bearing double blooms of deep blood-red. Early September.

CARLENE. A magnificent new variety, bearing huge double blooms of bright pure orange. Early August.

CORAL MIST. A new variety, coming into bloom early in August, and bearing a large single bloom of coral-pink.

DAWN PINK. The large single blooms are of a pure shell-pink colour with a lovely silver sheen. Mid-September.

DERBY DAY. The single blooms are of a rich shade of burnt orange. Early September.

EUGENE WONDER. Raised in America, the very double blooms are of a bright golden-yellow, flushed with orange. Mid-September.

HONEY POT. For the front of the border it makes a mound covered with small honey-coloured blooms. Mid-September.

MARGERY DAW. The large single blooms are of a rich cerise-red, like the Harold Robinson pyrethrum. Mid-September.

MOONLIGHT. The semi-double blooms are of a pale primrose-yellow. Early September.

POLLY PEACHUM. A new variety, bearing a large double bloom of deep peach-pink. Late August.

SPINDLE BERRY. The single blooms are of a rich shade of salmon-pink. Early September.

STARTLER. Outstanding, the double blooms being of a deep shade of claret-pink. Early September.

SUNNY DAY. One of the very best varieties, the double bloom being canary-yellow, with a rich perfume. Mid-August.

TAPESTRY RED. The fully double blooms are of deep crimson, with an attractive green centre. Early September.

TAPESTRY ROSE. The double flowers are of a soft rose-cerise colour, with an attractive green centre. Early October.

POMPONS

The habit of the plant is more compact than the Koreans, the button-like blooms being small, but produced in great freedom over a period of eight to ten weeks. The plants bloom at the same time, the latest bloom opening by early October. They grow to a height of about 2 ft. (60 cm) and are ideal plants for an exposed garden.

ANDY PANDY. The stems are almost devoid of foliage and are covered with tiny ball-like blossoms of brilliant yellow.

BABS. The dwarf, neat plants are covered with interesting, flat, shell-pink blooms over a long period.

BOB. Of compact habit, the plants cover themselves with tiny buttons of brilliant scarlet.

CREAM BOUQUET. The small button-like blooms are of a rich cream colour and are borne in profusion.

DENISE. The habit is very compact, the plant coming into bloom towards the end of August, when it covers itself with rich golden-yellow blooms.

FAIRIE. The tight button-like blooms are of a most attractive shade of strawberry-pink.

JANTE WELLS. Growing to a height of 2 ft. (60 cm) and bearing masses of golden-yellow buttons, this is an excellent cut-flower variety.

KIM. It makes a plant of densely-branched habit and bears masses of bronzy-scarlet blooms.

LEMON BOUQUET. One of the famous Bouquet series, which are excellent for cutting or for border decoration. The ball-shaped blooms are of a soft shade of lemon-yellow.

LILAC DAISY. The large button-like blooms are of an attractive shade of dusky lilac-pink.

LITTLE DORRIT. The neat, button-like blooms which are of a deep shell-pink colour, are produced in profusion.

LUSTRE. Of most compact habit, it bears tiny flowers of an unusual shade of antique lustre ware.

MASQUERADE. A recent novelty, the flat button-like blooms are of silvery rose, with a dark pinky-brown centre.

MORCAR GEM. Of very dwarf habit, making a plant as wide as it grows tall, and bearing masses of rosy-apricot flowers.

ORANGE LAD. The button-shaped blooms are of brilliant orange, flushed with red at the centre.

PAT. One of the most beautiful Pompons, the blooms being in the form of a double michaelmas daisy; in colour the petals are of a rich deep-pink colour.

TITANIA. A new variety, bearing large ball-like blooms of rich amber-bronze.

TOMMY TROUT. One of the most striking varieties, bearing flowers of rich coppery-amber.

WHITE BOUQUET. The first white pom, being of similar habit to 'Jante Wells' and just as free-flowering.

OUTDOOR SPRAY – DOUBLE VARIETIES

ALPHA. White with a lime-green centre. Withstands wet weather and is most prolific. October to November.

AURORA QUEEN. Ox-blood red with golden shading at the centre. August to September.

CLAUDIA QUEEN. Brilliant orange, shaded with gold. September to October.

DORIS JOICE. An upright grower with small foliage, the blooms are of rich brick-red. September to October.

GOLDEN ORFE. Extremely free and early, bearing large flowers of brilliant golden-yellow. August to September.

HUNSTANTON. Of dwarf habit, it bears flowers of a lovely shade of soft mauve-pink. August to September.

LILIAN HOEK. Pure orange with small foliage, it lasts well in water. August to September.

MADELEINE QUEEN. The reflexed blooms are of old rose, flushed with deeper rose. September to October.

MARION. The large reflexed blooms are of soft creamy-yellow. August to September.

NATHALIE. Late flowering, the large blooms are of a lovely shade of deep purple-pink. October to November.

PAMELA. The best of its colour which is deep golden-orange, shaded bronze. September to October.

PICCOLO QUEEN. Of compact habit, it bears flowers of lemon-yellow shaded gold at the centre. August to September.

RHEINGOLD QUEEN. The blooms are large and of rich golden-orange throughout. September to October.

SINGLE VARIETIES FOR OUTDOOR

ABEL MILES. Signal red with a striking golden centre. September.

ALAN PYE. Blood red tipped with green and a golden centre. Dark foliage. August to September.

APRICOT PYE. A deep apricot 'sport' of 'Alan Pye' and bearing masses of flowers. August to September.

BEN DICKSON. Orange-bronze with large flowers. September to October.

EVA JOICE. Bronze with a golden centre. Late August to November.

GOLDEN SUSY. The best single yellow spray. August to September.

PAT JOICE. Lilac-pink with star-like petals. September to October.

ROSE PYE. The most free flowering single, the rose-pink flowers measuring 2 in. (5 cm) across and with six rows of florets. Almost double flowering. August to September.

CASCADE CHRYSANTHEMUMS

For their grace and to provide a mass of colour in the greenhouse, there is nothing to equal the Cascades. A Japanese introduction of quite recent years, it makes an excellent pot plant for the conservatory or any high roofed glasshouse. A glass porch at home is also ideal and will provide a wealth of bloom during the early winter months for the minimum of attention. A well grown plant gives the impression of a waterfall tumbling with coloured blossom. Whilst its cultivation is easy, it is essential to nip out the early growth, leaving behind only two to three pairs of leaves. This must be repeated after three weeks for it is necessary to obtain a bushy habit at the base of the plant.

Cuttings are rooted in the usual way and potted into 3 in. (60) pots as soon as well rooted. When they have outgrown this size, re-pot into the size 48 and again into the size 32. When once they have reached this stage, a cane should be inserted into the pot as near the stem as possible which (the cane) should be bent or broken over the side of the pot.

To this, the stem should be tied as it makes growth and the pot then taken into the open and stood in a position which allows the cane to point towards the south. Do not stop again, but keep the side growths pinched out and pay regular attention to both watering and feeding.

When the plant is taken into the house again during September it will have sent its many shoots far beyond the end of the cane and when stood on a high shelf, the whole will have a most graceful appearance with the cascading habit clearly seen. The price of plants is the same as for most other types and of several excellent varieties, the best are:

HI-NO-HAKOMA, a vivid red.

MIKAGEYONIA, delicate salmon-pink.

SWALLOW, rich creamy-white.

YUSAN, attractive clear yellow.

CHRYSANTHEMUMS AS A POT PLANT

There is no more delightful decoration for the house or greenhouse than a well-grown pot plant and the chrysanthemum will provide a spectacular display. Though all types and varieties do well when grown in pots, there are several which are of dwarf habit and if well-grown, will provide a mass of bloom which will be at its best over several weeks. To obtain a spectacular show, two or three cuttings should be inserted in a size 60 pot towards the end of April. As soon as rooted and the roots are seen to be growing around the sides of the pot, re-pot into a size 48 and place in a cold frame at the end of May. Harden for two weeks before planting out in the usual way where they are grown-on during the summer months. The same attention should be given the plants as for all other pot grown chrysanthemums, except that the plants should be 'stopped' early in June and again towards the middle of July. In this way a bushy plant will be formed. Disbudding and keeping the stems free from all unwanted side growths will be necessary in the usual way, though too much disbudding may not be required, for a mass of small blooms which literally cover the plant will make for a more striking display than will larger blooms, fewer in number. Feed and water in the usual way and lift the plants into a greenhouse or living-room by early October. Careful staking will help to make a most effective display. The most useful varieties are the ever popular 'Blanche Poiteven', a pure white incurve; and 'Balcombe Pink', a very free-flowering deep shell-pink; but there are others, such as 'Marie Morin' and 'Pink Morin' and 'Pioneer'.

GROWING POT PLANT CHRYSANTHEMUMS FROM SEED

Growing from seed is recommended for the enthusiast who has available the necessary greenhouse space and some heat, for the seed must be sown in February to obtain best results. Many of our seedsmen and specialist growers offer seed saved from the best named varieties of each type, e.g. Japanese, decorative or single. Seed is not expensive, but it's irregular in its germination. Sow in a temperature of 55°F. (13°C.) and in a buoyant atmosphere. Sow thinly so that those seedlings which have made most growth may be lifted out without disturbing the others which are only just commencing to show. A sowing compost of decayed turf loam, with the addition of a liberal quantity of coarse sand and a handful of granulated peat per box, will give good results.

The young seedlings should be pricked out into small pots containing a similar compost when the house temperature may be dropped to 50°F. (10°C.) to ensure sturdy growth and the maximum amount of fresh air should be admitted. Potting-on and hardening will be the same as described for rooted cuttings; also disbudding. Regular watering and feeding with liquid manure water throughout the summer will be beneficial. During their growth the plants should be given considerable attention for to obtain one from a possible hundred plants which might prove worthy of addition to a specialist's catalogue, will be a fitting reward for ones' labours. A bloom of especially good form or unusual colouring should ensure that the plant be carefully marked for growing on the following year by means of cuttings. If the results come up to expectations at their second time of flowering they may be grown on for propagation on a commercial scale. Single varieties are the most easily raised from seed, seed of many of the decoratives often reverting to their original form with disappointing results. Cascade chrysanthemums come particularly true from seed.

SPORTS

Many of the best commercial varieties are those which have been obtained by way of a new colour break of a variety which have been produced previously from the same root stem. Thus we have flowers of two distinct colours growing from one root stock. When this happens the new variety is known as a 'sport' and may be a haphazard phenomenon or it may have its colour fixed and may be propagated and grown on as a new and distinct kind. The cause of this tendency of the chrysanthemum to 'sport' is because most new varieties are the products of hybridising, of crossing one distinct colour with another. When this happens the resultant bloom may be as expected from the crossing but in the cells of the plant, one colour may predominate more than the common colour. When this happens a new and distinct colour may appear.

To make use of this new colour break, the stem should be tied with raffia and the flower head cut off to encourage the stem to produce side growths which are used for propagation. These shoots are produced from the axils of the leaves. When large enough they are removed and rooted in a compost of sand and peat, in a temperature of 50°F. (10°C.) and are grown-on throughout the winter in the same way as described for rooting cuttings. The plant must be allowed to bloom the following season in order to discover whether it will have reverted to its normal colour or has remained 'fixed'.

CHRYSOGONUM

The little known *C virginianum*, a perennial, may be used for edging a large border for it grows only 12 in. (30 cm) tall and is in bloom from the end of May until mid-September. It has vivid green toothed leaves above which it bears its small blooms of gold which have five wide petals arranged like a star. Plant 9 in. (22 cm) apart in November, in ordinary soil and increase by division of the roots.

CHUSAN PALM, *see* Trachycarpus

CIMICIFUGA

A plant for the back of a border or for a shrubbery, its tapering spikes, like those of the eremurus, being borne during September and October. With their creamy-white blooms they make an excellent foil for the purple and red michaelmas daisies and would be more widely planted except for the unpleasant perfume of the flowers. The plants like ordinary loamy soil, containing a small quantity of humus and which is well drained in winter. Plant in November or in spring, allowing 3 ft. (1m) between the plants.

SPECIES AND VARIETIES

C. cordifolia. Quite happy in partial shade, it has attractive heart-shaped foliage and bears tapering spires of creamy-white 4–5 ft. (1.5 m). The form 'Elstead' bears a more refined bloom of deeper colouring.

C. racemosa. William Robinson described this as a 'handsome plant', its long feathery racemes borne during August and September on 5 ft. (1.5 m) stems. The form, 'White Pearl', bears a spike of pure white.

CINERARIA

It is a plant which is easy to grow, but which is just as easy to grow badly as it is to grow well.

Like the primula and calceolaria, it demands cool treatment throughout, a winter temperature of between 40°–45°F. (4°–7°C.) sufficient only to keep out frost and to prevent an excess of moisture forming in the atmosphere. In the south-west, all the plants require is to be wintered under glass without the need for artificial heat and to be given fresh air on all suitable occasions.

During recent years the cineraria has been much improved, the habit of the plant being more compact and whilst the leaves are smaller, the size of the flat individual blooms which go to make up the large flower truss, have greatly increased in size. The colour of the blooms, too, has been enhanced both in clarity and depth.

One of the finest strains is 'Morel's Intermediate', raised on the French Riviera. The variety 'Matador' bears flowers of rich crimson-red; 'Vieux Rose' of a lovely shade of dusky pink; and 'Bleu d'Orient' of bright deep blue. There is also a white form and each will germinate true to colour. The 'Hansa' strain is also excellent, the plants having a compact habit with small leaves, whilst it bears large flower trusses in rich, bright colours. 'Berlin Market' is also of similar habit and makes a large flower head.

Now comes the first yellow cineraria, introduced by Ernst Benary of Germany. It is named 'Citronella' and forms large flower heads of medium-sized flowers which open to a pleasing lemon-yellow colour, turning to deep cream with age. With the grey-green foliage, the flowers make a pleasing contrast.

SOWING THE SEED

The seed should be sown in two batches, one sowing being made in early April for the plants to come into bloom in the new year; another sowing is made in June to come into bloom during late spring and early summer the following year. The plants will take about ten months to bloom from the time the seed is sown, but forcing in any way will result in a 'drawn', weakly plant, which will make excessive foliage at the expense of bloom and will be prone to attack from aphis for which a constant search should be made throughout the life of the plant.

Though the seed is small it germinates more quickly than that of any other greenhouse plant, whilst almost one hundred per cent germination may be expected. The seed must, therefore, be sown thinly. Use the John Innes Sowing Compost and the seed should be just covered with a little of the dry compost. After giving a gentle watering, cover with brown paper to exclude light. This must be removed as soon as germination is observed. If the compost is kept in a comfortably moist condition this will take place about three weeks after sowing.

From the time the seedlings appear they must be grown as cool as possible, for every assistance must be given them to grow sturdy. They must be shaded from the sunlight, yet at all times must receive an abundance of light and they should be grown as close to the glass as possible. Also, provide the maximum amount of ventilation until the weather becomes cold and damp in late autumn. Where there is a cold frame available the plants will benefit from being placed in the frame where transplanted. The frame glass should be whitened to protect the young plants against strong sunlight, whilst the lights must be removed or partially opened during day-time until the end of September. The plants should be taken into the greenhouse again during October and from the month end artificial heat is employed so as to maintain a temperature of 40°–45°F. (4°–7°C.) throughout winter.

CARE OF THE PLANTS

The young plants will make rapid growth from the time of their appearance and watering will need to be done with care. They are copious drinkers during summer, but during dull weather moisture should be given only when the compost begins to dry out. Then give a thorough soaking so that the fibrous roots do not have to turn upwards in search of moisture which is all too often given only at the surface.

The seedlings will form large leaves at an early age and may be transplanted before they become overcrowded. They should be removed with the smooth, blunt end of a piece of cane, holding the leaves with one hand whilst the roots are loosened with the cane held in the other. The seedlings should be planted 2 in. (5 cm) apart so that they do not become 'drawn', and into a compost which should be similar to that used for seed sowing but to which is added 1 oz. of sulphate of potash per bushel. This will prevent the plants from growing 'soft' when they will be more liable to attack from aphis.

The young plants will soon be ready for moving into size 60 pots containing the John Innes No. 1 Potting Compost, but instead of the peat use leaf mould of good quality. This will provide additional food without giving an excess of nitrogen, for this would make the leaves grow too large and cause plant growth to become soft.

Ample ventilation and sufficient moisture is essential at this stage for the sturdy growth of the plants. By the end of autumn the plants may require moving to larger pots in which they will bloom, those from the first sowing coming into bloom in the New Year. During winter, water sparingly, remove the shading from the glass and employ just sufficient heat to exclude frost. It must be emphasized that over-watering when the plants are coming into bloom will prove disastrous, for if the plants begin to turn yellow and hang their heads there will be no hope of recovery. The plants should not suffer from lack of moisture, but should always be asking for it. At all times guard against aphis attacks and the spoiling of the handsome foliage by leaf miner. Spraying with quassia solution should be carried out as routine from the time the plants receive their first move, or dust with 'Lindex' to keep green fly (aphis) under control.

Leaf miner is more difficult to eradicate and here again, prevention is better than cure. The small grey insect lays its eggs on the underside of the leaves. The larvae work their way through the leaves making unsightly white lines which are clearly visble to the eye and when once they make their appearance little can be done to eradicate them.

One of the best deterrents is the introduction of a small amount of soot into the potting compost and watering the plants with soot water once each week from the time the plants are in the pots should prevent any serious outbreak.

When once the buds have formed, the temperature of the greenhouse may be raised to around 48°F. (9°C.) so that they may open together.

CINQUEFOIL, *see* Potentilla

CIRSIUM

Of the thistle family, *Cirsium japonicum* is a valuable plant for a sandy soil. A perennial, it is readily raised from seed and will bloom the first year if a sowing is made in spring. It grows 2 ft. (60 cm) tall, the variety 'Rose Crown' bearing small thistle heads of rosy-red above dark green spiny foliage. It remains in bloom for many weeks in the garden and remains fresh in water for several weeks when cut.

CLADANTHUS

Native of Spain, it is an annual and has the same degree of hardiness as tagetes. Sow in gentle heat early in the year and plant out in May, after hardening. *C. arabicus* grows 2 ft. (60 cm) tall and has feathery pale green foliage and pleasantly scented daisy-like flowers of brilliant golden-yellow. It blooms from July until October.

CLARKIA

A hardy annual, *Clarkia elegans* does not like transplanting and for this reason should be sown early in April where it is to bloom. Nor does it winter well outdoors except in the mildest parts. Plants from an autumn sowing will bloom at the end of May; those from a spring sowing, a month later.

To provide cut bloom, seed is sown in drills 12 in. (30 cm) apart and the young plants should be thinned to 3 in. (7.6 cm) apart. Under good cultivation the plants will attain a height of almost 3 ft. (1 m) and should be supported by twine on either side of the rows. If growing in the border, sow broadcast in small circles. Seed may also be sown in pots in a cool greenhouse in September to bloom in spring. Amongst the best varieties are 'Chieftain', soft mauve; 'Enchantress', satin-pink; 'Crimson Queen'; and 'Illumination', rose, suffused with orange.

CLARY, *see* Salvia sclarea

CLEMATIS

This colourful plant is useful to clothe a dead tree or trellis and is tolerant of cold conditions. The plants like a rich soil, well enriched with lime rubble, and are fond of some cow manure in their diet. Deep planting should be guarded against, and select a position away from winds. Plant in spring. In addition to the various species, the clematis is divided into several forms, which may be simplified if we group *Clematis jackmanii* and *C. viticella* together; and *C. patens* and *C. lanuginosa* for the former bear their blooms on the new wood; the latter on the old wood.

C. jackmanii:

Introduced a century ago, this is perhaps the best-known group, for the plants are of vigorous habit. They first bloom in July, and again in September, bearing their bloom on the new season's wood formed during spring and early summer. It is, therefore important to persuade the plants to make as much new growth each year as possible, and this means hard pruning. They may be cut back to whatever height required, possibly half-way would be right, removing the woody top growth either in November or in March. If in an exposed position, March pruning is to be preferred, but cutting must not be delayed until April, for by then the new growth will have begun. Where growth is rapid, such as in the south-west, the plants may be cut back to within 15 in. (38 cm) of the base. The new season's shoots will then quickly cover a trellis again, whilst the quality of bloom will be outstanding. So that the plants will provide bloom over as long a period as possible, use two plants of the same group together. One plant may then be left until mid-April before cutting back; this will encourage it to come into bloom later, and to continue until early November, thus extending the season.

COMTESSE DE BOUCHARD. A vigorous grower, and free-flowering variety. The blooms are of medium size, cup-shaped, and of a rich satin-pink colour.

GYPSY QUEEN. A lovely variety, bearing large, deep purple blooms of velvet-like appearance.

HAGLEY HYBRID. One of the best of all clematis. Introduced in 1956, it is of vigorous, free-flowering habit, bearing blooms of an unusual shade of dusky pink, with attractive brown stamens.

JACKMANII. The well-known form, bearing deep violet-purple blooms and which has given its name to the group.

JACKMANII RUBRA. One of the original hybrids and a fine variety, the blooms being rich crimson-red with attractive white stamens.

MME BARON VEILLARD. It bears rich lilac-rose blooms with green stamens. It is late to bloom, flowering August and September.

MME EDOUARD ANDRE. The blooms are star-shaped and are of deep petunia-red, with a velvet sheen.

PERLE D'AZURE. The blooms are large, of bright metallic-blue and are freely produced.

C. viticella

This group is comprised of a number of varieties, which may be described as being amongst the hardiest of clematis. For planting in an exposed garden, particularly where prevailing winds blow directly off the sea, the plants will prove capable of withstanding the severest weather. They may also be planted against a northerly wall. They are the latest to come into bloom, flowering about mid-July, and will continue until

October. They require the same methods of pruning as prescribed for the *jackmanii* group, though the plants may be left unpruned until early April. Plants of this group are valuable in that they are at their best when those of the *lanuginosa* and *patens* groups have almost finished blooming. They are rarely troubled by 'dieback', remaining healthy and vigorous for many years.

DANIEL DARONDO. Earlier into bloom than most in this section and not so long-lasting, but the semi-double blooms of bright purple are outstanding.

ERNEST MARKHAM. A fine old variety, bearing large velvety blooms of rich petunia-red.

LADY BETTY BALFOUR. The latest of all to bloom, the beautiful violet-blue flowers having striking golden stamens.

MARGOT KOSTER. A lovely variety, the large blooms being of a bright rose-pink, with a darker bar down each petal.

VILLE DE LYON. The bright carmine-red flowers are deeper at the edges, whilst they are attractively veined.

C. lanuginosa

This section requires quite different treatment, for the plants produce most of their blooms on the old wood and only a small proportion on the new wood. It therefore follows that for covering a high wall or an unsightly fence, this is the most suitable group for the plants may be allowed to attain considerable heights with the minimum of pruning. Regular feeding is important, for they make a large amount of wood and foliage which will tend to die back if starved. The only pruning necessary is to cut away dead wood and to thin out where there is overcrowding. The plants come into bloom before the end of May and continue until mid-July. Then, after a rest period, another flush of bloom is produced during autumn on the new season's wood. Any pruning should be done during March, so as not to harm the autumn display. The plants need great care in supporting, for when established they attain a considerable size and are heavy, especially in wet weather.

BEAUTY OF WORCESTER. An old favourite, bearing masses of rich violet-blue flowers.

BLUE GEM. The flowers are of a pleasing shade of soft blue, each petal barred white and with black stamens.

CRIMSON KING. The well-shaped blooms are of a striking shade of bright carmine-red, freely produced.

LADY NORTHCLIFFE. In bloom from June until November, it bears masses of large deep lavender-blue flowers with attractive white stamens.

MME LE COULTRE. Bearing masses of large pure white blooms, it is most striking when planted with one of the deep purple varieties.

MRS CHOLMONDLEY. A magnificent variety, coming into bloom towards the end of May, and bearing large flowers of clearest steel-blue.

PERCY PICTON. A fine variety, bearing beautifully shaped flowers of rich purple-blue.

THE PRESIDENT. The blooms are large and of rich violet-purple, with a paler bar down the centre of each petal.

W. E. GLADSTONE. The flowers are extremely large and are of a lovely shade of lilac, with white stamens and black anthers.

C. patens

Plants of this group bear almost all their bloom on the old wood, and so require almost no pruning. Simply thin out overcrowded wood and shorten any unduly straggling shoots. The best time to do this is in March, before new growth commences.

EDOUARD DESFOSSE. At its glorious best in the early weeks of summer, the mauve-blue flowers having a deep purple bar to each petal.

LASURSTERN. Extremely free-flowering, the large purple-blue flowers having attractive white anthers.

MISS BATEMAN. One of the best of white clematis, the beautifully shaped blooms having striking golden anthers.

NELLY MOSER. One of the loveliest of all, the delicate blush-pink blooms having a deep carmine-pink bar to each petal.

Several of the species will bloom during winter and early spring and will prove valuable for draping over an old wall or for covering a trellis. They will bloom when the large-flowered hybrids are devoid of colour:

C. tangutica is an unusual climber which should be more widely planted. It is a rampant grower, rapidly covering a trellis which will hide an ugly building. It bears its attractive, lantern-shaped, golden-yellow blooms from July until early November, and these are suspended from long wiry stems. The equally attractive seed heads, like silver tassels, are prominent during late autumn and winter. This is a plant which will flourish on a sunless wall.

C. armandii 'Apple Blossom' is one of the few evergreen species which bears small pale pink flowers very early in spring. It should be given a sheltered, sunny position where it will commence to bear its clusters of bloom in March. The dark green leathery leaves are as attractive as the bloom: the original form bears pure white bloom. *C. balearica,* the winter-flowering clematis, bears its creamy-white blooms during the winter months when planted in a warm, sunny corner. *C. macropetala* is a lovely plant for a low wall, for it is of neat habit and rarely exceeds a height of 6 ft. (2 m). It requires a warm, sunny wall, and will bear its pale blue flowers in late spring. These are followed by attractive tasselled heads in early autumn, which persist into winter. *C. montana,* an extremely hardy plant, bears small white, star-like flowers in great profusion during spring and early summer. The blooms are similar to those of the Wood Anemone. It is equally happy growing in full sun or on a cold, sunless wall. *C. montana* requires plenty of room, as it quickly covers a large trellis, and requires almost no pruning, for it blooms on the old wood. The variety *C. montana* Rubens is of similar habit, but bears beautiful purple foliage and deep rose-pink blooms.

CLEOME

A half-hardy annual, *C. pungens,* the Spider plant is suitable for border planting. 'Rose Queen' bears trusses of pink spider-like flowers, and 'Helen Campbell' pure white flowers on 3 ft. (60 cm) stems. Sow in gentle heat early in the year and after hardening, plant out 18 in. (45 cm) apart in May. It requires an open, sunny situation and a rich loamy soil. It blooms late in summer.

CLERODENDRON

A deciduous tree, native of China and Japan which reaches a height of 15 ft. (4.5 m) in a loamy soil. Valuable in that it blooms during August and September at the same time as the foliage begins to take on rich autumnal tints. Remove any dead wood in spring each year and propagate from suckers which grow from the base of the plant.

C. trichotomum is one of the most colourful of all small garden trees, its heavily scented flowers of purest white having a striking red calyx which is retained when the flowers are replaced by blue-black berries.

CLETHRA

Half-hardy deciduous and evergreen plants, they require a lime-free soil and a position sheltered from cold winds. *C. arborea,* native of Madeira is evergreen and against a sunny wall will attain a height of 10 ft. (3 m). In August and September, it bears its sweetly scented white flowers, like the lily-of-the-valley. *C. acuminata* is the North American Pepper Bush, and is also tall growing. It bears racemes of sweetly scented flowers in July, followed in autumn by the leaves turning rich yellow. *C. alnifolia* Rosea has also distinctive foliage and bears pinkish-white flowers.

Plant in April in a soil containing peat or leaf mould. Pruning consists of thinning out the old wood. Propagate by side shoots removed with a 'heel' and inserted in sandy compost under glass.

CLIANTHUS

C. puniceus is the Lobster Claw plant, its vivid scarlet blooms resembling a lobster's claw. The plant is evergreen and is of vigorous habit, bearing its blooms in clusters during early summer. October is the best time to plant and a position of full sun should be selected, for the plants will withstand dry, sun-baked conditions. They do, however, appreciate a little decayed manure in the soil. The plants are best grown up a lath frame or they may be supported by fastening to a wall, whilst they are kept in shape by the shortening of any long shoots after flowering.

CLIMBING PLANTS

INDOOR

Of recent years, indoor climbing plants have become popular again, due chiefly to contemporary architecture and oil-and-gas-fired heating which enables the modern home to be efficiently heated. Rooms are separated by semi-permanent partitions and it is for covering indoor trellis used for this purpose that the climbing evergreen is popular. Where space is limited in the small town flat, a corner of the kitchen may be converted into a delightful dining 'room' by the erection of trellis which may be quickly covered with plant growth to provide a continental-like appearance.

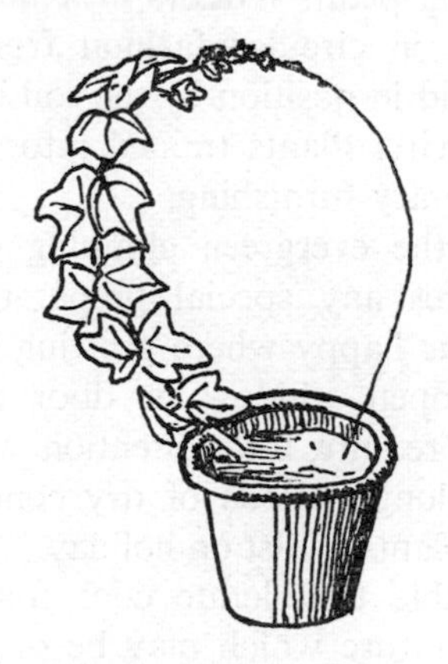

FIG. 2 *Training an ivy*

The plants are grown in pots, to be trained up the trellis as new shoots are formed. They may require help in climbing by tying the shoots to the trellis, or they will twine without assistance. Training plants in an upwards direction is by no means the only way to enjoy the beauty of a climbing plant for a pleasing effect may be obtained by fixing the pot at the top of the trellis, or to a wall bracket, and allowing the plant to cascade in the same way as plants of trailing habit growing in a hanging basket.

Most of the indoor evergreen climbers may also be used to 'frame' a window both in and outdoors. There are a number of suitable plants but the small-leaf ivies and geranium 'L'Elegante' are outstanding, the latter being used only indoors for it would not survive the rigours of the winter outside. For outdoors, the Virginian Creeper may be used or even a climbing rose which may be allowed to grow in a larger pot placed at one end of a window box and trained over a neatly made 'frame', made so that it does not interfere with the opening of the window. If a rose, which retains its glossy foliage for a long period after it has finished flowering, is used, it will be almost as evergreen as the ivies. So that the shoots may be regularly tied into position without difficulty, the plants should be used only around ground-floor windows, though where indoor plants are being used, almost any window will be suitable.

A pleasing effect may be obtained indoors by placing two plants one at either side of a window and training them up a section of trellis 9 in. (23 cm) wide and fixed to the wall at either side of the window and over the top. Or the trellis may be fixed to the inner wooden surrounds. There are few windows unsuitable for garlanding in this way and few that would not be greatly improved. The trellis would, of course, be painted to match the colour of the woodwork or walls and would be almost unnoticeable when covered with foliage, whilst it could be fixed so that it would not interfere with curtaining.

Another use of climbing plants indoors may be made by fixing a length of stout galvanized wire in circular fashion from one side of a pot to the other. This will be held in position by the soil into which the plant is set and trained around the wire. Plants trained into similar shapes will blend admirably with contemporary furnishing.

Of all indoor plants, the evergreen climbing plants are the easiest to manage for none requires any special preparation with their compost, whilst most are equally as happy where growing in shade as in light. Nor will draughts from an open window or door cause them the slightest worry. Again, most will require little attention with their watering, being able to withstand quite long periods of dry conditions so that one need give little thought to the plants whilst on holiday.

The plants are also able to tolerate cool conditions, several of them surviving a winter temperature which may be only a little above freezing. They have no vices in any way and as they may be used in the dullest corners of the home, merely growing them up a short central stake or up stakes placed in the pots in fan-shaped form, they may be considered amongst the most useful of all indoor plants.

OUTDOOR

These are plants which will cover a wall or trellis to which they are fastened for support; or they may be planted against a dead tree trunk about which they will twine and obtain the necessary support to enable them to pull themselves up to the sunlight. There are evergreen and deciduous climbers and those which will climb without artificial aid. These are known as self-clinging. This they do by means of suction pads present along the stems as in Ivy; *Campsis radicans; Ampelopsis veitchii; Hydrangea petiolaris; Decumaria barbara; Trachelospermum majus.* The plants will cling to any material and though they will quickly cover a wall, they will also attach themselves to the woodwork of a house such as window frames and gutters and will pull away the paint when removed. They should therefore not be allowed to grow where they may prove troublesome. The self-clinging plants usually make rapid growth and will cover the wall of home or outhouse more quickly than other plants.

Climbing plants may be divided into those which may be classed as being extremely hardy; those which are less hardy; and those which are tender. As a general rule those classed as 'less hardy' should be confined to gardens south of a line drawn across England from Chester to the Wash. As an example, the glorious powder-blue ceanothus makes an attractive picture covering the stone walls of the buildings which form the Great Court of Trinity College, Cambridge. In the colder north it rarely does well and makes only slow growth, though where given shelter it may be classed as being reasonably hardy if less so than many other climbers.

Those which demand the warm moist climate of the south-west, say west of a line drawn from Bristol to Bournemouth, may be classed as half-hardy and even here may demand a sunny wall to survive the winter. Here the campsis and the choisya, the magnolia and the abutilon grow well. Yet no hard and fast rule can be given as to what constitutes a favourable climate. For instance, Rothesay, on the west coast of Scotland, is renowned for its equable winter climate, being little different from that of Torquay, where all but the most tender climbers will survive.

Again, conditions may be improved by artificial shelter. Plants in a sunny courtyard will grow better than those growing in the same district which are more exposed. In such a situation the ceanothus might survive a hard winter, given protection, though its rate of growth would perhaps be slower than in the south.

Rate of growth too must play a part in the selection of plants. It may be necessary to cover an unsightly wall or building as quickly as possible;

then a rapidly growing plant must be selected, such as the climbing perennial nasturtium, *Tropaeolum speciosum,* which so loves a cool, moist climate and will make rapid growth, or *Clematis montana*. Many of the more vigorous modern climbing roses, such as Peace, will also quickly cover a wall.

For those who would wish to plant climbers which possess a vigorous habit these plants will be suitable:

Ampelopsis veitchii (Virginian Creeper)
Clematis montana; and *C. spooneri*
Climbing Rose: 'Caroline Testout', 'Elegance', 'Etoile de Hollande', 'Mme G. Staechelin', 'Mrs Arthur James', 'Peace', 'Souvenir de Claudius Denoyel'.
Hydrangea petiolaris
Lonicera caprifolium and *L. japonica* (honeysuckles).
Rambler Rose: 'Alberic Barbier', 'Albertine', 'Dr Van Fleet', 'Emily Gray', 'François Juranville'.
Vitis (Ornamental Vine), 'Miller's Burgundy.'
Wisteria sinensis

Where it is required to cover a low wall, possibly the front wall of a terrace or that part beneath a window, or where the fastening of the plants to a wall is not considered practical, a number of plants normally grown as shrubs will prove suitable. They will require almost no support and many will give a much better account of themselves where grown with the protection of a wall rather than in the shrubbery. One such plant is the wintersweet, *Chimonanthus fragrans,* which makes considerably more growth when grown against a wall. Other shrubs which do well against a wall and all of which are completely hardy are:

E – Evergreen

Aralia sieboldi (E)
Berberis (E)
Buddleia davidii (The Butterfly Plant)
Celastrus orbiculatus (Spindle Tree)
Cotoneaster
Cydonia japonica (Ornamental Quince)
Elaeagnus aurea (E)
Forsythia suspensa (Golden Bell Tree)
Garrya elliptica (E) (Catkin Plant)
Pyracantha (E) (The Fire Thorn)

These shrubs will more than reward one for their planting in that they will soon attain heights of from 4–8 ft. (1–2 m) and remain colourful for long periods. In addition to their flowers, several have attractive foliage.

A number of the less vigorous shrubs may also be planted to give protection to the less hardy climbers or where some colour is required at the base of the climber such as the clematis, especially plants of the *languginosa* group which tend to make plenty of top growth whilst leaving the base bare. Here the brightly-berried cotoneasters would be ideal for providing colour at the base, whilst the clematis would also benefit from the protection to its stem.

Some plants also enjoy covering at the base but for another reason: though the plants like a warm situation, they must have cool conditions at the roots and so should be covered by a dwarf-growing shrub. The interesting *Tricuspidaria lanceolata* is a plant requiring such conditions.

Again, it may take some plants such as the wistaria several years to come into bloom, yet whilst making growth, members of the jasmine family, planted nearby, could provide colour until the wistaria is ready to bloom.

SOIL CONDITIONS FOR OUTDOOR CLIMBERS

Soils, too, must play a part in the selection of plants, though it may be said that most wall plants will thrive in any good loamy soil. However, where the soil is of an acid, peaty nature, only those plants enjoying such conditions will thrive. Here the interesting *Tricuspidaria lanceolata* will flourish against a warm wall, likewise the magnolia and the arbutus (the Strawberry Tree). Where the soil is of a chalky nature, *Viburnum fragrans*, the buddleias, *Cydonia japonica*, the cotoneasters and clematis will flourish and all are extremely hardy. And where the soil is dry and sandy and lacking humus, the cotoneaster and the ivy, *Garrya elliptica* and *Fothergilla gardenii*, will succeed.

Wherever possible, a selection should be made which will provide colour the whole year round. However small the area to cover, this will be possible if the plants are kept in check and the less rampant growers are used.

Many plants will, however, not only provide glorious blooms but in autumn will give way to equally attractive berries whilst the foliage will take on autumnal tints. One such plant is *Celastrus articulatus*, which will attain a great height and prove hardy in almost any situation. Such a plant will provide colour the whole year round.

Not necessarily do all climbing plants make good wall plants; a number are happier when allowed to droop over a fence or garden wall and grow at random, needing little attention as to pruning or tying. The honeysuckles and the vines are better allowed to grow in this way than when trained against a wall. They are also valuable for covering a pergola or rustic screen. They prefer to grow over, rather than upwards.

CLIVIA

This is a plant of brilliance, likened to the 'Scarborough Lily' in its requirements. A native of Natal, the bulbs will bloom early in winter in a cool greenhouse. The clivia could be more widely grown for use in the home, for it is one of the few plants flowering indoors that enjoys shade. The bulbs hate disturbance when once established, and they will bloom better when they have become potbound. Planting should be done in April or May, the bulbs being planted 2 in. (5 cm) deep into a soil containing some humus and which is well drained.

The most striking species is *Clivia minata,* which produces large blooms of orange and apricot.

CLUSTER YEW, *see* Cephalotaxus

COBAEA

Native of South America, it is known as the Cup and Saucer Vine and is a plant of rapid growth, quickly covering a trellis. Indoors, it will grow well in a temperature of 50°F. (10°C.) and is perennial. If growing outdoors, treat as a half-hardy annual and plant in a sunny position about mid-May. The plant climbs by means of tendrils and will bloom profusely throughout summer, bearing urn-shaped flowers, about 1 in. (2.5 cm) across the top and which sit on a 'saucer' of small pale green leaves.

It is raised from seed which should be soaked for twenty-four hours before sowing two to a small pot in January in gentle heat. As the plant resents root disturbance, move to a larger pot rather than transplant.

COCKSCOMB, *see* Celosia

COCKSPUR THORN, *see* Crataegus

CODIAEUM, *see* Croton

COLCHICUM

This is generally referred to as the Autumn Crocus, to which it is not even botanically related. It is also confused with the Saffron Crocus, *Crocus sativus,* for it is known as the Meadow Saffron. They bloom from early September and through October, most of the species being at their best before the true Saffron Crocus, *C. sativus,* comes into bloom. Perhaps it is their rank foliage, produced during spring and early summer that has told against the popularity of these plants, for this gives the rockery and lawn a most untidy appearance. The true autumn-flowering crocuses are blessed with the dainty grass-like foliage of the spring-flowering species and so may have suffered in confusion with the colchicums, when considering planting the 'autumn crocus'. The colchicums are delightful plants and they should be used in the shrubbery, dell, on a grassy bank or by a stream where they may be planted in quantity. There they will produce their attractive cup-like flowers when the days are shortening and will be a delight in autumn. Or plant several species along a low wall with several mossy saxifrages for a soil covering and the effect will be delightful; or plant the yellow *Colchicum lutea* amongst plants of *Primula* Romeo. This is the spring-flowering colchicum and the only yellow species. Mossy saxifrages will be of the greatest value not only as a background to accentuate the colour of the flowers, but also to act in saving the plants from being splashed by rain.

Although the colchicums do well in almost any well-drained soil, even of poor quality, for they will flourish in a shrubbery of a town garden, they thrive best in one containing supplies of leaf mould. Where this is not present, fork into the soil a quantity of peat, spent hops or any available humus and do not forget to add some brick rubble. Coarse sand should be liberally sprinkled round the bulbs as they are set 3 in. (7.5 cm) below soil level. July is the best time for planting for the foliage dies down in late spring and the bulbs will be refortified and ready for lifting and replanting by midsummer. If left too long, they will commence to throw up their flower buds which may easily be damaged during planting whilst they should be established in their new quarters by the time they produce their flowers in September. Use the shrubbery or border rather than a rockery or lawn for their planting, for not only does the rather coarse foliage appear in the early new year, but it turns an unattractive brown colour in spring whilst dying back. As the foliage must be allowed to die back naturally to fortify the bulbs nothing can be done about its unsightliness. Except to cover it with nearby plants. The colchicums are not suitable for indoor culture; though *C. autumnale* may be flowered in a sunny window without soil or water.

SPECIES

Colchicum autumnale. This is the exception to planting the colchicum in the rockery, for its habit is so neat that its foliage should not prove too unsightly. The small lavender blooms appear in September. The white form (Album) is also attractive.

C. autumnale plenum. This is a most interesting double form for November flowering. The colour is deep rose and it is able to withstand adverse weather conditions. There is also a double white form but unfortunately both produce rank foliage.

C. bornmülleri. From the far Mediterranean area, it produces its large purple and white flowers during September.
C. byzantinum. In bloom throughout September and October, the lavender-rose flowers being most striking in the autumn sunshine.
C. luteus. Flowering from December until March, this is the only known yellow form of colchicum. It reached England from Afghanistan about 1875 and is really a species quite apart from the other colchicums. It produces its flowers from purple-tinted tubes which are a constant attraction to winter slugs, so it should be given the protection of a cool greenhouse where it will bloom undisturbed throughout winter.
C. parkinsoni. This is a lovely October-flowering species, bearing white flowers chequered with purple. It is of strong constitution.
C. speciosum. From Persia and as Farrar in *The English Rock Garden* says, it is 'one of the most beautiful plants in the world'. *The Century Book of Gardening* describes it as 'a noble flower'. Flowering late August and into November, when it produces its brilliant carmine-purple flowers in profusion, it should be planted in every shrubbery and border, and with it the expensive but pure white form to give an arresting display.

COLEUS

This, popular plant, known as the Ornamental Nettle on account of the shape of its leaves, is one of the most striking of all foliage plants. It is a plant of easy culture, though attention to detail is necessary if a bushy plant is to be grown and the plants are not to drop their leaves. The leaf colourings are quite amazing for they embrace almost every colour imaginable. Some bear leaves of bronze, orange and gold shades, others of various shades of pink and green. Some have plum-purple markings whilst others are a combination of all these colourings and with crimson included. They may also be obtained as self colours.

Seed is sown early in March in the propagating unit as used for most other foliage plants, for they require a temperature of around 60°–65°F. (16°–18°C.) in which to germinate. In such a temperature the seed will germinate rapidly. The John Innes Sowing Compost will be suitable, though the addition of a little more peat will enable the compost to retain the necessary moisture. A seed pan should be used, making the compost level so that all the seed will receive the same amount of moisture. The seed is small and must be sown with care, whilst it should be only lightly covered with a little peat dust. The compost should be kept moist, but not saturated and in about ten days the seed will have begun to germinate.

To prevent the young plants becoming 'drawn', they should be transferred to small pots as soon as possible, and here the John Innes Potting Compost, to which a little extra peat has been added, will be suitable.

Ample ventilation must now be provided, whilst the plants must never lack moisture or they will drop their leaves without warning. They will benefit from a daily syringing whilst they should also be shaded from the direct rays of the sun or scorching will occur. Firm potting is also essential or the plants will make weak, straggling growth. When several inches tall the growing point should be removed to encourage bushy growth. A winter temperature of 50°F. (10°C.) must be provided, taking care to protect the plants from draughts.

By early summer the plants will be ready for moving to size 48 pots containing a similar compost which should not be too rich otherwise the shoots will grow weak and 'leggy'. The plants will now require copious amounts of moisture and plenty of fresh air whilst they must be shaded from strong sunlight. As the plants enjoy a humid atmosphere it will be advisable to damp down the greenhouse whenever warm conditions prevail.

Should the plants make considerable growth they may require staking. This is best done by inserting the sticks or canes around the side of the pot and looping around them lengths of green twine.

The finest strains are those of T. Sakata and Co. of Yokohama, Japan and most striking is their Atlas 1970, bearing thick leaves of rich tangerine-bronze with a narrow edge of gold. 'Sunset Glory' has broad leaves of sunset-red whilst 'Festive Dance' has elegantly pointed leaves of scarlet with a broad margin of blackish-bronze and a wire edge of green.

COLLETIA

This interesting autumn-flowering shrub is a native of South America where it flourishes under desert-like conditions, and in Britain it enjoys the same dry sandy soil and sunny position. *Colletia armata* is an almost leafless spiny shrub which will most unexpectedly burst into a mass of tiny almond-scented white flowers in late September.

Plant in April and do little or no pruning. Propagate by cuttings removed with a 'heel' and inserted in sandy compost under glass.

COLLINSIA

C. bicolor 'Salmon Beauty' is an annual which grows well anywhere, producing masses of dainty salmon-pink flowers from June till the end of August on 12 in. (30 cm) stems. If the seed is sown thinly in early April, or in early autumn in a sheltered garden, no thinning will be necessary.

COLQUHOUNIA

Colquhounia coccinea is an interesting plant for a wall in the warmer parts of the British Isles. It bears heart-shaped leaves, which emit a refreshing fragrance when pressed, whilst the long umbel-like sprays of brilliant scarlet blossom are produced early in autumn. Plant in spring, into a soil containing some humus and support by wires. The plants will attain a height of 6–7 ft. (about 2 m) and are deciduous.

COLUMBINE, *see* Aquilegia

COLUMNEA

These beautiful plants are natives of Mexico and the West Indies and are grown more for their attractive flowers than for their foliage. One of the finest forms is *C. banksii* which bears tubular flowers of brilliant scarlet. Equally striking is *C. microphylla* which, though less free-flowering, bears crimson-scarlet flowers and very small dark green leaves. The flowers are borne at the leaf axils during the latter weeks of summer and early in autumn. After flowering, the trailing stems should be shortened so that they will be encouraged to form new growth which will bear bloom the following year. The plants have hairy stems and leaves, those of *C. gloriosa* being of a deep purple colour. It is, therefore, important not to splash the foliage when watering, which should be done more frequently than with most trailers. A minimum winter temperature of not less than 50°F. (10°C.) should be provided and a position of partial shade will suit the plants best. They like a compost made up of 3 parts fibrous loam, 1 part leaf mould and 1 part silver sand, but the secret of these plants achieving their greatest beauty is to 'stop' the shoots early in May and again early in July to encourage them to produce as many flowering stems as possible, for a plant well grown and in full bloom will be a sight worth seeing.

COMPTONIA

C. asplenifolia is the North American Sweet Fern, its handsome fern-like leaves releasing the scent of hay when handled. It grows 3–4 ft. (1 m) tall with narrow frond-like leaves and requires a deep, lime-free soil and an open situation. Plant in March. No pruning is required apart from the removal of dead wood. Propagate by cuttings removed from the ends of the new shoots and inserted in a sandy compost under glass. The plants may also be increased by division in March.

CONE-FLOWER, *see* Rudbeckia

CONIFERS

Mostly evergreen trees of which there are few apart from the conifers and with their hardiness and beauty of foliage and form, are indispensable for garden planting. They will flourish in almost any soil and owing to their wide variation in size and habit, may be used for all manner of purposes. Those like the Deodar Cedar and the Nootka Cypress with their pendent branches are most suitable for planting on a lawn or as a focal point in the garden. They may be used, together with other conifers to plant in groups, possibly in a sheltered corner to hide the junction of two walls or they may be planted about a shrub border. They require space to develop for if planted too closely, the branches will die back through lack of air and sunlight and where in contact with each other. They are mostly native of cool, mountainous regions and will not tolerate a stagnant atmosphere and over-crowding. They also require space to enable their beauty of outline to be seen to advantage. Some are so slow growing that they may be planted about the rock garden or in troughs, creating an effect of natural beauty. Others, such as Lawson's Cypress and the Monterey Cypress are valuable for a hedge, the latter for mild coastal areas, whilst Lawson's Cypress will be suitable for the coldest regions and will withstand hard clipping. Where planting in groups, a too sober effect may be prevented by the use of those conifers with silver or golden foliage and which will provide a striking contrast to the dark greens of most species. There are also several which take on bronze tinting in autumn, *Cryptomeria japonica* Elegans being one example. Besides those of semi-weeping form there are those of fastigiate habit, such as the Irish Juniper which makes a slender column and may be planted where space is limited. Others are grown for the beauty of their bark. *Picea rubens* with its reddish bark and *Metasequoia glyptostroboides* with its shaggy light brown bark are examples, the latter, like the larch and Swamp Cypress being amongst the few examples of deciduous conifers.

With the conifers, the male flowers are borne in catkins, the females in cones. The seeds are not enclosed in an ovary but lie naked on the surface of one of two scales, hence the name Gymnosperm given to those naked-seeded plants of the Order. Also, when the seed germinates, there are more than two seed leaves. The flowers are fertilised by wind for the conifers were amongst the first trees to inhabit the earth, before the advent of insects to act as pollinators.

Plant the conifers, as most evergreens, in April when they are starting into growth again and when the ground is in a friable condition after the frosts. Unless transplanted when young, conifers resent disturbance and must be moved with as large a soil ball as possible. Trees are usually transported with sacking tied tightly around the roots, to retain as much soil as possible. Young cupressus trees are usually sold from pots.

Young trees are intolerant of cold, draughty conditions and where planted in exposed ground, lengths of hessian or wattle hurdles should be fixed around them until established. It is suggested that where planting in colder parts, several of the more hardy species should be planted on the side of the prevailing wind to act as a shelter for those less hardy. *Abies veitchii* is fast growing with densely-arranged leaves but is not happy in a chalky soil. Instead, plant *A. cephalonica* which has pointed leaves and is fast-growing. Do not plant too deeply; cover the roots and give a thorough soaking. For the next few weeks when cold spring winds are often experienced, keep the roots moist and give the foliage a gentle spraying. All evergreens love a regular spraying during dry summer weather and will respond by making rapid growth and by the foliage taking on an attractive fresh appearance. Spraying is especially necessary where growing in a town garden where the trees suffer from soot deposits if not sprayed regularly.

CONIFERS FOR A CHALKY SOIL

Abies cephalonica
Abies pinsapo
Cephalotaxus harringtonia
Juniperus chinensis
Juniperus communis
Metasequoia glyptostroboides
Picea asperata
Picea omorika
Pinus montana
Taxus baccata
Thuya occidentalis
Thuya plicata
Torreya californica
Tsuga canadensis
Tsuga chinensis

CONIFERS OF DROOPING HABIT

Cedrus deodara
Cedrus libani 'Sargentii'
Chamaecyparis lawsoniana 'Bowleri'
Chamaecyparis lawsoniana 'Pendula'
Chamaecyparis nootkatensis 'Pendula'
Cupressus funebris
Cupressus sempervirens 'Fastigiata'
Dacrydium cupressinum
Ginkgo biloba 'Pendula'
Juniperus recurva
Larix leptolepis 'Pendula'
Picea omorika 'Pendula'
Picea smithiana
Picea spinulosa
Sequoia gigantea 'Pendula'
Taxodium ascendens
Tsuga canadensis 'Pendula'

Conifers

CONIFERS REQUIRING A LIME-FREE SOIL

Abies amabilis
Abies magnifica
Abies procera
Abies veitchii
Chamaecyparis thyoides
Picea abies
Picea rubra
Pinus banksiana
Pinus contorta
Pinus muricata
Pinus rigida
Pseudolarix amabilis
Sciadopitys verticillata
Taxodium ascendens 'Nutans'
Taxodium distichum
Tsuga heterophylla

CONIFERS REQUIRING A MILD CLIMATE

Agathis australis
Callitris tasmanica
Cunninghamia lanceolata
Cupressus lusitanica
Larix griffithii
Libocedrus formosana
Libocedrus plumosa
Pinus canariensis
Pinus insignis
Pinus patula
Podocarpus macrophyllus
Podocarpus spicatus
Tsuga dumosa

CONIFERS OF COLUMNAR HABIT

Cephalotaxus harringtonia 'Fastigiata'
Chamaecyparis lawsoniana 'Columnaris'
Chamaecyparis lawsoniana 'Kilmacurragh'
Cupressus goveniana
Cupressus sempervirens
Juniperus communis 'Hibernica'
Libocedrus decurrens
Thuya occidentalis 'Fastigiata'

CONIFERS WITH GOLDEN FOLIAGE

Cedrus atlantica 'Aurea'
Cedrus deodara 'Aurea'
Chamaecyparis hillieri
Chamaecyparis lawsoniana 'Lutea'
Chamaecyparis obtusa 'Crippsii'
Chamaecyparis pisifera 'Aurea'
Cupressus lawsoniana 'Stewartii'
Cupressus macrocarpa 'Donard Gold'
Cupressus pisifera 'Filifera'
Juniperus chinensis 'Aurea'
Juniperus communis 'Aurea'
Juniperus plumosa 'Aurea'
Pinus sylvestris 'Airea'
Pseudolarix amabilis
Thuya occidentalis 'Rheingold'
Thuya orientalis 'Elegantissima'
Thuya plicata 'Cuprea'
Tsuga formosana

CONIFERS WITH BRONZE FOLIAGE IN WINTER

Chamaecyparis formosensis
Chamaecyparis thyoides 'Ericoides'
Cryptomeria japonica 'Elegans'
Cunninghamia lanceolata
Juniperus communis 'Aurea'
Juniperus horizantalis 'Douglasii'
Juniperus virginiana 'Burkii'
Taxus x hunnewelliana
Thuya occidentalis 'Rheingold'
Thuya plicata 'Semperaurescens'

CONVALLARIA

C. majalis, the Lily-of-the-Valley is never happy in the mixed border, being rather too short stemmed to reveal its full beauty. It may be planted at the front of a shrubbery, but it may only be seen at its best when in a bed to itself. Like the montbretia, all too often this plant is confined to a sunless corner of the garden and to a soil which is almost barren. Here, even if it does bear a few spikes of hanging bells, they will be small and the spikes short-stemmed. Plant in a deeply worked soil enriched with decayed manure, to which is added some leaf mould or peat and some grit if the soil is heavy where the difference in quality will be most noticeable. Again, like the montbretia the roots are generally left year after year to become overcrowded and starved of food. They should be lifted every three or four years, split up and replanted into a well-prepared soil, enriched with manure and wood ash. Planting should be done early in November, young planting crowns as they are called in the nursery trade, coming into bloom eighteen months after planting. Set the crowns 12 in. (30 cm) apart and only just cover them with soil. Firm planting is necessary. A mulch of leaf mould and decayed manure should be given each autumn. Water with liquid manure during May before the plants come into bloom at the month end. This will greatly enhance the size of the waxy-white bells and extend the flowering season.

Early bloom may be obtained by making up the beds to the width of a frame. Around the beds side boards should be fixed in March and frame lights placed across them. The lights should be 12–15 in. (30–38 cm) above soil level so that the blooms are clear of the glass. Beds of lily-of the-valley in bloom before the end of April and early in May will prove a valuable source of income to sell in small bunches, backed by a fresh green leaf, to high class florists for ladies to use for evening wear. Flowers grown under glass will be free of rain splashings and command highest prices. But care should be taken in picking the sprays which should be gently 'tugged' from the rootstock whilst no more than one leaf should be removed from each 'crown' otherwise the plants will cease to bear flowering stems owing to its being deprived of nourishment.

Specialist growers produce the blossom from early November until May by using retarded 'crowns' usually obtained from Germany. Here, growers keep the crowns during summer and early autumn under refrigerated conditions in a temperature of just below freezing. The plants are dug up in March before coming into growth and are thus prevented from flowering until placed in a warm greenhouse. Upon receipt of the 'crowns' early in autumn, they are tightly packed side by side in strong wooden boxes, sterilised compost being packed around them and so that the plump points are just covered. They are placed in a frame or outdoors and covered with 6 in. (15 cm) of ashes and there they remain for four weeks to form roots. They are then taken to a darkened greenhouse or shed where the atmosphere is kept moist and where a temperature of 80°F. (27°C.) is maintained day and night for about two weeks. Pale yellow shoots will soon appear from whence the plants are allowed more light each day so that the leaves and stems will have turned a bright green colour by flowering time. This will be from three to four weeks after taking the plants indoors so that batches may be taken inside at monthly intervals. After flowering, forced 'crowns' are of no further value. It should be said that the plants will need large quantities of moisture during the forcing period.

Where a warm greenhouse is available outdoor 'crowns' may be lifted early in March to be planted into large pots containing a mixture of sandy loam and peat and if given a temperature of 50°F. (10°C.) and kept comfortably moist, they will come into bloom early in April. After flowering, replant the roots without delay.

CONVOLVULUS

The new dwarf strains are a welcome addition to the range of hardy bedding annuals, far removed in their habit from that obnoxious weed with the lovely white tubular flowers. The variety 'Royal Ensign', with its bright ultra-marine blue trumpets and the almost-prostrate 'Lavender Rosette', a lovely carpeting plant with its lavender flowers and grey foliage should be more widely grown.

The plants may be raised in a cold frame from an early March sowing or may be sown where they are to bloom and thinned out to 10 in. (25 cm) apart. They like a soil containing humus and an open sunny position.

COREOPSIS

This is a most valuable border plant in that it comes into bloom early in June and continues, whatever the weather, until September. For this reason, whilst never earning large returns it is a plant for the commercial cut flower grower. It likes a position of full sun and an ordinary well-drained soil, but not too much manure. It prefers peat and in a heavy soil, tends to be short-lived. As it so readily reproduces itself from seed, it is usually raised in this way. However, the new varieties are so outstanding that they should be in every border, and are propagated by root division. Plant in March, 20 in. (50 cm) apart. When marketing the bloom, cut with as long a stem as possible and make up into bunches of eighteen to twenty blooms. The bloom should be just fully open before it is cut.

VARIETIES

BADENGOLD. A glorious variety growing to a height of 3–4 ft. (about 1 m) and bearing clear golden-yellow blooms with attractively toothed petals.

MAYFIELD GIANT. An older variety but still a favourite on account of its freedom of flowering. The blooms are of a very deep guinea-gold colour, borne on 3–4 ft. (1 m) stems.

CORNEL, *see* Cornus

CORNFLOWER, *see* Centaurea

CORNUS

The Cornel or Dogwood makes a small tree and is valuable for windswept coastal gardens, and for planting in calcareous soils. It has silvery-grey leaves and bright red twigs and bark which is especially showy in winter. They grow well in a sandy soil and are extremely hardy. No pruning is necessary. Propagate from suckers or by layering the lower shoots in spring.

For a hedge *C. alba* with its rich red branches will form a dense thicket. The golden-leaf form *gouchaltii* is outstanding, the yellow leaves being splashed with pink. *C. hemsleyi* grows 18 ft. (5.5 m) tall and has leaves which are silvery on the underside. It bears white flowers. *C. macrophylla* grows 20 ft. (6 m) tall and bears its creamy-white flowers in large heads. The flowers are followed by blue-black berries. *C. mas*, the Cornelian Cherry, bears masses of small yellow flowers early in spring, followed by bright red berries. The form 'Aurea' has yellow leaves and 'Elegantissima' has leaves variegated with pink and gold.

CORYDALIS

A valuable race of bulbous plants, particularly suited to cool, shady positions. It does well in a town garden, where hedges, trees and shrubs will generally be found to supply the necessary shade. They will also bloom freely when growing in poor soils. In addition they have delightful fern-like foliage, blue-grey in colour, resembling the maiden hair fern. By planting several species, the flowering season may be extended from mid-March until the end of June.

The corydalis enjoys a soil containing some peat or leaf mould though it will flourish in one that contains little humus provided it is moist and cool. The bulbs are planted late in autumn; October is a suitable time, and they should be set 3 in. (7.6 cm) deep. The plants will increase naturally from self-sown seeds or they may be lifted and divided like snowdrops when flowering has finished, but before the foliage tends to die away.

SPECIES

Corydalis angustifolia. From the Caucasus, it must have a shady place and a light soil to produce its pale lavender sprays to advantage. Flowering to a height of about 8 in. (20 cm), it blooms from mid-March until May.
C. cava. Named because of its curious hollow bulb. A variety for the shaded rockery, producing its dainty sprays of lilac or white flowers above rich grey foliage.
C. densiflora. From Greece, it is of dwarf habit, covering itself with a mass of rosy pink flowers early in spring.
C. solida. Similar to *angustifolia* and though producing not quite so striking a bloom, it is of more compact habit.
C. wilsonii. Requiring more sunshine than the others, this is the most outstanding species. The bright yellow flowers, delicately marked with green are of more prostrate form. The foliage is of darkest green which accentuates the yellow flowers which are at their best during April and May.

CORYLOPSIS

There are two forms of this February and March-flowering shrub, which make compact plants 4 ft. (1 m) tall. *Corylopsis spicata* is first into bloom, bearing in February its pendant sulphur-yellow flowers which carry the delicious perfume of cowslips. Then in March blooms *C. pauciflora,* which also bears yellow, sweetly-scented flowers. In both cases the blooms appear before the heart-shaped leaves, which in autumn turn brilliant golden-yellow.

Plant in November in a well drained soil and in a sunny position and prune by cutting out the old wood after flowering. Propagate by layering or by cuttings removed with a 'heel' in July and rooted under glass.

COSMOS, *see* Cosmea

COSMEA

Also known as Cosmos, it is an annual and is a valuable cut flower. Like those other Mexican flowers, the dahlia and zinnia, it is at its best during late summer and autumn. As it does not easily transplant, it should be sown late in April where it is to bloom. Sow thinly and thin to 9 in. (22 cm) apart and later to 18 in. (45 cm). As the plants grow 3 ft. (1 m) tall, they may require support by surrounding them with twine or raffia fastened to strong canes. They require an open, sunny situation.

The two best for cutting are 'Goldcrest' which bears its semi-double blooms of glowing orange-scarlet over a long period; and 'Sunset' which bears flowers of vermilion flushed with gold.

COSTMARY, *see* Alecost

COTINUS

Syn: *Rhus cotinus*. It is the Venetian Sumach or Smoke Tree, so called because its plume-like inflorescences, produced in July, are fawn-pink and later turn to smoky-grey, whilst the round leaves change to rich shades of crimson and orange in autumn. They require an ordinary well-drained soil and no attention other than to remove dead wood in spring. Propagate from cuttings removed with a 'heel' in summer and rooted under glass; or by layering.

C. coggygria (*Rhus cotinus*) grows up to 15 ft. (4.5 m) tall, the best form being 'Atropurpureus', its plume-like inflorescences being of smoky-purple, whilst *foliis* 'Purpureis' has crimson-purple foliage which turns to lighter shades of red in autumn.

COTONEASTER

Deciduous or evergreen shrubs or small trees, valuable for the rich autumnal tints of their foliage and for their winter berries. They are lime-tolerant and flourish in the most exposed gardens. Several may be used as specimen trees i.e. *C. frigida*; to cover a wall; as a hedge; or in the shrub border for their habit is most diverse. Several may be planted on the rock garden, being of almost prostrate habit. To train them for whichever purpose they are required, carefully thin out overcrowded and dead and unwanted wood which should be done in spring. Propagation is by seed sown outdoors in drills or in a frame in spring; or by cuttings removed with a 'heel' and inserted in a sandy compost in July.

Of numerous species and varieties, *C. conspicuus* is a small leaved plant growing 4 ft. (1.2 m) tall, with gracefully arching branches, its white flowers followed by bright red berries. *C. franchetii sternianus* (Syn: *C. wardii*) has sage-green leaves, silver beneath and bright orange berries.

C. frigidus from the Himalayas is a handsome small deciduous tree with broad dark green leaves and bearing large clusters of crimson berries which it retains through winter. *C. horizontalis* will cover a wall horizontally, giving rich autumn colour with its red-tinted leaves and scarlet berries. *C. salicifolius* is the Willow-leaf Cotoneaster, a tall growing evergreen which is red with berries in autumn and winter.

COTYLEDON

This remarkable succulent, with round leaves serrated at the edges and depressed in the centre was known to cottagers as the Wall Pennywort for, like the House Leek, it grows on old walls. With its saucer-shaped leaves it was also known as the Navelwort. It takes its botanic name from the Greek, *kotule,* a dish, from the shape of the leaves which country children used to call 'Penny pies'. Gerard had a great affection for it, also for the Great Navelwort, a larger form which he described in detail, the flowers being of an 'incarnate' or flesh colour, the word Falstaff used to describe the colour of the carnation. Gerard has told that the Great Navelwort was to be seen growing from the walls of Westminster Abbey, 'over the door that leadeth from Chaucer's tomb to the old Palace' whilst he tells that the Wall Navelwort was native in the Alpine regions around Piedmont, though it was introduced into England at so early a date as to be classed (with *Sempervivum tectorum*) as a native plant.

The leaves are held on stalks 2 in. (5 cm) long whilst from the base arise the succulent flower stems which often attain a height of 12 in. (30 cm) or more. The greenish-white flowers, like tiny bells are borne in racemes and have the appearance of miniature Red (or green) Hot Pokers. It is a perennial, in bloom from June until the end of August when the plant takes on a pinkish tint, making it even more attractive.

CRAB APPLE, *see* Pyrus

CRANE'S BILL, *see* Geranium

CRASSULA

This native of South Africa enjoys cool conditions and plenty of light, apart from which the plant requires nothing fancy in its culture. The crassulas are of bushy, tree-like habit, and a number of them are capable of bearing a beautiful flower. *C. falcata* is one of the most outstanding, having interesting grey downy succulent leaves, formed one above another in vertical formation. It bears bright scarlet blooms in winter which have striking yellow stamens. They are produced above the foliage. The plant, like most of the crassulas, will eventually grow 2 ft. (60 cm) tall. *C. argentea* is also attractive, having oval leaves of shining silvery grey. *C. coccinea* (also known as *Rochea coccinea*) is most striking. Like all the crassulas it has horizontal pairs of leaves borne right up the stem and in early summer it bears small tubular flowers of bright cerise-pink.

A minimum temperature of 42°F. (5°C.) and watering sparingly will maintain the plant in good health through the winter.

CRATAEGUS

The Quickthorn, Hawthorn or May is the most useful of all plants for a hedge for it grows quickly, withstands clipping and always remains neat and efficient. It may also be grown as a standard tree and during early summer will smother itself in a mass of blossom. As a standard, it will form a large head on a 5 ft. (1.5 m) stem and may be kept in trim by the removal of unduly long shoots and branches. It will grow in ordinary soil and is propagated from seed and by layering. Ripe seed sown in early autumn will germinate within six to eight months.

Of many species and varieties *C. arkansana* grows 20 ft. (6 m) tall and bears berries the size of cherries whilst *C. calpodendron* makes a small compact tree and bears orange pear-like fruits. *C. crus-gallii*, the Cockspur Thorn makes a small spreading tree and bears trusses of white flowers followed by crimson fruits. *C. durobrivensis* bears the largest flowers, followed by large red fruits whilst *C. oxyacantha* Paul's Scarlet bears large trusses of double blooms of brilliant red. The variety 'Rosea Plena' is the Double Pink Thorn. *C. tenacetifolia* is the interesting Tansy-leaf Thorn, a downy-leaf species bearing fruits like golden crab apples.

CREEPING ZINNIA, *see* Sanvitalia

CRINODENDRON

The evergreen *C. hookerianum* requires a peaty soil and shelter from cold winds. It grows well against a sunny wall provided it is well supplied with moisture at the roots. It blooms in May and June, the flowers resembling crimson lanterns, and are evenly spaced along the branches. Under conditions which are favourable, it forms a dense bush.

Plant in spring and prune lightly after flowering, tipping back unduly long shoots. Propagate by layering or from cuttings of the half ripened shoots inserted in sandy soil under glass.

CRINUM

Those who can winter these striking bulbous plants in a greenhouse or frost-proof room will find them attractive subjects for tubs which may be arranged around a small courtyard or placed at the entrance to a house. When once established, the plants do not like disturbance, though they should be given a top dressing of peat every year in early spring. During summer they must never be allowed to suffer from lack of moisture. In favourable districts there are several species which may be grown entirely in the open beneath a wall and provided the plants are covered with litter, by way of straw or leaves or wood wool during early October, they should survive an average winter.

The bulbs produce their umbels of funnel-shaped blooms, like Regale lilies, during late summer, on sturdy stems 2 ft. (60 cm) tall. In tubs, or pots for the cold greenhouse, the bulbs should be planted in April, into a compost composed of good quality loam to which is added a little rotted manure, preferably cow manure, some peat and some grit to ensure good drainage. It is also necessary to place crocks in the tubs over holes which should have been drilled in the bottom. The bulbs should be buried with the neck just below the surface of the compost. After flowering, reduce watering to a minimum but not entirely for most are evergreen. Propagation is by offsets which are removed only when the bulbs are disturbed, divided and planted into fresh compost once every four to five years.

SPECIES

Crinum krelagi. It bears a beautiful flower, produced on 3 ft. (1 m) stems. The colour is deep rose pink and the plants will grow well outside given reasonable protection.

C. moorei. This species has been grown in English gardens for at least one hundred years, bearing its rich pink bloom during September. It is suitable for a sheltered garden, or for a cold greenhouse. The white flowered variety 'Album' is especially beautiful. The bulbs of this species are expensive but they will bloom for years if protected during winter.

C. x powellii. A hybrid and possibly the loveliest of the Crinums, bearing its blooms on 2 ft. (60 cm) stems and they are of an exquisite shade of rich pink. It is also one of the hardiest of all the species. A lovely variety, 'Album', which produces pure white flowers is also of easy culture.

CROCUS

INDOOR

The cultivation of crocus indoors presents few difficulties provided no forcing is done; indeed, like the snowdrop, the bulbs will resent anything but the coolest of temperatures throughout their days indoors.

For pot or bowl cultivation a top-size corm should be used, one at least 3½ in. (9 cm) in circumference and these should be in a sound, clean condition. The growing point may be seen at the top of the corm and so they should be handled with care. To a size 60 pot or a small bowl (and both the crocus and snowdrop look their daintiest when in fairly shallow bowls) six to eight corms should be planted 2 in. (5 cm) deep, spacing them about 1 in. (2.5 cm) apart each way. A compost similar to that recommended for window-boxes will be ideal, though to assist drainage a very small quantity of broken charcoal should be incorporated with the mixture. A good quality bulb fibre may be used instead of the peat, but I always recommend the use of a little loam and sand to give the compost some 'body'. A compost correctly made up of a handful of turf loam, one of a good quality horticultural peat, and one of rotted mushroom-bed compost to which is added a little coarse sand, some charcoal and a dusting of bone meal to a size 60 pot has always given excellent results. Those who live in a flat in the city must rely almost entirely on a prepared bulb fibre which will give good results provided it is of good quality. A compost which contains a small amount of turf loam will be more easily controlled as to moisture requirements than where a fibre only is being used. The crocus, flowering indoors in February or earlier, should be potted in early September so that it will have a prolonged period in which to form an adequate root system.

Where the pots can be stood in the open under a wall or in a cellar or attic and covered with deep ashes, soil or sand with an occasional watering for those indoors, this method will ensure a sound rooting system. At the end of November, the bowls may be removed from beneath their covering, watered if necessary and placed in a cool but light position where they will remain to bloom. Over watering and any degree of forcing will prove fatal to the crocus. Give water only when the compost appears to be dry.

OUTDOOR

When planting in grass, in orchard or lawn, a small amount of humus should be placed in the cavity formed when the turf has been removed. This should be done by making a firm cut with a spade on three sides. The turf should then be lifted and rolled over the side which has not been cut, exactly like a box lid on hinges. A turf 4 in. (10 cm) deep should be removed, which may easily be gauged with a little practice. Into the cavity should be placed some humus material and some coarse sand and into this the bulbs are pressed, flat side downwards. Six bulbs of 2¾ in. (7 cm) size should be placed under each turf which should be replaced into its original position and made quite firm by treading. The whole operation takes but a few minutes and a large area may be quickly planted.

Crocus planted in a rockery or along the edging of a path or between crazy paving are planted 3 in. (7.5 cm) deep by means of a trowel. Plant in clusters of self colours or species to obtain the best effect, spacing the bulbs 3 in. (7.5 cm) apart.

When planting in grass or in a special border, it will be advisable as far as possible to utilize a southern aspect for the spring-flowering species so that they will have access to every available hour of sunshine and come into bloom at the earliest possible date. The autumn-flowering species may be given a more northerly aspect for they come into bloom at a time when weather conditions are more favourable. But like the anemone, all crocus species will enjoy a position which receives at least a little protection from cold winds; this is one reason why they flower so well around the roots of large trees and in a rockery, sheltered by the stones.

No bulb is so valuable for growing in grass, either in a lawn or in rough grass. Flowering early, the foliage may be left completely undisturbed until it has turned yellow and died down. This will have taken place by early summer when the grass will need its first seasonal cut and so no untidy brown foliage will remain to spoil the appearance of the turf throughout the early summer weeks if the grass receives its first cut towards the middle of May.

A word should be said about planting in a heavy clay soil. Though the spring-flowering crocus may be planted until the end of November in ordinary soils and in favourable districts, those who garden in the north should plant not later than the end of October and a month earlier where the bulbs have to be planted in a clay soil, which so quickly becomes cold as the autumn sun dies down. The bulbs should be planted only 2 in. (5 cm) deep in a heavy soil and the month of September is the ideal time.

If the crocus has a drawback, it is that in some years of prolonged bad weather, the spring-flowering species may receive attention from birds which find the sweetened moisture at the base of the flowers much to their liking. The autumn-flowering species which are in bloom in a season of plenty, rarely suffer in this way and of the spring species, it is almost entirely the yellow varieties that are most liable to attack. Where growing in large numbers 'Glitterbangs', patented by Chase Cultivation Limited, prove valuable and they may later be used for protecting strawberries and again, for autumn-sown seeds. But for only small groups, the old method of placing black thread (rather than cotton) around the plants when in bud will prove effective. Do not wait until the flowers are fully showing colour before protecting, for the whole display may be destroyed in an hour.

The crocus which comes to Britain in such vast numbers each year from Holland has achieved much of its popularity from its ease of culture. Bulbs which have been planted a quarter of a century ago still produce a carpet of colour year after year and there seems little value in lifting them at any time and replanting the younger and most virile corms. But the crocus species, being more expensive and some being obtainable in only small numbers may be more rapidly increased, apart from those which are known to reproduce themselves easily from self-sown seed, by lifting every three years when the foliage has died down. On top of each parent corm will be found growing new corms of various sizes which should be removed, dried and replanted at the appropriate time. Those growing in a sandy loam which may have grown large may be used for pot culture.

SPECIES AND VARIETIES OF OUTDOOR CROCUS

AUTUMN-FLOWERING

C. asturicus. From the Pyrenees, it produces its lilac blooms, striped dark blue at the base, during October and November. As with *C. nudiflorus,* the leaves appear after the flowers.

C. cancellatus. The white form 'Albus' is one of the most beautiful of the autumn-flowering crocuses, the large silvery-white blooms being enhanced by the vivid orange anthers. The corms are sold as food in the markets of Damascus.

C. karduchorum. One of the finest of all the crocus species, and though discovered in 1859 is only now to be obtained commercially. It bears large flowers with broad petals, pointed at the tips, and which are of an exquisite shade of lavender-pink enhanced by the white throat. It blooms during October and has small, narrow leaves.

C. kashmiriana. A rare variety which may be found at high altitudes in Kashmir. It produces its rich pale blue flowers in October and makes an arresting display.

C. medius. May be seen at its best along the southern coast of France, but it grows equally well in the cooler climate of Britain. Its rich, deep violet blooms are marked at the throat with deep purple rays, and its orange-red stigmas makes it a most striking species. The leaves, too, are striking, having a wide central white line. It blooms in October.

C. nudiflorus. Much like *C. zonatus,* its lilac-mauve flowers appear before the leaves, in early October. It is a free and continuous-flowering plant, native of the Pyrenees. The plant increases by means of stolons which take about three years to grow into a flowering-size corm. The Yorkshire naturalist, William Crump, has described how the plant has become naturalised on the grassy hillsides around Halifax; and it was his opinion that at one time many of the hillside farms were the property of the Knights of St John, who brought the corms back with them from the Mediterranean countries.

C. pulchellus. It blooms during October, its lavender-blue flowers having white anthers and an orange throat. The variety 'Zephyr' is lovely, with large white blooms shaded on the inside with grey.

C. salzmannii. Though its natural habitat is southern Spain and Tangier, it is a vigorous, hardy crocus forming the largest corm of all the species, and bearing its silvery-lilac flowers, after its leaves, early in November.

C. sativus. The Saffron Crocus, in bloom during October and noted for the dense yellow deposit on the stigmas, of which the Elizabethan writer Gerard said: 'Moderate use of it is good for the head . . . it shaketh off drowsie sleep and maketh men merrie'. It would also be grown entirely for its lovely purple-pink flower, which is streaked a deeper purple and which remains in bloom for a long period.

C. speciosus. This is a magnificent late-September-flowering crocus remaining long in bloom and being happy in all soils and in any position in the garden. The large, almost globular, violet-blue flowers, are feathered with crimson and made brighter by their orange-red stigma. The form 'Aitchisonii' is especially fine, the long tapering blooms, which appear in October, being of a lovely shade of China blue; whilst 'Albus' has pure white blooms with the same bright orange stigma. The new 'Oxonian' bears a large globular bloom of dark blue, the interior being of pale blue.

C. zonatus. It blooms in September, the thin tubular lilac-pink blooms zoned with orange spots at the throat, and with their white stigmas, appearing before the leaves. It is also known as *C. kotschyanus* and is native to the Lebanon and Syria. It forms a large flat corm and needs a well-drained position.

WINTER-FLOWERING

C. boryi. Flowering in November and December and producing dainty blooms of creamy-white, with white anthers, this species is seen at its best when planted with *C. asturicus*. It is more tender than most and should be confined to southern gardens or to the alpine house.

C. imperati. One of the most handsome and valuable of all the species, it comes into bloom at the year's end and continues through the bleakest days until March. The outer petals are fawn or biscuit coloured, feathered with violet. There is also a pure white form, 'Montanus', found near Naples.

C. korolkowi. It comes into bloom early in January and continues until March, the flowers being star-shaped, while the brilliant yellow colour of the dainty blooms is much appreciated at this time. The exterior of the bloom is beautifully stippled with bronze and the inner surface has a varnished appearance.

C. laevigatus. A most valuable species from the Caspian Sea area, producing star-like flowers of rich violet during the bleakest days of December and January. It has a potent scent.

C. longiflorus. Though native to Sicily and Malta, it blooms during November when damp foggy weather frequently covers the land, and when colour in the garden is most appreciated. The long lavender tubes with their scarlet stigmas have a rich perfume.

C. niveus. A most outstanding variety, first recognized as a distinct species by the late E. A. Bowles in 1900. The flowers are large and globular, the pure white segments having a vivid orange throat. The vividness is accentuated by the scarlet stigmas. It is a hardy crocus, coming into bloom as the snow melts, towards the year's end, and continuing until March.

C. ochroleucus. A native of Galilee, it is a most outstanding November and December flowering crocus, the dainty creamy-white blooms with their white stigmas providing a pleasing foil for the purple and blue-flowered species, especially *C. laevigatus* and *C. longiflorus*, in bloom at the same time.

C. tournefortii. A rare crocus, it bears a large bloom of a unique shade of warm rosy-lilac with a rich orange throat, and is at its loveliest during the dark November days. It is found only in the Greek Archipelago and should be given a sheltered place facing south. Where happy, it will increase rapidly.

SPRING-FLOWERING

C. ancyrensis. The tiny bloom looks so delicate as it opens to the early February sunshine but these crocuses have a remarkable ability for standing up to severe weather. They should, however, be given some protection from the wind and are best planted in nooks about a rockery; on a grassy bank facing the sun; or as an edging to a path, where the blooms of deepest golden-orange will be produced in great profusion.

C. balansae. First discovered near Smyrna, it blooms during March and is one of the smallest and most fairylike of all crocuses. The best form is 'Zwanenburg Variety', its marigold-orange blooms suffused with mahogany on the outside, and stippled with bronze.

C. biflorus. From Greece, with attractive silvery white flowers, striped with blue. Very free-flowering, ideal under trees, and at its best in March. There are several forms of this species in which *C. biflorus* 'Weldeni', a pale grey, and 'Pusillus', white with an orange throat, from Yugoslavia, are outstanding and equally free flowering.

C. candidus. It blooms during April and comes from the Levant. The form 'Mountainii' is interesting. It bears freely, small globular flowers of richest orange. Seedlings raised by Van Tubergen have a yellow ground and are styled 'sub-flavus'.

C. chrysanthus. It is the large 'Golden Flowered Crocus' from Greece and Asia Minor and is one of the finest, coming into bloom towards the end of February and continuing until April. As to colour, it is the most variable species, an oustanding variety being 'E. A. Bowles', named by Van Tubergen, its raiser, after the English authority on the crocus. It has a large bloom of buttercup yellow, tinted with grey at the base. Plant with it the new 'Blue Pearl', its soft pearly-blue flowers having an attractive yellow centre. Equally desirable is the free-flowering 'Cream Beauty', its globular flowers being of a satisfying shade of soft creamy-yellow; and 'Peace', its huge snow-white blooms being veined with slate blue on the outside. The late E. A. Bowles himself named many lovely hybrids of his own raising, mostly after birds.

C. corsicus. A rare species but is most free-flowering and extremely handsome with its cream-coloured flowers feathered on the outside with mauve. It blooms during March.

C. dalmaticus. Lilac, yellow throat, flower tube about 2 in. (5 cm) long. Flowers February to March.

C. fleischeri. It should be planted on a grassy bank where the vivd red anthers may be seen through the almost transparent white star-shaped petals. Or plant it for contrast about clusters of Juliae primrose 'Wanda' or 'Marie Crousse', along the edge of a path or border. It is in flower from February until April.

C. minimus. One of the smallest species; the flowers are lavender, shaded buff on the outside, and only 1–2 in. (about 4 cm) high. It is a charming little plant, but is shy to increase.

C. olivieri. Found on the island of Chios in the Aegean, it is the latest to bloom, being at its best during April and into May and with *C. minimus* it is the smallest. For all that, its brilliant orange flowers are most striking beneath young evergreens.

C. sieberi. Similar to *C. fleischeri* but at its best throughout February. Its rich lavender-blue blooms of delicate golden throat make this a delightful crocus for the cold greenhouse when grown in pans, or for the window box. The new variety 'Violet Queen' has globular flowers of deep violet-blue.

C. susianus. The striking 'Cloth of Gold' crocus which reached this country from the shores of the Black Sea in 1587. It comes into bloom at the end of March, the glossy golden blooms being striped with brown on the exterior, while the narrow segments taper to a point. It is also known as *C. angustifolius.*

C. tomasinianus. Native of Bulgaria, it is in bloom during February and

March. It was named after the botanist Muzio de Tomasini, of Trieste, and is one of the best of all. The long thin tubes are of a brilliant shade of sapphire blue, and the new variety 'Taplow Ruby' has large flowers of deepest ruby-red. It increases rapidly both from its cormlets and from seed and is most valuable planted about a shrubbery, or as an edging to a path where it need not be disturbed.

C. vernus 'Albus'. The 'Snow Crocus' of the Alps which replaces the snow with carpets of creamy-white trumpets. Plant with it the new 'Haarlem Gem' which also blooms in March and bears lilac-blue flowers of great substance.

C. versicolor. It is to be found in numerous forms along the Riviera, the most outstanding being that known as the 'Cloth of Silver' crocus, its silvery-white blooms feathered by purple-crimson.

DUTCH HYBRIDS

A selection of the best modern hybrids, descendants of *C. vernus*, that are an essential element of the garden. Most of them produce larger blooms than the other species, so they are at their best during March and early April when massed under orchard or parkland trees or in the short grass of lawn or bank. They are excellent, too, for pot culture in a cold greenhouse or the home, in bloom when most of the spring-flowering species are over.

EARLY PERFECTION. Almost a pure navy blue and with a very large bloom. At its best early in March.

ENCHANTRESS. Bears a huge bloom of real Wedgwood blue.

JEANNE D'ARC. Produces a huge bloom of purest white.

KATHLEEN PARLOW. A magnificent white with a striking pale yellow throat.

LITTLE DORRIT. The dainty globular blooms are of a charming shade of silvery-blue.

MAMMOTH. Recognized as the finest of the yellow varieties, producing a huge cup-shaped bloom, and in flower over a long period.

NEGRO BOY. The deepest purple, to follow 'The Sultan'.

PALLAS. An old variety introduced in 1914, producing flowers heavily striped pale blue.

PAULUS POTTER. A striking variety, the huge blooms being of rich, glossy ruby-red.

QUEEN OF THE BLUES. Produces a large refined bloom of soft sky blue.

REMEMBRANCE. The colour of the old-fashioned lilac, and remains in bloom for a long period.

STRIPED BEAUTY. A popular novelty, grey-white, striped with porcelain blue.

THE SULTAN. Very deep purple of perfect shape.

VANGUARD. It is the first of the Dutch hybrids to bloom, the colour being pale lavender, flushed with grey on the exterior.

WHITE LADY. Possibly the largest white, and grows well anywhere.

CROSSANDRA

A greenhouse shrub growing 18 in. (45 cm) tall with evergreen gardenia-like leaves and bearing salmon-pink rhododendron-like flowers in terminal umbels. It remains almost perpetually in bloom.

Though of shrubby habit, the seed germinates freely if sown in a temperature of 70°F. (21°C.) and the compost does not dry out. Transplant the seedlings when large enough to handle to small pots and later to larger pots containing the John Innes Potting Compost fortified by some decayed manure and leaf mould. The plants should be given a temperature of 58°F. (14°C.), a room temperature being satisfactory and keep the compost comfortably moist.

Propagation is also by cuttings inserted in a sandy compost in a temperature of 68°F. (20°C.). They must not be allowed to lack moisture.

CROSS OF JERUSALEM, *see* Lychnis

CROTON

These tropical plants are by no means easy and are only for those houses which enjoy the unfluctuating warmth of central heating during winter. They require a winter temperature of at least 55°F. (13°C.) and they must be kept free from draughts.

The first croton to reach this country was *C. variegata,* which arrived from the East Indies in 1804. It is now often named *Codiaeum variegatum* and bears long glossy leaves in shades of yellow, green, purple and crimson, which are mottled and striped with yellow. The plants like some humidity, but they are never happy in sunlight and will drop their leaves if conditions are too dry for them. During summer give copious amounts of water and do not allow the plants to become too dry during winter. To keep the plants low and compact, nip out the leading shoot as the plant makes growth, so as to encourage side shoots to form.

The compost should be composed of 2 parts fibrous loam, 1 part leaf mould and 1 part coarse sand. To propagate, an electric propagator will be necessary to root the cuttings in a humid temperature of 65°F. (18°C.).

CROWN IMPERIAL, *see* Fritillaria

CRYPTOMERIA

The Japanese Cedar, an evergreen resembling the Wellingtonia in outline but slower in growth. The linear leaves are spirally arranged whilst the male catkins appear at the ends of the branches. The globular cones are prickly when ripe. It requires a deep well drained loam and a sheltered but sunny situation. Plant in April. Propagate by seed sown in spring or by cuttings from the tips of the branchlets.

C. japonica 'Elegans' bears foliage which turns bronze in winter and is the finest form, growing tall but bushy. *Globosa* 'Nana' makes a dense low tree with gracefully arching branchlets whilst *spiralis* is a tree of slender habit with leaves spirally arranged.

CUCKOO-FLOWER, *see* Cardamine

CUP AND SAUCER VINE, *see* Cobaea

CUPRESSUS

The Cypress or Cupress which has rounded branchlets and bears larger cones than chamaecyparis which require two years to ripen. The leaves are borne in whorls of four. Chamaecyparis is usually raised in the open ground but cupressus is grown in pots for it resents root disturbance. It is also less hardy and should be planted only in milder parts where *C. macrocarpa* makes a valuable hedge. Plant in April in a well-drained soil but which is retentive of summer moisture. Propagate from seed sown when ripe, in small pots; or from cuttings taken with a 'heel' in July and inserted in sandy compost under glass.

SPECIES AND VARIETIES

Cupressus arizonica. It makes a tall pyramidal tree with grey-blue foliage and has lax, spreading branches.

C. goveniana. The Californian Cypress, it resembles *C. marcrocarpa* but is of more columnar habit with shorter branchlets. The foliage is scented.

C. macrocarpa. The Monterey Cypress and a valuable tree for coastal planting where it is used as a hedge. Of rapid growth, it has bright green foliage which may be clipped in spring to maintain its shape. The variety 'Lutea' has soft yellow foliage, whilst 'Donard Gold' is of brighter yellow.

CURCURBITA

Under this heading come the ornamental gourds, the marrows and the squashes. But besides their edible value, a number bear small curiously shaped fruits which are interesting and colourful when grown against a trellis in a sunny position. *C. pepo* includes all the small fruiting gourds, and the raising of the plants should be as for marrows and squashes. Seed may also be sown in small pots, one seed to each, either in the warmth of a greenhouse or in a box which is covered with glass. They should be given a rich compost and must never lack moisture. If the seed is sown in spring the plants will be ready to be planted out towards the end of May. If the small fruits are removed when ripe others will take their place, and if those that are removed are polished and placed in a deep wooden bowl, they will make a colourful winter table decoration.

CYCLAMEN

HARDY

Some shade is essential to the requirements of the hardy cyclamen and for this reason they are generally found growing under tall trees and in a shrubbery, provided the soil is not acid. Cyclamen are lime lovers and grow best in those soils which have a high lime content. Where this does not occur, lime rubble should be worked in before the corms are planted. Ordinary hydrated lime may be given with equally good results provided drainage is completed by the addition of a small quantity of crushed brick or shingle. Leaf mould or peat should also be added if the soil is in poor condition. Planting preparations must be thorough for when once planted, the corms will remain in the ground for a lifetime, increasing in size each year, until they will, in about five years' time, be throwing dozens of dainty butterfly blooms from a single corm. Given a soil where conditions for reproducing themselves are suitable, they will produce small quantities of seed each year to carpet the ground for several feet around the original corm. For this reason they should not be planted too closely together. Plant in groups 5 ft. (1.5 m) apart, spacing three corms to each group 18 in. (45 cm) apart in the form of a triangle. Corms of the rare species being expensive, possibly no more than two could be obtained and these should be planted 2 ft. (60 cm) apart. It is surprising how quickly the self-sown seedlings will cover the ground and how attractive are their variegated leaves even before the corms are large enough to produce bloom. The foliage is an added attraction to the cyclamen of all groups. It cannot be over-emphasized that the ground should have been enriched and be cleared of perennial weeds before planting takes place for the corms hate to be disturbed when once they have settled down. This habit makes the hardy cyclamen an ideal plant for the garden where labour is at a minimum, and nowhere do they thrive better than under beech and oak trees, where the ground is bare of everything but the mould of decayed leaves. Here too, the ground may be slightly acid and the addition of some lime is important. A top dressing of lime rubble, finely crushed should also be given every three years and where the corms are planted in a border or shrubbery, they will appreciate a mulching with lime, peat and leaf mould in alternate years. Rarely is sufficient attention paid to the question of mulching bulbs with the result that they scarcely attain the full glory of which they are capable.

GROWING HARDY CYCLAMEN FROM SEED

Mr T. C. Mansfield, an authority on alpine plants, suggests sowing hardy cyclamen seed on a layer of lime stone chippings, and this has been found to be an excellent method. Boxes or seed pans may be used, pans being more suitable, for the seedlings are allowed to remain in the pans for twelve months where they will grow on slowly all the time. One method is to mix together a compost of equal parts peat and rotted turf loam. Over this is placed a layer of limestone chippings which have been pulverised but not too finely. Directly on to the chippings the seed is sown, thus from the beginning, the plant is provided with its lime requirements.

As with most alpine plants, the seed should be as fresh as possible when sown, and flowering as they do between September and April, early summer is taken to be the most suitable time. The seed should germinate readily if the pans are covered with a sheet of glass and allowed to stand in a cold frame or greenhouse shielded from the summer sunshine. Careful watering is necessary and especially during the following winter where the plants will remain in their seed pans in a cold frame, partially protected from severe frosts by covering with sacks or straw. The following spring should see the plants set out into small pots and in the cold frame they will be grown on for another twelve months. They will then have formed a small corm which will be ready for planting out early in summer. This may take another twelve months to become thoroughly established before flowering, after which in a soil they like, they will produce more and more bloom each season.

The species that flower during late winter and spring should be planted early in September and those that bloom before the new year should be planted during April. Rather more care should be taken in planting than with most other small bulbs. In the first place correct depth is important and like the more exotic greenhouse cyclamen, they must not be planted deeply. They should be pressed firmly into the soil with only the barest covering of soil over the top. Planting too deeply will cause failure. Again, it is essential in planting dormant corms to see that they are set the right way up, or smooth side downwards.

With shallow planting a mulch is important for with the wind and rain and the tendency of the corm to push itself out of the ground as it forms its root, the crown may become too exposed. A light mulch with leaf mould or peat and limestone chippings will correct this whilst at the same time supplying the plant with the food and humus it enjoys. And should there be any tendency of certain species to be tender, a light mulch will prevent damage to the corm.

HARDY CYCLAMEN FOR POT CULTURE

Many of the hardy species are equally lovely indoors and where a heated greenhouse is not available they will prove valuable by covering the entire winter period with a display of bloom. What they must be given is a compost containing not only lime rubble but a large quantity of peat or leaf mould. The compost should be made up of 2 parts peat (or leaf mould), 1 part loam, 1 part lime rubble. This should be thoroughly mixed and well moistened and for hardy cyclamen, an earthenware seed pan is the best container. Six corms of different species or whatever selection is required, should be pressed into the compost which has been made very firm. It should be firm, yet springy when the corms are pressed in, the top of the corms being level with the top of the compost. Planting should take place in early October and the pans may be stood in a cold shaded frame or placed directly into a cold greenhouse. They will require only occasional waterings otherwise the crowns will tend to rot should the weather be dull and misty. Depending upon the variety, they will bloom at Christmas and extend their flowering season right through the winter and spring. A single corm set in a small pot will also provide a charming addition to the dressing-table, for they will grow well in a window or they may be taken indoors from a frame or greenhouse as soon as the first buds are seen. Like the Persian cyclamen, they will remain longer in bloom and the colour will be enhanced if given an occasional application of liquid manure water. A charming effect is obtained if fresh sphagnum moss is placed around the corms as soon as they have made some growth; in fact a far greater use should be made of moss for all indoor bulbs and corms.

Two lovely varieties to plant together are *C. atkinsii* 'Album' and 'Roseum' for Christmas flowering, with *C. coum* and its white counterpart to follow.

SUMMER- AND AUTUMN-FLOWERING SPECIES OF HARDY CYCLAMEN

Cyclamen africanum. From North Africa, it requires a sheltered garden, preferably in the south of England and the west of Ireland, where

it will produce its daintily twisted pale pink blooms in October.
C. europaeum. From the European Alps, it possesses a strong perfume when in bloom during midsummer. It will produce its vivid crimson flowers to perfection either in full sun or partial shade and is a dwarf and dainty species. It is the exception to the rule of shallow planting for this species likes to be well down into the soil.
C. neapolitanum. Famed for its silver-mottled ivy-shaped leaves which precede its rich rose-pink flowers in September, this is a most valuable species to follow *C. europaeum* and also for flowering in a shaded position, under shrubs or conifers. There is also a pure white form which is scarce though is quite hardy and increases rapidly.

WINTER- AND SPRING-FLOWERING HARDY CYCLAMEN

C. alpinum. This hardy little species is usually in full bloom on Christmas Day, its white flowers attractively blotched with purple, making for a glorious midwinter display.
C. atkinsii. Valuable in that it will produce a succession of bloom from early December until the end of winter, this species produces deep carmine-pink flowers and has handsome deep green leaves. The white form, contrasting with the almost black leaves is particularly fine.
C. coum. Flowering like *C. atkinsii* over the entire winter, this is a species from Greece and Turkey and rapidly increases from self-sown seed. The flowers are vivid magenta, the leaves glossy green and it is the most dwarf of all.
C. hiemale. Producing its rich carmine flowers during the darkest winter months, when its white mottled leaves are also an attraction, this is a variety that seems to grow well in the town garden and in any soil, though it does like some lime rubble.
C. libanoticum. From Syria, and outstanding in every way. The delicately coloured pink flowers, crimson at the base, are larger than those of any other cyclamen, while they also carry a stronger perfume than all others. The flowering time is March and April when there are few other cyclamen in bloom until *C. repandum* comes along early in May.
C. repandum. Also from the Mediterranean regions from which most of the hardy cyclamen originate, this species produces a flat corm. The flowers being carried on long stems and at their best during May, this is an ideal variety for planting in short grass. The flowers are deep carmine-pink.

INDOOR

The most popular of all greenhouse plants to be raised from seed, for the plants may be enjoyed in the home during winter and after a summer rest may be brought into bloom again and again. Nor does it require a high temperature during winter. Where being grown in the greenhouse, a minimum winter temperature of 50°F. (10°C.) will be sufficient. This will be about the same temperature as an unheated room in the home in which the plants will be comfortable. Their removal, when coming into bloom from excessively high temperatures of the greenhouse to much lower temperatures of the home, often causes them to collapse when reaching their best. Grow the plants in an equable temperature of between 50°–55°F. (10°–12°C.) and this may be obtained by the use of a fumeless paraffin heater of the hot water type, or by electric heating. Cyclamen are not unduly tender plants and all that is necessary is to protect them from frost and excessive winter moisture which would cause the foliage to be troubled with mildew. Though they like moisture at their roots, a dry, buoyant atmosphere during winter and freedom from draughts is necessary.

The seeds, which are hard and slow to germinate, produce a scaleless bulb known as a corm which may be kept indefinitely. As the seed is hard-skinned it should be obtained as fresh as possible and from a specialist merchant, for seed of inferior quality will germinate unevenly, whilst those seeds which do germinate will most likely bear small flowers of pale colouring.

RELIABLE STRAINS OF INDOOR CYCLAMEN

The greatest name in the production of salmon-coloured cyclamen is Ferdinand Fischer of Germany, outstanding being his 'Pearl of Zehlendorf' which bears refined blooms of a deep shade of pure salmon. Also of German origin is 'Giant Baroque' which forms a plant of dwarf, stocky habit. The petals of the flowers are waved, the edges being frilled and they may be obtained in shades of pale pink, peach, salmon and white. Seventy-five per cent of the seeds bear bloom which is true to type.

Long famed for its cyclamen seed is the House of Blackmore & Langdon of Bath. Their 'Burgundy', which bears flowers of rich wine-red has received an Award of Merit from the Royal Horticultural Society. Others of great beauty are 'Afterglow', which bears large blooms of brilliant scarlet, and 'Bath Pink', the blooms of which are pure deep pink. The seed which is to be obtained from these specialists will have been saved from hand-pollinated plants and will produce about seventy-five per cent true to colour.

SOWING THE SEEDS OF INDOOR CYCLAMEN

The best time to make a sowing is early August. This will allow the seed several weeks of warm, sunny weather for germination. If sowing is delayed later than the middle of the month, it may not receive sufficient warmth to germinate before winter.

Make certain that the sowing receptacle is quite clean. Use a new box or seed pan which has been scrubbed, and as germination may take several weeks, the sowing compost must have been sterilised. It should also be neither too acid nor too alkaline, a pH of 6.5–7.0 being correct, whilst it should be made up to the following formula: 2 parts fibrous loam; 1 part leaf mould; 1 part silver sand and 1 oz. of superphosphate per bushel of compost.

This is a slight deviation from the John Innes Sowing Compost, chiefly in that leaf mould is used instead of peat, for cyclamen are lovers of leaf mould which must be sterilised with the soil. Make sure that the superphosphate is thoroughly mixed in and is used in fresh condition.

After filling the boxes (or pans) make the compost level and soak with clean water, leaving it twenty-four hours before the seed is sown. As the seed is not too small, it may be spaced ½ in. (1.25 cm) apart, staggering the rows which will allow the seedlings ample room to develop. With cyclamen it is important to prevent overcrowding. The seeds should be pressed about ½ in (1.25 cm) deep into the compost, the container being gently shaken to cover the seed and to make the compost level again. Water the compost lightly and cover with a sheet of clean glass over which should be placed a sheet of brown paper to prevent the compost from drying out too quickly, though in this respect it will be necessary to give the containers almost daily inspection.

Where relying on natural heat, germination may take from six to eight weeks, but if gentle night heat can be given germination should take only about five weeks. The ventilators should be closed during late noon to conserve warmth at night.

Once the seed has been sown, the compost must not be allowed to dry out though it must not be kept saturated. Never water to rule, and during dull weather give moisture only when necessary. The glass should be wiped clean of any moisture and the moment germination is observed, the paper must be removed. When the seedlings have formed their first leaf, the glass should also be removed.

During August and early September, whilst the seed is germinating, a day temperature will be about 60°F. (16°C.) possibly higher during sunny weather. At night the temperature will be around 50°F. (10°C.). As the seedlings make growth during the second half of September, temperatures will gradually fall and this will help the seedlings to grow sturdy, but by 1st October some night heat should be introduced so that the temperature does not fall below 48°F. (9°C.) and this should be maintained throughout winter.

When once the seed has germinated and for the rest of their life, the seedlings and young plants must be watered with care, otherwise they will readily damp off. Where the seeds have been spaced out at sowing time it will not be necessary to transplant as soon as they have formed their first leaf; they may be allowed to grow on until they have formed their second or third leaf. The corm will then have begun to form. The seedlings will be almost touching each other and may be transferred to 3 in. (7.5 cm) pots, supplied with drainage crocks and containing the John Innes Potting Compost No. 1.

Do not allow the seedlings to remain too long in the boxes, otherwise they will soon become 'drawn', from which condition they may not recover. Make the young plants comfortably firm in the pots but do not ram the compost down too tight. And do not over-water. It will now be the end of October and the sun will be rapidly losing strength; also, the usual winter fogs will tend to make the atmosphere damp inside the greenhouse. This means that artificial warmth will be necessary to prevent excess moisture causing the foliage to damp off. The compost will also be slow to dry out, so water sparingly taking care not to splash the foliage. To reduce watering to a minimum, stand the pots in trays containing a layer of moist peat or moss on the greenhouse bench.

Throughout the winter the young plants are kept growing on in a fairly dry, buoyant atmosphere and at all times they must be given the maximum amount of light by washing off any soot deposit as it forms on the outside glass. This is generally caused by foggy weather.

By the end of April, the plants will have formed four to five leaves whilst the compost will be filled with roots. The plants should then be transferred to size 60 pots containing a richer compost. This should be composed of 4 parts fibrous loam; 1 part leaf mould; 1 part decayed cow manure and 1 part coarse sand to which is added the John Innes Base in double strength per bushel of compost, plus ¾ oz. of ground limestone. Artificial heating may now be discontinued.

By early July, the plants will have formed good sized corms and will be drinking copiously. They will now be ready for their final potting into the size 48 pots, using a similar compost except that the sand is replaced by grit and the John Innes Base is now used in treble strength. When transplanting it will be found that the corm will have grown out of the compost and the top portion should be left exposed in its new pot so that about half the corm should be above soil level.

During July and August, the greenhouse must be well ventilated, the vents being opened both by day and by night except when the weather is wet and cold. Though the plants should be given plenty of light, they must be shaded from strong sunlight either by whitening the glass or by tacking brown paper on the inside of the roof on the side facing the sun. Also to prevent leaf scorching, the greenhouse should be damped down each morning so as to create as humid an atmosphere as possible. Surplus moisture should, however, not be allowed to remain at nightfall with the usual drop in temperature.

At this time the plants will be copious drinkers, but even so must never be over-watered and this is particularly important during autumn and winter. The safest way of discovering whether the plant requires moisture is to tap the pot and if a hollow, ringing sound is observed, this will denote that the plants need watering. Water should be given around the side of the pot, taking care not to wet the top of the corm and always using water of the same temperature as the room or greenhouse.

In addition to fluctuations of temperature between greenhouse and home and the possibility of draughts which will cause the foliage to collapse, over-watering may also cause harm to the cyclamen. Where growing in the home, which may be warmed by central heating, or by a gas fire, the plants will require more moisture than where the plants are grown cool. Likewise where growing in a warm greenhouse, but over-watering when the weather is cold and damp when the moisture evaporates slowly from the compost, will cause the foliage to turn yellow and the plants to gradually die back.

In the greenhouse whilst in their final pots, the plants should be placed on trays of gravel and should be allowed room to develop, for they make considerable leaf. The stems of both the foliage and the flowers being quite succulent, they will decay if there is not a free circulation of air around the plants to dry off excess moisture.

Cylamen are gross feeders and even though the compost will be rich, it will be advisable to feed the plants during August and September, whilst the flower buds are forming. Dilute soot water alternating with manure water, which may be obtained in bottles of concentrated strength, and applied once each week will ensure blooms of outstanding quality both as regards size and colour.

BRINGING INDOOR CYCLAMEN INTO BLOOM

The first buds will be observed in October, around the crown of the plant, half hidden by the beautiful deep green, mottled foliage. The night temperature of the greenhouse should then be kept at a minimum of 50°F. (10°C.), a little higher during daytime, and to achieve this with minimum of expense, the inside of the greenhouse should be lined with polythene sheeting. Cyclamen resent forcing conditions and even where they can be obtained, high temperatures must be prevented. The plants should be kept comfortably growing when the most advanced will begin to open their bloom above the foliage early in November, fifteen months after the seed was sown.

Those growing cyclamen for sale should pinch out any blooms which appear during November, for it will be required to have the plants at their best for Christmas. When buying a cyclamen, look about the base of the leaves for buds, for the more there are to develop, the greater will be the value of the plant. Plants which have bloomed too early will have few buds to come into bloom during that period between Christmas and

Easter when flowers are most expensive and when a well-grown cyclamen should provide many weeks of pleasure.

In the home, the plants must have a certain amount of light, but full light is not necessary. They should, however, be kept away from the source of heat as far as it is practical. Water only when necessary. Apply it around the side of the pot or stand the base of the pot in a bowl of water for several minutes. To prolong the display, dead blooms and decaying leaves should be removed as they form.

After flowering has finished in March, the plants still possess decorative value with their attractive dark green foliage and they may be allowed to grow on until April when water should be gradually withheld. The plants should then be placed outdoors, but the corms should not be allowed to dry out entirely. Just give them a little water whenever they become dry and during the first days of August re-pot into a size 60 pot containing a compost as previously described. Place on the greenhouse bench and start into growth again by keeping the compost comfortably moist.

CYDONIA, *see* Chaenomeles

CYNOGLOSSUM

C. nervosum is known as the 'Hounds-tongue' from the shape of its leaves whilst its rich mid-blue flowers, without any trace of mauve or magenta are borne on 20 in. (50 cm) stems through June and July. Like anchusa, it has a habit of dying back if not growing in a light, well-drained soil. Where it is happy it will send up its branching sprays, covered in a profusion of bloom and will prove one of the most colourful plants of the border. Plant 20 in. (50 cm) apart, only in spring. It will rarely survive autumnal planting. Propagation is by cuttings or by root division in spring.

CYPERUS

There are two forms which make attractive pot plants and are readily raised from seed. *C. alternifolius* known as the umbrella plant, grows up to 2 ft. (60 cm) tall and has narrow green leaves striped with white, and *C. natalensis* which is of easier culture.

Seed is sown in March in a temperature of 60°F. (16°C.) the John Innes Sowing Compost being suitable. As soon as large enough to handle the seedlings should be transplanted in clusters to a size 48 pot containing the J.I. potting compost. The loam should be of a heavy, greasy nature, a quality to be found in Kettering loam, whilst firm planting is essential.

From an early age, the plants require ample supplies of moisture and it is important to see that never at any time do they 'flag' or they may not recover. During summer the plants should be shaded and syringed daily during warm weather. Re-potting every second year in spring will keep the plants in a healthy condition.

CYPRESS, *see* Cupressus

CYPRESS PINE, *see* Callitris

CYRTANTHUS

A genus of about fifty species of bulbous plants, native of tropical east and of South Africa and taking their name from two Greek words, *hyrtos*, curved and *anthos*, flower, a reference to the curved perianth tube by which it is distinguished from crinum. The plants form large tunicated bulbs from which arise narrow strap-like leaves and drooping funnel-shaped flowers borne in umbels on hollow scapes. The perianth tube to which the stamens are attached is longer than the segments. The seed capsule is oblong and contains a number of flattened black seeds.

CULTURE

They require cool greenhouse or garden room culture away from the warmer parts, planting three to four bulbs of 3 in. (7.5 cm) diameter to a large pot and in March. There are several evergreen species which should not be dried off after flowering as necessary for those of deciduous habit. They require a compost made up of fibrous loam, leaf mould, decayed cow manure and coarse sand in equal parts and to be planted 3 in. (7.5 cm) deep. The flowering period is from mid-March until the end of summer, depending upon species.

After planting, stand the pots in a plunge bed outside or on a verandah in boxes and cover with ashes, also with a sheet of glass to retain the warmth of the spring sunshine. Then, towards the end of April move to a greenhouse, giving copious waterings as they make growth and whilst in bloom but gradually drying off the deciduous varieties after flowering. They should be wintered where frost is excluded but in a sunny sheltered garden may be flowered outdoors during summer. The plants will benefit from a weekly application of dilute manure water and a top dressing after flowering.

In a sheltered garden, the plants may be grown outdoors, planting the bulbs 4 in. (10 cm) deep and 12 in. (3 0cm) apart at the foot of a wall and lifting them after flowering, drying off and keeping them in a frost free room, to be planted out again the following April. The later flowering varieties are more suitable for outside culture.

When outdoor bulbs are lifted or indoor bulbs are re-potted, numerous bulblets will be seen clustering round the base. These are detached and grown on in boxes of sandy compost for twelve to eighteen months when they are potted and will have reached flowering size in another two years.

SPECIES

Cyrtanthus angustifolius. Syn: *C. ventricosus*. It makes a small ovoid bulb with a long tapering neck and it forms one to three narrow leaves 18 in. (45 cm) long. In its native South Africa it blooms from December until April; in July and August in North Europe and North America, bearing its bright orange-red drooping flowers in umbels of four to ten on a 12 in. (30 cm) scape. The flowers are about 2 in. (5 cm) long with spreading segments and the style longer than the stamens.

C. collinus. From the small ovoid bulbs appear two to three linear leaves 9 in. (23 cm) long and it bears on a slender 12 in. (30 cm) scape, six to ten bright red flowers 2 in. (5 cm) long.

C. huttoni. Native of Cape Province it is a most attractive species with broad strap-like leaves and bearing in May and June in a stout umbel, six to eight funnel or bell-shaped flowers of pale red with reflexing pointed segments.

C. mackenii. Native of the high mountainous ranges of Natal, it is the first to bloom, early in March where growing in gentle heat. It forms two to six long linear leaves and bears on a slender 10 in. (25 cm) scape, waxy white tubular flowers formed horizontally in umbels of six to ten. Varieties are also obtainable bearing flowers in shades of pink and apricot.

C. obliquus. The oldest known species, in cultivation from 1774 when it was discovered in its native Cape Province. It forms a large bulb 4 in. (10 cm) across and it has ten to twelve strap-like leaves arranged in two rows and produced after the flowers. In May and June it bears on a 2 ft. (60 cm) stem, an umbel of ten to twelve bright red drooping flowers, shaded yellow at the base and measuring 3 in. (7.5 cm) long.

CYTISUS

The Brooms are amongst the most valuable plants for a dry, sandy soil and an open, sunny situation. They resent transplanting and should be planted from pots whilst they dislike any form of pruning or clipping. For a wall, for it is not quite hardy, plant the Moroccan *C. battandieri*, its grey leaves covered in a silky sheen, whilst its bright yellow pineapple-scented flowers appear in July, in laburnum-like clusters.

C. albus, the White Portugal Broom and *C. praecox* also white-flowering, bloom early in summer and before, in a sheltered corner, being the first to bear flowers. *C. praecox* forms a drooping fountain-like plant with its stems smothered in bloom. In bloom shortly after are varieties of *C. scoparius,* the Common Yellow Broom, 'Cornish Cream' bearing pale creamy-yellow flowers and 'Dorothy Walpole', flowers of velvet-crimson. 'Lord Lambourne' has scarlet wings and a cream standard; and 'Donard Seedling' flowers of ruby-red streaked with yellow. Plant in spring, 4–5 ft. (1.5 m) apart and where they are sheltered from cold winds.

Propagation is by seed sown in small pots or by cuttings, rooted in pots under glass in August.

DABOECIA

Dainty heather-like plants in bloom from June until October and happy in full sun or partial shade. The most suitable for the rockery is *D. azorica* which bears large nodding heather bells of rich crimson. *D. polifolis* is charming, but flowers on 15 in. (38 cm) stems and is happier in the woodland garden. It bears handsome purple bells, and as a contrast *D. polifolia* 'Alba', pure white, is equally fine.

DAHLIA

No garden plant has been more greatly improved during recent years than the dahlia; to provide colour for the small garden from July until the appearance of the first hard frost blackens the foliage, there are a number of new varieties of dwarf compact habit which, for quality of bloom will rival those of more pretentious form. Those dwarf dahlias growing about 2 ft. (60 cm) tall and bearing flowers of decorative and cactus form, as against the older bedding dahlia growing only 15 in. (38 cm) tall and bearing single blooms, will produce as rich a display as the taller-growing varieties, and there will, moreover, be no need to provide support and adequate protection from strong winds.

SPROUTING THE TUBERS

A heated greenhouse is not necessary to grow dahlias, indeed a greenhouse is not necessary at all. The tubers may be started into growth in a cold frame, or even in deep boxes placed in a sunny position outdoors and covered with a sheet of glass. Additional protection may be provided by heaping soil, sand or straw around the boxes. Where the tubers are to be started without artificial warmth, they will be later into growth, though if the spring is one of more than average sunshine, they will have made sufficient growth to give early bloom when planted out in June.

The tubers should be planted about 1st April, not before, for until that time there will not be sufficient heat from the sun to stimulate them into growth. Also, as they should not be planted out until all fear of frost has gone, which will be during the first days of June, it is not advisable for the tubers to make excessive growth before that time. If the shoots are about 10 in. (25 cm) tall when they are planted out they will withstand the moving without any undue 'flagging', provided they are not allowed to lack moisture.

The tubers should be plump and clean when planted and, if they have to be stored for any length of time, they are placed in boxes of sand in a frost-free room away from a fire or any form of central heating which will cause them to shrivel. They should be planted on a layer of peat

FIG. 3 *Planting a dahlia root*

placed either in the boxes or in a frame. The tubers may be packed quite close together, a mixture of peat and sterilised loam being pressed about them so that there will be no air spaces, whilst they should be just covered with compost. Give a thorough soaking, place the glass in position, and all that is necessary during the next two months will be to keep them comfortably moist and to ventilate by day as the sun gathers strength.

Not until fear of frost has vanished should the tubers be planted out and they must be moved with care so that the shoots are not damaged.

The holes, large enough to accommodate the tubers, should be made before the dahlias are removed, so that they may be placed in their flowering quarters with the least possible delay. Do not bury the tubers too deeply. They should have about 1 in. (2.5 cm) of soil over the top after placing them on a layer of peat. To prevent damage from strong winds, the shoots should be given some support by means of stout twigs, and never must the tubers be allowed to lack moisture. Dahlias are copious drinkers and heavy feeders, so into the soil before planting, incorporate some decayed manure or bone meal in addition to giving a liberal dressing of peat. The plants will also benefit from a top dressing of peat given late in July.

Set out the tubers about 2 ft. (60 cm) apart or, better than planting each variety at the same distance apart, allow the same amount of space between the plants as their height.

During an average autumn, the plants will continue to bloom until the end of October, or to the middle of November, when the tubers should be lifted with care and cleaned of all surplus soil. After drying in a cool, airy room, they should be placed into boxes of sand or ashes to remain there until required to be started into growth again in April. To prevent mildew, it is advisable to dust the tubers with flowers of sulphur before storing.

A pleasing use of the dwarf dahlias may be made by growing them in small circular beds, using varieties of contrasting colours, with those of slightly taller habit planted towards the centre. Or a small border of dahlias may be made, planting those which are taller-growing to the back, with those of more dwarf habit to the front.

VARIETIES OF MERIT FOR A SMALL GARDEN

BROADACRE. It grows about 30 in. (76 cm) tall, its blunt-tipped flowers being of rich orange and borne on long stems.

CHARITY. It is a medium decorative dahlia, bearing almost globular flowers of salmon-pink flushed with carmine.

CHEERIO. It grows nearly 3 ft. (90 cm) tall and is one of the most decorative dahlias. It makes a plant of bushy habit and bears its cherry-red flowers on long wiry stems. The petals are thin and fluted, and are attractively tipped with white. (see plate 18).

COON. Well-named, for its flowers are darker than those of any other dahlia. It makes a bushy plant 28 in. (70 cm) high, and bears dark maroon flowers of rosette type.

DALHOUSIE PEACH. It forms a rounded bush-like plant 2 ft. (60 cm) tall and bears a succession of double blooms of the small decorative type. The flowers are of an attractive shade of peach-pink.

DOWNHAM. It grows 20 in. (50 cm) tall, and bears its large semi-cactus blooms of brilliant pure yellow in endless profusion. It comes quickly into bloom, its flowers having long stems which make them useful for indoor decoration.

DUNSHELT FLAME. The plants grow just over 30 in. (76 cm) tall and bear a profusion of bright scarlet blooms.

ECLIPSE. A very lovely variety, growing less than 30 in. (76 cm) tall and coming early into bloom. The small decorative blooms are orange, overlaid with a unique shade of rosy-vermilion.

FAME. It is one of the earliest varieties to bloom. It grows 3 ft. (1 m) tall, and bears on long, wiry stems blooms of bright apricot shaded with tangerine.

FREEMAN. A truly magnificent variety, growing about 28 in. (70 cm) tall, its deep carmine-red flowers being tipped with gold.

LANGSTEM. A variety growing 3 ft. (90 cm) tall and bearing globular flowers of a delightful combination of apricot and gold colouring. It is one of the longest and most free-flowering of dahlias.

LEMON HART. For best effect, plant with 'Market King', for it grows to a similar height and bears bright lemon-yellow flowers which have attractively quilled petals.

MARGARET HARRIS. It grows just over 30 in. (76 cm) tall, bearing flowers of small cactus form, cream at the centre, the petals being tipped and edged with lilac-pink.

MARKET KING. Though growing about 30 in. (76 cm) tall, it makes a compact, bushy plant, and bears a bloom of the colour of a ripe tomato, the petals being veined with gold to make it the brightest flower of the garden.

MARY BROOM. It grows 3 ft. (90 cm) tall and is one of the most spectacular varieties ever raised. The blooms are almost globular, the small but wide pillar-box red petals being tipped with white.

PEACH BEDDER. It makes a bushy plant almost 2 ft. (60 cm) tall, and bears masses of star-like blooms which are of a bright apricot-orange rather than of peach colouring.

RIVAL. It grows to a similar height to 'Market King', and bears a profusion of semi-cactus blooms of bright rose-cerise on long wiry stems.

ROTHESAY CASTLE. It grows 18 in. (45 cm) tall, comes early into bloom, and is most free-flowering. The flowers are a combination of cream overlaid with rose-pink.

STONYHURST. A fine dahlia for the small garden, growing little more than 2 ft. (60 cm) tall and bearing masses of exhibition-quality blooms of uniform cherry-red.

CLASSIFICATION AND VARIETIES

Section I. Single flowered

Divided into two classes: In Ia, the blooms must have smooth overlapping ray florets, giving perfect symmetry. In class Ib, the florets need not overlap so that the bloom does not reveal such a high standard of beauty. Most singles grow about 18 in. (45 cm) tall and are used for bedding:

COLTNESS GEM. An old favourite and still widely grown for its brilliant scarlet blooms freely produced.

NORTHERN GOLD. The blooms are large and of perfect symmetry and of pure golden-yellow.

PRINCESS MARIE JOSE. The large blooms are of a lovely shade of soft mauve pink.

Section II. Star-flowered

The small flowers have pointed florets which do not overlap. Varieties in this section are rarely to be seen.

Section III. Anemone-flowered

The bloom has an outer ring of flat ray florets surrounding a group of tubular florets:

VERA HIGGINS. The most popular in this section. It grows 3 ft. (90 cm) tall with the outer florets terra-cotta bronze, the disc petals being golden-yellow suffused with bronze.

Section IV. Collerette

The flowers are divided into three classes. Class IVa is for single Collerettes, those having a single row of ray florets and a single 'collar' half the length of the ray florets; class IVb is for Paeony-flowered Collerette dahlias, those having two or more rows of ray florets and 'collars'; and IVc is the Decorative Collerette dahlia, fully double all through. They grow 3 ft. (90 cm) tall.

BRIDE'S BOUQUET. All white with broad ray florets.

CLAIRE DE LUNE. The outer petals are pale yellow, the collar being of cream.

GOERLING'S ELITE. It has cardinal-red ray florets and a collar of yellow with the ray florets tipped yellow.

LA GIOCONDA. One of the most striking of dahlias with scarlet ray florets and a collar of pure gold.

SUNTAN. Deep apricot-orange throughout.

VERA LYNN. The ray florets are rose-pink with a contrasting cream collar.

Section V. Paeony-flowered

The blooms have two or more flattened ray florets and a central disc. There are three classes denoting size of bloom. Class Va is for flowers of more than 7 in. (17.5 cm) diameter; class Vb over 5 in. (12.5 cm) but not over 7 in. (17.5 cm); Vc not exceeding 5 in. (12.5 cm). The latter is the only class in which there is any interest. The plants grow 3 ft. (90 cm) tall.

BISHOP OF LLANDAFF. Popular for bedding, it has dark green, almost black leaves and bears crimson-red flowers.

SYMPHONIA. The orange-scarlet counterpart of 'Bishop of Llandaff'.

Section VI. Decorative

With the Cactus-flowered, the most important section, divided into five classes. Class VIa is for giant-flowered decoratives, the blooms exceeding 10 in. (25 cm) diameter; class VIb is for large-flowered varieties with flowers more than 8 in. (20 cm) diameter but less than 10 in. (25 cm); VIc is for medium-flowered with blooms over 6 in. (15 cm) but less than 8 in. (20 cm); VId has flowers more than 3 in. (7.5 cm) diameter but not more than 6 in. (15 cm). This is the most widely grown of all the twenty-four dahlia classes and there is often confusion between those varieties having blooms just over 3 in. (7.5 cm) diamater e.g. 'Jescot Jim' and the large flowered pompons of class VIIa which also have their bloom over 3 in. (7.5 cm) diameter i.e. 'Kokarde'. Class VIe is for miniature decoratives with bloom not exceeding 3 in. (7.5 cm) diameter.

Giant decoratives (*Class VIa*)

BONANZA. It bears flowers of peach-pink overlaid apricot-orange and is of exhibition form. 3–4 ft. (about 1 m).

COL. W. M. OGG. Of refined shape, the blooms are of a lovely shade of ivory-cream. 4 ft. (1.2 m).

CROYDON MASTERPIECE. The blooms are large and deep and of rich reddish-bronze. 4–5 ft. (about 1.5 m).

DORA RAMSEY. Of exhibition quality. The blooms are of a lovely shade of soft apricot tinted with rose. 3–4 ft. (about 1 m).

LADY LIDDELL. A suitable companion for the purple varieties for it has large flowers of deepest yellow. 3–4 ft. (about 1 m).

LAVENDER PERFECTION. The finest pure lavender-mauve decorative ever introduced, bearing flowers of exhibition quality throughout the season. 3–4 ft. (1 m).

MARGARET DUROSS. The blooms have enormous size and are of a beautiful shade of rich golden-bronze. 4–5 ft. (about 1.5 m).

WINIFRED ST REDWICK. It bears large white flowers of exhibition form and carried on long sturdy stems. 3–4 ft. (about 1 m).

Large flowered (*Class VIb*)

COVER GIRL. It blooms in profusion, the blooms being of clearest mauve, tipped with white. 3–4ft. (about 1 m).

FIRST LADY. Of a beautiful shade of brilliant yellow, the florets are long whilst the blooms are of exhibition form. 3–4 ft. (about 1 m).

MARGARET MORRIS. A beautiful variety, the deep orange flowers being shaded with rose-pink. 3–4 ft. (about 1 m).

MRS MCDONALD QUILL. Most striking in any company, the crimson-scarlet flowers are tipped with white. 3–4 ft. (about 1 m).

SILVER CITY. An ivory white of perfect exhibition form. 3–4 ft. (about 1 m).

Medium flowered (*Class VIc*)

ALBION. The best white dahlia for cutting for wreaths and bouquets, the blooms being held on long wiry stems. 3–4 ft. (about 1 m).

ARC DE TRIOMPHE. Of perfect form, the blooms are borne in profusion and are held on sturdy stems. 3–4 ft. (about 1 m).

BARROWFORD. The flowers have great substance and are of pure tangerine-orange. 3 ft. (90 cm).

EDITH. Cerise-flushed strawberry-pink with a golden-yellow centre. The flowers will keep fresh for ten days in water. 3–4 ft. (about 1 m).

GAWTHORPE. The blooms have perfect form and are of rich vermilion, shaded peach. 3–4 ft. (about 1 m).

Small flowered (*Class VId*)

ANGORA. Of ideal form, the pure white flowers are held on long wiry stems. Excellent for cutting, 3–4 ft. (about 1 m).

BOWLAND. Excellent for cutting and garden decoration, it bears rich salmon-orange blooms on long stems. 3–4 ft. (about 1 m).

GLORIE VAN HEEMSTEDE. The lemon yellow blooms of perfect form are borne on extra long stems for cutting. 3 ft. (90cm)

HAMARI FIESTA. A scarlet and yellow bi-colour, excellent for the show bench and in the garden. 3 ft. (90 cm).

KENDAL PRIDE. A first class cut flower, the blooms being of rich pink, suffused yellow and borne in profusion. 3–4 ft. (about 1 m).

LADY TWEEDSMUIR. An oustanding cut bloom bearing soft lilac-purple flowers of firmest texture and held on long wiry stems. 3–4 ft. (about 1 m).

MEIRO. A Japanese dahlia of excellent form, the blooms are of rich bluish-purple, held on long sturdy stems. 3 ft. (90 cm).

SHIRLEY WESTWELL. One of the most popular of all dahlias, it makes a bush almost as wide as it grows tall, bearing July to November a mass of brilliant scarlet flowers. 3 ft. (90 cm).

TRESOR D'OR. The blooms are of pure deep gold of ideal exhibition form. 3–4 ft. (about 1 m).

Miniature flowered (Class VIe)

DAVID HOWARD. A most decorative dahlia, the foliage being almost black, the blooms burnt gold. 3 ft. (90 cm).

DR GRAINGER. Of exhibition form, the blooms are of a rich shade of orange. 3 ft. (1 m).

MARY BROOM. The beautifully rounded blooms are of brightest scarlet, tipped with white. 3 ft. (90 cm).

MOTHER LISTER. The dainty white flowers are tipped and shaded at the centre with soft lavender. 3 ft. (90 cm).

NEWBY. A valuable cut flower dahlia, of a lovely shade of peach-pink with yellow shading. 3–4 ft. (about 1 m).

ROSE NEWBY. A pure rose-pink counterpart of the famous 'Newby'. 3–4 ft. (about 1 m).

SAFE SHOT. The flowers are produced in profusion and are bright orange, held on long stems. 3 ft. (90 cm).

Section VII. Double show and fancy dahlias

These have double blooms over 4 in. (10 cm) diameter with tubular florets, blunt at the end. The blooms are similar to the small decoratives which have superseded them in popularity.

BONNIE BLUE. A most valuable cut flower bearing bloom of soft bluish-mauve on long wiry stems. 3–4 ft. (about 1 m).

CHORUS GIRL. Sometimes classed as a small decorative, the blooms are almost globular and of rich mauve-pink. 3–4 ft. (about 1 m).

JEAN LISTER. The glistening pure white flowers are produced with freedom. 3 ft. (90 cm).

Section VIII. Ball or Pompon dahlias

They are divided into three classes. Class VIIIa, large-flowered poms or Ball dahlias between 3 in. (7.5 cm) and 4 in. (10 cm). diameter. Class VIIIb, for bloom between 2 in. (5 cm) and 3 in. (7.5 cm) diameter and known as miniature ball-type; Class VIIIc for flowers which do not exceed 2 in. (5 cm) diameter. These are the true poms.

Large pom or ball-type (Class VIIIa)

NELLIE BIRCH. The blooms are deep crimson-red. 3–4 ft. (about 1 m).

PRIDE OF BERLIN. An old favourite never surpassed in its colour, being of a lovely shade of lavender-pink. 3–4 ft. (about 1 m).

ROTHESAY SUPERB. It has flowers of brightest scarlet for cutting and exhibition. 3–4 ft. (about 1 m).

Medium pom or miniature ball (Class VIIIb)

CLARITY. Pale sulphur-yellow of perfect petal placement. 3–4 ft. (about 1 m).

EVELYN RICHARDS. An exhibitor's favourite, the creamy-white blooms being flushed with mauve. 3–4 ft. (about 1 m).

FLORENCE VERNON. Orchid-purple of ideal petal placement. A winner on the show bench. 3 ft. (90 cm).

HORN OF PLENTY. Flaming red with beautifully overlapping petals which make it an exhibitor's favourite. 3 ft. (90 cm).

LITTLE DAVID. The ball-shaped blooms are of clear orange. 3 ft. (90 cm).

TENDERNESS. The well-formed blooms are of a lovely shade of old rose. 3 ft. (1 m).

Pompon (VIIIc)

BURWOOD. Its ball-shaped blooms are of yellow tipped with red. An exhibitor's favourite. 3 ft. (90 cm).

CROSSFIELD EBONY. The best crimson pom for exhibition. 3 ft. (90 cm).

DIANA GREGORY. Lovely soft pink shading to mauve at the edge. 3 ft. (1 m).

JACKIE BROOKS. Golden yellow, tipped bronze. 3 ft. (90 cm).

JESSIE BUCHANAN. Pale pink throughout. 3 ft. (90 cm).

LITTLE PRINCE. Of exhibition form, the scarlet blooms are tipped with white. 3 ft. (90 cm).

TIPPIT. Yellow, tipped scarlet. 3 ft. (90 cm).

WILLO'S VIOLET. The most popular of all poms, the perfectly globular blooms being of violet-purple. 3 ft. (90 cm).

Section IX. Cactus dahlias

These bear double flowers with no central disc and with narrow pointed petals, straight or incurving. Raised from the original scarlet 'Juarezi', discovered in Mexico in 1872. Cactus dahlias are divided into eight classes, there being four classes for the true cactus and four for the semi-cactus. At one time there was a special class for the Giant cactus (over 10 in. 25.4 cm) but this is now merged with the Large cactus and Semi-cactus to form Class IXa for blooms measuring over 8 in. (20 cm) diameter. Class IXb is for Medium cactus and semi-cactus measuring 6–8 in. (15–20 cm); class IXc is for small cactus and semi-cactus, 4–6 in. (10–15 cm) diameter; and IXd is for the miniatures of less than 4 in. (10 cm) diameter. These classes are given in abbreviated form (e.g. G.S.C.=Giant Semi-Cactus) after the varieties which follow.

Large-flowered cactus (Class IXa)

COCARICO. Brilliant scarlet of exhibition quality. 3–4 ft. (about 1 m) (G.S.C.).

CROESO '69. An exhibitors' flower of pale pink shading to cream at the centre. 3–4 ft. (about 1 m) (L.S.C.).

HIGH STANDARD. The blooms are of perfect form and are of deep salmon, flushed with red. 3–4 ft. (about 1 m) (L.C.).

PAUL CRITCHLEY. Rich pink and of perfect form. 3 ft. (about 1 m) (L.C.).

NANTENAN. The large sulphur-yellow blooms are of ideal exhibition form. 3–4 ft. (about 1 m) (L.S.C.).

SUPERCLASS. With their narrow petals and excellent depth, the salmon-bronze bloom, overlaid with gold are ideal for garden and exhibition. 3–4 ft. (1 m) (L.C.).

Medium-flovered cactus (Class IXb)

AUTHORITY. National Challenge Cup winner, the blooms are as deep as they are wide, almost ball-shaped and of rich burnt-orange. 4–5 ft. (about 1.5 m) (M.S.C).

BEAUTY OF AALSMEER. The blooms are as deep as they are wide whilst the colour is carmine-pink overlaid salmon-pink. 4–5 ft. (about 1.5 m) (M.S.C.).

BEAUTY OF BAARN. Primrose-yellow of great depth and a Challenge Cup winner for the longest keeping dahlia when in water. 3–4 ft. (about 1 m) (M.S.C.).

CARNIVAL. Glowing scarlet with yellow tips. 3–4 ft. (about 1 m) (M.S.C.).

DENTELLE BLANCHE. The blooms are pure white with fimbriated petals which give it the appearance of a ball of white lace. 3–4 ft. (1 m) (M.C.).

FLEUR DE HOLLANDE. The flowers are of brightest scarlet shaded with gold at the centre and with great depth. 4–5 ft. (about 1.5 m) (M.S.).

PIQUANT. Blood red with white tips. 3–4 ft. (about 1 m) (M.C.).

ROTTERDAM. One of the outstanding dahlias of all time, the blooms having great depth and strong wiry stems and being of deep blood-red throughout. 3–4 ft. (1 m) (M.S.C.).

Small-flowered cactus (IXc)

GOYA'S VENUS. Wonderful for cutting, the blooms are of bright orange-pink, held on long cane-like stems. 3–4 ft. (about 1 m) (S.S.C.).

MURIEL GLADWELL. A most arresting dahlia, the well-shaped blooms of tomato-orange being held on sturdy 2 ft. stems. 3–4 ft. (about 1 m) (S.S.C.).

RITA RUTHERFORD. With its wiry stems 30 in. (75 cm) long and flowers of bronze, pink and yellow, this is on of the best of all cut-flower dahlias. 3–4 ft. (about 1 m) (S.S.C.).

SALMON RAYS. The flowers of deep salmon-pink have outstanding beauty, the petals being long and pleasantly twisted. 3–4 ft. (about 1 m) (S.C.).

SUPERMARKET. The finest yellow cactus for cutting, bearing almost globular blooms of brightest lemon-yellow. 3–4 ft. (about 1 m) (S.C.).

Miniature-flowered cactus (IXd)

JUDITH WHITE. Excellent for cutting, the blooms being of bright salmon-pink. 3–4 ft. (about 1 m).

PIROUETTE. Very free flowering, the blooms are neat and finely petalled and of pale sulphur-yellow. 3 ft. (90 cm).

Section X. Miscellaneous dahlias, not previously classified and not bedding dahlias

ANDRIES' WONDER. Of unusual formation, the blooms resemble a salmon incurved chrysanthemum. 3–4 ft. (about 1 m).

GIRAFFE. A double orchid-flowered dahlia of deep yellow, spotted and splashed pinkish-red. Valuable for floral decoration. 3 ft. (90 cm).

Section XI. Bedding dahlias

DOWNHAM. A dwarf cactus-flowered of brilliant yellow and which grows as wide as it grows tall. 18 in. (45 cm).

MAUREEN CREIGHTON. It bears its double decorative-type blooms of brilliant red in profusion. 18 in. (45 cm).

PARK WONDER. A dwarf cactus of pale sulphur-yellow and great freedom of flowering. 18 in. (45 cm).

PIPER'S PINK. The best pink bedder, the cactus-shaped blooms being of deep rose-pink. 18 in. (45 cm).

ROCQUENCOURT. Striking in its rich bronze foliage and orange coloured flowers. 20 in. (50 cm).

ROTHESAY CASTLE. The fully double blooms are of rich cream, overlaid pink. 20 in. (50 cm).

DAFFODIL, *see* Narcissus

DAISY BUSH, *see* Olearia

DAPHNE

A large genus of slow growing evergreen or deciduous shrubs, the daphnes are ideal small garden plants. Probably the best is *Daphne mezereum* a deciduous form which blooms during February and March, its leafless stems being covered with purple-pink blossom which possesses a sweet perfume. The flowers are followed by scarlet fruit. There is also an attractive white variety Alba, the grey-white flowers of which are followed by yellow fruits.

A hybrid, 'Somerset' should also be grown, for it bears its highly scented rose-pink flowers from the end of April to the end of May.

For the rock garden, *D. alpina* grows 18 in. (45 cm) tall and blooms during early summer. The white flowers, borne in terminal clusters have a spicy scent. It requires a soil containing lime rubble, likewise *D. blagayana* which like *D. alpina* is evergreen. It makes a low spreading plant, its leaves formed at the end of the branches where they act like a ruff for the clusters of fragrant milky white flowers, borne early in spring.

Almost no pruning will be advisable for any of the daphnes. Plant in November in a rich loam containing plenty of peat or leaf mould. Propagate from seed sown in small pots for daphnes do not like root disturbance, or by layering of the half-ripened wood.

DAVIDIA

The Chinese Dove Tree, *D. involucrata* is deciduous and grows up to 30 ft. (9 m) tall, its large heart-shaped leaves tapering to a point. It blooms early in summer and has large white bracts which surround the inflorescence, hence its name of Handkerchief Tree. It requires a rich deep loam and is propagated either by seeds or by cuttings of the half-ripened wood which are rooted under glass.

DAWN REDWOOD, *see* Metasequoia

DAY LILY, *see* Hemerocallis

DECUMARIA

A self-clinging semi-evergreen, *D. barbara* is native of the south-east United States and is closely related to *Hydrangea peteolaris*. Given a warm wall it will soon reach a height of 20 ft. (6.4 m) or will pull itself up a trunk of a dead tree. It has glossy green leaves and in midsummer bears large corymbs of creamy-white flowers. Plant in spring, into a soil containing decayed manure and peat or leaf mould. It requires a sheltered position.

DELPHINIUM

Known as the Queen of Border plants and rightly so for with their graceful tapering spires and growing to a height of from 6–7 ft. (2 m), no border is complete without them. They cannot, however, be considered as fool-proof as most other plants. They require care with their planting, whilst their densely flowered spikes require careful support. It was a century ago that James Kelway of Langport began his work of hybridising, making use of the Russian species *D. elatum* which reached this country during Tudor times and is the parent of modern strains. At the turn of the century, Messrs Blackmore and Langdon of Bath, took the hybridising of the delphinium a stage further and more recently the late Mr Frank Bishop made some magnificent introductions. In America, the firm of Vetterle and Reinelt of California have raised a series of lovely white-flowered varieties which possess a beauty all their own, and are certain to rival the better known 'blues' in popularity. And for the first time, at least one possesses perfume.

GROWING FROM SEED

Delphiniums may be propagated from seed; by division of the roots; or from cuttings. Where named varieties are not considered essential, raising plants from seed will give excellent results provided the finest of the modern strains are sown. Where there is a limit to one's outlay in the planting of the border, then raising a stock from seed will enable more to be spent on those plants which are not readily propagated in this way. A number of specialist growers may have young plants to offer in the spring of each year which will have been raised from seed of their own breeding. In this way enthusiasts are able to obtain plants of outstanding quality inexpensively, and without the trouble of sowing their own seed.

It has been found that where seed has been stored under refrigeration, it will germinate more rapidly, whilst at the same time retaining its original freshness over a period. In this way, seed saved by specialist growers in late summer will retain its vitality for sowing the following spring – the most suitable time. Refrigerated seed which will retain its freshness for several years, should, however, be sown immediately it is received. As the seed is small and fairly expensive, it should be sown with care, sowing in pans or boxes containing the John Innes Sowing Compost. Sow thinly and water thoroughly before placing a sheet of clean glass over the box or pan. Where possible place the container in a closed frame or in a warm greenhouse, shading the glass until germination has taken place to prevent moisture evaporation of the compost. Where gentle warmth is available, sow in March; where sowing under cool conditions, delay sowing until April. It is important for reliable germination that the seed is never allowed to become dry. The seedlings should be pricked out into a cold frame or into deeper boxes or better still into 3 in. (7.5 cm) pots as soon as they may comfortably be handled. They should then be placed in an open frame, where, kept moist they remain until the following spring, which as with most blue-flowered plants is the ideal time to plant. This is also the most suitable time to take cuttings or divide the roots.

TAKING CUTTINGS

To increase the stock as quickly as possible, specialist growers lift the roots in early April and pack them tightly together in a closed frame. The shoots or cuttings when 4 in. (10 cm) long are removed as they form. Remove them from just below the level of the compost, which should contain an abundance of peat, and into which they are packed. The cuttings are rooted in a mixture of peat (to retain moisture) and sand in a frame, and they should root in from four to five weeks. They are then moved to prepared beds outdoors and planted 6 in. (15 cm) apart in early July which is generally a humid month. This enables the plant to become thoroughly established before root action becomes less vigorous later in the season.

Another method is to lift roots growing in the border when the young shoots are about 4 in. (10 cm) high. This will be early May. Where two established plants of the same variety are growing together, one may be lifted without leaving too large a space in the border and this may be replanted with a rooted offset to come into bloom the following summer. After the root has been lifted, the soil is carefully dislodged and each shoot is then removed with a portion of root attached. The whole of the plant may be split up in this way, the woody centre being destroyed. The rooted offsets should be planted in a prepared bed, partially sheltered from strong sunlight and where they are kept growing on until replanted. They may also be immediately replanted in the border, but care should be taken to ensure that they receive their fair share of light and moisture which might be denied them by the established plants.

Another method of increasing stock is to divide the roots with two strong forks in the usual way. This should be done during March before the plant comes into growth, otherwise the new shoots might easily be damaged. A large root may be divided into four plants which will bloom the same summer if replanted at once, though later than usual.

PREPARATION OF THE SOIL

Thorough preparation of the soil is essential for success. The plants will never be happy in a cold, heavy soil and the crowns may die back during winter. If the soil is heavy, then give a good dressing of caustic lime to break up the clay particles. This should be done in November. Early in March the ground should have some sand or ashes incorporated, together with a liberal quantity of decayed manure. Cow manure is ideal for delphiniums where the soil is heavy. For a light, sandy soil incorporate some peat and plenty of decayed strawy manure to retain moisture.

The plants like a well-drained soil in winter and one retentive of summer moisture. Where the soil is of a good deep loam, only small quantities of peat and manure should be added. As slugs attack this plant in spring it is advisable to give a dressing of ashes round the crowns when the foliage has died down in autumn. For exhibition quality spikes, feeding with liquid manure once a week during May and early June will greatly enhance the quality of the spike.

When planting, allow 3–4 ft. (1 m) between the plants and select them with a view to obtaining bloom over as long a period as possible. For instance, the pale blue and mauve 'Jennifer Langdon' comes into bloom early in June, whilst the lovely rosy-mauve 'Nell Gwynne' does not come into bloom until the month end and remains colourful until late July. Where space permits plant two of the same variety together. Plant in April, spreading out the roots.

The shoots should be supported in plenty of time, first fixing hardwood stakes in triangular form round the plants, and from 2 ft. (60 cm) above ground level strong twine is fastened round the stakes at regular intervals. This will prevent the shoots being damaged by strong wind. At first the stakes may look unsightly but they will soon become obscured by the large pale green foliage of the plants and of those plants growing immediately in front. The flower spikes will be borne well above the stakes, but for the very large-flowered varieties such as the navy-blue and white 'Glamour Girl' which grows to a height of 6 ft. (2 cm) or more, and where growing for exhibition, it is advisable to stake each spike separately with a 6 ft. (2 m) cane. Where caught by the wind, the stems generally break away at the crown, thereby causing damage to the plant. Staking is more essential for the delphinium than for any other border plant, for not only is the size of their flower spikes equalled only by those of the eremurus, but of all plants, the delphinium must be given an open, sunny position. To plant too close to trees or a tall hedge will not only cause the plants to become 'drawn' with the result that the spikes will be loosely formed, but mildew, to which many of the older varieties are prone, will be encouraged.

After flowering, remove the dead heads and where the foliage has been removed in autumn, fork in a light dressing of manure around the plants and during winter surround the crowns with weathered ashes or sand.

There are four forms of the perennial delphinium: The large-flowering hybrids; the Belladonna or small flowering hybrids, having a more compact habit and bearing smaller spikes on branching stems; the Ruysii hybrids, the result of a cross between the large flowering strain and *D. nudicaule*, the Californian scarlet delphinium and *D. chinensis*, the Chinese Delphinium, which bears its dainty branched spikes on 12 in. (30 cm) stems throughout summer and with *D. nudicaule* is most valuable for bedding.

Delphinium ajacis, the Larkspur is one of the most popular of all annuals for it is useful for mixing with blue scabious and with other midsummer flowers on account of its dainty, feathery habit. It is extremely hardy and not liking disturbance, the seed is sown in September in drills 2 ft. (60 cm) apart where it is to bloom. Growing to a height of 3–4 ft. (1 m), it should be given a sheltered position though it enjoys ample sunshine. The plants require staking against strong winds but otherwise no additional attention. The bloom must be cut as soon as it shows colour, whilst many buds are still green, for not only is the bloom enhanced by the green tint, but larkspur will quickly drop its petals when fully open. The Giant Imperial and Double Stock-flowered strains are the most popular. Of these, 'Los Angeles', with its salmon, rose-tinted blooms; 'Rosamond', pure bright pink; 'Dazzler', rosey-scarlet; 'Blue Bell', bright mid-blue; and 'Sweet Lavender', are outstanding.

LARGE-FLOWERED PERENNIAL HYBRIDS (5–6 ft. about 2 m)

Early flowering

ARTIST. Mr B. J. Langdon of the famous Bath firm considers this one of the best varieties both for garden decoration and for cutting. The long, tapering spikes have pale blue florets and are produced with freedom.

BLUE RHAPSODY. A stately variety, the bloom being of a rich mid-blue with a small black centre.

HARVEST MOON. A lovely variety, the florets being pure silver-blue with a black eye.

JENNIFER LANGDON. The semi-double florets are pale blue and mauve with a small black centre.

LONDON PRIDE. A vigorous, free flowering variety, the florets are bright cornflower-blue with a white eye.

MRS NEWTON LEES. It forms a huge spike, the semi-double blooms being of mauve and pale blue.

SWAN LAKE. The spikes are well-built, the florets being white with a jet black eye. A superb variety.

Mid-season flowering

AGNES BROOKS. The spikes are of enormous size, the florets being of bright gentian-blue, its white eye being splashed with pale blue.

ALICE ARTINDALE. Without doubt the finest variety yet raised for cutting. It travels well and lasts long in water. The fully double florets are like ranunculus blooms and are of rich rose and pale blue.

BONINGALE GLORY. The tallest to bloom and suitable for a sheltered garden. The florets of the huge spikes are attractively waved, the inner petals being rosy-mauve, the outer petals pale blue.

CHARLES LANGDON. A most vigorous variety extremely resistant to mildew. The florets are mid-blue with a striking black centre.

EVENSONG. A superb variety, at its best planted near 'Swan Lake'. The deep violet florets have a jet black eye.

FRANZ LEHAR. An introduction of great beauty, bearing large tapering spikes over a long period. The florets are cobalt-blue with an unusual sepia eye.

GEORGE BISHOP. A glorious variety, the outer petals of the florets being bright mid-blue, the centre petals rosy-mauve with a brown eye.

GLAMOUR GIRL. The spikes are long, thin and graceful. The florets are of gentian-blue with a white eye.

ISOBEL. The dainty, tapering spikes are produced with freedom, the florets being of sky-blue flushed with pink.

LADY ELEANOR. An old favourite, the outer petals of the fully double florets being sky-blue, the inner petals purple.

LAURA FAIBROTHER. With 'Boningale Glory', the tallest. The semi-double florets are of pure mauve with a bright white eye.

ROYAL WINDSOR. The florets are sky-blue flushed with rose; the spikes long and tapering.

W. B. CRANFIELD. It bears a massive but refined spike, the florets being of a distinct reddish-purple colour with a large white eye.

WILD WALES. Quite outstanding, the bloom being forget-me-not blue, shaded with rose-mauve.

Late flowering

GARTER KNIGHT. Possibly the finest of all varieties, the spikes measuring up to 4 ft. (1.2 m) in length. The florets are of cobalt-blue flushed with purple.

JACK TAR. The florets are of similar colouring to 'Evensong' though without the violet shading. It is extremely resistant to mildew.

MERRIE ENGLAND. A fine new variety, the florets being a combination of deep French-grey-blue and rosy-mauve.

MRS FRANK BISHOP. With its long tapering spikes and rich gentian-blue it should be in every collection.

NELL GWYNNE. A fine old variety, the large semi-double blooms being of a lovely shade of rosy-mauve.

PURPLE PRINCE. Should not be omitted from any border. The massive spikes are of a pure purple colour whilst the plant is highly resistant to mildew.

NEW DWARF LARGE-FLOWERING VARIETIES

This new strain has been evolved during recent years for those with small gardens where they may be used as back row plants for a small border. They are also invaluable for planting in front of the tall flowering varieties. All are mid-season flowering and bloom at a height of 3–4 ft. (about 1 m). The flowers are proving long-lasting when cut.

BLUE PEARL. The florets of cornflower-blue have a rosette of the same colour at the centre. The spikes are thin and tapering.

CAPRICE. Of sturdy habit, the sky-blue florets have a striking ivory centre.

CINDERELLA. As compact as 'Janice', the florets are of a new shade of heliotrope with a sepia eye.

DAME FORTUNE. For such a compact plant the spikes are immense, the florets being pale mauve with a conspicuous white eye.

JANICE. A glorious glistening white of rather more compact habit than the others and later to bloom.

KINGFISHER. Most conspicuous, the florets being of a brilliant pure royal-blue colour.

KITTY. Slightly taller growing than the others, the florets are of pure gentian-blue with a tiny jet black eye.

PHYLLIS. The semi-double blooms are of a lovely shade of mauve-pink.

SOMERSET COUNTY. Growing slightly taller than the others, the florets are of a turquoise-blue flushed with pink and have a large white centre.

BELLADONNA HYBRIDS

Like *D. ruysii*, they continue to bear their short spikes of bloom on branched 3 ft. (1 m) stems from mid-June until late September. They are excellent for cutting and for planting with the large-flowered hybrids where their dainty habit is valuable in offsetting the stately formality of the large flowering varieties.

BLUE BEES. An old favourite on account of its earliness and freedom of flowering, from June until September. The colour is a clear sky-blue.

CAPRI. A lovely variety, bearing bloom of a delightful shade of powder-blue.

LAMARTINE. A striking variety, the blooms being single, of a rich purple-violet colour with a striking white eye.

NAPLES. A great favourite, the bloom being of a rich steely-blue, and semi-double.

PERSIMMON. Slightly taller growing than the others, the blooms are of a lovely shade of cobalt-blue.

THEODORA. Described as being electric-blue, the habit is excellent, the plants being long in bloom.

WENDY. Rich gentian-blue, early and free flowering.

RUYSII HYBRIDS

Growing to a height of just under 3 ft. (90 cm) they are the best of all delphiniums for cutting. The branched spikes should be cut when showing a delightful combination of pink and green. The bloom, cut at this stage, lasts well in water. For cutting, plant in beds 18 in. (45 cm) apart and at this distance staking should not be necessary. A new variety 'Rose Beauty', which bears a more rosy-pink bloom than the older 'Pink Sensation' is outstanding. The bloom of both varieties is similar in form to that of the Giant Imperial Larkspur, having a delightful feathery appearance and being ideal for small vases.

For those who would wish to plant a border devoted entirely to the delphinium, this plan would provide colour over as long a period as possible. The border could be edged with *Nepeta mussinii* which bears its mauve-blue spikes at the same time, its grey-green foliage blending admirably with the blues and pinks of the delphiniums.

DESFONTAINEA

This interesting shrub is almost hardy, but requires a sunny, sheltered position and a light, well-drained peaty soil and no pruning. *Desfontainea spinosa* grows 6 ft. (2 m) tall and has holly-like leaves. In August and September, it bears attractive tubular-shaped flowers of red and yellow.

Plant in spring and do no pruning apart from the shortening of any long shoots. Propagate from seed sown in small pots, or from cuttings made from the ends of the shoots and which are inserted in sandy soil under glass.

DEUTZIA

Native of the Himalayas and west China, it is a deciduous shrub with graceful arching stems and oval serrate leaves and is midsummer flowering. *D. shunii* has willow-like leaves and bears its panicles of pink and white flowers along the entire length of the stems. The hybrid *D. x rosea* makes a small, low shrub and bears bell-shaped pink flowers whilst *D. sieboldiana* is also compact and bears white mignonette-scented flowers with orange anthers.

Plant November to March and as they make new growth early, select a position sheltered from cold winds and plant in a soil containing plenty of leaf mould. Little pruning is necessary apart from the removal of dead wood. Propagate from cuttings inserted in a sandy compost under glass.

DIANTHUS

ANNUAL CARNATIONS

No annual is more pleasing as cut bloom about the home than the Chabaud Carnations of Raoul Martin raised in France. The blooms are double, have long stems and carry a delicious fragrance. They commence to bloom late in June and continue until autumn.

It should be treated as a biennial for best results, the seed being sown in a frame, preferably in boxes covered with glass, in July. Transplant the seedlings into the frames in September where they remain over winter and plant out into beds in early April. Or the seed may be sown in gentle heat in February and after the plants have been hardened, they are planted out early in May. The best method is to sow in midsummer for then the plants will come early into bloom the following season. For a sowing compost use one composed of loam, sand and a little peat and a small quantity of lime rubble. Where the plants are to be kept in frames through winter they should be dusted with sulphur once a month to prevent mildew (see plates 19–21).

The plants should be set out 9 in. (22 cm) apart in a position where they may receive full sunlight. They like a soil as light as possible containing plenty of lime and enriched with a small amount of cow manure.

There are at least thirty varieties, but those which are outstanding include 'Legion d'honneur', geranium red; 'Rose Pale', salmon pink; 'Dame Pointe', burgundy red; 'Marie Chabaud', primrose-yellow; 'Mikado, slate-lavender; and 'Etincelente', scarlet.

BORDER

Plants of extreme hardiness, border carnations do not require an abundance of manure and humus in the soil but they must have lime and a well-drained soil. For this reason they are best grown in a raised bed to which some shingle has been added if the soil is of a heavy nature. Borders will grow well in a heavy soil which is well drained, and some of the finest plants are to be seen growing in clay soil on a northerly slope which allows winter moisture to drain away. Should the soil contain a large proportion of clay particles, give a dressing in early winter of caustic (unhydrated) lime. This will heat up upon contact with moisture in the soil and a violent reaction will occur. At the same time the clay particles of the soil will disintegrate. The plants will, in addition, benefit from the lime which is an essential part of their diet. Where the soil is a friable loam of medium texture, give neither manure nor drainage materials. But work in some wood ash, for the potash content will enhance the colour of the bloom.

The ground should be brought to a fine tilth before planting whilst all perennial weeds should be removed. This is essential, for the best method of propagating borders is by layering and it is difficult to layer in weed-infested ground. Also, once established, borders resent disturbance for their roots are not as fibrous as those of many other hardy plants and to remove deeply rooting weeds growing near, may cause the carnations to die back. Border carnations are happiest where following a crop of potatoes which will clean the soil and bring it to a fine tilth. The residual manure will also benefit the plants and all that is necessary before planting will be to give a dressing with lime or to work in a quantity of lime rubble. They require an open, sunny situation.

Plant them early in spring, as soon as the ground is friable after the winter. The plants will then grow away rapidly and will not remain dormant for several months as they will if planted in the autumn. This may not be detrimental where the ground is very well drained, but if an excess of moisture remains about the roots, they will not be sufficiently active during winter to absorb this and will decay. In addition, to allow the soil to become weathered during winter will ensure a fine tilth. Plant during March and April, using pot-grown plants so that the roots will be disturbed as little as possible.

The ideal plant for setting out in spring is one from a 3 in. (7.5 cm) pot; a sturdy bushy plant which has been grown from a layer rooted the previous July. It will have been potted in September and wintered in a cold frame until March. By then, it will have become a hardy, vigorous plant capable of producing a number of choice blooms the first summer. When the ground is in the right condition the plants should be set out very firmly but should not be put in too deeply for this will cause rotting of the stem. They should be planted about 2 ft. (60 cm) apart each way to allow room for layering. During a dry spell of weather, each plant should be watered individually and this will be sufficient for them to become fully established. Carnations are able to stand up to dry conditions very much better than will, say scabious, sweet peas and chrysanthemums. For this reason the dianthus family probably require less attention than any other border flowers.

When growing in beds for garden decoration, plant 15 in. (38 cm) apart, so as to give a massed effect, for borders are at their loveliest when seen in this way.

STAKING AND DISBUDDING BORDER CARNATIONS

During summer, the plants will require little attention except for keeping the ground free of weeds. Regular hoeing will help towards the building up of a strong plant. By the end of April new growth will be observed, and by early May the flowering shoots will have formed. Almost daily the stems seem to lengthen and the buds become more swollen. Unwanted buds which appear clustered round the centre or main bud are removed. This is best done with the fingers and it is surprising how quickly disbudding may be done when once accustomed to the work. Should the main bud be removed by mistake, then allow one of the small buds to remain, for this will produce a flower almost as good. When the bud is taken, the stem should be secured to a bamboo cane by means of a wire ring placed over the top and allowed to rest on a leaf joint. Wire rings are inexpensive and not only may they be used over and over again but they are so much quicker than raffia or string, both of which may cut through the stem during windy weather.

When removing the surplus buds from around the main bud, care must be taken not to confuse this operation with thinning the side shoots which grow from the main stem. Each of these will produce excellent blooms for cutting or for garden decoration, and they should be disbudded in the same way as for the main stem bud. If the blooms are required for exhibition, the weaker shoots should be removed to allow the other to reach maturity by drawing on the food reserves of the plant. Throughout summer, stems will be branching from almost every leaf-joint and from the base of the main stem though possibly half must be removed. This will greatly extend the flowering season. Should it be the aim of the grower to have as much colour in his garden as possible, then little or no serious disbudding need be done.

LAYERING BORDER CARNATIONS

Though borders may be grown from cuttings and by sowing seed, plants from the former tend to weaken in constitution, whilst those from seed cannot be relied upon to come true to colour and form. Layering is the best method of propagation.

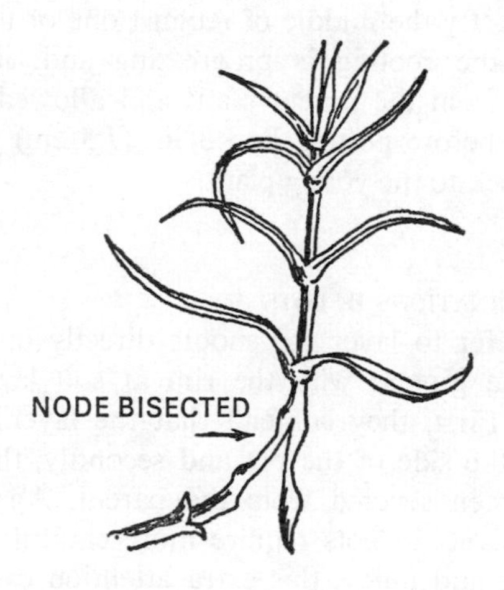

FIG. 4 *Layering a carnation*

This is done in July in the north, but the period may with safety be extended until the end of August in the south where the rooting period lasts until October. Too early layering should be avoided as the plants may make too much growth before the advent of autumn frosts. Layering frequently causes worry yet it is a simple operation.

Most plants will produce a number of shoots which are close to the soil for layering and by taking a sharp knife and cutting up to the joint which is nearest to the soil, many shoots can be layered in a day. The cut should be made at the centre of the stem and is made about 1 in. (2.5 cm) in length the cut must not be taken beyond the joint whilst care must be taken not to sever the cutting from the parent plant. The portion of stem which is not connected to the plant is then bent upwards and firmly pressed into the soil. A wire pin bent in the form of a hairpin is placed round that portion connecting cutting to parent and to keep it pressed into the soil.

Continuing to obtain its nutriment from its parent, the layer will quickly root, how soon depending upon the degree of moisture in the soil. The time taken is about four weeks under normal conditions compared with twice as long for cuttings severed from the parent. The layer will suffer little from either too wet or too dry conditions though, quickness in rooting depends upon a certain amount of help from the grower. Before any attempt at layering is made, the soil must be clear of all weeds and should have had a top dressing of peat and coarse sand in which it is to root. Careful hoeing round each plant before adding the top dressing will do much to retain the moisture content of the soil.

Many place cloches or sheets of glass over their layers. This may be a help during a cold wet period but is not usually necessary. During the rooting period, the only attention necessary is to keep the soil moist at the point of layering. By the middle of August one or two layers should be lifted to find how the rooting is progressing and, if satisfactory, each layer should be cut from the parent plant and allowed to remain in position for a few days before potting into 3 in. (7.5 cm) pots. This does not cause too great a shock to the young plant.

LAYERING BORDER CARNATIONS IN POTS

Some growers prefer to layer the shoots directly into small pots which are inserted into the ground with the rim at soil level. Reason for this method is twofold. First, they contend that the layers root more quickly if pegged down by the side of the pot and secondly, that the young plants receive no check when severed from the parent. Against these points it must be said that plants in pots require more careful watering than those in the open ground, and unless this extra attention can be given, layering should be done in the open ground.

When the rooted layers have been severed from the parent they are moved to small pots containing a mixture of well-rotted turf loam, peat and a little sand. They are then placed in a cold frame which is allowed to remain open during mild weather, or they may be moved to a sheltered place where they are allowed to remain in the open, resting on beds of rubble or ashes. Care must be taken to name the plants. Careful watering will be necessary after once the young plants have been placed in their new pots, and by late March they will be making bushy growth. Watering will require constant attention, and attacks from slugs are a menace. It may happen that a warm spell of weather may cause the soil in the pots to dry out quickly if the plants are not afforded some shade during day-time. Danger time is the night, when a severe ground frost follows hot sunshine and a pot wet with water. Shading is better than giving too much moisture at this time. By November, all the moisture needed by the plants will be obtained from the atmosphere.

Attacks from slugs are often a source of trouble during a wet autumn and winter. Preparations for killing these pests should be sprinkled between the rows of pots; common salt is an excellent deterrent.

If the plants are kept in closed frames during the winter, they should be given as much fresh air as possible as soon as the severe weather has passed.

VARIETIES OF BORDER CARNATIONS

Crimson and Scarlet Selfs

BONFIRE. The blooms, of beautiful form, are of a vivid cardinal-red colour, held on strong erect stems.

BOOKHAM GRAND. It is a plant of extreme vigour, the huge blooms being of rich wine-red which have a perfect form.

DAMASK CLOVE. A superb deep crimson with a powerful clove perfume and with petals like velvet.

FIERY CROSS. A fine scarlet self of perfect shape. A true exhibition variety, having all the good points of a first-class flower.

GIPSY CLOVE. One of the finest of all borders with a vigorous habit and bearing a massive bloom of glowing crimson.

OAKFIELD CLOVE. A huge flower of excellent formation and having the true clove scent. The colour is a bright crimson and the foliage a vivid green.

PERFECT CLOVE. A fine border, the glowing scarlet-crimson blooms being borne with freedom.

W. B. CRANFIELD. A magnificent scarlet self of vigorous habit. Bloom is very free, stem and calyx strong. Excellent both for exhibition and for cutting. An old favourite never surpassed.

Apricot and Orange Selfs

BRENDA BARKER. A grand pure apricot self of vigorous, free-flowering habit.

DALESMAN. A superb variety, bearing large perfectly formed blooms of a rich shade of apricot-orange.

ESMOND LOWE. The border counterpart of the market grower's perpetual carnation, 'Tangerine'. The bloom is small, but of a lovely golden-orange colour and its freedom of flowering is remarkable.

KING LEAR. In the author's humble opinion, one of the most perfect borders ever introduced. The huge blooms are of a rich tangerine colour and freely produced on strong stems.

LOYALIST. This is one of the finest of all, being extremely free-flowering and bearing a neat, medium-sized bloom of brilliant nasturtium orange.

OPHELIA. The blooms are large, of exhibition form and are of a lovely shade of deep apricot.

Pink Selfs

BOOKHAM PEACH. Of perfect exhibition form, the huge blooms are of a lovely shade of malmaison pink.

DAINTY CLOVE. The bloom is large and of perfect form, and is of an attractive shade of salmon-pink.

DAWN CLOVE. It is a strong grower and bears with freedom its perfect blooms of porcelain-pink.

HUNTER'S CLOVE. The colour may be best described as being of hunting pink or glowing rose-pink and of perfect form.

LUCY ASHTON. It possesses all the good points of 'Queen Clove' and bears a large bloom of a lovely shade of camellia-pink.

QUEEN CLOVE. One of the best of borders, of strong vigorous habit and with a rigid stem and perfect calyx. The colour is an exquisite shade of begonia-rose.

QUEEN MAB. An older variety, but the blooms are of such perfect symmetry and of such a lovely shade of old rose that it should still be grown.

White Selfs

AVALANCHE. A grand exhibition white of great purity, the bloom being held on long rigid stems.

BOOKHAM SPICE. Probably the best white ever introduced. The flowers are of perfect formation and possess great substance.

SNOW CLOVE. A pure white of exhibition form and having a powerful clove perfume.

SNOWY OWL. The blooms are large, of dazzling white and held on long, erect stems.

White Ground Fancy

BEVERLEY. The flowers are large and of excellent form for exhibition, the pure white blooms being edged and striped with garnet-red.

BOOKHAM DANDY. A grand border, the pure white blooms being marked with wide stripes of crimson-red.

BOOKHAM DREAM. A giant both in its constitution and in the size of its bloom, which is of pure white, heavily marked with scarlet.

BOOKHAM LAD. An outstanding variety, the large white blooms being heavily striped with scarlet to give a most striking effect.

DOROTHY ROBINSON. A fine exhibitor's variety, the perfectly shaped blooms being heavily splashed with rosy-red.

DUSKY MAID. A superb border, bearing a bloom of great substance, the white ground being heavily marked with deepest crimson, giving it quite a dusky appearance.

FANCY FREE. This has achieved fame on the show bench for the exquisite smoothness of its petals which are white, edged and marked with rose.

ISOBEL KENNEDY. The blooms are large and of perfect form and are heavily marked with crimson-red.

ROBERT SMITH. A new variety of vigorous habit, the glistening white blooms being lightly marked with rose-pink.

Yellow Selfs

BEAUTY OF CAMBRIDGE. An old favourite which should be grown for its smoothness of petal, perfect symmetry and glorious pure sulphur-yellow colouring.

BOOKHAM QUEEN. Outstanding in its section, the clear lemon-yellow blooms being held on rigid stems.

BOOKHAM SUN. An exhibitor's plant, the buttercup-yellow blooms radiating intense brightness and being of perfect form.

DAFFODIL. Of vigorous habit, its perfectly formed flowers are of bright canary-yellow.

KING CUP. The blooms are large, of perfect form and are of a deep clear mid-yellow colour.

Yellow Ground Fancies

ARGOSY. An old favourite, the clear yellow blooms being edged and marked with scarlet.

BOOKHAM FANCY. A fine novelty, the ground colour being clear, bright yellow, edged and ticked with purple.

BOOKHAM PRINCE. Of unusual colouring, the ground being of deep amber, heavily edged and marked with crimson.

CELIA. A beauty, the ground colour of moonlight yellow being marked with bright rose-pink.

GOSHAWK. A grand exhibitor's variety, the large blooms, borne on stiff stems, being of bright yellow, edged and barred with violet.

KATHLEEN BROOKS. Of perfect form and extremely free-flowering the ground colour is canary yellow, edged and striped with scarlet.

KESTREL. A striking variety, the vivid yellow ground colour being edged and striped with cyclamen-purple.

LIEUTENANT DOUGLAS. One of the best for the show bench, the blooms being large and of deep yellow marked with blood red.

THE MACINTOSH. An unusual variety, the ground colour is straw yellow, each petal being covered with spots and stripes of scarlet.

Purple, Grey and Lilac Selfs

ALBATROSS. A grey self of exquisite form, the petals having a velvet-like appearance and being overlaid with a silver sheen.

ANTIQUARY. Of sturdy, free-flowering habit it bears a large perfectly formed bloom of an unusual shade of lilac.

BODACH GLAS. It bears a large beautifully formed bloom of a lovely shade of lavender-grey.

CLARABELLE. Probably the best grey, having broad velvety petals, its blooms being held on long, sturdy stems.

GALA CLOVE. It is powerfully scented and bears a huge, perfectly formed bloom of bright purple.

GREYLAG. It makes a dwarf, compact plant and is striking planted with 'Highlander', with its blooms of pink and heliotrope. The dark grey blooms, held on sturdy stems, have an attractive silvery sheen.

LORD GRAY. A beauty for cutting and mixing with pinks, its medium-sized blooms being of a glittering heliotrope-grey colour.

Fancies

AFTON WATER. A magnificent border of bushy habit and possessing a sturdy stem. The blooms are of a bright rose, splashed with madder, producing a brilliant effect.

BONNIE LESLEY. It bears exhibition blooms of great beauty, being of a deep apricot colour, heavily suffused with brilliant scarlet.

BOOKHAM HEROINE. This is one of the best borders in cultivation bearing a profusion of large blooms of exhibition form. The base of the petals is shrimp-pink shading out to deep cherry-red.

CLANSMAN. A fine variety of robust constitution, the orange-scarlet blooms being striped with carmine-red.

DOUGLAS FANCY. The most striking of all the fancies. The habit is robust, the blooms of excellent formation and the colour a vivid rose-orange, edged rich crimson-red.

EBOR. An old favourite unsurpassed for colour. The ground is chocolate, striped vivid red and deep maroon, with the bloom perfect in form.

FANCY MONARCH. A brilliant introduction of huge size and perfect habit. The ground is dusky peach, speckled and edged with the nearest shade of blue possible.

HIGHLANDER. It bears an exquisite bloom, the buff ground being heavily marked with rose-pink.

LESLIE RENNISON. Raised by the late Mr R. Thain this is an excellent bedding variety. The flowers are small but of perfect shape; the colours, purple, overlaid pink.

PERPETUAL FLOWERING

Perpetual flowering carnations are the aristocrats of the dianthus family, and demand the best of conditions in which to produce their blooms. They must be kept from frost whilst they need the maximum amount of light and sunshine. They must have fresh air but no draughts; they must be kept warm in winter yet cool in summer; they need light and will not grow well in a smoke-ridden atmosphere. In other words, they must have their demands satisfied down to the last detail or they will not be worth the expense and trouble of growing them. This should not be taken to mean that they are 'hot house' flowers or difficult to grow. They will grow well in a temperature of 45°F. (7°C.) or even lower, and provided they receive the correct treatment of cultivation they will suffer little from pests and disease.

PROPAGATION OF PERPETUAL FLOWERING CARNATIONS

Almost the sole method of increasing stock is by means of cuttings which are freely produced from the stems. The most vigorous cuttings are those taken from halfway up the stem of a two- to three-year-old plant and they are neither too 'woody' nor too succulent for quick rooting. As with border cuttings, where layering is not carried out, it is a more certain method to procure a heel or joint to each cutting.

Layering is rarely resorted to for perpetuals grow in an upwards direction rather than in low clumps like borders. When taken, the cuttings are carefully trimmed by removing the lower pairs of leaves so that the element of 'damping off' will be reduced to a minimum. It is almost essential to use a propagating house for rooting the cuttings though certain growers situated in the more favourable positions in the south take their cuttings during mid-August and root them in a cold frame. By the end of September, they should be rooted and are moved to 3 in. (7.5 cm) pots and taken indoors into a temperature which is kept at a minimum of 45°F. (7°C.) during the winter months. The rooting medium should be of coarse sand and granulated peat, pressed firmly down and made moist before the cuttings are inserted; or sterilised soil may be used as for rooting chrysanthemums. The beginner will find that it is more difficult to root the cuttings in this way than it is to root borders by layering, for skill in controlling the moisture requirements is necessary. Seed pans or boxes may also be used for rooting but here again moisture control will be more difficult, for a hot sun will rapidly dry out the compost. The inclusion of the peat not only helps the cuttings to root but also prevents the sand from becoming baked in warm weather. Press the cutting very firmly into the compost and the distance for planting should be about 3 in. (7.5 cm) each way. Where a heated house is being used, the cuttings are generally taken during January and February when there is little fear of damage from too strong sunlight whilst rooting. Cuttings rooted in frames in August and September must at times be given protection from the sun. Only the strongest cuttings should be chosen, those between 5 to 6 in. (12–15 cm) in length and of sturdy appearance. They should not be set too deeply into the compost.

After about six weeks, the cuttings will have rooted under suitable conditions, and they will then be ready to move into 3 in. (7.5 cm) pots into a prepared compost made up of two-thirds turf loam, one-third peat and a little coarse sand. Care must be taken here and also throughout the life of the plant to see that the stem is not planted too deeply when it is moved. By early April the plants will be ready for transferring to large pots and during the following two or three weeks should be allowed as much fresh air as possible, in order that they may be ready for outdoor frames early in May. From April 1st until the end of the month it may only be necessary to put on the heating at night. No member of the dianthus family will tolerate excessive heat, though sunshine they must have, and it is this natural heat that should be encouraged. The sun will build up a sturdy plant which by careful hardening will be ready for the frames by early May. As the perpetuals are sun lovers, the glass of the greenhouse should be washed down each month and especially of those houses producing cut bloom. Perpetuals demand the maximum amount of light and sunshine and for that reason will never be at their glorious best in a smoky atmosphere or in the industrial regions of the north-east and midlands.

During summer, care must be taken to see that the plants are kept slowly growing by regulated watering. This means that the soil in the pots should be neither too dry nor too moist. Only experience will determine the correct moisture requirements.

STOPPING PERPETUAL FLOWERING CARNATIONS

Unlike borders or pinks, perpetuals must be 'stopped' to encourage a more sturdy bushy plant. Two 'stoppings' should be given, the first whilst the plants are in their first pots just before they are moved again; and a second time at the end of May when they have been thoroughly hardened off. All fear of frost damage will then have passed and the plants may be given only slight protection from cold winds. They will need a certain amount of feeding with weak manure water and with carnation fertiliser.

'Stopping' is done by removing with finger and thumb about 1 in. (2.5 cm) from the tip of the main stem. A full 1 in. (2.5 cm) must be removed, otherwise the shoot will grow again.

A final potting or planting into beds where they are to flower must be done during August and this is the most important stage in the life of the perpetual carnation for it will remain undisturbed for at least three years. Careful preparation of the compost and shallow planting is once again the secret of success.

MAKING UP THE BEDS FOR PERPETUAL FLOWERING CARNATIONS

Though in most cases the amateur will grow perpetuals in large pots, the cost of these will be prohibitive when grown on a large scale and they should be planted directly into floor beds. The plants should be placed about 1 ft. (30 cm) apart each way and not have the stem buried too deeply. The advantage of open beds is to provide unrestricted growth and as the plants will remain in their quarters for several years, this method calls for a saving in re-potting time.

To each plant, a 2 ft. (60 cm) cane should be placed in position and the main stem fastened to it by a wire ring in the manner of the borders but as the plants grow taller, they are held in place by wires stretched over the beds. This cuts out the necessity of tying and makes it easier to cut the bloom.

Feeding calls for attention to detail for too strong applications of farmyard manure or of artificials will make plant growth come weakly. Use a slow-acting fertiliser containing the correct proportions of phosphates and potash.

During the life of the plant, the bed surface must be raked over each month to aerate the soil and help to control moisture content.

VARIETIES OF PERPETUAL FLOWERING CARNATIONS

ALLWOOD'S CREAM. It is of a rich Jersey cream colour and possesses delicate scent. It bears a large bloom and is free flowering.

ALLWOOD'S MARKET PINK. A deep pink which lasts longer than any variety when in water.

ASHINGTON PINK. It is deep pink, easy to grow and bears a bloom of excellent shape.

CHERRY RIPE. A lovely cherry-red, it is considered to be one of the most free flowering of all perpetuals. A top class florists' variety.

CLAYTON CRIMSON. The brightest and most popular crimson, particularly attractive when under artificial lighting.

COUNTRY MAID. The plant is of vigorous constitution and the white medium-sized blooms are pencilled with bright red.

DAIRYMAID. An old favourite, the perfectly shaped medium-sized blooms being white, flaked deep pink.

DORIS ALLWOOD. The blooms are of a delicate salmon-pink, shaded and pencilled misty-grey whilst it carries a strong perfume.

GEORGE ALLWOOD. The finest white ever raised and one possessing a glorious clove perfume. The huge flowers are pure glistening white.

MARCHIONESS OF HEADFORD. A variety of unique colouring being of deep salmon with a cream edge.

MOORLAND'S PINK. A 'sport' from 'Ashington Pink', bearing bloom of delicate salmon-pink and of excellent shape.

No. 1 PINK. A rich salmon-pink grown by all commercial growers as their greatest standby whatever the season.

ROBERT ALLWOOD. Possibly the finest of all carnations, perfect for either exhibition or commerce. The bright scarlet bloom great petal count and is large in circumference.

ROYAL BUTTERFLY. The large bloom is of pale cream, overlaid deep claret. The habit of both plant and bloom are ideal for exhibition.

ROYAL CERISE. A superb grower of perfect shape and of a delightful shade of rich cherry-red.

ROYAL CRIMSON. The finest crimson, bearing a large bloom of perfect form and with a powerful perfume.

ROYAL FANCY. It bears the largest bloom of any carnation, the colour being of richest bronze, streaked with red.

SNOWDRIFT. A grand white for the exhibition or for home decoration as it does so well in pots and blooms continuously.

SPECTRUM SUPREME. A brilliant scarlet of great freedom of flowering.

TANGERINE. The bloom is not large but is of brightest tangerine, produced with freedom and is easily grown.

WYATT ALLWOOD. Deep yellow it makes an ideal pot plant. The blooms are large and of excellent shape.

MALMAISON CARNATIONS

This section is similar to the more popular perpetual flowering but though the subject of constant attention from the hybridist it has never reached the popularity of the perpetuals. When introduced in the 1850s, the plants grew leggy and produced a pale pink bloom but today it may be obtained in a wide range of colours. It should be propagated by layering where possible, but the taller shoots may be removed and inserted in sand and peat as for perpetuals. The cultural treatment is the same as for the perpetuals, though they will not tolerate as much heat. They are ideal for a room in the home where the temperature does not exceed 45°F. (7°C.) when they will bloom over a long period.

PESTS

RED SPIDER. These minute orange-coloured pests cause much trouble amongst carnations if allowed to get a hold. A too dry atmosphere will cause trouble. The mites attack the leaves which take on a shrivelled appearance. The pest may be eradicated by fumigating with azobazene at the rate of $\frac{1}{2}$ lb. to 1,000 cu. ft. of space. This will only be effective in a damp atmosphere, and is carried out as for nicotine fumigation. It must not be allowed to come into contact with either sweet peas or schizanthus.

THRIPS. These frequently give trouble by sucking the sap from leaves. They attack most glasshouse plants. Lindanc fumigations will deal with the pest, used in a similar manner as for nicotine shreds.

DISEASES

STEM ROT. Is caused by spores in the soil which cause the stem to become black and the plant to die. Too deep planting and over-watering can cause the trouble.

LEAF SPOT. This is a fungus which attacks the leaves causing ugly looking black spots which makes the plants unsaleable. The cause is generally too humid conditions but a cure may be obtained by spraying with a proprietary sulphur compound.

PINKS

To edge a border, especially where the soil is of a calcareous nature, there are no more valuable plant than the pink with its erect silvery-green foliage and extended flowering season. The plants may also be set out in beds to themselves; indeed, all members of the dianthus family are happier on their own, the exception being where used to edge a border. Here the plants will receive a greater amount of sunlight than when planted amongst front row border plants, and this they must have in abundance. For this reason pinks bloom more profusely when grown in beds to themselves or planted between paving stones.

Pinks prefer a dry, well-drained soil which is the opposite to the requirements of most border plants. Humus is not necessary, and where the soil is retentive of moisture, especially during winter or in areas of heavy rainfall, pinks rarely do well, tending to die back.

The plants can withstand severe frost; cold will not harm them in any way, as long as the soil is well-drained. A light, sandy soil suits them best and where a natural lime content is not present, incorporate plenty of lime rubble. Like the scabious and iris, the dianthus should be given a dressing with lime rubble or hydrated lime every year.

Where used for edging, the pink has an additional value in that its neat attractive foliage is evergreen, adding a touch of colour to the border at a time when there are few other flowers. Where used for bedding, too, the plants retain their compact habit and their foliage all the year round. But whereas the old garden pinks have a short season of flowering, at the end of which the blooms turn an unsightly brown colour, the *allwoodii* bloom throughout summer, the colour range being enormous. The refined blooms are borne on 12–15 in. (about 35 cm) stems without splitting their calyx. Where there is a grass pathway to a border, the dianthus with its compact upright habit is an ideal plant, for it does not foul the edges of the grass as does nepeta, whilst for all except the very smallest of borders, the pinks are more suitable than pansies or violas on account of their compact habit.

Early spring is the most suitable planting time, though near the dry east coast and where the soil is light and sandy, October planting is permissible. They should not be planted too deeply but plant firmly, spacing them 15 in. (38 cm) apart, whether used as an edging or where planting in beds.

Pinks are propagated by several methods – from slips, which are new shoots removed from the main stem with a 'heel' of the old wood; from pipings, which are new shoots removed at the socket; and by cuttings, the shoot being severed at a leaf joint when about 3–4 in. (7.5–10 cm) in length. If inserted in July in a compost of peat and sand which is kept moist, they will have rooted by early September, when they should be transferred to individual 2½ in. (6.5 cm) pots or to a frame containing a sandy compost where they remain through winter for planting out in April (see plates 22–25).

Only plants of the Perpetual Pinks require 'stopping', that is, they should have the growing point removed prior to planting out. To maintain the flowering period of modern hybrid pinks, the blooms should be removed as soon as the petals begin to fade. In this way the plants will begin to flower in June and will continue until the end of summer.

VARIETIES

ANNE. The blooms are held above the foliage on 12 in. (30 cm) stems and are double and of a lovely shade of salmon-pink.

AVOCA PURPLE. It is to be found in many Co. Wicklow cottage gardens. It bears a small purple flower, streaked with lines of darker purple and it is sweetly scented.

BLACK PRINCE. An old Irish variety now rarely seen and somewhat resembling Sops-in-Wine, its semi-double flowers being white with a large black centre or eye and with similar nutmeg scent.

BLANCHE. Very free flowering, the double blooms are of purest white with deeply fringed petals.

CHARITY. The habit is short and tufted whilst the plant is free flowering. The semi-double blooms have a white ground with clearly defined lacing of bright crimson.

DORIS. The double blooms of salmon-pink with a darker eye are sweetly scented and are of ideal exhibition form.

DUSKY. The result of back crossing the old Fringed pink with an *allwoodii* seedling and it is free and perpetual flowering. The blooms have fringed petals and are of a lovely shade of dusky-pink.

EARL OF ESSEX. One of those much loved of garden pinks which always splits its calyx but is always welcomed in the garden. The clear rose-pink blooms with their fringed petals have a small dark zone and sweet perfume (see plate 26).

ELIZABETH. A real treasure for the alpine garden, with erect silvery-green foliage and bearing, on a 4 in. (10 cm) stem a tiny double bloom of dusky pink with a brown centre. The scent diffused by so tiny a flower is truly remarkable.

EMILE PARE. One of the truly outstanding pinks, raised in 1840 in Orléans, France, by André Paré, probably having the Sweet William for one parent for it bears its double salmon-pink flowers in clusters and will survive only a few years so that it should be propagated annually.

ENID ANDERSON. A most striking pink, its semi-double clove-scented flowers of glowing crimson being enhanced by the silver-grey leaves.

ERNEST BALLARD. Raised at the Old Court Nurseries, Malvern, it forms a dense prostrate mat of grey foliage above which are borne on stiff stems, double flowers of crimson-red which have the old clove perfume.

FAITH. The first of the laced *allwoodii* (1946). The blooms are small but fully double with the petals broad and fringed. The ground colour is rosy-mauve with lacing of cardinal-red.

FRECKLES. An imperial pink, it bears a double bloom of unusual colouring, being dull salmon-pink flecked with red and with a penetrating spicy scent.

GRACE MATHER. A real gem for the trough garden forming a prostrate mat of neat blue-green foliage and bearing, on 5 in. (12.5 cm) stems, large double blooms of deep salmon-pink.

GUSFORD. An outstanding pink, bearing large double blooms of rosy-pink on 12 in. (30 cm) stems and which are deliciously scented.

HASLEMERE. Raised at the Ipswich nurseries of Thompson and Morgan, the large fragrant double flowers have a deep chocolate centre and fringed petals.

ICE QUEEN. A 'sport' from 'Dusky', it bears a highly scented double bloom of icy white which does not burst its calyx and which has fringed petals.

JOHN BALL. Raised and introduced by Turner of Slough about 1880. The bloom is large and double with a white ground and lacing and zoning of velvet-purple.

LILAC TIME. Raised by Mr C. H. Fielder of the Lindabruce Nurseries, Lancing, it is an imperial pink and bears fully double blooms of a lovely shade of lilac-pink with a powerful scent.

LINCOLNSHIRE LASS. It has been known since the beginning of the century and may be much older. The flowers are of an uninteresting flesh colour but the delicious scent makes it worthy of cultivation.

LONDON SUPERB. The large double blooms have a pale pink ground and are laced with purple. The fringed petals and perfume give it an old world charm.

MARS. One of the 'great' plants of the alpine garden, forming a dense tuft of silvery-grey above which it bears bright crimson-red flowers with the true clove perfume.

MARY. The first *allwoodii,* bearing double blooms of lavender-pink, zoned with maroon and with a delicate sweet perfume.

MISS CORY. Raised in Holland, it bears large double blooms of richest wine-red with the true clove perfume.

MISS SINKINS. A tiny replica of the more robust 'Mrs Sinkins' and found in a garden at Henfield, Sussex. The heavily fringed blooms of purest white have a delicious perfume.

MRS SINKINS. It was raised by a Mr Sinkins, Master of Slough Workhouse and named after his wife. The plant, which has the distinction of being incorporated in the Arms of the Borough of Slough, was introduced by the Slough nurseryman, Charles Turner, one of the great florists of the time and who had introduced the Cox's Orange Pippin apple to commerce. It is a pink of great character, its large white cabbage-like blooms, borne on 12 in. (30 cm) stems above a mat of silvery-green foliage possessing an almost overpowering perfume.

MONTY. The blooms are as large as an old five-shilling piece and are of solferino-purple, zoned with chocolate.

NAPOLEON III. Raised by André Paré in Orleans in 1840 from a Sweet William crossing and like 'Emile Paré' will flower itself to death in two years. It bears on 10 in. (25 cm) stems, large heads of double clove-scented flowers of a striking scarlet-cerise colour.

NYEWOOD'S CREAM. A delightful plant forming a hummock of grey-green above which it bears scented flowers of Jersey cream.

PADDINGTON. It was raised about 1820 by Thomas Hogg, a nurseryman of that part of London on which now stands Paddington Station. Of dwarf habit its double pink blooms have serrated edges and are richly scented.

ROSE DE MAI. It may be traced back to the beginning of the century and it is a beauty, the double blooms being of a lovely shade of creamy-mauve with fringed petals and glorious perfume.

RUTH FISCHER. Dating from the end of the last century, it is a most attractive variety of compact habit and bears small fully double flowers of purest white with a rich, sweet perfume.

SAM BARLOW. At one time it was to be found in every cottage garden though is now rarely seen. Like 'Mrs Sinkins' and so many of the old double pinks it splits its calyx but blooms in profusion, its white flowers having a maroon blotch at the centre and with penetrating clove perfume.

SHOW ARISTOCRAT. The plant has beautiful silver foliage and bears sweetly scented double blooms of flesh-pink with a deeper pink eye.

SHOW CLOVE. The rose-pink bloom is smaller than usual in this group but it carries the true clove perfume.

SHOW EXQUISITE. The bloom is almost as large as a Malmaison carnation and of a lovely shade of soft pink with a sweet perfume and with spicy undertones.

VICTORIAN. A laced pink of early Victorian times bearing huge blooms which often burst their calyx but is a most attractive variety. The white ground is zoned and laced with chocolate.

WHITE LADIES. To the grower of cut flowers, it is with Scabious 'Clive Greaves', the most profitable of all hardy plants, bearing its sweetly scented blooms of purest white throughout summer and they do not split their calyces. A variety of *D. plumarius* for it has the same fringed petals, it is a plant of neat habit and is tolerant of all conditions.

WILLIAM. One of the best of all white pinks with broad petals and outstanding clove perfume.

WILLIAM BROWNHILL. It dates from about 1780 and is one of the best of the laced pinks, the beautifully formed blooms being white, laced and zoned with maroon and they do not burst their calyces.

SWEET WILLIAM

Dianthus barbatus is treated either as a biennial or a perennial. It will however bloom early in summer the season after it is sown in May. Sow in shallow drills in a partially shaded position or in a frame during June after the summer bedding plants have been hardened and planted out. Sow broadcast or in drills keeping the seed at all times moist. Sow very thinly for this is a plant which will soon become straggly and weak if not given plenty of room to develop. Transplant to a sunny bed when the plants are large enough to handle, preferably where they are to bloom.

Besides the well-known Pink and Scarlet Beauty strains, the Giant Auricula-eyed and Sutton's Harlequin strains, are very attractive, ideal for bedding or for mixed borders. There is also a new dwarf strain, growing to a height of only 8 in. (20 cm), obtainable in pink and crimson and which is ideal for bedding.

There is also an annual form, Dwarf Red Monarch growing 9 in. (23 cm) tall and which will bloom in July from a sowing made under glass early in the year.

DIASCIA

An annual and like all South African plants, *Diascia barberae*, 'Salmon Queen', with its dainty flowers almost like short bells and in bloom during midsummer is ideal for bedding if given a dry soil and a sunny position. With its compact habit it also makes a charming greenhouse plant. The seed is sown early in March, the plants being grown on in small individual pots before planting out at the end of May.

DICENTRA

Known as the 'Bleeding Heart' from its heart-shaped flowers and though introduced only one hundred years ago, it quickly became a favourite in cottage gardens. It prefers a rich, moist, but well-drained soil and a position of partial shade. The plants come early into bloom and if it is to be grown in the shrubbery or border it should be planted close to those plants like the evergreen iris or the pyrethrum which makes plenty of early summer growth to provide it with protection from cold winds. It comes into bloom towards the end of May, for which reason it is invaluable in the border, and will continue to bloom until mid-July. Plant early in spring, propagation being by root division at the same time. If planted in a dry soil, frequent mulchings will be necessary.

SPECIES

D. eximia. Its nodding rose-pink flowers are borne along the 12 in. (30 cm) stems amidst attractive fern-like foliage. There is also a lovely pure white form Alba, now rare.

D. formosa Bountiful. This American hybrid bearing deep rosy-red flowers on 15 in. (38 cm) stems, has also the same attractive fern-like foliage as *D. eximia*. It does, however, remain in bloom until the end of August.

D. spectabilis. Its heart-shaped flowers are freely borne on slender 2 ft. (60 cm) stems and are of deep pink.

DICTAMNUS

It was named *D. fraxinella* because its leaves resemble those of the ash tree (*Fraxinus*). It is an interesting plant, hardy though a native of southern Europe. It is probably the bush that 'burned with fire' mentioned in Exodus. Like all members of the family, the flowers emit an orange-like fragrance and also exhale from rusty-coloured glands, an inflammable vapour which is particularly pronounced at night. As with the buds of the Balsam Poplar, it is possible to set alight the vapour if a match is held immediately above the plant on a calm evening, without causing it harm. When the leaves are pressed, they emit the refreshing aromatic perfume of balsam. Indeed, the whole plant is resinous. Tournefort wrote, 'the flowers and stalks are aromatic, balsamic and sweet', resembling the leaves and stems of rue when bruised.

D. fraxinella is a plant of beauty, forming a dense bush 2 ft. (60 cm) tall and bearing, in mid-summer, racemes of pale purple flowers, veined with deeper purple. There is also a pure white form, 'Alba'.

The plant does not take kindly to root division and is best propagated from seed, sown when ripe towards the end of summer. It requires a light soil to be long living.

DIEFFENBACHIA

Known as the 'Dumb Cane', for the juice is poisonous, and should the leaves be chewed, the tongue will swell and one may be deprived of speech for several days. For all that, this is a valuable house plant, being quick growing and soon reaching a height of 5–6 ft. (about 1.5 m) when it should be air-layered.

D. sequina is a native of South America, its oval leaves, which taper to a point, being 12 in. (30 cm) long and of bottle-green, beautifully marked with white. *D. bausei* has olive-green leaves, blotched with darker green and white, whilst *D. picta* is of more compact habit, its bottle-green leaves being splashed with white in much the same way as *C. sequina*.

E. J. Lowe in his *Beautiful Leaved Plants*, in which the illustrations are hand-painted, suggests a compost made up of 2 parts fibrous loam, 1 part leaf mould, 1 part peat and 1 part coarse sand, and the author has found this most suitable. Lowe also made special mention that no water should come into contact with the soft, fleshy stems during winter or they might collapse; this advice is well worth taking. During summer, however, the plants require plenty of moisture, whilst they should be given a position of partial shade. They require a winter temperature of not less than 50°F. (10°C.).

DIERAMA

For the back of a border *D. pulcherrimum* is a most delightful plant from South Africa. For eight weeks, from mid-July until mid-September it bears its arching sprays of drooping bells which are of a rich claret-purple colouring. Its tall montbretia-like foliage also gives added charm to the plant, whilst its spray of bloom waving in the late summer breezes have given it the name of the 'Wand Flower'. The bulbs resent disturbance and so should be left down ten years or more, during which time the plants will remain vigorous if given a yearly mulch (in June) with decayed manure. They like a well-drained, friable soil and should be planted in April, on a layer of sand and peat. This is an excellent plant to grow with the taller flowering irises. The plants may not prove completely hardy in a heavy soil.

VARIETIES

ALBA. The spikes or wands are of pure ivory-white and most attractive.

DWARF PINK HYBRID. This is quite a new break and like all hybrids is extremely vigorous and free flowering. The deep carmine-pink spikes are at their best during June and July, and are borne on 2 ft. (60 cm) stems above the evergreen foliage. A fine plant for mid-border planting.

NIGHTINGALE. A new introduction, the wands being of a lovely shade of fuchsia-pink.

RINGDOVE. The long pendulous spikes are of a delicate shade of shell-pink.

WINDHOVER. The spikes are thicker than most varieties, and are of a lovely lilac-pink shade.

DIERVILLA

The Weigela or Bush Honeysuckle is a hardy, deciduous shrub closely related to the honeysuckles. They have pointed oval leaves and bear long tubular flowers throughout summer. The plants grow well in sun or partial shade and in a well-drained soil, but they do appreciate some manure. They grow 3–4 ft. (1 m) tall, one of the best being 'Newport Red', which covers itself in clusters of ruby-red trumpets. For contrast, plant near it 'Alba', which bears snow-white flowers. 'Bouquet Rose' is also lovely, and well named.

Plant November to March. Cut back any long shoots after flowering and thin out where there is overcrowding. Propagate by cuttings of the half-ripe wood, removed in July and inserted in sandy soil under glass.

DIGITALIS

This hardy plant, common to Britain has, from earliest times, been revered both for its properties and because of the shape of the long finger-like flowers, those of a deep purple colour being reminiscent of priests' vestments. Whilst the plant is still used by the medical profession for its drug, digitalin, it is only occasionally to be found in the pleasure garden. The biennial hybrids raised by Rev. Wilks are of great beauty, but even more arresting is *D. mertonensis* which is fully perennial. This is an excellent plant for a shady position and is of more compact habit than the biennial form, bearing its rich strawberry-pink flowers on 3 ft. (1 m) stems. Equally lovely is *D. lutea* of similar habit and bearing bright yellow flowers; whilst for the front of the border *D. orientalis* bears small, dainty spikes of pure white on 18 in. (45 cm) stems.

Seed is sown in April in boxes or pans, the seedlings being transplanted to boxes of compost or to open ground beds when large enough to handle. The plants may be transferred to the border in November. These plants are readily raised from seed, and are difficult to obtain in any other way. Plant 12 in. (30 cm) apart and all are tolerant of some shade.

DIMORPHOTHECA.

An annual and though a native of South Africa, the Star of the Veldt is almost completely hardy and may be sown outside mid-April where it is to bloom or a month earlier in gentle heat. The plant grows to a height of 12 in. (30 cm) and blooms throughout late summer, *D. aurantiaca* bearing flowers in shades of orange and yellow. It needs a sunny situation and a light, sandy soil.

DISEASES (FLOWERS)

BASAL ROT

It frequently attacks narcissi bulbs and freesia corms as it does the gladiolus in a slightly different form, causing yellowing of the foliage and finally total decay. It begins at the base of bulb or corm and when the dry outer scales are removed, blackish-brown colouring may be seen around the base. Upon removal of the outer scales, pink spores may be seen whilst upon gentle pressure, the bulb (or corm) appears soft.

Fusarium bulbigenum is a soil fungus which may enter the bulb through root damage or through abrasions made in the bulb upon lifting and it may be undetected during storage, the yellowing of the leaves and unopened flower buds after re-planting being the first indication. Imported bulbs should be carefully inspected and stored in a dry frost free room but in a temperature not exceeding 60° F. (16°C.). Narcissus 'King Alfred' and 'Sir Watkin' and their progeny are rarely troubled by the disease whilst dipping bulbs and corms in a 0.5 per cent solution of formalin will give effective control.

BLACK DISEASE

It attacks lily-of-the-valley and is prevalent in Germany and Denmark and though unknown in Britain, it may occur on roots sent over from Europe for forcing. The covering scales of the buds become black and later, the buds fall off whilst the blackening extends to all parts of the plant, first as small grey spots on the leaves (later turning black) and also on the roots and underground stems. A careful watch should be kept on imported roots when being forced, so that where the disease is noticed, the plants may be removed and destroyed. After forcing, it is advisable to burn the plants so as not to introduce the disease to outdoor plants. Where older plants are lifted and divided outdoors, they should be replanted into freshly prepared ground.

BLACK MILDEW

It appears on roses growing under glass in the form of reddish-brown patches on the upper surface of the leaves, later as grey spots on the underside. It is often caused by condensation and may usually be prevented by correct ventilation. To control, spray with Fungex at the rate of 1 pint to 50 gallons of water and making up a fresh solution when required.

BLACK SPOT

It is the most troublesome and destructive of all rose diseases, appearing on the leaves first as small black spots, later spreading out to cover most of the leaves and also attacking the stems. The leaves turn yellow and fall off, giving the plants a bare appearance and without its foliage the plant will be unable to function and will soon die. As the fungus spores remain on the stems of prunings during winter, it is essential to remove all unwanted material as it is removed from the plants to be burnt.

Wintering on fallen leaves, the spores enter the plant tissues in spring so it is necessary to begin spraying from early May. Tulisan or Captan 50, applied every two to three weeks throughout summer will provide a protective coating to the stems and leaves. Or dust with Orthocide, containing 10 per cent Captan but it must not be used where a copper fungicide may have been used.

Weak Bordeaux Mixture containing 0.25 per cent copper sulphate will give satisfactory control but it must be used every ten days, a laborious task where spraying large numbers of plants.

It is found that rose leaves must be subjected to wet for at least six hours for black spot spores to germinate and where growing under glass, provide ample ventilation after spraying the plants.

BOTRYIS

B. gladiorum is one of several forms of botrytis which attacks many bulbous plants but is mostly troublesome on gladioli growing in areas of heavy rainfall and high humidity. As it attacks both foliage and corm, it may prove to be troublesome where the corms are planted in a badly drained soil or if lifting is too long delayed when flowering has ended. Its presence is observed by small greyish-brown spots which appear on the foliage of gladioli and also on the flowers, making them unsuitable for sale or for exhibition. Red flowering varieties may be troubled more than others and as spores are released in the wind or washed off by rain, the infected bloom and foliage must be removed and destroyed as soon as observed. Too close planting will encourage the disease and its spreading.

The disease may take the form of dry brown rot which penetrates down the neck to the centre of the corm, causing losses during storage, or the corms may decay from the base upwards whilst small black swellings may appear on the outside of the corms. Dusting the corms with flowers of sulphur or Botrilex when placing in storage will help to control the trouble. Commercial growers have found that dipping the corms for two hours in a 0.1 per cent solution of Mercuric chloride solution has prevented corm damage from the disease whilst keeping the corms after lifting and cleaning in a temperature of 84°F. (29°C.) for ten days to bring about the rapid healing of wounds caused in lifting has ensured almost trouble-free stock.

Corms infected by *Botrytis gladiorum* may fail to grow entirely or young growth will turn yellow, then brown and will die back at an early stage.

Botrytis known as Grey Mould disease which attacks *Anemone coronaria* and other species of anemone whereby the stems decay at ground level, may be due to careless removal of the bloom during periods of humidity, thus allowing the fungus to gain entry. Too close planting where growing under cloches or in frames and insufficient ventilation may encourage an outbreak and where growing under glass, the routine dusting of the foliage with the non-poisonous salicylanilide emulsified with Agral or in its proprietary form Shirlan AG, will prevent any serious outbreak.

The Grey Mould disease of the galanthus, *Botrytis galanthina* occasionally appears following alternating periods of frost and warmer weather. As soon as the shoots appear above ground they become covered with a grey mould which extends down to the bulb. Affected plants should be immediately dug up and destroyed and others dusted with flowers of sulphur.

The disease, in the form of *Botrytis cinerea* is one of the most troublesome of cyclamen diseases, especially of the indoor Persian cyclamen. It will attack seedling plants, also the new leaves and flower stems of older plants covering them with a grey mould and causing rapid collapse. It may also spread to the corm itself. It is caused by excessive warmth and humidity and careless watering. Ample ventilation is essential for the health of this plant whilst an over-rich soil will often cause its appearance. Routine spraying with Shirlan AG 91 oz. to 2 gallons of water has given encouraging results.

Botrytis hyacinthi is one of the most troublesome diseases of the hyacinth, in Holland often spreading through plantations at an alarming rate and it may be especially troublesome during a cold, wet spring. The tips of the leaves first become affected but the fungus quickly spreads downwards when the leaves become brown and decay near the base. The flowers may also be attacked with the familiar grey mould of botrytis which does not however, attack the bulbs. Spraying with Shirlan AG has given some control.

The disease appears on irises as *Botrytis convoluta* and is especially common to those species i.e. *I. pallida, I. germanica* forming a rhizomatous root. Its presence may be detected by the leaves turning yellow as they appear in spring and upon close inspection, the rhizomes will be seen to be covered with the purple mycelium of *B. convoluta.* Too deep planting of the rhizomes may encourage an outbreak, as will damage to the rhizomes when clearing the beds in autumn.

Upon inspection, the rhizome will be seen to have become shrivelled in which case it should be lifted and destroyed though where only one part of the root is attacked, it may be lifted and the diseased part cut away, treating the healthy cut part with a weak Lysol solution before replanting.

As *Botrytis elliptica,* it attacks lilies, especially *L. auratum* and its hybrids and to a less degree, *L. candidum* covering the leaves and buds with grey mould. Later deep spots may be observed on the stems though the bulbs are rarely attacked. Plants growing in badly drained soil or where they do not receive a free circulation of air will be those mostly troubled but spraying with a weak solution of Bordeaux Mixture will act as a safeguard.

Botrytis tulipae, another form, is the most troublesome of tulip diseases attacking all parts of the plant above ground and appearing on the leaves and stems as greyish-yellow spots. After a few days, the leaves and stems fall over whilst the bulb will also be affected in the form of black pits which are deeply embedded in the outer scales. During damp weather, the disease will make rapid headway causing complete disintegration of the bulb. The disease may originate in the soil or the spores may be carried by air currents and may also be spread by splashing by rain. The Murillo tulips are most susceptible.

Bordeaux Mixture should not be used on tulips but Shirlan AG as advised for *Anemone coronaria* will give almost complete control. The plants should be sprayed early in spring and at fortnightly intervals until the foliage begins to die back.

BROWN ROOT ROT

It is a disease of the roots of tuberous begonias causing them to turn brown and decay whilst it may also extend to the tuber. It may also attack cyclamen and primulas (of the same order). The foliage of affected plants turns pale yellow and falls away. It is a fungus of the soil and is common in leaf mould so that the almost sterile peat is a more suitable alternative to use for indoor plants. Outdoor plants that have become affected should be destroyed and the ground given a rest from begonias for at least two years.

COPPER WEB

It mostly attacks crocus corms and is a soil fungus, fortunately rare. Strands of purple mycelium cover the dry scales surrounding the corm and which later grow into the corm causing rapid decay. It attacks many vegetable crops and is especially a serious disease of sugar beet. There is no known cure and long rotational cropping must be practiced wherever it appears.

CROWN ROT

It is a disease which may attack *Anemone coronaria* and the bulbous irises especially during periods of excessive rain or where planted in badly drained soil. It attacks the leaf bases or flower stems, also the upper part of bulb or tuber with white mycelium. The routine dusting of bulbs and corms with Lindex at planting time and with flowers of sulphur upon lifting will usually prevent any serious outbreak.

DAMPING OFF

It affects seedlings from the earliest stage, causing them to wilt. It is due to a water-borne fungus which attacks the roots and stem at soil level, causing the stem to turn black and take on a wire-like appearance. It is the cause of Foot Rot in tomatoes. The fungus is present in most soils, hence the need for sterilised soil to be used in seed sowing composts but even so it may be introduced in watering. Conditions favouring an attack are over-crowding (sowing too thickly); over-watering and poor ventilation, each of which should be guarded against. Brassica and tomato seedlings raised under glass; stocks and asters are especially vulnerable, their stems becoming wiry when the plant falls over at soil level. There is no cure once the disease appears but an outbreak may be prevented by watering the compost after the seed is sown with Cheshunt Compound. This is especially necessary where using un-sterilised soil. Dusting the seedlings, and those of dahlia and chrysanthemum cuttings (where rooting) with flowers of sulphur or Orthocide dust will give some immunity. It is important to ensure that the containers used for watering are clean.

DOWNY MILDEW

It is a common disease of Cruciferae both in the wild and when cultivated. It is like a white powder which appears on the underside of the leaves, especially of seedlings and plants growing under glass. Seedlings raised in frames may become affected if not given sufficient ventilation; or plants set out in wet weather may be troubled. The disease may be prevented by dusting the plants with lime and sulphur (in equal parts) which should be a matter of routine under glass. If a serious outbreak occurs amongst frame-grown plants, the soil should be sterilised (by formaldehyde or Jeyes) before growing another crop.

DRY ROT

It is one of the most troublesome of gladiolus diseases, accounting for many losses, especially during storage. The first symptom is the yellowing of the leaves which decay near the base of the sheath before they fall away altogether. Upon lifting, the new corm will reveal brown markings which may also be seen to be scattered about the leaf sheaths. On the corm the markings turn black and become sunken. The disease begins inside the corm and works up into the foliage and as soon as any yellowing is observed, the plant must be dug up and burnt. Upon inspection, the corm will confirm the presence of the disease.

The fungus is unable to live in the soil, only in the corm so that when lifting the corms late in autumn, it is important to clean them as well as possible before inspecting each one separately. One badly infected corm may contaminate several hundred whilst in storage. It is also important when lifting to clear the ground of all dead foliage and stems upon which the fungus may remain during winter to contaminate the next year's crop. Better still, allow the ground to have a year growing some other crop before planting it again with gladioli.

The disease was first recorded in U.S.A. in 1909 and its presence was confirmed in Europe twenty years later. It also attacks the crocus, freesia and montbretia, indeed almost all corm-bearing plants. Planting in well drained land free from rank manure will do much to keep the trouble at bay whilst correct storage is all important for under damp or excessively warm conditions, the smallest outbreak will quickly assume dangerous proportions. It is also important to obtain clean stock from a reputable retailer whilst dipping the corms in a 0.1 per cent Mercuric chloride solution before planting has given complete protection. MC however is a poison and must be used with care. The disease will also attack freesias causing the foliage to turn yellow and the corms to shrivel and die. They should be treated in a similar way.

FIRE DISEASE

The disease is so called because of the rapidity with which it spreads, appearing on the foliage of narcissi as yellowish-brown spots which quickly cover the whole leaf. It may also appear on the flowers as greyish spots, making them unfit for market. The disease will not attack the bulbs. The removal of all dead foliage after flowering and spraying the young foliage in spring and early summer with Bordeaux Mixture (before and after flowering) will give a satisfactory measure of control. The disease may also affect muscari and hyacinths, reducing the plants to a black slimy mass.

FREESIA MOSAIC

It occasionally attacks freesias and is caused by a virus for affected plants have shown no sign of fungus activity. About a month after their planting, when they have developed three leaves, large brown circles appear on them causing them to wither. There is no known cure and affected plants should be dug up and destroyed without delay.

FUSARIUM YELLOWS

This disease, known also as 'Premature Yellowing' is one of the most troublesome of gladiolus diseases. Though infection enters the corm whilst in the ground, it is during storage that the disease is most noticeable. It also mostly attacks the lavender and purple flowering varieties whilst 'Picardy' and its offspring possess marked resistance. The disease is detected by the yellowing of the leaves between the veins, as if showing potassium deficiency but the veins remain green giving a striped appearance.

The disease enters the corm through the roots, first causing it to decay at the base. It then attacks the centre causing it to turn dark brown. The fungus may remain in the soil for as long as seven years during which time it is not advisable to plant gladioli in the same ground. To plant in the same place year after year will be an invitation to the building up of the disease for which there is no known cure. Severely infected corms may not reveal the disease until some time after they have been in storage when they may have been reduced to black mummy-like objects when upon handling, the corm will disintegrate. To prevent infected corms reaching this stage, it is advisable to inspect them frequently, removing any that show the first signs of disease. The fungus also attacks crocus corms in a similar way.

GREY BULB ROT

It is a common disease of indoor tulips and Spanish Iris and often shows itself immediately the boxes are taken indoors for forcing. Where present, the bulbs fail to bear a shoot and upon inspection, the tip will be seen to be brown. The disease extends inside the bulb causing the formation of grey patches and the decaying of the bulb. It was first discovered in Holland by Wokker in 1884 and in North America in 1925 by Whetzel and Arthur. It may also attack *Iris reticulata; Scilla sibirica* and *Fritillaria imperialis* causing complete destruction of the bulb. It is a soil fungus and where present, the ground should be soaked with a 2 per cent formalin solution at least two months before making fresh plantings whilst boxes in which tulips are to be forced should be subjected to the same treatment.

HARD ROT

Also known as Leaf Spot for on the foliage of gladioli and freesias, occasionally crocuses, the fungus causes small circular purple-brown spots whilst on the corms when the scales are removed, similar spots may be observed. Later, they turn black, sink into the flesh and become hard and woody. The fungus is similar to Dry Rot in that it winters in the corms and on the leaves so that all dead foliage must be cleared from the ground and burnt whilst the corms must be carefully inspected before placing into storage. The disease is most troublesome during a cold, wet summer and rarely affects freesias or gladioli grown under cloches or in a warm greenhouse. Where growing in the open, it is usually towards the end of July before the first signs appear on the foliage and where its presence is detected. The plants should be removed and burnt without delay. The disease is carried from one plant to another by splashing during heavy rain or by artificial watering. Too close planting should be avoided in districts of heavy rainfall.

Routine spraying with Bordeaux Mixture from midsummer onwards will usually prevent a serious outbreak. The Mixture is prepared by adding ½ lb. copper sulphate and ½ lb. hydrated lime to 6 gallons of water which is sprayed on the plants during a showery day. Repeat the treatment every fortnight and remember that a metal container should not be used for Bordeaux Mixture. Dipping the corms into a 0.1 per cent Mercuric chloride solution before planting will do much to control the disease which appears to attack Primulinus gladioli more so than the large flowered type. Planting in clean ground in alternate years will also do much to prevent an outbreak.

INK DISEASE

It attacks all the bulbous irises causing black ink-like markings to appear on the outer scales of the bulbs. Later, it will penetrate into the fleshy scales causing the bulb to mummify when it will disintegrate upon touching into a cloud of dust. The fungus may also appear on the leaves and will attack the leaves and corms of montbretia and tritonia; and the leaves and bulbs of lachenalia, causing brown spots to appear on the leaves. There is no known cure and where it is observed, the plants should be destroyed.

IRIS MOSAIC

This virus disease is rare on all but bulbous irises. It first appeared in England in 1928 and in North America six years later. It is carried to the leaves by aphides and is controlled by their extermination. The leaves become mottled or striped with pale green or yellow and dark spots may appear on the flowers. There is no cure and affected plants should be removed and destroyed.

LILY MOSAIC

It is similar to Iris Mosaic and affects most species of lilium causing spotting and streaking of the foliage and stunting of the foliage and flower stem. The virus is introduced by aphides which puncture the leaves to feed on the sap so that their extermination is the best means of prevention. There is no cure for diseased plants which should be destroyed.

LILY-OF-THE-VALLEY LEAF SPOT

This disease of lily-of-the-valley has proved troublesome in Europe but not in the British Isles. Dark brown spots appear on the leaves and may eventually cover the whole of the leaf surface causing entire defoliation. Spraying with a copper fungicide will usually prevent a serious outbreak if caught in time.

MILDEW

It attacks chrysanthemums, roses, apples and many garden and greenhouse plants, especially during humid weather, covering the leaves and stems with a white powdery deposit. Unrooted cuttings are also liable to attack if given insufficient ventilation and as a precaution, should be dusted with flowers of sulphur at regular intervals. Growing plants should be sprayed with Karathane Mildew Fungicide or with Mildew Specific at the first sign of the disease and at intervals of ten days until brought under control. Or dust with Karathane at regular intervals. In apples, 'Cox's Orange' and Lane's 'Prince Albert ' are susceptible whilst 'Worcester Pearmain' is highly resistant.

ROOT ROT

A disease of the soil, affecting plants of narcissi and lilium which may appear stunted whilst the foliage turns green. Upon lifting, the roots will be seen to be brown instead of white, especially near the tips and upon further inspection, the fungus may be seen to have attacked the basal plate. Formalin treatment as for Basal Rot will give some immunity but it is essential to plant new bulbs into ground which has not previously grown narcissi and lilies, whilst it is essential to plant into soil that is well drained.

RUST

It affects roses, chrysanthemums, antirrhinums (though the modern varieties show immunity), appearing as brown spots on the underside of the leaves. It can be most troublesome for the spores reproduce themselves upon reaching maturity in the same way as the mushroom. The disease is spread by excessive humidity and by planting too close so that a free circulation of air is prevented. For chrysanthemums, spray in July with a solution of 1 oz. potassium sulphide to 2 gallons of water or dust the plants with a mixture of slaked lime and flowers of sulphur in equal parts. This may discolour the foliage but will not do so if applied in liquid form. Half a pound of each should be boiled in $\frac{1}{2}$ gallon of water and when cool, use at the rate of a cupful to 1 gallon of water. For roses, spray with copper fungicide at a strength of 1 fluid oz. to 1 gallon of water, repeating if the attack is severe. For rust on asparagus which occurs on the fern, spray early in spring with weak Bordeaux Mixture.

SHANKING

Like *Pythium ultimum* it is a water mould which enters the bulb through the roots and attacks the base of the flower stem, preventing the shoot from developing. The flowering stem eventually decays completely or it may partially grow and be unable to bloom, a characteristic known to professional cut flower growers as 'shanking'. It is a disease of the soil on which Cheshunt Compound has no effect. Growing in heat sterilised soil or where the soil has been treated by Jeyes Fluid or a 2 per cent formalin solution will prevent a serious outbreak.

The closely-related *Phytophthara cactorum,* prevalent in North America attacks the base of the stems of tulips in wet, humid weather causing them to fall over before or as the blooms open, a condition known as 'topple'.

SMOULDER

This is one of the most unpleasant and troublesome of narcissus diseases attacking the foliage as it shows above ground in spring and causing it to decay at the neck. The fungus also occurs on the bulb reducing it to a black slimy mass and plants over the surrounding ground may quickly become affected. The fungus will affect the foliage from the bulb which will reveal its presence by numerous black spots beneath the outer skin. Where detected in the ground, bulbs should be dug up and burned whilst those bulbs in storage found to be unsound should be destroyed.

SOFT ROT

Closely related to *B. phyrophthorum,* the dreaded Leg Rot of potatoes, it is often present in turnips and carrots and other vegetable crops as well as in hyacinth and cyclamen. With hyacinths, it usually attacks the fleshy neck of the bulb but may attack all parts, secreting its poison into healthy cells.

The disease enters by means of wounds of the bulbs and attacks the lamella which binds the plant cells together. Soon the whole bulb becomes a decaying mass and emitting a most unpleasant smell. It is a soil fungus and so rotational cropping is essential with the hyacinth. It is also important to scrub down the trays, or boxes in which hyacinths are stored with a 1 per cent formalin solution a month before storing and again immediately afterwards. Hyacinths should not be grown near carrots or other vegetable crops.

The disease occasionally attacks the rhizomatous iris and is detected by the tips of the leaves turning yellow, then brown and finally becoming a slimy mass at the base before falling away. The rhizome will also be similarly affected, being a wet mass of evil-smelling bacteria. The diseased portion should be cut away at once and the healthy part, where cut, treated with a 1 per cent Lysol solution. It has been found that those plants given a light dressing with superphosphate of lime each year will rarely become affected.

SPOTTED WILT

A virus disease which attacks the cineraria, winter cherry (*Solanum capsicastrum*), and schizanthus, causing bronzing of the foliage or circular brown spots may be observed. The leaves turn down at the edges and the whole plant takes on a wilted appearance. There is no known cure but any affected plants should be destroyed whilst aphis, possibly the carrier, must be eliminated. Disinfecting all tools and pots and the knife used to remove side shoots from tomatoes should be routine.

STEM CANKER

It attacks roses first appearing as a red spot on the stems in winter. As the cells die the centre turns brown, whilst cracks appear later. The spots may encircle the stem causing that part above to die back. As the disease is killed by hard frost, it is rare in the north and rare on plants well supplied with potash. To control, spray with Orthocide Wettable at intervals of a fortnight.

STORAGE ROT

Lilies and narcissi suffer from the rotting of their bulbs during storage though *Lilium speciosum* and *L. longiflorum* are highly resistant. The disease enters through wounds and forms dark spots on the scales, later working its way down to the base and causing it to decay. The bulb emits an unpleasant smell. Care should be taken in lifting and transporting lilies especially, also in their planting for their scales are brittle and are easily damaged.

During the late 1930s, O'Leary and Guterman discovered that *Rhizopus necans* could be controlled by mixing bleaching powder with the lily potting compost or in the garden soil before planting, at the rate of 1 oz. to every 14 lb. of compost. It also gives control against Bulb Mite (*Rizoglyphus echinopus*). The disease in a slightly different form also attacks the corms of the gladiolus in storage and for which no cure is known.

STUMP ROT

It mostly attacks the centre part of the rosette of leaves formed by *Lilium longiflorum* and *L. giganteum* and is caused by soil splashing over the rosette during heavy rain where growing outdoors or by careless watering where growing indoors. It will also attack the leaves of sinningia (gloxinia) spreading down to the corm. It may be controlled by watering with Cheshunt Compound or Orthocide Fungicide, whilst 'topping' the pots with a layer of sand will help to prevent an outbreak.

TULIP ROOT ROT

It is a water mould introduced by the soil and causing grey areas at the base of the bulbs. The shoots fail to grow more than an inch or two and may be pulled from the bulb with gentle pressure. Sometimes the disease attacks only the roots through which it gains entry and the flower may appear normally before the fungus attacks the bulb. There is no known cure and diseased bulbs should be destroyed but as the disease attacks whilst the bulbs are rooting (in the dark), the use of clean compost or sterilised soil will prevent its appearance.

WHITE MOULD

It attacks narcissus and is most prevalent in the moister, warmer parts, appearing on the leaves as small grey spots. They gradually grow larger and become covered with a white powder and kill the foliage long before its time. The fungus seems unable to survive a severe winter in the ground whilst its appearance may be prevented by routine spraying of the foliage from early spring until it begins to die back, with a weak solution of Bordeaux Mixture.

WINTER BROWNING

It is a common trouble with *Anemone coronaria* due to the plants receiving little or no protection from cold winds and may wipe out an entire crop. Treading of the soil during wet weather and lack of potash will also bring about the trouble. The undue treading of the ground in tending or picking the crop thus causing consolidation of the soil and depriving the corms of the necessary oxygen will encourage Winter Browning. It may be kept in check by regular hoeing between the rows and by watering the foliage once every fortnight with a dilute solution of Cheshunt Compound. Vigorous young 'pea' size corms rarely suffer from the trouble and should be planted in preference to old 'jumbo' sized corms which have lost vigour.

YELLOW DISEASE

It is a disease of the hyacinth, prevalent only in Holland though it may spread to other parts by imported bulbs. The disease may enter the bulb through the foliage or flower stem and is difficult to detect during storage for it is present as tiny pale yellow spots between the scales. If however, a bulb is cut open, yellow slime exudes from between the scales. Stripping of the leaves may reveal its presence in newly planted bulbs when diseased bulbs should be dug up and destroyed and the surrounding ground treated with a 5 per cent formalin solution. The use of fresh manure or of artificials of high nitrogen content, causing the bulbs to become 'soft' may bring about an attack for which there is no known cure.

FLOWER DISEASES

FLOWER	DISEASE	TREATMENT
Anemone	*Botrytis cinerea*	Dust with Shirlan AG
	Puccinia pruni-spinosae	No known cure
	Sclerotium delphinii	Dust corms with Lindex
	Urocystis anemones	No known cure
	Winter browning	No known cure
Antirrhinum	*Puccinia antirrhini* (Rust)	Spray with weak Bordeaux Mixture
Begonia	*Botrytis cinerea* (Grey mould)	Spray with Bordeaux Mixture
	Spaerotheca pannosa (Powdery mildew)	Dust with flowers of sulphur
Chrysanthemum	*Phragmidium mucronatum* (Rust)	Spray with Potassium Sulphide or lime-sulphur
	Septoria chrysanthemella (Blotch)	Spray with Bordeaux Mixture
	Spaerotheca pannosa (Powdery mildew)	Dust with Karathane or flowers of sulphur
Colchicum	*Urocystis colchici*	No known cure
Crocus	*Fusarium oxysporum*	No known cure
	Rhizoctonia crocorum	No known cure
Cyclamen	*Bacterium carotovorum*	No known cure
	Botrytis cinerea	Spray with Shirlan AG
	Corticium solani	Water with Cheshunt Compound
	Thielaviopsis basicola	No known cure
Dahlia	*Sphaceloma* (Leaf spot)	Spray with Bordeaux Mixture
Delphinium	*Pseudomonas delphinii* (Black blotch)	Spray with weak Bordeaux Mixture
Dianthus	*Didymellina dianthi* (Ring spot)	Spray with weak Bordeaux Mixture
	Uromyces dianthi (Rust)	Spray with potassium Sulphide
Freesia	*Didymellina macrospora*	Spray with Bordeaux Mixture
	Fusarium bulbigenum	Dip corms in 0.5 per cent formalin

FLOWER	DISEASE	TREATMENT
(*Cont.*)	*Mosaic*	No known cure
	Sclerotinia gladioli	Dip corms in 0.1 per cent Mercuric Chloride Solution
	Septoria gladioli	Spray with Bordeaux Mixture
Fritillaria	*Sclerotium tuliparum*	Treat soil with 2 per cent formalin solution
Galanthus	*Botrytis galanthina*	Dust with flowers of sulphur
	Urocystis galanthi	No known cure
Gladiolus	*Bacterium marginatum*	Spray with Bordeaux Mixture
	Botrytis gladiorum	Spray with Bordeaux Mixture
	Fusarium bulbigenum	Dip corms in 0.5 per cent formalin
	Fusarium oxysporum	Dip corms in 0.5 per cent formalin
	Rhizopus necans	No known cure
	Sclerotinia gladioli	Dip corms in 0.1 per cent Mercuric Chloride Solution
	Septoria gladioli	Spray with Bordeaux Mixture
	Urocystis gladiolica	No known cure
Hemerocallis	*Didymellina macrospora*	Spray with Bordeaux Mixture
Hyacinthus	*Bacterium carotovorum*	No known cure
	Botrytis hyacinthi	Spray foliage with Shirlan AG
	Sclerotinia polyblastis	Spray with weak Bordeaux Mixture
	Xanthomonas hyacinthi	No known cure
Iris	*Bacterium carotovorum*	No known cure
	Botrytis convoluta	Spray with Shirlan AG
	Didymellina macrospora	Spray with Bordeaux Mixture
	Mosaic	No known cure
	Mystrosporium adustum	No known cure
	Puccinia iridis	No known cure
	Sclerotium tuliparum	Treat soil with 2 per cent formalin solution

FLOWER	DISEASE	TREATMENT
Lachenalia	*Mystrosporium adustum*	No known cure
Lily	*Botrytis elliptica*	Spray with Bordeaux Mixture
	Corticium solani	Water with Cheshunt Compound
	Cylindrocarpon radicicola	Dip bulbs in 0.5 per cent formalin solution
	Mosaic	No known cure
	Phytophthora parasitica	Water with Cheshunt Compound
	Rhizopus necans	Add bleaching powder to potting compost
Lily-of-Valley	*Dendrophoma convallariae*	Spray with copper fungicide
	Puccinia sessilis	No known cure
	Sclerotium denigrans	No known cure
Matthiola (Stock)	*Botrytis cinerea* (Grey mould)	Dust with flowers of sulphur
	Corticium solani (Black leg)	Water with Cheshunt Compound
Montbretia	*Mystrosporium adustum*	No known cure
	Sclerotinia gladioli	Dip corms in 0.1 per cent Mercuric Chloride solution
Narcissus	*Botrytis narcissicola*	Dip bulbs in 0.5 per cent formalin
	Cylindrocarpon radicicola	Dip bulbs in 0.5 per cent formalin
	Didymellina macrospora	Spray with Bordeaux Mixture
	Fusarium bulbigenum	Dip bulbs in 0.5 per cent formalin
	Ramularia vallisumbrosae	Spray with Bordeaux Mixture
	Sclerotinia polyblastis	Spray with Bordeaux Mixture
	Stagonospora curtisii	Spray with Bordeaux Mixture
Paeony	*Septoria paeoniae* (Leaf blotch)	Spray with copper fungicide
Pelargonium	*Botrytis cinerea* (Grey mould)	Dust with flowers of sulphur

FLOWER	DISEASE	TREATMENT
(*Cont.*)	*Phythium splendens* (Black leg)	Water with Cheshunt Compound
Rose	*Botrytis cinerea* (Grey mould)	Dust with Orthocide
	Diplocarpon rosae (Black spot)	Spray with Tulisan or Orthocide
	Phragmidium mucronatum (Rust)	Spray with Tulisan or Fungex
	Spaerotheca pannosa (Powdery mildew)	Spray with Karathane or Bordeaux Mixture
	Sphaceloma rosarum (Leaf spot)	Spray with Fungex
Scilla	*Sclerotium tuliparum*	Treat soil with 2 per cent formalin solution
Sinningia	*Phytophthora parasitica*	Water with Cheshunt Compound
	Botrytis tulipae	Spray or dust with
Tulip		Shirlan AG
	Phytophthora cryptogaea	Treat soil with 2 per cent formalin solution
	Pythium ultimum	No known cure
	Sclerotulm tuliparum	Treat soil with 2 per cent formalin solution

DIZYGOTHECA, *see* Fatsia

DODECATHEON

A small genus of perennial plants, native of North America and bearing their flowers late in spring in nodding umbels. Like all members of the family, they require a cool loamy soil which does not dry out during summer and a position of partial shade, such as an orchard. The soil should be enriched with either peat or leaf mould whilst the plants love a little well decayed manure in their diet.

The plants die back after flowering to reappear the following March, the most suitable time for planting and dividing the rosettes.

D. meadia was the first species to reach Britain, during the early years of the eighteenth century, where it was to be found growing in the gardens of Fulham Palace, introduced by the famous collector of plants, Bishop Compton. The plant carries the name of the celebrated Dr Richard Mead, in practice during the time of its introduction. In America its country name is Shooting Star, after the unusual and pretty reflexing of the petals which Henry Phillips likened to a half-opened parasol.

SPECIES

D. hendersoni. Its home is the Rocky Mountains and it grows 6 in. (15 cm) tall with smooth leaves of emerald green and it blooms in April. The flowers are crimson with a yellow ring around the centre and though they emit the clove perfume, it is not so pronounced as with the other species.

D. integrifolium. Readily distinguished from *D. meadia* by its fleshy bracts, it grows only 4 in. (10 cm) tall and bears cinnamon-scented flowers of outstanding beauty, being crimson, shaded white at the base and having orange shading in the throat.

D. meadia. Known as the Virginia Cowslip, it will attain a height of 15 in. (38 cm) in a cool, moist soil, bearing tufts of large, erect, toothed leaves above which arise tall scapes of rosy-purple or lilac-pink flowers with protruding yellow anthers. It is the most richly scented of all the species.

D. pauciflorum. It blooms later than the others, in June, bearing its flowers of rosy-lilac on 10 in. (25 cm) stems. The variety 'Redwings' is a superb plant, bearing large heads of deepest crimson with a pronounced scent.

DOG'S TOOTH VIOLET, *see* Erythronium

DOGWOOD, *see* Cornus

DONKEY'S EARS, *see* Statice

DORONICUM

One of the oldest of garden plants, introduced during the days of the Crusades, for at the time its root was widely used by Eastern apothecaries. The plant enjoys a well-drained soil containing some humus, and is propagated by division of the roots in autumn. Its value to the border is that it comes into bloom when there are few other plants showing colour, during the last days of April, or early in May in the North. It should, however, be protected from cold winds otherwise its flowers, with their rayed petals, may droop their heads and fail to open until the weather becomes warmer. It grows well in partial shade.

SPECIES AND VARIETIES

BUNCH OF GOLD. It bears an abundance of clear yellow flowers on 2 ft. (60 cm) stems during May and into June.

D. carpetanum. The bright golden-yellow blooms are borne on 3 ft. (1 m) stems and are excellent for cutting.

MISS MASON. A compact variety with a longer flowering season than the others, bearing its deep yellow flowers until nearly the end of June on 2–3 ft. (1 m) stems.

SPRING BEAUTY. The first fully double variety flowering in April on 20 in. (50 cm) stems and a valuable introduction for the cut flower grower, bearing its brilliant golden yellow flowers with freedom.

DOUGLAS FIR, *see* Pseudo tsuga

DRACAENA

In their native South Sea Islands they are known as Dragon Trees, and were at one time considered to be plants requiring the warm temperature of a greenhouse. They do appreciate rather more light and warmth than most foliage plants, but they are happy in a warm living-room, given a position close to a window where they will prove extremely tough and long-lasting. In fact *D. sanderiana* may be compared in this respect to the ivy or aspidistra. It grows upright and forms no stem branches and will attain a considerable height. For best effect, two or three plants should be grown together in a large pot, spacing them as far apart as possible. The plant has the appearance of maize, the leaves being about 6 in. (15 cm) long and attractively twisted. They are pale green, beautifully edged with creamy white.

If the plant grows too tall it may be air-layered. This sounds difficult but it is quite easy, and is the method used for propagating all the tall, single-stem plants such as the philodendron, ficus and deiffenbachia. When the plant has reached the required height the stem is partially cut immediately beneath a leaf joint in the same way in which carnations are layered. Do not cut right through the stem, merely make a nick in an upwards direction, and to keep it open, place in it a small piece of wood. Clean, damp moss is placed over the cut and around the stem, over which is placed a small piece of cellophane to maintain moisture. This is tied both above and below the cut to keep the moss in position. If the layering is done early in May, by the end of June the moss will have become a mass of roots. The stem is severed immediately below the roots and the layered plant potted. Cuttings may be taken of the remainder of the stem, or the plant may be allowed to grow-on at the reduced height to beautify the room.

Of quite different form is *D. terminalis,* its bright green leaves often reaching a length of almost 2 ft. (60 cm). They grow from the main stem in arching formation. The leaves are striped with pink and gold. Of similar habit is *D. ferrea,* a native of Indonesia. Its leaves are nearly 18 in. (45 cm) in length and about 3 in. (7.5 cm) wide and are of deep green, edged with crimson, whilst the underside is also of crimson. Another species, *D. fragrans,* makes rapid growth, its arching leaves being glossy green, stripped with olive-green, whilst the variety *Lindenii* has stripes of cream.

Closely related to the dracaena is the cordyline, so close, in fact, that since their introduction to Britain a century ago, they have never been correctly classified. The Club Pine of New Zealand, *Cordyline australis,* is in certain authentic books known as *D. australis.* It is an excellent room plant. Quite remarkable in its beauty is the variety 'Atropurpurea', which bears arching leaves of rich glossy purple.

One other of the dracaena which is worthy of growing is *D. godsefficina.* It makes a plant of spreading habit, its glossy broad bright-green leaves having large cream-coloured blotches. Like those plants of aspidistra-like form, this plant is best increased by means of the small shoots which appear around the base of the plant and which should be removed and inserted in small pots when 6 in. (15 cm) in length.

A room temperature during winter of 48°F. (9°C.) will be suitable for these dracaenas, and though they should be given copious amounts of water during summer, give moisture in winter only when the soil begins to dry out. If the plants are not given sufficient moisture during summer the lower leaves will turn brown at the edges and it will be necessary to remove them, thus spoiling the appearance of the plant.

A size 48 pot will suit the plants and a compost made up of 3 parts fibrous loam, 1 part peat and 1 part coarse sand.

DRACOCEPHALUM

A little-known border plant, *D. hemsleyanum* may be used for edging for it bears its short spikes of vivid blue on 9 in. (23 cm) stems from early July until September. It likes a sandy soil and a position of full sun. Plant in spring which is the right time to divide the roots.

Another species *D. prattii* blooms at the same time and is suitable for mid-border planting. It makes short, tufted growth but sends up its spikes of soft lavender-blue to a height of 4 ft. (1 m). It blooms June-August and requires an open, sunny position.

DRACUNCULUS

The Dragon Lily, *D. vulgaris* was at one time included with the Arum Lily family but is now given its own genus. It is known as the Dragon

Lily by its stem and leaves which are mottled with red. Native of the Mediterranean countries, it makes a large tuber from which arises a black spadix enclosing a spath of reddish-purple, shaded green on the outside. It has the appearance of decaying meat and has a similar unpleasant smell, necessary to attract flies and midges for its pollination.

If growing outdoors, plant beneath a warm wall when it will bloom in June. Indoors, plant one tuber to a 6 in. (15 cm) pot containing a compost made up of 2 parts turf loam, 1 part peat, 1 part sand and as it comes into growth early in the new year, water copiously. Indoors it will bloom in April.

DRAGON LILY, *see* Dracunculus

DRAGON TREE, *see* Dracaena

DUMB CANE, *see* Dieffenbachia

DUTCHMAN'S PIPE, *see* Aristolochia

ECCREMOCARPUS

For a warm wall, *E. scaber* is a striking plant of climbing form. It is perennial but owing to its tenderness and vigorous habit is best treated as a half-hardy annual. The seed however, must be sown in the early new year in gentle heat for it to bloom the same year – from August until October. A native of Chile, where it is known as the Glory Vine, it covers itself with clusters of rich orange-red tubular-shaped flowers, whilst there is also a golden-yellow form. Given the protection of a cool greenhouse, where bracken may also be placed over its roots, it will cover a large area of glass in a season and blooms from July until November.

ECHEVERIA

They require greenhouse culture and like the kalachoes, they bloom in winter. *E. fulgens* makes a beautiful pot plant, sending up its striking bright crimson sprays on stems nearly 2 ft. (60 cm) tall. *E. setosa* is of different habit, but equally colourful. It forms stemless rosettes from which arise sprays of red flowers which are tipped with yellow.

Another form which blooms during early spring is *E. gibbiflora*. Its pointed recurving leaves are marbled with white and are borne in rosette fashion. From the centre arise sprays of vivid orange flowers.

The plants, which enjoy a little lime in the compost, should be sown in the John Innes Sowing Compost in May and if covered with glass, the seed will germinate readily in an unheated greenhouse. By the end of summer, the plants will be large enough to handle and they should be transplanted into thumb-size pots containing the John Innes No. 1 Potting Compost and in which the plants will remain throughout the winter. During winter they should be given a minimum temperature of 50°F. (10°C.) and should be kept as dry as possible. In spring, the plants should be moved to size 60 pots containing the John Innes No. 2 Potting Compost and as with all succulents, fine shingle should be substituted for the sand.

During summer the plants should be grown as cool as possible, providing ample ventilation and watering copiously, but by September the amount of moisture and ventilation should be gradually reduced for winter flowering when the plants will bloom for weeks on end.

ECHINOPS

The Globe Thistles are amongst the best of all border plants and bloom in July and August. The plants prefer a well-drained sandy soil and are best increased by cuttings taken in early May and rooted in a closed frame. Division of the roots is possible, but the plants should not be disturbed more than necessary. The stems may be used for indoor decoration or they may be dried, cutting them just as the heads reach their best.

SPECIES

E. humilis 'Taplow Blue'. It makes a stout plant 3–4 ft. (1m) high and in addition to its silvery leaves and stem, bears a pretty globular flower of pale blue.

E. nivalis. This is a most colourful and interesting plant growing to a height of 5 ft. (1.5 m). Its stems are covered with a white meal, its leaves being of a metallic silver, whilst its globular flowers are as if covered with frost.

E. ritro. This is the best known member of the family. It has silvery stems and the thistle-like leaves are pale green on the top and silver beneath. During July and August it bears large purple globes on 3–4 ft. (about 1m) stems.

E. spaerocephalus. It grows to a height of nearly 6 ft. (2 m) and has silvery stems and leaves which are of reverse colouring to *E. ritro*. The large globular flowers are of a striking silver colour.

ECHINACEA

E. purpurea is a valuable plant for a heavy clay soil, being closely related to the rudbeckia. Plant 2 ft. (60 cm) apart and this is best done in November. Propagation is by division of the roots but the plants should be disturbed as little as possible. They grow 4 ft. (1.2 m) tall.

VARIETIES

ABENDSONNE. It blooms early in autumn, its large bright crimson-pink flowers having broad petals.

TAPLOW CRIMSON. It comes into bloom early in July, bearing rich crimson flowers with a striking black cone at the centre.

THE KING. Its flowers, nearly 4 in. (10 cm) in diameter, are of a deep crimson-purple colour with a black centre cone and they bloom from the end of July until October.

ECHIUM

An annual, at its best in a poor, dry soil. *E. plantagineum* 'Blue Bedder' makes a compact plant 15 in. (38 cm) high and is in bloom from June to September. It is so free flowering that it may be used for bedding. Sow the seed early April where it is to bloom, spacing the plants to 15 in. (38 cm) apart. It requires an open, sunny position.

EDELWEISS, *see* Leontopodium

EHRETIA

A genus of small trees, native of China and Formosa and of the Borage family. Growing 15 ft. (4.5 m) tall, they do well in most soils, including chalk and will flourish in all but the coldest parts. *E. dicksonii* has handsome leaves of shining dark green which measure up to 9 in. (22 cm) long whilst in July, it bears large panicles of small white flowers which are deliciously scented.

ELAEAGNUS

With its beautiful variegated foliage, this native of Japan is valuable for planting in a town garden, in partial shade, or in a sea-coast garden. Though slow to make a start after transplanting, it will later grow away rapidly and form a dense evergreen hedge. It makes an upright but compact bush, the stems having an attractive metallic silver covering. The plants require plenty of room to spread out in all directions and so should be reserved for the larger garden. They should be allowed 2–3 ft. (60–90 cm) in the row, and little or no clipping should be done for the first eighteen months after planting. The best planting times are April or early November, and when established the old wood should be systematically removed in autumn.

One of the best forms is *Elaeagnus macrophylla*, which bears silvery foliage in the early part of summer, later changing to grey-green. Equally striking is *E. pungens* 'Aurea Variegata', its foliage being of various shades of green and gold. It is at its best planted as a hedge in a sunless part of the garden. Clipping should be done during early spring, alternate years being perhaps sufficient. Another excellent evergreen form is *E. glabra*, the glossy leaves being blue-green and pointed. It is rather slower growing, yet makes an attractive compact hedge. Each of the species bears small bunches of creamy-yellow flowers throughout summer. The elaeagnus is more popularly known as the Wild Olive and should be much more widely planted.

ELDER, *see* Sambucus

ELM, *see* Ulmus

ENDYMION

The wood hyacinths or bluebells may be planted in the shrub border or beneath mature trees and will grow well under pine and fir trees, revelling in the acid soil caused by the fallen cones and 'needles'. They are also attractive when flowering against a background of silver birch trees, planted in grass which need not be cut short until the leaves of the bluebells have had time to die down in July. They also provide a pleasing display beneath ornamental cherries or in an orchard for they will bloom at the same time as the fruit trees whose pink and white blossom will present a delightful cover for the azure blue beneath.

Plant with them the Spanish hyacinths, like refined bluebells and which are also obtainable in shades of pink, blue and white.

The bulbs should be planted 3 in. (7.5 cm) deep and 4 in. (10 cm) apart in autumn, setting them in circular groups of a dozen or more. They will multiply rapidly and will seed themselves when established.

SPECIES AND VARIETIES OF THE SPANISH BLUEBELL

Endymion campanulata. Syn: *Scilla hispanica*. Growing to a similar height as the English bluebell, the Spanish hyacinths appear on more sturdy stems above the strap-like foliage. The flower spikes are more densely packed with bells and like our native bluebell, they are sweetly scented.

ALBA. The bells of purest white are borne on 12 in. (30 cm) stems and prove a pleasing contrast to the others.

BLUE QUEEN. It bears large pyramidal spikes of clear porcelain-blue.

EXCELSIOR. It grows 18 in. (45 cm) tall forming a dense spike, its large blue bells being of delicate lavender, veined with navy-blue.

FRANS HALS. Outstandingly beautiful with its large dangling bells of purest soft pink.

QUEEN OF THE PINKS. It bears its tightly packed spikes of enormous bells on 18 in. (45 cm) stems and the colour is dark rose-pink.

WHITE TRIUMPHATOR. The outstanding white with the huge bells borne on graceful arching stems.

ENKIANTHUS

Hardy deciduous shrubs of the heath family, they require a lime-free soil. Native of West China and Japan, the colour of the fading leaves is unexcelled by any other plant. The flowers are borne on stems which grow horizontally from the main branches and appear during May and June. *E. chinensis* makes a tall upright shrub, its leaves having crimson petioles whilst its yellow and red flowers are borne in large umbels. *E. perulatus* makes a dense leafy shrub 6 ft. (2 m) tall and it bears white urn-shaped flowers in May. The leaves turn brilliant red in autumn.

Plant October to November in a deep loam containing plenty of peat or leaf mould and do no pruning. Propagation is by layering in autumn or by cuttings of half-ripened wood, taken with a 'heel' and inserted in a sandy compost under glass.

EPIPHYLLUM

A genus of twenty-one species, native to Mexico, tropical South America and the West Indies. They are not parasitic plants but root on host plants and are usually found high above ground in the forks of trees, their roots growing in hollows which have become filled with soil and decayed vegetable matter. They are common plants of the tropical forests of Paraguay, their smooth flat succulent stems providing them with moisture to assist them over periods of drought. The stems replace leaves and are produced both as side shoots and from the base, often measuring up to 20 in. (50 cm) long. From the ends appear the flowers. The plants were named by Rudolph Hermann, Professor of Botany at Leyden University in 1689 from the Greek *epi*, upon and *phyllos*, a leaf for the flowers appear at the end of the leaf-like stems.

The plants require a humus-laden soil and one containing a high proportion of sand to encourage drainage though they should not be dried off like cacti. They require a minimum winter temperature of 42°F. (5°C.) and sufficient moisture to keep them alive, increasing supplies in summer. Give the plants a sunny window in winter and partial shade in summer.

Though the colours of the blooms range from scarlet and bright pink to pale yellow and white, it is only the latter which have perfume which is most pronounced in the glistening white flowered *E. cooperi* and especially at night.

SPECIES

Epiphyllum anguliger. It bears its strangely notched stems up to 18 in. (45 cm) long at the end of which appear small tubular flowers, like white tissue paper and they are sweetly scented both by day and by night. It blooms in October.

E. cooperi. The slowest growing species and possibly a natural hybrid, it bears most of its flowers on basal shoots. The flowers are pure white and tubular but open almost flat, like camellias. They remain fresh for about three days, scenting the air by day and by night with an exotic lily-like perfume which is most pronounced when first open.

A number of hybrids of the Cooperi type have great beauty but only 'London Sunshine', palest yellow, bears scented flowers.

E. oxypetalum. It opens its flowers by night and they are larger than most species, of ivory white, tinted with rosy-red on the outside and powerfully scented.

ERCILLA

Ercilla nolubilis is a hardy evergreen climber, suited to an east or north wall in the less exposed districts. It will grow to a height of about 15 ft. (4.5 m) and has strange leathery leaves. It bears spikes of small purple-black flowers early in spring. It is self-clinging and thrives in a loamy soil. Plant in April. Propagation is by cuttings taken in July.

EQUESTRIAN STAR-FLOWER, *see* Hippeastrum

EREMURUS

The Fox-tail Lily is a magnificent plant for a border or shrubbery where the elegant flower spikes receive shelter from strong winds. With its exotic pointed straplike foliage and huge flower spikes it rivals the delphinium as being the most stately of border plants. The fact that the plants are not completely hardy does not mean that they will not survive an English winter with the minimum of protection. South of a line drawn across England from Chester to the Wash there will be no need to give the tuberous roots protection provided the soil is well drained, likewise in sheltered parts of Scotland and Ireland. Elsewhere the crowns should be covered with ashes, coarse sand, or with bracken or heather, from December until March. Or dress the crowns with a forkful of decayed strawy manure, for the eremurus is a gross feeder.

The plants like a well-drained soil containing some peat and manure, preferably cow manure, and will only bear their huge spikes to the height and size of which they are capable, where the soil is deeply worked and enriched. That the ground should be thoroughly cleaned before planting is essential for all tuberous rooted plants. Plant in November 3–4 ft. (1 m) apart and select the varieties with care, so as to extend the flowering season over as long a period as possible. Though deep planting would give protection from frost, the eremurus resents it. Barely cover the crowns with soil, after planting on a layer of peat or sand. If the soil is heavy it is preferable to leave the crowns slightly exposed for this is the danger point. The crowns will rot away if any excess of winter moisture gathers about them. Where the soil is heavy, spread the tuberous roots over a small mound as for asparagus and spread out the roots with care. Planting will be best delayed until March. Propagation is by division of the crowns or by root cuttings, but the eremurus hates disturbance and the plants are best left untouched for five to six years. By that time they will bear up to a dozen spikes, a superb sight at the back of the border.

SPECIES AND VARIETIES

E. bungei. Growing to a height of 5–6 ft. (about 2 m) this is the most compact of all and should be the choice for a small border. The spikes of golden-yellow with their orange anthers, are borne during June and July.

E. himalaicus. It grows to a height of 8–10 ft. (about 3 m) and is in bloom from mid-May until late in June. Its massive spikes are covered with hundreds of white star-like flowers clustered together.

E. olgae. A late flowering species in bloom from late June until early

August. Of compact habit, the lovely apricot-pink spikes reach a height of 5–6 ft. (about 2 m) and are deliciously fragrant.

E. robustus. This is an excellent companion to *E. himalaicus*, growing to a similar height and bearing its rosy-pink spikes during June and July.

HIGHDOWN PINK. Bearing its lovely spikes of shell-pink on 6 ft. (2 m) stems during June, this is an excellent variety to accompany 'Sir Arthur Hazlerigg' at the back of a small border.

SIR ARTHUR HAZLERIGG. A lovely new hybrid, bearing compact spikes of copper-orange throughout July on 6 ft. (2 m) stems.

WHITE BEAUTY. A new hybrid bearing long tapering spikes of purest white on 6 ft. (2 m) stems during July and August.

ERIGERON

Growing to a height of 2 ft. (60 cm), the modern erigeron and its hybrid varieties is a great improvement on the older forms with their blooms of pale colouring and which would hang their heads as if in shame. The modern plant holds its bloom to the sun on stiff, erect stems and may be seen in rich tones of purple, lilac and rose. Those flowering on 18 in. (45 cm) stems may be planted 15 in. (38 cm) apart at the front of the border in groups of three or four; those growing to a height of 2 ft. (60 cm) and over should be used at the back of those of more dwarf habit. Modern varieties will bloom right through summer, providing plenty of bloom for cutting. Propagation is by division in spring, the most suitable planting time. Where growing for cutting do not cut the bloom until fully open or it may not open when in water.

VARIETIES

DARKEST OF ALL. A German variety, growing to a height of nearly 2 ft. (60 cm) and bearing dark purple blooms from mid-June until the end of August.

DIMITY. This is a glorious plant for the very front of the border, the bright pink flowers with their conspicuous yellow centre opening from copper coloured buds and borne on 9 in. (22 cm) stems.

INTEGRITY. Quite outstanding, making a compact free-flowering plant, the bloom being of a rich shade of rosy-pink with a striking golden centre.

SERENITY. The huge deep mauve flowers held on sturdy upright stems are of a rich shade of mauve-purple and are produced from mid-June until September.

UNITY. Excellent for the front of the border. The bright pink flowers are borne on 18 in. (45 cm) stems from June to August.

VANITY. Late to bear us pure pink flowers and late to finish. The blooms too, are as large as a ten penny piece and held on sturdy 2 ft. (60 cm) stems.

VIOLETTA. The semi-double blooms, borne on 20 in. (50 cm) stems are of a rich shade of violet-blue, and bloom throughout midsummer.

WIRRAL BLUE. Very free flowering and later into bloom than the others, the mid-lavender flowers being produced during July and August on 2 ft. (60 cm) stems.

WUPPERTHAL. An excellent variety, taller growing than the others, the violet flowers being borne on 2–3 ft. (60–90 cm) stems from early June until mid-August.

ERINUS

One of the loveliest plants for paving or a low wall is *E.* 'Dr Hanele', an evergreen of prostrate habit which bears sprays of glowing carmine flowers during early summer. Equally lovely is the hybrid, 'Mrs Charles Boyle', with its blooms of clear pink whilst both are enhanced by planting with the white form *E. alpinus* 'Alba'.

ERIOPHYLLUM

E. caespitosum is a plant for the front of the border which should be more widely grown. Its tufty leaves are of a striking silver-grey colour and one of woolly texture. From mid-June until early October it bears masses of yellow-rayed flowers on 12 in. (30 cm) stems. Because of its long flowering season and its attractive foliage it is a valuable plant for edging a border. Plant in November if the soil is well drained; in March if heavy.

ERODIUM

Heron's Bill and a member of the geranium family. Its value is in its long-flowering season, being in bloom from May until October. Of almost prostrate habit the plants may be used for paving. One of the best is *E. reichardi* 'Roseum', the green rosettes being studded with pink flowers and of which there is an attractive double form. Interesting is *E. guttatum* which bears white flowers on 6 in. (15 cm) stems, the blooms being blotched with chocolate colouring. Bearing stemless flowers of deep pink is the almost prostrate *E. corsicum* and its white counterpart, 'Album'.

ERYNGIUM

The Sea Hollies are amongst the best of all plants for drying for home decoration in winter. Not as tall-growing as the echinops, the eryngiums are suitable for planting towards the front of a border, 20 in. (50 cm) apart. Plant in November, a light sandy soil being suitable, provided it contains some humus to retain summer moisture. Propagation may be by root division, from seed sown early in summer, or by means of cuttings rooted in May. The latter is the best method to ensure rapid propagation of the named varieties.

SPECIES AND VARIETIES

E. oliverianum. It has long spiny leaves of silvery-green, like large holly leaves, whilst its small thistle-like flowers borne on branched stems are of deep amethyst-blue, the 3 ft. (90 cm) stems being of a similar colour. It is colourful from early June until late in August, when the stems are cut and dried.

E. planum. The branched stems reach a height of 2–3 ft. (60–90 cm) and bear sprays of pale silver-blue thistle-like flowers set in bracts of the same colouring. The stems too are blue, whilst the foliage is a striking silver-green. A new variety, 'Violetta', bears a deeper purple bloom which remains colourful into September.

E. tripartitum. It grows to a height of 3–4 ft. (about 1 m) and throughout summer bears its erect branching stems of steely-blue flowers and bracts.

JEWEL. This is a hybrid which has deeply cut leaves and bears its purple cone-shaped flowers on stems of the same colour.

ERYTHRONIUM

The Dog's Tooth Violets are amongst the best of spring flowering plants. Like the chionodoxas, they are at their best during March and April, whilst they increase both from their offsets and from self sown seed. The plant is known as the Dog's Tooth Violet on account of its long, pearly bulb or tuber which resembles a dog's tooth. Their garden value is in their liking for a cool, shady position. Plant the aconite in a slightly damp position; the crocus and chionodoxa in full sun; the snowdrop in partial shade; and in a position of almost total shade, sheltered from the prevailing winds but where the sun does not penetrate, i.e. a north border, plant the Dog's Tooth Violets which vary in height from 6–14 in. (15–35 cm). They grow readily if given a soil containing a high proportion of peat or leaf mould.

Not only will the plants flourish in a shrubbery which contains plenty of leaf mould (they are never at their best in the usual town garden with its soil devoid of humus), but they make a pleasing display in the garden beneath silver birch trees, planted on ground which slopes away from the sun. There against the silver-white bark of the trees and in grass which is kept closely cut, they bloom to perfection, shielded from the summer sun by the foliage of the trees. The Dog's Tooth Violet does well in pots. The long tooth-shaped bulbs are lifted during the last week of February, as soon as the tips are to be seen pushing through the soil. Great care is needed to lift and transfer the bulbs without exposing them to the cold longer than necessary. Into shallow bulb bowls they are placed, six to a bowl containing leaf mould and soil to which is added some coarse sand. Place in a warm room, in a window where they can enjoy the winter sunshine and water only when the soil begins to dry out. Within a week the bowl will have become a mass of the brilliantly marbled leaves and the dainty blossoms will hover above the foliage like butterflies. They will remain in bloom for three to four weeks. After flowering, replant outdoors.

PLANTING

Care of the bulbs when planting outdoors is important for they will quickly deteriorate if out of the ground too long or exposed to sunshine or wind. If allowed to become dry they will shrivel and no amount of care will nurse them back to health. September is the month to plant for then the weather is cool and if possible the bulbs should be lifted and planted within a few days, whilst still moist. They will then quickly become re-established. More so than most bulbs, they appreciate a regularly yearly mulch with peat given during August and to guard against conditions which may be too dry during summer the bulbs should be planted 4 in. (10 cm) deep.

If the soil is not disturbed, the plants will naturally seed or they may be increased by sowing seed in shallow pans or boxes during May in a cold frame which is shaded from strong sunlight. Sow the seed into a compost containing 2 parts peat or leaf mould, 1 part loam, 1 part coarse sand. Merely press the seed into the compost, water thoroughly and cover with a sheet of glass. The seed should germinate evenly and when large enough to handle, the seedlings are transplanted to boxes and allowed to remain there over winter, before being transferred to a shady position the following April. There they will bloom in two years' time. It is not advisable to lift and divide the clumps other than when they become overcrowded and to prevent damage to the bulbs when exposed to the air, any dividing should be done on a calm day.

SPECIES

Erythronium californicum. This is a most striking spring flower. The creamy white bell-shaped blooms, which are blotched with orange and brown are carried on stems 12 in. (30 cm) long. The foliage too, is attractive being deep green mottled with brown. The variety, 'Helenae', has fragrant petals of white, lined with deep yellow.

E. citrinum. A connoisseur's plant, and most lovely, the lemon-coloured flowers being carried on 9 in. (22 cm) stems. They remain in bloom over a long period.

E. dens-canis. Dog's Toothed it is named, not from the shape of the bloom, but of the bulb: The variety 'Franz Hals' is of a rich red-violet shade. It is carried on 6 in. (15 cm) stems which are also reddish coloured whilst the leaves are attractively mottled. It blooms mid-March to mid-April. There are other lovely varieties, 'Pink Perfection' bears flowers of a bright shell pink, and 'Snowflake', blooms of the purest white. Together with 'Franz Hals' they make a lovely trio for the shady garden or on a rockery facing away from the south (see plate 27).

E. hendersonii. Found growing naturally in the Oregon district of the U.S.A., this is a dwarf edition and bears its flowers of a pure lavender colour in great profusion during April. The flowers are almost identical with that of the hardy cyclamen.

E. tuolumnense. A plant for the connoisseur, producing its rich daffodil-coloured blooms on 12 in. (30 cm) stems throughout March. This is a hardy species bearing shiny unspotted green leaves and is happiest when planted in short grass under deciduous trees.

E. revolutum. Flowering very early in March it is the first to bloom and the most expensive to obtain. The white flowers are unique in that the petals roll back to reveal a crimson centre. It increases rapidly. From California, the main breeding-ground of the erythronium.

ESCALLONIA

Glossy-leaved evergreens which are hardy in coastal districts of Britain, though several may not be so if planted away from the sea. No plant will make a better hedge, for they provide all year round protection and in addition, bear large quantities of bloom through late summer and autumn when they form their pretty, waxy, tubular flowers. In the most exposed positions of a north-east garden, where in a severe winter they may require slight protection, sprays of beech or conifers may be pushed into the top of the hedge at the end of December to be removed early in March. Plant 3 ft. (60 cm) apart and they prefer a light, sandy soil enriched with some bone meal. No clipping or pruning should be done until early April, then remove any dead wood and clip carefully into shape. Planting is also best done at this time.

Of the many suitable varieties, none excels the new 'Glasnevin Hybrid', of vigorous upright growth, and which bears its rich red blooms over a long period. Plant with it for contrast 'Donard White', the first white-flowered escallonia.

Where a very hardy variety is required, plant 'Slieve Donard', of vigorous habit and which bears large pink flowers. Also good is 'Glory of Donard', which bears deep rose-carmine flowers. Carefully tended, the escallonias will make a hedge 6 ft. (2 m) tall furnished with foliage from top to bottom, there being no better hedge for a coastal garden.

ESCHSCHOLTZIA

Like the echium, this is a colourful annual for a poor, dry soil. Sown early in April, the seed germinates rapidly and the plants, growing to a height of about 10 in. (25 cm), bear a profusion of poppy-like flowers throughout summer. Amongst the best varieties are 'Dazzler', flame; 'Enchantress', rose and cream; 'Orange King'; and 'Crimson Queen'. Sown in groups about a shrubbery or border, the plants will add rich colouring.

EUCALYPTUS

There are several members of the eucalyptus family which may be grown under glass. They may be raised from seed which, like that of most foliage plants, should be sown in March in a temperature of 65°F. (18°C.). The eucalyptus is evergreen with aromatic leaves which are covered with an attractive grey 'bloom'. The best known forms are *E. citriodora,* the leaves having a rich lemon scent, and *E. globulus* which is the Australian Blue Gum, its grey-white leaves having a pungent fragrance.

Fresh seed must be used and it will germinate readily in the John Innes Sowing Compost if kept moist. As soon as large enough to handle, the young plants should be moved to 3 in. (7.5 cm) pots containing the John Innes Potting Compost. In these pots the plants remain over winter in a temperature of around 48°F. (9°C.), during which time they will require very little moisture.

Late in spring, the plants will be ready for moving to larger pots and throughout summer will require copious amounts of moisture and ample ventilation. The plants enjoy plenty of sunshine and fairly dry conditions, but will be quite happy growing under similar conditions as other foliage plants.

Soon after the plants have been moved to the larger pots, they should have the leading shoot pinched out to encourage them to 'break' and grow bushy.

EUCHARDIUM

An easily grown annual with a flowering season extending from mid-June until mid-September. It is best sown, either in autumn or spring, in the border where it is to bloom. Sow in small circles and in early summer thin to 6 in. (15 cm) apart. The 'Pink Ribbon' strain bears large rose-pink flowers with curious long ribbon-like petals, held on 15 in. (38 cm) stems. It will grow well in ordinary soil but requires an open, sunny situation.

EUCRYPHIA

Native of Chile and Tasmania, it dislikes lime and needs a soil well enriched with decayed manure and peat. Plant it with rhododendrons and with other acid lovers. It also needs protection from cold winds which may damage the foliage in spring. *E. glutinosa* is a deciduous form, its leaves taking on brilliant tints in autumn. It makes an erect shrub 12 ft. (3.5 m) tall, branching at the top and it blooms in July, its white flowers being enhanced by their tuft of golden stamens. *E. billardieri* is evergreen, making a dense leafy shrub and in June, bearing fragrant pendulous white flowers.

Plant both species in April which is the time to remove any old wood, otherwise no pruning will be necessary. Propagate by layering or by cuttings removed with a 'heel' and inserted in sandy compost under glass.

EULALIA

E. japonica, the Zebra-striped Rush, in its numerous forms is a most elegant plant, bearing narrow green leaves 3 ft. (1 m) in length and which are striped with yellow or white. It is almost hardy, requiring a winter temperature of only 42°F. (5°C.) and the minimum amount of moisture, whilst in summer the plants require copious amounts of moisture and shading from strong sunlight.

The plants are readily raised from seed sown in March in a temperature of 60°F. (16°C.). When large enough to handle, the seedlings are transplanted to a size 48 pot containing a mixture of 3 parts fibrous loam and 1 part each, leaf mould and coarse sand. They like a moist but buoyant atmosphere and during summer will appreciate a daily syringe.

EUONYMUS

The Spindle Tree which grows well in sunshine or shade and in town and coastal gardens. It is the deciduous forms which have attractively coloured foliage in autumn *Euonymus alatus* having branches closely set with tiny leaves which turn brilliant pink in September, whilst *E. latifolius planipes* has scarlet foliage in autumn, accentuated by its glossy crimson fruits. Both forms grow 4–5 ft. (about 1 m) tall and require only the minimum of trimming to train them into shape. They grow best in a lime-free soil.

The evergreen form withstands clipping and makes a splendid hedge. *E. japonicus* has glossy dark green foliage and grows 12 ft. (3.5 m) tall. The form 'Argenteo-marginatus' has silver-edged foliage. They may be planted alternately to make a handsome hedge.

Plant the evergreen species in early spring; the deciduous forms November to March. Propagate by cuttings or by layering in autumn whilst the deciduous kinds are increased by seed sown under glass in spring.

EUPHORBIA

These succulent plants are grown for their interesting forms rather than for their beauty, the single exception being *Euphorbia splendens*, Christ's Crown of Thorns, which produces long thorny stems of frightening appearance and which are clothed with bright green leaves. The plant bears, on short stems, clusters of scarlet flowers which have salmon-pink bracts, and they are almost perpetually in bloom. It grows 2 ft. (60 cm) or more in height and should be given ample support. It requires a sunny window, and, as it blooms throughout the year, always keep the compost comfortably moist, neither too wet nor too dry.

Of those forms of strange appearance, *E. pulvinata* resembles a cactus in that it is slow-growing and bears clusters of cucumber-like growths which are covered in spikes often up to an inch in length. *E. meloformis* and *E. polygona* are almost globular in form and covered in long spikes or thorns, whilst *E. caputmedusae* forms a short thick stem from which are produced snake-like branches, hence its name.

The plants like a compost as suggested for the aloe and other succulents, whilst they will require plenty of moisture during summer.

Propagation is always difficult with euphorbias, for when cut they eject a milk-like sap which will cause an unpleasant irritation if allowed to enter small cuts on the hands. The cuttings will not even begin to root until dried of the sap and even then take many weeks to form roots. During the rooting period they should be kept as dry as possible.

EVENING PRIMROSE, *see* Oenothera

EVERGREEN LABURNUM, *see* Piptanthus

EVERLASTING, *see* Helichrysum

EVERLASTING FLOWERS

As they are mostly native of South Africa, the so-called everlasting flowers require an open, sunny situation and a dry, sandy soil. Endowed by nature to withstand long periods of drought, they will retain their beauty and colour when removed from the plant and dried. They may be raised where they are to bloom, sowing the seed in April for as they are only half-hardy in the British Isles, they should not appear above ground until fear of hard frost has departed. Or as they will withstand transplanting, they may be raised under glass and planted out towards the end of May.

To dry, cut when at their best and hang the bunches in an airy room for two to three weeks to dry off. Do not hang them in a greenhouse for they would be subjected to excessive humidity and will not dry completely, tending to become mouldy. These everlasting flowers are valuable for winter decoration indoors:

Acrolineum roseum
Helichrysum monstrosum
Lonas inodora
Rhodanthe manglesii
Statice sinuata
Xeranthemum annuum

EXOCHORDA

The Pearl Bush is a hardy deciduous shrub, bearing in April and May, drooping racemes of round, pinky-white buds which open white. *Exochorda racemosa* makes a plant about 6 ft. (2 m) tall and requires a sunny situation and a rich loamy soil. Plant November to December and do no pruning apart from cutting back unduly long shoots. Propagate by cuttings of the half ripened shoots, taken with a 'heel' and rooted in a sandy compost under glass.

FAGUS

Where the soil is thin and overlying chalk, there is no more reliable hedging plant than the copper beech and whilst large, singly grown trees lose their foliage during late autumn, when used as a hedge, members of the beech family retain their leaves right through winter. That they are slow growing in comparison to a number of other plants possibly prevents them from being more widely planted. Retaining its foliage where used as a hedge, beech is valuable both as a windbreak and for privacy and in addition does not become too thick and occupy too much ground. Nor does it require clipping more than once every eighteen months. If not allowed to get out of hand, a beech hedge may be kept neat and compact and will never form excessive wood to the exclusion of foliage. The hedge may be kept low, possibly 3 ft. (90 cm) or less if it is required to surround a small border, or it may reach a height of 15 ft. (4.5 m) and still retain its neatness of form.

The plants, which should be obtained when 2 ft. (60 cm) high, are planted 15 in. (38 cm) apart in the row, or 18 in. (45 cm) if a double row is required. Plant firmly but not too deeply and where the soil is shallow, give a mulch during summer until the plants are established. November is the best time to plant.

The best form of the copper beech is River's Purple. The leaves are broader than those of the Copper Beech, smooth and of a deep crimson-red colour when the plants are established. The foliage is most attractive used for indoor decoration, and will remain fresh in water for several weeks.

Though making a valuable stock-proof hedge, the Common Beech, *Fagus sylvatica*, being slow growing, is too expensive to use anywhere but in a garden. For a thin, chalky soil there is no better hedge plant, for it will tolerate dry conditions and a hot soil. Retaining its red-brown leaves when planted as a hedge right through winter, the beech is as efficient as any evergreen and possesses the additional advantage in that it requires much less attention as to clipping and pruning than almost any other hedge. It is best planted as a double hedge, spacing the plants 16 in. (40cm) apart, 2 ft. (60 cm) plants being more readily established than older plants. Almost no clipping will be required for the first two to three years, and afterwards only in early April, to keep the plants in line. Long lasting and never becoming straggling nor making excessively thick woody stems however neglected, a beech hedge is warm-looking through winter, whilst there is no foliage to equal that of beech bursting into new growth in May.

A delightful mixed hedge may be made by alternative planting with copper beech, of similar habit and retaining its winter foliage in the same way.

Planting may be done between November and March, whenever the soil is in a suitable condition. And though well able to withstand a very dry soil, growth will be more vigorous if the plants are given a summer mulch.

FALSE ACACIA, *see* Robinia

FALSE CYPRESS, *see* Chamaecyparis

FALSE INDIGO, *see* Amorpha

FATSHEDERA

It is a plant of recent introduction, the result of a cross between the hedera (ivy) and the fatsia. It bears large five-lobed fig-like leaves which are glossy green and leathery to the touch and are borne on long footstalks from a single main stem. There is also a variegated form, the leaves being margined with gold. It is a fast-growing plant of upright habit and, as would be expected, it should be given a shaded position and the same cool conditions enjoyed by the ivy. It is a most valuable plant for an entrance hall, for it is not harmed by draughts.

As the plants are more attractive before they become too tall, when they tend to be rather bare at the base, they should be grown as slowly as possible, so use a pot no larger than the size 48 and always keep the plant on the dry side. The compost should be the same as suggested for the aspidistra and do not re-pot until absolutely necessary. The plants will greatly benefit from the regular wiping of the foliage and placing outdoors to enjoy gentle rain on all suitable occasions.

FATSIA

F. japonica, also known as *Aralia sieboldii*, is one of the largest-leaved of indoor plants. It is also known as the Japanese Aralia and the Dizygotheca, whilst it is frequently confused with the Castor Oil Plant, *Ricinus communis*. Like the aspidistra, the fatsia was a favourite house plant with the Victorians but has become less popular of recent years. In the south-west it may be grown outdoors with the minimum of winter protection. Outdoors it will bear globular heads of pure white flowers but rarely blooms indoors where the plant is grown for its large fig-like leaves. These are deep green and possess a leathery texture. There is also a variegated form, the leaves being splashed with white. The plants will grow in the darkest of rooms and require no special attention apart from ample supplies of water during summer. The compost should be composed of 2 parts fibrous loam, 1 part leaf mould and 1 part sand.

FELICIA

F. bergeriana, the Kingfisher Daisy, is a half-hardy annual and with its sky-blue star-like flowers borne on 6 in. (15 cm) stems, it is a plant deserving of greater popularity. Like all South African plants, it should be given a sunny position and a dry soil. Sow in April where it is to bloom which it will do from June until October. It makes an original edging plant or it may be used on the rock garden. The flowers have bright golden centres and in dull weather these are visible from a distance as the petals turn back.

FERNS

Herbaceous plants with fibrous roots or a rhizomatous rootstock and leaves (fronds) simple or divided, curling inwards towards the top. The leaf stalk is called a stripe; the midrib a rachis. They reproduce themselves by seed-like bodies known as spores which readily germinate under damp conditions, the small green body sending out tiny rootlets. On it are minute bodies which have similar functions to the stamens and pistils of flowering plants. When ripe, the males set free, moving bodies known as spermatozoids which pass down the neck of the bottle-shaped female organs so that fertilisation occurs from which the young plant arises.

Ferns are raised by sowing the spores of ripe fronds which are shaken into sterilised compost composed of soil and peat in equal parts and which is placed over a layer of moss. After sowing, water from the bottom of the container and cover with a piece of polythene to prevent too rapid moisture evaporation. If sown in a greenhouse or frame in spring, germination will take place during summer, depending upon the various species, their hardiness and readiness to germinate. When large enough to handle, the seedlings are pricked out into 3 in. (7.5 cm) pots and grown on in a moist atmosphere. At all times, ferns should be shaded from sunlight and grow best in partial shade. For this reason it is usual to make a fernery in that part of the garden where little else will grow.

The fern garden should be constructed of tufa stone which will protect the roots from excessive evaporation if planted between and about the stones with soil packed tightly about their roots whilst the plants will be cool during the summer. The bright green ferns will also be more attractive when displayed against the grey background of the stones. The soil should be mixed with peat, coconut fibre or leaf mould and some decayed manure to provide the necessary humus and food. There should be a considerable depth of soil between the stones into which the ferns are to be planted otherwise the roots will dry out during warm weather, during which time it is important to keep the plants well watered. As when constructing a rock garden, those ferns which require more moisture than others should be planted at the base of the fernery; those tolerant of dryer conditions being planted at the top. Early April is the best time to plant which is also the time to lift and divide those ferns which may be propagated by this method. This should be done during rainy weather, when cold winter winds have ceased.

Those ferns requiring a dry situation may be planted about old walls, in pockets of soil. These ferns do not usually require deep planting. Amongst those which require a dry elevated position are:

Asplenium adiantum – nigrum – Black Spleenwort
Asplenium ruta – muraria – Wall Rue
Asplenium septentrionale – Forked Spleenwort
Asplenium trichomanes – Common Spleenwort.
Asplenium viride – Green Spleenwort
Ceterach officinarum – Rusty-back
Cryptogramma crispa – Mountain Parsley fern
Cystopteris montana – Mountain Bladder fern
Lastrea montana – Hay-scented fern.
Polypodium robertianum – Limestone Polypody
Polypodium vulgare – Common Polypody.
Woodsia ilvensis – Oblong Woodsia

These ferns are also tolerant of chalk and limstone soils. No ferns like to have their roots perpetually waterlogged but several are more tolerant than others of moist conditions and may be planted by the side of ponds and streams. In dry weather, these ferns will require particular attention as to their watering:

Athyrium filix-foemina – Lady fern
Onoclea sensibilis – Sensitive fern
Osmunda regalis – Royal fern
Phyllitis scolopendrium – Hart's Tongue

FERN HOUSE

Ferns indoors will provide beauty the whole year round. They may be grown in pots in a room facing north or east or may be grown in a constructed house, preferably in a shaded corner formed in a courtyard by two walls adjoining at right angles or on the northerly side of a house where few other plants would grow. A small lean-to greenhouse may be used or a more simple construction of a wooden frame covered with polythene sheeting with the ferns arranged in pots on benches and shelves constructed of wood. The 'house' should be frost-proof but artificial warmth is not necessary except during severe winter weather when a modern portable heater may be installed. As an alternative to shelves, a fernery may be constructed at the back or highest part of the 'house' by the use of tufa stone as where growing outdoors so as to give a natural effect to the display. Here, the most dainty and delicate of ferns may be grown and will delight with their fresh green colouring and their cleanliness. If during summer, it is necessary to provide shade, whiten the inside of the glass or cover the roof with dark polythene. At all times, keep the plants comfortably moist whilst keeping them free of dead fronds by gently cutting them away at the base.

FERNS IN POTS

Ferns make attractive indoor pot plants. They may be grown by themselves or several used in a plant container in which may also grow other plants requiring the same conditions. If growing in pots, ensure that the pot is large enough to hold the roots without cramping and which would eventually cause them to decay at the centre through lack of food and moisture. Large plants should be divided into quite small pieces and in this way they will get away to a good start. Frequent re-potting into larger pots will keep them healthy.

A suitable compost should be made up of fibrous turf loam, peat and decayed manure in equal parts by bulk and to which is added a liberal quantity of coarse sand. Moorland soil containing the roots of heather and bracken may be used if chopped up. Make comfortably moist before placing in the pots over the crocks. The compost should have an open, friable appearance. Place the crown of the fern level with the rim of the pot and spread out the roots, pressing over and around them more compost so that it will be just below the rim of the pot to allow for watering. After making the plant quite firm in the pot, water in and stand in a cool room.

Ferns will not tolerate a stuffy atmosphere and will benefit from being placed outdoors in summer, away from the direct rays of the sun where they are watered and sprayed regularly. They will particularly benefit from night dews and coolness but should be removed into the home by early September. During winter, the plants will require watering only about once a week for at this time, the soil should not be kept in a saturated condition or the roots will decay. Water only when necessary – when the pot upon tapping gives a hollow ring.

Early April is the best time to re-pot when the plants may be placed in a frame or indoors to become established. They may be placed in the open towards the end of May when fear of frost has finished.

Besides the propagation of ferns from spores, established plants may be divided into offsets by 'teasing' them apart so that each crown has a tuft of roots attached. Those like the Common Polypody which forms thick fleshy or rhizomatous roots, may be pulled apart with the rhizomes attached. They require different potting from those with fibrous roots for the pots should be half filled with crocks and the fleshy roots placed over a thin layer of compost and covered in to a depth of 1 in. (2.5 cm). The method of growth of the rhizomatous species is to run for some distance just beneath the soil surface before throwing up a crown.

Another method of reproduction is by bulbils which with a few species grow on the fronds. When large enough they are readily detached and planted in pans 1 in. (2.5 cm) deep and 1 in. (2.5 cm) apart. The compost should be made up of sterilised loam, peat and sand in equal parts. Kept moist, the bulbils will soon grow into small plants when they may be moved to small pots.

HARDY FERNS FOR INDOOR AND OUTDOOR PLANTING

Adiantum capillus-veneris. The evergreen Maidenhair which is found about damp rocks in south-west England and Ireland. It grows 8 in. (20 cm) tall, its fronds being irregularly divided into alternate wedge-shaped leaflets.

Adiantum pedatum. One of the Maidenhair ferns, it grows only 9 in. (23 cm) tall and requires a warm, moist atmosphere. It flourishes in damp peat sheltered by stones or indoors in pots. It has whip-like branches and forked fronds.

Asplenium adiantum-nigrum. The Black Spleenwort may be planted with ericas and other acid-tolerant plants but it requires shade. Of tufted habit, it grows 12 in. (30 cm) tall and has shining black stalks whilst the ovate fronds are twice or thrice pinnate.

Asplenium marinum. The Sea Spleenwort, it is found in caves and on rocks by the sea in warmer parts of Britain, its rootstock being clothed in purple scales. It has lance-shaped fronds, pinnately divided into crenate lobes and with shining brown stalks.

Asplenium ruta-muraria. The Wall Rue, it grows 4 in. (10 cm) tall on old walls and in the crevices of rocks and is recognised by its ovate fronds twice-pinnately divided into rounded lobes.

Asplenium trichomanes. The Maidenhair Spleenwort, it grows 6 in. (15 cm) tall and has a shining black mid-rib on each side of which are deep green lobes regularly arranged.

Asplenium viride. The Green Spleenwort, distinguished by the green mid-rib and stalks. It makes a tufted plant 6 in. (15 cm) tall and requires a moist, cool situation.

Athyrium filix-foemina. The Lady fern which grows well in partial shade and in a well drained loam. From the sturdy rootstock arise pale green lance-shaped fronds, two to three times pinnate and 2 ft. (60 cm) in length with brown or yellow stalks. The form dissectum has fronds five-times pinnate.

Blechnum spicant. The native Hard fern which has a creeping root-stock and erect bright green leaves which grow 10 in. (25 cm) tall with glossy brown stalks.

Blechnum tabulare. A handsome evergreen fern growing 18 in. (45 cm) tall with leathery fronds of darkest green.

Ceterach officinarum. The Scaly Spleenwort which is to be found amongst

rocks and on old walls where it grows only 3 in. (7.5 cm) tall. The pinnately divided fronds have scaly stalks whilst the under surface is covered in rust-like scales, hence its name Rust-back.
Cryptogramma crispa. The Parsley fern which grows only 4 in. (10 cm) tall and forms a thick clump of finely divided fronds of brilliant green, like a clump of parsley. It is to be found on rocky outcrops in north England and in Scotland.
Cystopteris fragilis. The Brittle Bladder fern found on old ruins and about mountainous parts usually growing in shade. It has pinnately divided lance-shaped fronds 6 in. (15 cm) long and deeply toothed at the margins.
Cystopteris montana. The Mountain Bladder fern to be found about damp rocks in north Scotland and growing 9 in. (22 cm) tall. The triangular fronds are four times pinnately divided.
Dryopteris cristata. The Crested Buckler fern with fronds 18 in. (45 cm) long and pinnately divided, the short-stalked pinnae having short teeth.
Dryopteris dilatata. The Broad-leaf Buckler fern of native woodlands, growing 3 ft. (90 cm) tall with broad ovate lance-shaped leaves, twice or thrice pinnate.
Dryopteris filix-mas. The Male Buckler fern, with twice pinnate fronds 2 ft. (60 cm) long and lance-shaped in outline. The variety 'Grandiceps' has heavily crested fronds 18 in. (45 cm) long and which have a feathery appearance.
Matteuccia struthiopteris (Syn: *Onoclea germanica*). The large Ostrich Plume fern of north Europe with broad lance-shaped fronds tapering to the base and 3–4 ft. (1.2 m) in length. It requires moist, shady conditions.
Onoclea sensibilis. The Sensitive fern of North America, a neat and dainty species, its pale green twice-pinnate fronds are cut into lance-shaped pinnae, waved at the margins. It is lovely by the waterside.
Osmunda regalis. The Royal fern which requires marshy or bog-like conditions where it will form a large clump with fronds 6 ft. (2 m) high, the pinnae oblong and 6–12 in. (15–30 cm) long. The form 'Cristata' has the ends of the fronds and pinnae attractively crested whilst 'Purpurascens' has purple-tinted fronds.
Phyllitis scolopendrium. The Hart's Tongue which will grow almost anywhere but perferably in partial shade and in a moisture-holding soil. It forms tufts of bright green fronds 18 in. (45 cm) long with parallel rows of spawn cases at right-angles to the midrib. The variety 'Undulatum' has beautifully waved fronds, whilst 'Fimbriatum' has fronds which are finely divided. Syringe the plants often during dry weather.
Polypodium vulgare. The Common Polypody which is common about tree trunks and on ruined buildings. It has a creeping rootstock, the rhizomes appearing on the surface of the soil. It has alternate oblong fronds 9 in. (22 cm) long, pinnately cut into linear segments. They are deep green with golden spore cases. The polypodys do well in a limestone soil provided

they do not lack moisture. The variety 'Pulcherrimum' has broad, deeply divided fronds 12 in. (30 cm) long.

Polystichum aculeatum (Syn: *Aspidium aculeatum*). The Hard Shield fern, it prefers shade and a slightly acid soil. It is also more tolerant of dry conditions than other ferns. The rigid lance-shaped fronds grow 2 ft. (60-cm) long, the upper surface being glossy green, the underside covered in brown spiny scales. The twice-pinnate fronds have pinnules covered in bristles.

Polystichum lonchitis. The Holly fern which forms a dense tuft of oblong pinnate fronds 2 ft. (60 cm) long the pinnae having sharp spines at the edges, hence its name.

Polystichum setiferum (Syn: *P. angulare*). The Soft Shield Buckler fern which differs from *P. aculeatum* in having stalked and not sessile pinnules whilst they are less hard and stiff.

Thelypteris dryopteris. The dainty Oak fern, distinguished from the polypodys in its smooth fronds divided into three branches each with six or more pairs of pinnules. It grows 6 in. (15 cm) tall and likes a cool, shady nook.

Thelypteris oreofteris (Syn: *Hastrea montana*). The elegant Mountain fern, to be found on mountain sides and cliffs and which sends up its lemon-scented fronds to a height of 15 in. (38 cm).

Thelypteris phegopteris. The Beech fern, it is found in damp, shady places, usually woodlands where it produces its hairy 6 in. (15 cm) fronds of delicate green.

Thelypteris robertiana. The Limestones Polypody, it is found on limestone cliffs and on old walls. It makes a short-tufted plant with luxuriant foliage like that of the Oak fern but the fronds are covered with down.

Woodwardia angustifolia. Native of the United States, it has ovate fronds 6 in. (15 cm) long and borne on slender stalks and divided into eight to nine pairs of lance-shaped crenate pinnae. It grows best in sandy loam containing peat or leaf mould.

Woodwardia radicans. Native of south Europe, it needs a warm, sheltered garden in the British Isles. It sends up its graceful fronds 3–4 ft. (1m) long and 12 in. (30 cm) broad, divided into lance-shaped pinnae 12 in. (30 cm) long. The variety 'Cristata' has crested fronds.

INDOOR

Pteris argyrea, the Silver Brake Fern, was the first variegated leaf fern to reach Britain, which it did a century ago. Like most ferns, however, it is more suitable for the warm greenhouse than the home, and there are few out of the 10,000 or more species which will flourish under living-room conditions.

One of the best for this purpose is *Nephrolepis exaltata,* which has long drooping fronds, whilst *Pteris cretica* 'Albo Lineata', one of the Silver Brake family, has similar silver variegations to *P. argyrea* and grows well in the home. Probably the finest of the indoor ferns is *Asplenium nidus*, the Bird's Nest Fern, which forms a circle of elegant sword-shaped leaves which are of almost transparent green and have attractive black ribs. To see it at its best it should be viewed, like the caladiums, against light.

The ferns like moisture and should always be grown in the home on trays of moist sand or shingle.

Very different from *Asplenium nidus* is *A. bulbiferum,* which throws up masses of true fern-like fronds of palest green. If grown on a tray of moist sand and peat, plantlets will fall from its fronds and take root in the sand and peat.

The ferns like a compost made up of 2 parts loam, 2 parts peat or leaf mould and 1 part sand and they should never be allowed to lack moisture. They will enjoy being placed out of doors in summer during periods of gentle rain. Propagation is by root division in April.

FICUS

The plants are native of the tropical forests of Indonesia, Malaya and the Philippines, *F. elastica* being known as the India-rubber Plant. The ficus is related to the fig family, having oval glossy olive-green leaves. generally marked with crimson veins. They require rather more light than most plants of similar type and so should be grown near a window. They are quite hardy and a winter room temperature of not less than 42°F. (5°C.) will prove suitable. Great care, however, must be taken with their watering, for if given an excess, *F. elastica* will hang its leaves in a most depressing manner and may not recover. Later the leaves will fall off, leaving a bare stem. Give only just sufficient moisture to keep the compost comfortably moist, which will mean that the plants will require little during winter.

The large rubbery leaves are borne on short footstalks from a single, upright stem and they should be kept free of dust by regular wiping with a damp cloth. For potting use a size 60 pot and move to a larger size every two or three years. A compost made up of 3 parts fibrous loam, 2 parts peat and 1 part coarse sand will be suitable, but where a little decayed manure can be obtained, do use it for the ficus requires quite a rich diet. Propagation is by air-layering as described for the dracaena

There are other excellent forms suitable for a living-room, one of the best being *F. lyrata,* which requires a warmer temperature than *F. elastica.* Its leaves are shaped like those of the oak leaf. Excellent too, is *F. benjamina,* which has infinitely more grace than any form, the leaves being small and of dark, glossy green. For a large room, this is a striking plant, its huge leaves being covered in hairs and which should not be allowed to come into contact with moisture.

FIG TREE, *see* Ficus

FIR, *see* Abies

FIRE THORN, *see* Pyracantha

FITTONIA

Like many of the trailers and climbers, the plant is a native of the forests of South America, and though quite happy in a position of partial shade they must be given a winter temperature of not less than 48°F. (9°C.) whilst they must be kept away from draughts. Where grown in hanging baskets lined only with moss, the trailing shoots will root in the moss if kept moist, thus covering the outside of the basket with their pretty leaves. Or they may be grown in pots supported by wrought-iron brackets fastened to a wall.

F. argyroneura is most attractive, its pale green leaves being netted with silver veins, whilst *F. verschaffelti* has large deep green leaves which are veined with crimson.

The plants prefer an 'open' compost, composed of fibrous loam and peat in equal parts and to which is added a liberal amount of coarse sand or fine shingle. Guard against over-watering especially during winter but always ensure that the compost is comfortably moist. If the shoots become too vigorous or too woody, cut back in spring, when an abundance of new growth will follow during summer.

FITZROYA

A genus of a single species *F. cupressoides* is native of Patagonia. It is an evergreen shrub or small tree, in the British Isles requiring a mild climate. It has scale-like leaves and graceful drooping branchlets and where protected from cold winds will grow 30 ft. (9 m) tall. Plant in April in a well drained sandy soil. Propagate from cuttings removed with a 'heel' in July and inserted in sandy compost under glass.

FLAG, *see* Iris

FLAX, *see* Linum

FLEA-BANE, *see* Erigeron

FLOATING PLANTS

These are used to provide shade for fish and float on the surface of ponds and pools. It is usual to plant one to every 10 sq. ft. of surface area. They will also prevent the appearance of algae by shielding the sun's rays from the water. Plant May to September, on the surface of the water, into which they will send down their roots. In addition, several plants which provide surface shade but which have their roots in soil are mentioned here.

SPECIES

Aponogeton distachyum. The Water Hawthorn, native of South Africa and hardy in the British Isles in all but the coldest parts. It has floating lance-shaped leaves and bears spikes of white flowers with black stamens from May until autumn. Plant in containers 18 in. (45 cm) below the surface.

Azolla filiculoides. The Water Fern, a moss-like floating perennial with thread-like roots and bluish-green overlapping leaves which turn red in autumn. It bears its reproduction spores beneath the leaves.

Hydrocharis morsus-ranae. The Frog-bit, a floating perennial with thick heart-shaped leaves which turn bronze in autumn. The flowers, with their three white petals are held above the water and bloom in July and August.

Lemna gibba. The Fat Duckweed which grows over the water surface and is beloved by ducks. The roots grow down from the fronds which in this case are swollen. It is mostly found in stagnant water.

Lemna minor. The Common Duckweed, bearing tiny veined fronds which float on the surface and have a single root growing from each.

Lemna trisulca. The Ivy-Duckweed with lanceolate fronds and which moves continually from the surface to the bottom of the pond.

Stratiotes aloides. The Water Soldier, a bronzy-green perennial, submerged or above the surface when bearing its large white three-petalled flowers in June and July.

Villarsia nymphaeoides. The Floating Heart which has bright green heart-shaped leaves, spotted with purple and in July, bears crowded stems of bright yellow flowers. Plant with 18 in. (45 cm) of water over its roots.

FLOSS FLOWER, *see* Ageratum

FLOWERING CURRANTS, *see* Ribes

FLOWERING QUINCE, *see* Chaenomeles

FLOWER OF THE WEST WIND, *see* Zephranthes

FOAM FLOWER, *see* Tiarella

FOLIAGE PLANTS

INDOOR

Plants grown indoors almost entirely for their foliage possess one great advantage over flowering plants in that they will be in most instances fully evergreen, whilst they will demand far less attention to detail in their culture than is necessary to bring a plant into bloom and maintain its beauty for several months of the year. Foliage plants may also be grown in a more shady position than flowering plants and are thus more suited to living-room conditions, and whilst violent fluctuations of temperature may cause serious damage to the plants, they are better able to withstand changes of temperature and the conditions caused by the use of central heating, or a gas fire, than will flowering plants.

Most flowering plants require a window or a position to which bright light will penetrate, which in comparison with the area at one's disposal for the growing of indoor plants will be quite small. There is, therefore, considerably more scope for growing foliage plants with a liking for partial shade than flowering plants. Foliage plants, especially those having variegated leaves, have a beauty all their own and seem to fit into the scheme of contemporary furnishing much more suitably than the popular flowering plants. Again, they are able to withstand considerably longer periods without attention as to their watering, which, for those who work away from home, or who have to be away for lengthy periods, is an important factor in their popularity. But this does not mean that foliage plants will remain healthy and vigorous if left entirely to themselves, for the more care and attention they can be given, the more rewarding they will be, remaining healthy and attractive for many years.

REQUIREMENTS OF THE PLANTS

Though almost all foliage plants will respond better from individual treatment with their culture both in their soil requirements and with their preference for a certain temperature and position in the room, there are a number of conditions for maintaining the health of foliage plants which should be observed as a general guide to their culture.

Firstly, almost all foliage plants prefer partial shade and seem to be adverse to strong sunlight, so select a position away from the direct rays of the sun, the degree of shade being determined by the needs of the individual plants. The variegated leaf plants should, however, be given rather more light than those with green leaves, so that the leaf markings are accentuated.

Temperature also plays a part in the choice of a position, those plants requiring more warmth than others being placed closer to a fire or radiator and away from draughts. There are a number of indoor plants which will be quite tolerant of draughts and will in no way suffer harm where placed near a door which will be continually opened and closed, such as will occur in an entrance hall. Some plants, however, which will grow well in quite a low temperature will in no way tolerate draughts, and will suffer damage in a position whereby cold air reaches the plants upon the opening of an inside door. And guard against draughts from a closed window, for unless absolutely fool-proof in this respect – and few windows of the post-war home may be said to be completely draught-proof – it will be advisable to stand the plants some distance away. Cold air entering the room from a partially open window will also cause trouble where the plants are in a direct line. Careful selection of a suitable position is, therefore, of first importance, whilst a constant temperature will also help to maintain the health of the plants. In this respect certain plants should be given the warmth of a living-room throughout their life, whilst others will be quite happy where they receive no artificial warmth at all. For an entrance or corridor, or for a room which may be used only for 'best occasions', then those plants requiring less warmth should be used for decoration, having those of more tender habit in the living-room. But do not move the plants about unless they show signs of being unhappy. Those that are flourishing in a cool room should be left there and those growing well in shade should not be taken into the light. If they are not doing too well, that is another matter; then by all means change them around until they appear to be better suited to new conditions. Remember when making a selection of foliage plants that there are many of such easy culture that, at first, those which are more difficult to manage should be omitted.

Secondly, watering calls for care and there can be no definite rules, for so many factors must be taken into consideration. As a general rule, most plants will require very much less moisture during winter, when they are partially dormant, than during spring and summer, and it is during the winter months that great care must be taken not to over-water. Most indoor plants will require the very minimum of moisture at this time sufficient only to keep them alive; to soak the compost when the roots are not sufficiently active to utilize excess moisture will be for the compost to become sour whilst the roots will decay. As the roots become more active in late spring, with the natural rise in temperatures, the plants should be given more moisture which should be maintained through summer.

Room conditions will also determine moisture requirements for plants growing in a cool room will require considerably less moisture than where sitting-room temperatures are maintained and there is constant evaporation of moisture from the compost. Natural needs of the individual plants must also be considered and some will need more moisture throughout the year than others, and for these plants a compost capable of retaining the maximum amount of moisture should be provided. Plants growing in or near a window will require very much less moisture than where in a position away from sunlight, also where the room is centrally heated the plants will require more frequent waterings than otherwise. Those plants requiring rather warmer conditions, 'stove conditions' as they were usually called in the days of the hothouse or conservatory, will flourish where central heating is installed, for night temperatures in winter will never fall to dangerous levels. These rather more tender plants will be happier if the water they are to be given is allowed to stand in a warm room for an hour or so before use so that the chill may be off. Or add a little warm water to the cold before watering.

To maintain the glossy leaf plants in a healthy condition, it is advisable to wipe each leaf once every month with a damp cloth wrung from tepid water containing a few drops of vinegar. This will remove grease and smoke film which will block the pores of the leaves if not regularly removed, thus causing them to lack lustre and in time to die back. Any leaves which have become unsightly should be carefully removed; but the appearance of many of the foliage plants, especially those forming a stem of upright habit, will be spoilt if leaves are removed and everything should be done to preserve the foliage for as long as possible. The careful dusting of the glossy leaves as the routine dusting of the room is carried out will do much to preserve their health and beauty.

Finally, feeding and re-potting. Many plants will remain vigorous where growing for many years in the same pot, and all they require is a top dressing with fresh compost in spring each year, whilst to keep the foliage bright and healthy, watering the plants with ammonium carbonate solution, one teaspoonful dissolved in 2 pints of water, once in May and again in August, is all that is necessary. Foliage plants are not fussy as to their requirements, but they will respond to a little care by rewarding one with a richness of colouring that will compensate for any time spent on their culture.

FOLIAR FEEDING

Where growing in a soil of high alkaline content, such as often experienced where cultivating heavy land or where growing over chalk or limestone subsoils, plants will often suffer from a deficiency of iron and magnesium. This may be observed by the foliage turning yellow, a condition known as chlorosis, when the proper functions of the leaves cannot take place. If allowed to remain uncorrected, the plant will eventually die. Phosphorus and potash deficiency may be prevented by using a balanced fertiliser. A soil lacking phosphorus will produce a plant which shows blue colouring of the foliage, whilst those plants lacking potash will often fail to open their flower buds.

SOIL DEFICIENCIES

Iron deficiency may be corrected by the use of Sequestrene. It is the registered trade mark of J. R. Geigy, Basle, and is based on their iron chilate Sequestrone 138 Fe. It also contains salts of magnesium and manganese and is especially recommended for correction of iron deficiency (chlorosis) in roses growing in limestone soils. It may also be used for heaths, azaleas and rhododendrons which are growing in a soil with a too high lime content. Apply in spring at the rate of ½ fl. oz. to 1 gallon of water and during a rainy day, watering the foliage and the soil around the plants. Repeat in three weeks time. During dry weather, follow with a soaking of the soil with clear water to wash down the Sequestrene to the roots of the plants.

As an alternative to Sequestrene, water the foliage with ferrous sulphate and magnesium sulphate (separately) at fortnightly intervals, using ½ oz. dissolved in 1 gallon of water. This is best done in early morning when the sun is still off the plants and when the leaves are best able to absorb the solution.

Magnesium deficiency may be corrected by spraying the foliage with sulphate of magnesium (Epsom Salts) at a strength of 1 oz. dissolved in 1 gallon of water whilst phosphorus and potash deficiency may be quickly corrected by spraying with potassium phosphate at a strength of ½ oz. dissolved in 1 gallon of water. Potassium nitrate used at the same strength will correct nitrogen and potash deficiency.

A balanced foliar feed of proprietary make such as Murphy's or Welgro, containing salts of nitrogen, phosphorus and potash will quickly correct any deficiency caused by adverse soil conditions.

On rare occasions, the soil may lack boron, signified by the dying back of the stems of roses and herbaceous plants. This may be corrected by spraying the foliage with a solution of ⅛ oz. boric acid dissolved in 1 gallon of water.

Foliar feeding should commence in spring for the young leaves are better able to absorb the plant nutrients than older leaves. Spray both sides of the leaves, and the stems which are also able to absorb the nutrients. In order that the plants may be made thoroughly moist, a spreader or wetting agent should be used and which will already be present in most proprietary makes. An efficient spreader is Coverite, used at 1 fl. oz. to 12 gallons of water. The preparation Tulisan (based on Thiram) and used to prevent black spot in roses and mildew on roses and chrysanthemums may be included in the foliar spray so as to reduce the time taken in administering routine sprays.

FORGET-ME-NOT, *see* Myosotis

FORSYTHIA

There are several species of the Golden Bell Bush which bloom during March and April, bearing their bells along the stems before the leaves. Of great hardiness, it may be said that the plants will grow anywhere and rarely become too large for a small garden.

The first to bloom is *Forsythia giraldi,* which produces its yellow bells on arching stems. Its leaves are of a shade of olive-green and bronze. Excellent is *F. spectabilis* of upright habit, which bears its golden bells along the whole length of the stems; whilst *F. suspensa atrocaulis* forms arching sprays with the stems black, the bells being of primrose-yellow.

The plants will flourish in ordinary soil and in partial shade. Plant November to March and cut back the shoots after flowering to encourage the plants to form new shoots to carry next year's flowers.

Propagate by layering or from shoots of the partly ripened wood, taken with a 'heel' and inserted under glass.

FOTHERGILLA

It requires a position of partial shade and a soil containing plenty of peat or leaf mould. The 'flowers' appear as thick white stamens during April, before the leaves. *Fothergilla gardenii* is possibly the best form, growing only 3 ft. (1 m) tall, and besides its attractive 'flowers' its leaves turn brilliant scarlet in autumn.

Do no pruning and propagate by layering or from cuttings taken in autumn and rooted under glass.

FOUR O'CLOCK PLANT, *see* Mirabilis

FOXGLOVE, *see* Digitalis

FOX-TAIL LILY, *see* Eremurus

FRAXINUS

One of the most valuable of trees, growing well in any soil and in a wind swept locality, whilst it is tolerant of town conditions. The Common Ash, *F. excelsior* both in the weeping form, 'Pendula' and in the gold leaf form 'Aurea', are most handsome trees with their grey furrowed bark. To propagate, sow seed when ripe in autumn.

F. pensylvanica, the Red Ash of North America is a fast-growing tree with large leaves which provides shade like the mulberry and tulip tree. The form 'Variegata' has grey leaves mottled with white. *V. ornus* is the Manna Ash which makes a medium-sized tree and bears clusters of creamy-white flowers in June.

FREESIA

The freesia is one of the finest of all plants for growing under glass and like all the South African corm-bearing plants, it requires different treatment from bulbous plants. It does not require a period of darkness in which to form its roots. It requires the maximum amount of light from the time the seed is sown, for it is important that the plants do not become 'drawn'.

Freesia refracta reached this country from the Cape of Good Hope just a century ago, the sweetly perfumed white flowers having a deep orange throat. Until Armstrong introduced his deep pink freesia half a century later, little interest was taken in the flower. Crossed with *F. refracta* it produced a range of colourful hybrids which have been improved until the present time, Monsieur Morel's hybrids being the culmination of fifty years' work.

It is possible to obtain seed mixtures in each of the separate freesia colourings, the yellow to include shades of gold, primrose, orange and lemon; the red to include all shades from pale pink to deep crimson with the blue covering shades of mauve, blue and purple. The intermediate or pastel shades in which the modern hybrids may be obtained being surpassed in beauty by few other flowers. Another excellent strain is Sutton's New Hybrids; also that of Blackmore & Langdon. As with buying seeds of every description, it pays to obtain the best strains, saved and marketed by specialist growers. One thousand seeds will cost about 15 new pence.

The corms will bloom during winter and spring and all they require is a greenhouse with just enough heat to exclude frost. It is said that the freesia is difficult to grow but this is not so provided attention to detail can be given. Like so many greenhouse plants, however, it is just as easy to grow them badly as to grow them well.

SOWING THE SEED

The seed is small but not difficult and almost every grower sows in a different way as the seed is easy to germinate, but it must be remembered that the seedlings will not readily transplant. They must be sown where they are to bloom, hence deep boxes must be used, or the seed may be sown directly on to the greenhouse bed. When once the mature corms have been built up, they can be removed and re-planted, but this will not take place until after flowering and the foliage has been allowed to die back gradually.

The compost should consist of 2 parts good quality loam which has been sterilised, 2 parts peat and 1 part sand. Freesias are lovers of peat, but they must not be given a too rich soil otherwise the plants will grow lanky and produce foliage at the expense of bloom. The seed should be sown as thinly as possible and must be only lightly covered with dry compost. If sown too deeply, germination will be delayed.

As the plants will bloom where the seed is sown, deep boxes are essential, for the compost should be at least 4 in. (10 cm) deep. The boxes should be provided with drainage holes which must be 'crocked' before the compost is added. The compost should have been given a thorough soaking before the seed is sown about 1st May. The boxes are then placed outdoors and covered with glass to keep out heavy rain. The glass should be shaded on the inside with whitening, to protect the seedlings from strong sunshine but not to exclude light.

An open, sunny position should be chosen for the boxes, which should be watered only when the compost tends to dry out. By early June the seeds will have germinated, but so that the seedlings do not become saturated by rain, the glass should be kept in position until the young plants have made several inches of growth. Watering only when necessary they should be kept growing on until early October when the boxes should be moved to the greenhouse. Artificial heat will not be necessary until early November.

BRINGING ON THE PLANTS

Indoors, the plants will make rapid growth and to prevent them from becoming 'drawn' the boxes should be placed as close to the light as possible. The foliage and flower stems are supported by a few twigs or by large mesh wire-netting placed over the boxes.

At this stage, care must be taken in the watering of the plants, for freesias prefer dry soil conditions rather than a moist soil. Only just sufficient water is given to keep the blooms from flagging. This is one of those trade secrets which the experienced grower knows all about and which is not generally common knowledge. By late November, only six months from sowing the seed, the first buds will be observed and the temperature of the house should be brought up to 50°F. (10°C.) to hasten the opening of the buds. The blooms will continue to be produced through the winter and if a later batch is taken inside in early November these will prolong the supply throughout the spring.

AFTER FLOWERING

If the corms are to be saved and grown on another year, they must not be thrown outside or placed under the greenhouse bench, nor must they be knocked out of the pots or boxes until the foliage has definitely commenced to turn brown in colour, for whilst the foliage remains in a green condition, it is continuing to supply the corm with essential food for next season's crop. When the blooms have died down, remove the flower heads and the staking material and continue to give water just as before and until the foliage turns brown. My own method then is to place the pots or boxes in a shed as soon as the foliage dies back and there they remain until August when once again they are required for starting into growth. The corms are not re-potted, but the surface of the soil is carefully stirred and a little moist peat worked in. The boxes (or pots) are then placed outdoors in a sunny corner when they are watered and the cycle begins again.

If it is required to propagate where expensive new varieties are being grown, the corms should be shaken from the compost when the foliage has died down and re-potted after the offsets have been removed. The tiny offsets should then be placed in pans or shallow boxes containing a peaty compost, just covered and grown on during the summer in a cold frame, the same care being taken with the watering. Rather than grow on the corms for flowering in the following spring, it is often better to remove all bud stems and to allow the foliage to die back gradually, then to re-pot in August after a period of rest.

It should be said that where sterilised compost cannot be obtained for growing freesias from the seed sowing stage, the seed should be watered as soon as sown with Cheshunt Compound and again when the young foliage can first be seen.

GROWING OUTDOORS

Freesias for outdoors must open up a new popularity for this plant for not all possess a greenhouse or frame, but a hardy strain has been evolved from many of the best of the older-named varieties, which if given slight protection from heavy July rain appear to grow and bloom quite satisfactorily, especially in the south. They do better in the dry climate of south-eastern England than in the wetter west country. I have not yet tried out these plants in the exposed north, but from all accounts they seem satisfactory. They should be planted about 1st May when all fear of severe frost has departed. Place the corms 3 in. (7.5 cm) deep and into a soil containing plenty of peat or leaf mould. They also enjoy some cow manure which will tend to keep the soil cool during summer. But position seems more important than soil requirements. They enjoy full sunshine and yet must be protected from strong winds. Even where this protection is provided the foliage and flower stems will require the assistance of twigs to prevent damage. It has been suggested to me that the corms should be planted in clumps amongst heather or some other woody plant with a dwarf habit. This will not only provide some protection from rain, but some support for the plants. A barn-type cloche placed over the clumps at times of heavy rain would possibly be better but as yet there has been little experience with the outdoor freesias and more time must elapse before commenting in detail as to their possibilities as a commercial cut-flower crop.

VARIETIES

ALISON JOHNSTONE. The pure white blooms produced on short sturdy stems make this an ideal pot plant.

APOTHEOSE. A lovely variety producing large flowers of a deep mauve-pink colour with an attractive white throat.

APRICOT. The blooms are a rich apricot and tall growing, more suitable for cutting than for pot culture.

BUTTERCUP. The best pure yellow freesia in cultivation and which has received an Award of Merit from the R.H.S.

COTE D' AZURE. A superb variety producing an abundance of bright sky-blue flowers on long stems.

GOLDEN HARVEST. An outstanding variety, free flowering and of a bright golden yellow colour.

GWENDOLYN. A later flowering variety of a lovely shade of soft pink.

MARYON. The best sky-blue for cloche and greenhouse cutting. It flowers early and produces a large bloom of fine texture.

MAUVE QUEEN. Produces its rich purple-mauve flowers in abundance and on strong stems.

ORCHIDEA. The largest of all freesia blooms and produced on the longest stems. The colour is pale mauve with an attractive yellow throat.

ROBINETTA. A lovely rich crimson-red with a white throat and very free flowering.

ROSALIND. A grand variety for pots as it is very dwarf. The blooms are of a delicate rosy pink.

TREASURE. Its pale yellow blooms, shaded deep orange and lilac are enchanting and possess a rich fragrance.

YELLOW HAMMER. A lovely early flowering yellow for cutting, being large-flowered and a vigorous grower.

FREMONTIA

F. californica is semi-evergreen and will thrive in the most sun-drenched sandy soil totally devoid of humus. It is an ideal plant for a coastal garden in the south-west, or on the southern coast of Ireland. There it will reach a height of from 20–25 ft. (7 m) and remains in bloom from mid-May until early winter, being one of the longest flowering of all plants. The flowers are cup-shaped and of brightest gold which has given it the name of Californian Buttercup.

Plant early in spring, using pot-grown plants, placed into a soil containing some lime, best given in the form of lime rubble. The stems should be fastened to the wall and little or no pruning will be needed. Propagation is by sowing fresh seed in a heated greenhouse or frame and pricking out the seedlings as soon as large enough to handle into small pots.

FRENCH LAVENDER, *see* Santolina

FRIENDSHIP PLANT, *see* Pilea

FRINGE TREE, *see* Chionanthus

FRITILLARIA

A large genus distributed in northern Europe and Asia and North and South America with leafy stems and nodding flowers. *F. meleagris,* may be found growing in the British Isles in meadows and is in bloom late April and early May. It is known to country folk as the Snake's-head Lily. The fact that the fritillarias bloom throughout May makes them most valuable for the herbaceous border or rockery, where they should be planted in clusters, whilst several of them can be naturalised to good effect (see plate 28). Especially are *F. meleagris* and *F. pyrenaica* delightful when planted in an orchard where the grass is not cut before the beginning of July.

They enjoy a deeply dug soil and one containing plenty of humus. The bulbs appreciate any materials which will ensure a cool, moist soil during summer. Old mushroom-bed compost is ideal, also hop manure and peat. The plants must have plenty of moisture when making growth during late spring, but must have a well-drained soil otherwise the bulbs may decay during a wet winter. It is advisable to plant the large bulbs on beds of sand to encourage drainage while a mulch with rotted manure or peat after flowering will help to retain soil moisture during summer. When preparing the soil do not neglect the addition of lime rubble where this is not present in the natural form.

The bulbs are planted 6 in. (15 cm) deep in September or earlier if the soil is heavy and a strong root run is required before winter. Deep planting will prevent the bulbs from drying out during summer. They should be left down for several years for they do not like disturbance. When it is necessary to lift and divide the clumps, the small bulblets should be removed and planted into beds of peat and sand where they will attain flowering size in twelve months and be ready for their permanent quarters. This is a better method than growing from seed which takes five years for the plants to reach flowering size. If it is required to grow from seed, this should be sown as soon as ripe in pans of a pure peat and sand mixture, the seedlings being transferred to individual pots as soon as large enough to handle.

SPECIES

Fritillaria citrina. A delightful dwarf species from Asia Minor, a perfect rockery plant bearing lemon yellow flowers in May (see plate 29).

F. imperialis. The Crown Imperial, described by Parkinson in his *Paradisus* (1629) as being 'deserveth of first place in our garden of delight'. Its large bells of coppery-red and yellow are borne on sturdy 3 ft (90 cm) stems, each bell being surmounted by a whorl of feathery bright green foliage. Plant the bulbs in autumn, in a light, well-drained soil. Where the soil is heavy, plant on a layer of peat and sand and at an angle of 45° so that moisture can more easily drain from the bulbs. Plant in groups of three or four bulbs, spacing them 1 ft. (30 cm) apart and 4 in. (10 cm) deep. The bulbs will rapidly increase but should be disturbed as little as possible. It blooms in May.

F. meleagris. Producing its dainty drooping bell-shaped blooms on 12 in. (30 cm) stems during early summer, it is at its best in the orchard and wild garden planted in groups, the lovely pure white 'Alba', being enchanting when planted with the deep purple 'Nigra'. There is also a form, 'Praecox', which blooms early in April. Other than these of the white variety, the blooms are attractively chequered with pale mauve.

F. pontica. Another lovely rock garden plant producing its green, shaded rose-pink flowers during April and early May.

E. pudica. A delightful rockery plant bearing its charming little bells of purest golden yellow on only 4 in. (10 cm) stems through the springtime.

F. pyrenaica. The easiest of all the fritillarias to grow and perhaps the most beautiful. From the Pyrénées, and bears its purplish bell-shaped blooms, shaded green inside, throughout May on 2 ft. (60 cm) stems.

FUCHSIA

INDOOR

Where growing under glass, a minimum winter temperature of 50°F. (10°C.) is necessary. The plants require moisture and warmth. Propagation is by cuttings taken mid-March when established plants are cut back and started into new growth. Cuttings should be 3 in. (7.5 cm) long and trimmed to a leaf joint. A propagating frame should be used for their rooting at this time. Or remove shoots of the half-ripened wood in July and insert in sandy compost. They will root in about a month and are then planted into small pots containing a compost made up of fibrous loam, leaf mould, coarse sand and decayed cow manure in equal parts. Grow on under glass in a temperature of 50°–55°F. (10°–12°C.) and shade from strong sunlight in summer. Syringe when the weather is warm and transfer to larger pots late in summer. When established, pinch out the growing point to encourage bushy growth. The plants will appreciate occasional feeding with diluted manure water whilst they may be stood outdoors during the midsummer months, taking them indoors again early September for only a few species are frost hardy. Fuchsias are copious drinkers during summer but require only limited moisture in winter – just enough to keep them growing.

Indoors, fuchsias may be used for hanging baskets or make attractive plants grown in pots and placed in plant containers. They will require staking for they are of semi-drooping habit. They will bloom throughout the summer months, solitary or on axillary stalks.

VARIETIES OF INDOOR FUCHSIA.

BALLET GIRL. Outstanding with its rose sepals and double white corolla.

BRIDESMAID. The sepals are white flushed with rosy-pink, the double corolla being of soft lilac.

CHANG. The orange sepals are tipped with red, the single corolla being brilliant orange.

COACHMAN. It has salmon-pink sepals and a corolla of brilliant cerise.

DOLLAR PRINCESS. The sepals are scarlet, the double corolla being of royal purple.

EASTER BONNET. A symphony in pink.

GOLDEN DAWN. It has flesh pink sepals and a single orange shaded corolla.

H. DUTTERAIL. Of rich colouring with carmine sepals and double plum corolla.

KERNAN ROBSON. An all-red variety with double corolla.

ORANGE DROPS. Of unique colouring, it has orange sepals and a single corolla of darker orange.

TING-A-LING. Pure white throughout and with pale green foliage.

TORCH. It has broad pink sepals and a double corolla of pinkish-red flushed with orange.

VOODOO. The crimson-red sepals and double corolla of violet-grey combine to make this a most exciting fuchsia.

WINSTON CHURCHILL. The broad sepals are rosy-red with a double corolla of silvery-blue.

OUTDOOR

The indoor flowering or half-hardy varieties may be used for summer display outdoors and with the subtle colourings of the blooms are most attractive planted with *Cineraria maritima*, with its handsome grey fern-like fronds. After rooting the cuttings, grow on indoors in a temperature of 55°F. (13°C.) for if allowed to grow 'hard' they will bloom prematurely. Cuttings rooted in July will need moving to small pots in September and to larger pots early in March, pinching out the growing point and keeping them in a temperature of 50°F. (10°C). In April, they should be hardened in a frame before planting out in beds about June 1st. Plant firmly and space 15 in. (38 cm) apart.

Fuchsias may be used as standards for summer bedding display, beneath which geraniums, marguerites and begonias may be grown.

To build up a standard, allow the terminal shoot to grow on unchecked but removing the side growths. When the terminal shoot, supported by a cane has reached a height of 3 ft., (90 cm) side shoots are allowed to form to make up the head. The terminal shoot is then stopped and the side shoots when 6 in. (15 cm) long.

Those varieties mentioned for flowering indoors will make suitable standards and will bloom well where used for bedding, being tolerant of wet weather, like begonias.

The hardy fuchsias which grow wild in parts of west Scotland and southern Ireland may be used in the shrub border or as a hedge in those favourably situated gardens. They grow well near the sea and will reach a height of 6 ft. (2 m) or more with the minimum of attention. Plant 2 ft. (60 cm) apart in April. If cut down by frost or cold winds, clip hard back in April when they will make new growth from the base.

The same species may be used as a permanent surround to a flower bed. If the main shoots are pinched back when the plants are 18 in. (45 cm) tall, they will grow bushy without growing taller. The hardy fuchsias will grow well in partial shade. They require a sandy loam containing peat or leaf mould and some decayed manure. Water copiously during summer.

SPECIES AND VARIETIES (HARDY) OF OUTDOOR FUCHSIA

Fuchsia corallina. In the south-west it will attain a height of 20 ft. (6 m) with dark red stems and crimson-green leaves. The drooping flowers have crimson sepals and a plum-coloured corolla.

F. magellanica. Native of south Chile, it is a hardy species with finely toothed leaves, borne three in a whorl and bearing nodding scarlet flowers from June until October.

F. riccartoni. A hybrid of *F. corallina* and *F. magellanica,* raised at Riccarton, Edinburgh, in 1830. The hardiest fuchsia, it is laden with bright red flowers throughout summer and autumn. Valuable for coastal planting.

ALICE HOFFMAN. Hardy and free flowering, it has carmine sepals and a pure white corolla.

DRANE. The sepals are bright red with a double plum purple corolla.

DUNROBIN BEDDER. A very dwarf and hardy variety suitable for planting in a rock garden or for permanent bedding. It has red sepals and a purple corolla.

EMPRESS OF PRUSSIA. It bears large flowers with bright scarlet sepals and a single corolla of crimson-lake.

PHYLLIS. The sepals are rose-pink, the double corolla being of deeper pink.

PUMILA. A dwarf rock garden form having scarlet sepals and a purple corolla.

TOM THUMB. Dwarf and hardy, it has large flowers with reddish sepals and a bluish-mauve corolla.

FUNKIA

The Plantain Lily, also known as hosta, a fleshy-rooted plant which is suitable for a shady border or shrubbery for the showy variegations of its large handsome glossy leaves. The plants prefer a dry, sandy soil and should be planted in November. Propagation is by root division, but the plants should be left undisturbed for as long as possible. They are valuable for tub culture.

SPECIES

F. alba 'Marginata'. It forms a dense clump, bearing its pale mauve pendant blooms from mid-July until the end of September, its large pale green leaves being edged with white.

F. fortunei. Rather taller growing than the others, it has wide glaucous leaves, its pale lilac blooms being formed down only one side of the stem.

F. japonica 'Aurea'. It grows to a height of 2 ft. (60 cm), its flowers being pale mauve, its foliage beautifully veined with gold.

F. lancifolia 'Fortis'. This has narrow bright green foliage and bears mauve flowers.

F. medio 'Variegata'. Its wide glossy leaves are beautifully veined with silver.

GAILLARDIA

ANNUAL

The annual form *G. pulchella* 'Picta' is valuable for cutting, bearing flowers like those of the perennial form, on 18 in. (45 cm) stems, in shades of yellow, often with mahogany and bronze colourings. As it readily transplants, sow under glass early in March and plant out in May, 12 in. (30 cm) apart. The 'Lollipop' strain, bearing double ball-like flowers is outstanding. The blooms remain fresh for a week or more when cut and in water and do not drop their petals.

PERENNIAL

Though readily raised from seed, the best of the named varieties are propagated by offsets. Reaching Britain late in the nineteenth century, it quickly became popular on account of the bright colourings of its flowers and their extended season. The plants come into bloom early in July and remain colourful until early October, being excellent for small vases for they do not drop their petals. The gaillardia requires a light sandy soil enriched with humus in the form of peat and decayed manure. In a heavy soil the plants tend to die back after blooming for two years and so are best grown from seed and treated as biennials. Plant in early April,

15 in. (38 cm) apart and where marketing the bloom, cut with as long a stem as possible. The bloom should not be cut until fully open.

VARIETIES OF PERENNIAL GAILLARDIA

COPPER BEAUTY. The bloom is bright orange, shading to coppery brown and borne on 2 ft. (60 cm) stems.

CROFTWAY YELLOW. A variety bearing large refined blooms of golden-yellow on sturdy stems.

MANDARIN. It grows 3 ft. (90 cm) tall and bears large single blooms of rich tawny-red, tipped with gold.

MRS HAROLD LONGSTER. A fine variety bearing large golden-yellow blooms with a striking red centre.

TANGERINE. The medium-sized blooms are of rich tangerine-orange, freely produced.

TORCHLIGHT. Its bright golden-yellow blooms have an attractive maroon centre.

WIRRAL FLAME. Its blooms of blood-red with a striking centre are borne on 3 ft. (90 cm) stems and are amongst the most richly coloured flowers of the border.

GALANTHUS

Where planted in the south, the bulbs should be given a northerly aspect and some shade so that they may be kept as cool as possible. For this reason, the snowdrop is happiest where growing in grass which protects both bulbs and blooms from the sun. Or they may be planted around the base of mature trees, where the bulbs will receive shade and the blooms, shelter from March winds.

For massing, a 1¼ in. (3 cm) bulb will be large enough to ensure a display the first year. If smaller, they may not produce a flower until two years after planting. For window boxes and pots indoors use a 2 in. (5 cm) bulb, though the cost will be almost double. The bulbs should be light in colour and quite firm when received; a soft bulb may be a diseased bulb or one that has been incorrectly dried. The time for planting is around September 1st but not too late in the month, whilst those that flower before Christmas should be planted in May. All bulbs should be allowed plenty of time to make ample root growth before they come into bloom.

Snowdrops are not particular as to soil, provided it is well drained and does not contain too much clay. Ordinary loam is ideal, containing a small amount of leaf mould, whilst the bulbs appreciate a mulch with decayed manure and peat, given every other year in June, when the foliage has died down.

Though a bulb of good quality will quickly establish itself, division of the clumps immediately after flowering in April is more satisfactory. The bulbs will then produce a profusion of flowers in their first winter. If a light mulch is given early in June immediately after the foliage has died down it will help to conserve moisture in the bulb during a hot summer, which is vitally important, for the snowdrop dislikes excessive warmth and dryness.

Snowdrops may be planted in the shrubbery, as an edging to a path, or under any forest trees, but look at their best in short grass, especially in clusters about the lawn or in other grass which is kept short. Flowering and dying down before a lawn is first cut in April, unlike the daffodil the snowdrop does not suffer from the removal of its foliage before the sap has run back and fortified the bulb for next season's flowering.

They are ideal subjects for the rockery, planted with the winter-flowering crocus, in pockets containing peat or leaf mould and a little decayed manure. If possible they should be given a little protection to guard against cold winds which may retard flowering of the midwinter species. Or plant them around the trunks of large trees or near a hedge or wall, or in a part of the garden which may receive some protection from wind.

Like all the smaller bulbs, snowdrops should not be planted too deeply; 3 in. (7.5 cm) is the maximum. A trowel should be used; or, if planting on a lawn, it will be better to use a spade and to roll up a sq. ft. of turf to take about nine to ten bulbs, and firmly replace the turf by treading.

SPECIES AND VARIETIES

Galanthus allenii. A later spring-flowering variety, unique in that its leaves are the largest both in length and width of all the snowdrop species. The blooms are egg-shaped and of pure milk-white. 4 in. (10 cm).

G. byzantinus. A native of Turkey and besides its dainty habit it is one of the earliest to bloom, flowering from the end of November until mid-January, when *G. nivalis* succeeds it. Its large white flowers have an interesting green spot on each petal. 4 in. (10 cm).

G. colesbourne. Like the tall-growing Arnott's Seedling, it received an Award of Merit in 1951. Though dwarf in habit and ideal for window boxes, pots and cold house culture, the flowers are of great substance, frilled white and green. 4 in. (10 cm).

G. elwesii. One of the later flowering snowdrops, at its best early in March and one of the finest. The flowers are carried on long stems and are most attractive, with deep green markings on the petal tips and at the base. The foliage is of a delicate grey. 9 in. (23 cm)

G. imperati. Present in the Naples district of Italy, it produces its long-stemmed flowers at Christmas and is an excellent variety for cloche and pot work. The best form is 'Atkinsii'. 6 in. (15 cm).

G. latifolius. The best for a rockery, being very small and producing a thinly-petalled green-tipped flower late in March. 3 in. (7.5 cm).

G. nivalis. This is the common snowdrop, known and loved for its early spring-flowering. It does well in pots and under cloches. The form 'Viridipice' has green-tipped petals; 'Flore Plena' is the double variety which is not quite so early. The form 'Straffan' is outstanding, the blooms when fully open measuring 2 in. (5 cm) in diameter. The icy-white blooms are marked with green on the inner petals. The leaves are broad and strap-like, and each bulb will produce two blooms. 6 in. (15 cm).

G. olgae. From Greece, and valuable because it blooms in October. It prefers a dry position and is generally at its best on a rockery. It bears exquisitely shaped and entirely white flowers. 6 in. (15 cm).

G. plicatus. From the Crimea, it follows *G. nivalis* in its flowering-time. It bears huge pure white flowers and grey-green leaves and is most prolific in sowing its own seed. 8 in. (20 cm).

GALEGA

During medieval days, *G. officinalis* was used for feeding goats for it was said to increase their milk yield; hence its name Goat's Rue. It is not a striking plant, its pale green feathery foliage perhaps being more pleasing than its flowers. Its virtue lies in its hardiness and rather than use it in the large, sheltered border, it is a valuable back row plant for a small border where it will act as a wind break. The plants will also thrive in a dry, sandy soil and where exposed to salt winds. Plant in November and allow 3 ft. (1 m) between the plants. Propagation is by division of the roots.

VARIETIES

G. hartlandii 'Alba'. The pure white blooms are most attractive, more so than the mauve varieties.

HER MAJESTY. Similar to 'Lady Wilson', but the blooms are clear lilac-mauve.

LADY WILSON. The pale mauve pea-shaped flowers with their pretty silver keel are borne in clusters during July and August. In some soils the blossom is shaded rose-pink.

GALTONIA

Known as the Cape Hyacinth, *C. candicans* is an excellent border plant, indeed no other position in the garden will suit it so well. It bears its waxy pure white bells on leafless 6 ft. (2 m) stems during July and August and brings refreshing coolness to the border during midsummer. Though it is so tall growing, the plants never become coarse and occupy only a small area. In a well-drained ordinary soil, the bulbs are perfectly hardy. Plant 9 in. (22 cm) apart in March, the bulbs being placed on a layer of peat or sand. Propagation is by offsets every four or five years.

GARRYA

Native of California, *Garrya elliptica* which is fully evergreen, is one of the most valuable shrubs in the garden. It has handsome dark green leaves, grey on the underside, and from November until March bears clusters of long drooping greenish-yellow catkins. The plant does well in a sunless position or it may be grown against a wall. It requires little pruning, for the catkins are borne on the previous season's wood.

Propagate by layering, or by cuttings of the half-ripened wood rooted in a frame.

GASTERIA

The plants are similar in habit to the aloe and enjoy the same cultural treatment. If anything, their thick fleshy leaves which terminate in a spike are more prostrate and so the plants do not possess quite the same beauty. It is known as the Hart's Tongue for the leaves are long and tongue-shaped and are deep green with raised grey or white spots, being rough to the touch. *G. verrucosa* is perhaps the best known but *G. pulchra* is the most beautiful, bearing brilliant scarlet flowers during midsummer.

The plants are easy to grow and are propagated by means of offsets, which they freely produce. They require a compost made up of loam, leaf mould and sand in equal parts.

GAURA

G. lindheimeri is a little known plant from North America which will flourish in a dry, sandy soil. It grows 3 ft. (90 cm) tall and during July and August bears slender spikes of pale pink, tinged with purple above handsome narrow foliage. Plant in November and propagate by root division at this time.

GAZANIA

Though perennial, it is usually treated as an annual in Britain for if sown in gentle heat in January and planted out in early June after hardening, it will bloom late in July. Or it may be grown under glass, propagating by division or from cuttings. Low growing with attractive foliage, the large daisy-like flowers with their pointed petals are borne in profusion on 10 in. (25 cm) stems and are of rich colours with contrasting dark inner zones and markings.

Of named varieties, those raised by the Ebford Gardens at Topsham are outstanding. 'Boldness' is golden-yellow with a black centre; 'Fire King' is bright orange with a brown centre; and 'Strawberry' is soft pink. To grow from seed, the best strain is *G. longiscapa* 'Treasure Chest' which bears flowers in shades of pink, orange, red and gold.

GENTIAN

No garden plant bears a bloom of so intense a shade of blue and those who have grown *G. saxosa*, the New Zealand gentian, with its glistening white cup-shaped flowers and dark green leaves, will have come to appreciate this species even more than the glorious blue-flowered forms. Then there is the pink rambling form of *G. sino-ornata*, the trumpets being of a delicate shade of mauve-pink. Nor should it be thought that the gentian cannot be grown except where the soil is of an acid, peaty nature. True, the plants must have ample supplies of moisture about their roots at all times, and that means a soil well enriched with peat or leaf mould, but most of the species and hybrids will grow in ordinary garden soil so long as lime is not present. *G. acaulis* will flourish in a soil where lime is present in limited amounts.

Ordinary soil can be made to grow gentians by packing peat about the roots at planting time. Use pot-grown plants. Leaf mould is equally valuable and will keep the roots cool and moist throughout the year but this does not mean that the succulent roots will tolerate an abundance of moisture during winter. The plants like a well-drained soil wherever they are to be grown, and where the ground is low lying it will be better to grow instead, the Asiatic primulas, most of which will tolerate winter moisture. Gentians will bloom well in partial shade, whilst *G. asclepiadea* enjoys almost full shade, but as a rule they should be given an open, sunny situation. They will tolerate extreme cold and perhaps grow better in the cold, dry climate of north-east Scotland than elsewhere. Where grown cold, the plants show greater vigour and remain much longer in bloom.

Another cause of disappointment with gentians is with their planting. Firm planting is essential whether the plants have been grown in pots or are divisions from older plants. Planting time is also of importance for like most blue flowering plants, gentians are rarely happy if planted in autumn, even from pots. Plant either in spring or in June for the autumn flowering species and where the young plants have been grown-on in small pots. Gentians, though more readily established from pots, will resent being kept too long in the pots for they have a vigorous root run.

PROPAGATION

It is best to lift and divide those plants selected for propagation immediately after flowering, mid-October at the latest, and to pot each division separately. If the plants are kept under a frame light, which may be made of plastic material, during winter with care taken in their watering, they will be ready for planting out from April to June depending upon their flowering time. Spring and early summer so often being a period of dry weather, it is absolutely vital to keep the plants well watered until they are established. To allow the roots of recently planted gentians to dry out will only cause the plants to die back. Plant firmly, packing peat or leaf mould about the roots where the soil is lacking in those forms of humus, and to conserve moisture give the plants a mulching with the same materials. Keep the ground moist and remove all dead blooms as they form and the plants will be both long-lasting and long-flowering.

Gentian divisions, or offsets, are called 'thongs', and they are readily divided. If planted 12 in. (30 cm) apart in the open, either as an edging or about the rockery, alternate plants may be lifted and divided each year. This will prevent overcrowding, for where this occurs the plants do not bloom so freely. There will also be established plants to continue the display the following year, a two-year plant being capable of the most brilliant display.

Before lifting, the plants should be made moist at the roots. Shake away all surplus soil, then divide the plant into as many pieces as have roots. Each division should then be potted into a 3 in. (7.5 cm) pot using a compost composed of fibrous turf loam, coarse sand and either peat or leaf mould, in equal parts. Plant firmly and water well in, standing the pots in a frame where they will be sheltered from drying winds. As the compost in small pots will tend to dry out rapidly during warm weather, frequent waterings will be necessary, whilst to conserve moisture peat should be packed around the pots.

Where plants are to be sold, they should be made comfortably moist before removing from the pots to be separately wrapped. As they are not in any way brittle, the plants may be packed quite tightly together and sent through the post in a strong cardboard box.

It is not essential that gentians are grown-on in pots, but this does allow greater liberties with their planting. Where the 'thongs' are to be re-planted directly to the open ground, lifting and dividing should be done only in spring. To propagate in this way, when the plants are dormant in autumn, may cause the 'thongs to decay if too much moisture remains about the roots for long periods. To transplant during early summer may also cause trouble for at this time the plants may suffer from lack of moisture.

Those plants which are of short compact habit may be propagated by cuttings. These should be taken during August and inserted round the sides of a pot into a compost made up of equal parts peat and sand. They should be rooted under glass, the compost being kept moist. Rooting will take place in about six weeks when the young plants should be individually potted and grown-on in a frame for planting out in May.

Whilst the hybrids may only be increased true to form by vegetative methods, many of the species will grow well from seed. Those species which prove difficult to propagate by division should be raised from seed sown in autumn in pans containing a mixture of peat, sand and loam in equal quantities. Germination will be more reliable if the pans are covered with glass and are allowed to remain in the open where they will be exposed to hard frost. The seedlings will be ready for transplanting to small pots by early summer for growing on. The plants will be ready for planting out during autumn or preferably in spring.

SPECIES AND HYBRIDS

G. acaulis. It is the Stemless Gentian of Europe and one of the first to bloom in spring. It forms tufts of glossy leaves and bears, though sometimes not too freely, tubular flowers of pure deep blue on short stems. It prefers a heavier soil than most gentians.

G. altaica. A native of Siberia, it forms tufted rosettes from which appear navy-blue tubular flowers during May and June. It will not tolerate an excess of winter moisture.

G. asclepiadea. This is the Woodland or Willow Gentian, so called because of its willow-like foliage. It prefers almost full shade and a moist soil enriched with humus. It bears its bells of azure blue on 2 ft. (60 cm) stems, and there is also a beautiful white-flowered form, 'Alba'.

G. bernardii. A hybrid, the result of crossing *G. sino-ornata* with *G. veitchiorum* and is one of the best plants in the garden. It bears its rich Oxford-blue trumpets during August and September.

G. carolii. A hybrid from *G. farreri,* it comes into bloom in June and bears small trumpets of copper-sulphate blue.

G. dahurica. Of semi-prostrate habit, it blooms during midsummer, several 9 in. (22 cm) stems appearing from a central crown. From the leaf axils are borne tubular flowers of purple-blue.

G. depressa. Of prostrate habit, it forms a dense cushion studded during August and September with small greenish-blue flowers. It does not like winter moisture.

G. hybrida 'Devonhall'. A hybrid of vigorous constitution and bearing, during August, fat pale blue trumpets on 3 in. (7.5 cm) stems.

G. farreri. It enjoys partial shade and during July bears its electric blue trumpets on 3 in. (7.5 cm) stems. The blooms are made more attractive by their white throat.

G. freyniana. Much like *G. septemfida*, it is of easy culture and has similar glossy dark green foliage. Its blooms, however, are bright blue and of bell-like form.

G. gracilipes. Syn: *G. purdomi*, it is native of central China. The plants form a central rosette of dark green leaves from which arise numerous branching stems at the ends of which are borne bells of purple-blue during July and August.

G. hexa-farreri. Another hybrid, which forms rosettes like a sempervivum, from which are borne dark blue trumpets in July and August.

G. hexaphylla. A native of Tibet it forms a dense mat of small grey-green leaves, and during July bears six lobed flowers of a lovely shade of cobalt blue.

G. hybrida 'Inverleith'. A fine hybrid, having the same semi-prostrate form of its widely used parent *G. veitchiorum*, though the steel-blue trumpets are borne on 9 in. (22 cm) stems.

G. kesselringii. An interesting hybrid, it bears, during July and August, long trumpets of rich cream spotted with purple on 9 in. (22 cm) stems.

G. kochiana. Similar in habit to *G. acaulis*, it is valuable in that it bears its striking deep blue-and-green-spotted trumpets during May and June when few other gentians are in bloom.

G. kurroo. Found in the Himalayas, it forms a central rosette of long leaves from which arise numerous stems. From the leaf axils are borne small, clear blue trumpets, spotted with green and white. It blooms during early autumn.

G. lagodechiana. Like *G. acaulis* it does not object to lime if some peat is present in the soil. It bears its clusters of bell-shaped flowers during July on long prostrate stems, the blooms being of brilliant blue.

G. macaulayi. A hybrid from *G. sino-ornata* and *G. fareri*, it bears its trumpets of turquoise-blue during early autumn. The form 'Wells Variety' is more free-flowering, the blooms being of bright mid-blue and striped with green and purple.

G. newberryi. It forms a mat of rosettes like a sempervivum, and on 4 in. (10 cm) stems bears very pale blue flowers which have interesting bronzy-brown bands round the interior of the trumpets.

G. ornata. The dainty trumpets of Cambridge-blue with their attractive white throats are borne during August with great freedom.

G. orva. It makes a compact, prostrate plant, and during August is covered with small cups of cobalt-blue.

G. phlogifolia. From a central rosette arise 12 in. (30 cm) stems from the tops of which are borne numerous azure-blue flowers during midsummer.

G. prolata. From the Himalayas, it makes a compact plant and bears small narrow tubular flowers of vivid blue. It grows well anywhere provided it has plenty of moisture about its roots.

G. saxosa. The white-flowered gentian from New Zealand, the whiteness of the star-like flowers being accentuated by the glossy dark green leaves of the plant. Of easy culture it blooms during August with profusion.

G. sceptrum. It bears its blue-green bells on 15 in. (38 cm) stems above attractive shiny dark green leaves. It prefers semi-shade and a peaty soil.

G. septemfida. It is a plant of much interest from south Asia and which will flourish where other gentians prove disappointing. It blooms during August, its bright mid-blue flowers appearing on stems almost 12 in. (30 cm) high.

G. sinara. It is an ideal rockery plant, forming a compact mat above which are borne pretty bright mid-blue trumpets on 2 in. (5 cm) stems. In bloom during late summer.

G. sino-ornata. The last gentian to bloom, often flowering until the end of November. It does well in full sun or partial shade, its huge dark blue trumpets being borne in profusion on 3 in. (7.5 cm) stems. A hybrid form, 'Kidbrook Seedling', is very similar but blooms a month earlier.

G. stevenagensis. An excellent hybrid, the result of a cross between *G. sino-ornata* and *G. veitchiorum.* It bears large royal-blue trumpets late in autumn, which are held on 3 in. (7.5 cm) stems above grass-like foliage.

G. verna. The spring-flowering gentian liking a particularly well-drained soil and position of full sun when, during May and June, it bears its single flowers of vivid blue.

GERANIUM

ALPINE

If given a sunny position and a well-drained soil on the poor side, the geraniums will thrive in a calcareous soil. *G. farreri* is the first to bloom, in May, its large, flat flowers of pale lavender being most attractive. Flowering early and over a long season is *G. subcaulescens*, its deep pink flowers having black centres whilst the foliage is pale sage-green. Midsummer flowering is *G. argenteum* which has silver foliage and pink flowers, veined cerise-pink. *G. cinereum* 'Purpureum' bears rich pink blooms, whilst its white companion, 'Album', is even lovelier. They will grow true from seed.

HARDY

The Crane's Bills, the hardy geraniums, are natives of Britain and have been described by the gardening writers of Tudor and Stuart days from

Gerard's time. The plant likes very different conditions to the bedding geranium or pelargonium, in that it requires a cool, moist soil enriched with humus, whilst it is as happy in partial shade as in full sun, if provided with moisture at its roots. March is the best time to plant and as they make bushy growth when once established they should be given plenty of room. Allow at least 2 ft. (60 cm) between the plants. Propagation is by division of the roots in spring.

SPECIES AND VARIETIES OF HARDY GERANIUM

G. endressii. It makes a bushy plant and grows to a height of 15 in. (38 cm). It is one of the longest flowering of all border plants, coming into bloom towards the end of May and remaining colourful until October. Recommended varieties are 'A. T. Johnson', which bears flowers of silvery-pink and 'Wargrave Variety', which bears salmon-pink flowers.
G. grandiflorum. Of rather spreading habit, it bears its glorious blue flowers on 15 in. (38 cm) stems from June until the end of August.
G. macrorrhizum. Introduced from south-east Europe, it bears its large clear pink blooms on 15 in. (38 cm) stems during May and June. An additional beauty is its serrated foliage which takes on rich colourings of crimson and wine during autumn.
G. pratense 'Plenum'. Like most of the Crane's Bills this species grows as wide as it grows tall which is 18 in. (45 cm) for which reason it should not be given too close planting. It bears double mauve flowers of great beauty (see plate 30). Another lovely variety is 'Silver Queen' which bears its pale blue flowers with their silver sheen on 20 in. (50 cm) stems. Slightly taller is 'Alba', which bears large white cup-shaped flowers.
G. sylvaticum. Its deep blue flowers, veined crimson, are borne on 2 ft. (60 cm) stems during June and July. A pure white form, 'Album'; and 'Roseum', deep rose-pink are also delightful plants.

GERANIUM, *see also* Pelargonium

GERMANDER, *see* Teucrium

GESNERIA

The gesneria makes a plant of compact, bushy habit with leaves like those of the gloxinia and long tubular blooms like those of the streptocarpus, but borne in clusters. The plant grows about 12 in. (30 cm) tall, the flowers being yellow, orange or scarlet. *G. cardinalis* is a particularly fine plant, the blooms being of brilliant crimson-scarlet and the foliage glossy green. *G. exoniensis* is also beautiful, having large velvety leaves and bearing flowers of brilliant orange-scarlet.

The gesnerias require the same culture as the gloxinia and they bloom at the same time. They are, however, best wintered in their pots as for the achimene, to be started into growth again in spring, the pots being topped up with a little fresh compost. Like all members of the family, the plant requires a slightly acid soil.

GEUM

The first variety, 'Mrs Bradshaw', appeared almost a century ago as a seedling from *G. chiloense*, and with its long flowering habit quickly established itself as a valuable plant for bedding and later, for the front of the border. Ordinary soil suits it well and though a native of South America, the geum is an extremely hardy plant. Plant 18 in. (45 cm) apart, which is best done in November. Propagation is by root division or by sowing seed, several varieties coming reasonably true. Plant near the long flowering scabious, its rich blue flowers being enhanced by the vivid scarlets and yellows of the geums.

VARIETIES

DOLLY NORTH. The double blooms, borne on 2 ft. (60 cm) stems are of a vivid orange-flame colour, in bloom during June and July.

FIRE OPAL. Possibly the best of all geums. Of compact habit its intense flame-scarlet blooms are held on 18 in. (45 cm) stems and borne from the end of May until September. Could be used to edge a large border.

GOLDEN WEST. The large double blooms are of a deep golden-yellow colour borne on 18 in. (45 cm) stems from early June until August.

LADY STRATHEDEN. The blooms are of a pure lemon-yellow colour, fully double and produced through summer.

MRS BRADSHAW. Though a century old, it still remains a firm border favourite. The neat, small rounded flowers which are fully double and of a dark red colour are held on almost leafless 2 ft. (60 cm) stems and bloom from early June until the end of August.

PRINCESS JULIANA. A fine variety, its orange-rust blooms are held on rigid 2 ft. (60 cm) stems and are in bloom from early June until September.

RED WINGS. The blooms are rich crimson and borne in profusion on 2 ft. (60 cm) stems from May until September.

RUBIN. A new variety and growing to a height of almost 3 ft. (1 m) should be confined to the centre of the border. The signal-red blooms are flushed with crimson.

GIANT NEW ZEALAND REED, *see* Arundo

GIANT SUMMER HYACINTH *see* Galtonia

GILLA

Bearing heads of star-like flowers on 18 in. (45 cm) stems, this is an excellent plant either for border display or for cutting. *G. capitata*, bears blue flowers. The seed may be sown either early in autumn or in spring, the plants being thinned to 12 in. (30 cm) apart. Ordinary soil is suitable and the plant is completely hardy.

GINKGO

The Maidenhair Tree, *G. biloba* is a deciduous conifer which in a deep, moist loam will reach a height of 50 ft. (15 m). The leaves resemble those of the Maidenhair Fern and turn golden-yellow in autumn before they fall, whilst the female trees bear large yellow fruits. The variety 'Pendula' has weeping branches.

Plant November to March. Propagate from seeds sown under glass in August when fully ripe.

GLADIOLUS

INDOOR

Growing to a height of only 18 in. (45 cm) and flowering in spring, the *colvillei* gladioli require the same treatment as do freesias, and like freesias they must be in plenty of light. Again, like freesias they must be given no water until growth appears. The corms are inexpensive and are best planted late in August, one method being to set the corms 2 in. (5 cm) deep, placing three to a large pot. Use an acid-free soil for gladioli do not like acid conditions. If in doubt, add a handful of lime rubble to each pot and the same amount of peat or decayed mushroom-bed manure, and some sand. The pots are then removed to a plunge bed, under a wall is ideal, where they are covered with sand or weathered ashes and there they remain, covered with corrugated sheets or frame lights until early January. Gentle forcing only is required, a temperature of 50°F. (10°C.) being suitable and careful attention must be given to watering, giving only the minimum to keep the soil in a just moist condition. If there is any doubt about the greenhouse having sufficient light in midwinter it is better to leave the pots in the frame, protected from frost until mid-February when light conditions are better. The blooms may need support which should be provided by means of thin canes tied with raffia.

VARIETIES AND SPECIES FOR INDOOR FLOWERING

G. colvillei 'The Bride'. Perhaps the purest white flower in cultivation, for not only are the petals pure white, so are the anthers. The variety rubra bears flowers of carmine-red.

G. nanus 'Amanda Mahy'. It bears flowers of bright salmon-red and produces a spike of 12 in. (30 cm). It is valuable in that it may be planted early October both in a sheltered position in the open and in small pots for indoor flowering in May; or the corms may be grown on entirely in a cold frame. Outdoors it blooms from mid-June to mid-July.

OUTDOOR

The gladiolus is one of the easiest of all plants to grow provided the soil has been well prepared. It enjoys a soil which has been previously manured for a crop of peas or potatoes, for it should not be given any manure which is too rich in nitrogen. This will cause the spike to grow too tall and soft and the flowers will lack the much-desired brightness of colour. The exhibitor will go to any lengths in the preparations of the soil, but normally, one into which some humus-forming materials have been incorporated, is all that is necessary. Spent hops, peat, old mushroom-bed manure, decayed leaves, all are suitable and should be forked in during March, the ground having been dug over and cleaned and left to weather over winter. The often sour soil of a town garden should have a light dressing of lime. At planting-time, a 2 oz. per sq. yd. dressing with bone meal and liberal quantities of bonfire-ash raked in will supply the slow-acting fertilizers and the much-needed potash to build up sturdy growth and enhance the colour of the blooms.

Equally important as the preparation of the soil, is situation, for the gladioli will grow to a height of up to 5 ft. (1.5 m) and even the more dwarf *primulinus* and miniatures may be seriously damaged by winds when the stems are heavy with bloom. Delightful as they are in a mixed border, gladioli are at their best when planted in beds to themselves, for here they may be given the culture they deserve, and the brilliance of their blooms may be seen without distraction. But in this way they must be given protection against winds. They are essentially lovers of the sun and require an open aspect.

PLANTING THE OUTDOOR CORMS

For early bloom, those situated in the favourable districts of the south-west may plant about March 1st. In the south midlands, planting may be done at the end of March and in the colder north, mid-April is soon enough. It is important to allow the ground time in which to become warmed by the spring sunshine and if delayed it is better to wait a week or so after the usual planting-time. Corms, bulbs or seeds which are in soil too cold to allow any root growth will be better left in their containers. Again, it is a debatable point as to the hardiness of the gladiolus and growth should not be too advanced whilst frost is likely to be experienced.

Planting should be done to provide a succession of bloom, though the varieties themselves will naturally bloom at various times and in a cold exposed position it will be advisable to plant early and to grow quick-maturing varieties. This is more important than generally realized. For instance, the new small-flowered variety, 'Scotia', a lovely yellow with peach markings, will take up to one hundred days to reach maturity, compared with only eighty days taken by the ruffled miniature, 'Statuette'. Should the soil not be in suitable condition for planting the corms before mid-April, it will mean that 'Scotia' may not be in bloom until late in August and in a cold summer it could be well into September before it flowers in an exposed garden.

The size of the corm is a matter for consideration. The medium-sized high-crowned corm is the best and will produce a larger spike and of better quality than a flatter, but possibly larger corm. Not all varieties produce the same sized corm and a $4–4\frac{3}{4}$ in. (10–12 cm) corm of some varieties may give as good a flowering spike as a $4\frac{3}{4}–5\frac{1}{2}$ in. (12–14 cm) corm of another variety. Like the very large-sized anemone corms which have become acclimatized to the soil and climate in which they have been grown, larger size gladioli corms may have passed their best. For an exhibition spike, the $4\frac{3}{4}–5\frac{1}{2}$ in. (12–14 cm) size is most suitable. The cut-flower grower, growing for a market that is not quite up to exhibition standards, will find the $4–4\frac{3}{4}$ in. (10–12 cm) size is suitable. If growing on for a second season, the $4–4\frac{3}{4}$ in. (10–12 cm) size is a necessity for after cutting the spikes, even if several leaves are allowed to remain, the corm will produce a smaller spike in its second season. If in the first place the $3\frac{1}{4}–4$ in. (8–10 cm) corm is used, the spike produced the following season will be of little use for marketing (see plates 31 and 32).

The corm should be firm at the top when gently pressed with the fingers and it should not show any blemishes, neither should it be blackish brown in appearance which would signify 'fusarium yellows'.

The best way of planting is in trenches, for here the ground may be given more concentrated attention and staking and cutting is more easily carried out. As the gladiolus is essentially a cut flower, the beds should be of sufficient width to make cutting as easy as possible and trenches 2 ft. (60 cm) wide are suggested. Again and even more important, if the same distance is allowed between the beds or trenches, this ground can be used in alternate years so that the corms are not grown on the same soil for two successive years.

Plant the corms with a trowel or spade but never with a dibber, for no cavity must be allowed beneath the base where the roots are produced. When planting in trenches, a popular method is to remove a 4 in. (10 cm) depth of soil with a spade and to 'set' the corms rather than to 'plant' them. In clay soil, slightly shallower planting will be advised.

On ground that is well drained the trench bed system is to be recommended in preference to the more popular idea of planting in raised beds for not only is watering more effective in dry weather, but it is possible to earth-up the plants as top growth takes place; or a soil and peat mulch may be given around the plants. This will assist the plants during dry weather but more important, it may do away with the need to stake the plants, especially in the case of the smaller-flowering varieties. The commercial grower will plant in beds with a pathway between each and the corms will be set no more than 6 in. (15 cm) apart. In this way they will support each other and if earthed-up, little or no staking will be necessary. When planting in mixed borders, it is preferable to plant about 8 in. (20 cm) apart in groups of four. The flower spikes may be supported by tying green twine around the clump and if in a sheltered position it may then be possible to do away with a stake. If growing in a wind-swept position, beds of gladioli should be supported by extending twine along the rows fastened to stout stakes at each end and every 6 ft. (2 m). The exhibitor will stake each individual spike.

Planting the corms directly onto a layer of sand is to be recommended where the soil is heavy.

Corms planted early in April will produce shoots above the soil at the end of the month, but they will require no attention until early June when the first earthing-up may be necessary and spraying once each week with a weak solution of potassium permanganate which will make the leaves bitter to insects and so keep off fly attacks. As the stems become swollen, generally towards the end of June, gentle feeding should be commenced. This is the most vital stage in the life of the gladiolus, for on it depends not only length of stem and quality of the bloom, but also the vigour of the corm for next season and the formation of the cormlets for growing on. As it is necessary to provide the cormlets with both nitrogen and phosphates, this is given as diluted animal manure water, made by dissolving poultry, pig, sheep or horse droppings by suspending the droppings contained in a bag, into a barrel or tank of water. Bearing in mind that it is better to provide the corms with a continuous supply of dilute fertilizer rather than an occasional concentrated feed, the manure water should be used once a week. An excellent method of applying this is to make with the back of a rake, a drill 2 in. (5 cm) deep on either side of the bed or trench and to pour the manure water into the drills. It will gradually find its way to the corms.

Gladioli also love soot water which may be given once a month or even alternate fortnights in addition to the liquid manure. It cannot be stated too often that feeding should continue until long after the flower spike has been removed, until early autumn so that the old corm will be built up for the following season's flowering and the cormlets will continue to make growth.

The corms will appreciate a mulch of peat or hop manure in June for though the gladiolus enjoys a position of full sun, it will not tolerate a soil which is too dry. During a dry period and especially if the soil does not contain much humus, regular soaking in addition to the manure water application is essential to the formation of a first-class spike and good spawn growth.

GROWING UNDER CLOCHES

The gladiolus does well under barn cloches but it makes such rapid growth that though early flowers may be required, the removal of the glass before the often cold and windy weather of May has departed, may be to cause the spike to be damaged. Cloching and planting, is not advised before mid-March, so that the coverings may be removed towards the end of May, not before.

The trench method is most suitable, made to the width of the glass to be used and dug about 10 in. (25 cm) deep. It should be prepared as described, but 3 in. (7.5 cm) from the top of the trench should be allowed which will enable the stems to make maximum growth whilst covered with the cloches. The corms should be planted 6 in. (15 cm) below the top of the trench, being 3 in. (7.5 cm) deep in the soil. Plant in rows allowing 6 in. (15 cm) between each corm. Closer planting is not advised or mildew might be troublesome. It is advisable to prepare the trench during early winter to allow time for consolidation and the cloches should be placed over the trenches at the end of February to allow the soil to warm up before planting the corms. Feeding with liquid manure and soot water should continue from mid-May and as soon as the cloches are removed the top 3 in. (7.5 cm) of soil should be filled in to give the plants extra support, peat or hop manure being mixed with soil for this purpose which also acts as a mulch.

PROPAGATION OF THE CORMLETS

Though the gladiolus is not difficult to grow from seed, named varieties are increased by growing on the tiny corms or spawn to be found clustered at the base of the old corms when it is lifted in October. Some varieties are fairly shy with their cormlet production, others spawn in abundance, the shy varieties remaining more expensive even though they may be long-established varieties. During the course of a growing season the gladiolus produces a completely new corm, perhaps two, immediately above the corm originally planted which will have withered away. Upon lifting, the decayed corm should be removed and burnt. The cormlets, varying in size from little more than pin-heads to small corms will be found clustered round the base of the newly formed corm. Every one, however small, will be carefully removed, for all will eventually grow into flowering-sized corms. The very smallest will take up to three years to attain flowering size, those which are larger will bloom in two years. This being so, it is advisable to segregate them, using clean paper bags in which are punched several air holes and to store them during winter, having removed all soil particles at lifting-time.

The cormlets are grown on in beds, the soil first being brought to a fine tilth with the back of the head of a rake; drills 1 in. (2.5 cm) deep are made 6 in. (15 cm) apart and along each is sprinkled a mixture of peat and some coarse sand. The cormlets are then dropped into the drills in the same way as for planting peas, early April being the most suitable time. A position of full sun is advisable and it is essential to keep the ground free from weeds. Neither must the corms be allowed to suffer from lack of moisture. An aid to germination is to place the cormlets in a warm room for a fortnight before planting out and to immerse them in cold water the night previous to planting. So that they may be handled easily, they should be taken from the water an hour before required for planting and partially dried off in a warm room. They must not be allowed to flower during their time in the seed bed, as all their energies must be reserved for the formation of a substantial corm. The corms should be fed with dilute liquid animal-manure water during August and September and they are lifted and stored in early October in the same way as for mature corms.

Lift the corms early October, when the foliage appears to be turning yellow. Make a dozen or more into a bunch and string up in a dry, airy room to dry off. Then remove soil and the dead corm which has borne last season's flower, also the cormlets before cutting away the leaves to within 1 in. (2.5 cm) of the neck of the new corm. The corms are stored in trays in a frost free room until time for planting out in spring.

CLASSIFICATION OF GLADIOLUS

Grandiflorus. Covers all sections except the *Primulinus* group.

(a) Midget-flowered. Florets not to exceed 1½ in. (4 cm) diameter.

(b) Miniature-flowered. Florets must be over 1½ in. (4 cm) diameter but must not exceed 2½ in.

(c) Small-flowered. Florets must be over 2½ in. (6.5 cm) diameter but must not exceed 4 in.
(d) Medium-flowered. Florets must be over 4 in. (10 cm) diameter but must not exceed 5 in.
(e) Large-flowered. Florets must be over 5 in. (12.5 cm) diameter but must not exceed 6 in.
(f) Giant-flowered. Florets to be over 6 in. (15 cm) diameter.

Primulinus. The florets to be hooded and must not exceed 3 in. (7.5 cm) in diameter and carried on slender stems.

Primulinus grandiflorus. The florets to be hooded and over 3 in. (7.5 cm) in diameter. Of similar build to the *Primulinus* and loosely arranged on wiry stems.

GLAND BELLFLOWER, *see* Adenophora

GLECHOMA, *see* Nepeta

GLOBE FLOWER, *see* Trollius

GLOBE THISTLE, *see* Echinops

GLORY OF THE SNOW, *see* Chionodoxa

GLORY VINE, *see* Eccremocarpus

GLOSSARY

ACHENE. A hard, dry, one-seeded fruit as in clematis.

AERATE. To allow air to reach down into the soil.

ALTERNATE. Where the leaves are arranged on the stem one after another on opposite sides; or where the stamens of a flower appear in the spaces between the petals.

ANNULUS. The veil or tissue ring present around the upper part of a mushroom stem and which was joined to the cap enclosing the gills before the mushroom opened.

ANTHER. Pollen-bearing part of the stamen.

AQUATIC. Plants growing in water.

AXIL. The upper angle formed by stem and leaf base.

BEARDED. Long hairs formed like a beard, as present on the crests of certain irises.

BERRY. A fruit containing seeds embedded in its juice as in gooseberry.

BIENNIAL. A plant of two years growth; one which blooms the year following that in which the seed is sown as with stocks, Canterbury bells.

BOLT. To run to seed prematurely.

BRACTEOLES. Small bracts attached to the base of the pedicels.

BRACTS. Small leaves which differ from the others and which are present on the flower stalks.

BREAK. Development of new growth or side shoots.

BUDDING. The removal of a bud or eye from one plant to 'marry' or bring into contact with the cambium layers of another so that both may grow as a single unit.

BUD. It contains the flowers; the leaves; or is the growth 'bud'.

BULB. An underground leaf-bud with fleshy scales compressed together around a disc-like woody stem as in allium, tulip, hyacinth, squill. When the fleshy leaves are folded round each other, the bulbs are known as 'tunicated'. If a bulb is cut into two, at the centre will be found the flower stem with its incipient bloom packed tightly away and protected by the scales or leaves which not only protect but store up food and moisture. This they obtain from the green leaves above ground before they die back.

BULBILS. Tiny bulbs borne in the leaf axils and which may be detached when ripe for growing on to flowering size bulbs in possibly two to three years.

BULBOUS. A plant with bulb-like stems.

CALCAREOUS. Term used for a chalk or limestone soil.

CALYX. The green whorl of leaf-like organs of the flower situated below the corolla.

CAMBIUM LAYER. The layer of dividing cells situated between the wood and bark of a tree or of any Dicotyledon. Each year, by division of the cells, a layer of wood and a layer of bark is formed, causing the plant to increase in size (Exogenous). This provides the rings on the inner wood.

CAMPANULATE. Bell-like as in campanula, leucojon, galanthus.

CAPITATE. Growing in heads as in Compositae.

CAPSULE. A dry, many-seeded vessel.

CATKIN. Flowers of one sex closely crowded together, the perianths being replaced by bracts as in willow.

CLIMBER. A plant with long, straggling shoots.

CORDATE. Used to describe a heart-shaped leaf with two rounded lobes at the base and terminating to a point.

CORM. It differs from a bulb in that it is solid and has neither tunicated nor scaly leaves. A corm is a round or flat stem which dies back each year, leaving behind at the base a new corm developed by the action of the leaves, e.g. gladiolus, crocus.

CORMLET. A small corm often to be found clustering around the disc-like base of the mother corm. These are detached when the plant has died back and the corms are lifted. They are grown on in boxes of sandy compost until reaching flowering size.

COROLLA. A whorl of floral leaves, known as petals, situated between calyx and stamens.

CORONA. The crown or trumpet, situated at the centre of narcissus.

CORYMB. A raceme of flowers on pedicels which decrease in length as they approach the top of the stem, thus bringing them on to the same level.

COTYLEDON. The seed lobes usually forming the first leaves of the plant.

CRENATE. The small, rounded teeth at the margins of leaves.

CRUCIFORM. A flower having four petals arranged in cross-formation as in wallflower, aubrietia.

CUTTING. The name used for a shoot of hard or soft-wooded plant which has been removed for rooting.

CYME. A terminal inflorescence beneath which are side branches with a terminal flower.

DECIDUOUS. Plants which lose their leaves in autumn as in beech, maple.

DECUMBENT. Stems which lie flat on the ground but with growth rising up at the tips.

DEHISCENCE. The manner in which an ovary opens to shed its seed.

DENTATE. With triangular teeth at the margins of leaves.

DICOTYLEDONS. Plants with two seed leaves and afterwards having net-veined leaves.

DIGITATE. Leaves divided into finger-shaped leaflets all of which begin from the top of the petiole as horse chestnut.

DIOECIOUS. Plants with differently sexed flowers on different plants with stamens on one plant, pistils on another as in willow.

DISC. The surface from which stamens and pistils arise; or the round flat surface at the base of a corm from where the roots form; or the central florets of compositae as in daisy, helenium

DRUPE. A fleshy fruit with a hard stone as in plum.

EMBRYO. The germ of a plant in the seed. Every fertilised seed contains an embryo.

ENTIRE. Leaves not divided nor toothed at the margins.

EVERGREEN. A tree or shrub that retains its foliage all the year round.

EXTRORSE. Anthers which shed their pollen outwards.

EYE. A bud which is removed with a piece of bark attached (the shield) in the propagation of roses and other hard wooded plants.

FASTIGIATE. Used to denote the appearance of a tree when the branches are close and upright as in various forms of cupressus and juniperus.

FIMBRIATE. The petals of flowers which are fringed at the margins as with certain pinks and carnations.

FLORETS. The small rayed petals of compositae.

FROND. The leafy part of a fern.

GALBALUS. The fleshy cone of juniperus and cupressus.

GLABROUS. Smooth, glossy; without surface hairs as the leaves of laurel, camellia.

GLANDULAR. Resin secreting cells found on the leaves of certain plants as bay laurel.

GLANDULAR-HAIRS. Hairs which are tipped with glands and present on various (or all) parts of certain plants, e.g. the leaves of *Pelargonium tomentosum.*

GLAUCOUS. The leaves of those plants which have a bluish lustre, usually caused by minute hairs.

GLOBOSE. Globular, like the flowers of *Echinops ritro.*

GRAFT. Used for the uniting of scion to chosen rootstock, a method of propagation.

HALF-HARDY. Applies to those plants which are given protection during winter and early spring in all parts of the British Isles except where winter climatic conditions are favourable such as Cornwall and south Devon; south Wales; south-west Scotland. Depending upon their degree of hardiness, half-hardy plants are wintered in a cold frame or greenhouse with or without heat.

HARDY. A term applied to those plants which may be grown outdoors in all parts of the British Isles without protection through their period of growth, e.g. lupin, wallflower.

HEAD. A terminal inflorescence surrounded by an involucre.

HEELING IN. This is the term used for covering plants with soil and pressing in with the heel until such time as ready to plant in their permanent quarters. The method is to make a trench 9 in. (22 cm) deep and to place the roots at the bottom. Soil is then placed over them to a depth of about 6 in. (30 cm). This will ensure that the roots are kept moist and protected from frost until the ground is ready for their permanent planting.

HERBACEOUS. Plants with succulent green parts, not woody.

HERMAPHRODITE. Where stamens and pistil is present in a flower.

HIRSUTE. Generally used to denote leaves which are covered in long soft silky hairs.

HOARY. Plants with leaves covered in white down.

HUMILIS. Dwarf, compact, low growing.

HUMUS. The condition of the soil when various materials, e.g. leaf mould, peat, have been incorporated and bacterial action has broken down the soil to a condition conducive to supporting satisfactory plant growth.

The addition of organic manures will aid bacterial action and improve the soil.

IMBRICATE. Arranged over each other like the scales of a leaf bud.

INCURVED. Flowers with petals which curve inwards, like those of incurved chrysanthemums.

INFLORESCENCE. Arrangement of the flowers on the stem.

INTERNODE. The space between the nodes of a stem.

INTRORSE. Anthers which open inwards towards the pistil.

INVOLUCRE. The whorled bracts at the base of a single flower or flower head.

KEEL. The lower pair of petals of pea-like flowers, e.g. sweet pea.

LABIATE. Lipped; the corolla or calyx divided into two unequal parts as with lavender, rosemary.

LANCEOLATE. Lance-shaped, narrow, tapering to each end as with the leaves of *Lilium brownii*.

LAX. Loosely arranged, describing flower arrangement on the stem.

LAYERING. The term used for propagating a shoot by bending it so that it comes in contact with the soil. A cut is then made in the shoot in an upwards direction as far as the lower node or joint and so that a part of the stem remains joined to the parent plant. The flap or tongue is bent away and a small pebble inserted near the node to prevent it from closing again when planted in the ground. It is from the node that the cutting will form its roots in three to four weeks time when it may be severed from the parent plant and grown on.

LEAFLET. The sub-division of compound leaves.

LEGUME. A one-celled and two-valved seed vessel.

LINEAR. Used to describe leaves which are long and narrow.

LIP. A term used to describe the most conspicuous part of an orchid.

MARCESCENT. Dying back but still in its place as the calyx and corolla of certain flowers and the leaves of a beech hedge.

MIDRIB. The large vein extending down the centre of a leaf from petiole to tip.

MONOCARPIC. Flowering and fruiting once only like certain cacti and agaves.

MONOCOTYLEDON. Flowers with only one sheathing cotyledon or seed leaf as lily, crocus (bulbs and corms).

MONOECIOUS. Stamens and pistils on separate flowers but on the same plant.

MYCELIUM. The spawn threads of mushrooms and other fungi which run through the growing medium before branching upwards to fruit (the fungus).

NECTARY. A honey-secreting organ to be found at the base of petals as in fritillary; helleborus.

NETTED. Leaves (or corms) covered in veins or fibres to produce a net-like appearance.

NODE. The point on a stem from where a leaf is produced.

NUTANS. A term used for drooping or nodding flowers as *Ornithogalum nutans*.

OPPOSITE. A term used to denote leaves which are formed in pairs, one on either side of a stem, opposite to each other.

OVARY. The immature seed vessel.

OVULE. The name applied to the young seed before fertilisation.

PALMATE. Usually applied to leaves, the segments spreading out like the fingers of a hand from a central point.

PANICLE. Flowers borne in a raceme at the end of branching pedicels.

PARAPITIC. Used to describe one plant which lives on another.

PEDICEL. The stalk of an individual bloom.

PEDUNCLE. A flower stalk.

PELTATE. A term used to describe those leaves when the point of attachment is on the face and not the side as is more usual.

PERENNIAL. A plant of more than two years duration but usually continuing to flower and fruit each year indefinitely as paeony, primrose.

PERIANTH. The floral parts when calyx and corolla are indistinguishable as in tulip.

PERSISTENT. Usually for leaves remaining on the plant, not falling off as in evergreens.

PETAL. The division of the corolla.

PETAL-LIKE. A term used for sepals which have assumed the appearance of petals as in clematis.

PETOILE. The lower stalk of a leaf from where it joins the main stem and up to the first leaf or pair of leaves.

PILEUS. The cap or fruiting body of a mushroom or fungus.

PILOSE. Covered in stiffish hairs.

PINNATE. When leaflets are arranged in pairs on either side of a stem, opposite to each other.

PINNATIFID. A leaf cut almost to the mid-rib into a number of segments.

PIPING. An unflowered shoot of a pink or carnation which is removed by pulling away in an upwards direction so that the terminal leaves and a small piece of stem remain for rooting.

PISTIL. The ovary, style and stigma together.

PLICATE. Folded, like some leaves before becoming unfolded.

POLLEN. Dust on the anthers which fertilises the ovules.

POLLINATION. Where the ovules are fertilised by the pollen tube.

POME. A name used to describe the apple and pear.

PRAECOX. Early flowering.

PRATENSIS. Growing in pastureland or meadows as *Cardamine pratensis*, 'Our Lady's Smock'.

QUINATE. Arranged in fives as the leaves of *Akebia quinata*.

RACEME. Stalked flowers borne in a spike as in laburnum.

RADICAL. Arising from just above the root, used to describe the leaves of plants of tufted habit.

RAY-FLORETS. The outer flat rayed segments as in pyrethrum, daisy.

REFLEXED. Bent backwards as in certain varieties of chrysanthemum.

REGULAR. When all parts of a flower are alike.

RETICULATE. Forming a net of fibres as with the bulbs of *Iris reticulata*.

RHIZOME. The thickened base of a shoot or stem which grows horizontally usually beneath the soil from which roots are produced and from the end of which arises the flowering stem as in Solomon's Seal and *Iris germanica*. The corms of Crocosmia are joined by a thin rhizome.

RINGENT. An open two-lipped corolla.

ROOTSTOCK. A thick short rhizomatous root as in *Scabiosa caucasica*.

ROSETTE. Leaves radiating from a central underground stem, often overlapping to form a circle from the centre of which arises the flower stem.

RUNNER. A shoot growing along the ground and rooting at the end as with violet, strawberry.

RUPESTRIS. Growing about rocky formations.

SAGITTATE. Used to describe leaves shaped like an arrow.

SALVER-SHAPED. A corolla with a slender tube as in several campanulas.

SCALES. Used to describe the overlapping fleshy leaves of lily bulbs.

SCAPE. A leafless flower stem arising from below soil level as in tulip, hyacinth, galtonia.

SCION. The term used for a cutting which is to be united to a selected rootstock to produce a fruit tree or ornamental tree where it is desired to reproduce a variety.

SEGMENT. Used to describe a petal of a flower or parts of a leaf divided almost to the mid-rib.

SEPALS. The division of the calyx.

SERRATE. Saw-edged or toothed, used to describe leaves.

SESSILE. Stalkless, as for certain leaves.

SHEATH. The lower part of a leaf which forms a sheath or protection around the stem.

SIMPLE. Not branched, lobed or divided.

SINUATE. Bluntly lobed as in the leaves of the oak.

SOLITARY. Flowers borne singly, one on a stalk.

SPADIX. A succulent spike bearing multitudes of flowers tightly packed together as in Arum.

SPATHE. The large bract enclosing the spadix.

SPIKE. Like a raceme except that the flowers are stalkless.

SPINE. A persistent woody thorn.

SPUR. An extension to the lower part of a corolla as in aquilegia, tropaeolum.

STAMEN. The male organ of a flower composed of filament and anther.

STANDARD. The large upper petal of a pea-flower as in sweet pea.

STELLATE. Star-like; radiating from the centre.

STIGMA. The sticky cellular part at the top of carpel or style to which pollen adheres.

STIPUIES. Leaf-like appendages at the base of the petiole.

STOLON. An underground runner producing roots at intervals.

STRAP-LIKE. Used to describe leaves which are broad and long as in vallota, clivia.

STYLE. The termination of a carpel bearing the stigma.

SUCKER. A leafy stem produced from an underground shoot usually some distance from the main stem as in lilac, plum.

SUPERIOR. Used to describe a calyx when its tube is wholly attached to the ovary.

TAP ROOT. A root with a long tapering body as in parsnip, carrot.

TENDRIL. A wire-like organ developed from the end of leaf or at the axil by which the plant attaches itself to objects to enable it to climb.

TERETE. A round stem.

TERNATE. Arranged in threes as the leaves of *Choisya ternata*.

THROAT. The orifice of the tube of a corolla.

TOMENTOSE. The silky covering of leaves of certain plants as in *Pelergonium tomentosum*.

TOOTHED. Used to describe the edge of leaf or petal, divided into numerous small teeth.

TRIFOLIATE. Leaves composed of three leaflets as in clover.

TRIFOLIOLATE. Three leaflets proceeding from the same point.

TRIPARTITE. Divided into three almost to the base.

TRUNCATE. As if cut off at the end.

TUBER. A swollen underground stem furnished with buds or 'eyes' as in anemone, oxalis, begonia.

TUBEROUS-ROOTED. With the base of the stem similar to a tuber.

TUNIC. A thin membranous covering of a bulb as in narcissus, tulip.

UMBEL. Where numerous flower stalks arise from the same point and bear their flowers at the same level as in umbelliferous plants.

VERNALIS. Spring-flowering.

VERNATION. The leaf arrangement when in bud.

VERSATILE. Applied to anthers which have a back and forwards movement in the wind as with lilium.

WHORL. Usually applied to leaves arranged in circular fashion around an axis or stem as in asperula.

GLOXINIA

With its large trumpets, a well-grown plant will bear as many as fifty or more blooms in a single season to make this one of the most beautiful of greenhouse perennials. The long dark green leaves are deeply ribbed, whilst the blooms held on short erect stems are frilled at the margin.

SOWING THE SEED

Gloxinias require a slightly acid soil so do not use the John Innes Compost which contains lime. Instead, use a sowing compost made up of leaf mould, horticultural peat and silver sand in equal parts. The leaf mould must be genuine mould, not just partially decayed leaves. Use a small seed pan which has been well crocked and over the crocks place some leaf mould of rough quality and then the finely riddled compost. Make level and firm, and give a soaking with water several hours before sowing.

It is important to sow quality seed from a specialist grower or sow the 'Weidenhoff' strain of Weiser & Virnich which produces large refined flowers in brilliant colours. An ounce of seed is said to contain over half a million seeds which cost three times as much as an ounce of gold so they should be treated with respect. The Gloxinia is one of the few plants whose seed will germinate readily if more than three years of age. Surplus seed should be kept for sowing the following year rather than to sow an excess at one time.

Mix the seed with a small amount of dry silver sand before sowing which is done with finger and thumb and as evenly as possible, taking care not to sow too thickly. Cover with a little silver sand and water lightly before placing in a propagator, kept at a temperature of 70°F. (21°C.).

If the seed is sown about 1st March and a humid atmosphere is maintained, it will germinate by the month's end and by mid-April the seedlings will be ready for transplanting, into a compost made up of 2 parts fibrous loam, 1 part leaf mould and 1 part coarse sand. The young plants should be grown on in a similar temperature and great care must be taken with their watering, for if not given sufficient, they will die back, and if given too much, fungal diseases may cause them to wilt. Inspect the seedlings frequently and water when necessary.

By mid-May, the young plants will be ready to move to size 60 pots (60 to a cast) in which they will bloom. The compost should be made up as for the first transplanting, but this time add 1 part of well decayed manure, preferably cow manure. Also include a little charcoal to keep the compost sweet. Lime must not be used.

A minimum night temperature of 52°F. (11°C.) should be maintained. but the plants must be protected from strong sunlight by whitening the greenhouse glass. Though requiring warmth and humidity, all the Gesneriaceae have a dislike of strong sunlight which will cause burning of the foliage. If the plants continue to make vigorous growth they should be moved to a larger pot containing a similar compost, for they must not become root-bound. Take care not to plant too deeply, otherwise rotting of the crown may occur.

As the plants come into bloom early in July they will require copious amounts of water, and to prevent evaporation the pots should be placed in trays of moist peat to which water is given along with the plants. The plants will continue to bloom until the end of autumn when moisture is gradually withheld to allow the foliage to die back. The small tubers should then be shaken from the pots, cleaned and stored in boxes of sand in the cupboard of a living-room. The sand will prevent the tubers from shrivelling. They may be started into growth again in spring.

The plants should be sprayed for thrips before coming into bloom. Spray with 'Lindex' solution at a strength of ¼ fl. oz. to 1½ gallons of water soon after the first potting and repeat ten days later.

VARIETIES

Switzerland bears large scarlet trumpets with a frilled edge of white. Three Dutch introductions, 'Gerda Lodder' (scarlet), 'Lodder's Rose' (deep pink) and 'Glory of Utrecht' (violet), are equally lovely as are those of Messrs. Blackmore & Langdon. Beautiful is 'Bacchus', the blooms being of a rich shade of wine red, whilst 'Her Majesty' bears a flower of purest white. Lovely, too, is 'Pink Princess', the white bloom being edged with deep pink; and 'Grenadier', which bears trumpets of rich glowing scarlet.

GOAT'S BEARD, *see* Spiraea

GOAT'S RUE, *see* Galega

GOAT'S THORN, *see* Astralagus

GODETIA

Deservedly amongst the most popular of all hardy annuals which are sown where they are to bloom. There are tall flowering varieties like the orange, 'Kelvedon Glory', and the salmon-pink 'Sybil Sherwood', which attain a height of 2 ft. (60 cm); intermediate varieties such as the crimson, 'Firelight' and the snow-white, 'Purity', so lovely together; and the dainty dwarf 'Lavender Queen' and 'Crimson Glow' which make compact plants less than 12 in. (30 cm) tall.

No annual will provide a more vivid display if given a sunny position and a soil containing a little humus. Sow early in April and thin to 10 in. (25 cm) or more apart, depending upon the height of the variety.

GOLD DUST, *see* Alyssum

GOLDEN DROP PLANT, *see* Onosma

GOLDEN LARCH, *see* Pseudolarix

GOLDEN ROD, *see* Solidago

GORSE, *see* Ulex

GOURDS, *see* Curcurbita

GRAMMANTHES

An annual and similar in form and habit to the mesembryanthemums, *G. gentianoides* is almost of prostrate habit and bears flowers of scarlet, orange and yellow and intermediate shades. Like all South African plants it likes a dry soil and a position of full sun. Sow in heat mid-March, planting out early in June. The plant is most colourful from the end of July until November.

GRAPE HYACINTH, *see* Muscari

GRASS

ORNAMENTAL

Many of the hardy grasses have an ornamental value in the garden almost the equal of bamboos and other herbaceous plants grown for the beauty of their foliage. They may be used in the border where their foliage and graceful silvery spikelets will tone down the brilliant colour of other flowering plants; or they may be planted by the side of a pond or stream and here *Deschampsia caespitosa* which grows 3–4 ft. (1 m) tall and enjoys a moist position will enhance the site with its elegant panicles of silvery-green. Or again, *Miscanthus sinensis,* its tall arching leaves having a central white stripe, may be planted with other grasses in an island site on a lawn. Here may be grouped several grasses, those of more vigorous habit being planted at the centre with those less robust grouped around them, graduating the heights to the edge of the lawn where *Festuca glauca* with its upright glaucous leaves may be planted to round off the display.

Hardy and easy to manage, most grasses are intolerant of a shallow calcareous soil. They require one of considerable depth, into which is incorporated some leaf mould and a little decayed manure. Plant in spring, allowing those species of vigorous habit space to develop. Mostly perennial, the grasses increase by underground stolons and stems and quickly make large tufts. To maintain them in a healthy condition, lift and divide in spring every four to five years and replant into freshly prepared ground. During dry weather, keep them well watered at the roots whilst they will respond from a regular syringing of the foliage which will take on a much richer colouring.

The annual grasses should be sown early in April where they are to grow. Scatter the seed thinly and rake into the top 1 in. (2.5 cm) of soil. Keep the ground moist and thin out if overcrowded. They will bloom in August and September. The hardy species may also be sown outdoors in September when they will begin to bloom early July.

The ornamental grasses are most attractive when used for indoor decoration, either together or with other flowers. The stems should be cut when the flowers are at their best, before they seed.

SPECIES AND VARIETIES OF ORNAMENTAL GRASS

Agrostis nebulosa. The Cloud Grass, an annual readily raised from seed. It forms a tuft 12 in. (30 cm) tall and bears graceful panicles which are suspended over the foliage like a cloud.

Agrostis setacea. The Bristle Bent, a native grass growing 18 in. (45 cm) tall and forming a tight hummock of glaucous leaves.

Alopecurus pratensis 'Variegatis'. A pretty variegated form of the Meadow Foxtail, growing 12 in. (30 cm) tall and forming a tuft of narrow leaves, striped with yellow.

Arundo donax. The Great or Provence Reed of south France and which will attain a height of 10 ft. (2.5 m) in a moist soil. Its glaucous arching leaves are borne on long stems whilst the red-flowering spikelets are borne in a panicle 12 in. (30 cm) long. The form 'Variegata' grows 8 ft. (2.5 m) tall and has leaves striped with silver.

Briza maxima. The Quaking Grass, it is a hardy annual growing 18 in. (45 cm) tall and in August bears gracefully drooping cylindrical spikelets of creamy-white.

Cortaderia selloana (Syn: *Gynerium argenteum*). The Pampas Grass which grows 6 ft. (2 m) tall with glaucous green arching leaves and early in autumn, bears silvery plumes 12 in. (30 cm) long. In the milder parts, the plumes retain their beauty for many weeks both on the plants and when cut and used for indoor decoration. The form Pumila grows more compact whilst Rendalteri bears plumes of pale silver-pink.

Deschampsia caespitosa. The Tufted Hair-grass, native of damp meadows and woodlands. It has narrow leaves and bears panicles of silver spikelets in mid-summer.

Deschampsia flexuosa. The Wavy Hair-grass which likes completely opposite conditions to *D. caespitosa.* It is native of dry heathlands where it sends up its slender shining stems to a height of 18 in. (45 cm) at the end of which are browny-yellow spikelets.

Festuca glauca. Native of dry hilly pastures, it bears tufts of glaucous-blue bristle-like leaves and branched cylindrical stems 9 in. (23 cm) long, at the end of which are borne the brown spikelets or panicles.

Glyceria maxima 'Variegata'. A valuable waterside plant growing 2 ft. (60 cm) tall, it has strap-like leaves striped green, yellow and white and which are tinted rosy-pink in autumn.

Helictotrichon sempervirens. It grows 4 ft. (1.2 m) tall and has tufts of narrow glaucous leaves, above which are formed large silvery panicles.

Melica uniflora. The Wood Melic, it grows 2 ft. (60 cm) tall and prefers a shady situation. It has leaves of delicate green and bears graceful chocolate coloured spikelets in May and June.

Miscanthus sinensis. A Japanese grass growing 6 ft. (2 m) tall, its broad arching leaves having a central white vein whilst in autumn it bears plumes of dark valvet-red. The variety 'Zebrinus' has leaves transversely marked with golden bars.

Panicum violaceum. A hardy annual, it is one of the most beautiful of the grasses, growing 3 ft. (90 cm) tall with pale green lance-shaped leaves and in July, bears recurving plumes of green and violet.

Panicum virgatum. A perennial growing 4 ft. (1.2 cm) tall, it forms a large tuft of linear leaves 12 in. (30 cm) long and gracefully recurving plumes of green and red.

Phalaris arundinacea. Gardener's Garters, it grows 6 ft. (2 m) tall and is a native grass, found by lakeside and in marshy ground. It has handsome flat leaves striped with white or yellow and bears long purple spikelets which sway in the wind.

Stipa pennata. The Feather Grass which makes a dense tuft and early in summer bears gracefully arching stems 2 ft. (60 cm) tall, terminating in large feathery plumes.

GREVILLEA

A shrub *G. robusta* is of erect habit with attractive fern-like leaves and in summer it bears orange flowers. It is readily raised from seed provided this is fresh, for the seed quickly loses vigour. The large flat pointed seeds should be sown with the points placed downwards, for if placed flat, they may not germinate. They should also be well covered with compost. Seed pans should be used and the John Innes Sowing Compost will be suitable, slightly increasing the amount of sand, for the grevillea likes a very well-drained compost at all times. A temperature of around 65°F. (18°C.) must be provided and a high degree of humidity is necessary, otherwise germination will be considerably delayed. To provide this, the pans should be placed on a layer of peat which should always be kept moist. The compost itself should never be allowed to become dry, for even under ideal conditions the seed will be erratic in germinating, and may not germinate at all if it becomes dry. The seeds will take several weeks to germinate and as soon as the seedlings are large enough to handle they should be transferred to 2½ in. (65 cm) pots containing the John Innes Potting Compost. The loam should be used in a fairly rough condition whilst the sand must be of a coarse gritty texture.

If the seed is sown in March, the young plants will be ready for moving during the latter part of summer, during which time they should be shaded from strong sunlight, whilst the atmosphere should be kept as moist as possible.

During winter the plants must be watered as little as possible and they should be kept growing in a temperature of 50°F. (10°C.), otherwise they will tend to become too woody. The sun's rays will increase the temperature in spring when the plants may be given additional moisture and plenty of fresh air on all suitable occasions. By early summer the plants will be ready for larger pots, using the size 48 and the John Innes No. 2 Potting Compost, again making sure that both the loam and sand is in a rough condition.

Neither 'stopping' nor staking is necessary, but the plants will appreciate an occasional feeding with soot water to maintain the foliage in its attractive green condition and to prevent the lower leaves from turning yellow.

GROUND COVER PLANTS

Amongst the most valuable of plants are those evergreens that form dense growth low down and will cover the ground in the shortest possible time. Some will grow in shade, others prefer an open sunny situation. They will clothe an unsightly bank or area of ground where little else will grow, providing beauty both with their flowers and foliage. They will also choke out all annual weeds and provided the ground was cleaned of perennial weeds before planting, there should be no weeding necessary during the entire life of the plants. As most ground cover plants are hardy and durable, they may occupy the site for many years and all that is necessary is to cut them back where there is over-crowding.

Where planting in shade, it is preferable to grow those plants with variegated leaves so that their foliage will be seen from afar and will brighten the darkest corner. In shade, few plants will bloom to advantage and rely on their foliage to provide colour. Where growing in the shade of mature trees, it is advisable to work additional leaf mould or peat into the soil before planting for here the ground will usually be too dry for rapid plant growth.

PLANTS FOR GROUND COVER
(A) *acid loving;* (B) *lime tolerant.*

Arcotstaphylos nevadensis (A)
Aucuba japonica 'Variegata' (B)
Berberis calliantha (B)
Cyathodes colensoi (A)
Gaultheria procumbens (A)
Hypericum calycinum
Hypericum galioides
Ilex crenata 'Variegata'
Mahonia aquifolium (B)
Pachysandra terminalis 'Variegata' (B)
Pernettya mucronata (A)
Pernettya prostrata 'Pentlandii' (A)
Potentilla farreri (B)
Sarcococca humilis (B)
Vaccinium macrocarpum (A)
Vaccinium vitis-idaea (A)
Vinca major 'Variegata'

GUELDER ROSE, *see* Viburnum

GUERNSEY LILY, *see* Nerine

GUM TREE, *see* Eucalyptus

GYPSOPHILA

ALPINE

For the rock garden, the trailing *G. repens* 'Letchworth Variety' bears masses of deep-pink flowers from May to August. Another with trailing habit is the shell-pink flowered *G. fratensis,* whilst the new *G. nana* is an interesting plant, being of cushion habit and studded with almost stemless white flowers.

ANNUAL

An annual which likes a limestone soil or where this cannot be provided, quantities of lime rubble should be worked in before planting. A small sowing should be made in every garden, in any corner which receives the sun, for it is invaluable for mixing with other cut flower annuals, especially carnations and sweet peas. Seed is sown early September, the seedlings thinned to 12 in. (30 cm) apart in spring. The plants will produce a large quantity of bloom from mid-June until late September, which should be supported by twine fastened round the plants. The 'Monarch White' strain is excellent and plant with it *G. elegans* the attractive pink form.

PERENNIAL

Bearing its elegant sprays of grey-white or pale pink during July and August, *G. paniculata* is one of those indispensable flowers for 'mixing' with other cut flowers of stiffer habit, whilst in the border it forms a dense snow-like mound which will offset the more formal habit of nearby plants. It takes its name from the Greek meaning 'lover of lime', hence its value in a shallow lime-laden soil. Where the soil is of an acid nature, incorporate plenty of lime rubble, together with some grit, if the soil is at all heavy. The plants require no manure. Where the soil is light, plant in November; where heavy, planting is best delayed until March, allowing 3–4 ft. (1 m) between the plants. As with nepeta, it is a good idea to allow some of the foliage to remain on the plants over winter to give protection, then when this is removed in March, lime rubble should be placed round the roots.

To increase stock, this is done by taking and rooting cuttings. As they are difficult to root they should first be treated at the rooting end with beta indolyl butric acid, to be obtained from a chemist. The cuttings are removed in early May when 3 in. (7.6 cm) long, and after treating are inserted in a closed frame into a mixture of sand and peat. After rooting, the plants should be transferred to small pots for growing on.

Where marketing the bloom, liberal sized bunches should be cut with 2 ft. (60 cm) stems and the bloom should just be showing colour.

SPECIES AND VARIETIES OF PERENNIAL GYPSOPHILA

G. paniculata. This has single flowers, borne in profusion on 3–4 ft. (1m) stems.

BRISTOL FAIRY. Though not quite so free flowering, this variety has double flowers which give an attractive grey-white appearance.

FLAMINGO. The pale shell-pink form, the bloom being fully double. The plant requires rather more attention to detail in its culture.

HABRANTHUS

A genus of twenty species native of tropical and South America, closely related to hippeastrum, but from which it is distinguished by its stamens which are unequal in length and by the spathe which protects the pedicel. Native of Chile and Peru, the plants require winter protection in the British Isles and North America in all but the most sheltered gardens of the south-west. Here may be grown those species which bloom late in summer and early autumn for the winter and spring flowering species are best confined to a cool greenhouse or garden room. The bulb is globose with a dark brown tunic and from it arises narrow linear leaves, often deeply channelled. The flowers are funnel-shaped and are borne singly on a 12 in. (30 cm) stem.

Under glass, those flowering in early summer should be potted in September and those which bloom in autumn are potted in March, one bulb to a 6 in. (15 cm) pot in which it will remain for three to four years, during which time it will continue to increase in size. The plants require a compost made up of 2 parts fibrous loam and 1 part each leaf mould and decayed manure with a sprinkling of coarse sand and as they are lime lovers, the plants will benefit from some lime rubble (mortar) incorporated into the compost. The bulbs vary in size but the pot should not be too large, about twice the size of the bulb. Plant firmly with the neck and shoulder of the bulb just above the level of the compost. Stand the pots in a frame or in deep boxes in the greenhouse with bulb fibre or leaf mould packed around and over them and which is kept moist. In a temperature of 60°F. (16°C.) the bulbs will soon come into growth when the temperature is lowered to 52°F. (11°C.). The plants should be syringed regularly to maintain a humid atmosphere. Increase moisture supplies as the plants make growth and as soon as the flower stems appear, give a weekly application of dilute manure water, until the flowers begin to fade. After flowering, withhold water gradually as the foliage dies back and keep the plants almost dry and in a frost free room during the dormant period.

In the open, plant in April, at the base of a sunny wall, after working into the soil some decayed manure and leaf mould and a liberal dressing of lime rubble. To encourage winter drainage, incorporate some grit or coarse sand whilst the soil is being prepared. Plant 4 in. (10 cm) deep and 6 in. (15 cm) apart and to protect the plants from frost, cover with bracken during winter or with a thick mulch of decayed leaves.

Bulbs which have occupied the pots for several years should be re-potted into fresh compost and into a larger pot when numerous offsets will be seen to have formed around the base of the original bulb. These are detached and planted into small pots and after two years to larger pots in which they will bloom the following year.

Plants may be raised from seed sown in pans containing the John Innes Sowing Compost. Freshly harvested seed will germinate quickly and if the seeds are sown 1 in. (2.5 cm) apart or one to a 3 in. (7.5 cm) pot in a temperature of 60° (16°C.) they will germinate in ten to twelve days. The seedlings are kept growing on until winter when less moisture is given but not entirely withheld, whilst a night temperature of not less than 52°F. (11°C.) is maintained. Early in spring, the young plants are moved to larger pots and grown on for another year when they are moved to a larger size pot in which they will bloom the following year.

SPECIES

Habranthus andersonii. Native of Chile and Argentina, it has narrow pale green leaves 6 in. (15 cm) long which appear just before the flowers in June. The funnel-shaped flowers are yellow, shaded with copper on the outside and are borne singly on a 6 in. (15 cm) stem.

H. brachyandrus. From a large ovoid bulb are produced linear leaves 12 in. (30 cm) long. They appear before the flowers which are of pale lilac-rose, shaded claret at the base and are borne singly on a 15 in. (38 cm) stem. It blooms from July until September and may be grown outside where a warm winter climate is normal.

H. robustus. Native of the hillsides surrounding Buenos Aires, it has narrow grey-green deeply channelled leaves and which appear after the flowers. The flowers have funnels 3 in. (7.5 cm) long and are borne on a slender scape 9 in. (23 cm) tall. They are rosy-pink, shaded green in the throat and appear during August and September.

H. versicolor. Native of Uruguay, it has linear leaves 12 in. (30 cm) long which appear after the flowers. It blooms from December until February, the flowers having 2 in. (5 cm) long tubes and opening to nearly 3 in.

(7.5 cm) across. They are white, shaded red at the tips of the petals and green at the base and are most striking.

HAMAMELIS

Witch Hazel, *Hamamelis mollis,* likes a rich, deeply worked soil but will flourish under town garden conditions. From December until March it bears fragrant golden flowers along its branches, if in a sunny position. The flowers, which have narrow, twisted petals, possess a delicious fragrance. The plants possess an added beauty in that their hazel-like leaves turn brilliant crimson before they fall in autumn. Another form is *H. vernalis,* which during winter bears small pale yellow flowers, whilst *brevipetala* bears its straight-petalled golden flowers during January and February.

Propagate by seed which takes two to three years to germinate or by layering in August.

HANDKERCHIEF TREE, *see* Davidia

HANGING BASKETS

Suspended from the wall of a courtyard which would otherwise be devoid of colour, or from the eaves of a bungalow where the baskets are low enough to be tended with ease, or even hung around a small garden on stout wires held up by angle-iron stakes, hanging baskets add charm and colour to what may otherwise be drab surroundings. They are so easily made up and yet, they are rarely seen, except about hotels and places of public entertainment.

The strong wire 'baskets' may be obtained from most large sundriesmen for a small sum. They will, if carefully watered and tended, remain colourful from early summer, say the end of May, until well into autumn. And so that watering may be accomplished without trouble, the baskets should be hung where they may easily be reached. The regular removal of dead flowers, too, will make accessibility an essential.

The baskets are made of strong galvanised wire fastened closely together which ensures that the compost does not fall through, yet at the same time it allows any excess water to escape. This is very important, for during a warm, dry summer watering may be necessary twice a day, the foliage and blooms also being given a spraying at intervals.

In making up the baskets they should first be lined to a depth of 1 in. (2.5 cm) with new sphagnum moss which will not only restrain the compost from falling through the wires but will help to absorb the moisture, and will remain green all summer. Then place round the sides a layer of new turf, with the soil side to the centre, and carefully mould it to the exact shape of the basket. To within 2 in. (5 cm) of the top of the basket is placed the prepared compost. This should consist of turf loam to which is added some decayed manure or leaf mould, a sprinkling of lime to keep it sweet, and a little bone meal to provide a constant source of plant food right through the summer.

As an alternative to moss, polythene may be used, into which small holes have been cut to allow excess moisture to escape, but where moss can be obtained, it should be used.

To prepare the basket, it should first be placed between two bricks on the greenhouse or shed bench to prevent it from moving, or it may be placed over a large plant pot or washing-up bowl. Take care to mould the moss and the turves around the basket so that full use may be made of the basket's size.

After filling to the top with prepared compost, allow it forty-eight hours to settle before planting. The compost will have fallen 1 in. (2.5 cm) below the rim which will allow for watering and for additional moss to be placed around the plants.

The plants should be from small pots and only those with an informal trailing habit should be used, the varieties required being the opposite in habit to most of those used for window boxes. Instead of their being of dwarf, compact habit, those of trailing habit are more suitable. If there is somewhere to keep the plants growing on when moved to the basket, planting may be done early in May and the basket moved to the open at the month end, for most of the plants to be used for summer flowering will be only half-hardy. By planting early, the basket will be colourful and filled with greenery when placed in its permanent position.

As a centre piece, a short-jointed zonal pelargonium will be attractive, possibly Paul Crampal (scarlet) or Henry Jacoby, bearing blooms of turkey-red and with brilliant green leaves. Also yellow calceolarias and trailing mimulus, petunias and verbena; and around the sides, trailing lobelia and double nasturtiums raised from cuttings. Also of trailing habit is *Campanula frigalis* which will bear through summer, bells of clearest blue. The pendant or basket begonias will also provide brilliance of colour.

HARDENING

Hardy and half-hardy annuals, raised under glass, possibly in gentle heat, will require hardening before being planted out in spring or early summer. Where a cold frame is available, the pans or boxes should be moved to it about mid-March, as soon as hard frosts have gone but as a precaution, cover the frame at night with sacking until the month end. The frame lights should be kept in place at first, raising them only on calm sunny days but as the weather becomes warmer, raise the lights at the side and a little more each day until they may be removed altogether by day and early April, also by night when the plants will receive all the protection necessary from the frame boards. As the plants harden, they will take on a rich deep green colour and if grown well from the start, should be short and bushy when planted out.

HAREBELL, *see* Campanula

HART'S TONGUE, *see* Gasteria

HAWORTHIA

These are easily grown and readily propagated plants which bear their tubular greenish-white flowers well above their rosettes of leaves. Like the Christmas Cactus, the haworthias prefer partial shade and are never happy in a sunny window. If the plants are to be grown in a window, a northerly position suits them best.

The plants are of compact habit, the leaves being pointed and often encrusted with white, which is most pronounced in ***H. margaritifera*** and ***H. attenuata.*** An interesting plant is ***H. tessellata,*** the surface of the leaves appearing as if cracked all over.

A compost which is suitable for the aloes suits the plants well and during summer they should be well watered but given very little moisture throughout winter. Propagation is by means of offsets, which readily form.

HAWTHORN, *see* Crataegus

HEATH (Heather)

OUTDOOR

Amongst the most valuable plants for a labour-saving garden are the evergreen heaths and heathers. There are no plants better able to suppress weeds and none which require so little attention after planting. The calluna and its varieties are the true heathers or lings; the ericas being the heaths. The heathers have tiny overlapping leaves which enables them to conserve moisture and resist drought. The heaths have separated needle-like leaves, borne in whorls along the stem. Whilst heathers and heaths grow well in acid soils of a peaty nature, they will also flourish in any ordinary soil provided it contains peat or leaf mould, whilst *Erica carnea*, the winter-flowering heath and most of the rather less hardy Mediterranean species are tolerant of soils containing lime in which the summer-flowering *Erica stricta* will also do well.

All except the south European species are hardy in the most exposed parts of Britain but most plants are not happy if exposed to cold easterly winds and the heaths are no exception. They should be given the protection of each other's company by planting in groups and where growing in an exposed situation, they should be protected by low hurdles or evergreen shrubs, until established. Here, the low-growing species should be used, planting them on low mounds of soil so that the full beauty of their flowers and foliage may be seen. The taller flowering Mediterranean heaths should be planted in warm gardens and in those gardens which may be sheltered from prevailing winds.

There are heathers for all purposes. Some, like *Erica cinerea* 'Mrs Dill' which grows only 3 in. (7.5 cm) high and bears bright pink flowers in July and August, form tiny hummocks and may be planted in a trough with other plants requiring an acid soil. Others such as *Calluna vulgaris* 'Foxhollow Wanderer' grows 5 in. (12.5 cm) tall and quickly spreads over the ground to form a rich green carpet, handsome even when not in bloom. It may be planted in pockets of peat between crazy paving stones.

Those of less rambling habit. e.g. *Calluna vulgaris* 'Foxii nana', 'Mrs Pat', and 'Mullion' and which grow 6 in. (15 cm) tall are suitable for the rock garden. With them may be planted other plants requiring similar conditions and miniature flowering bulbs which will push themselves up through the heathers and give colour almost the whole year round as the heathers do, both from their flowers and foliage. Heathers are attractive where planted amongst grey rockery stones which provide a background for their purple, pink and crimson flowers. When established, they will drape themselves over the stones to give a natural effect.

A rock garden of heathers may be constructed alongside a pathway or at one side of a lawn. Indeed, where only a small garden is available, the heather garden could occupy almost the entire area, being surrounded by a path of shingle or paving stones which will allow the heathers to be visited whatever the weather. The garden may be constructed by making an undulating mound about 3 ft. (1 m) high as its highest point. Into the soil, peat and other humus materials are mixed and several large flat stones are placed into the soil to give a natural effect. Two or three stones are better than numerous small pieces and will give a more realistic appearance if flat at the top than if pointed. Those heathers should be selected to give all year colour and amongst them may be planted several cupressus of compact habit and slow growth to break up any uniformity.

Heathers require an open, sunny situation and fresh air. They do not grow well beneath trees or in shade. Sunlight brings out the fullness of colouring in the foliage and flowers and the plants should be seen with the sunlight upon them as they will when in a southerly or westerly aspect. Again, to plant them under deciduous trees will mean that they will become entangled in a mass of dead leaves which will prove difficult to clear and will entirely spoil the effect. Plant in groups of three or four for best effect and at any time between mid-October and mid-April, though pot (or container) grown plants may be set out at any time except during a drought or when the ground is frosted. If the soil is dry at planting time, keep the plants well watered until established. They will also derive great benefit from an occasional spraying during summer.

Planting distances depend upon species and variety and this will vary between 4 in. (10 cm) and 4 ft. (1.2 m) for the Tree heaths. Normally, plant 15 in. (38 cm) apart, setting the roots well into the soil and packing moist peat around them before filling in with soil. Plant firmly, treading round the plants if the soil is friable. When completed, the lower branches should rest on the soil and the plants will soon begin to spread out, preventing weeds from appearing. Where planted 18 in. (45 cm) apart, three plants to a sq. yd., they will have completely covered the ground in three years, producing a bold effect.

The plants will appreciate a liberal dressing of peat or leaf mould in spring and this should be worked right up to the crown of the plant. Early March is the best time to top dress and at the same time, cut away any long and straggling shoots. After four years, it will be advisable to take the shears over the plants at this time, cutting them back lightly to remove dead flowers and this will encourage them to form new wood whilst keeping the plants neat and bushy.

It may happen that where heaths are growing in an alkaline soil, they may suffer from iron deficiency with the foliage taking on a pale appearance. This often happens to varieties of *Calluna vulgaris*. It is quickly corrected by watering the plants with Sequestrene, a small teaspoonful to 1 pint of water per plant. This is watered into the soil around the plant, preferably in April, during wet weather.

To propagate, lift and divide whenever the groupings become overcrowded or it is desired to increase stock. Most heaths may be divided into numerous pieces each with roots and foliage. The plants are pulled apart in the same way as dividing a primrose and replanting the pieces as soon as possible. It is advisable to lift those plants at the centre of each group. This will prevent overcrowding yet will not harm the display. When other plants are removed, the display may be maintained by replanting several of the divisions in the same group.

Plants may also be increased from cuttings. These are shoots of the half-ripened wood removed with a 'heel' in July. They should be about 4 in. (10 cm) long and after treating with hormone powder, plant 2 in. (5 cm) apart in a frame containing sandy compost; or under cloches.

Shade from the sun and keep them moist when they will have rooted by the following spring and may be planted out.

Plants may also be increased by layering the lower shoots, bending them back at a leaf joint and pegging into the ground. They will often root naturally by this method, sending out roots from a leaf joint where the stems touch the ground.

SPECIES AND VARIETIES OF OUTDOOR HEATHERS

Calluna vulgaris. The Scottish Heather or Ling which requires a lime-free soil, preferably of an acid nature and which bloom July to November. Many varieties have strikingly handsome foliage in shades of silver, gold, orange and red. Amongst the best are:

ALBA. The lucky Scottish White heather, it has pale green foliage and bears white flowering spikes. August to September. 18 in. (45 cm).

AUREA. It has pale yellow foliage, especially at the tips of the shoots and bears white flowers. September to October. 4 in. (10 cm).

CARLTON. A very good white heather, for it also bears bloom on the lateral shoots and as if covered in snow. August to September. 18 in. (45 cm).

PILOSA. Its foliage is downy, as if covered in frost whilst the flowers are pure white. August to September. 12 in. (30 cm).

PLENA. Discovered in Germany, the long tapering spikes are made up of fully double flowers. August to September. 18 in. (45 cm).

PRAECOX. The earliest white heather and a plant of dwarf, erect habit. July to August. 9 in. (23 cm).

RIGIDA. Probably the best white heather, making a compact plant and covering itself in spikes of purest white. August to September. 6 in. (15 cm).

ALPORTII. A Derbyshire heather with sage-green foliage and bearing bright crimson flowers. August to September. 20 in. (50 cm).

ARGENTEA. In spring, its soft green foliage turns to silver at the tips. The flowers are mauve. August to September. 12 in. (30 cm).

BARNETT ANLEY. Its long slender erect spikes are heavily clustered with flowers of petunia purple. August to September. 18 in. (45 cm).

BLAZEAWAY. Raised by J. W. Sparkes, introducer of so many outstanding heathers, its foliage in winter takes on shades of orange, gold and flame. The flowers are of soft mauve. August to September. 16 in. (40 cm).

COCCINEA. One of the brightest for the alpine garden with rich crimson flower spikes which contrast well with the downy pale grey foliage. August to September. 8 in. (20 cm).

COUNTY WICKLOW. A pretty heather of almost prostrate habit and bearing double flowers of soft shell pink. It requires a warm garden to reach its best. August to September. 6 in. (15 cm).

CUPRAEA. Received a certificate from R.H.S. in 1873 and is still planted for its copper-coloured foliage which turns reddish-bronze in winter. It bears purple flowers. August to September. 12 in. (30 cm).

DARLEYENSIS. It has distinctive parsley curled foliage of reddish-brown and bears flowers of deep rose-pink. August to September. 12 in. (30 cm).

DAVID EASON. It continues the display after other varieties have finished, bearing deep reddish-purple flowers above pale green foliage. A good ground-cover plant. October to November. 15 in. (38 cm).

DURFORDII. Blooms until Christmas, it has dark bottle-green foliage and bears spikes of glowing pink. October to December. 16 in. (40 cm).

ELSIE PURNELL. It has grey-green foliage and bears flower spikes 12 in. (30 cm) long packed with double flowers of soft silvery pink. Lovely for cutting. September to October. 24 in. (60 cm).

FLORE PLENA. The earliest of the doubles to bloom, bearing flowers of soft lilac pink. July to September. 12 in. (30 cm).

GOLD HAZE. It should be in every collection with its brilliant golden foliage and long elegant sprays of purest white. August to September. 20 in. (50 cm).

HOOKSTONE. A charming heather of upright habit and excellent for cutting, the rose pink flowers appearing on long feathery spikes. August to September. 24 in. (60 cm).

J. H. HAMILTON. An outstanding heather bearing fully double blooms of brightest pink. August to September. 8 in. (20 cm).

JOY VANSTONE. It bears flowers of lilac-pink above golden foliage which later turns orange and then red. August to September. 16 in. (40 cm).

NANA COMPACTA. A carpeting heather forming a prostrate mat studded with tiny pink spikes. August to September. 4 in. (10 cm).

ROBERT CHAPMAN. Raised by J. W. Sparkes, it bears flowers of soft purple and has golden foliage which turns orange, flame and crimson during winter. August to September. 16 in. (40 cm).

RUTH SPARKES. A beautiful double white heather of compact habit and with bright golden foliage. August to September. 9 in. (23 cm).

SILVER QUEEN. A lovely variety, quite different in habit and colour, bearing flowers of soft silvery mauve above pale silvery foliage. August to September. 16 in. (40 cm).

SPITFIRE. Its golden foliage turns deep bronze in winter whilst it bears deep pink flowers. August to September. 10 in. (25 cm).

TENUIS. The Scarlet heather and a valuable ground-cover plant. It is the first heather to bloom and bears its crimson-red flowers over a long period. July to October. 6 in. (15 cm).

Daboecia polifolia. It is St Daboccia's heath, native of the bogs of Connemara, hence its liking for a moist peat soil. It will also grow well in partial shade though not beneath mature trees which will prevent the plants from receiving sufficient moisture. Most varieties grow about 20 in. (50 cm). tall and have glossy dark green leaves. They are easily recognised by their broader leaves which are white on the underside whilst the oval flowers are borne on long tapering stems. Cut off the dead flower heads after flowering. *D. polifolia* makes a bushy plant bearing long spikes of purple.

ALBA. One of the finest of the white heathers with deep green foliage and bearing masses of snow-white flowers on 12 in. (30 cm) spikes. June to October. 2 ft. (60 cm).

ATROPURPUREA. It makes a bushy spreading plant and bears elegant spikes of deep purple. June to October. 2 ft. (60 cm).

HOOKSTONE PURPLE. The flowers are large and of rich purple-mauve. June to October. 20 in. (50 cm).

PARTNER'S VARIETY. The most compact form, making a rounded bush, covered in tubular bells of deepest crimson. July to September. 6 in. (15 cm).

PRAEGERAE. Discovered in Connemara by Dr Praeger, it is compact and quite distinct with its flowers of pure salmon-pink borne on graceful arching spikes. June to October. 12 in. (30 cm).

Erica arborea. The Tree heaths, native of south Europe and where growing in a warm garden, attaining a height of 6–7 ft. (2 m) but up to 20 ft. (6 m). in their native lands. Their feathery foliage is an added attraction whilst they bloom from April until June. They should be used as background shrubs for they are of pyramidal habit, resembling cupressus trees. If damaged by frost, cut well back when they will come again. *E. arborea* bears white flowers which scent the air with their honey-like perfume.

E. alpina. The hardiest of the Tree heaths, it is a plant of upright habit with bright green foliage and it bears sweetly scented white flowers. March to May. 6 ft. (2 m).

E. australis. The Spanish heath and a suitable companion for *alpina.* It bears large bells of bright rosy-red and has attractive narrow leaves. March-June. 6 ft. (2 m).

RIVERSLEA. It has handsome pale green foliage and bears flowers of deep carmine-red. April-June. 6 ft. (2 m).

Erica carnea. The Mountain Heath, it is autumn and winter-flowering and was discovered in the European Alps. It is a plant of exceptional hardiness bearing deep pink flowers and will grow well in any soil, including chalk or limestone. The bell-shaped flowers are borne in clusters from early October until April, depending upon variety. They grow 8 in. (20 cm) tall and make bushy spreading plants 12–16 in. (30–40 cm) across. They may be kept healthy and in trim by clipping off the old flowers every third year.

ADA COLLINGS. The best white flowered winter heath of compact habit with stiff greenish bronze foliage. March to April. 6–8 in. (15–20 cm).

ATRO-RUBRA. It has glaucous foliage and bears flowers of rosy-crimson. March to April. 6–8 in. (15–20 cm).

AUREA. The only winter heath with golden foliage and bearing flowers of deepest pink. February to April. 6–8 in. (15–20 cm).

EILEEN PORTER. Slow-growing and requiring more shelter than the others, it smothers itself in flowers of glowing carmine over 7–8 months. October to April. 6 in. (15 cm).

KING GEORGE V. An outstanding heath of compact habit with dark green foliage and covered in rosy-pink flowers from December to March. 8 in. (20 cm).

LOUGHRIGG. It has pale green foliage turning bronze at the tips in winter whilst it bears long spikes of soft purple. February to April. 8 in. (20 cm).

PINK SPANGLES. Its flowers of rosy-red and lilac sepals give the plant an attractive bi-colour effect. January to March. 6–8 in. (15–20 cm).

PRAECOX RUBRA. Of vigorous spreading habit, it has dark foliage and bears flowers of deep glowing red. December to April. 6–8 in. (15–20 cm).

QUEEN MARY. Grows best in a lime-laden soil and has deep glaucous foliage against which the rich pink flowers are most attractive. December to March. 6–8 in. (15–20 cm).

RUBY GLOW. Of spreading habit, it has large bells of deep ruby red enhanced by the dark green foliage. February to March. 6–8 in. (15–20 cm).

Erica ciliaris. The Dorset heath, in bloom from mid-July until the end of October and requiring warmer climatic conditions than other native heaths. It requires a moist soil so pack plenty of peat around its roots when planting and handle with care for the stems easily break. Spray the foliage during dry weather and lightly clip the plants into shape after flowering. *E. ciliaris* grows 2 ft. (60 cm) tall and bears spikes of clearest pink.

AUREA. The only golden variety, it makes a compact hummock of quaint leaf formation whilst it bears pale pink flowers. July to October. 12 in. (30 cm).

GLOBOSA. It has rich grey-green foliage over which it bears erect spikes of deep pink. July to October. 16 in. (40 cm).

MAWEANA. An outstanding variety of shrubby habit with deep green foliage and bearing large flowers of rich crimson. July to October. 12 in. (30 cm).

MRS C. H. GILL. The best of the Dorset heaths with dark green foliage and bearing flowers of rich cerise-red. July to October. 10 in. (25 cm).

STAPEHILL. It bears long elegant spikes of creamy-white flushed with purple. July to October. 12 in. (30 cm).

WYCH. It has cream-coloured flowers attractively flushed with pink. July to October. 10 in. (25 cm).

Erica cinerea. The Bell heather which will retain its compact habit for many years with the minimum of clipping. It requires an open, sunny position and a soil containing plenty of peat. It will not grow in a limestone or chalk-laden soil. Water well after planting and syringe frequently during dry weather. It blooms mid-June to mid-September and rarely exceeds 12 in. (30 cm) high.

ALBA MINOR. It forms a small compact hummock of pale green foliage and bears its white flowers over a longer period than any *E. cinerea* variety. June to October. 6 in. (15 cm).

APPLE BLOSSOM. A pretty free flowering variety, its white flowers being shaded with soft pink. June to August. 10 in. (25 cm).

ATRO-SANGUINEA (Smith's Variety). The most intense scarlet of all heaths, accentuated by the dark green foliage. June to September. 6 in. (15 cm).

C. D. EASON. It has dark green foliage and produces its glowing carmine pink flowers with freedom. June to September. 6 in. (15 cm).

CEVENNES. Of compact, upright habit it bears flowers of an unusual shade of rosy-lavender. June to October. 9 in. (23 cm).

COCCINEA. A prostrate form, the first to bloom and bearing flowers of intense carmine-red. 3 in. (7.6 cm).

DOMINO. It has dark green foliage whilst its flowers are black and white; the sepals and stems black, the flowers white. June to September. 8 in. (20 cm).

EDEN VALLEY. Of low spreading habit, the bi-coloured flowers are soft lavender, shading to white at the base. June to October. 5 in. (12.7 cm).

FRED CORSTON. A beautiful variety with mid-green foliage and bearing flowers of salmon-pink. July to September. 10 in. (25 cm).

JANET. A dwarf with pale green foliage and bearing flowers of delicate candy-pink. June to August. 6 in. (15 cm).

JOHN EASON. The foliage is golden-bronze, the flowers deep salmon-pink. June to August. 10 in. (25 cm).

JOYCE BURFITT. Outstanding, with dark green foliage and bearing long spikes of maroon-red flowers. June to October. 12 in. (30 cm).

P. S. PATRICK. A vigorous upright grower, it bears long spikes of brilliant purple. July to October. 12 in. (30 cm).

ROMILEY. Raised in Cheshire, it is more vigorous and freer flowering than 'Coccinea', its flowers being of bright crimson-red. June to August. 8 in. (20 cm).

ROSEA. An old favourite with dark green foliage and bearing flowers of bright pink. June to September. 10 in. (25 cm).

VELVET NIGHT. Plant with the pinks for it has flowers of strikingly contrasting purple-black. June to August. 9 in. (23 cm).

VICTORIA. Neat and compact, it bears long spikes of deep royal purple. June to August. 6 in. (15 cm).

Erica hibernica. The Irish heath, native of Co. Clare, it is lime tolerant and blooms from late March until the end of May, bridging the gap between the winter and June-flowering heaths to provide all year round colour. It grows 2–3 ft. (1 m) high and should be planted together in groups, possibly as a centre piece for a bed of mixed heaths. *E. hibernica* has glaucous foliage and bears flower spikes of pale flesh-pink.

ALBA. The White Heath of Ireland, it makes a plant of dense upright growth. February to April. 3 ft. (1 m).

BRIGHTNESS. A magnificent tall heath with dark foliage and forming in winter, bronze buds which open in spring to deep rose-pink flowers. March to May. 3 ft. (1 m).

W. T. RACKLIFF. A compact form with rich green foliage and bearing slightly scented flowers of pure white. February to April. 2 ft. (60 cm).

E. lusitanica. It has a long flowering period, its flower buds being tinted pink and opening white. They are scented. January to April. 5 ft. (1.5 m).

Erica tetralix. The Cross-leaf heath, a moorland plant of north England which requires a damp, peaty soil and an open situation. Extremely hardy, they rarely exceed 12 in. (30 cm) in height and have much-branched stems. They bear their flowers in drooping terminal clusters between June and October.

ALBA MOLLIS. It has silvery-grey foliage, especially when young whilst it bears spikes of purest white. June to October. 8 in. (20 cm).

CONNIE UNDERWOOD. It has silver-grey foliage with which the rich crimson flowers contrast beautifully. June to October. 9 in. (23 cm).

DAPHNE UNDERWOOD. With its grey-green foliage and carmine-red flowers it is a striking heath. June to October. 9 in. (23 cm).

HOOKSTONE PINK. It has grey-green foliage and bears terracotta buds which open to spikes of clear pink. June to October. 12 in. (30 cm).

PINK STAR. Its soft grey foliage enhances the beauty of its star-like pink flowers. June to October 8 in. (20 cm).

Erica vagans. The Cornish heath, in bloom between August and October and growing to a height of 1½–2 ft. (45–60 cm). It grows best in a moist loamy soil to which has been added some peat but it is also tolerant of a slightly alkaline soil. It does not do well in a dry sandy soil.

ALBA SUPERBA. It blooms until Christmas, bearing long graceful spikes of creamy-white. August to December. 20 in. (50 cm).

BIRCH GLOW. A distinct and lovely heath with bright green foliage and bearing elegant spikes of bright rose. August to October. 16 in. (40 cm).

GEORGE UNDERWOOD. The flowers are shell-pink touched with cream. August to October. 18 in. (45 cm).

KEVERNENSIS ALBA. It makes a low bushy plant heavily covered in short white spikes. August to October. 10 in. (25 cm).

LYONESSE. A beautiful heath, the creamy-white flowers having golden anthers. August to October. 16 in. (40 cm).

MRS D. F. MAXWELL. One of the finest heaths ever introduced, it is of compact habit and bears its rose-pink flowers in profusion. August to October. 18 in. (45 cm).

MRS DONALDSON. A most choice variety bearing flowers of a lovely shade of creamy-salmon. August to September. 18 in. (45 cm).

PYRENESE PINK. The deep pink flowers shade off to cream, giving it a bi-colour effect. August to October. 16 in. (40 cm).

ST KEVERNE. Outstanding in its purity of colour, the bright salmon-pink flowers having no shading of blue. August to October. 16 in. (40 cm).

HEATHS UNDER GLASS

Winter-flowering heaths such as *Erica hyemalis, E. nivalis* and *E. gracilis* may be brought into bloom late autumn and will continue until the end of winter. One method is to lift plants growing outdoors, including *E. carnea* and its varieties and to put these in, in September. They may be brought into bloom in a frame or cool greenhouse or in gentle heat and depending upon variety will give colour until April.

Cuttings of the recognised pot heaths should be taken October to December. These should consist of young shoots 1 in. (2.5 cm) long, removed from stock plants and with a 'heel'. Do not remove any leaves before inserting in pots or boxes of silver sand and moist peat. Plant 1 in. (2.5 cm) apart, make firm and water in, then cover with a sheet of glass or polythene. They will soon root in a temperature of 50°F. (10°C.) and late in February pot separately, using 3 in. (7.5 cm) pots. Here they remain on the greenhouse bench, shaded from strong sunlight and kept comfortably moist until ready to be moved to larger pots containing a mixture of sand and peat. This is done in September and the plants occupy the greenhouse until the following May when they are moved to the open. During the next four months they should be copiously watered and every four weeks, given an application of sulphate of ammonia (½ oz. to 1 gallon of water). This will encourage them to make plenty of lush growth so that they will not bloom until winter. During the second winter, pinch out the shoots frequently, to build up a bushy plant.

Of shrubby, upright habit, the South African Cape Heaths make excellent indoor plants, possessing a pleasing feathery habit, and all they require is a position of partial shade and one which is free from frost. A winter temperature of 45°F. (7°C.) will be sufficient and this is the time of year when the plants are in bloom. The compost, which should be made up of 3 parts peat, 1 part loam and 1 part coarse sand, should at all times be kept comfortably moist. A compost too dry or one too wet for long periods will cause the spikes to drop their buds or blooms.

Propagation is from cuttings taken from the tips of the plants after flowering in spring. These are inserted in a mixture of sand and peat and should be kept warm and moist, otherwise rooting will be delayed. Rooted cuttings should be transferred to small pots and, after two to three weeks' growth, should be pinched back to encourage a bushy habit. If small plants are required for a narrow ledge or elsewhere they may be grown on in small pots, in which they make quite delightful plants. Plants are sent to market in their thousands in this size pot by specialist growers and are known as 'tots'.

Throughout summer, heaths may be grown-on outdoors, where they should be kept thoroughly moist, and are taken indoors in early October to come into bloom in a warm, but not too warm, room.

HEBE

The shrubby veronicas, now called Hebes, are amongst the most valuable plants for coastal planting. With their neat glossy foliage and bearing their short, fat flower spikes throughout summer and autumn, they are indispensable shrubs for a small garden. They are hardy and flourish in ordinary soil. One of the best is 'Bowle's Variety', which remains like a cloud of pale lilac from July until November and it grows only 3 ft. (1 m) tall. Also lovely is 'Warley Pink', which bears its rose-pink flowers in similar profusion. They require little pruning.

Outstanding is *H. pulkeana* which grows 6 ft. (2 m) tall and needs protection from cold winds. It bears its pale lavender-blue flowers in panicles 12 in. (30 cm) long. *H. salicifolia* has willow-like leaves and grows 10 ft. (2.7 m), making a valuable hedge for a coastal garden. Its white flowers tinted with mauve appear in long racemes. *H. brachysiphon* is also hardy and suitable for coastal planting.

No pruning is necessary until the plants become straggly when they are cut back. Propagate by cuttings taken in autumn and inserted in a sandy compost under glass.

HEDERA

INDOOR

Though of more trailing than of climbing habit, there are a number of handsome members of the ivy family which have a daintiness of form and colouring of foliage that makes them almost unrecognizable from the common ivy. They may be used in many ways in the home, withstanding cold temperatures and a shaded position, whilst they will survive many weeks without moisture. They are most attractive trailing from wall brackets or where allowed to twine up a trellis around a window, or over

A drift of Golden Gar!ic (*Allium moly*) page 26)

Artemesia schmidtii nana – a valuable foliage plant (page 50)

Asters – propagating Michaelmas Daisies (page 56)

(*Above*) Dark, self-coloured, show-type Auricula (page 70)

(*Above*) *Begonia metallica*, an excellent house plar (page 94)

Robust 'Himalaya' variety of *Begonia rex* (page 87)

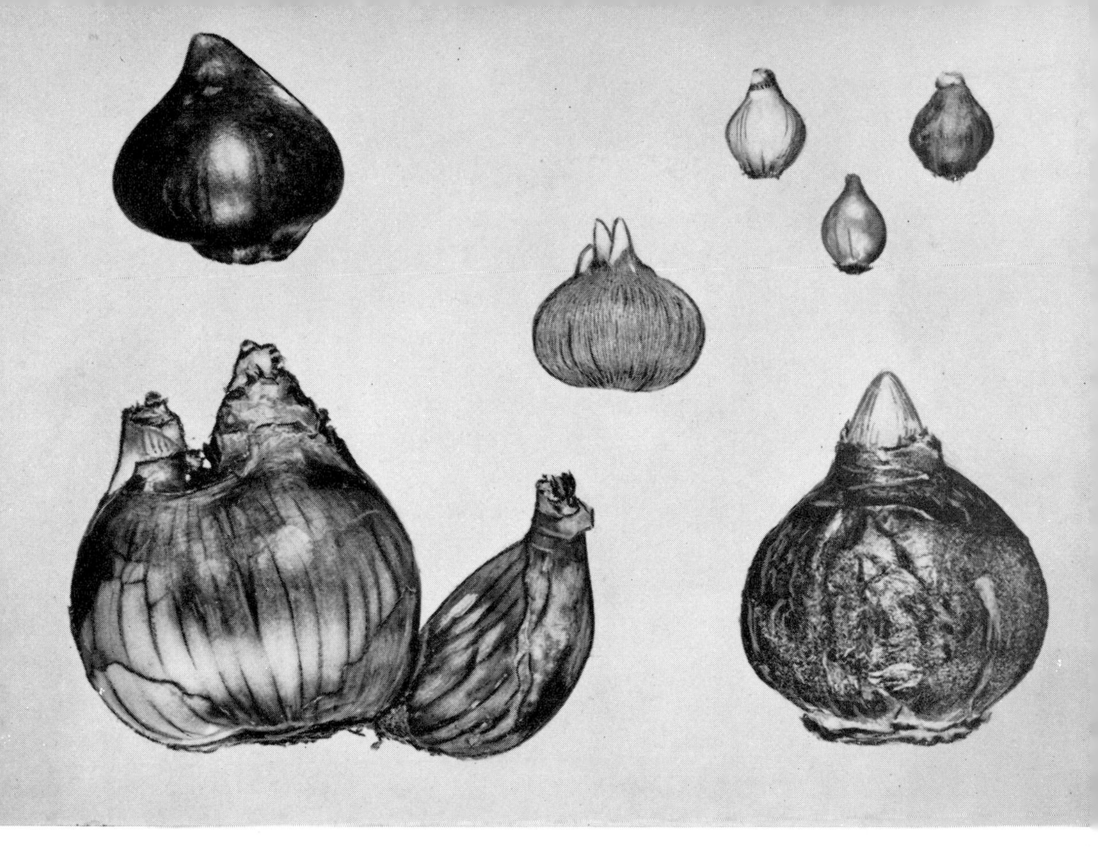

Bulbs: Top row: Tulip and Snowdrops. Centre: Crocus. Bottom: Daffodil and Hyacinth (pages 118–135)

Rooting bulbs outdoors in a plunge bed (page 119)

A plunge bed in a back yard (page 119)

Planting Crocus in a seed pan (page 216)

Tulips ready to bring into daylight (page 122)

A variety of Mamillaria cactus (page 138)

Campanula isophylla – beautiful in a hanging basket (page 151)

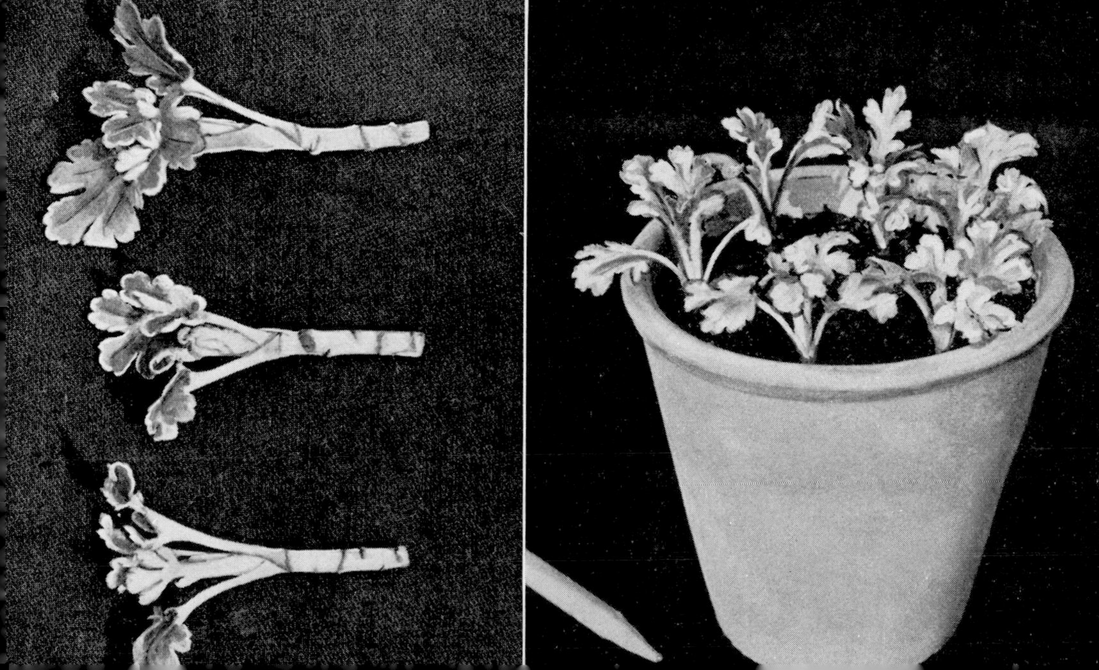

Propagating Chrysanthemums from cuttings, using a sharp knife and after trimming, planting around the edge of a four-inch pot (page 176)

'Cheerio', a small cactus Dahlia (page 238)

Growing Dianthus from seed. Glass is covered with brown paper to hasten germination. Seedlings can be lifted and 'pricked out' using a piece of split cane (page 255)

ropagating
'inks. 'Heels'
f cuttings
hould not be
oo long
elow). Label
uttings in pots
o avoid con-
usion. A
utting (*right*)
hich is
lready pro-
ucing a stem
nd bud should
e discarded. Plants received by post should be
horoughly soaked before planting, but soil should not
e washed off (page 268)

'Earl of Essex', a popular garden Pink (page 269)

Erythronium dens-canis, for shady garden or rockery (page 306)

Fritillaria meleagris, the Snake's-head Lily (page 332)

Fritillaria citrina, a dwarf species from Asia Minor (page 332)

Geranium pratense – grows as wide as it grows tall (page 346)

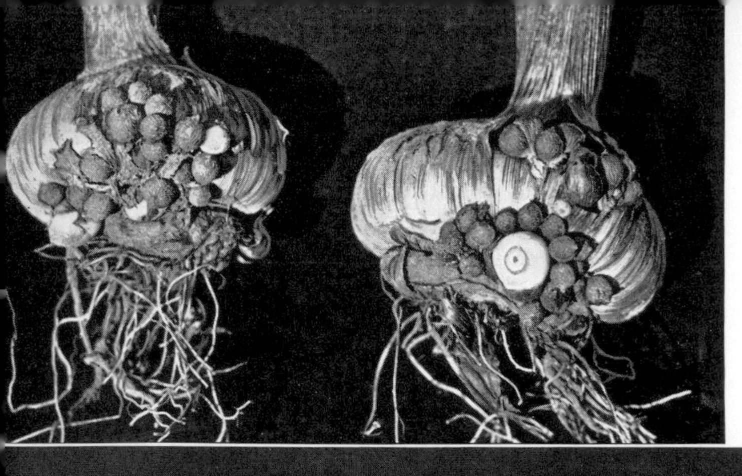

Gladioli in autumn. Note new corm, shrivelled remains of old one, and cormlets (page 352)

Gladiolus corms: 12–14 cm, 10–12 cm, and 8–10 cm (page 350)

Helianthus, 'Capenoch Star' (page 391)

Dutch Iris, 'Wedgwood' variety (pages 423-424)

Sweet Pea seeds should be carefully nicked to reveal part of the green layer beneath the outer skin, and sown as shown (page 442)

Lilium candidum – the Madonna Lily (page 458)

Muscari armeniacum 'Blue Pearl' (page 487)

Tazetta Narcissus 'Geranium' (page 500)

Miniature *Narcissus triandrus*
Superior in pots (page 502)

(*Right*) The Star of Bethlehem,
Ornithogalum umbellatum
(page 521)

Inexpensive, yet rarely seen –
Oxalis adenophylla (page 522)

Paeony 'Marie Crousse' (page 527)

Ivy-leafed Pelargonium 'L'Elegante' (page 540)

Ornamental leaves of Pelargonium 'Marechal McMahon' (page 542)

Pelargonium crispum 'Variegatum' (page 544)

Pelargonium capitatum – leaves used in perfumery (page 544)

'Carisbrooke', a show variety of Pelargonium (page 546)

Peperomia, a native of the Amazon valley (page 549)

Hose-in-hose Primrose 'Brimstone' (page 583)

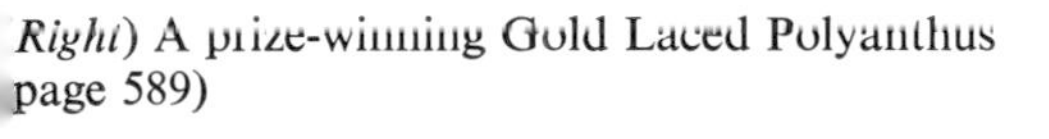

Right) A prize-winning Gold Laced Polyanthus page 589)

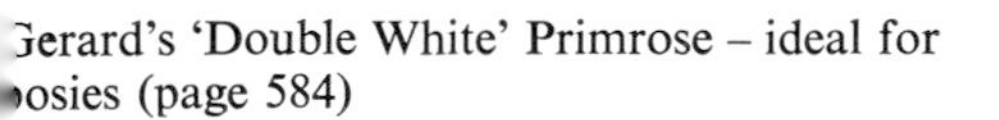

Gerard's 'Double White' Primrose – ideal for posies (page 584)

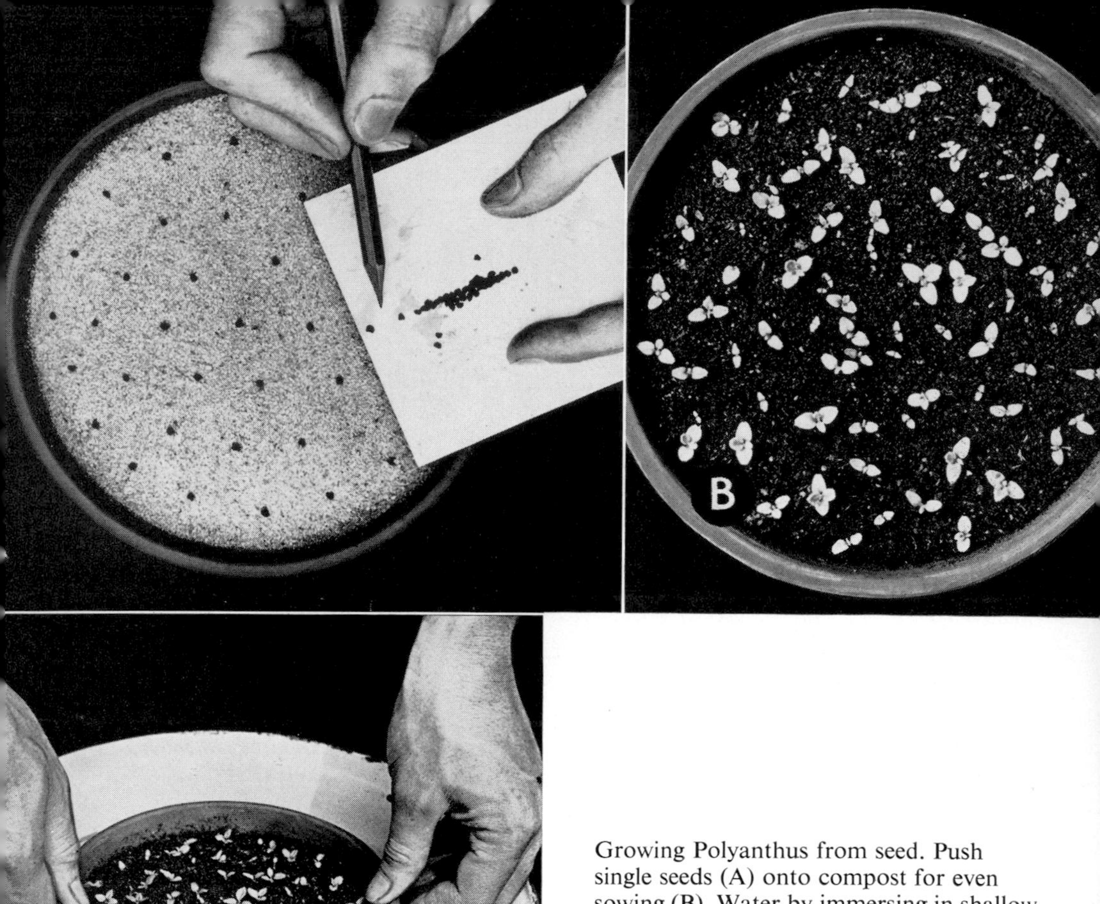

Growing Polyanthus from seed. Push single seeds (A) onto compost for even sowing (B). Water by immersing in shallow bowl (C). Roots develop rapidly (D) by the time that the seedlings are ready to transplant (E) (page 590)

'Teasing apart' Polyanthus roots
(page 590)

Puschkinia libanotica, the Lebanon Squill (page 602)

Planting a rose. Dig deep and wide enough (A) to take full spread of roots (B) and fill in (C) without disturbing them. Make the plant firm by treading (D) (pages 612–614)

Propagating miniature roses. Remove cutting with sharp knife (A) and moisten tip in hormone rooting compound (B). Using small dibber, insert cutting around edge of pot (C) and make firm (D) (page 639)

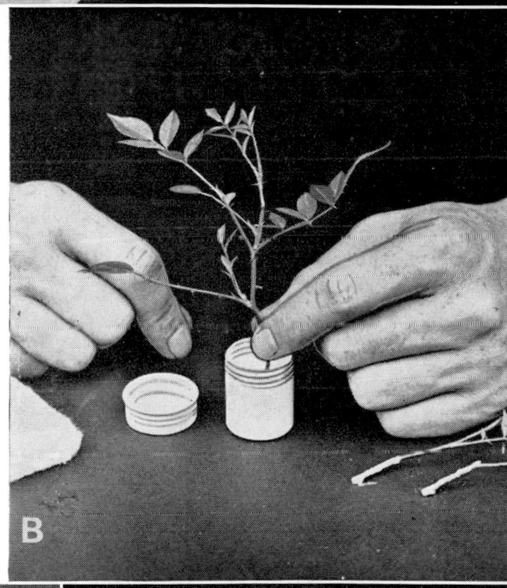

Miniature Rose 'Baby Masquerade'
(page 623)

Scabious 'Clive Greaves' – one of the
loveliest of all cut flowers (page 664)

Thunbergia alata – Black-eyed Susan
(page 696)

Tradescantia or Spiderwort (page 704)
(Below) *Tulipa kaufmanniana*, a native of Central Asia (page 728)

Show Pansies. Staked and tied (A) and with board nailed to stake to protect them from excessive rain and sunlight (B). Twelve fancy Pansies arranged for exhibition (C) (page 743)

Making a windowbox. The numerous holes for drainage are most important (page 760)

Zinnia 'Sombrero', an outstanding cutflower variety (page 768)

a pedestal in the same way as the climbing fig. Those species and varieties which have variegated leaves of dainty form will bring colour to a shaded room, whilst the plants will remain healthy for many years with the minimum of attention. For a cold, draughty entrance hall, or passage, there are no more useful plants, nor any more graceful.

Ordinary fresh loam will be suitable for ivies, but where a little decayed manure can be incorporated it is surprising how much brighter will be the leaf colourings, whilst growth will be more vigorous. Pot firmly and during summer water frequently. Where grown as pot plants and when ready for water, stand the pots in the kitchen sink so that the pots are half immersed and where the leaves may also be syringed. If treated in this way,

FIG. 5 *Ornamental ivy in hanging basket*

the plants will take on a new beauty, whilst there will be no sign of Red Spider.

A species of great beauty is *Hedera canariensis*, which has large golden leaves and appreciates more light. It must also never be over-watered. *H. angularis* 'Aurea' is also lovely and has large leaves, edged with gold. A most attractive form is *H. cristata*, the Holly Ivy, the leaves having frilled edges. Lovely, too, is the variety, 'Glacier', its leaves edged and veined with silver to give it a frosted appearance, whilst the variety 'Jubilee' has a small dainty leaf with a striking golden centre. These American ivies make plants of more bushy habit than the others. 'Silver Sheen' is also lovely, the glossy dark green leaves being edged with silver. Another ivy of leafy, bushy habit is *H. sagittifolia* which bears pointed leaves shaped like arrow-heads, and similar in habit is *H. cavendishii*, its small leaves bring edged with cream. Another attractive ivy is 'Marmaret Efeu'. It is a slow grower and makes a pleasing table plant, its dark green leaves being beautifully flecked with pale green. Very similar in appearance is the miniature Luzi, which is charming used for a tiny indoor garden. It is like a miniature tree and is very slow growing. Another for indoor gardens is *H. minima*, which with its tiny leaves has the appearance of a miniature Japanese tree.

OUTDOOR

The common ivy or hedera will provide handsome winter colour by its attractive foliage. It is a self-clinging plant, which often becomes troublesome and unsightly if neglected, but when kept under control and frequently clipped it will bring great beauty to a northerly wall, especially where the variegated-leaf varieties, with their neat habit, are grown. Of these, Silver Queen, with its neat grey-green leaves edged with silver, and *Hedera angularis* 'Aurea', the soft green leaves of which are edged with gold, are most attractive plants, far superior to the Common Green Ivy, *H. helix*. Also with variegated foliage is *H. dentata* 'Variegata', which has quite a different habit to the other ivies. It forms large olive-green leaves, mottled and edged with bright yellow, and it is not self-clinging. Support may be given by means of wires, or the plant may be grown up a trellis. It is evergreen, and there is no more colourful plant to cover a trellis situated in a sunless part of the garden.

Of the green ivies, *H. caenwoodiana* has small, beautifully shaped leaves, whilst *H. hibernica*, the Irish Ivy, has large deep green leaves and is not so rampant a grower as the common ivy. One green ivy which is quite outstanding is *H. purpurea*, the purple-leaf ivy, the foliage taking on the most glorious crimson and bronze tints during winter. It is very like *H. ampelopsis*, but has the added advantage in that it will retain its foliage through winter.

Pot-grown ivies may be planted at any time, but preferably in April, when they will get away to a rapid start. Though the plants will thrive in any soil, the difference will be surprising if they are given some decayed manure or garden compost and a yearly mulch; the plants then take on a richness of colouring rarely believed possible.

Established plants should be clipped each year with a pair of sharp shears, and any shoots which may be growing into a window-frame or spouting should be removed. Young plants will also appreciate an occasional syringe during summer.

HEDGING PLANTS, THEIR PROPAGATION

Arbutus. As it does not readily transplant, sow the seed in individual pots, or layer shoots into small pots inserted in the ground.

Aucuba. Seed may be sown, or remove cuttings of the new season's wood in September and root in a frame.

Azalea. Seed may be sown in late summer, or by layering. Or remove cuttings with a heel in July.

Bay (*Laurus nobilis*). Insert cuttings in a frame in July.

Beech. Sow ripe seed in March in drills.

Berberis. Sow seed in March; layer in September; or take cuttings with a heel in August.

Birch. Sow seed, which germinates quickly.

Blackthorn (*Prunus spinosa*). Sow seed.

Box. Either layer, or lift and divide plants in October.

Caryopteris. Insert cuttings under a bell-jar in July.

Ceanothus. Insert cuttings under a bell-jar in July.

Cistus. Insert cuttings in a frame in August.

Cornus (Dogwood). Sow seed in autumn, or by layering during early summer.

Cotoneaster. Sow seed in March or take and insert cuttings in September, preferably under glass.

Cupressus macrocarpa. From seed.

Cupressus, other species. By rooting short cuttings containing a heel, in a frame or under cloches.

Cydonia. By sowing seed in autumn, or by cuttings or layers.

Daphne. By sowing seeds in summer; from cuttings; or by layering in early autumn.

Elaeagnus. From cuttings or by layers in August.

Escallonia. From cuttings, which root quickly, and which may be inserted any time between April and October.

Forsythia. From layers or cuttings in autumn.

Fuchsia riccartonii. Root cuttings in pots in a frame.

Garrya elliptica. Best increased by rooting layers in small pots.

Gooseberry. Take cuttings in September or layer.

Gorse (Ulex). Insert cuttings in small pots in April as it does not readily transplant.

Hazel (Corylus). By removing suckers.

Hippophae rhamnoides (Sea Buckthorn). From seed, suckers, layers or cuttings.

Holly. From cuttings in a frame, taken with a heel.

Hornbeam. Sow seed.

Hydrangea. Insert cuttings in a closed frame in May.

Kerria japonica. From cuttings inserted under a bell-jar.

Laurel. From seed sown in March; by layers; or by cutting inserted in a frame in autumn.

Lavender. Cuttings inserted in a frame in July.

Lonicera nitida. Cuttings inserted in July.

Olearia haastii. By cuttings inserted under a bell-jar in early autumn.

Osmanthus delavayi. Layer or insert cuttings into a frame containing plenty of peat in September.

Osmarea burkwoodii. Insert cuttings under a bell-jar or frame in August.

Pittosporum. By rooting cuttings in July in a frame or under a bell-jar or cloche.

Poplar. By inserting cuttings in a frame at any time from May until October.

Potentilla. Insert cuttings in a frame early in autumn.

Privet. Take cuttings and root in the open or in a frame any time from April until October.

Pyracantha. Sow seed in spring or insert cuttings under a frame or bell-jar in August.

Quickthorn (Crataegus). Fresh seed will quickly germinate and this is the best way to propagate.

Rhododendron ponticum. By layering.

Ribes (Flowering Currant). Layer, or take cuttings in September and insert under cloches.

Rose. For a hedge, roses are increased by taking cuttings in August and inserting in a frame.

Rosemary. From seed or by inserting cuttings in a frame or in the open in August.

Santolina (Cotton Lavender). Insert cuttings in a frame in July.

Senecio greyi. Layer in autumn or insert cuttings under glass in August.

Spartium junceum (Spanish Broom). It resents transplanting; in spring into individual pots.

Tamarisk. From cuttings inserted in the open.

Thuya. Sow seed in spring or insert cuttings in a sandy soil in autumn.

Weigela. Insert cuttings from July to autumn under glass.

Yew. From seed or by cuttings inserted under a bell-jar in July.

HEDYSARUM

H. multijugum, native of Mongolia is a hardy deciduous shrub growing 3–4 ft. (about 1 m) tall. It has bright green pinnate foliage and bears spikes of purple vetch-like flowers throughout summer. It requires a sunny position and a well-drained sandy soil. Little or no pruning is necessary. Propagate by layering in autumn or by cuttings, removed in July and inserted in a sandy compost under glass.

HELENIUM

ANNUAL

The variety 'Sunny Boy' is the best annual form of the popular autumn-flowering perennial. Growing about 20 in. (50 cm) tall, it blooms at the same time, from late in July until mid-October, with attractive fern-like foliage and bearing yellow daisy-like flowers with a central cone of deeper yellow. Sow in gentle heat under glass in spring or in a frame and plant out early in May. Of upright growth, the plants may be supported in exposed gardens by a few twigs inserted about the foliage.

PERENNIAL

A valuable plant for late summer and autumn colour, for no plant bears more bloom from a single root at the same time. It is also extremely hardy, flourishing in a heavy soil and proving almost indestructible. Planting is done during November, allowing 2 ft. (60 cm) between the plants. Propagation is by division of the roots after flowering, the smaller outer pieces being the most vigorous. The woody centre part of old plants should be discarded. Space out the plants along the back row with care, so that they will provide a succession of colour when the midsummer display has ended.

VARIETIES OF PERENNIAL HELENIUM

BAUDIREKTOR LINNE. A new variety bearing very large flowers of a deep orange-red colour. Growing to a height of 3–4 ft. (about 1 m), it is an ideal back row plant for a small border. It is in bloom during autumn and persists until early November.

CHIPPERFIELD ORANGE. An old favourite growing to a height of 5 ft. (1.5 m) and bearing masses of orange-yellow flowers, striped and tinged with scarlet.

COPPER SPRAY. A fine variety of branching habit, its coppery-orange flowers are borne on 3–4 ft. (about 1 m) stems from mid-July until September.

GOLD FOX. A fine new variety of erect habit. Its large tawny-orange flowers are produced on 3–4 ft. (about 1 m) stems over a very long season, from mid-July until October.

KARNEOL. A valuable late flowering variety, bearing its rich bronze flowers on 3–4 ft. (1 m) stems during late autumn.

MOERHEIM BEAUTY. An old favourite, but one of the finest of all border plants, forming dense round bushes of bloom, like Pumilum magnificum, which flowers at the same height. The blooms are of a rich crimson-bronze colour and are in bloom from late July until September.

PUMILUM MAGNIFICUM. An old favourite bearing a mass of bright golden-yellow flowers during July and August at a height of almost 3–4 ft. (1 m). It remains one of the very best of border plants.

RUBRUM. Also tall growing bearing coppery-crimson flowers on 5 ft. (1.5 m) stems during August and September.

RUBY THORNLEY. The blooms, the size of a ten-penny piece are orange, flecked with crimson and borne on 5–6 ft. (2 m) stems. They are in bloom from mid-August until late October.

SPATROT. A recent continental introduction, which bears its rich chestnut red flowers on 4–5 ft. (about 1.5 m) stems during autumn.

THE BISHOP. Growing to a height of 3–4 ft. (about 1 m) its warm orange-yellow flowers have an attractive brown centre. Valuable because it flowers during late autumn.

WALTRAUD. Like all the new heleniums, its golden-bronze blooms are very large and are borne on 3–4 ft. (about 1 m) stems during July and August.

WYNDLEY. The growth is erect and bushy, the blooms being large and of a deep golden colour, suffused with orange. It blooms from late in July until early September.

HELIANTHEMUM

Now grown in almost as large quantities and in as great a number of varieties as the aubretias, *H. vulgare* is a valuable plant for a hot, dry soil being in bloom from mid-June until late in August. The plant has an added attraction in its shiny, dark-green or grey leaves which clothe the strong wiry stems and which are evergreen. It requires a position of full sun when its blooms will open in long succession. Some of the loveliest varieties are:

BEN ATTOW. Lemon-yellow.

BEN LAWERS. Tangerine, shaded lemon.

BEN LEDI. Rich crimson.

FIREBALL. Double scarlet.

JUBILEE. Double yellow, flushed orange.

SNOWBALL. Double pure white.

ST JOHN'S COLLEGE. Deep golden-yellow.

WATERGATE ROSE. Deep salmon-pink.

WISLEY PRIMROSE. Grey foliage, pale yellow.

Of the other species, *H. alpestre* with prostrate foliage, covers itself with yellow blooms, whilst *H. umbellatum* is taller growing and bears icy-white flowers early in summer. All are readily increased from cuttings which when rooted should be grown on in small pots until planted out between October and March.

HELIANTHUS

In bloom from the end of July until late in October, the perennial sunflowers are most stately plants, but they require a sheltered garden and a large border, for they are of vigorous habit and grow tall (see plate 33). The plants prefer a moist soil, one enriched with plenty of humus so that they will bear an abundance of bloom. November is the best time to plant and divide. Like the helenium and solidago, frequent division will maintain the vigour of the plants. Allow 3–4 ft. (about 1 m) between the plants and it will be advisable to stake securely.

VARIETIES

CAPENOCH STAR. The clear yellow single flowers are borne in profusion on 5–6 ft. (2m) stems, through August and September.

LODDON GOLD. An outstanding plant, bearing fully double flowers of bright yellow from early August until late October on 5 ft. (1.5 m) stems.

MONARCH. This is the tallest growing variety, the large deep yellow blooms being borne on 6 ft. (2 m) stems from early August until November.

SOLEIL D'OR. A very old variety and the most compact, bearing its semi-double golden flowers on 3–4 ft. (about 1 m) stems. It is a fine plant for a small border.

HELICHRYSUM

Annual or perennial plants often with handsome silver foliage whilst others bear everlasting flowers and are given half-hardy treatment. Seed is sown in gentle heat early in spring and after transplanting and hardening, the plants are set out 8in. (20 cm) apart in their flowering beds early in June. The 'Dwarf Spangle' strain grows 12 in. (30 cm) tall and in August and September bears long-stemmed flowers of 1–2 in. (2.5–5.0 cm) diameter with the papery petals arranged in numerous rows. The flowers are pink, red, orange, yellow and when harvested and dried remain fresh throughout winter. 'Monstrosum' grows 30 in. (76 cm) tall and bears flowers with a similar colour range.

Of the perennial forms *H. alveolatum* makes an upright shrub 3 ft. (1 m) tall and retains its silvery leaves throughout winter. In April it should be cut hard back to prevent it becoming leggy. *H. fontanasii* has the whitest leaves of all the helichrysums which it retains through winter. The leaves are long and pointed and have a felt-like feel and appearance. *H. italicum,* the Miniature Curry plant grows 12 in. (30 cm) tall and has brilliant silver foliage. It may be clipped in spring to form a low hedge.

HELIOPSIS

Closely allied to the helianthus, it is a plant for the middle of a border, in bloom from mid-July until the end of September. It makes a plant of dense but compact habit and bears a profusion of bloom. November is the best time to plant and a cool, moist but well-drained soil suits it best. Allow 2 ft. (60 cm) between the plants. Propagation is by division of the roots.

SPECIES AND VARIETIES

H. gigantea. Rather taller growing than the others, it is free flowering and bears rich orange-coloured, semi-double blooms.

GOLD FEATHER. An excellent cut flower, growing 3–4 ft. (about 1 m) tall and bearing in August and September, a profusion of double blooms of brilliant golden-yellow.

GOLD GREENHEART. A magnificent plant in every way, and which should be in every border. A continental introduction, the fully double primrose-yellow blooms have an attractive green centre, whilst the habit is all that could be desired. 3–4 ft. (about 1m).

LIGHT OF LODDON. The semi-double flowers are of a deep golden-yellow colour, freely produced.

H. scabra 'Incomparabilis'. The golden-orange semi-double, zinnia-like blooms are borne from mid-July until autumn.

SUMMER SUN. An excellent new variety of compact habit and bearing a profusion of orange-coloured bloom.

HELIOTROPE, *see* Heliotropium

HELIOTROPIUM

The Cherry Pie, which bears corymbs of mauve, purple or white flowers and which carry the fragrance of ripe cherries. It is a plant of shrubby habit requiring a winter temperature of 50°F. (10°C.) but is hardy enough to be used for summer bedding if planted out in June.

If the seed is sown in March in a temperature of 65°F. (18C°), it will quickly germinate and the plants will come into bloom early the following spring, continuing through summer and autumn. The John Innes Sowing Compost will prove suitable, sowing in pans and transplanting the seedlings to small pots as soon as large enough. During summer, the plants should be kept growing on, before being transferred to larger pots early in autumn.

The plants resemble the geranium in their water requirements. Very little should be given during winter and only limited amounts in summer, just sufficient to stimulate the plants into new growth. To keep the plants bushy, side shoots must be pinched back continually, whilst plants which have become old must be cut well back. Regular syringing will keep the plants healthy and enable the woody stems to 'break into' new growth more easily.

There are several varieties which may be expected to come true from seed and 'Lemoine's Giant', which bears flowers of rich heliotrope, is strong-growing and free-flowering. 'Vilmorin's Variety' bears deep purple flowers and 'White Lady', flowers of purest white.

HELLEBORE, *see* Helleborus

HELLEBORUS

Planted in beds to themselves, beneath a wall, or in the mixed border, the *Helleborus niger* or Christmas Rose bears its waxy-white blooms from Christmas until the end of March, to be followed by *Helleborus orientalis*, the Lenten Rose. The Christmas Rose may be used anywhere about the border for it is in bloom and leaf at a time when other plants are dormant. The plants will tolerate full shade and though they require an abundance of summer moisture, they like a soil which does not retain excess water during the winter months. They should be planted at one end of a border or in the shrubbery, so that they may remain undisturbed and the bloom may be conveniently picked without undue treading of the soil during winter. Plenty of grit should be incorporated if the soil tends to be of a heavy nature, whereas a light soil should receive plenty of decayed manure and some peat to retain summer moisture when the buds are forming. Christmas Roses will never bloom well if the soil is starved. Do them well and they will respond with a copious amount of bloom. The end of March, immediately after flowering is the time to plant, always retaining at least three or four buds when the roots are divided. Take care to ensure that the roots do not dry out at planting time and set them 18 in. (45 cm) apart. During summer a dressing of decayed manure will make an enormous difference to the quality of the bloom, and in November when the border is tidied, place clean straw around the large green leaves of the plants to prevent soil splashing the white blooms.

SPECIES AND VARIETIES

One of the loveliest has the unusual name of 'Coombe Fishacre Purple', the petals being plum coloured and shaded deep green.

Helleborus niger 'Maximus'. This is the first to bloom, its earliest flowers appearing mid-December and are white tinged with pink, crimson or purple. The true helleborus is the 'Bath Variety', for in that city it was at one time grown in vast quantities as winter decoration for the fashionable ladies. This variety comes into bloom rather later, from early January and they will continue to bloom until March. A superb variety introduced from Ireland and producing pure creamy-white blooms during the same period is called St Brigid. It is the finest of all the Christmas Roses.

Helleborus orientalis. Bearing its bloom on 18 in. (45 cm) stems from early March until the end of April, the plants should be set 14 in. (35 cm) apart at the front of the border, preferably near *Anchusa myosotidiflora* which will bloom almost at the same time. In such a position the plants will obtain the dappled sunshine they require, never being happy in full shade, like the Christmas Rose. Otherwise the plants require the same cultural conditions as *H. niger*.

HEMEROCALLIS

The Day Lily, so called because its tubular lily-like flowers last only for a day and are then replaced by others, is amongst the best of all plants to use towards the front of a border. Plant where the flowers cannot be hidden and with a view to enjoying their colour from mid-May until the end of August. The flowers are borne on branched stems above a wealth of strap-like foliage. The plants should be left undisturbed for as long as possible, the dahlia-like roots being divided and replanted in November. They like a soil containing plenty of moisture, a medium loam enriched with humus in the form of decayed manure or peat, rather than a clay soil for they do not like an excess of winter moisture.

VARIETIES

APRICOT. Through June and July it bears its blooms of deep apricot, flushed with gold.

CITRINA. The long yellow trumpets are of a pure lemon-yellow and carry a delicate fragrance.

HYPERION. A lovely variety, the large canary-yellow trumpets are borne throughout July on 3–4 ft. (1m) stems.

IMPERATOR. Outstanding, the large blooms being of a rich golden-orange colour.

MARGARET PERRY. Almost as late flowering as 'Radiant', the blooms are large and of an intense scarlet-orange.

RADIANT. The last to come into bloom in mid-July, when it bears its rich orange-red flowers until the end of August on 2 ft. (60 cm) stems.

Royal Sovereign. Coming into bloom early in June and continuing until early August, this variety remains longer in bloom than any other. Its lily-like trumpets are of a rich buttercup-yellow, borne on 2 ft. (60 cm) stems.

Tangerine. The first into bloom in May and the most dwarf. The tangerine blooms appear until the end of June on 20 in. (50 cm) stems. A front row variety.

Viscountess Byng. A lovely variety, bearing flowers of a soft rose-pink colour.

HEMLOCK, *see* Tsuga

HEPATICA

It is a delightful plant and will bloom with the snowdrops if the weather is not too severe when it bears its dainty anemone-shaped flowers on only 3 in. (7.5 cm) stems which enable it to survive strong winds. It enjoys, like most spring flowers, partial shade and may be planted in the alpine garden or at the margin of a border. It is attractive planted near an entrance to home or garden where the dainty blooms may be observed as soon as they open.

Hepaticas love a well-drained loam and like snowdrops are best planted in March, when in bloom, spacing them 6 in. (15 cm) apart. They should not be lifted and divided, except when it is necessary to increase stock for the plants resent disturbance, and blooms only after occupying the same place for two to three years. Hence they have never achieved the popularity with the modern gardener that they did with those of old who exercised greater patience with their plants. Care is also needed when planting, and not to bury the crown below soil level. Hepaticas are readily raised from seed sown in July, in small pots where they remain over winter until planted out in April where they are to bloom eventually.

SPECIES AND VARIETIES

Hepatica angulosa. Native of the Caucasus, it differs from *H. triloba* in that it has five-lobed leaves and its flowers come rather later, in March and April. The blooms are sky-blue with striking black anthers and are held on 6 in. (15 cm) stems. It grows well in a chalk-laden soil and is almost evergreen.

Lodden Blue. The finest form, bearing large flowers of the most brilliant blue on 5 in. (12.7 cm) stems.

H. triloba. Known to Elizabethan gardeners as the Trinity Flower from its three-lobed leaves. The flowers, which are pink, white or sky-blue are held on 4 in. (10 cm) stems. Sometimes, like the primrose, the flowers will double when they take on the attractive camellia-shape of the double white primrose.

HERON'S BILL, *see* Erodium

HESPERIS

'Amongst the most desirable of hardy flowers', wrote William Robinson. *H. matronalis* is a lovely old-fashioned plant for the border. Known as the Sweet Rocket, it will flourish in ordinary soil where it may be left undisturbed for years. Plant 18 in. (45 cm) apart in November and increase by division of the roots, or from cuttings taken in May. It grows 3 ft. (90 cm) tall and bears its white or purple flowers in midsummer. The double form, 'Flore Plena', is the best and bears its deliciously fragrant purple sprays throughout June.

HEUCHERA

New hybrids, in bloom from mid-May until August, have brought about a greater popularity for this graceful flower. Several may be planted to make an original edging to a border, such as the compact 'Freedom', which bears its arching sprays of carmine-pink through summer. The plants prefer a light, dry soil, though I have found them equally long lasting and free flowering growing in a heavy clay soil. Plant in November and bury the woody rootstock deeply or it will eventually push itself out of the ground.

VARIETIES

CARMEN. The flowers of a deep carmine-pink colour are large and freely produced.

LADY ROMNEY. Rather taller growing than the others, its coral pink flower sprays growing to a height of 2 ft. (60 cm) so it should be planted just behind the front row plants.

OAKINGTON JEWEL. An older favourite, which with its long sprays of coppery-rose and bronzy foliage still holds its own in any border. It grows to a similar height as 'Lady Romney'.

ORPHEE. As a contrast to the rich crimson of 'Red Spangles', this dainty variety with its sprays of pure white should be planted nearby, for it grows to a similar height.

RED SPANGLES. A superb garden plant, the bright crimson-red sprays are held on 18 in. (45 cm) stems and are borne in profusion right through summer.

RHAPSODY. Growing to a height of 2 ft. (60 cm), the bright rose-pink sprays are produced with Freedom throughout summer. A lovely flower for cutting.

SHERE VARIETY. Ideal for edging, its brilliant scarlet sprays, composed of dozens of tiny flowers, being held on 15 in. (38 cm) stems.

SNOWFLAKES. Growing to a height of 2 ft. (60 cm), this variety, with its large pure white flowers provides a happy contrast to 'Rhapsody' and 'Carmen'.

HIBERTIA

H. dentata is a charming climber to plant where a temperature of 50°F. (10°C.) can be maintained in winter. The large yellow blooms are borne during the first four months of the year when there are few other plants in bloom. The long deep green leaves are tinged with bronze.

The thin stems will reach a height of 8 ft. (2.5 m) and may be grown against trellis or inside the roof of greenhouse or garden room, the plants being either in the open ground or in pots. Pruning consists of cutting back straggling shoots immediately after flowering in spring.

HIBISCUS

Closely related to the mallow, it is one of the fastest growing of all plants attaining a height of 5–6 ft. (about 2 m) in a single summer and it may be treated as an annual or perennial.

The variety of *H. grandiflorus* named 'Southern Belle' bears one of the largest of all flowers, measuring 12 in. (30 cm) across and they are white or of various shades of crimson and pink.

If the seed is sown in January in gentle heat, the young plants will be ready to set out in April and will grow so quickly that they will bloom before the end of July. To help the seed with its germination, place the seeds in a bowl and over them pour water heated to 160°F. (71°C.) Leave for an hour, drain off and partially dry before sowing two to a small pot, retaining the strongest seedling.

HIMALAYAN LILY, *see* Cardiocrinum

HIPPEASTRUM

A genus of about seventy species, native of tropical America and known as the Equestrian Star-flower. It takes its name from *Hippeus*, a knight on horseback and *astron*, a star. They have large tunicated bulbs from which arise strap-like leaves, whilst the flowers are borne two to six on a hollow scape about 12 in. (30 cm) tall. Where a temperature of 60°F. (16°C.) can be provided, they will bloom in winter and where 'prepared' bulbs are obtainable, will come early into bloom. Where there is no heat but protection from frost can be given, they will bloom in spring. Several species and hybrid varieties may be grown outdoors in those gardens of the British Isles and North America which enjoy a mild winter climate. In America it is known as *Amaryllis hippeastrum*.

Where growing indoors, their culture commences in October when 'prepared' bulbs are potted. A bulb of not less than 11 in. (28 cm) in circumference should be used for it will bloom well in its first year under glass. For outside, 10 in. (26 cm) is suitable. To encourage the bulbs to make early growth, place the base of the bulb in a tray containing a 1 in. (2.5 cm) depth of water and allow it to remain for three to four days. The pots should be not larger than twice the diameter of the bulb and here they will remain until the following January when they are re-potted into a larger size for by then the bulbs will also have grown larger. An 11–12 in. (28–30 cm) bulb may be expected to bear two flowering spikes.

Hippeastrums like a rich well drained soil and one which is sufficiently porous to admit air to the roots. The compost should be composed of fibrous loam, decayed manure (cow manure or old mushroom bed compost is suitable) and grit in equal parts. Pack the compost firmly around the bulb, leaving the top exposed and with the compost just below the rim of the pot to allow for watering. To start them into growth, give them some bottom heat either by placing them on a shelf above a central heating radiator or on a mantlepiece above an open fire and where they will remain for several weeks, removing them to a window as soon as the flower spikes appear. Where a warm greenhouse is available, place the pots in deep boxes and pack moist peat around them. Stand the boxes above the hot pipes and if possible maintain a temperature of 65°F. (18°C.) though this may be 10°F. (i.e., down to 13°C.) lower without ill effect, when the bulbs will be later into bloom. As soon as the flower stems appear, which will be about five weeks after potting, the temperature should not exceed 60°F. (16°C.).

Whilst making growth, syringe the foliage frequently and water at the top as often as necessary for when growing, Hippeastrums are copious drinkers. To enhance the quality of bloom and to build up the bulb, give a weekly application of dilute manure water when once the flower stem appears and until the plants are rested. It is also advisable to wipe the leaves with clean water as most are evergreen and in the home, the leaves collect dust. The leaves may grow to 3 ft. (90 cm) in length and will be 4 in. (10 cm) broad. So that the flower stem remains straight and where the plants are growing in a sunny window in the home, turn the pots several degrees each day. By regulating the temperature, it will be possible to have the bulbs in bloom from February until midsummer but always provide them with brisk bottom heat to start them into growth. Keep the bulbs supplied with water until the early autumn when the foliage begins to turn yellow and die back. Moisture is then withheld until early in the new year when the bulbs are re-potted. With care in their cultivation, and this means removing the dead flowers before they can seed, the bulbs should remain healthy and free flowering for many years, increasing in size until they attain a circumference of 14½ in. (36 cm) or more.

Where growing outdoors, plant the bulbs in April, 4 in. (10 cm) deep and in a sunny position such as at the base of a wall where they may be protected from winter frost by covering them with bracken and a frame light reared against the wall. *Hippeastrum pratense* is possibly the best species for outdoor culture when it will bloom during June and July.

The plants are readily increased both from offsets and by seed. Upon lifting the bulbs for re-potting in January, small offsets will be found clustering about the base. These are detached and grown on in pans or boxes in a sandy compost for twelve months until large enough to plant individually into small pots. Here they remain for another year when they are again moved to a larger pot. They will bloom in three years from the time of removal from the mother bulb.

Plants are readily raised from seed sown one to a small pot. Sow when ripe in late summer and into a sandy loam and if the greenhouse (or room) temperature does not fall below 60°F. (16°C.), the seed will germinate in about twelve days and within a month the seedlings will have formed tiny bulbs which with careful watering will begin to swell. During winter, they should be given only sufficient moisture to keep them alive and in spring will be ready to move to larger pots. They will bloom within three years from time of sowing.

SPECIES AND VARIETIES

Hippeastrum advenum. Native of Chile, it has narrow leaves of blue-green 15 in. (38 cm) long and as it flowers towards the end of summer, it may be planted in a warm border. The flowers with their long narrow petals, are borne horizontally and are scarlet shaded with yellow.

H. aulicum. Native of Brazil, it has large strap-shaped leaves and bears flowers with a tube up to 6 in. (15 cm) long. It is bright crimson, shading to green at the base of the tube.

H. bifidum. Native of Argentina, where it grows on hillsides around Buenos Aires, it is almost hardy but as it blooms in April is best given greenhouse protection in the British Isles and North America. Of easy culture, it bears large trumpets of clear orange-scarlet on a 15 in. (38 cm) stem.

H. candidum. Native of Argentina, it is one of the hardiest species and blooms during midsummer, bearing its elegant trumpets of purest white on 2 ft. (60 cm) stems. The flowers are sweetly scented.

H. equestre. Like *H. reginae,* it is found in Mexico, South America and in the West Indies where it is known as the Barbados Lily. It is the oldest known species, discovered in 1698 and in its native haunts, it grows with abandon from its stoliferous bulbs and blooms almost the whole year. It has strap-like leaves of brilliant green and bears bright red flowers on a 12 in. (30 cm) scape. In the British Isles, it blooms during summer. The variety 'Splendens' bears larger flowers of vivid scarlet.

H. pratense. Native of Chile, it makes a smaller bulb than most species and has narrow leaves 12 in. (30 cm) long. It bears two to three bright scarlet blooms, shaded yellow at the base on a 12 in. (30 cm) scape during spring and early summer.

H. reginae. It is native of Mexico, South America and of several islands of the West Indies. It first flowered in 1728, on the birthday of Queen Caroline. It has leaves 2 ft. (30 cm) long which appear after the flowers which are at their best during midsummer. They have a long tapering trumpet and open to 4 in. (10 cm) across, being of brilliant scarlet with a white star in the throat.

H. reticulatum. Native of Brazil, it is fully evergreen with broad strap-like leaves and it bears its mauve-pink flowers in autumn. Their tubes measure 4 in. (10 cm) long. The variety 'Striatifolium' has a distinctive white mid-rib.

H. rutilum. Native of Venezuela, it has narrow leaves 15 in. (38 cm) long and is the most free flowering species, bearing in autumn several large blooms to each 12 in. (30 cm) stem. They are bright crimson with pointed petals, the long elegant tube being shaded with green.

H. vittatum. Introduced from the Chilean Andes in 1769, it is a vigorous species with large strap-like leaves and bearing its flowers on 2 ft. (60 cm) stems. Each flower measures up to 6 in. (15 cm) long and up to six appear on each scape. They are white, striped with magenta whilst there is also a pure white form, 'Album'.

This species was used by a Prescot watchmaker, Arthur Johnson, who crossed it with *H. reginae* and raised the first hybrid in 1799. It was named *H. johnsonii.* It makes a large bulb from which arises a 2 ft. (60 cm) scape bearing three to four tubular flowers of brilliant scarlet, streaked with white. From this plant, de Graaf of Leiden raised a number of outstanding hybrid varieties known as the Dutch hybrids and bearing flowers of almost every known colour except blue. Of the named varieties, the following are worthy of a place in every collection:

ANNA PAVLOVA. Brilliant scarlet of outstanding form.

APPLEBLOSSOM. The large tubular flowers are of softest pink shading to white down the throat.

BELINDA. The broad petalled flowers are of deep crimson shading to scarlet at the base of the petals.

BORDEAUX. Raised by W. S. Warmerhaven, the large solid blooms are of deep wine-red, exquisite under artificial light.

BOUQUET. A Ludwig hybrid, the large refined blooms being of an attractive blend of salmon and begonia-pink.

DURANGO. The flowers with their long tubes are of clearest orange throughout.

GLORIOUS VICTORY. A variety of outstanding form, the long elegant trumpets being of salmon-orange, paler at the petal edges, darker in the throat.

HALLEY. The flowers open wide and are of vivid scarlet with a glistening frost-like appearance.

KING OF THE STRIPES. The large blooms with their long elegant tubes are white with broad stripes of velvet red.

LEGION STANDARD. The large blooms of rosy-red open wide to reveal a glistening white throat.

MARGARET ROSE. The blooms open wide but have a long slender tube and are of an unusual shade of deep shrimp-pink.

MINERVA. It bears a flower of great beauty being white edged brick red with veining of similar colouring from the end of the tube to the tips of the petals.

MONT BLANC. The blooms are white, shaded green in the throat which gives an appearance of icy-whiteness.

QUEEN OF THE WHITES. The large refined flowers are of pure velvety white with a distinctive sheen.

QUEEN SUPERIOR. Raised by van Meeuwen, it is one of the finest hybrids ever raised, the large handsome blooms being of pure ox-blood red.

HOLLY, *see* Ilex

HOLLYHOCK, *see* Althaea

HONESTY, *see* Lunaria

HONEY PLANT, *see* Hoya

HONEYSUCKLE, *see* Lonicera

HORNBEAM, *see* Carpinus

HORNED VIOLET, *see* Viola

HORSE CHESTNUT, *see* Aesculus

HOSTA, *see* Funkia

HOT-WATER PLANT, *see* Achimenes

HOUNDS-TONGUE, *see* Cynoglossum

HOUSELEEK, *see* Sempervivum

HOYA

The variegated leaf form of *H. carnosa* is one of the most beautiful of indoor plants, its leaves being thick and glossy, oval in shape and of a bottle-green colour, edged with gold and pink. It is fully evergreen and is suitable for room culture provided the winter temperature does not fall below 45°F. (7°C.). It is an easily grown plant and does best in partial shade. Apart from the beauty of its leaves, it bears waxy blush-white flowers during summer, each of which produces a drop of pure honey, hence its name of the Honey Plant, a native of Australia which reached Britain in 1850. The plants will attain a height of 6–7 ft. (2 m) and will not bloom profusely until several years old.

It will bloom best in a poor soil, when the leaf variegations will be more distinct, so use a size 60 pot containing fresh loam, a little peat and some crushed lime rubble or broken brick to keep the soil 'open'. With this plant it is important for the roots not to become pot-bound, so the plants should be re-potted every third year which, with care, may be done without removing the foliage from the trellis. Guard against giving too much moisture during winter, in fact after October the plants may require none at all until the following spring. An excess of moisture about the roots at this time may cause them to die back. Although generally described as a stove plant, the *Hoya carnosa* is more suited to the partial shade of a living-room than to the greenhouse and presents no difficulties in any way.

New plants may be readily raised by striking cuttings in summer around the edges of a pot in a compost of peat and silver sand kept quite moist.

Other species suitable for living-room conditions are *Hoya australis,* of similar habit to *H. carnosa* and bearing the same leathery leaves but with creamy flowers; and *H. bella,* which is similar but of less robust habit and is perhaps more suitable for a hanging basket, or for trailing from a pedestal.

HYACINTHUS

It makes a large bulb, up to 8 in. (20 cm) in diameter, one of which will fill a size 60 pot. Yellow varieties, whilst as vigorous, produce a smaller bulb, 5 in. (18 cm) being the top size. For forcing, a 7.8 in. (18–20 cm) bulb is suitable; for outside bedding, a bulb of 5½–6½ in. (14–16 cm) may be used. A correctly ripened bulb should possess a silvery sheen and should be firm when pressed at the base. It will be suspect if any degree of softness is noticed.

Bulbs of Roman hyacinths which come into bloom before Christmas are smaller, a forcing-size bulb being 5½ in. (14 cm) . It is not advisable on the matter of economy to try to force smaller bulbs, though where being grown under ordinary room conditions a 6½ in. (16 cm) bulb will be satisfactory and one of 4¾–5¼ in. (12–13 cm) size for the Roman hyacinths.

Bulbs specially prepared in Holland, to flower a fortnight earlier than usual may be obtained. This is done by earlier lifting and ripening of the bulbs and then placing them in cold storage. They are more expensive but it is possible to have them in bloom by Christmas. These bulbs are not sent out until required for immediate planting, which is at the end of September.

Hyacinths enjoy a moist soil. Unlike the freesia they must not be allowed to suffer from dry conditions or the flower spike will be stunted. Bulb fibre can be used with satisfactory results, but as with all bulbs, where it is possible to make up a compost to the bulb's requirements, better results are obtained. A light loamy soil with which has been incorporated some coarse sand and moist peat in equal proportions will prove ideal. It is important where planting hyacinths to have the peat thoroughly moist before mixing, as dry peat is difficult to bring into a moist condition and if too dry the peat will take up moisture from the soil needed by the bulbs. The bulbs also appreciate an eggcupful of bone meal worked into the compost to a larger bowl but this is not essential. Potting should not be too firm or there will be a tendency for the bulbs to push themselves out of the pots as they form their dense rooting system. The bulbs should be so planted, that just their tops are showing above the soil. As a rule, Roman hyacinths which reach us from France and are smaller flowering, should be planted about 1st September. As many as five or six bulbs may be planted in the same size bowl that would hold only three Dutch hyacinths. If the Romans are planted late in August or early September a succession of bloom may be enjoyed from mid-December until early spring, if this first planting is followed by another of the 'prepared' Dutch hyacinths later in September, and with yet another planting of forcing-size bulbs early in October. If at this final planting 6–7 in. (16–18 cm) bulbs are also planted these could be grown in a cold-house and would extend the flowering season right into spring. An outside planting, also in October, would carry the display into summer, with, if required, a cloche covering to connect the cold-house blooms with those growing in the open – a six months' display of the easily grown hyacinth.

When planting is completed the pots should, where possible, be placed under the protection of a wall and covered with either sand or weathered ashes for as long as five weeks. Those living in a flat cannot use this method of root formation and must be content with placing the bowls in a dark cupboard or in any dry place which is quite cool. A cellar is ideal for here the compost will not dry out and the bulbs, like those in the open will require no watering after they have been given a soaking immediately after planting. It is essential with all bulbs and none more so than with hyacinths, that they should form a heavy rooting system before any attempt at forcing. If not, either a thin or a stunted spike will result. Cool conditions while the roots are being formed is essential. Nor will the hyacinth stand up to hard forcing as soon as taken indoors. This must be done gradually. The Romans and the prepared bulbs will be taken indoors at the same time, about 1st November and they should at first be partially shaded with sacking or brown paper. Some, having knocked the ash from the pots, first place them under the benches of the greenhouse for several days until they become accustomed to the light whilst the house temperature should be no more than 45°F. (7°C.). Ventilate thoroughly until the house temperature is gradually increased, then give the pots all the light possible and copious amounts of water. The temperature should now be kept at a steady 60°F. (16°C.) and growth will be rapid from the end of November. But never at any time allow the bulbs to suffer from a dry compost. They will require water most days and without splashing the flower spikes they will appreciate some damping down of the floor at midday. Hyacinths, grown in the home will not be subject to such a high degree of forcing, but the same rules will apply. When the roots have formed, introduce them gradually to the warmer room temperature and stronger light. Full exposure to strong light too soon may cause the leaf tips to turn brown and some stunting may result.

Pots should be placed in as bright a position as possible as soon as acclimatised. They will not require such copious amounts of water as where high temperatures are being employed, but never let the soil become dry.

As growth advances and the flower spike makes headway, it will be advisable to strengthen it by means of a wire support, obtained with the pots from any sundriesman. If the plants have been grown as described they will be sturdy and with a compact flower spike and staking may not be necessary unless it appears that the bloom will be extra large.

GLASS CULTURE

Hyacinths are one of the easiest of bulbs to grow in jars or in special glass containers which permits the base of the bulb to be just above the level of the water into which it sends its roots. Rain water should be used if possible and a few pieces of charcoal placed at the bottom of the container to keep it 'sweet'. No soil is used but the bulbs should be treated in the same way as when growing in pots or bowls containing soil or fibre. They require the same amount of time in total darkness and to be kept in a cool place until a strong rooting system has formed. The bulbs will use up the water fairly quickly and every few days the containers should be topped up so that the water is level with the base of the bulb or just below. Only when the jars are filled with roots and top growth has begun, should they be moved to a warm room to flower.

PLANTING OUTDOORS

It is said that the hyacinth should be grown only indoors where its fragrance, its earliness and the long life of the blooms make it an ideal indoor flower; that outdoors it is too stiff, too formal for an interesting display. Yet like the geranium, in its formality lies its charm for it is different in habit from almost all other spring-flowering bulbs. A bed of hyacinths massed together with a background of flowering cherries or against the silvery bark of the birch, or interplanted with aubretia or arabis, make a display of the utmost charm. Two-colour schemes are most attractive, using pink and blue, or purple and white hyacinths. Or the yellow-flowering variety, 'City of Haarlem', interplanted with purple aubretia, 'Dr Mules', is a sight of the utmost brilliance. Or plant the pure white variety, 'L'Innocence', amidst a bed of the pale pink arabis; or carpeted with aubretia 'Russell's Crimson', it is outstanding.

Yet another lovely combination is to use a white hyacinth with a red Juliae primrose 'Joan Schofield' or 'Mrs Frank Neave' is suitable; or plant the delicately coloured 'Pink Pearl' amidst a bed of primrose, 'Romeo', which bears huge rich purple blooms. The rich crimson-red variety, 'Garibaldi', will give a startling display if carpeted with a pure white primrose,

'Craddock's White' or 'Juliae Alba'. Carpeting will reduce the required number to at least half. Beds of hyacinths are too expensive to plant at random, but small beds on either side of a porch where they can enjoy the late spring sunshine will be long lasting, bright and fragrant.

The hyacinth's love of water is equally applicable to outside plantings and a sandy soil containing plenty of humus, rotted manure or peat, will prove suitable. A dry, clay soil containing no humus will rarely produce a large, well-proportioned and sturdy bloom. Chopped seaweed is also excellent for hyacinths and so is cow manure. Heavy feeding is not necessary, the important point is to supply the moisture-holding humus.

Mid-October seems to be the most suitable time for planting both bulbs and the carpeting plants. Plant the bulbs 3 in. (7.5 cm) deep, and should the soil be on the heavy side, place around the bulbs a sand and peat mixture. Planting distance will depend upon the carpeting plants. If no under-planting is being done, plant the bulbs 9 in. (23 cm) apart. In the mixed border a few clumps of hyacinths will add colour and fragrance, and here they may be left in the ground year after year. If given a top dressing of peat and rotted manure in midsummer they will continue to produce a spike of reasonable size for a number of years.

LIFTING AND DRYING THE BULBS

With hyacinths being as expensive as they are, this is all important. Plants that have flowered outdoors should have the dead blooms removed immediately they have completed their flowering. This will be when the blooms begin to turn brown. If they are allowed to form seed, they will exhaust the bulbs. Do not remove the stems, only the flower heads, for the stems contain sap that is required to be put back in the bulbs.

When the leaves have turned yellow is the time to lift the bulbs. This will be some time in May to make way for the summer bedding plants. Select a dry day for lifting and leave the bulbs on the bed exposed to the air for several hours. Then remove as much dry soil as possible and place the bulbs in a dry, open shed for several weeks until correctly dried. The bulbs should be turned weekly as they are liable to sweat. If bulbs of $6\frac{1}{4}$ in. (16 cm) size have originally been used, they may be used again the following year and after that should be planted in a mixed border. It is more economical to purchase the large bulbs which will produce a first-class bloom for two years. Bulbs which have been forced and possibly subject to high temperatures should be allowed to die down after flowering and then should be knocked from the pots and allowed to dry out in the same way as those planted outside. They will be of little use for indoor planting again, nor will they give a bloom worthy of an outdoor display. They are best planted in a border the following October where they will produce spikes of indifferent sizes and shapes but will still be useful for cutting for the house.

VARIETIES

ARENTINE ARENDSEN. Early, it forces well and bears a large spike of purest white.

CITY OF HAARLEM. The best exhibition yellow, the colour being primrose yellow, but like all of this colour, it will not tolerate any heavy forcing.

DR LIEBER. A variety of great merit, producing a huge spike of pale porcelain blue, flushed violet and is good for forcing.

GARIBALDI. The best bright red for early forcing, forming a very compact spike.

GRAND MAITRE. An attractive variety on account of its deeply coloured stem and rich lavender-blue flowers. Medium late.

JAN BOS. An early-flowering variety bearing a bloom of deep crimson.

KING OF THE BLUES. Very late and of compact habit, the bloom being of darkest blue.

KING OF THE VIOLETS. The colour is attractive rich violet-mauve, the bloom very large.

LADY DERBY. An older variety and perhaps the best for individual pots. It is mid-season flowering, producing a bloom of bright shell pink.

LA VICTOIRE. Best described as deep carmine-rose, the bloom being of large exhibition proportions.

L'INNOCENCE. Very early; a very easy and robust grower and produces a refined bloom of purest white.

LORD BALFOUR. Produces a bloom with very large bells, the colour being lilac, tinged with violet.

MARCONI. A very late-flowering variety, making a large bloom of a bright rose colour.

OSTARA. A recent introduction, bearing a large, compact spike of pale forget-me-not blue.

PINK PEARL. Produces a most attractive bloom of a deep rosy pink colour and is both early and stands quite hard forcing.

PRINS HENDRIK. Produces a very strongly perfumed broad spike of a deep shade of yellow.

QUEEN OF THE BLUES. A lovely old variety of a pale shade of azure blue and is an excellent forcer.

QUEEN OF THE WHITES. Later flowering than most whites and ideal for indoor or outside cultivation.

TUBERGEN'S SCARLET. The dainty small-flowering spikes are early and of almost a vermilion colour.

YELLOW HAMMER. A variety known to our gardens during the last century but still the best garden yellow. The colour is deep cream, the spike being compact and very early.

HYDRANGEA

INDOOR

This ever-popular plant is grown in thousands for functions and special occasions which demand a floral display of the highest class. To commence, cuttings should be obtained early in August and rooted in frames of sandy soil or singly in small pots. The shoots should be prepared by having the lower leaves removed. There is little risk of hydrangeas suffering from excess moisture and from the beginning they may be given ample supplies. By late October the cuttings will have rooted and in order that they will not be damaged by frost, they should be re-potted in size 60 pots and taken to a greenhouse which is frost-proof though high temperatures are not required. Here they remain throughout winter whilst forming strong plants and a heavy root growth. By spring the plants should again be re-potted into size 48 pots and allowed to stand in the open throughout summer. Place the pots in a bed of ashes so that they will not be subject to excessive drying out. By autumn the plants will have reached maturity and be large enough to carry six heads of bloom in spring. They should be lifted into a house having a temperature of 50°F. (10°C.) and given a weekly application of liquid manure together with their daily requirements of ordinary water. Of the many lovely varieties, perhaps Marèchal Foch is the most useful for it will respond quickly to watering with copper sulphate solution when the flowers will turn a brilliant blue colour. To encourage this, the compost should be of an acid nature.

Hydrangeas are pruned by removing the old flower heads in early summer with a few inches of stalk. Rather will the plants suffer from being root bound and to prevent this condition should be re-potted each year. Cuttings are taken from the short side growths and rooted as described. In many districts the hydrangea is sufficiently hardy to withstand a winter in a cold house or even in the open ground but a severe frost may prevent a prolific display and may even cause the plant to have no bloom at all that spring.

OUTDOOR

No plant has a longer season of flowering; from August until November there will be a succession of bloom until they finally fade on the approach of winter. Hydrangea blooms are as colourful in autumn as in summer. The plants appreciate some protection when young and are best planted out from pots. The blooms will turn blue only in an acid soil and where this is desired, peat should be packed about the roots. Plant in May. One of the best varieties of *Hydrangea macrophylla* is 'Carmen', coming early into bloom and bearing deep crimson flowers which are of an attractive shade of wine-purple where growing in a peaty soil. For late flowering, King George V is excellent. Normally the flower heads are of deep rose-pink, but when the plants are in an acid soil and watered with copper sulphate solution, a deep clear blue is obtained.

H. paniculata 'Grandiflora' is also a beautiful plant, bearing large panicles of sterile florets during summer and autumn. The flowers are white, later fading to pale and then to deep pink.

After flowering, cut away the dead blooms with about 6 in. (15 cm) of stem but where there is plenty of new unflowered shoots, cut back the old wood near to the base. Propagate from shoots of the new wood, inserted in sandy compost under glass. When rooted, move to small pots and plant out in May.

H. petiolaris is one of the finest of all wall plants, not nearly so frequently planted as it should be. Though hardy and quick-growing, it is rarely seen, even though it will cover the entire north wall of a house in a few years. Planted in a cool, northerly position, in a well-drained soil containing a little decayed cow manure, it will eventually work its way around the house, covering the four walls with its glossy ivy-like leaves, and bearing flat creamy-white flowers 6–8 in. (15–20 cm) across. A wall clothed with its dark green leaves and contrasting white flowers during early summer is a never-to-be-forgotten sight.

Planting may be done at any time between November and March, and the plants will require supporting by means of trellis or wires.

HYPERICUM

The shrubby St John's Worts which grow well in all soils, in full sun or partial shade. Low growing and hardy, they may be used to cover a bank or to hide unsightly ground when the evergreen species will, with their attractive foliage, give colour all the year. They are particularly handsome when in bloom, the bright yellow rose-like flowers having a tuft of golden stamens. *H. calycinum,* the 'Rose of Sharon', is the best ground-cover plant for dry places, making a low spreading plant and is evergreen. So too, is *H. patulum* 'Henryi' which bears, through summer and autumn, golden yellow flowers nearly 3 in. (7.5 cm) across. Hidcote variety is equally free and bears even larger flowers. The semi-evergreen *H. kouytchense* which has leaves which colour in autumn and which is in bloom July to October, makes a spreading shrub 3–4 ft. (about 1 m) tall.

Plant November to March and cut back in spring each year to maintain their bushy habit. Propagate by division or by cuttings taken with a 'heel' and inserted in sandy compost in a frame.

IBERIS

An annual, candytuft grows about 9 in. (23 cm) tall and may be sown in April, as an edging to a bed or border, as it was in cottage gardens of old for it comes up uniformly and always retains its compact habit.

The original strain bears flat or slightly cone-shaped flowers in shades of purple, mauve, rose and pink, also white. The flowers have no perfume but retain their colour for a long period.

Two outstanding strains for the border, or for sowing where the plants may be grown to provide cut blooms for the home is the hyacinth-flowered 'White' and 'Rose Cardinal', the latter bearing flowers of brilliant cardinal-red. They grow 15 in. (38 cm) tall and bloom from June until September.

The evergreen perennial form, *Iberis sempervirens*, was first grown in the Chelsea Physic Garden in 1740, having been sent from Persia, its native land. It forms a bushy plant 12 in. (30 cm) tall and where long established, is often to be found some 20 in. (50 cm) in diameter. It has small dark green leaves above which it bears multitudes of flowers of glistening white, as pure as driven snow. It is a plant of great endurance and hardiness.

The variety, 'Little Gem', flowering in May, makes a compact little bush, whilst *Iberis jordani* of prostrate habit, is suitable for crazy paving. Both these bear clear white blooms, but *I. gibraltarica* bears large mauve flowers and blooms a month earlier.

ILEX

Hollies are mostly evergreen trees which grow well in partial shade and in a heavy soil. *I. aquifolium* with its glossy prickly leaves will make a good hedge and will need clipping only occasionally. It is unisexual, i.e. male and female flowers are borne on different trees, the female bearing berries but requires a nearby male for pollination. Propagation is by seed sown in spring and which may take eighteen months to germinate; or by cuttings of half-ripened wood removed with a 'heel' and rooted under glass. First treat with hormone powder. To make a hedge, plant 15 in. (38 cm) apart, using two-year-old plants.

I. aquifolium has many varieties, including 'Argenteo-marginata', the Broad-leaf Silver Holly; 'Flavescens', the dark green leaves being shaded with gold; 'Golden King' with its wide margin of yellow; and 'Handsworth Silver', its narrow leaves having a clear white margin. The form 'Scotica' has smooth spineless leaves. *I. intricata* from the Himalayas makes a small bush, suitable for a low hedge and has bright green crenate leaves and scarlet berries.

IMPATIENS

The Busy Lizzie of cottage windows was introduced from Zanzibar as recently as 1896 and named *Impatiens sultanii* in honour of the Sultan.

It makes a compact plant, growing less than 6 in. (15 cm) tall and is readily raised from seed sown in gentle heat in March, or by cuttings removed immediately beneath a leaf joint and inserted in a glass of water. Rooting will take place in two to three weeks when the plant may be potted and brought into bloom so maintaining a succession.

The variety 'Scarlet Baby' bears flowers of brilliant red whilst 'Salmon Princess', a plant of low spreading habit, bears flowers of a luminous shade of salmon rose.

The Common Balsam, *I. balsamina* was known to gardeners of early Tudor times although a native of tropical Asia. An annual, it is today grown almost entirely indoors yet during Gerard's time was widely used for summer bedding, planting as he advised 'in the most hot and fertile place in the garden'. It grows 18 in. (45 cm) tall with lance-shaped serrated leaves and it bears its rosy-red flowers during July and August when during periods of drought and intense heat, it will retain its freshness and beauty. Phillips reported having seen plants of shrub-like proportions growing in the Tuilleries Gardens and with flowers of crimson, scarlet, purple or white 'as large as a moderate rose'.

It takes its name 'Touch-me-not' from the characteristic of the capsule which releases the seed when ripe at the slightest touch. The plants are readily raised from seed sown in pans or boxes in gentle heat in March and transplanted to small pots when large enough to handle. At all times, the impatiens need a well-nourished soil and where growing in pots indoors, needs regular watering with dilute liquid manure to prevent its leaves from falling. Nor should the plants be allowed to dry out at the roots.

The finest strain is the double camellia-flowered, the large flowers being closely set along the whole length of stem. The 'Double Scarlet' bush-flowered strain is also excellent for bedding or as a pot plant.

INCARVILLEA

It may not be completely hardy in a low-lying garden, where the soil is heavy and winter moisture proves troublesome. But where it can be given a light, sandy soil enriched with decayed manure, or peat to retain summer moisture, and is given a mulch in November to provide winter protection, it will be long living. The thick, fleshy tuberous roots are best planted in March. If planted in November they may decay before root action commences. Spread out the roots on a layer of peat and sand, 2 in. (5 cm) below the surface of the soil. Propagation is by root cuttings but established plants should be disturbed as little as possible.

SPECIES

I. delavayi. Introduced from north China about a century ago, it is one of the finest border plants in cultivation. Its gloxinia-like blooms are borne in long racemes on 2 ft. (60 cm) stems during June and are of a deep shade of mauve-pink with an attractive yellow throat. The variety, 'Bees' Pink' bears a bloom of delicate shell-pink flushed with cerise.
I. olgae. Slightly taller growing, it bears its racemes of rosy-purple during June and has pretty fern-like foliage. All the incarvilleas are long lasting in water if cut when the bloom is just opening.

INCENSE CEDAR, *see* Libocedrus

INDIAN BEAN TREE, *see* Catalpa

INDIA RUBBER PLANT, *see* Ficus

INULA

Members of the inula family are amongst our most valuable border plants, from the bushy little *I. ensifolia* to the tall *I. afghanica* which is an excellent plant for a large border. It bears its shaggy sunflower-like blooms of bright golden-orange from mid-June until August, and like all the inulas it likes a position of full sun and an ordinary soil. Plant 3–4 ft. (about 1 m) apart in November and propagate by root division.

SPECIES AND VARIETIES

GOLDEN BEAUTY. A new hybrid, similar in habit to the others, having leathery foliage and bearing shaggy bright yellow bloom.

I. glandulosa. The plants, which grow 2 ft. (60 cm) tall remain compact throughout summer and are ideal for a small border. The shaggy golden-orange flowers are borne in profusion and are excellent for cutting.

I. oculis christi. So called because the yellow rayed blooms, borne on 2 ft. (60 cm) stems have a striking green centre

I. royleana. This is a most striking plant, growing only 20 in. (50 cm) tall and bearing right through summer bright golden-orange flowers which have a pin-chushion centre and from which protrude its shaggy rayed petals.

IRIS

A genus of hardy perennials which for convenience may be divided into two main groups, those having a rhizomatous root stock and those which grow from a bulb. They are amongst the most beautiful of all garden plants, entirely at home in the mixed border but perhaps happier in a border to themselves. The plants require a deeply worked soil and appreciate some lime in their diet, which is usually given in the form of lime rubble or mortar whilst they like a soil that is well-drained in winter. They also require a position where the summer sunshine can ripen the rhizomes, so essential if they are to bear a full complement of bloom each year.

The Flag Iris is an excellent town garden plant being tolerant of deposits of soot and a sulphur-laden atmosphere which will in no way harm the sword-like leaves. Usually, however, the plants are confined to a shady corner or to the shrubbery where the sun rarely reaches them and where the soil is devoid of nourishment and so are never at their best as when planted in a specially prepared border to themselves.

Plants should be divided every four years as they quickly exhaust the soil. The most suitable time is late July, after flowering which commences mid-May. Or they may be lifted and divided in October or in March though they will not bloom the same year.

The roots may be divided into sections by cutting with a sharp knife, but each piece must have an 'eye' from which the leaves arise and from which the new plant can develop.

Set the pieces 2 ft. (60 cm) apart, laying them just below the surface with

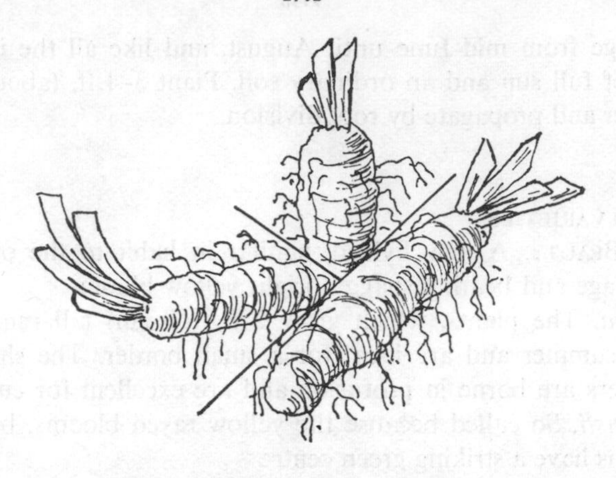

FIG. 6 *Dividing a rhizomatous iris*

the fibrous roots downwards and the top of the rhizome exposed to the sun to encourage ripening.

The rhizomatous irises may be divided into three groups:

(a) The Bearded Iris, descended from *I. germanica.* They have a creeping rootstock from which arise dark green sword-like leaves 18 in. (45 cm) long and an erect scape with several flowers attached at the end. The flowers are like large orchids in form and colour and have 'fall' petals which are bearded or crested.

(b) The Beardless Iris. Into this group come the difficult *I. stylosa,* the winter-flowering iris and several other species whose large, handsome blooms are free of any beard.

(c) The Cushion or Oncocyclus Iris. Here the bud appears at the end of a short stolon whilst the scape bears only a single flower, usually of great size and beauty. The Regalia irises of this section bear more than one flower to a scape. Each of them likes a well-drained gritty soil and shallow planting.

SPECIES AND VARIETIES

Iris acutiloba (c). Native of the Caucasus, it has a slender creeping rootstock and curving narrow leaves. It is one of the Cushion irises growing 3 in. (7.5 cm) high with broad falls of pale lilac with dark purple veins and black hairs at the base, the standards being erect and of pale lilac, waved at the margins.

I. alberti (a). Native of Turkestan, it has sword-like leaves 2 ft. (60 cm) long and in May and June bears its flowers in loose panicles on stems 2ft. (60 cm) in length. The standards are lilac; the falls white, veined with lilac and bronze and densely bearded.

I. albicans (a). At one time classed as a variety of *I. florentina* but now classed as a distinct species. Native of the Yemen, it is now to be found around the Mediterranean where it is used to bind the sandy soil against erosion. It bears a flower of purest white and may have been introduced into Britain by the Romans.

I. atropurpurea (c). Native of Syria and Iran, it has sikle-shaped leaves 6 in. (15 cm) long and flowers with narrow falls, bearded with yellow and with black tips; the standards are large and round and deep purple-black.

Iris aurea (b). Native of the west Himalayas, it grows 3–4 ft. (1 m) tall and is readily raised from seed. Its flowers are borne in two sessile clusters during June and are of brilliant golden-yellow throughout, the fall being waved at the edges.

I. barnumae (c). It is to be found amongst the hills of Kurdistan and grows only 4 in. (10 cm) tall, bearing flowers of port wine colour, the falls being narrower than the standards and with a yellow beard. A yellow variety Sulphurea emits the delicious scent of lily-of-the-valley.

I. bartoni (a). Native of Afghanistan, it has sword-like leaves 18 in. (45 cm) long and 2 in. (5 cm) broad and it blooms in June, bearing its sweetly scented flowers in clusters of two or three. The creamy-white falls and standards veined with green and purple and with an orange beard combine to make this a most arresting species.

I. biflora (a). Syn: *I. fragrans*. Native of south Europe, it has glaucous sword-like leaves and bears its flowers in April at a height of 15 in. (38 cm). They are of bright purple, the standards being erect, the falls having a yellow beard and they are deliciously scented.

I. biliotti (a). A rare native of Asia Minor, it has sword-like leaves, distinctly striped and 2 ft. (60 cm) in length and it bears its heavily scented flowers in May and June. The falls are 3 in. (7.5 cm) long and are of reddish-purple, veined with black and with a pronounced white beard; the standards being of purple-blue with navy-blue veins.

I. bracteata (a). Native of Oregon, it has leaves 2 ft. (60 cm) long, glaucous on one side only and it bears its flowers on an angled stem rather shorter than the leaves and with purple sheathing bracts. The flowers are palest yellow, the falls veined with blue.

I. chamaeiris (a). A southern European species bearing its flowers on 6 in. (15 cm) stems amidst a tuft of pale green leaves. It blooms in April, the long spoon-shaped falls being of brightest yellow, veined with brown and with a yellow beard; the narrow standards are of pale yellow.

I. cristata (b). Native of east United States, it grows only 6 in. (15 cm) tall, its flower stem appearing in April from a rosette of linear leaves. The flowers are of amethyst-blue with a yellow crest on the falls and with attractively crisped margins.

I. douglasiana (b). Native of California, it forms a tuft of linear leaves from which arises in June, flowers of primrose-yellow, the falls veined with lilac.

I. duthiei (a). Native of north-west India, it has a knotty rhizomatous rootstock from which arises a tuft of yellowish leaves 2 ft. (60 cm) in length though the solitary flowers appear in May when the leaves are only making their appearance. The flowers with their horizontal falls are purple-lilac with darker veins and a white beard; the standards also being of purple-lilac.

I. flavescens (a). Native of south-east Europe and Afghanistan, it resembles *I. germanica* in its habit and sword-like leaves whilst it bears its lemon-yellow flowers bearded with orange on 2 ft. (60 cm) stems.

I. florentina (a). The Florence Iris whose roots, when dry, possess the fragrance of the violet and in medieval times were used as powder to be placed amongst clothes and linen and to perfume the hair. The dry roots when burnt will perfume a musty smelling room whilst chewed, it will sweeten the breath.

I. gatesii (c). A rare native of the lower mountainous regions of Armeria and Asia Minor, it is known as the Prince of Irises for it bears one of the most beautiful flowers of all cultivated plants. It has narrow dark green foliage and bears its flowers on a 20 in. (50 cm) stem in June. The standards are silvery-white, dotted with violet; the falls being cream, splashed with brownish-mauve. The flowers measure 5 in. (12.7 cm) across and should be protected from wind for they bruise easily. It should be planted at the base of a wall in full sun and where it may be dry in winter.

I. germanica (a). The Common Flag or German iris is a native of central Europe and is the oldest iris to be given garden culture. It is believed to have been grown in the ninth century, in the monastery garden of Reichenau by its Abbot, Walfred Strabo. The plant is one of the hardiest and toughest in cultivation, well-nigh indestructible but though by its rugged constitution persisting through the years, it received little attention from breeders until the present century. Though the modern flag irises are always classed as being the offspring of *I. germanica,* this species has played little part in their raising in comparison with the scented *I. pallida.* Under this heading however may be listed a number of varieties of outstanding beauty and which are richly scented.

ALINE. It is an older variety but one of great beauty, being a pure azure blue self and carrying a more powerful perfume than any other variety, perhaps obtaining its scent from *I. pallida* which bears flowers of similar colouring. 3–4 ft. (about 1 m).

BLUE SHIMMER. A most handsome iris, bearing a large crisp white flower, feathered with blue and sweetly scented. 3–4 ft. (about 1 m).

CHRISTABEL. One of the few irises to bear copper-coloured blooms and to be scented. It is an iris of exquisite texture with coppery-purple fall petals and blooms with freedom. 3–4 ft. (about 1 m).

CLEO. Described by its raiser as chartreuse-green in colour, it is a most interesting iris in any company and in addition it has the scent of orange blossom. 3–4 ft. (about 1 m).

EBONY QUEEN. Like a number of the darker coloured irises, this one is also deliciously scented. The blue-black flowers have great substance and come into bloom before all others. 3–4 ft. (about 1 m).

FASCINATION. A gorgeous iris to plant near those of darkest colouring for its flowers are of a lovely shade of dusky lilac-pink with a sweet perfume. 3–4 ft. (about 1 m).

HARIETTE HALLOWAY. A new iris of outstanding form, the ruffled flowers of great substance being of a lovely shade of medium blue with a powerful perfume. 3–4 ft. (about 1 m).

INSPIRATION. Though introduced in 1937 no iris bears a flower of the same rosy-cerise colouring and none has a sweeter fragrance. 3–4 ft. (about 1 m).

IVORY GLEAM. The huge refined blooms are of solid ivory with touches of gold at the edges of the falls and diffusing a perfume like lily-of-the-valley. 3–4 ft. (about 1 m).

LAGOS. Valuable in that with 'Coastal Command', it is the latest iris to bloom, its large cream and gold flowers with their soft sweet perfume, opening about mid-June to extend the season by several weeks. 3–4 ft. (about 1 m).

MAGGADAN. One of the most unusual and sweetly scented varieties. The standards are slate-blue, the ivory-white falls being flushed with slate. 3–4 ft. (about 1 m).

MANYUSA. One of the outstanding pink irises, the flowers with their exquisitely ruffled petals, being of soft orchid-pink with a sweet orange perfume. 3–4 ft. (about 1 m).

MATTIE GATES. A most attractive variety and surprisingly, one of the few yellows with pronounced perfume. The standards are of soft, almost primrose-yellow with the falls of brightest gold, blazoned with white. 3–4 ft. (about 1 m).

MOONBEAM. With 'Mattie Gates', it is the most richly scented of all the yellow irises, the large blooms of clear sulphur-yellow having the scent of the lily-of-the-valley. A variety for the front of the border. 2 ft. (60 cm).

RADIANT. This fine iris will be a valuable addition to the front of any border for its bright apricot-orange standards and terracotta falls ensure that it receives the attention it deserves. In addition, it is free flowering and richly scented. 2 ft. (60 cm).

ROSE VIOLET. A front of the border bi-colour with rose-pink standards and violet falls, the whole being a bloom of great substance. Valuable for its lateness of flowering and its rich gardenia scent. 2 ft. (60 cm).

I. graminea (b). It is to be found about the lower alpine regions of central Europe and has linear leaves 18 in. (45 cm) long which tower above the two-edge flower stems of only half that height. The flowers which measure

2 in. (5 cm) across are borne in twos and emit the delicious smell of ripe plums. The falls are lavender, veined with violet and tipped yellow, the standards being of brightest mauve. The long stamens are protected by the petal-like stigmas. This iris grows well in full sun or partial shade and is hardy anywhere.

Iris hexagona (b). Native of southern United States, it has sword-shaped leaves 3–4 ft. (about 1 m) in length and bears its flowers on forked stems 3–4ft. (about 1 m) tall during April and May. They are of soft lilac with spoon-shaped standards and obovate falls.

Iris hookeriana (a). A handsome native of Bengal with fleshy rhizomes and pale green leaves which appear when the flowers have died back. The flowers are borne two to a stem, the purple falls being heavily bearded with white hairs whilst the narrow standards are of bluish-purple.

Iris kaempferi (b). They are beardless, producing flowers with the appearance of the clematis and they vary considerably in size. They may be hybrids. It has sickle-like leaves 4 in. (10 cm) long and large broad flowers, the standards of pale lilac are spotted with purple; the creamy-white falls being spotted with black. The flower was widely used for hybridising by Sir Michael Foster early in the century.

Iris kaempferi (b). Are beardless, producing flowers having the appearance of the clematis and they vary considerably in size. They may be propagated by division immediately after flowering or by sowing the seed in the moist soil where the plants are to bloom. The ripened seed is gathered in September when it may be sown in pots of peat and loam and wintered under glass, or it may be sown in the open ground where the plants will appear the following spring. The plants must be kept free of lime; they enjoy an acid soil which is another way in which they differ from their near relations.

DARK CLOUDS. A lovely variety bearing double flowers of dark violet-blue with a bright yellow blotch.

GEI-SHOO-UI. A strong grower, the white blooms are attractively edged with rose.

HERCULE. Bearing a very large bloom of lavender-blue, flushed purple.

MOONLIGHT WAVES. Bears a huge double-white bloom with striking yellow blotch.

PURPLE EAST. A taller-growing variety, the deep purple blooms being single and most outstanding.

TIGER DANCE. The pale blue flowers are veined violet and crimson.

I. lortetii (c). Native of the Lebanon, it resembles *I. gatesii*. It has sword-like leaves and in June bears large flowers, the standards being of a beautiful shade of soft pink veined with mauve; the falls of palest blue spotted with crimson, heavily at the centre.

I. lutescens (a). Native of south Europe, it has glaucous sword-like leaves and stems on which it bears in May, large yellow flowers, the falls being veined with purple-brown.

I. missouriensis (b). Native of the Rocky Mountains, it forms a tuft of linear leaves 12 in. (30 cm) long which taper to a point and it blooms in May, the large lilac-blue flowers being veined with purple, the falls delicately shaded with yellow.

I. pallida (a). A Flag Iris similar to *I. germanica* and which has had a considerable influence on the raising of the modern hybrid varieties. It is native of south Europe, arriving in the British Isles during Elizabethan times. Gerard mentions that it grew in his garden in Holborn with 'leaves much broader than any other (iris) and . . . with fair large flowers of a light blue or (as we term it) a watchet colour'. And he adds, 'the flowers do smell exceeding sweete, much like the orange flower'. To some, the perfume nearly resembles vanilla; to others, it is likened to civet.

I. pseudacorus (b). It is the Yellow or Water Iris to be seen growing by the side of rivers and in marshlands throughout the British Isles and in France, whose King Louis VII took the flower as his blazon during the Crusades and gave it his name, 'flower of Louis'. It is the yellow 'Vagabond Flag' of Shakespeare's *Antony and Cleopatra* and it became the national symbol of medieval France, in heraldic language: 'Azure powdered with fleurs-de-lis or'. In 1339 when Edward III made claim to the throne of France and began hostilities against Philip VI (Philip of Valois), he took for his arms the three Plantagenet Lions and the fleur-de-lis of France.

It is a delightful plant with a scented flower some 3 in. (7.5 cm) across and of a soft shade of golden-yellow. The flowers are produced in succession from May until August amidst sword-like leaves. From the dried rhizomatous roots, a delicately scented essential oil is obtained which at one time was used to adulterate oil of *Acorus calamus*.

I. sari (c). It grows on the banks of the River Sar and it blooms in May, its large handsome flowers resembling those of *I. gatesii*. It has sword-like leaves 6 in. (15 cm) long and bears flowers of soft violet, spotted and veined with deeper violet, with the falls rather darker than the standards and with a dark brown beard.

I. sibirica (b). It was grown in Elizabethan gardens and in spite of its name, is native to all parts of Europe. It makes a pleasing waterside plant but will flourish in the border provided lime is not present in the soil. It has elegant grass-like foliage and bears two or three flowers together of brightest lilac-blue.

Two outstanding varieties are 'Helen Aster', a rose-red and 'Nottingham Lace', introduced in 1960, the wine-red flowers being laced with white.

I. stylosa (b). Syn: *I. unguicularis*. Native of Algeria, it has a thin rhizomatous rootstock from which arises a tuft of bright green linear leaves 12 in. (30 cm) high. As it blooms in January and February, it should be given as warm a position as possible such as the foot of a wall facing south and beneath the eaves of a house where the plant will be protected from excess winter moisture. It requires a poor, dry soil well supplied with lime rubble. The sweetly scented flowers are of a most beautiful shade of sky blue, the variety speciosa being veined with white. The partially open buds if taken indoors will open in water and remain fresh for a week or more.

I. susiana (c). The Mourning iris, 'so fit for a mourning habit', wrote Parkinson. 'I think in the whole compasse of nature's store, there is not a more pathetical . . . among all the flowers I know, coming neare into the colour of it.' It takes its name from the ruined city of Susa in Persia where it was discovered, the enormous flower being white, veined and marked with black. It must be given a dry, sunny situation.

Gerard said, 'it doth prosper well in my garden' and he likened its bloom to the Ginny Hen. He also called it the Turkey Flower de Luce, it having reached England from Constantinople early in the sixteenth century.

I. tectorum (b). Native of Japan, it has pale green sword-shaped leaves 12 in. (30 cm) long and it blooms in May and June, on 12 in. (30 cm) stems. The flowers are of brightest lilac with the falls attractively crisped and veined with purple.

I. variegata (a). Native of central Europe, it forms tufts of sword-like leaves 18 in. (45 cm) long, shaded purple at the base and it blooms in May and June, bearing its flowers on a glaucous stem 18 in. (45 cm) high. The falls are of deep claret-red with a yellow beard whilst the erect standards are of contrasting golden-yellow.

I. versicolor (b). Native of east United States and Newfoundland where it is common by the side of streams and in damp meadows. It has glaucous sword-shaped leaves and in May and June, bears clusters of deep purple flowers variegated with yellow and green on round smooth stems 2 ft. (60 cm) in height. The standards are usually of paler colouring than the falls though there is considerable variation in colour.

VARIETIES OF THE REGELIO-ONCOCYCLUS SECTION

ANCILLA. A free-flowering form growing to a height of 12 in. (30 cm) with a standard of white veined soft blue and with the falls netted with grey and with a dark brown flake.

CHIONE. It grows 15 in. (38 cm) tall the standards being white, veined with pale blue and the falls netted greyish-brown with a blackish brown flake.

CLARA. Similar to *I. susiana* and growing 20 in. (50 cm) tall, the white ground standards being netted and veined with blackish-purple, the falls being of dark brown with a black blotch.

DARDANUS. An *I. iberica* hybrid, the creamy-white standards shaded and veined with lilac, the falls veined with purple on a cream ground. Of vigorous habit, it grows 2 ft. (60 cm) tall.

MERCURIUS. An *I. barnumae* seedling growing 20 in. (50 cm) tall, the large rounded flowers having deep purple standards and broad, spreading falls of bronzy-purple.

SYLPHIDE. It has large refined flowers borne on 18 in. (45 cm) stems, the white standards being mottled with grey; the falls being of greyish-brown with a black blotch.

THESEUS. A free flowering iris of outstanding beauty, the standards being of deep violet with black veins; the falls veined and blotched violet on a cream ground.

THOR. An *I. sari* hybrid growing 10 in. (25 cm) tall with a pearl-grey ground throughout veined with dull purple and with a bright purple blotch on the falls.

BULBOUS

The bulbous species may be divided again into three main groups:

(a). Those of the reticulata group which are distributed in Asia Minor and are characterised by the handsome netted tunics of the bulbs; by their dwarf habit and earliness to bloom. They have rush-like leaves and bear their flowers singly. They are usually scented. They make admirable pot plants and are delightful on the rock garden.

(b). This section includes those of the xiphium group, native of the Iberian Peninsular and north-west Africa. They have smooth tunics and bear their flowers on stems 2 ft. (60 cm) tall. Of this group belong the English and Dutch irises, grown by the million each year for their cut bloom during spring and early summer.

(c). This section includes the Juno irises of Bokhara, south Russia, Turkey, Persia, and Afghanistan. They are bulbous but also have thick fleshy roots and deeply channelled leaves which enclose the stem at their base. They are the most difficult of all irises to maintain, requiring a rich diet and plenty of moisture whilst growing but need to be kept almost dry during their rest period.

CULTURE OF BULBOUS IRIS

Iris of the reticulata group will come into bloom outdoors early in February and are suitable for naturalising in short grass and for planting on the rock garden where they prove valuable on account of their earliness to bloom and neat foliage. They will be seen at their loveliest planted in small groups near a rock, particularly if it is a piece of weathered Yorkshire stone, greyish white in colour, which shows off the rich purple, crimson and pale blue colours of the bloom to the greatest advantage.

On the rockery or as an edging to a border it will come into bloom in mid-February, at a height of only 6 in. (15 cm). Planted in short grass where it looks delightful, it will reach a height of 9 in. (22 cm). In the semi-shade of trees, particularly among silver birch, with their brightly coloured winter bark, the bulbs may be left untouched for several years, after which they are best lifted, divided and replanted during September. Plant the bulbs only 2 in. (5 cm) deep and 2 in. (5 cm) apart outdoors, for their habit is neat and upright and they make only a few rush-like leaves. Though they like a soil containing some peat or leaf mould and a little decayed manure, lime rubble is, to nearly all members of the iris family, the most important part of their diet; and plenty of mortar should be worked into the soil before the bulbs are planted in autumn. A dressing with lime rubble each autumn will keep the bulbs healthy and free-flowering. When the blooms begin to die back, they should be removed, so as to conserve the energies of the plant.

It is as pot plants that the dwarf irises are at their best and the bulbs should be planted in pots or bowls in September. Plant them only 1 in. (2.5 cm) apart, and as many as a dozen bulbs may be planted in an earthenware bulb bowl. After two months in the plunge bed, or in a cool, dark place they may be slowly introduced to the living-room. In early spring, their richly coloured blooms will scent a large room.

Irises of the xiphium section may be planted in groups in the border and in prepared beds for cutting. Planting should be carried out from October 1st until mid-November, though October planting will ensure the formation of roots and therefore anchorage against hard frost.

One method of planting is first to prepare a bed 3–4 ft. (1m) wide which will enable picking of the blooms to be done easily. As irises like dry conditions, especially during winter, a raised bed will suit them best. This should be made up in September, incorporating some sand and leaf mould. If the soil is light in texture, some well-rotted manure, either of horses', cows' or pigs' will give it the necessary 'body', for although the iris enjoys dry conditions, some humus is essential to the formation of a tall, well-formed spike. A raised bed will ensure that the bulbs are in no way waterlogged should a long, wet period be experienced in the early part of the winter and before the bulbs have formed their roots. If the ground lacks lime, rake in some lime rubble or a small quantity of hydrated lime, but the bed must be given 3 weeks in which to settle down, for bulbous irises are like onions and shallots in their demanding of a firm planting bed. To make quite sure that the bed is compact, it is as well to tread it all over a day or so before the bulbs are planted. Some writers on the subject suggest planting in drills made 3 in. (7.5 cm) deep and placing the bulbs 1 in. (2.5 cm) apart. One method is to plant as for shallots, merely pressing the bulbs into the soil with their noses at ground-level. They are planted 2 in. (5 cm) apart in rows the same distance apart so that they will hold up each other when coming into bloom and will require no staking other than placing a cane at the four corners of the bed and tying around a thick piece of twine. So as to ensure a firm bed, run the garden roller over as soon as planting has taken place, and of course provided the soil is in no way sticky in which case planting will wait until it becomes friable.

The blooms will be ready for picking early in May, six months after planting the bulbs, and they must be removed before the flower heads are fully opened. Several days may be saved, and every day means a more profitable crop, if the blooms are removed as soon as showing colour and allowed to stand in buckets of cold water for twenty-four hours before dispatching to market or shop. The flower stems, being of a high water content, will continue to supply the blooms with moisture and they will continue to open all the time. A bloom which is too open will bruise and rapidly deteriorate.

If a good-sized bulb is planted, one $2\frac{3}{4}$–$3\frac{1}{4}$ in. (7–8 cm) as against the $2\frac{1}{4}$–$2\frac{3}{4}$ in. (6–7 cm) size, and the beds watered with liquid manure every fortnight, it will be possible to grow the same bulbs for a second season. In order to help the bulbs as much as possible it is advisable to leave a piece of stem and a leaf rather than to cut the bloom at ground-level. Bulbous iris grown in the border for decoration should have the flower heads removed after flowering and the stem and foliage should be allowed to die down as for all other bulbs.

Bulbous iris used in the border should be planted in clumps of six bulbs. As when planted in beds for cutting plant in separate colours. But do not neglect to plant the Spanish and English types which will prolong the flowering season into July, and whilst the Spanish and English iris may not have much commercial value, they are invaluable for early summer cutting. They require much the same treatment as described for Dutch iris, though the English iris enjoys a heavier and slightly moister soil.

All these irises should, where possible, be planted in a position sheltered from the prevailing winds on account of their tall habit. A wall or hurdle screen is ideal; the shelter of trees is not suitable for the plants must have full sun.

DUTCH IRIS UNDER GLASS

The early outdoor forcing of the Dutch hybrids under frames and barn cloches is widely carried on by commercial growers to supply the early spring markets. The same methods may also be used to provide early bloom for the home. All the Dutch varieties will bloom well under outdoor glass and will come into bloom three weeks in advance of uncovered bulbs.

The varieties 'Wedgwood' (see plate 34) and 'Yellow Queen' are also suitable for indoor cultivation in gentle heat and like all the others may also be used for cool glasshouse flowering. By using a slightly heated house, followed by the early-flowering varieties under glass outdoors, it is possible to have Dutch iris for cutting from late February until the Spanish varieties come into bloom in June, a six months' display. As these iris are essentially flowers for cutting rather than for pot culture in the home, no word will be made of home cultivation under this heading. That is reserved for the dwarf-flowering species.

Where cloches and frames are to be used, the beds should be planted to the width of the glass covering. Where possible, a trench should be prepared, a 4 in. (10 cm) space being left at the top which will give the plant the extra height, for it is little use covering the bulbs and then having to remove the glass before they have come into bloom unless Dutch lights are available with which to cover the plants when the cloches have ceased to play their part. If the lights are placed on their sides and held in place by wooden stakes and over the top are also placed lights, a miniature greenhouse will have been made which will provide the correct height for Dutch irises.

The bulbs are planted early in October and are covered early in February, the plants coming into bloom in April when advantage may be taken of the Easter market, if it is a late Easter. If a cool or heated greenhouse is to be used, the bulbs are planted either in large pots or deep boxes. They should be planted about September 1st into a mixture of loam, coarse sand and well-rotted manure. Old mushroom-bed manure is ideal, so are spent hops. Some lime rubble should also be added. As many as six bulbs may be planted into each large pot, or if growing in boxes, plant them 2 in. (5 cm) apart each way. A 4–4¾ in. (10–12 cm) bulb should be used. After planting and watering, the pots and boxes are stood outside in a plunge bed of ashes or sand until thoroughly rooted, which will take about two months. The first batch may then be taken indoors. 'Wedgwood' will be the first and a temperature of 50°F. (10°C.) will be detrimental to the production of a top-quality flower. Only when the buds are showing should the temperature be increased to 55°F. (13°C.). At this period, the bulbs will need ample supplies of water and a dry, well-ventilated atmosphere. 'Wedgwood' taken indoors about November 1st should be ready for cutting when grown in a temperature of 50°F. (10°C.) by March 1st. Later batches of 'Yellow Queen' and 'Imperator' will be taken indoors before the year end and will be in bloom from the end of March. Those outdoors under lights or cloches will bridge the gap until the unprotected 'Wedgwood' is ready.

Iris may also be grown in a quite cold house by planting in pots or boxes as described which are taken indoors when the tomato crop has ended in late October, or the bulbs may be planted directly into the greenhouse-floor border in October. In this case, they should be covered with sacking to keep out the light until early December, when they should be rooted. They should at no time be over-watered and will be in bloom early in April. Owing to their occupying the greenhouse for some considerable time when planted in this way, planting in pots or boxes is preferable.

SPECIES AND VARIETIES OF DUTCH IRIS

Iris alata (c). Known as the Scorpion Iris, it is native of south Europe and north Africa and is winter-flowering, coming into bloom early in November and though perfectly hardy in all but the most exposed gardens, its exquisite beauty is best enjoyed in the shelter of the alpine house. Outdoors, it requires a well-drained soil and a sun-baked situation and where sheltered from cold winds, will continue to bloom until the new year. It has lance-shaped pale green leaves 12 in. (30 cm) long and bears its flowers on a 6 in. (15 cm) stem. They are of bright lavender, the falls having a golden crest or keel whilst the spoon-shaped standards spread out horizontally.

I. bakeriana (a). Native of Persia and Iraq, it comes into bloom a week or so before others in this section and is one of the most handsome of all irises. Like *I. reticulata* the flowers have the same delicious scent of violets. It has ovoid bulbs with an eight-ribbed tunic of pale brown and sharply pointed leaves 9 in. (21 cm) in length. The flowers have standards of pale ultramarine-blue and broad spoon-shaped falls of white, blotched with deep blue and edged with violet. The flowers are borne on 6 in. (15 cm) stems and are hardy and tolerant of adverse weather though for the bulbs to be long living, they should be planted where excess winter moisture can readily drain away and where they may be baked by the summer sunshine.

Iris boissieri (b). Native of the Jerez Mountains of Portugal and Spain, it is a handsome species but difficult to manage in the open and is best grown under glass where it can be almost dried off after flowering. It has linear leaves and bears its flowers in June on a 12 in. (30 cm) stem. The standards are purple, shaded reddish at the base; the falls being of reddish-purple with a golden-yellow beard.

I. bucharica (c). Native of Bokhara, it has pale yellow leaves like those of the leek in shape and from the axils arise in April, white and yellow flowers on 12 in. (30 cm) stems. The falls are tipped with yellow and are beautifully frilled at the margin. In a well-drained soil it increases rapidly.

I. caucasica (c). A delightful little iris, native of the Caucasus Mountains of north Persia and growing 6 in. (15 cm) tall with bright green leaves which taper to a point. It is one of the earliest of the Junos to bloom, flowering towards the end of February or early March, the pale lime-yellow flower being 3 in. (7.5 cm) across. The variety 'Major' bears larger flowers than the type.

I. danfordiae (a). Native of the Cilician Taurus, it requires a light, well-drained soil containing plenty of lime rubble, and a sunny situation. Once established, do not disturb it for it is from the newly formed bulblets that next season's flowers will arise. It bears its golden-yellow flowers, speckled with brown, during February, and is almost leafless at flowering time. It blooms on a 4 in. (10 cm) stem and is untroubled by the severest weather, for the flowers will emerge through frozen snow.

I. filifolia. (b). Native of south Spain and north Africa, it resembles the Spanish iris in miniature when in bloom as well as in its bulb and foliage. It has slender grass-like leaves and it bears its flower of bright uniform purple in June on a 12 in. (30 cm) stem.

I. fosteriana (c). It is to be found about rocky hillsides in south Russia and Afghanistan. It more nearly resembles (b) in its bulb for it forms few fleshy roots whilst its flowers and linear leaves resemble the Spanish iris. It blooms in March, the standards being of bright mauve and almost horizontal, the falls pale yellow, waved at the margin. The flowers are borne on a 12 in. (30 cm) stem.

I. graeberiana (c). A rare native of Turkestan bearing sheaths of leaves 12 in. (30 cm) high and from the axils, small flowers of silvery-mauve with cobalt-blue shading on the falls.

I. histrio (a). Native of Syria and the Lebanon, it is probably the first of the reticulata group to bloom, in January if the weather is not too unkind when it bears its lilac-blue flowers with a white crest on the falls, on a 6 in. (15 cm) stem.

The best form is 'Aintabensis' which bears a larger flower of soft pale blue with an orange flash down the centre of the falls and is long lasting when in bloom.

I. histrioides (a). One of the best of all winter flowering bulbs, coming into bloom as soon as the snow begins to melt and bearing its flowers before the leaves. It is tolerant of all weathers and growing only 4 in. (10 cm) tall, it is at its loveliest in pans in the alpine house or planted in sunny pockets about the rock garden. But wherever it is planted, it should be in a position where the bulbs can be ripened (baked) by the summer sunshine. It is lovely planted in a narrow border beneath the eaves of a house and facing due south. It bears flowers of brightest ultramarine, spotted with black with a white crest on the falls. The form 'Major' bears larger flowers than the type.

I. juncea (b). A delightful little Spanish iris of south Spain and north Africa which blooms during June and July followed by rush-like leaves which grow 12 in. (30 cm) long. The flower is of deepest golden-yellow, borne on a 12 in. (30 cm) stem and it has a powerful scent which is retained after it is cut and placed in water.

I. kolpakowskiana (a). Native of Turkestan, being rare in the wild and in cultivation for it is difficult to manage. It bears its violet-scented flowers in March before the deeply channelled linear leaves have made much growth. The flowers are borne on a 3 in. (7.5 cm) stem, the standards being of soft lilac-purple, the falls of rich purple with a golden crest veined with purple and they are sweetly scented.

I. magnifica (c). Syn: *I. vicaria.* Native of Turkestan, it is one of the finest of the Junos and one of the tallest, attaining a height of 2 ft. (60 cm) with as many as ten flowers appearing in succession from the axils of the leaves. They appear in April and are palest lavender with an orange crest on the falls.

I. orchioides (c). The beautiful Orchid iris, found amongst the lower hill-sides of Bokhara and which has bulbs larger than a hen's egg. It has broad bright green leaves from the axils of which are produced flowers with pale yellow standards and deep yellow falls with an orange crest at the centre. The variety 'Sulphurea' is pale uniform yellow and 'Coerulea', palest blue with a yellow ridge on the falls.

I. persica (c). Depicted on the first plate of the *Botanical Magazine* in 1787 since when it has been growing in gardens but usually with difficulty. It makes a bulb the size of a bantam's egg from which arises a stem 6 in. (15 cm) high with two or three narrow leaves. The flowers appear in February and measure 3 in. (7.5 cm) across. They are violet scented and of an unusual shade of greenish-blue with a purple blotch at the tips of the falls and a yellow crest. To be long living and free flowering, the bulbs must be grown where the summer sunshine can ripen them fully but where the garden is exposed and the soil badly-drained, it is best grown in pans in the alpine house.

I. reticulata (a). It is to be found on lower mountainous slopes from the Caucasus to Persia and as a pot plant and in the garden is outstanding, being hardy and bearing flowers of exquisite daintiness and perfume from mid-February until early April whilst the leaves will be no taller than the flowers. The plants increase rapidly and bloom year after year, the dark purple flowers, like tiny Dutch irises appearing on a 6 in. (15 cm) stem. The flowers are enhanced by an orange flash on the falls. There are many lovely varieties:

CANTAB. It bears amazingly lovely flowers of brilliant Cambridge blue with a brilliant orange crest on the falls.

CLARETTE. The result of a cross with *I. bakeriana*, it has standards of sky blue and falls of deep blue with a white flake.

HARMONY. The flowers have both substance and durability and are of rich pansy purple with a central orange ridge on the falls.

JEANNINE. Received an Award of Merit in 1966. It bears flowers of sky blue of great clarity of colour.

JOYCE. Obtained from an *I. histrioides* x *I. reticulata* cross, the flowers being clear blue with an orange ridge on the falls.

J. S. Dijt. The earliest to bloom, the powerfully scented blooms being of reddish-purple.

Pauline. It bears a large bloom of brightest purple with a large white crest on the falls.

Purple Gem. Early to bloom, the standards are of ruby purple, the falls purple black spotted with white.

Royal Blue. The blooms are large and of uniform Oxford blue.

Violet Beauty. The standards are of bright Parma violet, the falls having a conspicuous orange crest.

Wentworth. Sweetly scented and early, the blooms are of brightest royal purple, the falls crested with gold.

I. sindjarensis (c). Syn: *I. aucheri*. A beautiful species, native of the Djebel Sindjar Mountains of Mesopotamia with long pear-shaped bulbs and leaves narrowing to a point. From the axils are borne early in March, vanilla-scented flowers of a unique shade of slate-blue, the falls having a creamy-white crest veined with green.

I. vartanii (a). It would be found growing about the rocky hillsides around Nazareth in the time of Christ and has heavily ridged bulbs from which arise horny-tipped leaves and flowers which measure 4 in. (10 cm) across, being the largest in this section. Following a warm, dry summer, the first flowers often appear in October and continue through winter. They are of a lovely shade of lavender-blue, the falls having a white crest whilst the claw is spotted with black. The flowers are borne on a 6 in. (15 cm) stem and are deliciously scented of almonds. There is a white form, 'Alba'.

I. willmottiana (c). Native of the higher mountainous regions of Turkestan, it is rare in cultivation and in the wild. It has a large round bulb and blooms before the end of February at a height of 6 in. (15 cm). The flowers are about 2 in. (5 cm) across and are palest blue, the falls being flecked with darker blue.

I. winogradowii (a). It is found close to the Black Sea and is similar to *I. histrioides* in its habit and flowering time but its flowers are of a lovely shade of creamy-yellow, the falls having a central ridge of brightest orange.

IRON-BARK TREE, *see* Eucalyptus

IRON TREE, *see* Parrotia

ISOLEPSIS

I. gracilis forms a tuft of foliage whose elegant rush-like leaves droop around it, completely hiding the pot. If placed to the front of the greenhouse bench, the foliage will eventually droop over the side and reach the floor. Or the plants may be used in hanging baskets.

Seed is sown in early March in a temperature of 60°F. (16°C.), using a seed pan containing the John Innes Sowing Compost. The seed is small and care must be taken with its handling, but provided the compost is not allowed to dry out, it will germinate in about three weeks, the seedlings being moved to $2\frac{1}{2}$ in. (6.5 cm) pots when large enough to handle.

During summer, water freely and shade the plants, whilst during winter provide a temperature of 52°F. (11°C.) and give the plants little moisture. The plants should be transferred to size 60 pots about twelve months after the seed is sown.

IVY, *see* Hedera

IVY ARUM, *see* Scindapsus

IXIA

The Africa Corn lily, in appearance resembling montbretia and freesia and possessing the qualities of both. In warmer parts they may be planted in beds during September and covered with bracken to keep out winter frosts. They will bloom during June, the long graceful stems making them a desirable cut flower. Or they may be wintered in a cold frame, planting the bulbs in September and withholding water in the same way as for freesias, until spring growth commences, when they may be watered and brought into bloom during May. Again, they may be planted in deep boxes, wintered in a cold frame and moved to a cool greenhouse during February when they will commence to bloom during April.

Wherever they are planted they enjoy a light, sandy soil, containing no manure but a little peat. They are also lime lovers and should be given ample supplies of lime rubble.

The corms should be planted 4 in. (10 cm) deep and 4 in. (10 cm) apart in a sunny position if grown in the open. They may be grown to perfection in narrow beds under a high wall facing due south in sheltered garden or under lights reared against a wall. This will not only keep out frost, but excessive moisture.

Marketing and cutting ixias calls for some judgment for if cut before the blooms open they may never do so, and those that may have opened too far may close up and not recover if remaining too long out of water. It is for this reason that the ixia is not more widely grown.

SPECIES

Ixia azurea. Bears its blooms of pure sky-blue during June and blooms well in the open in a sandy soil.

I. bucephalus. An excellent cut-flower species, bearing its vivid rose-coloured flowers on 2 ft. (60 cm) stems.

I. gigantea 'Alba'. It bears a striking bloom of purest white, which carries a delicious perfume.

I. lutea. A lovely form, producing its rich yellow star-like flowers on long sprays.

I. rosea 'Plena'. Produces its almost double flowers of rich pink on 15 in. (38 cm) stems.

I. scariosa. The pale lilac blooms, carried on 18 in. (45 cm) stems, appear in June.

I. viridiflora. This is included, though difficult to obtain, being one of the most striking flowers in cultivation. It is easy and bears in profusion, flowers of vivid green, flushed deep blue and purple.

JACOBAEA

This hardy annual, also listed as *Senecio elegans*, is now obtainable in a double form, in crimson, purple, white and rose. The seed is sown at the end of March and as the plants attain a height of 18 in. (45 cm) they should be thinned to 12 in. (30 cm) apart.

Senecio arenarius, is the annual cineraria, of similar height and which bears its flowers through summer.

JACOBINIA

Native of tropical Brazil, *J. pauciflora* is a valuable winter-flowering plant, bearing conical heads of loosely petalled flowers of a lovely shade of soft salmon-pink. It requires a minimum winter temperature of 48°F. (9°C.) and should be allowed to bloom in a sunny window, though it prefers some shade when resting in summer. Water with care at all times and cut back in March after flowering. Propagate from cuttings inserted in a sandy compost around the side of a pot.

JAPANESE ARALIA, *see* Fatsia

JAPANESE CEDAR, *see* Cryptomeria

JAPONICA, *see* Chaenomeles

JASMINE, *see* Jasminium

JASMINUM

From the flowers of the White Jasmine, the perfume 'oil of jasmine' is obtained whilst the ancients captured the evanescent odour of the blossoms by means of enflouage, embedding the fresh flowers in fat from which they made odoriferous ointments.

It is a genus of more than 120 species of evergreen and deciduous shrubs, all indigenous to the east, to be found from Persia to central China, though *J. azoricum* is native of the Azores and *J. odoratissimum* of Madeira. Many are completely hardy in the British Isles and are successful in ordinary soil whilst with their small glossy leaves, like those of the privet and other plants of the family they are tolerant of town garden conditions. They are readily increased from cuttings of the half-ripened wood, removed in August and rooted under glass. Any pruning should be done in April after flowering.

SPECIES AND VARIETIES

Jasminum azoricum. It is a twining species, vigorous and evergreen but in Britain is suitable only for the mildest localities in which it will bloom almost throughout the year. The flowers are white and heavily scented.

J. beesianum. It rarely exceeds a height of 6 ft. (2 m) but may be planted against a low wall. It makes a slender plant and blooms during July and August, its small deep rosy-red flowers having a powerful spicy scent.

J. humile 'Glabrum'. Native of Nepal, it is a plant of vigorous habit growing 9 ft. (3 m) in height and from May until August bears clusters of deliciously scented yellow flowers.

J. nudiflorum. This native of China did not reach English cottage gardens until quite recent times when it came to be appreciated for its winter blossoms which appear in opposite pairs along the twiggy stems, devoid of leaves at this time of year. Any pruning should be done in April, after flowering so that the new shoots produced during summer will be able to carry the full complement of blossom. It is not scented but its primrose-yellow flowers are always appreciated when the days begin to lengthen and the cold intensifies.

With its flexible stems and small narrow leaves, it may be trained about trellis or against a wall and always remains neat and tidy, whilst no degree of cold will harm it.

J. odoratissimum. Native of Madeira, it is suitable only for the mildest parts of Britain where it will bear its primrose-yellow flowers in threes from the tips of the branches. The flowers have a heavy sweet perfume.

J. officinale. The Common White Jasmine of Tudor gardens, a strong-growing twining plant which will attain a height of 9–10 ft. (about 3 m) and is evergreen in all but the coldest localities of the British Isles. It is a plant now rarely to be found though with its deliciously scented blossoms, borne from June until October, it is a charming companion for the Winter Jasmine which comes into bloom early in November and continues until late in March when *J. revoltum* comes into its scented loveliness.

JERUSALEM SAGE, *see* Phlomis

JEW'S MALLOW, *see* Kerria

JONQUIL, *see* Narcissus

JOSEPH'S COAT, *see* Scindapsus

JUNIPER, *see* Juniperus

JUNIPERUS

A genus of hardy evergreens, mostly native of the north Hemisphere with needle-like leaves, scattered or imbricated; the male flowers borne solitary or in crowded catkins; the cones round and berry-like. The junipers are amongst the finest of all evergreens for a cold, windswept garden and for a calcareous soil, though they will grow best in a deep, heavy loam. Plant in April. Propagate from seed which takes two years to ripen. First remove the pulp by immersing the fruits (used to flavour gin) in water for two days and then shake up with sand. This will hasten germination. Or propagate by cuttings of the half-ripened shoots removed with a 'heel' and inserted in sandy compost under glass.

SPECIES AND VARIETIES

Juniperus chinensis. The Chinese juniper, the male plant having graceful drooping branches and glaucous foliage; the female being of slender pyramidal habit and bearing purple fruits. Of many forms, 'Aurea' ('Young's Golden') has handsome golden foliage whilst 'Fortunei' has grey foliage and is of more loose habit. 'Variegata' is a slow-growing spreading plant, its glaucous foliage being marked with silver.

J. communis. The Common Juniper of the British Isles, which makes a dense bush 20 ft. (6 m) tall. It has stiff awl-shaped leaves and bears fleshy glaucous cones (fruits). The variety 'Hibernica', the Irish Juniper, grows upright, making a dense slender column of grey-green. The form 'Aurea' has golden foliage which turns bronze in winter whilst 'Compressa' makes a dense dome-shaped plant of glaucous-grey and rarely exceeds 12 in. (30 cm) in height.

J. horizontalis. The Creeping Juniper of prostrate habit for the rock garden, the form 'Douglasii' having foliage of glaucous blue changing to bronze in winter whilst that of 'Plumosa' is purple-tinted in winter.

J. packyphloea. The Alligator Juniper of south United States, so called for the strange square markings of the bark. Its foliage is bright silvery-blue but it requires a mild part for it to succeed.

J. pfitzeriana. Pfitzer's Juniper, it makes a spreading shrub with scale-like foliage, the branchlets recurving at the tips. The variety 'Armstrongii' is of more dense habit and has foliage of apple-green tinted with gold which is more accentuated in the variety 'Aurea'.

J. recurva. The Drooping Juniper of the Himalayas, with feathery re curved branches and sharply pointed grey-green leaves arranged in threes. The male form, 'Densa' is of more compact habit.

J. sabina. The Savin, an upright branching shrub with scale-like pointed leaves which makes a valuable hedge. The form 'Cupressifolia' has plumose branches clothed with dark green leaves.

J. virginiana. The Red Cedar of the United States which makes a tall tree of pyramidal form. From its fragrant wood, pencils are made. The form 'Burkii', makes a dense pyramid, its blue-grey foliage turning bronzy-purple in winter.

KAFFIR LILY, *see* Schizostylis

KALANCHOE

Closely related to the crassula, the kalanchoes are to be found in their natural state in Madagascar and along the east African coast. *K. bloos-*

feldiana, like all the kalanchoes, bears dark green leaves which are covered with minute hairs. These are used by the plant as protection against strong sunlight. The leaves of *K. blossfeldiana* are strikingly edged with scarlet, whilst in winter it bears beads of orange-red tubular flowers on 15 in. (38 cm) stems which last a long time on the plant, or when cut and placed in water.

K. flammea is also grown for florists' use, for it bears orange-scarlet flower heads and has grey-green leaves whilst *K. marmorata* has green leaves marked with purple. In early spring it bears glistening white flowers. *K. tomentosa* has leaves which are covered with down and attractively tipped with chocolate brown.

The kalanchoes are propagated from cuttings taken during summer and rooted in a sunny window around the side of a pot containing a sandy compost. The plants will come into bloom in eighteen months' time.

At no time should the plants be over-watered; give sufficient to keep them growing and after flowering give only enough to keep them alive, increasing the amount again during summer. They like a compost which is both rich and well drained, that suggested for growing aloes being suitable. They also like more winter warmth than most succulents, requiring a minimum temperature of 48°F. (9°C.) and protection against draughts in any form.

KANSAS FEATHER, *see* Liatris

KEPHALARIA

The tall pale yellow scabious *K. tartarica* will grow almost anywhere, and though providing little colour in the border is a useful cut flower. But for the amount of growth the plants make their flowers are most disappointing, whilst their support requires attention. The blooms are produced throughout July. Plant 2 ft. (60 cm) apart in November, at the back of the border.

KERRIA

K. japonica is the Jew's Mallow and is one of the most pleasing of small shrubs, making a spreading bush 4 ft. (1 m) tall. Though completely hardy, it is also valuable to plant against a wall where it will attain a height of 6 ft. (2 m), its arching branches clothed in pale green serrated leaves and in April and May, rich golden yellow flowers. The double form is more attractive. Plant November to March in a sandy loam and in an open, sunny position. Prune in June after flowering, shortening the flowering stems to healthy side growths.

As the plants make basal growths, they may be propagated by root division in autumn or by cuttings rooted under glass.

KINGCUP, *see* Caltha

KINGFISHER DAISY, *see* Felicia

KNAPWEED, *see* Centaurea

KNIPHOFIA (Tritoma)

The Torch Lily, *K. uvaria,* was introduced this country 250 years ago from South Africa, but due to its short season never achieved the popularity the brilliance of its spikes would seem to have assured for it. During recent years the introduction of new varieties will, if chosen with care, provide a continuation of colour from mid-July until mid-October, and they may be numbered amongst the most colourful of all autumn plants. Greatly in their favour is their extreme hardiness, their ability to thrive in ordinary soil and the fact that they require no staking. There are varieties flowering at varying heights, from 20 in. (50 cm) to 6 ft (2 m) and which may be planted from the front to the back of the border. Plant in November and increase by root division, though they should be disturbed as little as possible.

VARIETIES

GRACE SAMUEL. It blooms during August and September on 4 ft (1 m) stems, the spikes being of a lovely rich amber colour.

JAMES. Very early to bloom, by mid-July, the spikes being of an unusual carmine-red with a primrose-yellow base and borne on 5 ft. (1.5 m) stems.

JOHN BENARY. A new introduction bearing large spikes of deep orange-red and growing to a height of 5 ft. (1.5 m).

MAID OF ORLEANS. With the introduction of this lovely plant no longer can this flower be called the Red Hot Poker, for its spikes are of a lovely creamy-white, tinged with green and borne on 3–4 ft. (1 m) stems. Its bloom provides a valuable contrast to the rich scarlet and yellow colours, whilst it blooms during August and September.

PAULINE SAMUEL. A robust and long flowering variety, the bloom being large and of a vivid vermilion scarlet.

THEODORE. One of the Wrexham hybrids raised by Mr Watkin Samuel. It grows to a height of 5 ft. (1.5 m) and blooms late, from mid-September until November. The spikes are of a rich crimson self colour.

KNOTWEED, *see* Polygonum

KOCHIA

This shrub-like plant is attractive during summer with its feathery pale-green foliage, lovely when planted with geraniums and salvias, whilst during late summer and autumn its foliage turns a rich mahogany colour, equally as attractive. Seed is sown in early May where the plants are to remain, but they are best raised in gentle heat from an early March sowing. Growing to a height of 2 ft. (50 cm) the plants should be given the same distance in the beds. *K. childsii* is an annual and is the best species to grow.

KOLKWITZIA

The Beauty Bush, *K. amabilis* is a graceful hardy bush, growing 3–4 ft. (1 m) tall and bearing throughout June, small pink trumpets with yellow shading in the throats. It will only bloom freely if given a sunny situation where its wood will ripen, and it is worthy of some care in this respect.

Plant November to March in a rich loam. The only pruning necessary is to cut back the flowering shoots in July. Propagate from cuttings of the half-ripened wood inserted in a sandy compost under glass.

LABURNUM

It is one of the most colourful and valuable of ornamental trees, never growing too large and when in bloom early in summer, the bright golden yellow cascading flowers present a picture of great beauty against the green bark and twigs. *L. vossii* is the finest form, bearing its flowers in long racemes and they are sweetly scented. The laburnum has an added value in that it will grow and bloom well in partial shade and under town conditions. It is also not particular as to soil. A warning should be given in that the seeds, which are borne in long green pods like peas (it is of the same family), are poisonous if eaten, as they often are by children. Little or no pruning will be necessary apart from the regular removal of suckers. It is propagated from seeds sown in autumn and by grafting.

LACHENALIA

It is the Cape Cowslip, a deciduous bulbous-rooted plant which is valuable in that it will bloom during the first four months of the year in a temperature of 55°F. (13°C.). The plants have dark green strap-like leaves like those of the hyacinth, whilst its golden-orange blooms are borne in stiff spikes. With the form *L. glaucina* the flowers are creamy-white tinted with red.

The seed is sown in pans in January for the plants to come into bloom in twelve months' time. The John Innes Compost may be used and a temperature of 60°F. (16°C.) is required for quick germination. Also, the seed must be quite fresh. As soon as large enough to handle, the seedlings should be transferred to small pots containing the John Innes Potting Compost in which they are kept growing until ready for size 60 pots towards the end of summer. The plants should be kept as cool as possible, damping down the house and shading from the direct rays of the sun.

As the flower spikes appear in December, the plants should be given as sunny a position as possible and they will require ample supplies of moisture. The plants will also benefit from an occasional application of dilute liquid manure.

The bulbs should be dried off during summer so that they may ripen. They are re-potted during September and by gentle watering are brought into new growth to bloom again in the New Year.

LAD'S LOVE, *see* Artemesia

LANTANA

A genus of woody downy plants, native of Jamaica and mostly of semi-trailing habit. The sweetly-scented flowers are borne in stalked flat heads, the best known species being L. *camara* which should be grown under glass in the British Isles or in a room with a winter temperature of not less than 48°F. (9°C.). During Victorian times, a number of hybrids were raised from *L. camara* and *L. nivea* and which were used for summer bedding with heliotrope and pelargoniums.

The plants require an open soil enriched with a little decayed manure and to encourage a bushy habit, should have the tips of the side shoots pinched out as the plants make growth.

Propagate from cuttings rooted in a sandy compost.

LARCH, *see* Larix

LARGE-LEAVED YEW, *see* Cephalotaxus

LARIX

A deciduous tree with long shoots and tufted linear leaves which open palest green and turn deep yellow in autumn. The cones are ovoid and consist of a few woody scales. Plant November to March in a well drained soil. Propagate from seed sown outdoors in spring in a prepared bed and allow the seedlings to remain two years before planting out 15 in. (38 cm) apart. Here they remain for two more years before moving to permanent quarters. *Larix europea* (*L. decidua*) has horizontal spreading breaches and bears its leaves in grass-like tufts. The cones are erect and remain long on the tree. The Japanese larch, *L. leptolepis*, possesses similar quarters. *Larix europea* (*L. decidua*) has horizontally spreading branches hardiness and also makes a tall, upright tree. It is remarkable for its purple-red twigs. The form 'Pendula' is of weeping habit. *L. griffithii* requires a mild climate and is a tree of beautiful form with long needle-like leaves and pendent branches.

LARKSPUR, *see* Delphinium

LATHYRUS

One of a genus of more than one hundred species of annual or perennial plants, only *L. odoratus*, named by Linnaeus, having perfume. It is an annual, discovered in Sicily early in 1697 by a monk, Father Cupani, and he described the plant in his *Hortus Catholicus* of the same year.

It took someone with foresight to consider this flower worthy of development. But in 1870, Henry Eckford, a Midlothian man, when gardener to Dr Sankey at Sandywell in Gloucestershire began cross-fertilizing sweet peas. When he realized the commercial possibilities, he left his employment to devote his time to the raising of new varieties, working in his garden at Wem, Shropshire. So successful was he that at the Bi-centenary Sweet Pea Exhibition held at the Crystal Palace in 1900, out of the 274 varieties exhibited, half had been raised by Eckford. His variety 'Lady Eve Balfour', described in Robert Bolton's catalogue as being 'pale lavender, shaded grey', was chosen as the most outstanding variety yet raised and was used for the presentation bouquet at the Exhibition. The flowers were deliciously scented.

In the same year, in the garden of the Countess Spencer at Althorp Park, Northamptonshire, appeared the first sweet pea with frilled petals. It appeared amongst plants of 'Prima Donna', raised by the head gardener, Silas Cole, who named it after the Countess and almost overnight, Silas Cole and his new sweet pea earned universal fame. From a single seed pod came the world's most beautiful flower which had a perfume above all others. But 'Countess Spencer' had one great fault, it did not breed true to type. Nature, however, has a way of correcting these things for in the following year, there appeared in the garden of a Cambridge grocer, Mr W. J. Unwin, a deeper pink form of Eckford's 'Prima Donna' which also had waved petals and which he named Gladys Unwin, after his eldest daughter. As soon as Unwin realized that it would breed true, he sold his business and set about raising new sweet peas for the commercial cut-flower market and quickly achieved success. His son, Mr Charles Unwin has told that 'Prima Donna' had been grown for at least eight years before producing its waved form, which it did at almost the same time in three different places, and never did so again, this being one of the enigmas of horticultural history.

The sweet pea is a hardy annual but like so many hardy annuals, it responds better to half-hardy or biennial culture, being sown in a cold frame either in August or in March or in gentle heat early in the year. The plants are moved to their flowering quarters in April when they will come into bloom in June. The seed should be sown individually in small pots or in boxes and spaced 1 in. (2.5 cm) apart so that the young plants will have space to develop from the moment of germination. When about 4 in. (10 cm) high, the growing point should be removed to persuade the plants to develop a more bushy habit. Sweet peas require an open, sunny situation and will not tolerate shade.

A plant of climbing habit, it requires to be grown up canes or against a trellis or netting about which it will climb by means of tendrils. It may also be grown up twiggy branches when it will make a delightfully scented 'hedge' some 6 ft. (2 m) tall. If the dead blooms are continually removed before they set seed, the display will continue into autumn. For exhibition, the plants are grown against canes and supported by wire rings for they have their tendrils removed so that the total energy of the plant is concentrated in the blooms.

VARIETIES

BALLERINA. The blooms are large with outstanding petal texture and are of rich cream with a deep picotee edge of rose. They have pronounced fragrance.

CREAM GIGANTIC. Raised from 'Gigantic' and likewise introduced by Messrs Robert Bolton, the blooms are of similar substance but of a rich shade of cream with pronounced perfume.

CRIMSON EXCELSIOR. The most richly scented crimson, the enormous blooms being of a striking shade of crimson-red.

ELIZABETH TAYLOR. The best of its colour which is a lovely rich deep mauve with an equally rich perfume. An exhibitor's favourite.

EVENSONG. One of the most beautiful and sweetly scented of all sweet peas, shades of soft blue and lilac merging with delightful results whilst it is unsurpassed by any variety in vigour and perfume.

GIGANTIC. Introduced in 1932 when it received the Gold Medal of the National Sweet Pea Society, it is a white and has never been surpassed. It also received the Abol Trophy 'for the greatest advancement since the First World War in any one species or strain of plants'. The florets are enormous and the petals so heavily frilled as to give rise to a 'double' effect, whilst none has a richer perfume.

JOHN NESS. Raised by the late Mr John Ness, it is an outstanding exhibitor's variety, the flowers which are borne in fours and fives being of an attractive shade of clear mid-lavender with a spicy scent.

LEAMINGTON. Raised by Rev. Kenneth College and introduced in 1958, no variety carries a more pronounced perfume nor is there one which has received more honours, including the Award of Merit from the R.H.S. and one from the National Sweet Pea Society. The frilly-petalled flowers are large and of a lovely shade of deep, clear lilac with ideal placement.

MABEL GOWER. A long established favourite being introduced in 1949, the blooms are of a most attractive shade of clear medium blue with a sweet vanilla perfume.

PATIENCE. The large frilled blooms are of a lovely shade of deepest lilac with outstanding petal texture and a lovely sweet scent.

PICCADILLY. A variety with outstanding scent, the large frilly blooms being of deep rose-red, suffused with salmon.

PIXIE. The most heavily scented of the Unwin striped sweet peas for which the firm are famous. The deep cream ground is veined and marbled with salmon-orange.

ROSE FONDANT. Raised by Messrs Unwins of Histon, it has the richest perfume of any sweet pea of its colour, and is one of the most beautiful. The frilly blooms are of a subtle shade of soft rose-pink suffused with salmon.

ROSE FRILLS. A picotee of great charm, the huge frilly blooms, the largest since 'Gigantic', having a white ground and a wide edge of rose-pink with a delicious perfume.

FOR BEDDING

The 'Colour Carpet' strain, in which the plants grow 9 in. (22 cm) tall and have a spreading habit, is an excellent plant for bedding, or to use at the front of a border. Seed may be sown either in August or early in April outdoors, or in frames or boxes in March, the plants being set out in April where they are to bloom. But though hardy, winter protection is preferable for sweet peas. The plants come into bloom in early June, and continue until the end of summer. The blooms are large, fully frilled, and borne three to a stem, whilst they may be obtained in all the popular sweet pea colours. The seed will germinate more quickly if it is chipped before sowing, taking care not to damage the 'eye' or germinating point (see plates 35 and 36).

UNDER GLASS

With their cleanliness, delightful art shades and delicate fragrance, sweet peas are always assured of an all-year-round market provided they are grown well.

This means the production of long sturdy stems each containing several well-formed blooms of large size. Only blooms approaching exhibition quality will demand a worth-while market price and so professional sweet pea growing is not considered easy. Almost all the earliest bloom is grown in the Scilly Isles and reaches the markets early in February. This is followed in late March, April and May by choice bloom from Essex and parts of East Anglia and Kent where early spring sunshine is at a maximum. For greenhouse bloom, only pink and mauve colours are in demand and there is a large choice of varieties from which to choose. For the very earliest flowering, February and early March, the special winter-flowering varieties must be grown of which the best are 'Memory', a pure lavender and 'Shirley Temple', rich deep pink. To follow, there is a wide choice though some will grow better under glass than others. Of these, suitable are:

AMBITION, a lavender purple

LIME LIGHT. Lavender, flushed violet, and carrying four large blooms per stem.

MRS C. KAY, another lovely lavender

MONTY, a large blush pink and cream ground

RECONNAISANCE, a large cream, edged pink

Of the older varieties, 'Magnet', a cream pink overlaid salmon and 'Goldfinch', orange standard and pink wings are still hard to beat.

CULTURAL REQUIREMENTS UNDER GLASS

Light is the most important factor in growing sweet peas under glass. If this is not obtained, there will be trouble from bud-dropping which will make those blooms left to reach maturity quite unfit for marketing. Where sweet peas are being grown, the glass of the greenhouse must be kept clean and though ample ventilation is needed to prevent mildew, draughts must be excluded for this is also a cause of bud-drop. Too much water can also contribute to the trouble, also insufficient moisture, so it will be seen that it is not easy to grow sweet peas to perfection under glass.

Seed is sown in individual thumb-size pots in a cold frame in August where they will soon begin to 'shoot'. The plants must be taken indoors in early October and allowed to stand near the glass. They should be moved to a larger pot for best results and transferred to beds in late January.

Sweet peas like manure and they like lime and in November when the tomato crop is removed, the beds should be trenched and a 6 in. (15 cm) layer of manure placed at the bottom. If already trenched and manured for a tomato crop so much the better. A layer of soil should be placed on top and over this should be added a 1 in. (2.5 cm) dressing of hydrated lime. Another 2 in. (5 cm) of soil should complete the trench.

An even temperature of 50°F. (10°C.) should be continually maintained and whatever ventilation necessary should be provided; this will depend upon the outside weather conditions. The plants should be placed 6 in. (15 cm) apart in the trenches which will have been made 18 in. (45 cm) apart. Give a heavy watering when planting is completed and this should provide the plants with sufficient moisture for a considerable time.

No 'stopping' is done for the plants are required to bloom early and are grown on the single shoot system up long canes to which they are fastened by means of a wire ring. All tendrils must be removed when observed. In order to encourage the young plants to come into bloom early, the soil should be kept on the dry side though this does not mean it should be too dry. As soon as the plants show in bud, the air temperature may be lowered to just under 50°F. (10°C.) and as the weather outside becomes warmer, a daily syringing with luke-warm water is beneficial. The bloom should be cut with as long a stem as possible and made up into bunches each containing a dozen stems. The market salesman does not favour mixed bunches or boxes of mixed colours, and only one variety to each box. For that reason possibly only two varieties should be grown. The plants should require little or no feeding until they have been in bloom about a month when a weekly application of liquid manure will be beneficial. Given a correct even temperature, careful applications of water and freedom from draughts, sweet peas will give both pleasure and profit but it is not always easy to provide just the ideal conditions.

LAUREL, *see* Laurus

LAURUS

The laurel is one of the best plants to make a screen in a town garden, but it is necessary to keep it under constant attention for if it becomes neglected and begins to form thick woody growth, this will not only be made at the expense of foliage, but will leave unsightly gaps in the hedge when cut away. A laurel hedge should not be trimmed with shears, for this will leave behind a large number of leaves which have been mutilated and which will turn an unsightly brown colour. Rather should the shoots be shortened with the secateurs in autumn each year.

A thick shelter or hedge of brilliant green may be formed by planting laurel in three rows, allowing 3 ft. (1 m) between each plant and pruning in such a way that the back row is allowed to grow 12 in. (30 cm) taller than the plant in front, whilst the middle row plants are allowed to grow 12 in. (30 cm) taller than the front row. This will form a dense screen 10 ft. (3 m) thick and could be planted as a dividing hedge between two gardens, or as a shelter belt in a large garden which is exposed to strong winds. It is an idea which may even be used for a hedge to a small front garden, substituting such a hedge for the usual straggling privet, often so tall that it shuts out considerable light from the front (or back) rooms. Laurel planted in this way would be a substitute for those tiny unkempt lawns, for such a hedge would take up considerable room. Or a similar planting may be made to cover an unsightly bank and at the same time give privacy and also act as a windbreak. The plants will require but little attention apart from the removal of any unduly long shoots each autumn. What is more, the plants will be completely oblivious to frost and snow. Like all town garden hedges, the plants will appreciate the spraying of their foliage with a horse-pipe or water-can during summer. This will remove any deposits from their leaves and freshen up the plants in every way.

The laurels grow well in the shade of trees or buildings, whilst they are tolerant of draughts and cold winds. This is also an excellent plant for a chalky soil, for it is not generally realised that the common laurel, *L. rotundifolia*, is a member of the great prunus family and not of the true Bay Laurel family, *Laurus nobilis*, hence its liking for a chalky soil or one containing plenty of lime.

LAVENDER, *see* Lavandula

LAVENDER, COTTON, *see* Santolina

LAVANDULA

With its attractive silvery-grey foliage, lavender will be colourful and aromatic even where perhaps not bearing a large amount of bloom, whereas no plant makes a more attractive low evergreen hedge. Used to surround a bed of scented crimson roses such as Ena Harkness, or a bed of fragrant pinks, the dwarf lavender remains neat and compact and may be clipped into shape when the plants are young. The most dwarf variety is *L. nana* 'Compacta', its habit being erect and with, careful clipping in spring, which is the best time to trim all lavenders, the hedge may be kept at a height of no more than 12 in. (30 cm). It bears pale lavender-mauve flowers and is one of the latest to bloom. Even more compact is the white flowered counterpart which will not grow more than 8 in. (20 cm) tall.

Little more vigorous is the Munstead lavender, famed for its deep purple flowers. It may be kept at a height of 15 in. (38 cm) without trouble. Another of dwarf habit is 'Hidcote Variety', its dark violet blooms being the richest coloured of all the lavenders. The charming 'Loddon Pink', slow growing and of compact habit, should be in every collection if only for its shell pink flowers. Yet another dwarf lavender is 'Folgate Blue' which grows to a height of 18 in. (45 cm) and bears pale blue flowers. When describing the heights of these different varieties, it does not include the flower spikes though they grow more dwarf than will the Old English or large-flowering lavenders which will form a hedge 3–4 ft. (1 m) in height where the plants find conditions to their liking. In comparison with the dwarfs, they produce a much larger amount of bloom and for this reason are planted commercially where a large quantity of bloom is required.

The best form of the Old English lavender is the 'Seal Variety', which is extremely free flowering, the blooms being rich in oil where grown well. For making a hedge, however, the habit is less compact than that of the other robust lavenders. It has green foliage and on a matured bush more than a thousand flower spikes may be removed. Almost as vigorous and free-flowering is 'Grappenhall' which bears a dark purple flower, whilst the original Dutch lavender makes a large spreading bush with attractive silver-grey foliage and bears a pale mauve flower. The leaves are broader than with those of other lavenders. An excellent variety to plant by the side of a path is the 'Wildersee' lavender. It is slow growing and compact and has small, narrow green leaves. It bears its dark lavender-mauve flowers very early and is a good variety to grow in the north.

There is an interesting white Dutch lavender, which is not so hardy and tolerant of adverse conditions as the mauve flowered varieties, but for all that, is very lovely planted with the 'Grappenhall' or 'Wildersee' lavender and it possesses a strong perfume. More tender than the white

Dutch is *L. stoechas,* which should only be planted in more favourable climates. It makes a compact plant 12 in. (30 cm) high and bears spikes of deep violet-blue.

HARVESTING

In order to retain their maximum perfume, the spikes should be removed just before the bloom begins to fade, towards the end of July. Where growing commercially for oil extraction, it is advisable to leave the spikes on the plants for another fortnight or until the bloom has faded. Almost the whole of the stem is removed, the spikes being severed with a sharp knife or with scissors and placed on pieces of canvas. The lengths of canvas are then placed on large trays or on shelves in a dry, airy room, but away from the direct rays of the sun, where they will dry in about two weeks. The blooms may then be rubbed from the stems between the hands and if the spikes have been placed in the same direction, this will facilitate the work. When growing for oil extraction, the spikes should be removed to the stills intact and immediately after cutting.

Where growing for use in the home, the dried blooms may be used in pot-pourris but are most often made up into small muslin bags to be placed amongst clothes. After removing the bloom do not discard the stems for they may be made up into bundles and if lighted, the sticks will burn slowly like incense.

Like all herbs except the mints, lavender prefers a light, sandy soil and will always be more fragrant growing along the drier eastern side of Britain than elsewhere. Soil of a chalky nature suits it most of all, bringing out its powerful fragrance as nowhere else. It also likes plenty of sunshine which is essential where grown for oil extraction. It requires neither manure nor humus, though a little top manure may be forked round the plants in autumn. The more robust varieties should be planted between 3–4 ft. (about 1 m) apart, slightly closer where planting a hedge. For a dwarf hedge, plant the compact lavenders about 15 in. (38 cm) apart, for the plants will always grow as wide as they grow tall. Planting is best done early in spring and the younger the plant the better it will transplant. When moved after two years, the plants may die back. Plant firmly and into clean ground.

Propagation is by cuttings or slips which should be removed with a heel. They should be obtained in midsummer and inserted in boxes containing a sandy compost or in a similar compost in a frame. If the compost is kept moist and the slips have been given firm planting they will root within a month. Small numbers may be grown on in small pots to be set out in spring or the rooted cuttings may be allowed to occupy the frame until that time. In the south, the slips may be inserted in beds of sandy soil in the open and may be planted in their permanent quarters in autumn.

The plants should be kept in shape by judicious clipping, first for the removal of any long, straggling shoots, later they may be clipped with a small pair of shears. The plants should not be allowed to form too much old wood for they will then be impossible to cut back without causing serious damage. For this reason purchase young plants and where planting for a hedge, trim them lightly from the beginning, doing so early in spring.

LAVATERA

L. trimestris is the annual mallow, the deep rose-pink, Loveliness, growing to a height of 3–4 ft. (1 m) and forming a dense bush-like plant. Requiring a moist soil, the seed is sown early in April and thinned to 2 ft. (60 cm) apart. The mallow is at its best during August and September, a delightful plant for the back of the border.

LAYIA

This annual is native of California and is at its best in a dry, sandy soil. It should be given a position of full sun. Seed is sown early in April to the front of the border for the plants grow only 12 in. (30 cm) high and bear most of their bloom during August and September. *L. elegans*, 'Tidy Tips', bears dainty yellow flowers edged with white, and is a most attractive plant.

LEBANON CANDYTUFT, *see* Aethionema

LEBANON SQUILL, *see* Puschkinia

LENTEN ROSE, *see* Helleborus

LEONTOPODIUM

The edelweiss or *Leontopodium alpinum* is native of the higher alpine regions of central Europe and is a hardy perennial growing 6 in. (15 cm) tall. It forms rosettes of downy grey foliage and bears pale yellow flowers surrounded by white star-shaped flowers which, like the foliage are also covered in down. It blooms July to September and requires an open, sunny situation and a well-drained soil. It does well on the rock garden.

It may be raised from seed sown under glass early in spring, the plants being set out towards the end of summer. Or seed may be sown outdoors in boxes early in the year and will germinate more quickly if subjected to hard frost.

LEOPARD'S BANE, *see* Doronicum

LEPTOSIPHON

For edging, this is a charming little annual plant, coming quickly into bloom from an early April sowing. The new French hybrids, reaching a height of only 4 in. (10 cm), cover themselves in star-like flowers embracing all the colours of the rainbow. Coming into bloom early in June, they remain colourful until mid-September. Do no thinning.

LEPTOSYNE

An annual, it is valuable both for cutting and garden decoration, with its semi-double marguerite-like flowers of golden yellow ('Golden Rosette') and growing to a height of 18 in. (45 cm). The bloom is held on wiry stems and borne from mid-June until September from a late March sowing. Thin to 10 in. (25 cm) apart.

LEUCOCORYNE

At the foot of the Andes Mountains, this delightful plant bears its long wiry, arched stems of brilliant blue and white blooms which possess a delicious fragrance. The small freesia-like bulbs are of such easy culture, growing to perfection in a cool house temperature of around 45°F. (7°C.) that they should be grown by all flower lovers. They require the cultural treatment of freesias and watsonias, keeping them dry after the bulbs have been planted in pots or boxes and until growth commences. The bulbs are planted in August and placed in a cold frame or shed until rooting takes place. They may be taken to a warm room in the early new year or to a cold greenhouse in February, where they will bloom in spring.

Those who would like an original display should try the leucocoryne in the long boxes or containers specially made for the window display of plants. The container may be placed in an attic or cellar until growth commences towards the year end, and it may then be removed to a cold frame or greenhouse whilst the window is being used for geraniums or other plants. In early March, if the container is placed in the window and the bulbs are kept comfortably moist, the reward will be graceful arching sprays of richly coloured blooms which will remain colourful for five to six weeks. The delicious fragrance will be an additional attraction.

If the bulbs could be obtained inexpensively, this is a cut-flower plant that would find a ready sale in the markets. The flower sprays remain fresh in water for fully a week and it is 'something different'.

LEUCOJUM

Its name, derived from the Greek means 'white violet', though the flower more closely resembles a snowdrop. We know it as the Snowflake and it is a delightful plant, more streamlined than the snowdrop whilst its petals are more evenly spaced, forming a bell of perfect symmetry. Of the three best known species, one blooms during October, another in February, whilst the summer snowflake is at its best in May.

The Summer Snowflake is also known as the Loddon Lily, for it was on the banks of the little Hampshire and Berkshire river that William Curtis discovered it growing wild, towards the end of the nineteenth century. Those who know Ireland may have picked the same flower alongside of the Shannon. It is an ideal plant for growing along the banks of a stream or pool.

All three species should be planted for they are quite inexpensive and easily obtained. If they have a fault it is the length of time they take to come into bloom, often two years after planting and they do not like disturbance when once established. Increasing relatively slowly both from self-grown seed and bulblets there is no need to divide the clumps after flowering as frequently as with snowdrops. The leucojums favour a soil that is moist yet well drained, and one containing liberal quantities of leaf mould. Like snowdrops they also appreciate a peat mulch after flowering. They make ideal pot plants, especially those that bloom in February. They should be potted into bowls or pots during late August and stood in cold frames or under a wall, well covered with ashes or sand until early December when they should be taken to the cold-house or indoors to a cool room. They should be watered sparingly for they enjoy soil conditions which err on the side of dryness. Early in January they will come into bloom and remain fresh and slightly fragrant for a month at least.

In the open ground the spring and early summer species should be planted during early September so that they will have settled down before the winter weather. Plant 2 in. (5 cm) deep, and if a little peat and coarse sand is sprinkled over the bulbs before the soil is filled in around them, this helps to get them away to a good start.

The autumn-flowering species should be planted in April or early May so that they too may have ample time to form their roots before coming into bloom.

SPECIES

Leucojum aestivum. The Summer Snowflake. The best form is 'Gravetye' which produces small drooping bell-shaped flowers on stems up to 18 in. (45 cm) in length. Grown in a sheltered position this variety could well become a popular cut flower for the difficult May period. Plant it near a clump of Solomon's Seal and the green and white effect will be enchanting. It will grow to perfection on the banks of a stream.

L. autumnale. As its name implies, it is October flowering when it produces its grass-like foliage and snowbells tinged with pink. This species should be given a more shaded position than the spring-flowering leucojums. Plant with *Gentiana sino-ornata,* the brilliant blue trumpets enhancing the white and pink colouring of the leucojum.

L. vernum. Similar to *L. autumnale,* but here the rounded petals are tipped with vivid green which makes them most attractive as pot plants or on the rockery. They flower throughout February and into March and carry a pleasing perfume.

LEWISIA

The plants are amongst the most beautiful of the rockery and given a sun-baked position and a good loamy soil will produce their striped flowers during June and July. The leaves are succulent, the roots fleshy, and though liking a hot sunny position, they must have a soil which will retain moisture during summer. The most outstanding variety is *L. cotyledon* 'Howell's Variety', which bears rosettes of thin long leaves and on 9 in. (23 cm) stems its large blooms of rich apricot, striped with salmon-pink. Another lovely species is *L. heckneri*, its blooms being pink, striped with white. All are natives of Columbia and Oregon, and as a contrast plant *L. columbiana* which bears sprays of richly striped purple flowers and is of dwarf habit.

LIATRIS

The Snakeroot, so-called on account of its long, woody tuberous roots, and one of the best of border plants. From the end of July until late in September, it bears spikes of rosy-mauve which open from the top downwards and which remain fresh in water for a considerable time. It likes a well-drained soil thoroughly enriched with peat and a little decayed manure. The roots should be given shallow planting. Spread them out just below the surface of the soil, pack peat around them and plant in March, 20 in. (50 cm) apart. Propagation is by root division, though *L. pycnostachya*, the 'Kansas Feather' which blooms during September and October, is readily increased from seed.

SPECIES AND VARIETIES

L. callilepsis 'Kobold'. This is a striking new variety bearing long spikes of deep lilac-blue, deeper in colour and of more compact form than the species.

SEPTEMBER GLORY. Also of excellent cut flower habit, the long poker-like spikes being covered with tiny button-like flowers of a deep rosy-mauve colour.

GOBLIN. A hybrid, bearing tapering spikes of deep lavender-mauve and growing to a height of only 2 ft. (60 cm).

L. spicata. The long spikes of deep reddish-purple flowers are freely produced through autumn.

LIBOCEDRUS

Closely related to thuya, it is the Incense Cedar of California with flattened branches and small spreading leaves. The cones are oval and composed of four to six flat scales. Plant in April in a deep loam and in a warm, sheltered situation. Propagate from seed sown in spring or from cuttings of the side shoots removed with a 'heel' and inserted in sandy compost.

L. decurrens makes a tall column-like tree with small bright green glossy leaves borne on flattened branchlets. In the form 'Aureovariegata', golden leaves appear with green leaves. *L. chilensis*, the Chilean cedar, makes a heavily branched tree with drooping branches whilst the cones, held on short stalks, are also drooping.

LIGULARIA

L. clivorum is a valuable plant for a moist, shady border, having large glaucous foliage and bearing tall, erect flower stems from mid-July until the end of September. It is possibly not a plant for the select border, but several new varieties may bring added popularity to the plant. Plenty of peat should be incorporated into the soil together with as much compost as possible if the ground is in any way of a light, dry nature. Plant in November and increase by root division.

VARIETIES

GREYGNOG GOLD. The bright golden flowers with their attractive bronzy centres are borne in long spikes above its glaucous foliage. 4 ft. (1.2 m).

OTHELLO. It has attractive purple glossy leaves and bears a spike of orange-yellow flowers. 5 ft. (1.5 m).

LIGUSTRUM

Privet is the most commonly planted of all hedging plants. Its good points are its tolerance of the worst of town conditions, whilst it quickly makes a dense screen and withstands considerable clipping. Against this, it makes too rapid growth, demanding clipping on numerous occasions throughout the year, whilst it is a gross feeder and a robber of the soil. Its value lies in its vigour, for young plants cut back to 12 in. (30 cm) of ground level after planting will grow into a thick hedge 5 ft. (1.5 m) tall within two years, and where privacy is required, this is much in its favour. It is evergreen, keeps its colour well and if correctly trimmed will prove efficient against tresspass though it does not possess such a neat, compact appearance as does *Lonicera nitida*. Much could be done to make a privet hedge more attractive by clipping into an original form when the hedge has attained its required height. One is the battlement formation. A hedge cannot be cut and kept in this shape until it has reached its required height, but it is an alternative to the rather monotonous straight top. To make a battlemented hedge, first make the top level with the shears, then with the secateurs cut away a section the length of a board previously cut to the required dimensions of a lower line (B). This is repeated along the hedge, the length of each section removed corresponding with that which remains level with line (A). For several weeks, the hedge where cut away, will appear woody and devoid of foliage, though it will quickly heal over, when the hedge may be trimmed with a pair of small shears.

Ligustrum ovalifolium is the green evergreen privet; the golden form being *L. ovalifolium foliuis aureis*. The latter is most attractive when planted with green privet to form a mixed hedge and it is surprising that it is not more commonly seen. It is slower growing but the evergreen form may be cut back to keep the hedge level.

Where planted as a windbreak or for privacy, a double row hedge will prove more satisfactory, and will form a hedge of much better appearance having greater depth and being more easily shaped. Plant 12 in. (30 cm) apart or 16 in. (40 cm) apart when a double row is planted.

LILAC, *see* Syringa

LILIUM

No plant is so versatile as the lily. The list of species and varieties covers a flowering period from May until September; some are tall-growing, some dwarf; some suitable for lime-free soils while others thrive in a calcareous soil. Some species enjoy a shady situation, others a position of full sun. Certain lilies are happier growing in a cool greenhouse, others bloom to perfection in the more exposed garden.

For filling in spaces in the shrubbery or herbaceous border, there is no more showy plant and there are species for all positions. From the tall-growing *L. henryi*, which reaches a height of almost 6 ft. (2 m) to the dwarf-growing *L. chalcedonicum*, which flowers at a height of 2 ft. (60 cm). The border is an ideal place for lilies, provided it is not lacking in humus, for they love to have their roots in shade and their heads in the sunshine. But in a lily border, they are even more striking and their perfume will be enjoyed throughout the whole garden especially after a shower of rain or in the early evening. They may be planted in autumn or spring and except for staking, and a top dressing in autumn, will need no further attention for years. They require some shelter from strong winds and particularly do they look attractive when planted against a stone wall or wattle hurdles.

For the small courtyard or for the side of an entrance or porch, lilies look delightful when growing in tubs containing a compost into which is incorporated plenty of peat to retain moisture. Those growing to a height of no more than 3–4 ft. (1 m) should be used for this purpose, *Lilium speciosum* or *L. tenuifolium* being most suitable, also *L. davidii* Maxwill when there should be at least one in bloom from mid-May until September, five to six bulbs being planted to a large tub.

PREPARATION OF THE GROUND

Lilies, like most bulbs enjoy a soil well supplied with humus but not an excess of manure. A mixture of old mushroom-bed manure and peat is invaluable, but peat and leaf mould or spent hops will be satisfactory. They enjoy a moisture holding compost, rather than a rich soil and they like an annual top dressing or mulch in early autumn. Not only will this help to retain moisture in the soil and keep down annual weeds, but it must be remembered that many of the lilies are stem-rooting plants and thus a top dressing is essential to its healthy growth. Should the soil be heavy, as much peat, coarse sand and decayed leaves as possible should be worked in. If growing in a shrubbery in the town garden where lilies will do well, the soil will generally be of an acid nature and plenty of lime rubble should be worked in. It will be advisable to refrain from planting those species that do not enjoy a calcareous soil such as *L. brownii*, but many enjoy a slightly acid soil and provided the shrubbery is not over-acid, lime need not be given. As much coarse sand or grit, and plenty of peat worked in to a depth of 8 in. (20 cm) will be all that is necessary for healthy root formation. A dry, hard soil, so frequently seen in shrub borders will never grow good lilies.

When growing in a border entirely to themselves, it will be advisable to raise the bed to help with drainage before working in the humus materials. A raised bed will also display the blooms at their best. It is not advisable to plant beneath large trees or too close to a privet hedge, which will consume large quantities of moisture from the surrounding ground. For this reason wattle hurdles or a stone wall are better for it is not advisable to plant beneath large trees or too close to a privet hedge. Planted in clumps in odd beds against the side of a house, lilies look in their right setting and the taller varieties which make rapid growth during early summer, might be used to hide the wall of a garage or shed. *Lilium regale* is ideal for such a purpose, so are the Tiger lilies. The bulbs are quite inexpensive, and where it may be thought that they are less hardy than other plants, this is not so, for nowhere do they grow better than in Scotland. But they do require individual attention, for the species are distinct in their habits. This makes the lily the more interesting for if each variety is given that little attention, it will reward one with a display of the utmost beauty. It should not be forgotten that most lilies are equally lovely when cut and used in the home as they are in the garden, and those that are inexpensive should be planted for this purpose. They will provide colour and fragrance throughout the summer.

PLANTING

Each variety should be studied for planting depths. Some species, the stem rooters like to be planted as deep as 8 in. (20 cm) such as *L. davidii* 'Maxwill'; others like only 3 in. (7.5 cm) of soil over them, the average depth being about 4 in. (10 cm) for a number of varieties are often planted too deeply.

Where the soil tends to be in any way heavy or sticky, the bulbs should be planted on a base of sand and covered with a sand and peat mixture. Lily bulbs are received packed in dry peat and it is unwise to remove them or expose them to winds. Keep them in the containers until planting time. As a rule, the early summer-flowering bulbs should be planted in the early autumn and those that bloom during July and August may be planted early in spring. Again, if the soil tends to be cold and sticky, spring planting is preferable. It depends on the soil rather than on the position, for in an exposed garden in the north, provided the soil is sandy and humus laden, the bulbs may be planted during autumn and covered with bracken in winter.

Careful study of the individual requirements of the different species as to the need for shade or full sun will mean obtaining the best from each bulb.

FOR SHADE

	Height	In Bloom
L. amabile	3–4 ft. (about 1 m)	July
L. candidum	3–4 ft. (about 1 m)	June
L. croceum	6 ft. (2 m)	June-July
L. henryi	9–10 ft. (about 3 m)	August
L. leucanthum	7–8 ft. (about 2.5 m)	August
L. michiganense	7–8 ft (about 2.5 m)	July
L. pardalinum	6–7 ft. (2 m)	August
L. pyrenaicum	3–4 ft. (about 1 m)	May
L. speciosum	3–4 ft. (about 1 m)	September
L. umbellatum	3–4 ft. (about 1m)	June

FOR A DAMP SITUATION OR HEAVY SOIL

L. michiganense	8 ft. (2.5 m)	July
L. pomponium	3–4 ft. (about 1 m)	June
L. pyrenaicum	3 ft. (1 m)	May
L. roezlii	3–4 ft. (about 1 m)	June-July

LIME LOVERS

L. candidum	3–4 ft. (about 1 m)	June
L. chalcedonicum	3–4 ft. (about 1 m)	July
L. henryi	9–10 ft. (about 3 m)	August
L. leucanthum	7–8 ft. (2.5 m)	August
L. martagon	5–6 ft. (1.5 m)	July
L. pomponium	3–4 ft. (about 1)	June

FOR SUNNY POSITION

L. brownii	3–4 ft. (about 1 m)	July
L. chalcedonicum	3–4 ft. (about 1 m)	July
L. dauricum	3–4 ft. (about 1 m)	August
L. hansonii	5 ft. (1.5 m)	June-July
L. martagon	5 ft. (1.5 m)	July
L. davidii maxwill	6 ft. (2 m)	July
L. pomponium	3–4 ft. (about 1 m)	June
L. pumilum	1–2 ft. (about 0.5 m)	June
L. regale	5 ft. (1.5 m)	July-August
L. tigrinum	5 ft. (1.5 m)	August-September

REQUIRING A LIME-FREE SOIL

	Height	In Bloom
L. amabile	3–4 ft. (about 1m)	July
L. auratum	6 ft. (2 m)	August
L. hansonii	5 ft. (1.5 m)	June-July
L. pardalinum	6–7 ft. (2 m)	August
L. superbum	5 ft. (1.5 m)	July-August
L. tenuifolium	2 ft. (60 cm)	May

Though this may clarify the requirements of the individual bulbs, it must be said that many are most versatile, *L. regale*, for instance, thriving in partial shade and full sun and in a calcareous soil or one of an acid nature. When planting, care must be taken to note the heights that the plants will attain, so that tall varieties may be placed at the back of the border and those with a dwarf habit to the front. When planting, remove the soil to a depth of 6 in. (15 cm) and into the aperture is sprinkled a mixture of sand and peat. The bulbs are then firmly placed on to this, spacing them 6 in. (15 cm) apart and the soil is then carefully filled in around them and made firm with pressure of the hands. This is a better method than planting with a trowel.

As the bulbs make growth, a mixture of peat and soil should be placed round the stems. Staking should be done before the plants are tall enough to be damaged by winds but those of dwarf habit may require no staking.

After flowering, and the stem has died down in autumn, it is removed and a mulch of peat placed over the bulbs. If there is fear of hard frost, bracken or straw may be placed over the bulbs and removed early in April.

PROPAGATION

There are various methods of increasing stock. Several species, *L. regale,* and *L. croceum,* for example, will rapidly increase in a soil which suits them, and if lifted in alternate years the bulbs may be separated and replanted just as with the daffodil. The Tiger Lily, *L. tigrinum* and L. *umbellatum* are increased from the tiny bulbils which form at the point where the leaves join the stems. They will be found to have ripened by early September when they should be removed and planted into drills 2 in. (5 cm) deep and into which has been added some sand and peat. They will reach flowering size in two years if not allowed to lack moisture. Or most varieties may be increased from scales carefully removed in autumn and planted into boxes of loam, peat and sand, covered with 1 in. (2.5 cm) of compost and placed in a cold frame until they have formed bulbs. They are then planted out and will bear flowers in three years. But the easiest method, that most practised by amateurs, is to sow seed which with a number of varieties will germinate quickly and evenly, and will grow into

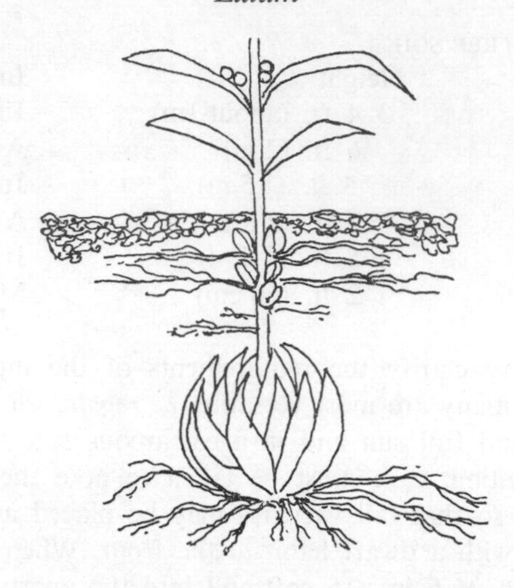

FIG. 7 *Stem-rooting lily, showing bulbils*

flowering-size bulbs within two years; *L. longiflorum* will flower within fifteen months. The seed should be sown in a greenhouse or frame in early April into boxes 8 in. (20 cm) deep and containing a mixture of loam, peat and sand in equal quantities. The seed is sown thinly and is lightly covered with peat. When germination has taken place, the containers are placed in a well-ventilated frame, shaded from strong sunlight and there they will remain until the following April, protected from severe weather by covering the lights with sacking. After twelve months the bulbs may be planted into trenches of peat and sand and they should remain there for another twelve months. They will by then have formed flowering-size bulbs, ready for planting early in April in the border where they will come into bloom a few weeks later.

Many of the stem-rooting varieties will produce dozens of tiny bulbils on the stems beneath the soil. These will root themselves and may be grown on in boxes of sand and peat for two years before planting in the border.

GLASSHOUSE CULTIVATION

Most of the hardy lilies are suitable for cold-house culture in large-size pots. Here they will find a ready sale with florists for church decoration and wreath-making, and they will be equally useful in the home in pots or as cut bloom. In the cold-house during early summer, they will give of their charm and fragrance. The best varieties and those most profitable for cutting are *L. longiflorum, L. speciosum* and *L. elegans. L. longiflorum* is grown commercially as the Easter Lily. While *L. speciosum* grows equally well in the open as indoors, *L. longiflorum,* is better suited to warm-house treatment. *L. regale* and *L. tigrinum,* both so easily grown outdoors, bloom to perfection in a cold-house. The blooms remain clean and in a fresh condition for some time. Grown three or four bulbs to a large pot, they are ideal subjects for a hall or porch.

PLANTING

Large pots are necessary for not only are most lilies tall-growing, but being stem-rooting, ample space must be allowed for top dressing as growth progresses, and rather than plant one bulb to each pot, it is better to use a large pot and to plant three or four bulbs in each. Deep planting and constant top dressing is the secret of success with *l. longiflorum.* Correct drainage and a moisture holding compost go hand-in-hand with indoor lily culture. The pots should be well 'crocked' and 4 in. (10 cm) of compost placed over the crocks. This should be composed of decayed turf loam passed through a sieve, to which is added 1 part of peat and 1 part of coarse sand. It is better not to use manure with lilies. If the bulbs are just covered with compost, there will be 4 in. (10 cm) to be filled up as growth is made. For Easter and early summer-flowering, the bulbs will be potted in October and will stand in a cold frame until the turn of the year. The lights should be covered with sacking to keep out light and frost. In introducing all lilies to a warm greenhouse or room, this must be done gradually with no undue forcing. The temperature of the greenhouse should be just sufficient to keep out frost and the bulbs growing slowly. *L. longiflorum* will stand a higher temperature when growth is almost complete and the buds are forming during early March, when it may be increased to 60°F. (16°C.). *L. regale* or *L. speciosum* are not satisfactory when forced; allow them to take their time. Both are better suited to a cold greenhouse, *L. regale* being potted in early March and allowed to stand in a frame or under the greenhouse bench until the roots have formed. They will bloom during June and throughout summer if taken indoors in batches. Likewise, will the flowering period of *L. longiflorum* be spread over the spring and early summer months. *L. speciosum* will come into bloom during late August and from later pottings may be had in bloom up to Christmas. The slower-growing *L. candidum* should be potted in late August and taken to a cold-house in early March and like *L. regale,* it will come into bloom early in summer.

After flowering, the bulbs should be slowly dried off, bearing in mind that more bulbs are spoilt through not giving them sufficient care after flowering than for any other cause. Water should be gradually withheld until the foliage has turned yellow. The bulbs should then be removed and placed in boxes of peat in a cool, dry room until spring, when they may then be planted in the border. An 8 in. (20 cm) bulb should be used for indoor cultivation.

MARKETING

This calls for the greatest need for care, especially when marketing white lilies. In order to retain the spotless white condition of the petals, first remove the stamens with the greatest care so that the pollen does not smear the throat when travelling or when exposed to winds. Then give the blooms a long drink in rainwater and when boxing, pack cotton-wool carefully around each throat after covering the box with a layer of blue tissue paper to enhance the vivid whiteness of the blooms. To hold the blooms in position, place a thin strip of wood across the box just below the heads.

SPECIES AND HYBRIDS

Lilium amabile. Growing to a height of only 2 ft. (60 cm) it is a stem-rooting lily which should be planted in a gritty lime-free soil. It enjoys best a position to the front of a shrubbery. A native of Korea, it bears its deep red flowers in early July.

L. auratum. Introduced from Japan in 1862, and though possibly the most stately lily in cultivation has a habit of not flowering after its second year where conditions do not suit it, though it is difficult to determine the most satisfactory soil. It is not partial to lime and as it is not quite hardy and is stem-rooting, it should be planted 6 in. (15 cm) deep. Known as the Golden-rayed Lily, the large white petals carry a distinct yellow marking down the centre. The crimson anthers enhance the beauty of the bloom.

L. brownii. It grows well in any soil. Its large funnel-shaped blooms are purple on the outside and white within. It reaches a height of 3–4 ft. (1 m) and is at its best in July. It is particular neither as to soil nor to position.

L. candidum. A lime-loving lily, the Madonna Lily of the Real East is base-rooting and should be planted with the soil only just covering the bulb. It enjoys a sunny position where it should be left undisturbed. In bloom June and July (see plate 37).

L. cernuum. A dainty little lily growing to a height of 2 ft. (60 cm) and producing its lilac-pink blooms throughout June and into July. Plant 4 in. (10 cm) deep.

L. chalcedonicum. The old Scarlet Turk's Cap lily, producing its brilliant scarlet blooms during July. It enjoys a sunny position and some lime rubble in the soil. Once established, leave undisturbed.

L. croceum. A variety known to Elizabethan gardens, it produces cup-shaped flowers of vivid orange during June. It grows to a height of 5 ft. (1.5 m) in a heavy soil. It will grow in an acid or calcareous soil and should be planted 6 in. (15 cm) deep.

L. davidii 'Maxwill'. A Canadian lily of extreme vigour and hardiness, flowering in July in a sunny position and in a soil containing plenty of

grit. The flowers are of a brilliant orange-red and of great substance.
The variety *Willmottiae* is at its best during early August, its vivid orange-red blooms being carried on 3–4 ft. (1m) stems. It does best in partial shade and in a soil containing plenty of peat or leaf mould.

L. elegans. Also known as *L. thunbergianum.* It is of dwarf habit and may be obtained in a number of forms ranging in colour from orange to deep mahogany. At its best in July.

L. hansonii. It loves partial shade and a gritty soil and is July flowering, the orange flowers spotted brown carrying a delicious fragrance. Since receiving an Award from the R.H.S. in 1878, it has been a most dependable lily.

L. henryi. August flowering, it grows to a height of 9–10 ft. (3 m) and enjoys a position of partial shade and plenty of moisture at its roots. It is hardy and does well in a calcareous soil.

L. martagon. One of the Turk's Cap lilies, which grow well in any soil and in any position. It is base-rooting and enjoys some lime rubble in the soil. The quaint drooping rose-pink flowers are spotted with purple and black. The white Turk's Cap lily, *L. martagon* 'Album', is one of the most handsome of all lilies.

L. pardalium. The Panther Lily of California, flowering in August. The petals of the flowers are attractively turned back, the colour being vivid orange-red, yellow at the base. Base-rooting, it enjoys a moist, peaty, lime-free soil.

L. phildauricum. Bears its rich apricot blooms in clusters on sturdy stems 2 ft. (60 cm) tall. Requiring a lime-free soil, it is a lovely lily for massing at the front of a border.

L. pomponium. It blooms profusely in a limy soil, but loves a stiff loam with some lime rubble worked in. Its blooms possess an unpleasant perfume, but this is compensated by their brilliant scarlet colouring. Base-rooting, it requires shallow planting.

L. pyrenaicum. The old yellow Turk's Cap Lily and useful in that it comes into bloom in May. It is base-rooting, requiring shallow planting. The lovely lemon-green coloured wax-like blooms are carred on 3 ft. (1 m) stems. Likes a shady position.

L. pumilum. A lily suitable for the rockery or a window-box as it grows only 15 in. (38 cm) tall and produces its vivid scarlet blooms in profusion during June. Enjoys a position of full sun.

L. regale. The most popular of lilies and rightly so, for it is not particular as to soil, is free-flowering and carries a rich perfume. It is stem-rooting and requires planting 8 in. (20 cm) deep. Its white blooms of great substance look attractive in the shrubbery or amongst evergreens.

L. roezlii. A valuable lily in that it blooms to perfection near water, in the bog garden or on the banks of streams or ponds, where the brilliant orange-red blooms remain long in flower.

L. speciosum. Formally *L. lancifolium,* and one of the loveliest of all lilies. It should be planted in clumps, preferably by itself where it will bloom in profusion year after year. It carries a pleasing perfume and is lovely planted in large tubs near a house or in a small courtyard. It is stem-rooting and requires planting 6 in. (15 cm) deep. In the border it should be planted near the Tiger lily for both bloom during August and September.

L. superbum. A lover of a moist, leafy, lime-free soil, this lily grows to a height of 8–9 ft. (about 3 m), bearing many crimson flowers, spotted with mahogany.

L. tigrinum. A very old easily grown lily which enjoys a rich soil and a position of full sun. The bulbs are stem-rooting and like deep planting. September flowering. The deep orange flowers are spotted with black.

L. tenuifolium. From Siberia, it bears its brilliant scarlet blooms late in May in a peaty, lime-free soil and in a position of partial shade. Suitable for the front of a border.

L. umbellatum. A dwarf lily, base-rooting and June flowering. It is one of the easiest of all for the beginner's collection and blooms to perfection beneath tall trees and shrubs. Perhaps the best variety is 'Golden Fleece', with its attractive golden blooms, tipped with scarlet.

A simple classification of flowering times may help in selecting the varieties.

MAY

L. pyrenaicum
L. tenuifolium

JUNE

L. candidum
L. cernuum
L. croceum
L. tenuifolium
L. phildauricum
L. pomponium
L. pumilum
L. roezlii
L. umbellatum

JULY

L. amabile
L. brownii
L. chalcedonicum
L. davidii Maxwill
L. elegans
L. hansonii
L. regale
L. superbum

AUGUST-SEPTEMBER

L. auratum
L. davidii Willmottiae
L. henryi
L. pardalinun
L. speciosum
L. tigrinum

LILY OF THE FIELD, *see* Sternbergia

LILY-OF-THE-VALLEY, *see* Convallaria

LIME, *see* Tilia

LIMNANTHES

An annual, it should be used like the leptosiphon, for edging. The plants are sweetly scented and attractive to bees. *L. douglasii*, grows to a height of 5 in. (12 cm) and bears yellow and white flowers. Do no thinning and sow the seed late in March.

LIMONIUM, *see* Statice

LINARIA

ANNUAL

The flowers of *L. maroccana* 'Fairy Bouquet' are like miniature antirrhinums, produced in profusion on 6 in. (15 cm) stems throughout midsummer. The colour range is enormous, white, ruby, rose, purple, yellow, the spikes being neat and compact. Sow in any soil towards the end of March, no thinning being necessary if the seed is sown thinly.

PERENNIAL

Liking a dry, sandy soil, few alpine plants are more free flowering than the dwarf flaxes, nor of more easy culture. *L. œquitriloba*, almost the smallest of all alpines, bears tiny lavender-blue flowers, but more colourful is *L. alpina*, the violet flowers having a vivid orange lip. The variety *L. rosea* bears bloom of a clear pink. *L. organifolia* covers itself with tiny violet flowers and *L. hepaticifolia* bears its shell-pink blooms on stems which often appear several feet away from the plant.

LINDEN, *see* Tilia

LINUM

An annual, growing to a height of 15 in. (38 cm) and bearing dainty flowers of the most brilliant colours. Continuity may be obtained if the first sowing is made towards the end of March, with a second sowing early in May. Thin to 6 in. (15 cm) apart. *L. grandiflorum* 'Coeruleum', bears bloom of brightest blue; 'Rubrum', vivid crimson, and borne in great profusion.

LIQUIDAMBAR

Natives of North America and east Asia and closely resembling the maples in habit, also with their large leaves which take on brilliant autumnal tints, the liquidambars are fast growing trees requiring a rich, deep soil. No pruning is necessary apart from the shortening of unduly long shoots. Propagation is best done by layering the lower branches or by cuttings taken of the half-ripe wood in July and rooted under glass.

L. styraciflua, the Sweet Gum of North America, is a handsome tree with red cork-like bark and maple-like leaves which take on rich crimson colourings in autumn. *L. orientalis* is a slower growing tree with smaller leaves, attractively tinted in autumn.

LIRIODENDRON

The Tulip Tree of North America, a fast-growing shade bearing tree with three-lobed leaves which turn clear yellow in autumn. It requires an open sunny situation and a rich loamy soil. In summer it bears beautiful tulip-shaped flowers of green and yellow. No pruning will be necessary. Propagate from seed sown outdoors in spring under glass.

L. tulipifera 'Aureo-marginata' has leaves edged with gold, whilst Fastigiatum grows erect and narrow and is suitable for a confined space.

LITHOPS

They are the Living Stones, plants of the arid lands of South Africa and so amazingly camouflaged as to be almost unrecognizable from the stones amongst which they grow. They are most interesting plants of rubbery texture and growing no more than 1 in. (2.5 cm) tall have the appearance of pebbles. At the top of the plant is a thin slit, as if two portions are divided. This is actually the case, for each year the plant divides to reveal another tiny plant between the two portions. The following year it, too, will divide. And so the original plant soon grows into a rock-like cluster. They will bear either white or yellow flowers towards the end of summer which appear from the centre of the plant.

Lithops grow from April until September during which time they should be watered about once each week. During winter the plants will completely shrivel up and appear as if dead and during this time they should on no account be given water. Then in spring, as the plant comes to life again, watering should recommence but at all times this must be done as sparingly as possible.

Use a small pot which has been well crocked, and the compost should be made up of 2 parts silver sand and 1 part sterilized loam to which is added a small quantity of lime rubble. Plant firmly but do not cover the body of the plant. They should be placed in as sunny a position as possible and do not worry if you have to be away from home for longer than expected, for it is rarely that the plants are troubled by lack of moisture, though they are easily killed by too much.

These interesting little plants are an ideal hobby for those living in a confined space but where there may be room for dozens of pots to be left in a sunny position. Besides the flowers, the tops of the plants are covered with the most exquisite patterns in almost every colour, outstanding in this respect being *Lithops karasmontana* which has yellowy-brown markings and bears a pure white bloom. There are nearly 100 species, all of them worthy of growing, and no plant is more labour-saving nor less troubled by pest and disease.

LIVINGSTONE DAISY, *see* Mesembryanthemum

LIVING STONES *see* Lithops

LOBELIA

Though, like the antirrhinum really perennial, the lobelia is always given half-hardy annual treatment, the seed being sown in heat in February. Cover lightly with sand and place a sheet of clean glass over the box to hasten germination. Whilst hardening, it is advisable to pinch back those shoots which tend to become straggly, to encourage a bushy plant. The plants should be set out at the end of May or early in June when the spring bedding display has finished. They are used entirely for edging and should be spaced 6 in. (15 cm) apart.

Though the navy-blue, 'Mrs Clibran', with its striking white eye is the most widely planted of all lobelias, the paler 'Blue Stone', is more attractive planted with pink flowering plants; whilst the dark sea-blue 'Crystal Palace'; the white, 'Snowball'; and the uncommon crimson-red, 'Prima Donna', should be grown by way of originality. Or three or four varieties planted alternately will produce a colourful effect. The Trailing variety Sapphire is suitable for hanging baskets and window boxes.

L. cardinalis. This tuberous-rooted perennial lobelia gives a magnificent splash of colour to the border during August and September and provided it is given suitable soil conditions, is hardier than generally believed. The plants prefer partial shade, a well-drained soil in winter and plenty of summer moisture. A light soil is essential, summer moisture being provided by the addition of plenty of decayed manure and peat or leaf mould. Plant early May, preferably near plants having silver foliage, which will provide the necessary contrast to the scarlet spikes and bronze leaves of the lobelia. Propagation is by root division in April, the crown being 'teased' apart in the same way as when dividing the eremurus or hemerocallis. Where the garden is not favourably situated it will be advisable to cover the crowns with ashes in November, or lift and store in a frost-free room.

PURPLE EMPEROR. A distinct variety having copper coloured foliage and bearing a long spike of violet-purple. 3–4 ft. (1 m).

QUEEN VICTORIA. The neat foliage is a rich copper-red colour, the bloom brilliant scarlet and borne in long graceful spikes. 4 ft. (1.2 m).

RUSSIAN PRINCESS. The foliage is a dark bronzy-green, the flower spikes a pale shell-pink. 3 ft. (90 cm).

THE BISHOP. A truly magnificent plant having purple-green foliage and bearing a bloom of vivid vermilion-red. 3 ft. (90 cm).

LOBSTER CLAW, *see* Clianthus

LONAS

An everlasting, it is almost hardy and requires an open, sunny situation and a dry, sandy soil. An annual, it blooms July to October, bearing small golden buttons on 12 in. (30 cm) stems and they may be dried and used for winter decoration. Sow in April where the plants are to bloom.

LONICERA

CLIMBING SPECIES

The honeysuckles are best planted against a west wall, where the evening sunshine will bring out the rich fragrance of the flowers to the full.

There are over 200 species in cultivation, ranging from the tall-growing climbers, which often reach a height of 20 ft. (6 m) or more, to the dwarf bush honeysuckles. There is no better plant for growing against a trellis, for draping over an old tree trunk, or covering a wall, for it possesses the same informal habit as the clematis, and remains long in bloom. When honeysuckle plants are covering a house, their dense twiggy growth makes it necessary to support them, either by wires or by a trellis, over which the stems will twine. Many of the honeysuckles are evergreen, others deciduous, whilst varieties may be planted to provide a succession of bloom.

Natives of the British Isles, North America and the borders of Russia and China, the honeysuckles are extremely hardy. They will flourish in all parts of Britain and they will thrive in any good garden soil. The plants will resent any hard cutting when once established, and may die back altogether if heavily pruned. They are sturdiest when allowed to grow unrestricted, having only the dead wood cut out each year after flowering. Propagation is by cuttings of the new season's wood. These should be about 9 in. (23 cm) long, and inserted in sand under a bell cloche. November is the best time to plant, and remember that they do like a deeply dug soil. As with the clematis, when given a soil enriched with a little farmyard manure and summer mulch, they will respond with greatly improved flowers and a longer flowering season.

The first of the climbing honeysuckles to bloom is *Lonicera periclymenum* 'Belgica', the early Dutch variety, which bears its yellow tubular blooms in May and June. Then follows the late Dutch honeysuckle, 'Serotina', which bears its reddish-purple flowers until the end of summer. Both possess an almost overpowering fragrance, delightful in a walled garden where the perfume does not easily escape. Also early to bloom is *L. caprifolium*, possibly the loveliest of all, bearing long cream-coloured tubular flowers, flushed with shell-pink and more freely produced in this plant than in any other species. Like all the deciduous honeysuckles, the blooms are followed by vividly coloured berries in autumn, those of *L. caprifolium* being brilliant orange. *L. henryi*, with its long oval leaves and yellow and red blooms, is valuable in that the blooms produced during late summer are followed by striking black fruits in winter. It will reach a height of 12 ft. (3.5 m).

Semi-evergreen is the Japanese Honeysuckle, *L. japonica*, which is of vigorous habit, quickly reaching a height of 25 ft. (7 m). It bears its sweetly-scented white flowers at the leaf axils and blooms throughout summer. The variety 'Haliana' bears primrose-coloured flowers.

Another evergreen climbing species is *L. americana*, a hybrid of *L. caprifolium; L. americana* is a really fine climber of neat habit. The tubular blooms are pink and yellow and possess strong perfume.

Lonicera tragophylla is a little known yet one of the best of all the climbing honeysuckles. It is of neat habit and bears large orange-yellow tubular flowers often up to 4 in. (10 cm) long, several being borne together. The plant will make up to 8–9 ft. (2.5 m) of growth in a cool soil but where not unduly exposed to cold winds. It remains in bloom longer than any honeysuckle, from early summer until autumn, and has a continuous mass of bloom of great beauty. To encourage the formation of large bloom, the new season's shoots should be shortened back several inches, whilst the plants will benefit from applications of decayed manure in spring and again after blooming. A small quantity of cow manure and 2 oz. of bone meal should be worked into the soil before planting

SHRUBBY SPECIES B

The evergreen shrubby honeysuckles are ideal hedging plants for the small town garden, being neat and compact with their small dark glossy green leaves, whilst the plants form little coarse wood. Growth is rapid though a hedge will rarely exceed 4 ft. (3 m) in height and for this reason is possibly the best of all hedges for a small garden, withstanding hard and continual clipping and forming a thick hedge in all soils. It is equally valuable for coastal planting.

Plant early spring, spacing 15 in. (38 cm) apart, though slightly wider if a double row is to be planted. Clipping should also be done at this time, to enable the new season's wood to bear the fragrant creamy-white flowers during midsummer, which are followed by brightly coloured berries in autumn. Excessive clipping will prevent the appearance of both flowers and berries.

The most commonly known of the bush honeysuckles is *Lonicera nitida,* a native of China, which bears purple berries in autumn. Another is *L. yunnanensis,* of more upright habit, and which should be planted 12 in. (30 cm) apart. It is also quicker growing and makes a hedge up to 5 ft. (1.5 m) tall. Both are hardy, but *L. fragrantissima* requires a more sheltered position, and is seen at its best in the south and western side of Britain. This is an interesting species and though only semi-evergreen, bears its heavily fragrant creamy bells through winter. It should be planted and pruned in April. Of vigorous habit and more suitable for a large garden is the Russian honeysuckle, *L. tartarica,* which quickly makes a large bush, as deep as it grows tall. It requires constant clipping to retain its shape and to prevent the formation of woody growth.

LOOSESTRIFE, *see* Lysimachia

LOVE-IN-A-MIST, *see* Nigella

LOVE-LIES-BLEEDING, *see* Amaranthus

LUNARIA

It may have reached Britain during Tudor times but it is more likely to have arrived at an earlier date for Gerard describes the White Satin flower as he names it, in detail and said that it was found growing wild 'in the woods about Pinner and Harrow, on the hills about 12 miles from London, and in Essex'. It may therefore be a native plant and may be Chaucer's 'lunarie'. Honesty is its more common name for its seed is clearly (honestly) seen through the transparent skin of the seed pod and appears as a piece of white satin.

'The stalks are charged or laden with many flowers like the common stock gillofloure, of a purple colour', wrote Gerard and these are followed by the flat seed cases which are produced in sprays like tiny moons, hence its name lunaria. They are of the size of a ten penny piece. Parkinson called it the Pennyflower. If removed when the seeds have formed and dried, they provide lasting indoor decoration for the winter.

The plant is biennial, the seed being sown where it is to bloom, in August, and thinning the seedlings to 6 in. (15 cm) apart. The plant grows 2 ft. (60 cm) in height and its flowers, which appear in May, are of a most brilliant purple colouring. There is also a pure white form which is a striking contrast when in bloom with the type.

LUPINUS

ANNUAL

Hardy and growing strongly in the poorest of soils *L. hartwegii*, grows to a height of 2–3 ft. (60–90 cm) and may be obtained in white, deep blue and rose. Seed is sown in the open at the end of March and is sufficiently large to be planted individually about 15 in. (38 cm) apart. *L. nana*, is the Tom Thumb lupin, growing to a height of only 15 in. (38 cm) and is useful for the front of a border or shrubbery.

PERENNIAL

Introduced in 1826 from America, the lupin with its loosely packed spikes of blue or white remained in obscurity until the advent of George Russell's wonderful hybrids which appeared in 1936. In bloom throughout June and July, the Russell lupins bear tall spikes, densely packed with florets which give the blooms a richer appearance then provided by any other plant in the border. Many colour combinations have been evolved, each of the utmost beauty. The lupin will grow well and prove to be long lasting in ordinary soil, but not in chalk. Plant in November and allow 3–4 ft. (about 1 m) between the plants. Propagation is by root division, or either from root or stem cuttings taken early in May. The plants will not require staking unless growing in an exposed position. After the main flower spikes have begun to die back they should be removed to prevent the formation of seed. Smaller spikes will then continue to form throughout summer and autumn.

VARIETIES OF PERENNIAL LUPIN

ALICIA PARRETT. Both bells and standards are of a pale cream colour, the spikes being large and refined. A lovely contrast to the crimsons. 3–4 ft. (about 1 m).

BETTIE EVERETT. It forms a large handsome spike with broad florets of a lovely shade of soft pink, the centre of the standards being shaded with ivory.

BETTY ASTELL. Of compact habit, the spikes are thick and of great size, both bells and standards being of a pure dusky-pink colour. 3–4 ft. (about 1 m).

CANARY BIRD. The tall, tapering spikes are of bright canary yellow.

CHARMAINE. One of the most richly coloured plants of the border. The huge tapering spikes have flame coloured bells and gold standards. 3 ft. (90 cm).

CHERRY PIE. The spike is of great length, the bells being of a rich pure cherry-red, the standards bright carmine-red. 3–4 ft. (about 1 m).

GLADYS COOPER. One of the most stately plants of the garden. The huge tapering spikes, often 2 ft. (60 cm) in length are tightly packed with smoky sky-blue bells; the standards being rose-pink. 4 ft. (1.2 m).

MRS GARNET BOTFIELD. A fine small border variety of sturdy compact habit. Both bells and standards are of rich buff-yellow. 3 ft. (90 cm).

MRS MICKLETHWAITE. The most compact variety, growing to a height of only 2–3 ft. (60–90 cm), the bells being salmon-pink, the standards gold.

ORION. The large spikes have bells of crimson and standards of bright terracotta, both blending to produce a bloom of great beauty. 3 ft. (90 cm).

PLATO. The tall spikes are of a rich plum-purple colour. 4 ft. (1.2 m).

RAPTURE. The huge tapering spikes remain longer in bloom than most lupins. The bright rose-pink bells and ivory standards combine to make a spike of great beauty. 4 ft. (1.2 m).

RHAPSODY. A rose-pink lupin of charm, the standards being flushed with crimson. It grows only 2 ft. (60 cm) tall.

RITA. The spikes are smaller than most, the habit of the plant being compact making it ideal for a small border or for planting in beds. Here its crimson spikes are most showy. 2 ft. (60 cm).

RIVERSLEA. This is a crimson self-colour, the long tapering spikes being produced with great freedom. 3–4 ft. (about 1 m).

ROYAL DRAKE. The immense spikes are of a rich purple-bronze self colour, remaining in bloom over a long period. 3 ft. (90 cm).

SERENADE. Growing less than 3 ft. (90 cm) tall, it is one of the most striking of all lupins, with orange-red bells and crimson standards.

THUNDERCLOUD. The spikes are shorter than most, but borne in great profusion. Both bells and standards are of a deep purple, flushed with rose. 3 ft. (90 cm).

TOM REEVES. The spike is perhaps shorter than the other yellows, like that of 'Thundercloud' and produced with the same freedom. Its rich yellow self colour makes it a fine border plant. 3 ft. (90 cm).

TORCHLIGHT. A striking variety, the bells being deep salmon-red, the standards brilliant gold. 4 ft. (1.2 m).

VISCOUNTESS COWDRAY. A superb variety somewhat slow to increase but with its long, tightly-packed spikes of pure crimson-red it is outstanding. 3 ft. (90 cm).

LUPIN TREE

The perennial Tree Lupin, *L. arboreus* was introduced from California in 1800. It is a plant of shrubby habit, growing 5 ft. (1.5 m) tall and in July, bears multitudes of small spikes of laburnum-yellow. There is also a white and a mauve variety. The plants, which have become naturalized in parts of south England, require a light sandy soil and resent moving. Seed should be sown where the plants are to bloom or in small pots so that the young plants may be set out without disturbing the roots. The flowers have the sweet mossy perfume of the sweet pea and laburnum of the same family.

All lupins appreciate a well-drained sandy soil, devoid of manure and other fertilizers nor do they like lime but one in which their tap roots can penetrate to a considerable depth. They like the sun for they are native of the warm temperate regions but are of extreme hardiness. For the plants to be long living, the flower spikes should be removed before they can set seed.

The plant takes its name from the Greek, *lupe*, sadness or grief, for it is said that the seeds if not boiled in several waters before consumed, are so bitter as to cause contortions of the face and tears in the eyes. Virgil called it *Tristes lupinus.*

LYCHNIS

In bloom from mid-June until the end of August, this lovely plant, with its upright, compact habit has been grown in English gardens since the days of the Crusades. With the cross-like form of the tiny blooms which make up the flower head, it was known as the Cross of Jerusalem and in a well-drained soil it is an extremely hardy plant.

The plants come into bloom early in June and continue until the end of August, the vivid crimson-scarlet flower heads being enhanced by the dark green glossy foliage. Planting is best done in November and with their stiff upright habit, 18 in. (45 cm) only need be allowed between the plants. Propagation is by division of the roots which readily split up into small pieces.

SPECIES AND VARIETIES

L. chalceodonica. An indispensable and almost indestructible border plant, its flat heads of crimson-scarlet being held on erect 3–4 ft. (1 m) stems. There is an interesting double form, 'Rubra-plena'; and one which bears blooms of a unique shade of salmon-red, 'Salmonea'.

L. coronaria 'Atrosanguinea'. A valuable border plant having white woolly foliage and bearing deep crimson flower heads during June and July. The variety 'Abbotswood Rose' bears magenta-pink flowers on 20 in. (50 cm) stems.

LYCORIS

A genus of ten species, to be found in west China and Japan and named after a lady well known in Roman history. Closely allied to nerine, it has a large tunicated bulb with a short neck and strap-like leaves. At the end of a stout scape are borne funnel-shaped flowers with waved petals. In the British Isles and North America it should be grown in the open only in those gardens enjoying a frost free climate, though *L. radiata* and *L. squamigera* have been known to survive several degrees of frost outdoors.

Out doors, select a border where the bulbs will be baked by the summer sunshine and where they may be kept as dry as possible when forming their flower buds, and during winter. A bed at the base of a wall, in full sun and beneath the eaves of a house is suitable. The soil should be well drained and have incorporated some decayed manure, together with peat or leaf mould, in addition to a liberal sprinkling of sand or grit. Plant the bulbs (as for nerine) early in August before the flower stems appear, placing them on a layer of sand to assist with drainage and ensure that the neck of the bulb is just above soil level. Plant 12 in. (30 cm) apart and do not disturb for several years when each year the bulbs will increase in size and form numerous offsets which will attain flowering size in about four years. The leaves die down before the flower buds appear in August when moisture should be withheld. After flowering, place decayed leaves over the bulbs and where possible cover with a frame light reared against the wall though the lycoris is more tolerant of winter moisture than amaryllis and hippeastrum. New leaves will begin to form early in the new year.

Indoors, plant one bulb to a 5 in. (12.5 cm) pot, using a compost made up of fibrous loam, peat and decayed manure in equal parts and to which has been added a liberal sprinkling of sand or grit. Plant in July or early August, with the neck of the bulb just above the level of the compost. Water sparingly until the flowers have appeared then give an occasional application of dilute manure water. After flowering, the leaves will begin to form and during winter and spring, give only sufficient moisture to keep the plants growing whilst frost should be excluded by placing the pots in a living-room of the home (if the greenhouse is unheated) where the night temperature does not fall below 40°F. (4°C.). High temperatures must be avoided. As with nerine, it is important to allow the plants a two months' rest period from the time the leaves fade in June until the flower stems appear in August. At this time and until after the flowers have formed, moisture should be withheld.

The plants should remain in the pots for several years, giving a light top dressing of fresh loam and dehydrated cow manure each year when the leaves have died back.

The bulbs readily form offsets which are detached when the bulbs are ready for re-potting after four to five years and which should be done early in August. The offsets are planted one to a small pot, containing a mixture of fibrous loam and sand. They arc grown on for two years, giving them the necessary rest when they will be ready to plant into larger pots in which they will bloom in two more years.

It is readily raised from seed (like nerine) sown in April and which should be scattered over the surface of the John Innes Sowing Compost. Keep in a moist condition and transplant to small pots (one to each) in about eighteen months. Here they will remain for twelve months when they are moved to larger pots in which they will bloom in two more years.

SPECIES AND VARIETIES

Lycoris aurea. Syn: *Nerine aurea.* Native of west China, it is known as the Golden Spider Lily, the bright golden-yellow flowers appearing in an umbel on an 18 in. (45 cm) scape. The funnel-shaped flowers have narrow wavy segments and measure about 3 in. (7.5 cm) across.

L. incarnata. Native of central China and north Japan, it bears its funnel-shaped flowers of soft salmon-pink veined with blue in a large umbel of six to twelve and on a 15 in. (38 cm) scape. The segments are only slightly waved but have a soft sweet perfume.

L. radiata. Outstanding in its beauty, it is the hardiest species. The flowers are bright orange-red and scentless, attractively reflexed and borne on a 15 in. (38 cm) stem. The variety 'Alba' (Syn: *L. albiflora*) bears creamy-white flowers whilst 'Carnea' has white flowers tinted with pink.

L. sanguinea. It is native of Japan and blooms three to four weeks earlier than other species. It has bright crimson-red flowers which measure about 2 in. (5 cm) across and are borne in a small umbel on an 18 in. (45 cm) stem.

L. squamigera. One of the finest forms, being almost hardy and bearing large umbels of sweetly scented flowers of rosy-lilac pencilled with blue on a 2 ft. (60 cm) scape and which appear after the narrow strap-like leaves. The handsome flowers measure nearly 3 in. (7.5 cm) across.

L. sprengeri. Native of south Japan, it has long stalked flowers of rich mauve-pink without a distinct tube above the ovary. They are borne on 12 in. (30 cm) stems.

L. straminea. Native of west China, it is closely related to *L. aurea* and bears, on a 2 ft. (60 cm) scape, straw-coloured flowers shaded with pink.

LYSIMACHIA

The perennial Loosestrife is a plant for a moist soil and a partially shaded border, each of the species growing to a height of around 2 ft. (60 cm) and flowering during July. Plant in spring 18 in. (45 cm) apart and propagate by root division.

SPECIES

L. clethroides. Its long sprays of pure white bell-shaped flowers are borne above its foliage which turns an attractive gold and crimson in autumn.

L. punctata. Its yellow flowers are produced in whorls during July and August on long tapering stems.

LYTHRUM

The Purple Loosestrife is one of the most showy perennials for the middle of the border and though making a dense bush when established, the masses of flower spikes remain rigid and upright and the plant rarely needs staking. It is a plant for the back rows of a small border flowering from June until September. It prefers a moist soil, the plants flourishing in heavy clay provided it is reasonably well drained. Plant in autumn. For a cold garden it is one of the most reliable plants.

SPECIES AND VARIETIES

L. salicaria, 'The Beacon'. Well named, for its spikes of deep crimson-red may be seen from afar. 3 ft. (90 cm).

BRIGHTNESS. The spikes of vivid rose are colourful well into autumn. 3 ft. (90 cm).

L. virgatum, 'Dropmore Purple'. Similar in habit to *L. salicaria,* but the spikes are shorter and not quite so compact. The blooms are of a rich rose-purple. 3 ft. (90 cm).

MARDEN'S PINK. Of compact habit, the rose-pink spikes are produced through the summer. 30 in. (76 cm).

ROBERT. Very compact, the dainty spikes of clear pink grow to a height of only 2 ft. (60 cm).

ROSE QUEEN. Ideal for a small border to accompany 'Robert', its compact flower stems bearing spikes of pale rose-pink. 2 ft. (60 cm).

MAHONIA

Evergreen shrubs, distinguished from berberis by the compound leaves and spineless stems. Tolerant of all soils and partial shade, they are used for open woodland planting and for game coverts or they may be used to cover an unsightly bank. *M. aquifolium* is the best known, forming a dense low shrub 3 ft. (90 cm) tall, its glossy dark green leaves turning bronze and red in winter. Early in spring it bears golden-yellow flowers in dense racemes and they are followed by blue-black grape-like fruits, hence its name of Oregon Grape. Equally striking is *M. japonica* which has palmate leaves and in winter bears clusters of sweetly scented pale yellow flowers. Also valuable in that it blooms in autumn is *M. fortunei* which grows 5 ft. (1.5 m) tall and bears its bright yellow flowers in erect racemes.

Plant October to March and if the plants become leggy, cut back hard after flowering when new growth will appear near the base. Propagate by division or from suckers; also by cuttings, taken of the new wood and inserted in sandy compost under glass.

MAIDENHAIR TREE, *see* Ginkgo

MALCOMIA

Growing to a height of 6 in. (15 cm) and obtainable in shades of crimson, mauve, white and blue, the Virginia Stock, *Malcomia maritima* is a pretty plant with which to edge a border, sown at the end of March. No thinning is necessary if sown sparsely. The most compact form is 'Nana compacta' but the variety 'Crimson King', bearing bright red flowers, is the most showy.

MALLOW, *see* Lavatera and Malva

MALOPE

An annual, it is similar in form and habit to the lavatera. *M. trifida* makes a bushy plant 3 ft. (90 cm) tall. The bloom is of various shades of crimson and rose, at its best from August until late October. Sow at the back of the border early in April and thin to 20 in. (50 cm) apart. The variety 'Tetra Red' bears large trumpets of carmine-red with deeper red veining.

MALVA

M. alcea is perennial and bears its sprays of vivid rose-coloured flowers on 4 ft. (1.2 m) stems from mid-July until September, its beautifully cut leaves being an added attraction. This is an excellent plant for the back of a small border. It does well in any soil provided it is well drained and the plants should be allowed 3 ft. (90 cm) in which to make their bushy growth.

MAMMOTH TREE, *see* Sequoiadendron

MANETTIA

It is a native of Brazil and Peru and like the maurandia is an evergreen twining plant of slender, neat habit but which requires warmer conditions than most plants, a minimum winter temperature of 56°F. (14°C.). It should be given a position which is light and sunny, otherwise it will not bear its small tubular flowers of scarlet and gold during summer. Where it obtains the necessary light it will bloom through autumn too, being one of the longest flowering of all plants. Whilst flowering the plant requires ample supplies of moisture, which should be reduced to a minimum during winter. The manettia likes a compost made up of 2 parts loam, 2 parts peat and 1 part silver sand and a size 48 pot should be used for a flowering plant as they make plenty of roots. As the plants begin to twine at an early age, small twigs should be placed about the pot for the shoots to obtain a hold until ready to climb up the trellis.

MAPLE, *see* Acer

MARANTA

The Arrowroot plant of South America, from which the foodstuff is obtained, makes a bushy plant and bears large palm-like leaves. A number of them, such as *M. fasciata*, from Brazil, require more warmth than can be provided in the average home, but where a winter temperature of 50°F. (10°C.) can be maintained, which should be possible in a centrally heated home, a number of species and varieties make most attractive plants. The plants are named after the Venetian physician Maranti, who lived in the sixteenth century, and one of the most striking is *M. vittata*, for the creamy-white markings on the pale green tapering leaves are said to be the most perfect of any plant in cultivation. The leaves have double bars at regular intervals from the mid-rib almost to the leaf margin. *M. arundinacea* 'Variegata' grows 3–4 ft. (about 1 m) tall and bears oblong leaves mottled with various shades of green. A plant of more compact habit is *M. leuconeura* which rarely exceeds a height of 12 in. (30 cm), the pale green leaves being blotched with white and with purple markings on the underside.

The plants prefer as shady a position as possible and a rich compost made up of 2 parts fibrous loam, 2 parts leaf mould, 1 part sand and 1 part decayed manure. During summer water freely but the plants will require very little moisture during winter.

Propagation is by division of the tuberous roots early in spring after the plants have spent several years in the pots or by means of rooted offsets.

MARIGOLD, *see* Tagetes

MARSH MARIGOLD, *see* Caltha

MASK FLOWER, *see* Alonsoa

MASTERWORT, *see* Astrantia

MATTHIOLA

It is the Night-Scented Stock, so much appreciated if sown beneath a window which may be left open during a warm summer evening. *M. bicornis*, grows to a height of 10 in. (25 cm), covering itself in tiny flowers of palest lilac. Sow early April, no thinning being necessary if the seed is sown sparingly.

The garden stocks are descended from a single species, *M. incana* and its variety *M. annua*, annual (or biennial) plants which are native of south Europe.

BROMPTON STOCKS

Stocks may be divided into four main groups, one for each season of the year. First come the spring-flowering or Brompton Stocks, developed at the Brompton Road Nurseries of Messrs London and Wise (who laid out the gardens at Blenheim Palace) early in the eighteenth century. The plants are true biennials, obtainable in white, purple, crimson and rose and should be sown early in July removing the dark green seedlings (as with all stocks) as these will give only single flowers. The seedlings should be transplanted in August and they may either be set out in their flowering quarters in October or wintered under a frame and planted out in March. Or they may be grown on in pots to bloom under glass early in the year. 'Lavender Lady' and the crimson 'Queen Astrid' are outstanding varieties.

SUMMER FLOWERING OR TEN-WEEK STOCKS

They are so called because they may be brought into bloom within ten weeks of sowing the seed in gentle heat early in March. Hansen's 100 per cent Double is a recommended strain for bedding, making branched plants only 9 in. (23 cm) tall and obtainable in all the stock shades including apple-blossom pink, light blue and yellow.

The 'Excelsior Mammoth' strain in a similar colour range, including blood red, grows 2 ft. (60 cm) tall, the large densely packed spikes having a most majestic appearance in the border whilst they are valuable for cutting so that their delicious clove scent may be enjoyed indoors.

AUTUMN FLOWERING STOCKS

These are the 'Intermediate' or 'East Lothian' stocks, discovered in a cottage garden in East Lothian. They should be sown in March for they take several weeks longer to come into bloom than the Ten-week type. They are at their best during the autumn and until the arrival of the November frosts. The 'Kelvedon' strain in all the rich stock colours is outstanding.

WINTER FLOWERING STOCKS

They are the finest of all stocks, known as the 'Winter Beauty' or 'Beauty of Nice' strain. The seed should be sown in July for winter flowering under glass. Making plants of vigorous branching habit, they grow 18 in. (45 cm) tall and produce their large dense spikes throughout winter under glass. Amongst the finest varieties are 'Mont Blanc' (white); 'Queen Alexandra' (rose-lilac); 'Crimson King' and 'Salmon King'. Under glass in a temperature of 50°F. (10°C.) the perfume of the flowers is almost overpowering.

MAURANDIA

It should be given a position near a window, for it seems to be tolerant of draughts and does not require a winter temperature of more than 42°F. (5°C.). During the summer the plant requires copious amounts of water but very little in winter and an ordinary compost composed of 3 parts fibrous loam, 1 part peat or leaf mould and 1 part silver sand suits it well.

The plants are evergreen and are of slender neat habit. They will twine around a trellis or stake and are suitable to garland a small window.

Natives of Mexico, one of the best forms is *M. barclayana*, which has glossy heart-shaped leaves and during summer bears tubular flowers of violet and white. *M. semperflorens*, rather more vigorous, bears rich purple flowers.

MAY, *see* Crataegus

MEADOW FOAM, *see* Limnanthes

MEADOW RUE, *see* Thalictrum

MEADOW SAFFRON, *see* Colchicum

MECONOPSIS

Most members of this family, natives of the lower Himalayas are monocarpic, that is, they die after flowering once. The beautiful *M. baileyii* and *M. cambrica*, however, are true perennials and like the others may readily be raised from seed. The plants should not, however, be allowed to bloom the first season, for if they are allowed to concentrate on forming flowers before they become fully established, they may die back in the same way as the others. Plant in March 18 in. (45 cm) apart and both plants require quite different treatment.

SPECIES

Meconopsis betonicifolia 'Baileyii' (3–4 ft. 1 m). Known as the Blue Himalayan Poppy, no plant is more beautiful. It bears its pure sky-blue flowers with their bright golden anthers during July above its rosette-like foliage. The plants, which are best raised from seed, require a rich leafy soil and a position of partial shade for the roots should never be allowed to dry out.

M. cambrica (2 ft. 60 cm). The Welsh poppy bears its single pale yellow blooms from June until September and will flourish in ordinary soil and in a position of full sun. Though not so beautiful as *M. baileyii*, it is a valuable plant for the front of the border on account of its long flowering period.

MELIANTHUS

A half-hardy evergreen shrub, *M. major* has glaucous pinnate leaves almost 12 in. (30 cm) long whilst the crimson-brown flowers are borne in erect terminal racemes of similar length. Growing up to 10 ft. (3 m) tall, it requires the protection of a warm wall in the British Isles away from the milder climate of south-west England, Ireland and Scotland. It is known as the Honey Plant from the honey-like scent of its flowers.

Plant in April in a deep loamy soil and prune only to maintain its shape. Propagate from cuttings of the new season's wood, taken with a 'heel' and inserted in a sandy compost under glass.

MELITTIS

A hardy plant from southern Europe, it grows readily from seed sown in May and will bloom in twelve months, at a time when the border is lacking in colour. It bears its white and purple lipped flowers from the leaf axils on stems 12 in. (30 cm) high. While the flowers are visited by bees, the leaves carry a pleasant musk-like perfume. Closely related to the salvia family, plants of *M. melissophyllum* will grow well in shade and may be planted in the shrubbery.

MERTENSIA

An interesting hardy perennial, in bloom during April and May, when it dies back almost completely. It should be planted about the border, close to the later flowering plants which will hide the spaces it leaves when it dies down. Plant 2 ft. (60 cm) apart in November, in ordinary soil.

SPECIES

M. paniculata. Its tiny flowers, like forget-me-nots, are of the same colour and are borne on graceful arching stems during April and May.

M. virginica. The nodding clusters of sky-blue flowers are borne from long arching sprays. Its glaucous grey-green foliage adds to its beauty.

MESEMBRYANTHEMUM

This dwarf growing succulent, native of South Africa is an annual and should be given a sunny position and a dry, sandy soil. It will survive almost desert-like conditions and is ideal for bedding in the sandy soils of coastal gardens, remaining in bloom throughout summer and autumn, the daisy-shaped flowers opening only when the sun shines on them. *M. criniflorum,* is also known as the Livingstone Daisy, and no annual has a more brilliant colour range with its blooms of crimson, pink, apricot and yellow and numerous art shades.

For the plants to come early into bloom, seed is sown under glass, either in heat, or under a frame or cloches early in March, the plants being set out towards the end of May. Or seed may be sown early in May where plants are to bloom, though by this method flowering will be several weeks later.

METASEQUOIA

The Dawn Redwood of west China – 'a living relic of a fossil genus'. It is deciduous, with foliage which turns a lovely shade of soft salmon-pink in autumn. Extremely hardy, *M. glyptostroboides* is quick growing and in the wild will attain a height of 100 ft. (30 m). Of dense pyramidal habit, it has shaggy brown bark which adds to its beauty. Plant November to March, in a well-drained soil but this tree is lime-tolerant though is slow growing in a shallow soil. It is hardy though should be given a sheltered, sunny situation.

Propagate from seed which germinates readily or from cuttings taken with a 'heel' and inserted in a sandy compost under glass. When rooted, it will attain a height of 20 ft. (6 m) in ten years.

MEXICAN BREAD FRUIT, *see* Monstera

MEXICAN ORANGE FLOWER, *see* Choisya

MEXICAN SOAP PLANT, *see* Agave

MEXICAN TIGER FLOWER, *see* Tigridia

MICHAELMAS DAISY, *see* Aster

MICONIA

M. magnifica is an evergreen bearing large velvety leaves which are tinted bronze on the upper side, crimson-purple beneath. It is an extremely beautiful plant, requiring stove conditions.

Seed should be sown in April in a temperature of 70°F. (21°C.), using for a compost a mixture of leaf mould and sand. The seed should not be covered. A humid atmosphere must be provided and as soon as the seedlings are large enough to handle they should be moved to small pots containing the John Innes Potting Compost in which leaf mould is substituted for the peat.

During winter the plants will require a temperature of 55°F. (12°C.) and very little moisture, though in summer they should be watered freely and given a syringe whenever conditions are warm.

MICROMERIA

M. corsica is a most uncommon little plant, like a tiny heather and growing only 3 in. (7.5 cm) tall. It has minute grey foliage and pretty purple flowers, very like those of heather, whilst the plant can tolerate dry conditions. The foliage, which is excellent for using in sachets and pot-pourris when dry, carries a strong smell of incense. The plants may be trodden upon and will withstand cutting, so may be planted about a chamomile 'lawn' or between crazy paving stone.

MIGNONETTE, *see* Reseda

MILKWEED, *see* Asclepias

MIMOSA

The Sensitive plant is *Mimosa pudica*, native of tropical America and Asia and which is readily raised from seed sown in John Innes compost in spring. It is grown as an annual and is a most handsome pot plant for a warm room with its reddish-brown stems and fern-like leaves formed into numerous leaflets. When gently touched with finger or pencil, the leaves curl up and they do so in the evening. The plant requires a sunny window and copious amounts of water in summer, when it will bear fluffy pink flowers like those of the more familiar yellow flowering mimosas.

MIMOSA, *see also* Acacia

MIMULUS

A perennial, where it is happy it is long-lived, where not, it is a short-lived plant. Though it likes a moist soil, it will only be long-lasting where the soil is well-drained in winter. A light loam enriched with peat and decayed manure is ideal. Plant in spring 6 in. (15 cm) apart and increase by root division.

VARIETIES

A. T. JOHNSON. The large orange trumpets are spotted with brown.

CANARY BIRD. The buttercup-yellow trumpets have a crimson spotted throat.

CERISE QUEEN. The large trumpet-shaped blooms are of a brilliant cerise-red, in bloom from mid-June until August.

RED EMPEROR. Glowing crimson-scarlet of compact habit and being longer living than the older 'Whitecroft Scarlet'.

MIRABILIS

It reached England from Spain early in the sixteenth century and by the publication of his *Herbal* (1597), Gerard said that it had long been growing in his garden. The generic name was given to the plant on account of the diversity of colours in the flowers which are striped with yellow, white and red. They are so numerous that the whole plant is covered in bloom from early July until late October. On a warm day the flowers refuse to open until late afternoon for which reason it is also known as the Four O'clock plant. During autumn, the flowers are open throughout the day but in a warm climate, so exacting are the flowers in their opening at four o'clock that the plant is grown solely as a time keeper.

It is a half-hardy perennial with a tuberous root and in the British Isles it should be lifted after it has finished flowering, late in October and the roots stored during winter in moist sand. However, it grows so readily from seed that from a sowing made in gentle heat in March, the plants will come into bloom early in August. The seed is sown in boxes or pans and the seedlings transplanted to small pots from which they are set out early in June. The plants grow 3–4 ft. (1 m) tall and grow bushy and so should be allowed 3–4 ft. (1 m) in which to develop.

It was often to be seen in many a cottage garden, left to take care of itself for unless the winter was harsh, the roots received all the protection necessary from plants growing nearby and survived all but the most severe conditions.

MISCANTHUS, *see* Eulalia

MISTLETOE, *see* Viscum

MIST PROPAGATION

The method of rooting cuttings in artificially heated soil (*see* Soil heating) on a greenhouse bench and under a controlled water spray has revolutionised plant cultivation during recent years. The bench is converted into a trough 8 in. (20 cm) deep and the soil warming cable laid as described for Soil heating, taking care to provide adequate drainage. The misting equipment consists of a 'sensing' element or artificial 'leaf', overhead water lines and misting nozzles.

The sensing element is placed amongst the cuttings so that it is exposed to the same conditions. The element is fitted with two carbon electrodes connected to the controller and the supply. The controller is connected to a solenoid valve so that when a film of moisture crosses the exposed ends of the electrodes, the current flows through to the controller and cuts off the circuit to the solenoid valve and at the same time, the water supply to the mist nozzles. The water connecting the electrodes will evaporate as it does from the plants with the result that there will be a reversible action, the water flowing into the nozzles and creating a mist-like spray. The solenoid valve continually opens and closes and on a warm, sunny day may do so many times in each hour, though may do so only once or twice during dull weather. The misting will enable the cutting to form its roots before it loses its moisture content and where the rooting medium is warm and moist, rooting will quickly take place. Even hard wooded cuttings will root in half the time normally taken.

It should be said that to obtain a fine mist which is essential in plant propagation, a water pressure of 50 lb. is necessary and this should be checked before commencing the operation.

MOCK ORANGE, *see* Philadelphus

MOLUCCELLA

M. laevis or Bells of Ireland is an interesting annual, much in demand by flower arrangers. It grows 3–4 ft. (1 m) tall, its tapering stems being covered in green sheaths or flowers, arranged all the way up and which if dried, may be kept through winter. Sow the seed in a temperature of 60°F. (16°C.) early in the year and transplant when large enough to handle, planting out 12 in. (30 cm) apart towards the end of May. This plant requires a higher temperature for its seed to germinate than most other annuals.

MONARDA

The Bergamot, *M. didyma* was introduced to this country from Canada two centuries ago. It grows wild in the Oswego district of Ontario where its leaves were appreciated for making a beverage. The pungent lemon perfume of its leaves, its compact habit, its hardiness and the interesting shape of its colourful blooms make this one of the best of all plants for the border. The monarda will thrive in any soil and in partial shade. Plant in November and propagate by root division. The plants are in bloom during July and August and are ideal for a small border.

VARIETIES

CAMBRIDGE SCARLET. This is the best form of *M. didyma*, its foliage being extremely pungent, its vivid scarlet whorls being held on stiff stems.

CROFTWAY PINK. The blooms are of a clear rose-pink colour, the habit being compact and free flowering.

DARK PONTICUM. Growing to a height of 3 ft. (90 cm) its purple blooms are interesting rather than beautiful.

MAHOGANY. The dark crimson blooms are a pleasing contrast to the pink shades. 2 ft. (60 cm).

MELISSA. Growing to a height of 3 ft. (90 cm) the blooms are of a lovely soft shell-pink colour.

PRAIRIE GLOW. A new variety which should have a great future. The bright brick-red flowers are held on 2 ft. (60 cm) stems and remain colourful until September.

MONKEY PUZZLE, *see* Araucaria

MONK'S HOOD, *see* Aconitum

MONSTERA

M. delicosa is an interesting plant from Mexico and a member of the philodendron family. Its large yellow flowers are replaced by fleshy fruits from which it takes its name of the Mexican Bread Fruit. It is evergreen, the plants requiring a winter temperature of not less than 48°F. (9°C.), and whilst it requires light, the direct rays of the sun should be avoided, so grow it in or near a window which does not face due south. The John Innes Potting Compost No. 1 suits it well and during summer the plant will require plenty of moisture, though only very limited supplies during the winter months.

It bears large deeply cut leaves with serrated edges which are leathery to the touch. Its flowers, too, are handsome, whilst there is the fruit to follow. An additional attraction is that as new growth breaks from the back of the previous leaf, it causes the plant to grow at an angle of about 60°. Though the plant has aerial roots which cling to the trellis up which it will climb, it may also be grown as a table plant.

Propagation is by cutting the stems into 4 in. (10 cm) lengths immediately below a leaf joint. If inserted into a sandy compost around the side of the pot the cuttings will root in a month if taken during summer.

MONTBRETIA (CROCOSMIA)

Once, the montbretia was planted in those odd corners of the garden where nothing else seemed to grow. The introduction of many lovely hybrids which are hardy given the minimum of protection, has, with their large flowers in the most exciting colour range, made this a first-rate plant deserving of some care in its culture. The earliest varieties come into bloom towards the end of July and continue to the end of autumn, the sprays being delightful when used with michaelmas daisies for indoor decoration.

Planted in light, sandy soil enriched with some humus in the form of peat or leaf mould and a small quantity of decayed manure the flower sprays will take on a quality never believed possible with these flowers. Only where the soil is light and well-drained will the corms survive a wet winter; in a heavy soil they should be lifted each November and re-planted in spring. This, however, will mean that the plants will not bear the amount of bloom of which they are capable and the clumps are best left undisturbed for three or four years, after which time they should be lifted and divided. Plant in groups of six corms, spacing them 6 in. (15 cm) apart and 4 in. (10 cm) deep to provide protection against frost. March is the most suitable time to plant and if given a peat mulch in early December after their strap-like foliage has died back, the plants should come safely through the severest winter. The plants will also appreciate a mulch of decayed manure, old mushroom bed manure is ideal, given in mid summer for they form a mass of cormlets and fibrous roots which will greatly exhaust the soil if not continually enriched.

VARIETIES

Red and Orange Shades

COMET. The blooms are star-shaped, of a rich shade of golden-orange with an unusual band of crimson. Mid-season.

FIERY CROSS. The huge broad petalled blooms are of a fiery orange with a primrose centre and dark stems. Mid-season.

GLADIATOR. A magnificent early flowering variety, the deep crimson bloom being shaded white at the centre.

INDIAN CHIEF. Early to flower, the bloom being coppery-orange with a golden throat. Very free.

LORD WILSON. The deep crimson-scarlet blooms are of exquisite shape and are formed along the branching stems. Late.

MEPHISTOPHELES. A new variety and a beauty. The plant is compact, the bloom borne on 2 ft. (60 cm) stems and of a vivid flame colour with a striking golden centre. It is early into bloom.

PROMETHEUS. A richly coloured variety of great beauty, the dark orange flowers being flushed with mahogany.

RED KNIGHT. Mid-season, the vermilion bloom shaded gold at the centre being very striking.

SIR MATTHEW WILSON. The intense scarlet flowers are shaded carmine at the centre and borne on tall branching stems.

STAR OF THE EAST. A new introduction, the huge clear orange blooms being early into flower.

Lemon and Golden Shades

APRICOT QUEEN. Very early to flower, being of a pleasing shade of golden-apricot.

CITRONELLA. The star-shaped bloom is of a pure primrose-yellow, attractively blotched maroon at the centre.

J. CROSS. An older variety, late flowering and for its bright self apricot colouring is still in demand.

LADY OXFORD. An unusual and distinct variety, the pale yellow bloom being attractively shaded peach-pink.

LEMON QUEEN. The blooms are huge and of an attractive lemon colour, shaded cream at the centre. Mid-season.

Pink Shades

E. A. BOWLES. Early and robust, the bloom being of a deep rose-pink with a striking yellow throat.

JESSIE. Extremely early and free flowering, the bloom being of a delightful shade of shrimp-pink.

MOUNTAIN ASH, *see* Sorbus

MOUNTAIN CRANBERRY, *see* Vaccinum

MULLEIN, *see* Verbascum

MUSCARI

The Grape Hyacinths are at their best when planted with the miniature hybrid daffodils, those which bloom at a height of from 8–12 in. (20–30 cm) or plant them with Juliae primroses or with the old 'Double White' primrose, or 'Cloth of Gold', a double yellow, for they all bloom at the same time. *Muscari botryoides* 'Album', with its dainty white bells, is particularly well placed in a bed of dark red primroses. The bulbs are most inexpensive, costing in many instances, only ten pence per dozen, so that they could well be planted with much greater lavishness and there are many species and hybrid varieties to choose from, quite apart from the well-known 'Heavenly Blue'; $3\frac{1}{4}$ in. (8 cm) bulbs will give the best results, and these are planted towards the end of September. After several years, the bulbs will have formed large clumps which should be lifted and divided, otherwise there will gradually be a reduction in quality and quantity of bloom. The most suitable time for lifting and dividing is early in summer, just before the foliage has died down. They may then be easily located. Before replanting, work into the soil any humus-forming material, together with a small quantity of fine shingle or coarse sand. Muscari enjoy a sunny position under a wall, but they are most tolerant of shade and even of adverse soil conditions.

SPECIES AND HYBRID VARIETIES

M. argaei 'Album'. This is a distinctive plant which blooms during May and June, producing tiny spikes of chalk white which last for many weeks. 4 in .(10 cm).

M. armeniacum 'Cantab'. This is one of the loveliest of all the muscari, its sturdy spikes of a good shade of Cambridge blue produced with freedom. It makes a pleasing contrast with the cobalt-blue spires of 'Blue Pearl'. 6 in. (15 cm) (see plate 38).

M. botryoides. This is the Italian Grape Hyacinth, its tightly packed spikes of dark blue globular flowers resembling a bunch of black grapes. It flowers throughout April and into May and to see it at its best, plant with it the white form, 'Album'. 6in. (15 cm).

M. comosum. Described by Louise Wilder as a 'quaint monstrosity', the Tassel Hyacinth grows 12 in. (30 cm) high. The spike, produced during April and May, descends from top to bottom in shades of brown to green and blue, and it is excellent for indoor decoration. It has a pleasing fragrance.

M. conicum 'Heavenly Blue'. The best-known of all the muscari. It is the Starch Hyacinth and bears many conical spikes of gentian-blue, agreeably scented. The tiny globular bells are packed tightly together on 9 in. (22 cm) stems, making this an excellent form for planting in grass and for cutting. It is in its prime during April.

M. latifolium. It has broad, strap-like leaves like those of the Arum lily, from which the dark blue bud pushes up to open pale blue at the top, dark blue at the base. An interesting and free-flowering species for April. 10 in. (25 cm).

M. moschatum 'Flavum'. The bloom of the Musk Hyacinth, from which the whole family takes its name, is anything but beautiful, the purple 'grapes' turning first grey and then to an uninteresting shade of browny-yellow. The perfume, however, is pungent and incense-like, a few blooms scenting a large room. Plant it in those odd corners about the garden where it may be left undisturbed to distil its fragrance during March and April. 6 in. (15 cm).

M. neglectum. This is a most attractive little chap, with blue-black spikes on short sturdy stems above greenish-yellow foliage. It is free-flowering and blooms through April and May. 6 in. (15 cm).

M. paradoxum. The bulbs are cheap enough, and the blooms quite beautiful, large and of a unique shade of peacock-green. The foliage is neat and tidy, so this is an excellent species for the rockery, at its best during April. 6 in. (15 cm).

M. plumosum. This is the Feather Hyacinth. Its blooms are sterile, and drawn out like fine threads, so that it has a feather-like appearance. Its rich violet 'feathers' are borne in May and are most useful for 'mixing' with other miniature flowers for indoor decoration. 9 in. (22 cm).

M. polyanthum 'Album'. A native of Greece with, in April and May, bears dainty spikes of rich creamy white flowers which are in delightful contrast to the blue-flowered forms. 6 in. (15 cm).

M. tubergenianum. Known as the Oxford and Cambridge Hyacinth and was introduced from northern Persia during recent years. The top half of the bloom is of Cambridge blue, the lower portion of Oxford blue without any trace of purple colouring. The large blooms are borne from the end of April until well into June and are excellent for cutting. 9 in. (22 cm).

MUSK, *see* Mimulus

MYOSOTIS

This brilliant blue flower should be given a position of dappled shade, like that of the woodlands where it grows naturally. It is thus happy when used for planting at the base of apple trees or in beds which are partially shaded, and also when used to under-plant tulips. Though perennial, it is usually treated as a biennial. The seed is sown into a soil well enriched with some moisture-holding humus, such as peat or leaf mould and never at any time must the plants be allowed to suffer from lack of water. Sow in shallow drills in June and line the drills with peat before sowing.

For edging, Sutton's Miniature strain is charming, the plants making little mounds only 4 in. (10 cm) high and obtainable bearing white, pink or blue flowers. Growing to a height of only 8 in. (20 cm) is the bright

'Blue Perfection'; the rosy-pink, 'Carmine King'; and the well-known, 'Royal Blue'. The early flowering mid-blue, 'Express', grows to a height of 6 in. (15 cm).

For the scree garden plant *M. uniflora* which bears unusual lemon-coloured flowers and has attractive grey foliage. *M. nummularia* is an almost prostrate shrub bearing white flowers, followed by pink berries late in summer. Delightful for a trough or for growing in pans is *M. rupicola* which makes a tiny hummock and bears heads of brightest blue. *M. alpestris* 'Ruth Fischer' is a moisture-loving plant and is never happier than in the bog or woodland garden where it bears its sky-blue flowers early in summer.

MYRICA

The Bayberry of North America, evergreen or deciduous trees, requiring an acid soil and a cool, damp situation. With their handsome aromatic leaves, they may be planted in the wild garden. *M. pensylvanica* is deciduous but its white fruits hang through winter. *M. californica* is evergreen, growing 10 ft. (3 m) high and in June, bears green flowers.

Plant the deciduous species October to March; the evergreen species in spring. The only pruning necessary is to remove the old wood. Propagate by layering in autumn.

MYRTLE, *see* Myrtus

MYRTUS

Myrtles are small evergreen trees which should be confined to the warmer parts of Britain, where they will grow 15 ft. (4.5 m) tall. Both the small glossy dark green leaves and the white flowers are scented. They enjoy a soil containing peat which will help to retain summer moisture and maintain it in a slightly acid condition. No pruning is necessary. Propagate by cuttings of half-ripened wood inserted into sandy compost under cloches.

M. communis, the Common Myrtle bears a profusion of white flowers in July and blooms best where growing against a sunny wall, whilst the form 'Microphylla' has smaller leaves and is equally free-flowering. *M. luma*, native of Chile is worthy of planting for its bark which peels off to reveal a bright cream inner layer.

NARCISSUS

INDOOR

There is no happier flower than the daffodil where it is given cool treatment. True, more forced daffodils are sold in the early new year than any other flower, forced in a temperature of about 56°F. (14°C.), but it is only under cool conditions, where a temperature of between 40°–48°F. (4°–9°C.) is maintained to keep out frost and prevent mildew, that the daffodil reaches perfection. Only under such conditions can the delicate whites, pinks and creams of the modern varieties be seen at their best, shielded from soot deposits, soil splashings and strong winds. The daffodil is at its best under conditions somewhere between those required for the snowdrop and the tulip; a cold, damp atmosphere it does not enjoy, neither is it happy in considerable heat. In the temperature of a living-room or a slightly warmed greenhouse, the plants will bloom to perfection from February until late in April if varieties for successional flowering are selected. For home culture the bulbs are best planted in size 48 pots, four bulbs to each pot; or earthenware bowls are effective. A compost composed of a good fibrous loam to which is added a small quantity of peat and some coarse sand is ideal and as the bulbs have a tendency to decay at the base, ample drainage must be provided.

Now comes the crucial point. More often than not, daffodil bulbs are then placed in a darkened cupboard in a room which is much too warm for them. For from twelve to sixteen weeks, absolutely cool conditions are essential from the time the bulbs are planted about the first week of September. Daffodils require a long period for the formation of their rooting system and while this is taking place, cold conditions are necessary. Where possible, the home grower should place the pots outdoors in a position away from the sunlight; or a cellar will be as good and there the bulbs must remain until they have rooted and made about 2 in. (5 cm) of top growth. This will mean taking them indoors about mid-December, depending upon the variety. All too frequently we are impatient and not only place the bulbs in too warm a room, but bring them from the darkness before the bulbs are sufficiently well-rooted, with the result that the foliage turns yellow and no bloom may appear or they may be stunted.

When placing the pots outdoors, a covering of 8 in. (20 cm) of ashes or soil is placed over to keep them as dark as possible. Remember to plant the bulbs with the tops just above soil-level and plant firmly. When growing for cut bloom, it is usual to use strong wooden boxes and to place the bulbs almost touching each other. The boxes are placed outdoors as described. As to moisture, if the bulbs are watered at planting-time and placed in a cold, dark room, no further watering should be necessary. This again brings us to the folly of placing all newly planted bulbs in too warm a position while rooting, for the compost will dry out too quickly and serious damage may be done to the bulbs before the too-dry condition can be corrected. For those living in flats where space is restricted to the use of a cupboard or position under the kitchen sink, the pots should be stood in deep boxes and around them packed peat, which should be kept moist. More peat, too, should be used in the compost for this will also retain moisture. Peat packed round the pots will keep the bulbs as cool as possible. A garage or garden shed is better for rooting the bulbs than a room in the home.

Early in January, or (for certain varieties) just before Christmas, if a warm greenhouse is available, the soil or ashes are carefully shaken away and the pots or boxes placed in a position of partial shade – under a greenhouse bench or in a semi-darkened position in a living-room, where they remain for ten days to become accustomed to the light. The compost should be kept moist and more water given as the bulbs are brought on either by the heat of the sun or from the artificial warmth of a room or greenhouse. As they reach flowering stage, almost copious amounts of water will be required.

The blooms should be supported by inserting thin sticks or canes around the pots or boxes as soon as the buds show the first signs of colour. Staking is essential for all but the dwarf-growing species, which must be an additional reason for using pots instead of bowls, which are often too shallow to permit efficient staking.

Where the blooms are to be marketed, they should be removed before being fully open and where they have to travel a distance for exhibition or sale, they must be allowed to stand in cold rainwater for twelve hours before packing. Flowers growing in the open and which are required in bloom at the earliest possible moment should be cut the moment the buds show colour and placed in water in a slightly warmed room or greenhouse where they will open in forty-eight hours, often a saving of at least two to three days on the time taken if left to open outdoors. This may enable the bloom to be marketed on a Friday, which is the best selling day of the week, rather than on the Monday or Tuesday, which would otherwise have to be the case if the blooms were to be marketed whilst at their best.

PREPARATION OF THE GROUND FOR OUTDOOR NARCISSUS

A well-drained soil, deeply worked and containing some humus, is the ideal for the narcissus. The plant will in no way tolerate a water-logged soil, nor one of an acid nature. The bulbs flower in abundance in a soil which contains leaf mould, such as is to be found in woodlands, though the best bloom is from bulbs planted away from the hungry roots of tall trees. Dappled shade will enable the blooms to remain in a fresh condition for as long a period as possible and will prevent fading should the weather be unduly warm. The ideal position is between the rows of orchard trees, planting in soil of fibrous loam which has received an occasional top dressing of peat or well-rotted manure. In no way should animal manure be allowed to come in contact with the bulbs though wool shoddy, decayed leaves, peat, hop manure and bean or pea haulm worked into the ground will help to retain the necessary moisture for the bulbs. Wet, clay soil should be lightened and made suitable for daffodils by incorporating a quantity of grit before opening up the texture with humus materials. Woodland soil will be suitable for planting without any special preparation. Land which is of an acid nature should be given a light dressing of lime though this should not be over-done, for the daffodil prefers a soil which is slightly acid to one too alkaline.

When planting in the border or beds the ground should be prepared early in August, so that it is allowed time to settle before the bulbs are planted early in September. Before planting, the ground should be given a light dressing with wood ash and a 2 oz. per sq, yd. dressing of steamed bone flour which is raked into the soil. Established beds, or bulbs planted in short grass will also benefit from a yearly dressing with bone flour and a peat mulch given early in February before the foliage appears above the ground. When planting with members of the primrose family, such as blue primroses or blue polyanthus, which are so enchanting when planted with 'Golden Harvest' or 'King Alfred' daffodils, the plants should be placed into position immediately after the bulbs are planted 4 in. (10 cm) deep and 8 in. (20 cm) apart. If the ground is used for summer

FIG. 8 *Planting depth for narcissus*

bedding plants, these may be left in position and planting of the beds delayed until early October, though September planting is preferable. The advantage of planting in an orchard or a shrub border is that the bulbs will in no way interfere with the plants or trees. They may be planted at the correct time and left undisturbed. Another valuable point is that the blooms will be afforded protection from cold and often severe winds during spring.

When planting in an orchard, it will not be possible to cut the grass until the foliage has died down and this will be mid-July. The ground will then be clean and short to facilitate gathering of the fruit. The scythe should be used again early in December when the fruit has been removed and this will keep the grass short and tidy whilst the bulbs are in their spring glory. Cutting should not take place after December 1st or there may be the chance that the young shoots of snowdrops, daffodils or other early flowering bulbs will be harmed, though this does not apply to bulbs planted round the trees in the circles made by the grass having been cut away. Here the soil should be lightly forked over and top dressed early in November. Here too, primroses and polyanthus may be planted, also in the woodland garden where they may be left untouched for several years.

Nor is it advisable to plant daffodils on the lawn which is to be kept frequently cut and tidy all the year round. Where possible plant the bulbs along the edge of a lawn if there is no other place available and this strip can be left uncut until the foliage has died down.

When planting for profit, the bulbs will either be planted in prepared beds or in orchards, but in either case the beds should be made 3–4 ft. (1 m) wide to enable picking to be done easily. And so that staking will not be necessary, plant the bulbs 4 in. (10 cm) apart each way.

SIZE OF OUTDOOR BULBS

Daffodils and narcissi are advertised as 'Double-Nose 1', 'Double-Nose 2', 'Double-Nose 3 and Rounds'. There is approximately fifty pence per 100 difference in the price. 'Double-Nose 1' bulbs are of enormous proportions and each will produce several flowers. The 'Double-Nose 2' size are fifty pence per 100 cheaper and generally used for cool-house or home culture in pots. 'Rounds' are suitable for garden culture while large 'Rounds' are the best for forcing. As with all bulbs, the firm, virile, round bulb of good average size is the most suitable for planting under cool conditions indoors and for open-air culture in frames. Often the top-size bulbs have lost their vitality and have become too acclimatized to their original surroundings to be satisfactory. Mr Guy Wilson has suggested that even a good-sized offset will produce a top-quality bloom if given cool treatment. The skin should be clean, light brown and smooth and the bulbs firm when lightly pressed with the thumb. Smaller-sized rounds will be ideal for outdoor planting if they measure up to this standard.

Bulbs planted in borders or in beds which may remain undisturbed, should be lifted and divided every three years. This is not vital if one is pressed for time during September, but it should be the rule where possible and where bloom of exhibition quality is required. The offsets should be removed and planted in a nursery bed and the bulbs which will have now formed clusters of good-sized bulbs, should be carefully pulled apart and replanted into fresh beds if possible or into the same ground which has been sweetened and enriched. This work of lifting, dividing and replanting may be done any time after the leaves have turned yellow and died down, which may safely be taken to be from 1st July. Indeed many growers growing under cloches or frames plant the bulbs early in August to allow them as long as possible in the ground before covering them early in the new year.

Most of the well-known exhibitors lift and divide the bulbs in July before new root action commences, for they contend that should this take place, the check will seriously impair the ability of the bulbs to produce a bloom of top quality. But the daffodil is one of the most accommodating of all plants, and may even be planted as late as December without ill-effect. Though the blooms are later to appear in spring, they may be only a little below exhibition standard. If the bulbs are to remain out of the ground for any length of time they should be placed in shallow trays in a dry but cool room. If sending the bulbs through the post, they should be packed in dry peat to prevent bruising.

GROWING UNDER CLOCHES

Daffodils and narcissi are ideal plants for growing under barn-cloches or frames. The bulbs should be planted in August if possible into prepared beds of the required width. If cloches are being used, two rows may be placed side by side with a path between each double row. A method is to cover the beds with straw and soil or ashes during early autumn in order that moisture may be retained and strong rooting system formed. The covering is removed towards the year's end when the rows are covered with glass.

As soon as the plants have finished flowering, the glass is removed; and when the foliage has died down, this is removed and the bed given a top dressing with peat and old mushroom-bed compost. When growing daffodils in the home garden, down the sides of paths or in beds near the house, much untidiness of the leaves may be overcome while they are dying down if they are tied loosely together in knots. Or again where time for gardening is limited, the leaves may be pegged down at ground level with wire sprigs and the summer bedding plants set out between the rows of leaves early in June. The leaves may be removed in a month's time and before the bedding plants have grown too large.

R.H.S. CLASSIFICATION AND DESCRIPTION OF SOME OF THE BEST ALL-ROUND VARIETIES

I – *Trumpet Narcissi*

Distinguishing characters: one flower to a stem: trumpet or corona as long as, or longer than, the perianth segments.

(a) Perianth coloured; Corona coloured not paler than perianth.

BRANDON. Deep golden-yellow self. The flower is large and the trumpet very deeply frilled. It is exceptionally vigorous; free flowering and a splendid forcer, flowering under glass at Christmas. Recommended. Excellent for naturalizing.

CROMARTY. Award of Merit, R.H.S., 1938. One of the most perfect exhibition trumpets ever sent out, of faultless form and beautiful quality; broad, smooth, flat perianth standing at right angles to the neatly flanged and serrated trumpet; colour, deep self-gold; very free flowering and splendid habit. A grand town garden plant.

GOLDEN HARVEST. A first-class yellow trumpet of good colour, large size and fine form. The trumpet is deep golden-yellow set off by an overlapping perianth of clear yellow. Fine for show and excellent for garden decoration.

HUNTER'S MOON. Award of Merit, R.H.S., 1943. A lovely trumpet arrayed in graded tints of clear shining, cool, luminous lemon. Vigorous, free-blooming and extraordinarily durable. It is a sheer delight for cutting, a most beautiful garden plant and exquisite in pots.

KING ALFRED. A superb all-round trumpet daffodil growing 24 in. (60 cm) tall, and of great substance. The colour is deep golden-yellow, the flower of very large size. Excellent for forcing for cut flower.

KINGSCOURT. A large flower of faultless form, superb quality and uniform rich deep golden colour. Flat, velvet smooth perianth of immense breadth and noble perfectly balanced bell-mouthed trumpet. Has for some time been rated in the R.H.S. *Year Book* Ballot as the best exhibition yellow Trumpet.

LEINSTER. It is one of the best yellow trumpets. It bears a large flower of ideal show form and substance having very broad quite flat perianth standing at right angles to the beautifully balanced trumpet which has a well-flanged and frilled mouth; colour a particularly pleasing clear deep self-lemon. A tall and vigorous grower it is grand in pots.

UNSURPASSABLE. One of the largest flowered trumpet daffodils, of deep golden-yellow throughout. A striking flower, and first-class garden plant.

(b) Perianth white; corona coloured.

FORESIGHT. A flower of first-class form and good quality with broad, flat, erect milk-white perianth and perfectly proportioned neatly flanged golden trumpet. The first to open outdoors. Vigorous and free flowering.

PRESIDENT LEBRUN. Pure white perianth with large creamy trumpet. A very fine bicolour variety of good substance. An ideal variety for pots and bowls.

TROUSSEAU. One of the very finest daffodils in existence, of superb quality and perfect form. Very broad, flat, pure white perianth and well-proportioned straight trumpet which opens soft yellow.

(c) Perianth white; corona white not paler than the perianth.

Beersheba. One of the finest and largest white trumpets yet raised. The flowers are of perfect form and texture. The flower stems long. The flower is of the purest white; very free and vigorous.

Broughshane. A glorious giant white trumpet of perfect form and balance, the trumpet being widely flanged and frilled. The great flower, which is of immense substance and is quite exceptionally durable, is carried on a strong stem over 2 ft. (60 cm) high, while the foliage measures as much as 2 in. (5 cm) across.

Cantatrice. A flower of smooth texture; with a pointed even perianth, standing out from the well-balanced perfectly smooth, rather slender trumpet. It is pure white throughout.

Mount Hood. A large white trumpet variety with well-proportioned blooms of great substance The growth is tall and vigorous, does well in pots. Award of Merit, R.H.S.

Mrs E. H. Krelage. It bears a creamy white trumpet of perfect form, the mouth of the trumpet being beautifully reflexed.

Silver Wedding. A medium-sized absolutely pure white flower of the utmost charm and refinement, smooth, clear-cut, sharp-pointed perianth, and perfectly proportioned slender trumpet.

Vigil. It has a broad sharp-pointed perianth and graceful well-proportioned well-flanged trumpet. A flower of fine parchment-like texture and ice-white throughout. A strong grower with beautiful blue-green foliage.

II – *Large-Cupped Narcissi*

Distinguishing characters: One flower to a stem. Cup or Corona more than one-third, but, less than equal to the length of the perianth segments.

(a) Perianth coloured; corona coloured, not paler than the perianth.

Adamant. Well-formed smooth perianth of a deep saffron-yellow colour, medium-sized, rich saffron-orange crown frilled at margin. A flower of unique and striking colouring. Award of Merit.

Aranjuez. A flower of the very highest quality with beautiful smooth, clear yellow, round perianth of exceptional texture, shallow expanded deep yellow crown, widely margined deep orange-red.

Carbineer. A magnificent flower of great substance with broad, flat, bright, rich yellow perianth and deep, bright orange-red cup.

Carlton. Broad, overlapping, flat perianth of 5 in (12.5 cm) diameter, and a large expanded cup, nicely frilled at the mouth; the whole flower is clear, soft yellow. Excellent for forcing, garden decoration and cut flower. Award of Merit.

Fortune. A giant Incomparabilis of largest size, perfect form and gorgeous colour, borne on a very tall, strong stem. Perfectly flat, well-overlapping perianth of great substance and of a clear brilliant yellow colour, with a very large and bold crown of glowing coppery red-orange. Very early flowering, it is, outstanding variety for garden, cutting or forcing.

HAVELOCK. Deep yellow throughout, a flower of most perfect form. Perianth broad and overlapping, of 4 in. (10 cm) diameter. One of the finest exhibition or garden varieties; tall and early.

HELIOS. Perianth deep primrose yellow, cup deep yellow, darkening to orange as it expands; flower of fine form, free flowering; one of the earliest; very vigorous. Splendid for bowls and forces easily.

KILLIGREW. Bright yellow perianth and a large, bright orange-red cup, a flower of wonderful quality.

LEPRECHAUN. A small flower of jewel-like brilliance; deep, clear lemon-gold pointed perianth, and small goblet-shaped, clear ruby-red cup; good stem; free and vigorous.

RUSTOM PASHA. One of the aristocrats of the Incomparabilis section with large flowers of intense colouring. The large flat perianth is deep golden yellow and the large crown is of the most brilliant orange-red which is abolutely sunproof.

SCARLET ELEGANCE. A splendid red and yellow Incomparabilis of distinct merit for garden and forcing. The perianth is clear yellow and the cup of rich red. Tall and vigorous.

(b) Perianth white; corona coloured.

BRUNSWICK. Broad, flat white perianth and finely formed well-balanced crown of charming pale lemon colour, shading to a paler colour at centre, of lasting quality.

COVERACK PERFECTION. A flower of charm with a broad white perianth and wide shallow saucer crown. The ground colour is white, while it is edged and flushed pale gold and salmon and has a faint tinge of green behind the anthers.

FERMOY. A magnificent flower of great size and fine quality; very large pure white perianth of substance and well-proportioned beautifully frilled bowl-shaped crown, bright orange-red at the mouth, fading to gold in the base; very vigorous.

GREEN ISLAND. Headed the list in the R.H.S. Daffodil Ballot several times in the 1950s, as the best exhibition II (b). It bears a flower of very large size, great substance and waxen smooth texture, with smooth rounded white segments of such immense width that they form an almost complete circle; well-proportioned, shallow bowl-shaped frilled cup, greenish white at the base inside, passing to white, which in turn passes to a band of clear, cool, greenish lemon at the margin.

JOHN EVELYN. It has a large solid white perianth of great substance, 4 in. (10 cm) broad and overlapping, with flat, soft yellow cup. An outstanding cut flower. Award of Merit.

KILWORTH. An outstandingly fine large dark red and white flower with broad white perianth and perfectly proportioned bowl-shaped crown of intensely vivid dark solid orange-red with a touch of dark green in the eye; tall and vigorous.

LINGERING LIGHT. A good-sized flower of refinement, with a spreading pure white perianth of delicate texture, and well-proportioned shallow bowl-shaped crown, flushed pale apricot pink.

MRS R. O. BACKHOUSE. A lovely variety with perianth of ivory-white and beautifully proportioned. Slim, long trumpet of apricot pink, changing to shell-pink at the deeply fringed edge; early.

RED HACKLE. A magnificent and brilliant red and white of fine habit and form. Broad, rather pointed, very slightly reflexing, pure white perianth; frilled bowl-shaped crown of solid, intense, deep orange-red.

ROSE OF TRALEE. The best coloured of a remarkable series of pink-crowned seedlings bred from self-fertilized White Sentinel; flower of beautiful form and quality, having pure white perianth pointed at the tips; long, nicely flanged crown of rosy apricot pink right down to the base, passing off white before the flower dies.

WILD ROSE. One of the brightest pinks to date, as the cup is rosy pink to the base, the colour being retained till the flower dies. Not a large flower, but of attractive form and balance.

(c) Perianth white; corona white, not paler than the perianth.

CARNLOUGH. A constant winner of championships at overseas shows. A distinct flower of strong sturdy habit, fine size and superb quality. Broad, firm, flat, pure white perianth; on first opening the crown is faintest citron with a frill of soft coral pink.

GREENLAND. A lovely flower of perfect proportion and true Leedsii character. Of splendid substance it is of pure ice-white throughout, both the broad perianth and medium-length crown, and has an entrancing tint of cold sea-green at the base, a vigorous plant with strong foliage and a sturdy, short-necked stem.

KILLALOE. Award of Merit. A most striking immense pure white flower of grand quality. The perianth segments are pointed, very broad, flat, and of fine substance. The large cup is beautifully flanged and frilled at the mouth.

III – *Small-Cupped Narcissi*

Distinguishing characters: one flower to a stem; cup or corona not more than one-third the length of the perianth segments. (a) Perianth coloured; corona coloured, not paler than the perianth.

BALLYSILLAN. An early and brilliant cupped flower of fine quality with clear yellow perianth of very smooth, firm texture and shallow vivid deep red cup, good stem and neck.

CHUNGKING. The finest red and yellow small-cupped yet sent out. It is a tall-stemmed flower of fine quality, having very broad, clear, rich golden perianth and perfectly proportioned intense deep vivid red shallow crown. A plant of vigour, increasing rapidly.

(b) Perianth white; corona coloured.

BLARNEY. One of the most distinct and charming of this type. Large flower, having firm, satin smooth, snow-white perianth and flat, salmon-orange crown with a narrow primrose rim.

MISTY MOON. An exquisite flower, having large pure white perianth and large eye, with grey-white centre and the outer half a halo of soft pale salmon-orange.

(c) Perianth white; corona white, not paler than the perianth.

CHINESE WHITE. A very large flower of faultless form and quality; pure white throughout, except for a faint touch of green in the eye. Very broad, circular, satin smooth perianth of great substance, fully 4 in. (10 cm) in diameter, and perfectly proportioned shallow-fluted saucer crown. A superb flower, quite unique, it created a sensation when winning the Engleheart Cup.

FOGGY DEW. A most beautiful large small-crowned flower, half-sister to 'Chinese White'. Very broad, rounded, much overlapping, large pure white perianth of fine substance and quality, smallish frilled white crown, having a deep sage-green centre.

PORTRUSH. A lovely late-flowering variety, with broad, flat, pure white perianth of great substance and almost flat, white crown with deep green eye. Tall, vigorous plant with good stem, very free of bloom and increase; should make a good market flower, coming at the end of the season.

IV – *Double Narcissi*

Distinguishing character: double flowers.

FEU DE JOIE. A distinct flower. The perianth petals are long and pure white, the short petals are of a brilliant orange-scarlet. The flower is fully double, which makes it lovely for decoration. Tall, strong, free and early.

INGLESCOMBE. Fully double flowers of a buttery primrose yellow, with broad rounded petals. The finest double-yellow daffodil yet introduced.

MARY COPELAND. Regular pure white perianth with segments glowing orange-scarlet; large, of fine form and tall. One of the most beautiful of all doubles.

TEXAS. Large, fully double flowers, round form, petals rich yellow, interspersed with fiery orange segments. Ideal for forcing.

V – *Triandrus Narcissi*

Distinguishing characters: characteristics of *Narcissus triandrus* clearly evident.

NIVETH. A triandrus hybrid of the greatest beauty and more vigour than is usual in this strain; purest stainless white of perfect grace and quality; comes particularly fine and very durable in a cold greenhouse.

SILVER CHIMES. Pure white perianth with a delicate pale primrose cup, each stem produces six or more flowers; a lovely exhibition variety and beautiful when grown in pots or bowls. Sweetly scented.

THALIA. A very lovely triandrus hybrid; late flowering, bearing three to four drooping, glistening, snow-white flowers. Vigorous grower.

VI – *Cyclamineus Narcissi*

Distinguishing characters: characteristics of *Narcissus cyclamineus* clearly evident.

BERYL. Charming drooping flower of primrose-yellow colouring and with a globular orange cup. Award of Merit, R.H.S. Very dwarf habit.

FEBRUARY GOLD. Outstanding in every way, having a lemon-yellow reflexed perianth and a bright yellow trumpet frilled at the edge.

LE BEAU. With its drooping flower of a soft shade of yellow throughout and long graceful trumpet, this is the ideal pot or alpine plant.

VII – *Jonquilla Narcissi*

Distinguishing characters: characteristics of any of the *Narcissus jonquilla* group clearly evident.

GOLDEN PERFECTION. Golden yellow, broad overlapping perianth, large golden cup of good form. Two to three flowers on each stem.

TREVITHIAN. A beautiful variety carrying a large cluster of lovely pale self-lemon-yellow flowers, early flowering and forces well. Bears several sweetly scented flowers on a stem.

VIII – *Tazetta Narcissi*

Distinguishing characters: characteristics of any of the *Narcissus tazetta* group clearly evident.

CRAGFORD. A notable new Poetaz variety; creamy white perianth, rich orange eye, three to four flowers on a stem. A splendid variety to grow in bowls. Can be had in flower by Christmas.

GERANIUM. A splendid variety, pure white perianth, and geranium red cup. An excellent variety for growing in bowls for late display (see plate 39).

MARTHA WASHINGTON. Pure white, well-overlapping perianth with clear orange crown, two to three flowers on a stem. The flowers are much larger than those of any other Poetaz.

SCARLET GEM. A showy and attractive Poetaz for growing in pots. Perianth of primrose yellow enhanced by a deep orange cup.

IX – *Poeticus Narcissi*

Distinguishing characters: characteristics of the *Narcissus poeticus* group without admixture of any other.

ACTAEA. Broad snow-white perianth with very large yellow cup, broadly margined dark red. Very early, tall, large flower. Vigorous. Splendid for pots, bowls and garden.

MINIATURE

Miniature daffodils are generally neglected, but for planting around the roots of young trees or as an edging to a path or border, on the rockery, for a window-box, and especially in pans for the alpine house or cold frame, the dainty species which bloom on no taller than 10 in. (25 cm) stems, make a delightful display during April. They are inexpensive and so easily grown in any fibrous loam to which has been added a little peat and some coarse sand or shingle. Small pockets should be prepared on the rockery to take four or five bulbs, which are planted in September after the rockery has been cleaned and the established plants stripped of their straggling growth. When using a window-box or tub it is not always convenient to plant before mid-October, which will not be too late, as the miniatures do not seem to require so long a season in which to make root growth.

SPECIES

Narcissus bulbocodium. The lovely little hoop-petticoat narcissus which takes a year to become established in the open ground before it blooms. The variety 'Citrinus' bears flowers of a paler, lemon-yellow shade; whilst 'Teniufolium' is dwarfer and earlier to come into bloom. This latter variety bears thin rush-like leaves which are of almost prostrate habit.

N. cyclamineus. At its best by the side of a stream for it likes a peaty, moist soil and the protection of grass in summer. The loveliest variety is that of recent introduction called 'Snipe', which grows to a height of 9 in. (23 cm) and makes a delightful pot plant. The reflexed perianth is of pure white, the trumpet deep yellow.

N. jonquilla. In its original form, a plant of the utmost charm, growing to a height of 9 in. (23 cm) and in bloom during May. It enjoys a warm, sunny border or a rockery facing south.

N. juncifolius. The rush jonquil, growing to a height of only 4 in. (10 cm) and flowering late in May. It is happy in the shelter of rockery stones and is charming in pans in the alpine house, but does not like open ground planting. It bears a rich perfume and a unique large trumpet. An additional value is that after flowering its leaves die back almost out of sight.

N. lobularis. The earliest of all to bloom, appearing in February in a sheltered corner. It reaches a height of only 6 in. (15 cm).

N. W. P. MILNER. A hybrid of charm, bearing its palest sulphur trumpets on 9 in. (22 cm) stems. It is most attractive in the rockery, under trees or for pot culture.

N. minimus. This is the smallest of all the trumpet daffodils, bearing its dainty fringed trumpets on stems only 4 in. (10 cm) in length. Though lovely planted alongside the edge of a path or border, this fairylike daffodil is at its best in pans in the alpine house.

N. nanus. It follows *N. lobularis*, in bloom during March and so keeps continuity from February to May with these dainty rock garden species. It bears an exquisite yellow bloom, the exact replica of a 'King Alfred' daffodil, on 6 in. (15 cm) stems.

N. odorus. The Spanish campernelle, possessing a rich perfume and should be planted at the edge of a border beneath a window to allow its fragrance to enter the house during May. The star-like blooms are borne in clusters of three or four on 8 in. (20 cm) stems. There is also an attractive double form known as Queen Anne's Irish jonquil, which is of a warm gold colour and deliciously scented.

N. triandrus. Of this species, *Robinson in The English Flower Garden* says: 'as a pot plant it has no superior for delicate beauty. Mr Rawson of Windermere grew it in pots which bore 50–100 blooms'. He tells us that after the leaves have faded the pots should be rested, top dressed, but neve rre-potted and this is the most successful way to grow 'Angels' Tears' daffodils, so called on account of the corolla hanging like a teardrop beneath the perianth (see plate 40). There are several lovely varieties, the white, 'Albus', flowering in April on 6 in. (15 cm) stems.

NASTURTIUM, *see* Tropaeolum

NEMESIA

A half hardy annual and like the marigolds, the modern nemesia has been improved out of all recognition during recent years, the habit being more compact, more free flowering and longer in bloom. Though classed as being half-hardy they are almost hardy, and in sheltered gardens in the south, the plants may be set out after hardening, at the end of April, with a second planting early in June to carry the flowering period through summer. For an early planting seed is sown in gentle heat in early February, a month later for an early June planting; or seed may be sown in a cold frame early in March.

Both 'Blue Gem' and 'Dwarf Orange' may be used for edging for they grow only 8 in. (20 cm) tall, the large-flowered or *Strumosa suttonii* strain being used for the main bedding display. Their freedom of flowering and rich colours is difficult to equal and like the dwarf marigolds they are not troubled by wet weather. To make a bushy plant the growing point should be pinched out whilst the plants are being hardened.

NEMOPHILA

It enjoys a cool, moist soil and a position of partial shade. There is a semi-climbing species *N. aurita,* bearing deep mauve flowers; and *N. menziesii*, 'Baby Blue Eyes', an annual which bears pale blue flowers with a striking white centre. Growing to a height of only 6 in. (15 cm) and coming into bloom mid-May from an early April sowing, the latter is an attractive edging plant.

NEPETA

With its attractive grey foliage, its rich purple flower spikes and informality, combined with a long flowering season, makes the nepeta an outstanding perennial.

Like most blue flowers, the nepetas prefer a well-drained, sandy soil and spring planting. Where used to edge the border, plant 12 in. (30 cm) apart. The plants are propagated by root division or by cuttings of the ripened wood removed in July and rooted in sandy soil in a frame or under cloches.

SPECIES AND VARIETIES

N. macrantha. Growing to a height of 2 ft. (60 cm) this plant is not suitable for edging, but with its attractive foliage and stiff spikes of violet, it is a pleasing plant for the border.

N. mussinii. More plants of this species must be raised each year for border edging than any other. It bears its deep blue-mauve flowers in sprays from mid-May until September, its grey-green foliage acting as a striking foil for those white and yellow flowering plants placed immediately behind. It grows to a height of only 12 in. (30 cm) and as with most blue flowers it appreciates some winter protection. This is usually given by allowing the dead flower stems to remain on the plant until early spring. If the untidy appearance offends the eye, cut back half-way. A new form, 'Violacea' bears bloom of a deep purple-mauve.

SIX HILLS GIANT. A hybrid growing to a height of about 20 in. (50 cm) and bearing its branched spikes of violet-blue above grey-green foliage. *N. glechoma*. At one time a popular plant for hanging baskets when it was known as the trailing nepeta, for it is of the catmint family, its leaves having the same pungent smell. It is a native plant and is perennial, considered by some to be a weed but if so, it is a charming one, trailing over the ground in the manner of ivy, by roots formed at the joints of the dainty kidney-shaped leaves which are variegated white and green. Like all the 'mints', it enjoys a rich moist soil and a cool, semi-shaded position. It bears its dainty whorls of purple-blue lavender-like flowers during June and July when they are often mistaken for violets. In medieval times, the plant was known as Ale-hoof for its leaves were used to clarify ale, as a substitute for hops whilst in rural districts, the leaves were dried and made into 'tea'.

NERINE

A valuable greenhouse subject for it comes into bloom in October and must be given a little heat not so much for the production of bloom, but to enable the foliage to ripen during winter and spring. The bulbs should be potted in large-size pots early in August, the neck of the bulbs being above the surface of the compost which should be of fibrous loam, coarse sand and peat. They should not be given a rich compost, but one which is well drained and contains some humus. Unlike most bulbs, no darkened position should be given for them to root. The pots should be placed in the full sun of frame or greenhouse, sparingly watered, and by early autumn, they will produce their leafless stems of brilliantly coloured blooms held in tight clusters.

There are three outstanding species:

Nerine bowdenii. Which bears its lovely soft pink blooms late in September.
N. filifolia. Which produces its neat, rose pink flowers during November.
N. sarniensis. The true Guernsey Lily, grown in large quantities in the Channel Islands, the flowers being rich carmine-red.

After flowering appear the strap-like leaves, which continue to grow until early spring. From late April, all water must be withheld so that the bulbs can ripen off thoroughly. They should be left in the pots and placed in a covered, but ventilated frame so that they may remain dry and can enjoy the sunshine. Early in August they may be watered and started into growth again. No repotting is done. They prefer to become pot-bound and do not like disturbance.

GROWING OUTDOOR

In favourable districts of the south and west country, nerines may be cultivated, as they are in Guernsey, in the open in a sheltered position under a south wall. There they will bloom during September and throughout October, the species *N. bowdenii,* which is early flowering, being the most suitable. They must be given a dry, well-drained soil, one containing plenty of grit and sand and no manure and should be left undisturbed.

NERIUM

Native of the Mediterranean regions, it contains poisonous milky juices used by the natives of Central Africa to poison their darts. *N. oleander* should be grown indoors in the British Isles, in a minimum winter temperature of 50°F. (10°C.). It is evergreen, with narrow pointed grey-green leaves and it bears its pale pink flowers, like those of the wild rose, in terminal clusters. They are deliciously scented. The plant will grow up to 3–4 ft. (about 1 m) tall and should be kept well supplied with moisture at its roots, especially during summer. The plants may be stood on the floor near a sunny window if they grow too tall for a table.

Propagate from cuttings, removed in spring and inserted with the base of the stem in a jar of water when the roots will quickly form. Nip out the growing point and pot on when they will bloom the following year.

NEW ZEALAND DAISY, *see* Olearia

NICOTIANA

This annual used to be confined to the mixed border, for growing to a height of 4 ft. (1 m) and opening its blooms only at night, it was planted solely for its perfume. The modern nicotianas not only make dwarf, compact plants, but are free flowering and more important, the flowers remain open through the daytime, a compensation for their having lost some of their perfume.

The tobacco plants are native of South America and should be given half-hardy treatment. They may be planted out from mid-May, spacing them 12 in. (30 cm) apart and planting firmly. The varieties 'Crimson' and 'White Bedder' make compact, bushy growth and are seen at their best when planted together and edged with dwarf antirrhinums.

Even more compact is 'Idol' which grows 9 in. (22 cm) tall and bears crimson-red flowers on well-branched plants from July until September.

For the border, the taller 'Lime Green' is a most unusual flower for floral decoration, its blooms being of soft lime-green. It grows 2 ft. (60 cm) tall.

NIEREMBERGIA

Native of the mountain ranges of South America, the tiny shrublets cover themselves with campanula-like blooms from May to October. The loveliest is *N. cærulea* which bears deepest violet flowers with a bright yellow centre. More dwarf is *N. rivularis* which bears white gentian-like blooms with a golden throat. Both are delightful plants for the rock garden, requiring a sunny position and a soil retentive of summer moisture.

NIGELLA

One of the loveliest of annuals, its country name Love-in-a-Mist, so well describes the flowers of misty blue half hidden in a frill of linear foliage, like that of fennel and of brilliant green. It takes its botanical name from niger, black, from the colour of its seeds, those of *N. sativa* having a nutmeg-like fragrance and from early Tudor times were imported from Egypt to use for flavouring cakes. The aromatic seeds were put in muslin bags to place amongst linen and clothes whilst Gerard advises warming the seed over hot ashes to scent a musty room or to be inhaled, to assist with the breathing from a cold in the head. Tournefort said that apothecaries distilled from the seed, an oil used as a substitute for spikenard.

N. damascena, native of the near east, was first grown in English gardens during early Elizabethan times. It is illustrated in Gerard's *Herbal,* exactly as it is today and he writes, 'nigella is both fair and pleasant'. He also tells us that it had the old Anglo-Saxon name of Gith and was also called St Catherine's flower. The seed crushed into powder and mixed with vinegar, 'taketh away freckles and is a most excellent remedie'. The seeds of *N. damascena* are not aromatic.

By the end of the sixteenth century double-flowered forms were grown and every cottage garden grew the single form, seed being sown in the open early in April and thinning the seedlings to 4 in. (10 cm) apart. Miller recommends making a sowing in August when the soil is well-drained. The plants will come into bloom fully a month earlier than those from a spring sowing. They grow 18 in. (45 cm) tall and are suitable for indoor decoration, looking most attractive in a pewter tankard.

The variety 'Miss Jekyll', bearing flowers of cornflower-blue; and Oxford-blue, are both lovely whilst equally fine is the new 'Persian Rose', bearing flowers of antique rose-pink.

NOLANA

The Chilean Bellflower is a valuable annual plant of easy culture for rock garden, window box or for edging. It makes a compact plant 6 in. (15 cm) tall and the same across and covers itself in bell-shaped flowers of brightest mauve with zones of white and yellow. It requires a dry, sunny situation. Sow the seed in gentle warmth early in the year and plant out 8 in. (20 cm) apart early in June, after hardening.

NOMOCHARIS

A genus of sixteen species, closely related to lilium with bulbs which are without tunics and with thin narrow leaves arranged in whorls along the stem. The segments of the flowers which are held on long pedicels, are divided to the base so that they open flat and face downwards. Native of the Himalayas and west China, it was discovered by Mr Reginald Farrer growing at a hight of 10,000 ft. or more but is rare in cultivation.

It requires similar conditions as in its native land, i.e. a long cold winter followed by long periods of sunshine to ripen the bulbs. An open situation, preferably in the northern part of Europe and America is therefore essential. It also requires a well drained soil, containing some humus and plenty of grit. Plant the bulbs in September, 9 in. (23 cm) apart and 4 in. (10 cm) deep, but the best method of obtaining a stock is to raise plants from seed for at all stages of growth they resent transplanting.

Seed of most species now in cultivation is obtainable from Messrs Thompson and Morgan of Ipswich and it will germinate quickly. Sow in April, the contents of a packet in a pan filled with the John Innes Sowing Compost. Sow thinly, only just covering with compost and keep comfortably moist but giving little or no moisture in winter. During this time, the pans are best kept in a frame to which plenty of fresh air is admitted. In two years' time, carefully knock from the pan the seedlings and the compost which will be a mass of fibrous roots and without disturbing, plant where it is intended they should bloom. Keep moist during their first summer outdoors until established when they may be expected to bloom the following summer. Where conditions are suitable to them, the plants will readily seed themselves to replace those which die off each year.

SPECIES

Nomocharia aperta. Native of Tibet and west China, it grows 2 ft. (60 cm) tall and bears in June, four to six pale pink pendulous flowers held on long pedicels. Both the inner and outer segments have a crimson blotch at the base.

N. farreri. Native of north Burma, it grows 3–4 ft. (1 m) tall, the stems being clothed in whorls of narrow leaves whilst the white flowers, blotched with purple at the base of each segment open to 4 in. (10 cm) across.

N. mairei. Native of west China, it grows 2 ft. (60 cm) tall, its large white flowers being heavily spotted with purple whilst the inner segments are heavily fringed. The variety 'Candida', rare in cultivation has pure white flowers devoid of markings.

N. meleagrina. Native of east Tibet and Yunnan, its large white flowers, borne on a 3–4 ft. (1 m) stem are heavily marked with crimson-purple spots resembling the Snake's-head Fritillary, hence its botanical name.

N. pardanthina. Native of west China, it resembles *N. mairei* except in the absence of fringing to the inner segments. The pale pink flowers are blotched with crimson at the base of each inner segment and are borne eight to ten on a 3–4 ft. (1 m) stem.

N. saluenensis. Native of west Himalayas, it grows 2 ft. (60 cm) tall and bears five to six large flowers of deep rose-pink which open saucer-shaped, the three inner segments being blotched at the base.

NOTHOFAGUS

Known as the Southern Beech for it is native of South America and Australasia. Those of South America are hardy in the British Isles for they come from the southern extremities of the American continent. They are fast growing, attaining a height of 100 ft. (30 m) and require a deep loamy soil. They will not grow in chalk. They require no pruning.

Of the two hardy deciduous species, *N. antarctica* has small glossy heart-shaped leaves, whilst the branches grow twisted. *N. procera* has attractively veined leaves like those of the Hornbeam.

NOTHOLIRION

A genus of six species, closely related to lilium and fritillaria, the bulbs being distinguished from those of lilium by the smaller number of scales, whilst it also forms a long basal leaf which protects the dormant flower spike. The flowers resemble those of nomocharis, being funnel-shaped and nodding and are borne on long pedicels though they do not open flat as with nomocharis. Like that genus, notholirion is native of the Himalaya region, extending from Afghanistan to west China. The plants require similar cultural conditions but seem to be of more tender habit and produce their basal leaf in midwinter so should be confined to those gardens where frost and rain are not excessive.

Outdoors, the bulbs should be planted 4 in. (10 cm) deep and 12 in. (30 cm) apart in September, in an open sunny situation and in a well-drained sandy soil containing plenty of humus. It is advisable to plant near to low shrubs which will provide the basal leaf in winter with protection from frost and cold drying winds for if the foliage is harmed, the plants may fail to bloom as the dormant flower spike is enclosed within the broad basal leaf.

Where the garden is unfavourably situated, the plants are best grown in pots under glass, one bulb to a 5 in. (12.5 cm) pot containing a sandy compost. *N. macrophyllum* is the most suitable species to grow under glass for it grows only 15 in. (38 cm) tall and blooms during April and May. The bulbs should be planted in September 2 in. (5 cm) deep, giving only sufficient moisture to maintain plant growth during winter but increasing supplies as the flower spike forms early in spring. After flowering the bulb dies away but leaves behind numerous bulbils which form inside the scales and in the wild, serve to propagate the species.

Upon lifting the old bulb late in summer when most species have finished flowering, the bulbils are detached and planted in boxes containing a sandy compost in which they are grown for two years. They are then moved to individual pots in which they remain for two more years to attain flowering size.

SPECIES

Notholirion bulbuliferum. Native of Tibet and Nepal, it was so named for the multitudes of bulbils it produces. Its funnel-shaped flowers are of a lovely shade of pinkish-mauve tipped with green and as many as twenty to thirty appear on a 5 ft. (1.5 m) stem. It blooms during July and August.

N. campanulatum. Native of north Burma, it forms a basal leaf almost 2 ft. (60 cm) in length and bears its crimson bell-shaped flowers in an umbel of twenty or more at the end of a 3–4 ft. (about 1 m) stem. Each segment is tipped with green whilst each flower is more than 2 in. (5 cm) long. It is midsummer flowering.

N. macrophyllum. Native of Nepal, it is too tender to grow outdoors in the British Isles and should be confined to the alpine house where it will bloom during April and May. The funnel-shaped flowers of lilac-mauve are borne six to eight on a wiry stem 15 in. (38 cm) tall, the basal leaf being of similar length.

N. thomsoniana. Syn: *Fritillaria thomsoniana.* Native of the alpine meadows of the Himalayas, it is one of the most handsome plants in cultivation, bearing its sweetly scented mauve flowers in a large umbel of thirty or more at the end of a 3–4 ft. (1 m) stem. But as it blooms in April and May, it is best grown under glass in the British Isles and North America in all but those gardens enjoying a warm winter climate as it is so readily spoiled by frost and rain. It produces a large irregular bulb.

NOTOSPARTIUM

The Pink Broom of New Zealand. *N. carmichaeliae* is a half-hardy deciduous shrub growing 4 ft. (1 m) tall with graceful arching stems which in July and August are clothed in small lilac-pink pea-shaped flowers. Plant in April from pots and into a sandy soil. It requires an open sunny situation and requires no pruning. Propagate from seeds sown as soon as ripe in small pots.

NYMPHAEA

A genus of fifty species present in British Isles; south-east Asia and the tropical water of the Amazon. The Common White Water-lily, *N. alba,* which floats upon the surface is the only species native to the British Isles. Nymphaea should be planted in spring, in small heaps of soil at the bottom of a pond or in sunken containers. For most lilies, the water should be 15–18 in. (38–45 cm) deep. A greater depth of water will check growth and retard flowering. The larger the leaves and flowers, the greater should be the depth of water. Water lilies require an open, sunny situation.

SPECIES AND VARIETIES

Nymphaea alba. Native of the British Isles, it bears a flower about 6 in. (15 cm) across with yellow stamens and which open only in sunshine. They bloom during July and August and have a soft, delicate scent. The variety 'Pygmaea Alba' bears tiny flowers and may be planted in 6 in. (15 cm) of water.

N. caroliniana. Native of North America, the large globular flowers being a delicate shade of salmon-pink with a soft, sweet perfume. The form 'Nivea', bears large flowers of purest white and rosea, flowers of bright rose-pink with golden stamens. Both are pleasantly scented.

N. froebelii. One of the best water-lilies for tub culture on a veranda or roof garden. It requires only 12 in. (30 cm) of water and bears its scented wine-red flowers with freedom.

N. gladstoniana. One of the most beautiful of all hybrid varieties, the huge white paeony-line blooms, with touches of green on the sepals possess outstanding perfume.

N. odorata. This attractive North American species introduced in 1786, resembles *N. alba* in almost all respects but its more pronounced scent. It comes into bloom before the end of June and from the fleshy rootstocks arise heart-shaped leaves 9 in. (23 cm) across and deliciously-scented white flowers, tinted with rose. The flowers open early morning to some 6 in. (15 cm) across when they are most fragrant for by afternoon they have closed again.

The varieties of *N. odorata* are numerous but none has a more powerful perfume than 'Grandiflora', nor is there one with greater beauty, its enormous flowers of golden yellow having the same sweet perfume as Paeony, Duchesse de Nemours and much of its beauty. The form minor, found in shallow swamps in many North American states, bears small star-like flowers with golden anthers and they are scented.

The variety 'Rose Arey' is one of the most beautiful of all the water-lilies, the pointed petals giving the flowers a star-like appearance whilst they are of deep rose-pink with a pronounced sweet perfume. Equally fine is 'Helen Fowler', with small dark green leaves and bearing flowers of richest pink with golden anthers. It has a sweet perfume.

N. stellata 'Coerulea'. The Blue Lotus of the Nile and requiring the protection of a greenhouse in northern latitudes. It more nearly resembles the nelubium in that its flowers, of soft sky blue stand several inches above the water. They have a hyacinth-like scent.

N. virginalis. It is early into bloom (April) and continues until mid-October, its pure-white flowers having broad petals and exquisite perfume.

N. zanzibariensis. The Royal Purple Water-lily of Africa which can be grown in a small tub indoors. Its dark green leaves are blotched with brown whilst its star-like flowers of richest blue have conspicuous golden anthers and a powerful perfume. The form 'Jupiter', deepest purple-blue, is equally fine and strongly scented.

HYBRIDS OF N. LAYDEKERI

N. fulva. The leaves are blotched with brown and crimson whilst the creamy yellow flowers are lined with red against which the golden stamens shine like torches. The blooms have a soft, delicate perfume.

N. lilacea. Like all the members of this group, the flowers are small and dainty, measuring 3 in. (7.5 cm) across and in this case they are of a lovely shade of lilac-rose with a delicate tea rose scent.

HYBRIDS OF N. MARLIACEA

N. albida. It is the largest of all water-lilies, the large globular flowers with their wax-like petals being as white as driven snow and with an exotic heavy fragrance. The flowers are enhanced by the bottle-green leaves.

N. carnea. It is a most free flowering form, its flesh-tinted flowers having a vanilla-like perfume.

N. chromatella. It has enormous flowers of clearest primrose yellow and which have a rich sweet almond-like perfume. The blooms are borne from May until the end of autumn. The leaves when young, are attractively mottled with brown.

OAK, *see* Quercus

OBEDIENT PLANT, *see* Physostegia

OCYMUM

The Sweet Basil, *Ocymum basilum*, is a perennial plant which is only half hardy and so is best treated as a half hardy annual. The seed is sown in gentle heat or under cloches early in spring. The leaves possess a distinct clove flavour and used in salads with discretion will impart a pleasant taste. In a good summer the plant, which grows 2 ft. (60 cm) tall, will form plenty of leaf which may be dried and used in stuffing and in pot-pourris.

Bush Basil, *O. minimum* grows only 6 in. (15 cm) tall and is given the same culture. Where there are no facilities for sowing early, make a sowing outdoors in April, thinning the plants to 12 in. (30 cm) apart. They require a rich soil.

OENOTHERA

BIENNIAL

The Evening Primrose, so called on account of its large pale primrose coloured blooms and the delicate perfume they give off on summer evenings. The plants grow to a height of 4 ft. (1 m) and remain colourful through the latter part of summer, the blooms soon dying but being replaced by new flowers almost daily. They should be planted towards the back of a border and about a shrubbery. The seed is sown in early

July and if sown thinly, transplanting is not necessary. The plants are set out into their permanent quarters in October.

PERENNIAL

Present in their natural haunts across the United States of America, from Virginia to California, there are no more colourful, nor more free flowering plants for the front of a border. Coming into bloom before the end of June, they bear a profusion of clear yellow or white cup-shaped flowers and thrive in almost all soils, especially where sandy. They must, however, be given an open sunny position in which to open their blooms. Plant 18 in. (45 cm) apart in November. Propagate by division of the roots.

SPECIES AND VARIETIES OF PERENNIAL OENOTHERA

FIREWORKS. So called because of its interesting red buds, which combining with the trusses of yellow flowers produce a most colourful effect. In bloom from June until the end of August.

O. fruticosa 'Youngii'. The best known species, bearing a profusion of pale yellow flowers on 18 in. (45 cm) stems from June until October.

O. glaber. A colourful addition to its rich golden cup-like flowers is its bronzy foliage. 2 ft. (60 cm).

O. speciosa. This is one of the loveliest plants of the border though so little planted. Its first blooms appear in a sheltered position before the end of May and are borne right through summer and autumn.

YELLOW RIVER. Similar in every way, except that the blooms are of a richer gold and are larger.

OLEANDER, *see* Nerium

OLEARIA

Extremely hard, several of the New Zealand Daisies make a satisfactory hedge. Being highly resistant to salt winds, the plants make a valuable windbreak for a coastal garden. They bear white daisy-like flowers. Particularly for a windbreak, plant *Olearia oleifolia*, with its glossy olive-like leaves; *O. ilicifolia*, which has foliage like that of holly; *O. forsteri*, like pittosporum; and *O. haastii*, with its oval, mealy-green foliage.

Plant in April, 2 ft. (60 cm) apart, and little clipping should be necessary, though if a hedge of these plants is found to be neglected, having made an excess of thick wood, it will be safe to cut hard back with the saw and pruners. The wood will quickly break into new growth again. This cutting back is best done in April.

OLIVERANTHUS

O. elegans makes a bushy little plant about 6 in. (15 cm) tall and grows to a similar width. In habit, it is like the echeveria, forming loose rosettes of pointed succulent leaves of blue-green. From the centre of the rosettes arises a stem 6 in. (15 cm) tall at the end of which are borne, in pairs, drooping bells like small chinese lanterns which are coloured orange on the outside and yellow inside. They come into bloom in June and continue until the end of September, a well-grown plant having as many as twenty flowers open at the same time. The plants may be grown on in a warm greenhouse and taken indoors when in bloom, where in the window of a warm room they will bloom through summer with the minimum of attention.

Whilst the plants are readily propagated by means of cuttings, they also grow easily from seed which should be sown in a propagator during February. As the plants prefer a neutral or slightly acid compost, the John Innes Compost should not be used. Instead, make up a compost of 2 parts sterilised loam and 1 part each of leaf mould (which should also be sterilised), silver sand and peat.

Use a small pan which should have been scrubbed quite clean and which should be 'crocked' before the compost is added. As the seed is small, it will be advisable to mix with it a little dry silver sand before sowing. As with most Gesneriacea a very light covering of silver sand will help to prevent the seedlings from damping off, but on no account must the covering be overdone. If the compost has been made moist before the seed is sown, it will require only a very light syringing afterwards. The pans should then be placed in the propagator.

Seed may also be sown in April, relying on the natural warmth of the sun for germination, and here the seed pans should be covered with glass to help with germination. By sowing later in the year, however, the plants will not be as large and as free-flowering when they come into bloom the following summer as where they have been sown early in the year and have the whole of the spring and summer to develop.

Whether the seed is to be germinated in a propagator or not it will require a temperature of between 60°–65°F. (16–18°C.) if germination is not to be delayed, and whilst a humid atmosphere will be necessary for rapid germination, moisture should not be allowed to drip on to the compost. Condensation should be wiped from the glass as it forms. Should the surface of the compost dry out, it will be advisable to water by immersing the base of the pan in water until moisture can be seen to have reached the surface.

If the seed has been kept comfortably moist and the temperature constant at about 60°F. (16°C.), the seed will germinate in from three to four weeks; as soon as the seedlings begin to make growth the glass may be removed and the temperature reduced to 55°F. (12°C.). After another month, the seedlings will be ready for transplanting. This should be done when large enough to handle.

Very small pots should be used, those known as 'thumb' size, and in these pots the plants will remain from the time of the first transplanting until the following spring. The John Innes No. 1 Potting Compost may be used, but omitting the chalk or limestone, so that if anything the compost will be slightly acid. The sand to be used should be coarse and gritty.

After potting, the greenhouse should be kept closed for several days until the plants are firmly established, or, if they have been moved to a frame, the lights should be kept in position for a week or so during which time the plants should be watered sparingly.

As soon as the plants are re-established in the pots, they should be given the maximum amount of fresh air and watered freely until October when watering should be reduced and by closing the ventilators during early afternoon, sufficient heat will be stored to prevent the night temperature from falling too low. From the beginning of November, however, artificial heat should be provided so that throughout winter, the temperature will not fall below 50°F. (10°C.), otherwise the plants may drop their leaves. During this time only sufficient moisture should be given to keep the plants alive whilst careless splashing of the foliage should be avoided.

Early in spring the plants will start to bear lateral shoots, and it will then be time to pot on into larger pots in which they will bloom. The John Innes No. 2 Potting Compost should be used, again omitting the lime and substituting fine shingle for sand. Nitrogenous manures must not be used for they will cause the plants to make too much lush growth at the expense of the bloom.

The plants must not be 'stopped' but should be grown on as cool as possible, watering freely and giving ample ventilation. They will be in bloom by mid-June and after flowering should be given the same treatment each winter. The plants may remain in the flowering pots for at least two years before repotting.

OMPHALODES

ANNUAL

O. linifolia is an attractive plant for the front of a border, its tiny cream coloured flowers and grey foliage providing a striking contrast to the more richly coloured subjects. Sow at the end of March and thin to 6 in. (15 cm) apart.

PERENNIAL

It likes partial shade and a cool moist soil. It is a useful plant for the front of a border in that it blooms during April and May. Planting is best done in June, immediately after flowering. Propagation is by division of the roots.

SPECIES OF PERENNIAL OMPHALODES

O. cappadocica. It bears its sprays of dark blue forget-me-not flowers on 12 in. (30 cm) stems during May, its deep green foliage adding to its attractiveness.

O. verna. In a sheltered border it comes into bloom before the end of March, bearing tiny bright blue flowers which have a striking white eye.

ONOSMA

Biennial or perennial plants with grey, hairy foliage and bearing long tubular flowers with the delicate fragrance of almonds, the onosmas are amongst the loveliest plants of the alpine garden, in bloom during May and June. Perhaps the best is *O. echioides,* a biennial, known as the Golden Drop plant, the pale yellow tubes hanging in bunches and remaining quite indifferent to wind and rain. It is readily raised from seed sown in July, the plants being set out early the following spring. Similar, but an evergreen perennial is *O. tauricum* which bears its flowers on 8 in. (20 cm) stems early in summer. *Onosma cassia* bears flowers of similar shape but of purest white and on hairy stems, whilst *O. siehiana* bears pink and crimson blooms through June and July.

OPUNTIA

Known as the Prickly Pear, the opuntias form flat oval-shaped pads which are stems, not leaves, and which grow one above another. They are difficult to handle, having barbed hairs which grow in tufts and which readily come away, penetrating the hands, but they have few spikes or spines, whilst their flowers are extremely beautiful. One of the best for flowering in Britain is the yellow-flowering *O. salmiana*, the others being rather shy in flowering. An additional interest is that an edible fruit follows the flowers and, if left on the plant, the fruit will later bear bloom and so on indefinitely.

ORCHID, *see* Orchis

ORCHID CACTUS, *see* Epiphyllum

ORCHIS

HARDY

There are a number of orchids of easy culture and which are sufficiently hardy to be planted outdoors in the British Isles whilst several may be grown indoors in pots in ordinary room temperatures. Outdoors, they should be planted where they may be sheltered from cold northerly winds but where they may be protected from the direct rays of the mid-summer sunshine. Semi-woodland conditions and in the company of ferns is ideal for hardy orchids and the soil should be made moisture-holding by the addition of peat or leaf mould. They should be planted in groups of three or four and left undisturbed for many years. Plant early March, whilst the buds are still dormant and cover the roots with only 1 in. (2.5 cm) of soil and so that the buds are just above soil level. They must be kept moist whilst becoming established and will appreciate an occasional mulch with leaf mould and finely screened loam. Several may be used about a shaded rock garden and will benefit from the cool root run provided by the stones.

HARDY ORCHIDS FOR OUTSIDE PLANTING

Bletilla striata. Native of Yunnan, it is of easy culture but requires protection from hard frost and cold winds. It requires a moist soil and a semi-shaded situation and does well on the rock garden, protected from winds by the stones. It has handsome ribbed leaves and sends up its flower stems to a height of 12 in. (30 cm). It bears several rosy-red flowers to each stem, the blooms having star-like outer segments and a long pentstemon-like tube shaded white at the base. Plant 1 in. (2.5 cm) deep and 6 in. (15 cm) apart.

It does well in pans in a cool greenhouse or frame when it will bloom in July, or it may be flowered outdoors, placing the pans in the open in May. It requires large amounts of moisture during the summer.

Calapogon pulchellus. A tuberous rooted orchid, native of North America, with grass-like leaves and bearing its flowers in loose racemes during July and August on 18 in. (45 cm) stems. The flowers are purple with a tuft of yellow hairs on the 'crest', like *Calypso borealis.* Like that plant, it requires a moist soil and partial shade.

Calypso borealis. Native of North America, it is a handsome orchid, requiring a damp almost bog-like situation and a lime-free soil. It has a stem which forms a bulbous base and bears a single heart-shaped leaf. The flowers are borne solitary and have sepals of rosy-purple and a blush-white lip, blotched brown and with a tuft of yellow hairs at the mouth. It is sweetly scented.

Cypripedium calceolus. The Lady's Slipper Orchid, a British species, bearing its flowers early in summer, usually two to a stem and at a height of about 12 in. (30 cm). The leaves are ribbed and downy whilst the flowers have the lips chocolate coloured and the pouch which inflates to form the slipper, is bright yellow. It is the only hardy orchid to appreciate some lime in its diet.

Cypripedium montanum. Native of north-west America, it grows 12 in. (30 cm) tall and to each stem bears two to three large flowers with brownish-purple sepals and a white slipper. It is deliciously lemon-scented.

Cypripedium pubescens. A hardy North American species which sends up its leafy stem to a height of 20 in. (50 cm) and blooms in June. It is the easiest and most free flowering of all garden orchids once established. The lips and petals are brownish-yellow whilst the pouch is clear yellow.

Cypripedium spectabile. The Moccasin Flower of west-North America. It bears its flowers in May at the end of leafy stems 2 ft. (60 cm) tall. The leaves are long and pointed and palest green whilst each stem bears three to four large flowers with blush-white petals and a pouch of rose-pink.

Epipactis gigantea. A hardy orchid of the moist woodlands of north-west America which spreads by underground stolons. It blooms in July and August, bearing six to twelve purple mottled flowers on a 12 in. (30 cm) stem.

Orchis foliosa. Native of the higher mountainous regions of Madeira, it will survive an average winter in British gardens if given the protection of decayed leaves or is planted amongst evergreens used as ground cover. It is a handsome orchid to grow under glass, bearing in June large spikes of lilac-mauve flowers on 2 ft. (60 cm) stems. It requires a cool, moist root run.

Orchis spectabilis. A charming little plant, to be found from New Brunswick to South Carolina and bearing in May and June on a 6 in. (15 cm) stem, a dense spike of purple and white flowers which arises from twin glossy green leaves. It requires a cool, moist soil but one which is well drained in winter.

Pleione limprichtii. The Windowsill Orchid, so called because it will bloom to perfection in a north or east window in the home. Use the John Innes Potting Compost fortified with additional peat or leaf mould and plant late February, two to a 4 in. (10 cm) pot and covering only the base of the bulbous root. Place in a temperature of 50°F. (10°C.) and give more water as the plant makes growth.

Outdoors, plant in partial shade and where the flowers are protected from cold winds for it blooms in May, in April indoors. It is one of the hardiest of all orchids and is rarely troubled by frost.

It has swept-back petals of carmine-pink and a long elegant tube of creamy white, flecked with scarlet inside. In form, the flower resembles bletilla and several are borne in each 6 in. (15 cm) stem. It is a pretty plant for the alpine house or rock garden.

ORCHIDS UNDER GLASS

There are a number of orchids which may be cultivated in a greenhouse or garden room with a temperature no higher than 50°F. (10°C.) during winter. Into this category come the cymbidiums, odontoglossums and oncidiums which frequent high altitudes and require no more heat than necessary to exclude frost and keep the plants growing. They also require a buoyant atmosphere. Under such conditions they will require no yearly rest as do the hot-house orchids but will continue to bloom almost continuously.

Early September is the time to plant them. They have large fleshy roots and require ample space to develop. Use a 5 in. (12.5 cm) pot which must be well crocked and a mixture of Osmunda fibre from Japan and turf loam in equal parts and adding a small quantity of sphagnum moss. Hot house orchids, the cattleyas and miltonias favour a compost composed of 3 parts Osmunda and 1 part moss. They require a winter temperature of 60°–65°F. (16°–18°C.) and each plant will produce about six blooms during winter. Do not use too large a pot for orchids. Like many bulbous plants, they bloom better where the roots when once established, are somewhat restricted. Plant so that the crown is just above the rim of the pot with the compost just below the rim to allow for watering. Do not over-water, especially immediately after planting and when the plants are established, give only sufficient that the plants can utilise, especially with odontoglossums. They may be given more water during spring and summer. Rain water is softer and better for orchids and preferably where raised to the temperature of the house. Shading from strong sunlight in spring and summer is essential and this is done by fixing hessian canvas to the inside roof rafters. This may also be used in winter to keep out frost. At no time give the plants manure though they will benefit from an occasional application of dilute manure water. Cymbidiums especially, are heavy feeders Give plenty of fresh air but guard against draughts.

SPECIES OF EASY CULTURE

Cymbidium giganteum. It bears elegant sprays of yellow and brick red blooms and is most free flowering during the winter months.

Cymbidium lowianum. At its best in spring, bearing sprays of maroon and cream coloured flowers.

Odontoglossum nobile. Spring flowering, its large handsome blooms being of red, white and cream.

Odontoglossum pulchellum. It bears heavily scented flowers of purest white and is free flowering in winter.

OREGON GRAPE, *see* Mahonia

ORIENTAL BELLFLOWER, *see* Ostrowskya

ORNAMENTAL NETTLE, *see* Coleus

ORNAMENTAL QUINCE, *see* Chaenomeles

ORNITHOGALUM

Stars of Bethlehem they are called, for it is in Palestine that the white-flowered *O. umbellatum* is to be found. Flowering from April until the end of June, they will hold their umbels of star-like flowers (see plate 41), in almost perfect condition for nearly three months. Not only is their flowering range considerable but so is their variation of height, so that some are suitable for the rockery and for massing in short grass; others do well in a border or shrubbery; whilst those that grow taller are useful for planting at the back of a border and in any sunny corner protected from the wind. Also in their favour is that several species will bloom in partial shade. *O. nutans* and *O. umbellatum* are suitable for planting in grass or plant them in the shrubbery or between cracks in crazy paving. They will increase rapidly from self-sown seed or by the formation of bulblets.

Though they thrive in any ordinary garden soil they are lovers of woodland conditions and grow best if some peat or leaf mould is worked into the soil at planting-time, while they also appreciate a mulch in October. They may be lifted, divided and replanted in October possibly once in four years.

The bulbs should be obtained in as fresh a condition as possible and are best planted 4 in. (10 cm) deep, spacing them 6 in. (15 cm) apart.

SPECIES

Ornithogalum balansae. Growing to a height of only 8 in. (20 cm) and bearing its snow-white blooms striped grey-green on the outside early in April, it is suitable for a rockery or for the alpine house where it will bloom late in March.

O. narbonense. Similar in form and habit to *O. balansae,* this species bears in May a sturdy spike of milky-white blooms, striped green on the outside. It looks most attractive towards the front of a shrubbery. The broad leaves are shorter than the scape.

O. nutans. A woodlander, it is at its best when naturalised and so freely does it seed, that within two years a few bulbs will have become a mass of milky-white blooms, flushed with green on the outside. The umbels are held on 16 in. (40 cm) stems and are in bloom throughout May.

O. pyramidale. It bears its tall spires of rich creamy white flowers during May and June on 2 ft. (60 cm) stems and is one of the most outstanding plants for the water-side garden, flowering at the same time as many of the candelabra primulas. It has bright green lance-shaped leaves.

O. thyrsoides. The Chincherinchee, native of the hedgerows of Cape Province, an onomatopoeic word to describe the sound of wind as it blows through the papery textured flowers. It bears its white flowers with brown centres in large spikes on 2 ft. (60 cm) stems and they remain fresh in water for many weeks.

O. umbellatum. The Star of Bethlehem, it is an excellent species for orchard planting, the dainty white flowers being borne on 6 in. (15 cm) stems during May. They open at noon and close at sundown.

OSMANTHUS

An evergreen which makes a neat, compact hedge 3–4 ft. (about 1 m) tall. It is hardy, and withstands clipping though it is slow growing and this is rarely necessary. *O. delavayi* has deep green, box-like foliage and bears fragrant white flowers in May, making a handsome hedge with its arching sprays, and black fruit late in summer. Another species, *O. forrestii*, also a native of China, is of robust, erect habit, bearing serrated leaves and creamy-white flowers. It is perhaps not as hardy as *O. delavayi*.

Planting should be done in late autumn, allowing 2 ft. (60 cm) between the plants, and any pruning should be done in midsummer after flowering.

OSMAREA

With its neat box-like foliage, crowded with fragrant white flowers during early summer, *Osmarea burkwoodii* is an excellent plant for a hedge for towns and coastal gardens and in all soils. It will make a neat hedge 5–6 ft. (2 m) tall and requires little pruning other than cutting out any dead wood in autumn. Plant 2 ft. (60 cm) apart and allow it to grow away for eighteen months before clipping.

OSTEOMELES

The Chinese Hawthorn, *O. subrotunda* is evergreen, its fern-like leaves being covered in silky hairs whilst its clusters of white hawthorn-like flowers appear in June. North of the Thames, it needs the protection of a warm wall or sheltered border. Plant in April, into a well-drained soil containing plenty of peat or leaf mould. Do no pruning apart from thinning out dead wood. Propagate by layering in autumn or by cuttings of the half-ripened wood removed with a 'heel' and inserted in sandy compost under glass.

OSTROWSKYA

Its fleshy, tuberous roots, like giant parsnips reach down into the soil to a depth of 2 ft. (60 cm) or more and for this reason this hardy border plant must be given a deeply-worked soil. It also likes a soil containing plenty of sand or grit to drain away winter moisture, and a position of full sun. *O. magnifica,* the Oriental bellflower is a most handsome plant, producing its leaves in whorls and its huge bell-shaped flowers of silvery-lilac during early summer on 3–4 ft. (about 1 m) stems. Plant towards the centre of the border, 2 ft. (60 cm) apart and in groups of three. As the plants make growth early in the season it is advisable to plant close to other plants which make early growth so as to afford some protection. The evergreen *Iris germanica* is excellent for this purpose.

OSWEGO TEA, *see* Monarda

OXALIS

Several species of this tuberous-rooted genus have immense beauty. They have a flowering season extending from May until mid-November and are of the utmost value on the rockery and in a fairly shady position in the shrubbery or under woodland trees. They enjoy best a cool, leafy soil, moist but by no means water-logged. If sufficient leaf mould is not present, work in some peat or decayed manure. September is the time to plant, the tubers being set just beneath the surface. The dainty November-flowering *Oxalis variabilis,* a plant for the rockery, should be planted in April. All the species enjoy a liberal mulching with peat or leaf mould early in autumn.

SPECIES

Oxalis adenophylla. So inexpensive and yet so rarely seen. From South America, and bearing its rosy pink flowers on 2 in. (5 cm) stalks throughout early summer. This species appreciates some limestone in the soil (see plate 42).

O. bowieana. Like a rock-rose in form and bearing its bright rose flowers with their attractive golden centre throughout June and July. Found in its natural state on Table Mountain, in South Africa, and requires a warm, sunny position and a soil containing plenty of grit.

O. braziliensis. From Brazil, and so requires a warm pocket in the rockery where it will receive full sunshine and some slight protection. Bears most arresting little blooms of deep wine red. Grow it with *Ramonda nathaliae*, with its bright lavender-blue flowers. The combination is delightful throughout June.

O. chrysantha. From Chile, and an amazing plant, setting no seed but increasing by its tuberous roots like alstroemeria. Not entirely hardy so should be planted near a path or stone where its roots can run beneath and so obtain protection in this way. The dwarf blooms are of pure golden yellow.

O. deppei. From Mexico, and the exception to the rule of shallow planting. The tubers should be planted 6 in. (15 cm) deep in beds of leaf mould. It bears flowers of an unusual brick red colour from June until late in August. The foliage is of rich shades of green and purple.

O. enneaphylla. Native of the Falkland Isles, the large waxy white cup-shaped blooms have a distinctive green centre. In bloom throughout the summer.

O. lobata. Losing its leaves in June when it should be divided, it produces its bloom in the manner of a colchicum, in September when it bears masses of golden yellow flowers. It needs the same treatment as *O. bowieana.*

O. variabilis. It produces its clear shell pink flowers during November and planted with *Crocus speciosus* or *C. asturicus*, it will provide a most pleasing combination during the dullest days.

OXYGENATING PLANTS

In every pond or pool it is necessary to use ten oxygenating plants for every 20 sq. ft. of surface area. They are essential to absorb carbon dioxide whilst they give off oxygen which is required for the survival of fish for which they are also a source of food. They also provide shelter from strong sunlight and the cold. They will use up mineral salts from the water and help to keep the water clear and in a suitable condition for fish and other water plants survive. They are planted at the bottom of the pool between April and October, usually in containers which hold six plants (see Water gardens).

SPECIES

Callitriche palustris. The Water Starwort, common in mud and ponds, the leaves borne in floating whorls in which are borne the minute green flowers during summer.

Hottonia palustris. The Water Violet, so called because of its violet-coloured flowers which are borne above the water during early summer. A member of the primula family, it has leaves which are deeply segmented.

Myriophyllum spicatum. The Spiked Water-Milfoil, its pinnate leaves borne in whorls of five, its crimson flowers borne in an erect spike.

Myriophyllum verticillatum. The Whorled Water-Milfoil, a trailing submerged plant with its pinnate leaves borne in whorls all the way up the stem. During July and August it bears greenish flowers in spikes.

Potamogeton crispus. The Curly Pondweed which has toothed lettuce-like leaves, four-angled stems and bears its tiny petalless flowers in stalked spikes.

Potamogeton pectinatus. The Fennel Pondweed with thin leathery leaves and tiny flowers borne in dense spikes.

Ranunculus aquatilis. The Lodewort, which is found floating on fresh water pools, the leaves divided into three lobes. The white cup-shaped flowers, borne on the surface of the water during summer, resemble water lilies.

PAEONIA

The paeony is one of the oldest plants known to civilization, for the bloom was used for simmering, to provide the women of ancient China with a lotion for their complexions. For the modern garden, where saving of labour is the first consideration, the plants require no staking, whilst they may be left undisturbed indefinitely. It is suitable for a partially shaded border where few other plants would flourish. A border made entirely of the paeony will provide a lavish display during the month of June. Possibly its short flowering season, for the earliest varieties come into bloom about June 1st, and the later varieties will have finished blooming by early July, has contributed to its neglect. However with its richly coloured foliage, bronze at first in spring and again in the autumn, the oriental beauty of its bloom, its complete hardiness and freedom from disease, should do much to counteract its short flowering season, in addition to its labour-saving qualities. The plants may be confined to the shrubbery or border; or used in beds to themselves and in the open woodland whilst they will be happy in sunshine or in shade. In full sun, however, the blooms will open more quickly and their great beauty will too soon be over. Some protection from the hot June sun will not only prolong the display but will also enhance the colour of the bloom, the flesh-tinted varieties tending to fade if exposed to strong sunlight.

A border devoted entirely to the paeony may be said to be a flower bed rather than a border, for there will be little variation in the height of the plants. This may be accentuated by raising the centre of the bed when it is being prepared, and especially is this desirable where it is possible to walk along each side. Or where planting against a background of wattle hurdles or a hedge, the ground should be gently sloped from the back towards the front. In this way the full beauty of the flowers will be revealed. If one has the choice, select a northerly aspect for making the bed, for no amount of frost will harm the plants, whilst the flowers will appreciate the shade. As the plants will be permanent, choice of position and preparation of the ground needs careful consideration.

PREPARATION OF THE SOIL

Paeonies are gross feeders and for this reason are rarely seen flowering so profusely in town gardens as they do in the country, where manure is more readily obtainable. If the plants cannot be grown well they should not be planted. Like the phlox, the paeony enjoys a rich, moist soil preferably a heavy loam, provided it is well-drained in winter. For this reason, and remembering that the bed will, if required, remain down for as long as fifty years, it will be beneficial to remove the top soil from badly-drained land to a depth of 2 ft. (60 cm) and at the bottom place a layer of crushed brick. In replacing the soil, as much manure as can be obtained should be incorporated, well-decayed cow manure, old mush-room bed compost, spent hops and decayed garden compost all being suitable. Paeonies are also peat lovers, indeed any material which will help to retain summer moisture and provide plant food should be used. A 2 oz. per sq. yd. dressing with bone meal as the soil is replaced, will also benefit the plants. Where the soil is light and sandy even larger quantities of humus should be added so that the maximum of summer moisture will be retained. In this respect wool shoddy and bark fibre are excellent substitutes for farmyard manure. Little will be gained by planting in the inert, acid soil of a town garden, for under such conditions the plants may not even bloom.

PLANTING

October is the best time to plant, the ground being prepared a month before, to allow it time to settle down. Where the soil is friable, planting may be done at any time, when not frosted, from mid-October until early

March, but not later, for the 'eyes' are very brittle and may be broken off if moved when coming into growth.

Paeonies like a firm soil and should be set out 3 ft. (1 m) apart with the crowns just below soil level. They resent too deep planting; 2 in. (5 cm) is deep enough but they must be made firm whilst it is better to plant young roots with just two 'eyes' than old roots which have lost vigour. The young roots will not bloom the first summer, but will do so the following June, and from then onwards will form stout bushy plants and give a profusion of bloom which should be cut in the bud stage. From then until the time when the blooms commence to drop their petals will be almost three weeks in a cool room and if half an inch of stem is removed from each bloom every five days.

The blooms will travel well if picked in the bud stage just showing colour. Immature buds may not open, let them be just bursting and no more. More mature than this, they will expand too rapidly and may be too advanced by the time they reach the florist.

Where making up a border of paeonies, the choice should be made so as to give as long a flowering period as possible. In warm districts, south of a line drawn across England from Worcester to Cambridge, the early flowering varieties will come into bloom during the last days of May. In the north, the late flowering varieties will continue to bloom until mid-July, so make the most of the early and late flowering varieties.

After flowering, remove all dead bloom but do not cut back the foliage. This is needed to build up a strong crown for the following season, besides providing rich autumn colour. Then cut back to 4 in. (10 cm) of soil level when the border is tidied in early November and give a dressing of manure, peat or leaf mould around the crowns. To provide the plants with as much humus as possible is the secret of success with paeonies, and this will also help them to overcome adverse conditions, such as a hot shallow soil.

DOUBLE PAEONY SINENSIS 2–3 FT. (60–90 CM)
(S)=Scented
Early Flowering

DUCHESSE DE NEMOURS (S). The large incurved bloom is of pure white.

FESTIVA MAXIMA. Taller than most, the white blooms have a red blotch on the petals.

F. KOPPIUS. An old favourite, the large double blooms are of a rich ruby-red colour.

JAMES KELWAY (S). Best described as milky-white, shaded yellow at the centre. Tall growing and long flowering.

JULIE ELIE. The very large incurved flowers are of a beautiful shade of satin-rose. Free-flowering and tall.

KELWAY'S LOVELY (S). The large, handsome blooms are of a bright rose colour, freely produced.

KELWAY'S ROSEMARY (S). So named because of its perfume. The refined bloom is rose-pink with a silver sheen.

LAURA DESSERT. The nearest to a yellow paeony, the sulphur-yellow petals shading to deeper yellow at the centre.

LORD KITCHENER. Brilliant cherry-red of great substance.

MADAME CALOT. Worth planting for its coloured foliage in autumn. The large blooms, freely produced, are of an exquisite creamy-pink colour.

MRS F. DAVIDSON. A most attractive variety, the petals being coral-pink on the outsides, cream on the insides.

PAOLA. The beautifully formed incurved blooms are of a delicate lavender-pink shade. Tall.

PHILLIPE RIVOIRE (S). Magnificent. The crimson blooms have a black sheen and possess the fragrance of an 'Ena Harkness' rose.

ROSE QUEEN. The first to bloom, the large flowers have salmon-pink outer petals, with ivory-white inner petals.

SARAH BERNHARDT (S). One of the best, the apple blossom pink blooms, tipped with silver are produced throughout June.

THERESE. The large white blooms flushed with pink are of handsome form and are freely produced.

Mid-Season Flowering

ADOLPHE ROUSSEAU. The large bright maroon flowers are borne on tall stems.

ALICE HARDING. The white blooms are beautifully tinted with amber.

BARONESS SCHROEDER. The blush-white blooms are globular and are produced with freedom.

CLAIRE DUBOIS. The bloom is satin-pink flushed with silver. Tall and strong growing.

ELIZABETH STONE (S). A new early mid-season variety, the bloom being of a charming combination of rose and lilac.

GERMAINE BIGOT (S). The semi-double blooms are white, shaded with salmon-pink and carry a spicy scent.

HER GRACE (S). The outer petals are pale pink with the centre packed with golden petaloids.

KELWAY'S GLORIOUS (S). Often described as the finest of all paeonies. The huge full blooms being glistening white and strongly scented.

KELWAY'S SUPREME. The blooms are blush-white and produced over a long period. The foliage is crimson in autumn.

MARIE CROUSSE (S). The large blooms are of coral-pink, shaded salmon-pink (see plate 43).

PETER BRAND. A fine new variety, the bloom being deep glistening crimson.

PRIMEVERE (S). The outer petals are white and reflexed, the inner petaloids being rich yellow. Tall growing.

Late Flowering

AUGUST DESSERT. The blooms are not large, but are of a bright salmon-pink, edged with silver.

BRIDAL VEIL. The outer petals are shrimp-pink, the inner petals creamy-white and partially incurved.

DR H. BARNSBY. The large flowers are of deep glistening crimson.

FELIX CROUSSE. Free flowering, the blooms are of carmine-rose, flushed with silver.

KARL ROSENFIELD. Very late to bloom, bearing flowers of deep crimson-red. A superb variety and the last to fade.

MME EDOUARD DARIAT (S). The large globular cream coloured blooms are shaded carmine at the centre.

PURE DELIGHT (S). Late flowering, the small shell-pink blooms are borne in profusion.

SIR WILFRED LAURIER. One of the latest of its colour, the blooms being pure dark crimson.

SOLANGE a distinct variety, the outer petals are buff-amber, the inner petals salmon-orange. Very tall.

SOUVENIR DE LOUIS BIGOT. A fine paeony, the blooms are large and globular and are of a rich salmon-rose colour.

IMPERIAL PAEONIES

This is a new race, known as the cup-and-saucer paeony and bearing a bloom of great beauty. It has a wide outer rim, the centre being tightly packed with hundreds of petaloids in rosette formation. They bloom at the same time as *P. sinensis* and require the same culture. The plants are more expensive, but one or two at least should be in every border for no bloom possesses greater beauty.

ADMIRAL HARWOOD. The outer rim or saucer petals are blush white, the rosette-like petaloids being pale lemon tipped brown. Early to bloom.

BOWL OF BEAUTY. The outer petals are pale pink, the tightly packed petaloids being creamy white. Late.

CRIMSON GLORY. The outer petals are ruby-red, the petaloids being red, tipped with gold. Mid-season.

EMPEROR OF INDIA. Amazingly beautiful, having crimson outer petals and a centre rosette of brilliant gold. Late.

EVENING WORLD. The outer rim petals are peach-pink, the centre petaloids being flesh-pink. Mid-season.

GLOBE OF LIGHT. One of the most beautiful and amazing blooms in cultivation. The rim petals are pure deep rose, the centre petaloids being of pure gold, in bloom over the whole of June and into July.

GREAT SPORT. The rim petals are bright cherry-red, with pale rose petaloids tipped with gold. Late.

KELWAY'S UNIQUE. The outer petals are shell-pink, the cushion of petaloids being creamy-buff, in bloom through June.

KING GEORGE VI. Bright rose-pink throughout, with the centre petaloids tipped with gold. Mid-season.

KING OF ENGLAND. Having crimson rim petals and a centre cushion of ruby petaloids, edged with gold. Early.

QUEEN ALEXANDRA. The outer petals are glistening white, the inner petals being soft lemon-yellow. Early.

WINSTON CHURCHILL. The broad outer petals are rose-pink, with a centre rosette of golden-yellow. Early.

VARIETIES OF SINGLE PAEONIES

CHOCOLATE SOLDIER. The maroon-red flowers are as if varnished. Mid-season.

DAYSPRING. The beautiful pale pink blooms are borne in trusses. Early.

DISPLAY. A new variety having crimson petals, and yellow petaloids, edged with red. Late.

EVA. The blooms are deep salmon-pink, with the foliage taking on rich autumnal tints. Mid-season.

LADY WORSLEY. The trusses of deep rose come into bloom before the end of May. Early.

MISTRAL. The blooms are of deep cherry with golden stamens. Early. A most distinctive variety.

PERFECT DAY. The blooms are bright rose-pink with an attractive silver sheen. Mid-season.

WHITLEYI MAJOR (S). Also known as 'The Bride'. The richly scented pure white blooms with golden anthers are borne in clusters on tall stems. Very early.

PAEONY SPECIES

May Flowering

P. lobata. The single goblet-shaped blooms are of a bright, salmon-scarlet colour. 2 ft. (60 cm).

P. mlokosewitchii. A glorious variety bearing globe-shaped lemon-yellow blooms with coral-pink stamens, and having bronzy foliage. 3 ft. (90 cm).

P. officinalis. This is the bright green-leaved scentless paeony found in cottage gardens, *P. officinalis* 'Rubra Plena' being the Old Double Crimson Paeony. There is also a white form, 'Alba Plena'; and a pink, 'Rosea Plena'.

P. peregrina. The handsome blooms are produced during May and are of crimson-scarlet with golden anthers. 2 ft. (60 cm).

P. willmottiae. The single pure white blooms make a delightful contrast with the plum-coloured foliage.

FLOWERING PERIOD OF PAEONIES
(Late May to Early July)

Early	*Mid-season*	*Late*
'Admiral Harwood'	'Adolphe Rousseau'	'Auguste Dessert'
'Dayspring'	'Alice Harding'	'Bowl of Beauty'
'Duchesse de Nemours'	'Baroness Schroeder'	'Bridal Veil'
'Festiva maxima'	'Chocolate Soldier'	'Display'
'F. Koppius'	'Claire Dubois'	'Dr H. Barnsby'
'James Kelway'	'Crimson Glory'	'Emperor of India'
'Julie Elie'	'Elizabeth Stone'	'Felix Crousse'
'Kelway's Lovely'	'Eva'	'Great Sport'
'Kelway's Rosemary'	'Evening World'	'Karl Rosenfield' (Very late)
'King of England'	'Germaine Bigot'	'Mme Edouard Dariat'
'Lady Worsley'	'Globe of Light'	'Pure Delight' (Very late)
'Laura Dessert'	'Her Grace'	'Silver Flare' (Very late)
'Lord Kitchener'	'Kelway's Glorious'	'Sir Wilfred Laurier'
'Madame Calot'	'Kelway's Supreme'	'Solange' (Very late)
'Mistral'	'Kelways' Unique'	'Souvenir de Louis Bigot'
'Mrs F. Davidson'	'King George VI'	
'Paola'	'Marie Crousse'	
'Phillipe Rivoire'	'Perfect Day'	
'Queen Alexandra'	'Peter Brand'	
'Rose Queen' (Very early)	'Primevere'	
'Sarah Bernhardt'	'Sir Edward Elgar'	
'Therese'		
'Winston Churchill'		
'Whitleyi Major'		

TREE PAEONY

It is *P. moutan* or *P. suffruticosa* of the Imperial Gardens of Peking, and is a hard wooded deciduous shrub. As it comes early into bloom and also produces its new shoots early, it is important to give it protection from cold winds by planting it amongst evergreens. It is also necessary to plant in a position where the early morning spring sunshine is unable to reach them for if the buds become frosted, they will be damaged should the sun shine upon them before the frost has had time to thaw. They should therefore not be planted facing east or south but rather in a westerly position, where the sun does not reach them until later. No amount of cold will harm the plants it is only when coming into new growth early in spring that they may be damaged.

Tree paeonies make considerable growth, eventually growing to 8 ft. (3 m) in height and the same across, so must be allowed room to develop. A fully grown plant will bear up to one hundred blooms, each as large as a dinner plate and so will require a rich soil and a yearly top dressing of decayed manure to maintain its vigour. Plant 8 ft. (2.5 m) apart at any time between mid-October and early March when the ground is in suitable condition. Obtain pot-grown plants, usually two years old which will begin to bloom in two years' time.

Propagation is by layering the lower branches which should be done in autumn; or by grafting either on to seedling rootstock of *P. moutan* or on to a root of *P. lactiflora*. Plant 6 in. (15 cm) below the graft so that the plant will eventually form its own roots.

VARIETIES

BELLE D'ORLEANS. The blooms are fully double and of glowing garnet-red.

BENI-KIRIN. One of the finest, the large double flowers being of purest orange.

COMTESSE DE TUDER. The fully double blooms are of delicate shell-pink tinted with rose-red.

DIAMOND JUBILEE. The semi-double blooms of deep crimson-red grow to an enormous size.

JEANE D'ARC. The large semi-double blooms are of an attractive shade of salmon-pink with darker shading at the centre.

REINE DES VIOLETTES. The fully double blooms grow to great size and are of soft lilac-pink.

PAGODA TREE, *see* Sophora

PANCRATIUM

A genus of fifteen species of bulbous plants, native of the Mediterranean coastline, central Asia, tropical Africa, and India, and closely related to hymenocallis. It takes its name from *pan*, all and *kratys*, powerful, the reference being to its medicinal qualities. From the extended neck of the bulb arises a rosette of strap-like leaves and umbels of sweetly-scented white flowers, remarkable for their central staminal cup which resembles the corona of a narcissus whilst the six narrow outer petals converge into a long tube. Most species require warm greenhouse culture but those which are native of the Mediterranean region are sufficiently hardy to be grown outdoors in those gardens of the British Isles which enjoy a mild winter climate.

Probably the best for outdoor culture is *P. illyricum* which blooms early in summer. It requires an open, sunny situation where the bulbs will ripen thoroughly, without which they will not bloom. Plant in October, 6 in. (15 cm) deep and 12 in. (30 cm) apart in a well-drained soil enriched with some decayed manure and to which some peat or leaf mould has been incorporated. The bulbs should be given a covering of decayed leaves to protect them against frost but as they bloom before the end of May, plant only where late frosts are not to be experienced. To assist with winter drainage it is advisable to plant the bulbs on a layer of sand. The plants will require ample supplies of moisture as they make growth in spring and should not be disturbed for at least four years.

Indoors, plant one bulb to a large pot using a compost made up of fibrous loam, leaf mould, dehydrated cow manure and silver sand in equal parts, leaving the neck of the bulb exposed. Pot firmly, *P. illyricum* in October, *P. maritimum* in March and provide a minimum night temperature of 55°F. (13°C.). Water sparingly during winter but increase supplies in spring and early summer as the sun gathers strength. After flowering, gradually withhold water whilst the foliage dies back and place the pots on their side for the bulbs to ripen. They should be given a top dressing in autumn and slowly started into growth again. Indoors, *P. Illyricum* will bloom towards the end of April and *P. maritimum* during July and August.

The usual method is by offsets which will have formed around the mother bulb after three years and detached when the bulbs are lifted in October. They should be planted into 3 in. (7.5 cm) pots containing a sandy compost and in which they are grown for two years before being planted out or moved to larger pots. They will bloom in two years' time.

SPECIES

Pancratium canariense. Native of the Canary Islands, it has broad strap-like glaucous leaves and bears in autumn, pure white flowers in a large umbel of eight to twelve on a 2 ft. (60 cm) scape. The staminal cup is short, the petals being narrow and reflexed. In the British Isles it should be grown under glass except in the warmest parts.

P. illyricum. It is found amongst the sand dunes of Mediterranean bays of southern Europe, being especially prominent near St Tropez and has a large pear-shaped bulb with a neck almost 12 in. (30 cm) long, encased in brown scales. The lance-shaped leaves are covered in a glaucous 'bloom' whilst the powerfully scented snow-white blooms appear in June in an umbel of eight to twelve on an 18 in. (45 cm) scape.

P. maritimum. It is found growing along the shores of the Mediterranean where it is known as the Sea Lily or Sea Daffodil on account of its staminal cup. It has brown-coated pear-shaped bulbs and lance-shaped glaucous leaves which are evergreen. The large powerfully scented translucent white flowers are borne in umbels of four to eight on a 15 in. (38 cm) stem. They have a long perianth tube whilst the staminal cup is longer than in other species.

P. sickenbergerii. It is found in furrows of the desert lands of Egypt and Arabia and has a bulb 1 in. (2.5 cm) through with a long neck. It blooms in September in an umbel of four to eight, the white flowers being quite small with the perianth segments short. They are borne on a 12 in. (30 cm) scape which is followed by spirally curled leaves.

P. tortuosum. It is present about the coral rocks along the coast of Egypt and North Africa and around the Isle of Elba where it forms a small globose bulb with a cylindrical neck. The leaves are twisted and appear in a tuft of eight to twelve at the same time as the flowers in January. The white flowers are larger than of other species and are borne two to four in an umbel on a 12 in. (30 cm) scape.

P. verecumdum. Native of north India, it has a large bulb 2 in. (5 cm) through with a cylindrical neck and strap like leaves 18 in. (45 cm) long. It blooms in autumn, two to six pure white flowers appearing on a 15 in. (38 cm) scape.

PANSY, *see* Viola

PAPAVER

ANNUAL

P. rhoeas, the Shirley Poppy, is an annual which is sown where it is to bloom. The double flowered strain embracing shades of plum, slate-blue, crimson, scarlet add a richness of colour to a border or shrubbery equalled by few plants. Seed is sown early in April to bloom from July until October, and a sowing may also be made in a sheltered border in early September to bloom mid-May until August. It may be sown in a clay or sandy soil, but for an autumn sowing the ground should be well-drained. Sow thinly and thin the plants to 12 in. (30 cm) apart.

A variety of brilliance is 'Ladybird' which bears its scarlet flowers, with a black spot at the base of each petal, at a height of only 20 in. (50 cm). Slightly taller growing is 'Pink Chiffon' which forms a large ball of twisted petals like a paeony and of a shade of brightest pink.

PERENNIAL

For planting behind the front row border plants there are no plants more colourful for early summer flowering than the oriental poppies. That its period of flowering is short, covering no more than the six weeks of midsummer should not go against its planting, two plants being sufficient to provide a splash of rich colour. Where the soil is well-drained, planting may be done in November; where the soil is heavy, it is better planted in spring, 2 ft. (60 cm) apart. As the hairy flower stems tend to fall over when coming into bloom, they should be given support by inserting twiggy sticks about the plants as they make growth. Propagation is from root cuttings, and to a less extent by division of the thick fleshy roots, but once established the plants should be disturbed as little as possible. There are a number of varieties of *P. orientale*.

VARIETIES OF PERENNIAL PAPAVER

BARR'S WHITE. The huge blooms are of pure white with a striking black centre.

COLONEL BOWLES. The huge scarlet blooms are held on erect 2½ ft. (75 cm) stems.

CRIMSON BROCADE. The large blooms are of a deep crimson colour and come into bloom before any other variety, during the latter days of May.

ENCHANTRESS. A new variety of great beauty, the huge blooms being of a luminous carmine-pink.

INDIAN CHIEF. The deep crimson-brown blooms make a striking contrast to Barr's White.

MRS STOBART. Growing to a height of over 3 ft. (1 m) and flowering later than the others, its soft salmon-pink flowers are quite beautiful.

ORIFLAMME. A new variety, the blooms being fully double and of brillant scarlet-orange.

PETER PAN. The best variety for a small border. The small salmon-scarlet blooms are held on erect 12 in. (30 cm) stems.

SALMON GLOW. Long flowering. The blooms are double and of a glorious salmon colour, flushed with glowing orange.

PARROTIA

It makes a small, shrub-like tree, and is occasionally found in a shrubbery, but used for a hedge it is highly efficient and is one of the most attractive of all shrubs. It is deciduous, but in winter its bare twigs take on a silvery appearance and its beech-like foliage in autumn is striking in shades of rose, pink and orange.

If cut back after planting to 12 in. (30 cm) from ground level, it makes a spreading, bushy plant and will withstand shaping with the shears. Any dead wood should be removed in autumn after the leaves have fallen. *Parrotia persica* enjoys a rich, loamy soil and an open, sunny position when growth, once established, will not be too slow. Plant November, 3 ft. (1 m) apart.

PASSIFLORA

Natives of South America, *Passiflora caerulea* is the best known species of passion flower. It is a twining plant, climbing by tendrils and it may be grown against a sunny wall in favourable districts outdoors or in a greenhouse or garden room. The plant was so named by early Spanish missionaries to South America for the parts of the flower are representative of the Crucifixion. The three stigmas represent the three nails of the Cross; the five anthers, the five wounds. The ten petals represent the ten apostles and the corona, the Crown of Thorns. They bear edible fruits known as Passion fruits.

P. caerulea enjoys rich well-drained soil and when in bloom from June until October, should be kept well watered. It may be trained against a trellis and will quickly reach a height of several feet.

The flowers are amongst the most beautiful of all, with petals of pale blue and purple styles, whilst the filaments of the corona appear as two perfect circles and are purple at the base, white in the centre and pale blue at the tips. The variety 'Constance Elliot' is pure white and is sweetly scented.

PASSION FLOWER, *see* Passiflora

PEACH, *see* Prunus

PEARL BUSH, *see* Exochorda

PEARL EVERLASTING, *see* Anaphalis

PEA TREE, *see* Caragana

PELARGONIUM

For a prolonged display, both in the garden and later as a pot plant for the living-room, no plant is more labour-saving or more colourful. But for all those who grow the large-flowered zonal geraniums, few know of the charms of the fancy leaf varieties which bear a long profusion of dainty flowers, which hover above the plants like butterflies. A small bed planted with the attractive 'Lass O'Gowrie', which has tricolour leaves and bears dainty scarlet flowers on short wiry stems, will command attention from all who see it. With the brilliant 'Paul Crampel' plant the silver-leafed 'Caroline Schmidt', which also bears red flowers, and which will accentuate the beauty of the plants with its silver foliage. The bed may be edged with the almost black-leafed 'Black Vesuvius', to give a display of outstanding beauty. It is only when a bed contains nothing but geraniums that the plants are seen to the best advantage.

Geraniums are propagated from cuttings, taken at any time of the year except in the midwinter months, but so that they will have made substantial plants for bedding out the following May and June, the cuttings are best taken in August, when they will root in about fifteen days, without fear of damping off. The cuttings should be short jointed and 3 in. (7.5 cm) long, the lower leaves being removed so that the shoot has not to support too much foliage whilst at the same time forming roots. It will be found that they will root more quickly and with an almost one hundred per cent regularity, if placed round the sides of size 48 pots (i.e. pots with 48 to a 'cast') filled with a compost of sand and loam. If unsterilized soil is used, the cuttings may be troubled with Black Rot before they are rooted. The cuttings will root in a cold frame at this time of the year, which should be shaded to give protection from the summer sun. Or a greenhouse may be used. Nurserymen, who may have rooted cuttings in this way in midsummer, will find that the tops of the plants, where wintered

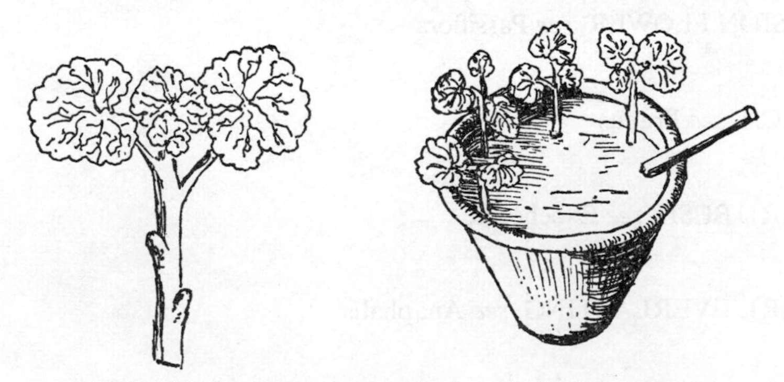

FIG. 9 *Rooting geranium cuttings*

in gentle heat to keep out the frost, may be removed in early April to encourage bushy plants; these shoots may also be rooted in gentle heat. Only enough heat is needed to exclude frost; geraniums are never happy in a warm, stuffy atmosphere.

The cuttings will root more readily if allowed to remain away from the rays of the sun for an hour before inserting in the pots. This will remove excess moisture which geranium cuttings find detrimental to easy rooting. Whilst rooting, a shaded frame or greenhouse should ensure that the cuttings retain all the moisture necessary, for at no stage in its life does the geranium enjoy excessively moist conditions. As soon as rooted, the cuttings should be moved to individual pots containing a mixture of equal parts of well-rotted manure (mushroom-bed compost is ideal), turf loam, and coarse sand, with a sprinkling of bone meal. Though geraniums, being natives of South Africa, enjoy dry conditions, they thrive in a reasonably rich compost which, if provided from the beginning, will enhance the quality of the bloom and size of the plant, though an excess of nitrogenous manure in the flowering beds will cause the plants to make too much leaf at the expense of the bloom.

If the plants are potted in September, they may remain undisturbed through winter, being given only the minimum of moisture, of which the fibruous roots are able to take in but the very smallest quantities during the winter period. By mid-April the plants should be removed to a cold frame to harden off gradually, so that they will be ready for outdoor beds and window-boxes by the end of May. If the plants are not to be used until mid-June – and in a northerly district a late display of tulips may not warrant their lifting before that time – they should be moved to larger pots containing a mixture of fibrous loam and decayed manure, otherwise they may become starved, pot-bound, and drawn. At all times it is desirable to build up a sturdy, short-jointed plant.

When planting out in beds, choose a damp, calm day if possible, so that the plants get away to a good start. If dry conditions prevail, see that they are kept moist until they become established, when they will take care of themselves, thus showing their worth as bedding plants in those dry borders beneath the walls of house or terrace, and their need for so little attention also makes them ideal plants for parks and large gardens, where labour is often required elsewhere. Space the plants 15 in. (38 cm) apart and plant firmly; geraniums are never happy in a loose soil. If sending geraniums away, the nurseryman should, after removing them from the pots, wrap each plant separately in paper and stand them upright in deep boxes. In this way, with the top of the box left open, the plants are put on rail, and should reach their destination free from any serious damages.

The bed to accommodate the plants should receive some preparation, for, though geraniums will flower well in the impoverished, dry soil frequently found in town gardens, they bloom better in a soil enriched with decayed manure. Particularly does the geranium, in its numerous forms, enjoy decayed manure more than any other material.

Lime is also of the greatest importance to geraniums; they may, in fact, be classed as real lime lovers, and town soils especially should be given a liberal dressing with lime whilst the beds are being made up. Equally essential is potash. This will build up a 'hard' plant, one able to withstand excessive rains, and will correct any tendency for the plant to make too much soft leaf. At the same time, potash in the soil will bring out richness of colouring in the blooms, and increase their size. A 2 oz. per sq. yd. dressing is sufficient, supplied in the form of sulphate of potash. In the south-west, where the plants may remain in their beds for long periods, often from April until December, a 2 oz. per sq. yd. dressing of bone meal, a slow-acting fertilizer, should be forked into the beds. This will keep up the flowering qualities of the plants when the food value of the manure will be diminishing.

After planting, geraniums will require little attention other than the removal of the dead blooms, but they will appreciate a weekly application of dilute manure water from the end of July; given if possible during a wet day.

The geranium is such an accommodating plant that it may be planted in almost any position of the garden. Though happiest in full sun, especially in the dry soil of a border beneath a wall, the plant will bloom abundantly in a position of partial shade. One of the loveliest beds that it has been my privilege to see was oval in shape, situated in the corner of a garden, and almost surrounded by tall evergreens, judiciously planted so that the rays of the sun from the south could reach the bed unhindered. The bed, filled with nothing but the glowing pink 'Queen of Denmark', was one of great beauty, and showed that the zonal geranium could be incorporated with more natural surroundings.

After the bedding season is ended, the plants should be lifted before the frosts arrive. In my own garden overlooking the North Sea I have allowed the plants to continue to bear the odd bloom or two until the end of November, for early frosts are rarely experienced close to the sea. Those inland would be advised to lift at the end of October, and in any case the beds may be required for a spring display of tulips, wallflowers, or forget-me-nots, which should be in the ground by the first week of November.

One of the difficulties in using geraniums for bedding is in retaining the plants for flowering another season. Those who possess a heated green-house will be presented with no trouble, and already cuttings may have been removed, and will by late October, have become rooted. Plants from the beds should be lifted and placed in dry, sandy soil, either in pots or boxes. In a frost-proof house the plants will winter without any trouble, and may be potted up to provide more cuttings in April. These may be utilized for bedding for another season, being re-potted in April after the cuttings have been removed. Where there is no heated greenhouse available, a frost-proof room may help the plants to survive the winter. Here the plants should be cleaned of all soil and stripped of most of their leaves. The stems should be tied together in small numbers, and the plants suspended from a beam or large nail. But the room must be dry and frost-proof; or the plants, with most of the foliage removed, may be kept in boxes of dry soil.

Plants may also be lifted and potted, and after shortening any unduly long shoots, the plants may be grown on in a warm room where they will continue to bloom throughout the winter, to be planted out again in May.

ZONAL GERANIUMS FOR BEDDING

AUDREY. It is fully double and one of the loveliest of all geraniums, the flowers being of a most attractive shade of pure phlox-pink.

BANBURY CROSS. A new variety, the red-brick-coloured bloom being of large proportions, and most striking with its pronounced white eye.

BEATRIX LITTLE. A valuable variety owing to its dwarf, compact habit, making it an ideal bedder for a small garden or for a window-box. The colour is intense scarlet.

COLONEL DRABBE. Most striking with its large well-formed bloom, being fully double and of a deep shade of crimson with a white centre.

DORIS MOORE. Of good bedding habit, the large blooms are of a lovely shade of cherry-red.

DOUBLE HENRY JACOBY. An old favourite of dwarf habit, bearing fully double flowers of rich Turkey-red.

DR ROSS HADDON. A plant with a compact habit, and bearing double blooms of brightest crimson.

EDWARD HOCKEY. An old variety of most compact habit, bearing vermilion flowers, veined with a darker shade of vermilion.

GUSTAVE EMICH. Not a new geranium, although it has only recently become popular. The semi-double blooms are of striking orange-scarlet.

JOHN CROSS. A variety of beauty, the salmon-pink flowers possessing a bright silvery sheen.

KING OF DENMARK. The ever popular semi-double, bearing flowers of a delicate shade of shell-pink.

LADY ELLENDON. Vigorous and free-flowering, the rose-coloured blooms retain their colour through sunshine and rain.

LADY WILSON. Of dwarf, compact habit and most free-flowering, the white blooms are strikingly edged and veined with Turkey-red.

LORD CURZON. An arresting variety, the rich purple blooms having a clear white eye.

MAURETANIA. The large blush-white blooms are ringed at the centre with rose-pink.

MRS E. HILL. Its lovely shade of clear salmon-pink has made it a most popular bedding plant of recent years.

NOTTING HILL BEAUTY. Of true bedding habit, the bloom is of an attractive shade of soft geranium-lake.

OYSTERSHELL. The habit is excellent for bedding; the name describes well its unusual colouring.

PAUL CRAMPEL. Possibly more have been propagated than of any other plant for summer bedding. Of excellent habit, its rich colouring has given to art the colour known as 'geranium-red'.

PIERRE COURTOISE. A magnificent variety, the huge blooms are of a rich shade of velvet-crimson.

IVY LEAFED

They are so called from the ivy-like appearance of their leaves and are of trailing habit with long-jointed stems which gives them a dainty, informal appearance. With some, the leaves are zoned but usually they are without. They may be used in window-boxes, tubs and urns where they will trail over the side or they may be trained up a trellis, cane or wire and fastened by raffia or plastic-coated twist ties. In the garden room they will cover a wall or trellis in two to three years. They may also be pinched back and made to form a more compact plant to be used for bedding. 'Galilee' and 'Lilac Gem' are varieties of compact habit. In the more exposed gardens, they should be wintered indoors and planted out mid-May when they will come into bloom almost at once. They are ideal for covering a low wall or bank.

Propagation is simple for the thin wiry stems root more readily than do the thicker stems of the zonal pelargonium. October is the best time to take and root cuttings, keeping them as dry as possible and shading from the direct rays of the sun.

VARIETIES OF IVY-LEAFED PELARGONIUMS

ALICE CROUSSE. It blooms with freedom, bearing masses of large flower heads of bright cerise-pink.

BEATRICE COTTINGTON. Lovely with 'Alice Crousse', it bears large double blooms of soft purple.

GALILEE. An old and still popular ivy of great charm, the flowers being of bright clear pink.

L'ELEGANTE. It is well named for it is a most elegant plant, and is one of the most useful for a living-room, being quite happy in a sunless room and in a temperature little above freezing. It is an ivy-leaf geranium which bears insignificant little mauve-pink flowers and is grown entirely for its attractive foliage, the ivy-shaped leaves being of bottle-green, edged with cream, whilst the veins are also cream coloured (see plate 44). Where the plant is grown with the minimum of moisture, and it will withstand long periods without water, the leaves take on a purplish tint which adds to their beauty.

This geranium may be used in so many ways in the home. It may be grown in a size 48 pot and trailed up a trellis dividing one part of a room from another, for the plants will flourish with the minimum of light, or it may be used for hanging baskets, for its shoots have a lightness to be found in few other plants. Again, it may be used as a specimen plant, the shoots being trained up a cane.

The plants require a soil which is devoid of peat, a compost made up of 3 parts fibrous loam and 1 part decayed manure, and they appreciate firm potting. Water with care, for, like all members of the geranium family, the plants prefer to be grown on the dry side, especially during winter. They may be left for several weeks without moisture.

It is readily propagated by removing the shoots during summer and autumn and rooting them round the side of small pots in a sandy compost. Geraniums love to have contact with the side of a pot, probably because there will be adequate drainage for the cuttings, and they will root much more quickly than if inserted into the centre of a pot or into boxes. An additional aid to rapid rooting is to allow the cuttings to remain on a sheet of paper for two to three hours before inserting into the pots. This will allow excess moisture to evaporate. Give only the minimum of moisture whilst the cuttings are rooting, after which they should be removed to small pots for growing-on.

LEOPARD. A most handsome 'ivy' with large deep green leaves and flowers of soft orchid pink, blotched with red and which more resemble the flowers of an orchid than a pelargonium.

LILAC GEM. An interesting dwarf 'ivy' with pungently scented leaves and bearing small flowers of palest purple.

MADAME CROUSSE. Of tall, vigorous habit, it blooms in profusion right through summer, the flowers being of a lovely shade of soft apple-blossom pink.

MRS W. A. R. CLIFTON. Superb in a white window-box or against the washed walls of a house for its foliage is darkest green, its flowers brilliant red.

SOUVENIR DE CHAS. TURNER. An old favourite bearing large double blooms of deepest pink feathered with purple.

THE DUCHESS. The large handsome blooms are white with clear white edges and striped and feathered Tyrian purple.

ORNAMENTAL LEAFED

These interesting plants are grown almost entirely for the rich colouring of their foliage – for their blooms are mostly insignificant. They are valuable not only for providing a contrast to other bedding plants, but should be used where soil is poor, and the garden shaded. Apart from the removal of any yellowing leaves, they will require little or no attention throughout summer. There are many lovely varieties, all retaining their foliage in the home throughout winter. They make excellent indoor pot plants.

A HAPPY THOUGHT. It is well named for the soft green foliage, each leaf having a pale yellow 'butterfly' in the centre, makes a plant of the utmost charm.

CAROLINE SCHMIDT. Bearing a silver-green leaf, edged with palest gold and the largest bloom of all the ornamental-leaf geraniums, this is a variety which should be more widely grown. The blooms are fully double and of deep blood-red.

LASS O'GOWRIE. A striking variety, the rich cream-coloured leaves being marked with crimson. The plant has a dainty, compact habit, its single brick-red flowers hovering above the foliage.

MARECHAL MCMAHON. A fine plant for grouping in a shrub border or for providing a contrast to a bed of carpeting plants. The almost symmetrical leaves of 'Maréchal McMahon' are of a rich golden colour, heavily zoned with bronze and edged with palest yellow (see plate 45).

MASTERPIECE. Striking in that its almost black leaves are edged with gold.

MRS BURDETT-COUTTS. Rare and expensive, it is for the connoisseur; the leaves are marked with zones of cream, pale green, crimson and bronze.

MRS HENRY COX. One of the best in this section and a real exhibitor's tricolour, the exotic leaves being zoned with gold, silver, and red.

MRS POLLOCK. Of vigorous habit, it propagates readily and is a most arresting plant, the green leaves being zoned yellow, bronze, and orange.

VERONA. An old variety, valuable in that its pure golden-yellow leaves provide a marked contrast to the darker-foliaged varieties.

SCENTED-LEAFED

They require more light than begonia rex and even less moisture but otherwise are similar in their requirements, being happy in ordinary room temperatures and able to survive long periods without moisture. They are essentially indoor plants, which are not only of fascinating appearance but will prove a never-ending source of interest. The plants are easy to manage.

When potting-on the plants, use 2 parts fresh loam, 1 part decayed manure and one part grit or sand. Geraniums do not like peat but are great lovers of decayed manure. Your local nurseryman will make up a box of suitable compost, though only specialist growers will be able to supply a wide variety of plants. Their comprehensive catalogues are well worthy of detailed study before deciding upon what varieties to order.

It should be remembered that the plants may occupy the same pots and the same compost for many years so it will be advisable to go to a little trouble in providing them with a suitable compost at the beginning.

As the plants make growth a number, such as *P. crispum* 'Variegatum' (see plate 46), will require staking for though most are of compact habit some grow tall. At first a centre stake, which may be concealed by the main stem, should be inserted and when the plants are forming numerous side shoots these may also require supporting though this will not be necessary for two or more years, during which time large specimen plants will have been formed. Those such as *P. tomentosum*, which are slower growing and are of more prostrate habit, will not require support.

To maintain the plants in a healthy condition, any leaves should be removed as they die back and any unduly long shoots should be pinched back to maintain the shape of the plant; occasionally stirring the surface soil, and top dressing with a little fresh soil and decayed manure, will maintain the plants in a vigorous condition. But over-watering must be guarded against, for whereas begonias will enjoy an abundance of moisture, geraniums, being natives of South Africa, prefer a dry atmosphere in addition to dry conditions about their roots. Give only sufficient moisture to keep the plants alive and it is surprising how little they require during all but the three midsummer months. In prolonging one's holiday away from home, there will be fear in that the geraniums will be requiring moisture if the plants are placed away from the direct rays of the sun before leaving. As with all plants grown for the fragrance of their foliage, the drier they are grown the more pronounced will be their perfume.

The fact that the scented-leaf geraniums possess a pungent, aromatic fragrance rather than a sweet perfume is greatly in their favour, for one never tires of the fragrance of peppermint, sage, lemon and orange, whereas a sweet, sickly perfume indoors can become oppressive.

Yet in addition to their perfume, the scented-leaf geraniums would be worth growing for the beauty of their foliage. With some varieties it is small and waved, with others it is large and hairy, whilst some have foliage which is attractively edged with gold, all of which contributes in making them the most interesting of all indoor plants.

SPECIES AND VARIETIES OF SCENTED-LEAFED PELARGONIUMS

Pelargonium abrotanifolium. Interesting in that the leaves have the appearance of and carry the aromatic scent of southernwood (Lad's Love) whilst it bears masses of small white flowers.

P. andersonii. Of dwarf, compact habit, it bears dark mauve flowers whilst the rose-scented leaves have a deep purple zone.

P. aspericum. Its leaves are beautifully 'cut', like those of the oak leaf, and appear as if dusted with gold. They are oily to the touch and emit a strong smell of nutmeg.

P. capitatum. This plant is important in commerce, for its leaves, which smell strongly of roses, are now used for the essence, to replace the more expensive attar of roses in perfumery (see plate 47).

P. citrodorum. The attractive pale green foliage is deeply serrated like that of the oak leaf, whilst it carries a strong lemon perfume. It bears mauve flowers.

P. clorinda. It makes a bushy plant of excellent shape, the leaves having a pleasing eucalyptus perfume. The large rose-coloured blooms are feathered with purple and are equal in size to those of the Show pelargoniums.

P. crispum 'Variegatum'. It is a plant of pyramidal shape and covers itself in a mass of small crimped cream-edged leaves which carry the pungent aroma of lemons. The form *P. crispum* 'Minor' bears even smaller leaves which have the aromatic scent of verbena. All members of the *P. crispum* section are excellent for making pot-pourri and scented bags.

P. denticulatum. It has attractive large toothed leaves which carry a strong lemon perfume.

P. endsleigh. It is a plant of almost prostrate habit, the leaves having a dark centre whilst they carry a sharp, but not unattractive, peppery scent.

P. filicifolium. Cannell's in their famous catalogue of 1910, suggest that with its fern-like foliage 'it is well adapted for bouquets and button-holes', whilst it carries the pungent smell of wormwood.

P fragrans. It forms a plant of beautiful shape, its foliage being silvery green, and though the catalogues describe its fragrance as like that of nutmeg, to some it is more reminiscent of featherfew.

P. graveolens 'Lady Plymouth'. One of the finest plants in this section, for not only are its leaves beautifully serrated but are variegated with cream and carry the perfume of the old cabbage rose.

P. 'Joy Lucille'. It is a hybrid having deeply cut foliage of bright green and which have the soft, downy texture and peppermint fragrance of *P. tomentosum*.

P. 'Lady Mary'. It is a hybrid of the *P. crispum* section, having nutmeg-scented leaves and bearing a profusion of purple flowers. Listed in Cannell's catalogue of 1910.

P. 'Little Gem'. It is well named for it makes a dwarf, compact plant with tiny, waved leaves which carry the delicate fragrance of roses.

P. odoratissimum. It is a rare plant, the leaves being small and velvet-like and with a most pronounced apple scent which is quite delicious.

Along the shores of the Mediterranean the leaves are used when making apple jelly.

P. 'Pretty Polly'. Of dwarf habit its flowers are of vivid cerise, whilst its small leaves possess the scent of almonds.

P. 'Prince of Orange'. It makes a neat, compact plant, its tiny leaves being waved and serrated, whilst they emit a powerful orange perfume and should be included in all pot-pourris.

P. 'Purple Unique'. It is one of the famous Uniques, popular during the early eighteenth century and possessing a strong, vigorous habit so that they make large specimen plants. The leaves possess the perfume of absinthe, whilst it bears trusses of large magenta flowers.

P. quercifolium. This is the true oak-leaf geranium of which there are major and minor forms, the leaves being deeply serrated and possessing an oily, pungent scent, whilst they are sticky to the touch.

P. 'Scarlet Unique'. It is also known as Moore's Victory, its leaves having a pungent smell like incense, whilst the vivid scarlet blooms are more showy than any in this section.

P. stenopelatum. This is a true ivy-leaf geranium and the only one having scented foliage. It is a lovely plant for a hanging basket or window-box, for apart from its leaves which carry the aroma of wormwood, it bears bright crimson flowers.

P. tomentosum. Its leaves are unique in that they are large and flat and covered with tiny hairs which gives them a thick, velvet-like appearance. The hairs are to protect the plant from the sun's rays. The leaves were used to make mint scented jelly.

SHOW

Also known as the Regal Pelargonium, it has bright green leaves free of zoning. The flowers are larger than those of the zonals and borne in looser trusses. For this reason they are more liable to damage by adverse weather and are usually grown indoors where they are amongst the most beautiful of plants, in bloom throughout summer. They require a temperature sufficient to exclude winter frost, 45°F. (7°C.) being suitable and as dry an atmosphere as possible. Between October and April, the plants will require the minimum of moisture which should not be allowed to come into contact with the foliage. During summer, ventilate liberally and shade the plants from the direct rays of the sun. At this time the plants will require ample supplies of moisture. To obtain winter bloom, it is advisable to remove the flower buds during the latter weeks of summer to conserve the energies of the plant for winter flowering. After flowering, cut back the plants if they have grown leggy. In spring, re-pot using a compost made up of 2 parts fibrous turf loam; 1 part each decayed manure and coarse sand. To each barrowful of compost, mix in 2 lb. of mortar and 4 oz. of sulphate of potash to give the blooms intensity of colour. Plants in bloom will benefit from an occasional watering with dilute liquid manure during summer.

VARIETIES OF SHOW PELARGONIUMS

ADMIRAL BERESFORD. Most striking with its bright crimson red flowers and contrasting white centre.

CARISBROOKE. The finest pink show, bearing large refined trusses filled with large florets of pale camelia pink, feathered with crimson and purple (see plate 48).

CHERITON. The flowers are of an unusual shade of cherry red with the upper petals marked with maroon.

CONSPICUOUS. Most striking with its flowers of ox-blood red with black markings and ruffled petals.

DR MASTERS. The large flowers are of beetroot-purple with orchid-purple markings on each petal.

GAY NINETIES. A variety of charm, with ruffled petals of ivory-white with purple splashing at the centre.

GRAND OPERA. The very large flowers are of clearest pink, flushed with crimson on the reverse.

GRANDMA FISCHER. A most attractive variety from U.S.A., bearing blooms of bright orange-pink with ruffled petals.

HOUSE AND GARDEN. Named in honour of the American magazine, the Turkey-red flowers shade off to pink in the throat.

MRS W. J. GODFREY. The large refined blooms are of phlox pink shading to white at the centre and with maroon markings on the upper petals.

PRINCE JOHN. Most effective in its colourings of cardinal-red and maroon.

SUE JARRETT. The neat, compact flowers are of a lovely shade of salmon-pink with a maroon spot on the upper petals.

WALTZ TIME. The blooms have attractively ruffled petals and are of pale orchid purple with red feathering.

PELARGONIUMS, FROM SEED

The 'F1 Carefree' hybrids are readily raised from seed and some interesting plants may result. Sown in January in a temperature of 60°F. (16°C.) they will bloom in their first summer and though there will be some variation in colour, this is an inexpensive method of raising geraniums for bedding or greenhouse display.

Sow thinly, using the John Innes Compost and water sparingly. When the seedlings are large enough to handle, transplant to small pots containing a compost as recommended for show pelargoniums and after hardening, plant out early June.

MINIATURE SHOW PELARGONIUMS

Delightful additions to the range of flowering pelargoniums are the miniatures, which have the true show 'blood' and for their charm rely on the size and beauty of their flowers rather than on their foliage, though one or two varieties have scented foliage which adds to their interest. The bloom is like that of the show pelargoniums but in miniature form, whilst it is produced without a break right through summer and early autumn; it is amazing how new blooms replace those that drop their petals in one steady succession. They are also the first of the pelargonium family to come into bloom, early in April.

The plants rarely exceed a height of 12 in. (30 cm) remaining bushy and compact and never becoming straggling as do many of the shows. They are charming plants for a small room, constantly in bloom and requiring almost no attention the whole year round other than an occasional watering in warm weather. Neither gas nor central heating will trouble them in any way, though the blooms will drop their petals if the atmosphere is too dry.

Cuttings may be taken and struck during August and September. They will root more readily than most pelargoniums, for the shoots are short-jointed and quite hard. Three or four should be inserted around small pots and kept comfortably moist, when they will root in three weeks. Nothing could be easier. Potted into individual pots early in October they should be kept almost dry through winter, then given more water to bring them on in March. For a potting compost, use a fresh fibrous loam to which is added a small amount of sharp sand, and some very well-decayed manure which all pelargoniums love.

As soon as the newly rooted cuttings begin to grow in spring it is advisable to pinch out the growing point, when it will be found that new shoots will appear from soil level, making a most compact plant.

Of a dozen or so varieties, though there may be more, one with richly scented foliage is 'Catford Belle', with its large orchid-mauve flowers, blotched petunia-purple, whilst the foliage is pleasantly rose-scented. This plant will grow in a window for several years and remain colourful and fresh.

A race of hybrid miniatures, known as the Langley Smith miniature pelargoniums, with bloom like tiny azaleas and beautifully marked, includes 'Mrs H. G. Smith', white, with attractive mauve upper petals; 'Newcourt', pale mauve, veined rich purple; 'Kerlander', soft shell-pink; 'Langley', salmon-pink, flushed purple; and 'Spring Park', lavender-mauve. There is also a hybrid having scented foliage and beautiful cyclamen-pink flowers called 'Floris'. Others of interest and charm are 'Petit Pierre', of perfect miniature form and bearing a bloom of rich Tyrian rose; and 'Sancho Panza', with its bloom of rich purple.

PELLIONIA

Of the nettle family, they are trailing plants which are at their best in partial shade whilst they are happy in a winter temperature of 50°F. (10°C.). During summer, or where a temperature of around 60°F. (16°C.) is provided, the plants will be copious drinkers.

The better known form is *P. devoreana* which has olive-green leaves marked with bronze, but *P. pulchra* is more striking, with longer leaves of bright green, the colour being accentuated by striking black veins and purple stems.

The plants require a soil which is retentive of moisture, one made up of fibrous loam and leaf mould in about equal quantities and containing a little coarse sand.

PENNYFLOWER, *see* Lunaria

PENTSTEMON

Similar to the bedding strains in form, there are a number of species and varieties suitable for planting in the border where they will prove hardy if given a well-drained, sandy soil. A light peat mulch will certainly bring them through a hard winter, but the soil must be well-drained. The compact habit of the plants, together with their long flowering season should have brought them greater popularity, but the general belief that they are tender seems difficult to overcome. They are best planted in spring, about 18 in. (45 cm) apart and propagation is either by root division or from cuttings inserted in a closed frame in August.

SPECIES AND VARIETIES

GARNET. In a well-drained soil may be considered completely hardy. It forms a thick bush 2 ft. (60 cm) high, and throughout summer is covered in tubular flowers of garnet-red, its pale green foliage providing additional colour.

P. barbatus 'Coccineus'. This is the tallest pentstemon, reaching a height of 2 ft. (60 cm), and bearing spikes of brick-red tubular flowers from June until September and even until November in a sheltered position.

P. campanulatus 'Evelyn'. Growing to a height of 20 in. (50 cm) it provides a charming contrast to the blue flowered species. The slender spikes are almost entirely covered with blooms of pale pink which have an attractive white throat. They flower from mid-June until August.

P. deutus. Growing to a height of almost 2 ft. (60 cm), the sturdy spikes are of clear blue, in bloom during May and June.

P. heterophyllus 'True Blue'. The vivid Cambridge-blue tubular flowers are borne on graceful 18 in. (45 cm) stems from June until August. Another variety, 'Blue Gem', bears spikes of similar height and which are of a brilliant sky-blue colour.

P. schonholzeri. Possibly the hardiest species, its striking crimson-red blooms being borne on 2 ft. (60 cm) stems from June until October.

STAPPLEFORD GEM. A lovely new introduction, its graceful spikes covered in small violet-blue flowers reach a height of nearly 2 ft. (60 cm) and bloom from June until August.

PAEONY, *see* Paeonia

PEPEROMIA

To be found growing along the Amazon valley, these plants require a draught-free room and rather more warmth than most other trailing plants. They have fleshy stems and small heart-shaped or melon-shaped leaves, beautifully variegated.

P. sandersii is one of the most beautiful of indoor plants. It makes a bushy plant of branched and trailing habit. The silver-coloured leaves have green veins whilst the stems are crimson. This is the Pepper Elder of Brazil. Of similar habit is *P. magnoliaefolia* which has bright green fleshy stems and leaves outlined with cream. These species may be used as pot plants about the home in the usual way.

Of more trailing habit is *P. scandens* which, with its yellow and green leaves and red stems, is a most colourful plant. A pleasing contrast is *P. obtusifolia* which is taller growing than most peperomias and has fleshy dark green leaves, edged with purple. Pleasing, too, is *P. hederaefolia* which makes a low-growing plant of bushy trailing habit, its heart-shaped leaves being extremely corrugated and of a lovely pearly-grey colour; whilst *P. caperata* has attractive small dark green leaves.

Where baskets are being used, two or three plants can be grown together for a most pleasing effect and if the trailing shoots become untidy they should be cut well back in spring when they will break anew.

Propagation is by pieces of the shoots removed at a leaf joint and inserted around the side of pots containing a sandy compost. They must not be over-watered or may decay. They will root more quickly in a propagator.

Towards the end of May is the best time to take cuttings and by early July they will be ready for individual pots. Peperomias like a compost made up of 3 parts fibrous loam, 2 parts peat or leaf mould and 1 part sand. The plants should be kept comfortably moist but they must not be given an excess of moisture during winter, when a minimum temperature of 52°F. (11°C.) should be maintained. And guard against draughts.

PEPPER ELDER, *see* Peperomia

PERILLA

An annual, renowned for its rich purple foliage which makes an attractive centre-piece for a circular bed. *P. nankinsensis*, makes a compact plant 18 in. (45 cm) tall. It is raised from seed sown in heat early in February, for it is slow to germinate. Plant out early June, allowing 15 in. (38 cm) between the plants.

PERIWINKLE, *see* Vinca

PERNETTYA

The Prickly Heath, *P. mucronata* is a hardy evergreen shrub from the southernmost tip of South America. It makes a valuable hedge as it forms a dense thicket of small serrated leaves which terminate in a prickly point. Its flowers are white and heath-like and appear in June. They are followed by large round berries which range in colour from white to deep purple-red. The variety 'Alba' bears white fruits; 'Bell's Seedling', crimson; and 'Lilacina', fruits of lilac-pink.

Plant in November or in March, in a well-drained soil containing plenty of peat for it requires an acid soil. Do not prune except to shorten unduly long shoots. Propagate from seed sown in spring or by cuttings removed with a 'heel' and inserted in sandy compost under glass.

PERUVIAN LILY, *see* Alstroemeria

PESTS

BULB MITE

A microscopic pest which mostly attacks tulip and lily bulbs, entering the bulbs where damaged upon lifting. During storage, the pest does its work of destruction, its presence being detected by small red marks on the outer skin of the bulbs. It is difficult to eradicate as it collects on the underside of the scales but as a precaution, dip the bulbs before planting into a solution made up of 1 oz. sulphide of potassium dissolved in 3 gallons of water.

CHAFER

There are three forms of the chafer beetle which will attack roses: (a) the cockchafer; (b) the rose chafer; (c) the ordinary garden chafer. The larva lives in the soil which should be treated with Gammexane (which will also kill wireworm) as routine when the ground is prepared or which should be applied whenever the pest are known to be present.

Whilst the grey-coloured grubs will attack the roots, the beetle-like insects will swarm on to the stems, buds and blooms, so devouring the bud that the blooms will be malformed beyond recognition. The damage to the roots will often be so severe as to bring about the total collapse of the plant and as the eggs are laid in the soil, it is here that the pests should be exterminated (by a soil fumigant). The pests may be removed from the plants by spraying with liquid Derris containing a spreader or with Sybol (based on Gamma-BHC and derris) and repeated at intervals of three weeks.

EARWIG

These reddish-brown creatures, about 1 in. (2.5 cm) long and with their numerous legs will cause similar damage to the succulent stems of chrysanthemums and dahlias as do woodlice whilst they hide by day in the multi-petalled blooms. Spraying the plants with Lindex and the soil around will eliminate the pests or use Lindex dust or Gammexane at fortnightly intervals.

EELWORM

It mostly attacks narcissus bulbs, affected bulbs becoming distorted whilst the leaves become swollen and turn yellow almost as soon as they appear above ground. Upon opening, the bulbs reveal large brown areas, soft to the touch and it is here that the pest exists. The flower stems grow stunted. The pest is most troublesome in heavily manured ground and rank manure must be avoided. The worms (nematodes) are pointed and enter the bulb by piercing a minute hole. Here the females lay their eggs in large numbers and upon hatching, feed upon the living plant cells.

To prevent an attack, narcissus bulbs should be placed in a clean sack and immersed in a tank of water for three hours, heated to and kept at a constant temperature of 110°F. (40°C.). Only narcissus will withstand such a high temperature and for so long a period.

Eelworm will also attack chrysanthemums, irises, potatoes and tomatoes and where the pest has proved troublesome, the ground should be treated with Jeys Fluid at a strength of two tablespoons to a gallon of water before planting.

GREEN FLY

These insects, which cluster about the stems and buds of roses and other flowers, feeding on the juices of the plant, will be either green or brown in colour. The eggs are laid on the plants, spring being the most troublesome time, for during mild winters, the eggs are not destroyed. As routine, roses and other perennial plants should be sprayed in May with Abol–X (1 fl. oz. to 2 gallons of water). This preparation contains menazon which is absorbed into the stems and will give protection for two months. A second application in July will give complete immunity during the season.

MICE

Mice may attack the corms of anemones and crocus, also peas and beans when newly planted and may devour large numbers, usually at night, before being detected. Before planting, it is advisable to dust the seeds (or corms) with carbolic powder or red lead, 1 oz. of which will treat 100 corms of crocus or anemone. It is however, poisonous and must be kept away from children and animals. It should be used just prior to planting, washing the hands afterwards. Or mouse traps may be used along the rows, to be kept baited.

MILLIPEDE

It is almost black in colour and should not be mistaken for the centipede which is yellow with numerous legs and which is a friend of the gardener, living on harmful pests. The millepede moves slowly and when touched, rolls itself into a ball. Like the wireworm it attacks the roots of young plants. It will be exterminated by treating the soil as it is prepared with Aldrin or Gammexane.

NARCISSUS FLY

The pest lays its eggs during May in the necks of narcissus bulbs and may also attack other Amaryllidaceae in similar fashion. The fly has a hairy body and is black with a yellow ring around the base of the head and also at the end of its body. It resembles a small bee but has only two wings. The eggs are white and hatch out to white grubs which work their way down into the bulb, often devouring the whole of the inside. So destructive is the pest that commercial growers of new varieties treat the bulbs each year before planting as for eelworm. Where growing on a small scale, dip the bulbs when dormant in August, into a solution made up of 4 pints liquid BHC; 1 pint spreader (Agrol LN) to 100 gallons of water or its equivalent. Immerse the bulbs for three hours before spreading out to dry and planting.

ROSE LEAF CUTTER BEE

In appearance it resembles the honey bee and the female will employ a cutting action to remove circular portions of a rose leaf for the purpose of lining the cells where they lay their eggs. The female carries the piece of leaf away to line the cell where only one egg is laid and where the larva feeds on honey stored for its use. Disfigurement will be caused to the plant, whilst disease will be likely to enter at the severed parts. No control is known except to destroy the nests which are to be found in decayed wood, such as dead tree trunks.

SLUGS

They are especialy troublesome during damp weather, attacking seedlings and young plants and succulent plants such as bedding begonias and delphiniums, severing them at soil level when rows of young plants (or shoots) may be devoured on a single night. To prevent, water the plants and the soil before and after planting with Slugit. 1 oz. dissolved in 1 gallon of water will be sufficient to treat 10 sq. yd. Repeat the treatment four weeks later. This is safer and more effective than using metaldehyde.

THRIP

They are small steely-black thread-like insects which attack chrysanthemums, gladioli, roses and lilies. They attack all parts of the plant, especially the flower buds causing the blooms to open malformed. It was first discovered on gladioli in Ohio in 1930 since when it has become the most troublesome pest of this plant for they feed upon the corms during storage, covering them with a brown tissue. Their presence may be detected as in gladioli and lilies by the leaves turning yellow, then brown or there may be stunting of the flower spike. To control, dust all plants from early June with Lindex and as the pests reach adult form and begin to lay eggs within twelve days of hatching, repeat the dusting every ten to twelve days. Or spray the plants every tenth day with Sybol. As an additional precaution with gladioli, dip the corms for two hours before planting in a solution of Lysol or Jeyes Fluid and plant whilst still wet.

WHITE FLY

It is a small white moth-like creature which attacks greenhouse plants, especially tomatoes, auriculas, cinerarias and primulas, clustering about the leaves in large numbers and sucking the sap, greatly reducing the vigour of the plant and providing an entry for disease. A white fly parasite, obtainable from horticultural stations will keep the pest under control. Spraying flowering plants with malathion will keep them free or use malathion or Lindex dust. Increasing the humidity of the greenhouse (as for red spider) will help to eradicate the pest.

WIREWORM

The bright orange wire-like pest, the grub of the click beetle, will attack and devour the roots of most plants, often severing them completely whilst the pests will enter the corms of gladioli and bulbous plants, feeding upon them and causing their complete collapse. Treating the corms of gladioli and montbretia whilst in storage will do much to prevent an attack. The ground should be treated with Aldrin dust (or Gammexane) at 1 oz. per sq. yd. before planting, or with naphthalene at double the rate. These preparations will also exterminate leatherjackets and millipedes as well as carrot and onion fly.

WOODLICE

It is a surface pest which hides by day beneath stones and amongst rubbish, also in crevices in tree trunks and in greenhouse walls. At night, the pests attack the succulent stems of pansies and violas; chrysanthemums and dahlias as do slugs, often severing them completely. They are eliminated by watering cuttings and plants with Lindex ($\frac{1}{2}$ fl. oz. to 1 gallon of water). Under glass, they may be trapped with half turnips or potatoes.

PETUNIA

It is one of the more tender bedding annuals requiring to be raised in heat for planting out early in June. With their ruffled trumpets of the most intense colouring, great freedom of flowering and compact habit, no plant will make a more exotic display, nor one as colourful over so long a period.

Seed is sown in heat in early February, or for a later display towards the end of March in a cold frame, the plants being set out 9 in. (22 cm) apart early in June. The plants may also be grown in pots under glass to bloom in early summer.

For outdoor planting the more compact dwarf bedding strains should be used, the more robust varieties being given greenhouse culture. Outstanding for bedding is 'Comanche', the trumpets being of fiery scarlet whilst the 'American Alldouble' strain with its fringed petals is equally valuable for bedding. Others of merit are the bright rose-coloured, 'Rose of Heaven', and 'Silvery Lilac', well described; also 'El Toro', bearing blooms of deep velvety-red.

PHACELIA

Producing its pinnacles of Hairbell-like flowers of rich mid-blue, on 9 in. (22 cm) stems and coming into bloom early in June from an April sowing, *P. campanularia* is one of the most colourful of annuals for edging for a dry, sandy soil. It will bloom within eight weeks of sowing. Thin to 6 in. (15 cm) apart.

PHELLODENDRON

Native of China and Japan, it is of the same family as the rue, the large aromatic leaves releasing a pungent scent when pressed. It grows 20–30 ft. (8–9 m) high, the leaves taking on striking autumnal tints, whilst the greenish female flowers are followed by jet black berries. It grows well in all soils but is happiest in one of chalk or limestone. Prune only lightly, to maintain the shape. Propagation is usually by cuttings removed with a 'heel' and rooted under glass.

P. amuriensis has large compound leaves 12 in. (30 cm) long and is a tree of graceful habit, known as the Amur Cork tree because of its cork-like bark.

PHILADELPHUS

The Mock Orange, so-called because of the orange-flower scent and appearance of its blossom. No shrubs are of easier culture than the hybrid varieties with their long arching sprays of rich orange-scented blossoms. They are completely hardy and will grow well in all soils. They make an abundance of cane-like growth, which should be thinned where there is overcrowding or where there is an accumulation of dead wood.

One of the finest is 'Belle Etoile', the arching branches being crowded with single white flowers which have a distinctive pineapple perfume. Of the doubles, 'Dame Blanche' makes a small, neat bush covered with clusters of tiny double flowers, whilst 'Virginal', of more vigorous habit, bears flowers like tiny double Bourbon roses.

Propagate by cuttings 9 in. (23 cm) long of half-ripened wood which are inserted in trenches of sandy compost in August. Cover with cloches in winter and plant into permanent ground in autumn.

PHILODENDRON

Native of the West Indies and tropical South America, there are several species of this attractive plant which are of climbing habit but all require support either by means of a cane, or they should be allowed to grow up trellis. In their native haunts they climb trees for support. The plants require more light than most indoor climbing plants and they will not tolerate draughts, so give them a position away from the direct line of an open window. They will also require a minimum winter temperature of 52°F. (11°C.).

Their potting compost should be loose and 'open'; 3 parts fibrous loam, 1 part peat or leaf mould and 1 part silver sand; or, as an alternative to the sand, use a handful of crushed brick to a size 48 pot. Do not plant too firmly and never allow the plants to suffer from lack of moisture. During summer they will need copious amounts and in winter they will require more than is usual for indoor plants.

Propagation is by stem-cuttings, removed at a leaf joint during summer and inserted into a sandy compost around the side of a pot. They will root reasonably quickly if enclosed in polythene and placed in a sunny window, but where possible use a propagator. Afterwards the rooted cuttings should be moved to small pots for growing on.

The philodendrons have long pointed glossy leaves and outstanding in its beauty is *P. verrucosum*. Its leaves are of various shades of green with striking bands of maroon.

P. fenzlii is perhaps the most vigorous for it has aerial roots and pulls itself up rapidly. It has interesting glossy three-fingered leaves. *P. lacinatum* also has a twisted stem and aerial roots, whilst its leaves are attractively ribbed. Of more bushy habit is *P. erubescens*, which has very short leaf stalks, the glossy green heart-shaped leaves being crimson on the underside. Similar is *P. scandens* though the leaves are smaller and inclined to be pendant. There is another form with beautifully variegated leaves.

PHLOMIS

A genus of shrubby plants of the lavender family with woolly leaves and which flourish in ordinary well-drained soil.

P. russelliana is an interesting plant for the back of a border. It has large hairy leaves, densely powdered on the underside and sticky on top. It bears its clusters of attractive nettle-like flowers of canary-yellow during July and August. Plant in spring 2 ft. (60 cm) apart and propagate by root division. *P. fruticosa* is the Jerusalem Sage, a distinct shrub growing 3–4 ft. (about 1 m) tall with ovate leaves, woolly on the underside. It bears its yellow flowers in June and July in showy whorls. Propagate by cuttings which root readily in sandy compost in a frame. Plant out in May.

PHLOX

ALPINE

They are charming plants with their semi-trailing habit and freedom of flowering and should be on every rockery. They are happy anywhere but never more brilliant than in full sun. There are two main species *P. douglasii* and its hybrids, 'May Snow', producing sheets of brilliant white, and 'Violet Queen', with its purple-blue flowers. Their habit is almost prostrate and they may be planted about crazy paving. Slightly more erect in growth is *P. subulata* and its numerous hybrids, their brilliantly-coloured starry flowers being held well clear of the mat-like foliage. One of the loveliest is 'Betty', rich salmon-pink; 'Sampson', is rich rose-pink; whilst 'G. F. Wilson' is a lovely clear lavender, and 'Temiscaming' brilliant crimson of compact habit. A recently introduced species, of branching habit in bloom early in spring, *P. stolonifera* 'Blue Ridge', which produces spires of sky blue flowers.

BORDER

Until Captain Symons-Jeune developed this plant in the 1950s, it was only from German breeders that the best varieties of *Phlox decussata* reached this country. Though widely planted during the latter years of the nineteenth century, the short flowering season of the older varieties combined with the susceptibility of the stock to eelworm, tended to bring about the almost total eclipse of the phlox. The 'Symons-Jeune' phlox and the 'Windsor' hybrids have ensured a new popularity for this plant, though, due to dry conditions, so many failures in the border continue. Above all things the phlox must have copious amounts of moisture. The plants will grow best in a heavy soil; where it is light and dry, work in as much decayed humus as possible in addition to providing a midsummer mulch. Plant in November, 2 ft. (60 cm) apart and propagate by root division or from cuttings. The latter method ensures cleaner stocks. The plants come into bloom mid-July and continue until mid-September, scenting the evening with a delicate perfume.

VARIETIES OF BORDER PHLOX

Balmoral. The large trusses, with their refined individual flower pips are of a new shade of orchid-lavender. 3 ft. (90 cm).

BORDER GEM. An old favourite and one of the few to hold its own with the modern varieties. The colour is deep violet-blue. 3–4 ft. (about 1 m).

BRIGADIER. The well formed trusses are of a vivid orange-red. 3–4ft. (about 1 m).

CECIL HANBURY. Best described as glowing salmon-orange with a crimson eye. 2 ft. (60 cm).

CHARMAINE. A variety of arresting beauty, the large elegant trusses bearing 'pips' of bright cherry-red with a striking white centre. 3 ft. (90 cm).

CHERRY TIME. One of the first to come into bloom in July. The bright cherry-pink trusses, flushed with orange are borne on 2 ft. (60 cm) stems.

DRAMATIC. A place should be found for it in every garden, the large refined trusses being of an arresting shade of deep salmon-pink. 3 ft. (90 cm).

DYNAMIC. It grows 3–4 ft. (about 1 m) tall and bears large conical trusses of bright salmon-orange over a long season.

EVENTIDE. The trusses, produced in freedom are of a pastel blue, flushed with lilac. 3 ft. (90 cm).

FAIRY'S PETTICOAT. A glorious new variety, the long flower heads being of a lovely shade of mulberry-pink. 3–4 ft. (about 1m).

FANAL. The most brilliantly coloured phlox, being of an intense flame-red. Ideal for a small border. 30 in. (75 cm).

GAIETY. Growing only 2 ft. (60 cm) tall, it is one of the earliest to bloom, making a large truss and bearing blooms of cherry-red flushed with orange.

GLAMIS. A fine small border variety. The deep pink blooms have an attractive lavender eye and are borne on 30 in. (75 cm) stems.

HOLYROOD. Its huge trusses are of a deep shade of carmine-pink with a striking purple eye and borne on 2 ft. (60 cm) stems.

ICEBERG. Remaining in bloom almost ten weeks, the beautifully formed trusses are of white, flushed with violet. 3–4 ft. (about 1 m).

INDEPENDENCE. One of the brightest plants of the border, bearing large handsome trusses of vivid geranium scarlet with a crimson eye. 3 ft. (90 cm).

RED INDIAN. The refined, compact heads are of a delicious shade of port wine. 3 ft. (90 cm).

SANDRINGHAM. The cyclamen-purple pips build up into conical flower heads of great beauty. 30 in. (75 cm).

SCHNEERAUSCH. The habit is compact, the flowers pure white with a long flowering season. The best white. 3 ft. (90 cm).

SPATROT. Late into bloom, the heads being of a deep crimson-scarlet 2 ft. (60 cm).

SYMPHONY. Tall and late flowering, the individual pips are large and of a colour combination best described as strawberries and cream. 3–4 ft. (about 1 m).

TOITS DE PARIS. It makes a bushy plant 2 ft. (60 cm) tall and bears large trusses of soft lavender-blue flowers early in the season.

VINTAGE WINE. A variety of outstanding merit, the large shapely trusses being of a striking shade of rich claret-red. 3–4 ft. (about 1 m).

WINDSOR. The huge trusses are of a brilliant rosy-carmine colour, with a deeper carmine eye. 3 ft. (90 cm).

PHOTINIA

Native of the Himalayas, there are both deciduous and evergreen species, the deciduous taking on the glorious tints of autumn before the leaves fall. The plant is related to the Hawthorn family, and bears small hawthorn-like flowers during May and June. They grow best near the coast, several reaching a height of 18 ft. (5 m) given the protection of a wall where they receive some shelter. They should not be in a position of full sun. The evergreen species should be planted at the end of April, the deciduous species whenever the weather permits between Novembr and March. They like a soil enriched with humus.

The best of the deciduous species is *P. beauverdiana*, which has long pointed leaves and bears tiny white flowers at the tips of the shoots. These are followed by hawthorn-like clusters of red berries in autumn. The plant will reach a height of about 15 ft. (4.5 m) Of the evergreens, *P. davidsoniae* bears long saw-edged leaves and the same clusters of white flowers, followed by scarlet berries. It is of more vigorous habit, and will reach a height of 20 ft. (6 m). Pot-grown plants should be planted and pruning consists of cutting back and fastening the shoots to the wall as required.

PHYGELIUS

An evergreen, *P. capensis* is native of South Africa and will make vigorous growth if planted against a sunny wall. It is known as the Cape Figwort on account of its large scarlet buds which are shaped like a fig. When open, they form tubular flowers which are coloured vivid yellow on the inside. They appear from early June until mid-October, the plant remaining in bloom almost as long as the fremontia. It likes a sandy sun-drenched soil aided by a sunny wall to ripen its shoots.

If the plant is grown away from the south the roots and base should be given winter protection, for should the plant be cut down by hard frosts it will survive and send forth new shoots when protected at the base. Plant early in spring and prune at the same time, though in this respect all that is necessary is to cut out any decayed wood and shorten back long shoots. The plants are best grown up a trellis or with the shoots fastened to a wall. Propagation is by means of cuttings inserted under a bell cloche in July.

PHYSALIS

Better known as the Cape Gooseberry or Chinese Lantern, the dried blooms with their 'lantern' appearance being popular for Christmas decoration. It is *Physalis franchetti* which bears the blood-red 'lanterns' almost 2 in. (5 cm) in length and almost as wide. Inside is an edible berry. The plant has a creeping rootstock like mint and in the same way the roots are generally sold by the bushel for planting early in spring.

This is a plant which likes to air its head in the sunshine and its roots in a moist soil (again like mint) and liberal quantities of manure should be worked in. It is a difficult plant for a border owing to its creeping roots and is best planted in a bed to itself. Early April or as soon as the soil is friable, trenches 4 in. (10 cm) deep are made to the width of a spade. The roots are spread out so that they do not touch, in the same way as mint is planted. Allow 18 in. (45 cm) between the trenches. As the soil is replaced, rake in a 1 oz. per sq. yd. dressing of sulphate of potash. As soon as the shoots can be readily seen, the hoe should be used between the rows and a peat mulch given in June to retain moisture, and to keep down annual weeds. As the plant is surface rooting, any hoeing or weeding from mid-June should be done with care.

Well grown physalis should bloom on 2 ft. (60 cm) stems, but if the soil is dry and devoid of humus, the stems may only reach a height of 12 in. (30 cm) which will take away much of its value. The stems are cut just above ground level and towards the end of September. Plants may also be raised from seed which should be sown in a frame or under cloches in April. The plants will produce plenty of bloom the following year.

PHYSOSTEGIA

Apart from the michaelmas daisies, this is one of the few autumn flowering plants which bears pink or lilac flowers. The almost square spikes are borne from early September until November where they provide welcome colour to the border. The plants flourish in any ordinary loamy soil, and should be moved immediately after flowering. Propagation is by root division, the rooted offsets being readily pulled apart. Plant 20 in. (50 cm) apart.

VARIETIES

ALBA. Produces its spires of pure white from early August until October on graceful 3 ft. (1 m) stems.

BOUQUET ROSE. It grows 2 ft. (60 cm) tall and blooms from early August, bearing handsome spikes of a lovely shade of soft lavender-pink.

SUMMER SNOW. A variety of beauty growing less than 2 ft. (60 cm) tall and coming into bloom before the end of July, bearing long tapering spikes of ivory white enhanced by the dark green foliage.

SUMMER SPIRE. A new hybrid, bearing its slender spires of rose-pink on 30 in. (76 cm) stems and remaining long in bloom.

VIVID. Its short spikes of rosy-red flowers are borne on 18 in. (45 cm) stems. Plant 15 in. (38 cm) apart.

PHYTOLACCA

A native of the eastern coastline of America, *P. decandra* the Pokeberry is an interesting plant. Its flowers are borne in large round spikes and though first pure white, later take on a rosy-red colour. The individual blooms give way to dense spikes of purple fruits, almost like small blackberries which are retained through autumn, the foliage also taking on rich autumnal tints. The plants prefer an ordinary soil, enriched with peat or leaf mould to retain summer moisture. Plant early in December 3–4 ft. (about 1 m) apart. The plants grow readily from seed.

PICEA

Evergreen coniferous trees with four-sided leaves and pendulous woody cones ripening in their first year. Less hardy than the pines, spruce require a sheltered situation and a deeply worked soil retentive of summer moisture. They will not flourish in a shallow, calcareous soil. Plant in April, allowing each tree ample space to develop for the lower branches will die back if there is too close planting. Propagate by seeds sown outdoors in a prepared bed in which the seedlings remain two years and two more years after transplanting before planting out permanently.

SPECIES AND VARIETIES

Picea abies (Syn: *Abies excelsa*). The Norway Spruce or Christmas tree which in a sheltered situation will add 12 in. (30 cm) to its stature annually. It makes a handsome pyramidal tree, its erect branches clothed with dark green four-angled leaves. The form 'Pyramidata' is almost fastigiate whilst 'Repens' has branches which cover the ground to a distance of

4–5 ft. (about 1.5 m) in width though the tree rarely exceeds 12 in. (30 cm) in height.

P. asperta. The most lime tolerant of the spruces, it is a hardy Chinese species with sharply pointed blue-grey leaves and handsome cones. Of vigorous habit, it will attain a height of 100 ft. (30 m) in the wild.

P. breweriana. Brewer's Weeping Spruce of Oregon with horizontal branches which attain a considerable length and with slender pendent branchlets covered in flat linear leaves. The cylindrical cones are 3 in. (7.5 cm) long.

P. glauca (Syn: *P. alba*). The White or Silver Spruce of North America, it is the hardiest of the genus, suitable for planting in exposed situations. It makes a large spreading tree and requires a soil retentive of moisture. It grows well by the side of a pond or in low lying ground. The leaves are pointed and pale glaucous green whilst the brown cones are cylindrical. The form 'Nana' makes a low spreading plant suitable for the rock garden.

P. nigra (Syn: *P. mariana*). The Black Spruce of North America where it grows equally well in dry rocky terrain as well as in swamp-like ground. It has branches densely arranged and with its dark green foliage, the tree has a sombre appearance and should be planted with conifers of brighter green.

P. omorika. The Serbian Spruce from mountainous regions of south-east Europe with blunt glossy green leaves and small ovoid cones like those of the larch. It makes a tall slender tree and is lime-tolerant. The form 'Pendula' has twisted, drooping branches, the leaves being silver on the underside.

P. pungens. The American Blue Spruce, fast growing and hardy and one of the most handsome of all conifers with sharply pointed leaves and long drooping cones which are of a lovely pale silvery green colour. The tree is enhanced by its orange bark and aromatic scent. Of extreme hardiness, it grows well in swampy ground and also in a dry, calcareous soil.

P. rubra. The Red Spruce of North America, it makes a tall slim tree of pyramidal form and has red bark. Its bright green leaves are incurving, its cones being cylindrical. It requires a moist, deep soil and is not suitable for calcareous soils.

P. sitchensis. The Sitka Spruce of California, tolerant of all soils, its whorled branches clothed with stiff, sharp-pointed leaves which twist into all directions. The cylindrical drooping cones are 3 in. (7.5 cm) long. It is a rapidly growing tree, especially in a moist soil.

PIERIS

Members of this delightful family must have a lime-free soil containing plenty of peat whilst they enjoy partial shade. *Pieris forrestii* is perhaps the best form, but it is not completely hardy and requires shelter away from the south-west. Its foliage opens a brilliant scarlet whilst, during the last weeks of spring, it bears panicles of fragrant white heather-like flowers. *P. taiwanensis* is also striking; its foliage opens as a coral-red in April, which is in contrast to its pure white flowers borne in drooping panicles.

Plant in November and do no pruning apart from the removal of dead flowers in early summer. Propagate by layering in autumn or by cuttings inserted in a mixture of sand and peat under glass.

PILEA

P. cadierei is mentioned in few books on indoor plants yet it is a most charming little plant, which has become known as the Friendship Plant. It grows only 18 in. (45 cm) tall and requires a pot no larger than the size 60. Its compost should contain fibrous loam and peat or leaf mould in equal parts, into which a small quantity of grit has been incorporated.

The leaf markings are unusual looking as if a whitewash brush has been lightly taken across each leaf for the irregular white 'splashes' do not run into the deeply set ribs. The rest of the leaf is bright green and glossy. The plants enjoy a partially shaded position and plenty of moisture during summer. However, only the minimum quantities of water should be given during winter when a minimum temperature of 42°F. (5°C.) will keep the plants healthy.

Propagation is by means of cuttings inserted into a sandy compost around the side of a pot, May being the best time for taking the cuttings.

PINE, *see* Pinus

PINK, *see* Dianthus

PINK BROOM, *see* Notospartium

PINUS

Evergreen trees of the northern temperate regions and whilst the spruces require a moist soil, the pines flourish in dry, open situations. They are tolerant of calcareous soils but not of town garden conditions. Plant them away from smoky parts and in an open, sunny situation. Their needle-like leaves are borne in tufts of two, three or five; the cones solitary or in clusters and ripening their second year. Plant in April, preferably in a dry sandy soil. Propagate from seed sown in April in outdoor beds. Space the seeds 6 in. (15 cm) apart and allow the young plants to occupy the ground for two years.

SPECIES AND VARIETIES

Pinus aristata. The Bristle-cone Pine of Arizona which is slow growing and is known to attain an age of 4,000 years. The leaves are marked with white resinous dots whilst the cones have bristle-like scales. It may be grown on the rock garden.

P. ayacahuite. The Mexican White Pine which should be confined to warm gardens. It makes a tall, slender tree and is one of the most handsome of all conifers with long three-sided leaves and bearing cylindrical cones which droop down.

P. contorta. The Beach Pine which has contorted branches and is used for maritime planting in exposed positions and in dry, sandy soils. It does not take kindly to a chalky soil. It has grass green leaves and ovoid cones.

P. coulteri. The Big-cone Pine of California which makes a large spreading tree, the branches drooping at the tips. It has stiff glaucous leaves and bears yellow cones 12 in. (30 cm) long, the thick scales ending in a spine and often weighing 4 lb. or more.

P. densiflora. The Red Pine of Japan, used in Bonsai. It is distinguished from Scots Pine by its red bark whilst it bears its twisted dark green leaves in pairs.

P. griffithii. The Bhutan Pine which requires a sheltered garden. It forms a broad head but retains its lower branches better than any other pine. It has drooping glaucous leaves and long slender cones covered in wedge-shaped scales.

P. insignis (Syn: *P. radiata*). The Monterey Pine of California, noted for its pale green twisted leaves which are borne in threes and its long orange cones. Fast growing, it does best in milder parts and especially near the sea.

P. jeffreyi. A handsome tree to plant in gravelly soils. It has glaucous leaves 8 in. (20 cm) long, borne in threes and drooping brown cones of similar length. The tree is enhanced by its bright red bark.

P. nigra. The variety 'Austriaca' is the Austrian Pine; 'Calabrica' being the Corsican Pine, both being densely branched with dark foliage and suitable for maritime planting. The former is the best for a calcareous soil and transplants better.

P. parviflora. The White Pine of Japan, used in Bonsai. It makes a small glaucous tree with the short leaves borne in fives. They are silvery on the underside. The spreading tree is closely furnished with horizontal branches

P. pinaster. The Maritime Pine of south-west Europe which grows well in a sandy soil. It has broad dark green leaves and bears yellowish cones in dense clusters.

P. sabiniana. The Nut Pine, prized by North American Indians as food. The greyish leaves are 12 in. (30 cm) long, whilst the chocolate-brown cones with their pointed scales remain on the trees for several years.

P. sylvestris. The Scots Pine, a flat-topped tree with rough red bark and glaucous leaves. The cones are borne one to three. Extremely hardy, it will grow in gravelly soil but prefers a deep loam. The form Aurea has golden foliage in winter.

P. thunbergii. The Japanese Black Pine, useful to plant in maritime parts or in gravelly soil. It has dark leaves but bright yellow branches which are an attraction to the winter garden.

PIPTANTHUS

P. nepalensis, because of its large clover-shaped leaves and vivid golden blooms – like those of the laburnum – this strong-growing semi-evergreen shrub is known as the Evergreen Laburnum. The plant likes a west wall in the southern counties, but should be given a south-westerly position where planted away from the south coast.

During May and June, the blooms appear at the end of the shoots and pruning should be done immediately after flowering, shortening any long shoots and tying back. Early spring is the best time to plant, into a sandy soil containing some leaf mould. Propagation is by cuttings taken with a 'heel' in early August, and inserted under a bell cloche. Young plants should not be allowed to suffer from drought and must be artificially watered until established. When open the blooms have the appearance of the potentilla.

PITCHER PLANT, *see* Sarracenia

PITTOSPORUM

It is suitable only for the warmer parts of Britain and grows well on the sheltered west coast of Scotland; in south Wales; and in the south-west but is always at its best near the sea, where the salt-laden atmosphere protects it from hard frost.

Native of the Canary Islands and New Zealand, one of the hardiest forms is *P. tobira* 'Variegatum', which bears the neat glossy green foliage of the species, with clusters of cream flowers which are orange-scented. That most widely planted for garden use and for cutting is *P. tenuifolium* 'Mayi', which produces long sprigs of pale green glossy leaves which are attractively waved and striking with their shining black stems. Another form of the same species, 'Silver Queen', has soft grey leaves, margined with white.

Towards the end of April is the best time to plant, when the frosts have departed. The plants will then be able to become thoroughly established before the winter. A sandy soil is preferable but it must contain some peat or a little decayed manure. Plant 20 in. (50 cm) apart. After planting, the hedge should be lightly cut back to encourage bushy growth, but should require no other cutting where the foliage is used for market. The sprigs, cut when 18 in. (45 cm) long, are marketed during autumn and the early winter months, being used for mixing with spray chrysanthemums.

Along the south-west coast the plants will attain a height of 10 ft. (3 m) but may be kept to half that height by systematic cutting. Should the weather be severe, or if planted away from the coast, the hedge should be given a mulch with long straw during December to protect the roots from frost damage. Planted only as a hedge for garden use and colour, the hedge may be clipped at both top and sides during May of each year, and it should not need further attention until the following year. In every way a delightful plant for a hedge.

PLANE, *see* Platanus

PLANTAIN LILY, *see* Funkia

PLANT POTS

Clay pots are numbered by the quantity to a 'cast' and nurserymen know them as 'small 60s' rather than for their inside diameter of $2\frac{3}{4}$ in. which is also the depth of a 'small 60' size pot. Clay pots have both advantages and disadvantages. They are heavy to handle whilst they are easily broken by careless handling. Against this, they are porous and allow the roots of a plant to breathe and the moisture of the compost to evaporate so that the compost does not become sour.

No. to cast	Diameter	No. to cast	Diameter
Small 72	$1\frac{3}{4}$ in. (4.5 cm)	Large 32	$6\frac{1}{4}$ in. (16 cm)
Middle 72	2 in. (5 cm)	Large 28	7 in. (18 cm)
Large 72	$2\frac{1}{2}$ in. (6.5 cm)	Large 24	$7\frac{1}{2}$ in. (19 cm)
Small 60	$2\frac{3}{4}$ in. (7 cm)	Large 16	$8\frac{1}{2}$ in. (21.5 cm)
Middle 60	3 in. (7.5 cm)	Small 12	9 in. (23 cm)
Large 60	$3\frac{1}{2}$ in. (9 cm)	Large 12	10 in. (25 cm)
Small 54	4 in. (10 cm)	Large 8	11 in. (27.5 cm)
Large 54	$4\frac{1}{4}$ in. (11 cm)	Large 6	$12\frac{1}{2}$ in. (32 cm)
Small 48	$4\frac{1}{4}$ in. (11 cm)	Large 4	14 in. (35 cm)
Large 48	5 in. (12.5 cm)	Large 2	$15\frac{1}{2}$ in. (39 cm)
Large 40	$5\frac{1}{2}$ in. (14 cm)	Large 1	18 in. (45 cm)

PLASTIC POTS

These are now in wide use for they are almost unbreakable, light to handle and are easily stored. Each pot has a minimum of three drainage holes protected by a base rim, making the pot suitable for a capillary bench. The pots may be used in an outside plunge bed and may even be ploughed up with little likelihood of breakage. They are usually sold in cases when buying in quantity:

Size	No. in case	Size	No. in case
3 in. (7.5 cm)	950	4 in. (10 cm)	450
$3\frac{1}{4}$ in. (8 cm)	720	$4\frac{3}{4}$ in. (12 cm)	280
$3\frac{1}{4}$ in. (tall) (8 cm)	780	$5\frac{1}{4}$ in. (13.5 cm)	230
$3\frac{3}{4}$ in. (9.5 cm)	500	$4\frac{1}{4}$ x $5\frac{3}{4}$ deep (11 x 14.5 cm)	250

PROPAGATING POTS

These are made of black whalehide and have a life of twelve weeks with normal watering. They may then be planted out without removing the plant for the pot will eventually disintegrate. They are obtainable in sizes from $2\frac{1}{2}$ in. (6.5 cm) to $6\frac{3}{4}$ in. (17 cm) and with either a wide or a narrow base. They are sold in lots of one thousand, a thousand of the smallest size costing about two pounds.

Red whalehide pots have a life three times as long. They are obtainable in the same sizes and cost twice the price of the black whalehide pots. They are also made in $7\frac{1}{2}$ in. (19 cm) and 9 in. (23 cm) sizes suitable for growing a crop of tomatoes or chrysanthemums throughout the season, inside or outside.

Black whalehide pots $2\frac{1}{4}$ in. (6 cm) diameter and 5 in. (12.5 cm) deep, known as 'tubes' are used for sweet pea culture, one seed being planted in each 'tube' and the plant grown on until ready for planting out.

Pea pots as manufactured by Jiffy Pots and Root-o-Pots of Milton-of-Camps, Scotland, are also suitable for growing sweet peas as well as most other plants. Made of compressed peat into which the plants form a heavy rooting system, they are easily handled and produce plants of outstanding quality, supplying them with a suitable rooting medium and with food and humus when set out in their flowering quarters. The 'pots' are obtainable in sizes from 1½ in. (4 cm) (for sweet peas) to 4½ in. (11.5 cm); also as trays of six 2 in. (5 cm) squares for sweet peas and bedding plants.

WOODEN POTS

Laconeil pots are made of resin-bonded sawdust, contoured for even rooting all round the pot. They are light to handle and are long-lasting, being almost unbreakable with normal use whilst they do not absorb nitrogen from the compost as do clay pots.

SELF-WATERING POTS

These are manufactured by Messrs Marmax Ltd, and are known as 'camel' pots. They are composed of an inner pot made up of resin-bonded wood-fibre with an internal ridged surface which permits an easing of pressure on delicate roots. An outer plastic pot permits for a thirty-day supply of water under normal room temperatures so that house plants require the minimum of attention and will survive five to six weeks when one is away from home.

PLATANUS

Deciduous trees growing up to 100 ft. (30 m) tall and do well anywhere and in any soil. They are adaptable to town conditions. With their large leaves, they are valuable as shade-giving trees, the London planes being a familiar and much loved part of the scene. Whilst the leaves take on rich golden tints in autumn, the bark is also colourful, peeling off to reveal areas of yellow inner bark. The leaves are like those of the maple whilst it bears a ball-like fruit to each stalk. The trees need no attention apart from occasional thinning of shoots to keep the centre open. Propagate from seed sown in autumn outdoors or from cuttings taken with a 'heel'.

The hardiest and best form is that known as the London Plane, thought to be a cross between *P. occidentalis* and *P. orientalis*.

It has the good points of both parents. There is an erect form, 'Pyramidalis' which is suitable for planting in a restricted area and for street planting.

PLATYCODON

This, the Chinese Balloon flower is a most interesting plant. It prefers a light, sandy soil for its horse-radish-like roots and likes a sunny position. It should be planted at the front of the border, in November and 15 in. (38 cm) apart.

SPECIES

P. grandiflorum 'Mariesii'. Its pale blush blue flowers with their deeper blue veins first appear as tiny balloons or lanterns, opening to large star-like flowers during August and September. There is a double form, 'Flore Plenum' which may also be obtained in the pink form, 'Roseum'. They grow 20 in. (50 cm) tall.

PLECTRANTHUS

P. fruticosus is a dainty plant of trailing habit, having purple stems whilst its pale green leaves have deep pink veins. The leaves are fleshy whilst the plant also bears tiny pink flowers. It may be said to be of more creeping habit than trailing and is most pleasing where used to creep round the edges and trail over the sides of indoor plant tables. It grows quickly and is perfectly happy in partial shade, though it should not be placed at the back of the room. Fibrous loam containing a little peat and some silver sand will suit it well and care must be taken not to over-water, especially during winter, when a minimum temperature of 42°F. (5°C.) will prove suitable.

PLUM, *see* Prunus

PLUME POPPY, *see* Bocconia

PODOCARPUS

Evergreen trees or shrubs, native of Australasia, China and Chile and which should be confined to milder parts of the British Isles. The best species is *P. andinus*, the Chilean yew which greatly resembles the English yew. It makes a dense upright shrub but has brighter green foliage. *P. macrophyllus* of China is the hardiest species, of similar habit, its leaves being pale green above, glaucous on the underside.

Plant in spring, in a soil containing peat or leaf mould. Propagate from seed or from cuttings removed with a 'heel' in July and inserted in sandy compost under glass.

POKEBERRY, *see* Phytolacca

POLEMONIUM

A neglected border plant though it has been cultivated in Europe for two thousand years. It will flourish in ordinary soil, and is increased by root division after flowering. Plant in November, 12 in. (30 cm) apart; or in March if the soil is not too well-drained.

SPECIES AND VARIETIES

BLUE PEARL. A new introduction bearing bells of deep blue and growing to a height of only 12 in. (30 cm).

P. coeruleum. With its divided leaves it is known as Jacob's Ladder. It grows to a height of 2 ft. (60 cm) and comes into bloom early in June and continues through summer bearing its sprays of bright blue bells with their attractive gold stamens.

SAPPHIRE. This is a new variety bearing bloom of bright sky-blue on 20 in. (50 cm) stems and which comes into bloom in May.

POLIANTHES

A genus of a single species, native of Mexico and taking its name from *poly*, many and *anthos*, flower for its abundance of flowers. It is the poet Shelley's 'sweet tuberose, the sweetest flower for scent that blows'. During Victorian times the tuberous rooted bulbs, covered with the old leaf bases were imported from southern Italy, to be grown by specialists in heated glasshouses almost the whole year round, whilst few conservatories adjoining large houses were without its pleasantly scented flowers. The linear leaves are channelled and spotted with brown on the underside. The flowers are white and funnel-shaped and appear in a terminal raceme on a 2 ft. (60 cm) stem during July and August where growing outdoors or in unheated greenhouse.

The tubers may be brought into bloom in a temperature of 60°F. (16°C.) during the winter months or where there is no heat, they will bloom during summer depending upon planting time. From an early spring planting, they will bloom during the latter weeks of summer and in autumn. Plant one tuber to a large pot, using a compost made up of fibrous loam, coarse sand, leaf mould or peat and dehydrated cow manure in equal parts. Plant firmly, just covering the nose of the bulb and keeping the compost moist. Where a temperature of 65°F. (18°C.) can be given them and a moist atmosphere maintained, retarded bulbs may be brought into bloom within eight weeks. Whilst growing, they require copious amounts of moisture and after flowering will usually take a year before they bloom again. The tubers may be grown in a cool greenhouse or garden room and will bloom late in summer, the flowers remaining fresh for several weeks when they will scent the room with their lily-like perfume.

Plants may also be brought into bloom outdoors. They should be started into growth in small pots under glass and from a March planting may be set out in their flowering quarters early in June when there is no fear of frost damage to the foliage. Plant 9 in. (22 cm) apart in a warm border, leaving the top of the tuber exposed and where the tubers may be ripened by the sun. They require a soil enriched with some decayed manure and containing peat or leaf mould. During summer as it grows, it should be given ample supplies of moisture. It will bloom during August and September and after flowering, the tubers should be lifted and stored over winter in boxes of peat or sand until started into growth again in March. In a sunless summer however, the tubers will not ripen sufficiently to be worth keeping for another year and fresh tubers should be obtained in spring.

SPECIES AND VARIETIES

Polianthes tuberosa. Native of Mexico, it makes a large bulbous-like tuber and has leaves which are channelled. The funnel-shaped flowers open star-shaped and are borne in an elegant raceme of twenty or more at the end of a 2 ft. (60 cm) scape. The variety 'Gracilis' has narrower leaves and bears flowers with longer tubes whilst 'The Pearl' which bears double flowers of pearly-white is that most often cultivated for the high-class florist trade. The flowers are larger than those of the type.

POLYANTHUS, *see* Primula

POLYGALA

Though they love some peat or leaf mould in the soil, the polygalas are tolerant of a chalky soil, especially *P. calcarea* which bears its rich blue flowers early in summer. The flowers are shaped like those of the broom and are most attractive amidst their glossy-green leaves, like those of the box. The species *P. chamaebuxus* is known as the dwarf box. It likes a sunny position where it will soon make a large plant, its cream and bronze flowers being most attractive. Rarely exceeding a height of 3–4 ft. (about 1 m), *P. vayredae*, a native of Spain, is of similar habit and in May, bears flowers of rich crimson colouring.

POLYGONATUM

The Solomon's Seal, *P. multiflorum* is a plant which likes a moist soil and some shade. The rhizomes should be planted almost at the front of the border where it is most shaded or in the wild garden. There the arching sprays with the glossy leaves on the uppermost side and the dangling waxy-white bells beneath, provide a coolness rarely to be found in any other plant. The flower sprays, leafless at the base appear almost like asparagus from the creeping rootstock, so do not plant too closely to other border plants. Plant the rhizomes in November and make certain that the ground has been thoroughly cleaned and that plenty of leaf mould has been incorporated. It blooms in June on 2 ft. (60 cm) stems.

POLYGONUM

Native of the Himalayas, they are interesting plants of easy culture, and several planted in a border or used to cover a trellis, will provide colour from May until October. Plant 18 in. (45 cm) apart in November, and propagate by root division or from cuttings.

SPECIES

P. affine. It has oval leaves which turn bright yellow during early winter and in autumn it bears dozens of neat spikes of clear pink flowers on 1 ft. (30 cm) stems. The variety, 'Darjeeling Red' bears spikes of rich scarlet-red and is one of the best of all front row plants.

P. amplexicaule 'Atrosanguinea'. In spite of its name it should be in every small border. The plant is of compact habit and throughout autumn bears upright spikes of crimson, like a miniature lythrum.

P. baldschuanicum is a valuable climber of vigorous habit for covering a trellis or fence. It will attain a height of 12 ft. (3.5 m) in two years and eventually reach a height of 30 ft. (9 m) or more. It likes a cool, shady position and is deciduous. Early spring is the best time to plant, pot-grown plants proving the most reliable. The large bottle-green leaves are heart-shaped and during late summer and early autumn the plant bears long drooping sprays of creamy-pink flowers. To prune, cut out dead wood each year.

P. bistorta 'Superbum'. A charming plant and so valuable in that it comes into bloom mid-May and continues until the other members of the family come into bloom in August. It grows to a height of 2 ft. (60 cm) and bears sprays of tiny pure pink bells packed closely together along the stem. The sprays are lovely for cutting.

POMEGRANATE, *see* Punica

POPLAR, *see* Populus

POPPY, *see* Papaver

POPULUS

A genus of fast-growing deciduous trees, poplars can form a valuable wind-break or they may be planted to shut out an unsightly view. Poplars require a moist loam and will not grow well in a dry, sandy soil, moisture being necessary for their rapid growth. All have handsome, heart-shaped leaves and bear their male and female flowers on different trees. The Balsam poplars which are amongst the largest of all forest trees have their buds covered in a sweetly smelling resin whilst the unfolding leaves release a rich balsamic odour. Poplars will tolerate hard clipping to keep them to the requisite shape. Propagate by seeds or by cuttings 12 in. (30 cm) long and the half-ripened wood inserted in trenches of sand outdoors.

P. alba is a handsome species. It is known as the White Poplar because of the brilliant white under-surfaces of the leaves. The variety 'Paletskyana' has toothed leaves. *P. canescens*, the Grey Poplar, which has leaves of deep sage green, is a long living species and is the only poplar to do well on chalk. *P. nigra betulifolia* is the Manchester Poplar, one of the best of all poplars for a town garden for it makes a compact head and has downy shoots. *P. nigra italica* is the Lombardy Poplar, tall growing and columnar in habit and which may be planted to form a wind break. *P. tremuloides* is a pleasing species with pale yellow bark and small, evenly toothed leaves which tremble in the wind. The form Pendula is of weeping habit. Of the Balsam Poplars which should be planted for their fragrance, the best for a small garden is *P. candicans*, the Ontario Poplar which has broad leaves and which carry the scent of Balm of Gilead. *P. tacomahaca* grows tall and erect whilst *P. trichocarpa* is valuable for an exposed garden and is also balsam-scented.

PORTUGAL LAUREL, *see* Prunus

PORTULACA

Purslane's pale green propeller-like leaves give it the appearance of mistletoe, whilst it has attractive crimson-red stems which are quite succulent and may be used for boiling. The leaves are sometimes used in salads. *Portulaca oleracea* is a tender annual and seed should not be sown until late in spring. The plants require ordinary soil and a sunny situation.

POTENTILLA

With their strawberry-like flowers, the potentillas are valuable plants for a dry, sandy soil, making compact bushes 2 ft. (60 cm) tall and remaining in bloom through summer and autumn. In most instances the foliage is silvery-grey, making a striking contrast to the dark green foliage plants of the shrubbery. Almost no pruning will be necessary apart from the occasional removal of dead wood.

Amongst the best are *moyesi*, having grey leaves and large golden flowers; 'Moonlight', bearing pale primrose flowers; and 'Purdomi', with masses of lemon-yellow blooms.

Propagate by division in March or by cuttings of half-ripened wood inserted in sandy compost under glass.

POTERIUM

For a moist soil and especially for a small border, *P. obtusum* is a charming plant. A native of Japan, it has attractive bright green yew-like foliage and its rich pink blooms are well-described by its name of the Bottlebrush plant. They are indeed exactly like those brushes used for cleaning babies' bottles, bending over in an attractive way at the ends. There is a pure white form, 'Album', which also blooms from mid-July until the end of September. Plant 2 ft. (60 cm) apart in November and propagate by root division. It grows 3 ft. (90 cm) tall.

POT MARIGOLD, *see* Calendula

PRICKLY HEATH, *see* Pemettya

PRICKLY PEAR, *see* Opuntia

PRIMROSE, *see* Primula vulgaris

PRIMULA

ASIATIC

Amongst the most beautiful plants of the garden, the asiatic primulas should be more widely grown, but the belief persists that they are 'difficult', demanding special conditions which the average garden is unable to provide. It is thought that they will grow only in full shade and in boggy land. By the side of water they will flourish, and beneath mature trees where few other plants will grow, but they will also prove equally satisfactory elsewhere about the garden, provided the soil is well enriched with humus. There the asiatics will grow well even in full sunshine, though artificial watering may be necessary during dry periods in summer. This, however, should not be necessary where the plants are growing in a shrubbery or in partial shade, an orchard or wild garden being ideal places for growing them.

Like most members of the extensive primula family most of the asiatics die back completely during winter and so are excellent plants for a town garden where continual deposits of soot and sulphur, particularly during winter, may cause the death of many of those plants which do not shed their leaves.

To grow the asiatics well, so that they will be vigorous and long lasting, the soil must contain an abundance of humus, for the roots must at all times be kept in a moist condition. Nor will they grow well in a chalky soil. Humus materials of any form may be used, including decayed material from the garden compost heap, well-rotted manure, spent hops, shoddy or peat. For the townsman, peat augmented by some used hops or old mushroom bed compost which are readily obtainable, should be used as freely as possible, working well into the soil, for primulas are deep rooting plants even though they form their roots at, or above, soil level. The more humus that can be incorporated, the less moisture by artificial means will the plants require.

Most of them being long living and seeding themselves with freedom it is also necessary for the ground to be made completely clean before any planting is done. A regular mulch of peat, which is most inexpensive, will suppress all annual weeds in addition to providing an excellent means of maintaining soil moisture.

Many of the asiatic primulas will flourish in soil which remains damp almost the whole year round, in what may be described as semi-bog conditions, being one of the few plants to grow well under such conditions. An area of ground which is unduly damp, possibly where moisture drains from surrounding ground, could well be made both profitable and beautiful.

The ease with which these primulas fertilize each other has led to the introduction of numerous hybrids whose offspring bear bloom of extremely varied colouring, seed of the true species being somewhat difficult to obtain. It is usual to raise plants from seed, for there are few named varieties of the asiatics, apart from those named varieties of *P. denticulata,* the seed sowing is an inexpensive and rapid method of building up a large stock of plants. Propagation by division is also possible, but an established plant attains such large proportions and such beauty when in bloom that it is neither easy nor advisable to disturb it when it is known that it will provide a large number of self-sown seedlings, or seedlings may be readily raised from hand-sown seed.

GROWING ASIATIC PRIMULA FROM SEED

The seed should be gathered as soon as ripe, just when the capsules are on the point of opening. This will be about a month after the blooms have died away, the exact time depending upon the flowering period of the plant. Or, of course, the seed may be allowed to fall and germinate naturally, this being the method demanding least attention.

The seed should be sown as soon as possible after it has been saved, and whilst still in a fresh condition, when it will germinate rapidly. The seed of many asiatics will be sown in July or August using boxes or pans which should be covered with small sheets of glass to encourage germination. Or the boxes may be placed in a frame or cold greenhouse until they have germinated and until the plants are ready for transplanting.

The John Innes Sowing Compost may be used, or equal parts of sterilized loam, peat and sand. The seed, which must be sown thinly, should be pressed into the compost after it has been made level, but it should not be covered or germination will be delayed. The seed should always be kept comfortably moist and should be covered with a piece of brown paper to hasten germination. This must, of course, be removed as soon as the seed has germinated, but to prevent the seedlings from being scorched by the sun they should be given some shading.

The seedlings should be transplanted as soon as they are large enough to handle. They may be transferred to a frame or to boxes containing a similar compost, or directly to prepared beds outside. A border where the soil has been enriched with humus and brought to a fine tilth, and where a partially shaded position may be found, will prove suitable. The seedlings are set out 6 in. (15 cm) apart and made quite firm, and so as not to compact the ground unduly, a wide board should be provided for kneeling. Alternate plants may be removed for sale, the others being allowed to bloom and seed themselves. The seedlings should be well watered-in and kept in a continually moist condition.

The seedlings may either be kept in a frame over winter, which is preferable where growing north of the Thames, or they may be moved early in October to allow the plants to become established before winter. Plants moved at this time will bear bloom the first summer, though several species will not do so until the following year.

Plants propagated by root division should be lifted and divided in autumn or during March, as soon as they begin to make new growth.

P. denticulata and several other asiatics may also be increased by root cuttings which are thick and fleshy. A plant may be lifted after flowering and portions of roots cut into lengths of about 1 in. (2.5 cm). Use a compost made up of equal parts sterilized loam and sand, and where using a size 60 pot, six to eight root cuttings may be accommodated. The top of the root cutting should be just below the level of the compost which should always be kept moist. Within a few weeks plant growth will appear from the top of the cutting, and as soon as sufficient growth has been made, each plant should be re-potted into a 2½ in. (6.5 cm) pot for growing-on, or they may be planted directly into the open ground.

SPECIES AND VARIETIES OF ASIATIC PRIMULAS

P. alpicola. Discovered by Kingdon Ward in Tibet in 1925, it likes similar conditions to *P. aurantiaca*, a cool, moist soil and some shade. It is a dwarf counterpart of *P. florindae*, bearing its pale primrose flowers, likc cowslips, on fifteen-inch stems during June. The flowers are deliciously fragrant and have an attractive coating of meal.

P. asthore. A hybrid having *P. beesiana* and *P. bulleyana* for parents. Like so many of the asiatic primulas, the flowers are borne in candelabra form, in whorls or tiers one above the other and on 18 in. (45 cm) stems. The colours range from almost golden-brown to violet, whilst there are lovely shades of pink and yellow. The colours are of pastel tints and will tend to fade where exposed to strong sunlight, for the plants bloom at the height of summer.

P. aurantiaca. It grows to a height of 18 in. (45 cm) and bears its neat candelabras of coppery-orange during midsummer. It likes a cool, leaf-mould soil and a shaded situation. A position by the side of water and where shaded by mature trees suits it well.

P. beesiana. The plants will flourish in ordinary soil and make a pleasing effect in a shrubbery, where they should be confined on account of their rather coarse growth. The blooms are borne in candelabra fashion on 18 in. (45 cm) stems during early summer.

P. bulleyana. One of the best of the asiatics, it bears its whorls of buttery-yellow flowers on 20 in. (50 cm) stems during June and July, each stem carrying as many as six tiers of bloom. Like most in this section, it forms a large plant and should be allowed ample room to develop.

P. burmanica. Not so well known as it should be, for it is a lovely plant, bearing as many as six tiers of plum-coloured flowers on 2 ft. (60 cm) stems and makes a plant of compact habit. It grows well in ordinary soil and is a useful plant for the border or shrubbery. It blooms during early summer and planted in bold groups makes a striking display. There is also a lovely form known as 'Ascot Hybrid', which has scarlet flowers with a striking yellow eye.

P. capitata. It may be likened to a dwarf *P. denticulata,* for it bears its loose flat heads on 18 in. (45 cm) stems but comes into bloom when *P. denticulata* has ended. It is a most attractive plant with mealy grey-green foliage but is only short-lived. A better form is *P. mooreana,* which is stronger and longer living, the foliage being more hairy whilst the blooms are of a richer shade of purple-mauve. It should be planted near or with *P denticulata* whilst its roots must have ample moisture.

P. chionantha. It likes a low-lying situation, for it must have its roots in almost constant moisture and, where this is provided, it will be quite happy if its head is in full sunlight. Its narrow leaves are covered with farina or down, whilst it bears its umbels of creamy-white flowers on 2 ft. (60 cm) stems during the early weeks of summer.

P. chungensis. A most beautiful plant, which should be given almost full shade. Its brilliant orange blooms are borne in candelabra fashion in whorls as many as five to a 2 ft. (60 cm) stem, making a most arresting sight in summer.

P. denticulata. This must surely be one of the most useful of all garden plants, for it is one of the first to bloom, bearing its ball-shaped heads from early March in a sheltered corner. It is also perfectly happy either in full shade or full sunlight, and is of such hardy constitution as to be almost indestructible. Their foliage turns a rich orange-yellow in autumn before dying back. The plants will be most vigorous in a heavy, moist soil where the foliage will be of a rich green, whilst the stems will be thick and up to 2 ft. (60 cm) in length with the flowers almost as large as a tennis ball. The named varieties are readily increased from root cuttings or from rooted offsets which form around the main crown. They should be detached and replanted in October. The blooms last long in water. There are a number of early varieties:

ALBA. This is the pure white form so effective when used as a contrast to the richly coloured named varieties. The foliage is of a paler, more downy, colour than the others.

BENGAL ROSE. As its name implies, the blooms are of a soft shade of deep rose-pink.

CRIMSON EMPEROR. Also known as 'Red Emperor', the large pure crimson-red balls which are free of any purple colouring, being borne on stems 18 in. (45 cm) long.

HAY'S VARIETY. The blooms are the largest of all and are of a deep purple colour, flushed with crimson.

PRICHARD'S RUBY. This is the deepest red form, the large blooms being of a deep ruby-crimson colour.

PURPLE BEAUTY. It bears its drumstick blooms of rich purple-mauve in great profusion.

SPRING MAID. A superb variety, the large blooms being of a lovely dusky pink colouring.

TAYLOR'S VIOLET. New, the blooms being of an intense royal purple colour held on long, sturdy stems.

P. florindae. This, the Giant Cowslip of the Himalayas, is one of the most stately plants of the garden, growing to a height of 3 ft. (1 m) and bearing clusters of scented pale yellow flowers. It is the latest of the asiatics to bloom, for it is at its loveliest during August and September. It is readily raised from seed, the plants coming into bloom the second year from a late autumn or early spring sowing, or it may be increased by root division after flowering. This plant really does require a damp situation and will flourish even where its roots are continually submerged in water. It will grow well in a heavy soil where it is low lying.

P. helodoxa. The Bog Primula of China and a beautiful plant, its golden-yellow flowers appearing in tiers on 2 ft. (60 cm) stems during midsummer. It likes a damp situation but is happy in full sunlight. The plants come quite true from seed, sown by hand or naturally, for it does not readily cross with nearby plants as do most of the other asiatics.

P. japonica. It is tolerant of drier conditions than the other asiatics and should be planted beneath mature trees where the plants will obtain the shade they enjoy. They make large plants and bear their tiers of bloom in shades of pink and crimson on 2 ft. (60 cm) stems during early summer. There is a fine named form called 'Miller's Crimson' and one bearing pure white whorls called 'Postford White'.

P. pulverulenta. The best forms are so beautiful as to warrant propagating by offsets, which it forms in abundance. Its blooms are of rich crimson with attractive brown centres and those of less rich colouring should be eliminated. There is also a lovely strain known as the Bartley Strain, the blooms being of pastel shades of pink and salmon. The late Mr Dalrymple of the Bartley Nurseries also introduced 'Lady Thursby', bearing flowers of bright rose-pink with a yellow eye. The plants make vigorous growth and bear their whorls on attractive powdered stems at a height of 2 ft. (60 cm).

Crossed with *P. cockburniana,* which is only a short-living plant, *P. pulverulenta* has given several lovely named hybrids which must be propagated by root division. Fortunately they are quite as perennial as *P. pulverulenta.* 'Aileen Aroon' bears whorls of vivid orange-red, whilst 'Red Hugh' bears flowers of brilliant flame on 18 in. (45 cm) stems. Both are amongst the brightest plants of the garden.

P. rosea. Like *P. chionantha* it must be given a damp situation. It does well with *P. denticulata,* in a heavy soil which is low lying. In a wet, shady situation it will quickly form a dense clump of glossy, succulent leaves. The best form is known as 'Micia Visser de Geer', the blooms being twice the size of the species and of a rich rose-pink colour borne on 8 in. (20 cm) stems throughout spring. It is easily propagated by division.

P. sikkimensis. Whilst *P. florindae* prefers an open, bog-like situation, the Sikkim Cowslip enjoys the cool leafiness of the woodlands, partial shade and a soil containing plenty of leaf mould. It makes a much more compact plant from which arise on 15 in. (38 cm) stems the loose heads of fragrant pale yellow flowers. If given a position beneath trees which is free from grass and bracken, the ground will soon be covered with a mass of self-sown seedlings. It may also be increased by division. It blooms in May and June.

AURICULA

A race of plants descended from *P. auricula* and *P. pubescens* and with their foliage and bloom covered with farina. They were known as Mountain Cowslips or Bear's Ears from the shape of the leaves and the covering of farina (hairs). They are the hardy border auriculas though with them must be included several which are entirely free from farina but which, with their vigorous coarse habit cannot be classed with the more refined alpine auriculas. They are plants of great antiquity and with their hardiness, freedom of flowering, the rich 'old master' colouring of the blooms and their exotic fragrance, are amongst the loveliest of flowers, in bloom from April until mid-June. So powerful is their scent that from a small bed, it will be almost overpowering when the early summer sun shines down upon the plants whilst they impregnate the sun-baked soil with their perfume.

Not having been confined to pots under glass as have the show auriculas, the older border varieties have in no way lost their vigour and if they can be given a sunny situation and a soil which does not readily dry out during summer, they will prove adaptable to almost every garden, quickly growing into large vigorous clumps above which they bear on 9 in. (22 cm) stems, irregular heads of large velvety blooms. This irregularity adds to their old-world charm. There is nothing stiff about them and when cut and placed in small vases indoors, they will remain fresh for days and scent the house with their warm sweet honey fragrance. Plant in spring or in autumn and top dress each year.

VARIETIES

ADAM LORD. A variety of sturdy habit with serrated foliage which forms into numerous rosettes above which in large trusses appear the navy-blue flowers with their creamy-white centre.

AMETHYST. It forms a large truss of bright wine-purple flowers with a clearly defined white centre and is densely covered in farina.

BLUE MIST. Now rare but an outstanding variety and so well-named, for the medium-sized blooms are of pure sky blue with the farina providing a silvery, mist-like sheen.

BLUE VELVET. Of vigorous habit, it forms a symmetrical head of purple-blue flowers with a clearly defined creamy-white centre and emits a rich honey perfume.

BROADWELL GOLD. A superb variety found by Mr Joe Elliott in a Cotswold cottage garden but it is now rare. It bears large blooms of brilliant golden-yellow with beautifully waved petals and they diffuse a musk-like fragrance. All parts of the plant are covered with farina.

CELTIC KING. A strong grower, its foliage is without farina whilst its flowers are of a lovely shade of lemon-yellow with attractively waved petals.

CRAIG NORDIE. The flowers are of burgundy-red with a golden centre and are almost free of farina whilst the foliage is grey-green.

GOLDEN QUEEN. It forms a large truss of pale sulphur-yellow flowers with a white centre and which carry a rich sweet perfume.

LINNET. An old favourite which now seems to have been lost. It is late into bloom and is well-named for the blooms are of a combination of green, brown and mustard, the colours of the now almost equally rare bird of that name.

MCWATT'S BLUE. Raised by the late Dr McWatt in Scotland, the rich mid-blue flowers with their white centre carry a delicious honey perfume. The foliage is heavily mealed.

MRS NICHOLLS. A most attractive variety, the pale yellow blooms have a golden centre around which is a circle of white. The blooms have a pronounced musky scent.

OLD IRISH BLUE. A most beautiful auricula, with serrated foliage above which it bears large trusses of rich mid-blue. The flowers have a white centre and are heavily mealed whilst the perfume is outstanding.

OLD PURPLE DUSTY MILLER. Also known as 'Blue Dusty Miller' but the flowers are really of purple colouring, with pronounced perfume and are held on 12 in. (30 cm) stems. All parts of the plant are heavily mealed.

OLD RED DUSTY MILLER. Of great antiquity, it is so heavily mealed as to appear as if covered in flour. The flowers are not large but the colour is unique, being of crimson-brown, wallflower colour and with a heavy scent.

OLD SUFFOLK BRONZE. A most interesting variety more alpine than border for it is free from farina. It makes a plant of robust habit and over a period of at least ten weeks bears large trusses of flowers of dollar size and of shades of gold, bronze and buff, almost impossible to describe. The scent is intoxicating.

PRIMULA VULGARIS (PRIMROSE)

The plant takes its name from primaverola, a diminute of *Prima vera*, meaning the first flower of springtime. From this word, the obvious development was primerole which soon became 'prime-rose' which in Shakespeare's time was a plant held in so great esteem as to be the word most often used to denote excellence.

It was not until Turner used the word 'prymerose' did the name we know today, come to be used. Spenser and Shakespeare writing shortly afterwards, used the modern spelling. Sir Francis Bacon in *The Making of Gardens*, suggests planting 'prime-roses' for February flowering which is early for all but the most sheltered gardens in Devon and Cornwall.

It was during the reigns of the Tudors that primroses first came to be widely grown for the beauty and interest of their flowers, for by then many variations of the yellow wild primrose had become known. Tabernaemontanus, writing in the year 1500, described the double yellow primrose which may still be found in its double form growing wild in various parts of Britain. Variations of the primrose *Primula acaulis* or *P. vulgaris*, are extremely numerous and quite apart from the lovely double forms, there are the dainty Hose-in-Hose forms and the Jack-in-the-Greens, which originally had been found growing in the wild state. By Elizabethan times, these lovely primroses were widely grown in the dainty knot gardens of both manor house and cottage, for besides being of extremely perennial habit, the compact form of the plants made them ideal for this form of gardening.

JACK-IN-THE-GREENS

Here, the blooms are single and are backed by a ruff-like arrangement composed of tiny replicas of the primrose leaf, which provides a pleasing contrast to the clear colourings of the blooms. They were widely planted during the sixteenth century, the arrangement of the leaves behind the blooms greatly appealing to the Tudors as could well be imagined. The plants later became known as Jack-in-the-Pulpits.

Yet another form is the 'Gally Gaskin'. This is a single bloom which has a swollen, distorted calyx and is now rarely to be found. It has also a frilled ruff beneath the bloom. A picture attributed to Henri Coiffier de Ruse in the National Gallery, aptly illustrates this form, a frilled ruff appearing beneath the knees of the gentlemen which appear large, like the swollen calyx of the 'Gally Gaskin'.

CUP AND SAUCER PRIMROSES

Perhaps the loveliest of all forms of the primrose is that of the Hose-in-Hose, where one bloom grows from another to give a plant in full bloom a most dainty, feathery appearance, the flowers seeming to dance in the spring breezes in a fairy-like manner. They are also known as Duplex, Double-decker or Cup and Saucer primroses, though Parkinson's Hose-in-Hose beautifully illustrates the form. 'They remind one of breeches men do wear,' he writes. Gerard called them Two-in-Hose and included an illustration in his *Histoire of Plants*. The hose worn by men at that time were knitted with a much stronger wool than used today and it was the custom for one stocking to be placed in another before being passed to the wearer.

HOSE-IN-HOSE VARIETIES

(p) denotes Polyanthus form

ABERDEEN YELLOW. This is a very old variety bearing small clear yellow blooms, with the duplex clearly defined.

ASHFORT (p). It is tall growing and bears an umbel of brownish-red flowers above pale green foliage.

BRIMSTONE (p). May be said to be of semi-polyanthus form, for its large, clear sulphur-yellow bell-shaped blooms are borne from a main stem 3 in. (7.5 cm) long. An outstanding variety, remaining long in bloom (see plate 50).

CANARY BIRD. It is very early flowering, its bright, pure canary-yellow blooms being borne with great freedom.

CASTLE HOWARD. A lovely hose of true primrose habit, discovered in the grounds of the great Yorkshire house bearing that name. The blooms are of true primrose-yellow colour.

ERIN'S GEM. It bears a most interesting and dainty cream-coloured flower, the lower one having an attractive bright green stripe.

FLORA'S GARLAND. This is a lovely old Hose-in-Hose, bearing well-defined flowers of deep pink.

GOLD LACED HOSE (p). The Hose-in-Hose form of the florist's gold laced polyanthus and a most delightful plant of perfect cup and saucer form. The flowers resemble the gold laced in that the ground colour is of light or dark red (almost black) which accentuates the brilliant golden edge and centre. The flowers are deliciously scented.

GOLDILOCKS (p). It bears its deep yellow blooms in umbels on long stems making it an excellent cut flower.

IRISH MOLLY. A hose of great charm and of true primrose form, bearing a profusion of large flowers which are of a soft shade of mauve-pink. Also known as 'Lady Molly'.

IRISH SPARKLER (p). This striking variety is frequently confused with 'Old Vivid'. Its blooms are similarly formed but are of a crimson-red colour without the conspicuous yellow centre. Like most of the Hose-in-Hose, it makes an excellent cut flower.

LADY DORA (p). I have seen this lovely variety in an Irish garden, but have not yet had the pleasure of having it in my own garden. The colour of the small, dainty blooms is brilliant golden-yellow, whilst they possess a powerful perfume.

LADY LETTICE. In his delightful book, *Old Fashioned Flowers,* Sir Sacheverell Sitwell so rightly says that this plant should be in every garden. Planted along a path, its fairy-like flowers provide a delightful effect. It comes early into bloom in March and continues until the end of May, bearing masses of apricot-yellow flowers which are tinged with salmon-pink. Sadly, it is now almost extinct.

OLD SPOTTED HOSE (p). Of polyanthus habit, the large crimson blooms are held on sturdy stems, the centre of each petal being quaintly spotted near the edge.

WANDA HOSE. This is the Hose-in-Hose form of the ever-popular Juliae primrose, 'Wanda'. It has the same habit, the bright purple-red blooms being borne on short stems.

DOUBLE PRIMROSES

'Our garden Double Primrose, of all the rest is of the greatest beauty,' wrote Gerard in 1597, and he makes special mention of the 'Double White'. Thirty years later, Parkinson gave a full description of double primroses in which he mentioned that the leaves were larger than those of the single primrose, 'because it groweth in gardens'. In other words, the double primroses were then popular garden plants. Parkinson describes the flowers as being 'very thick and double and of the same sweet scent with them' (as the common single primrose). Indeed, the double primroses are among the most beautiful and interesting plants of the garden, and though they are considered by modern gardeners to be difficult to grow, the fact that Gerard's 'Double White' and the 'Double Sulphur' primrose of Tabernaemontanus are still with us and are still vigorous must surely do much to refute that charge.

The earliest writers repeatedly mention the ease with which the plants were grown. Phillip Miller, Curator of the Chelsea Physic Garden about the year 1720, wrote in his *Gardener's Dictionary* (1731), 'They will grow in almost any earth provided they have a shady situation,' and John Rea in his *Flora* published in 1665 said, 'Were it (the double primrose) not so common in every Countrywoman's garden, it would be more respected, for it is a sweet and dainty double flower . . .' Thus, through the years, the garden writers seem to have taken it for granted that those who grow the double primrose, and this would appear to be most gardeners, found it amongst the easiest of all plants to manage. Why, then, have these quite charming plants become little more than collector's pieces with present-day gardeners? The town garden with its acid and inert soil, due to continual deposits of soot and sulphur and complete lack of humus has now taken the place of the country cottage garden, where the soil was continually being revitalized with humus-forming manures. These took the form of night-soil, decayed farm-yard manure, sheep and poultry droppings, decayed leaves, supplying not only humus but food in the form of nitrogen, which the double primrose loves so well and without which the plants will not survive for long.

Though the double primrose will grow well in a town garden, given the right conditions, it has survived only in the cottage garden.

VARIETIES OF DOUBLE PRIMROSES

ARTHUR DE SMIT. Now rare but it is one of the loveliest of the doubles, and was raised in Germany. The blooms, borne on single footstalks, being of rich purple edged with yellow.

ARTHUR DU MOULIN. It goes by many names and is believed to have been introduced from Ireland towards the end of the nineteenth century, for de Moleyns is the family name of Lord Vertry. It is also known as 'Dumoulin' and 'des Moulens'. It is the most important double, being the first to bloom and being the only variety to yield pollen in quantity, with the exception of the less double blooms of 'Prince Silverwings'. The flowers are of beautiful formation and are of deepest violet, borne on short polyanthus stems.

BON ACCORD BEAUTY. The blooms are large and of deep purple-blue, the petals being edged with white and spotted with white towards the edges.

BON ACCORD BLUE. The largest of all the Bon Accords, the blooms are of a lovely shade of rich blue borne on decidedly polyanthus stems. Though now scarce it is one of the easiest to grow.

BON ACCORD BRIGHTNESS. Believed to be extinct, this must surely be the variety now called 'Crathes Crimson', for it answers the description in every aspect.

BON ACCORD CERISE. The petals are flat and perfectly rounded, forming a near button-like rosette of a lovely shade of clear cerise-pink, the bloom being sweetly perfumed. Of strong constitution.

BON ACCORD ELEGANS. Outstanding amongst the Bon Accords but is the most difficult to grow well, requiring a rich diet. The attractive orchid-pink flowers are edged and flecked with white.

BON ACCORD GEM. The blooms are of bright rosy-red, shaded with mauve and are produced with freedom. The attractively waved petals accentuate the beauty of the blooms. The easiest and most vigorous in this section.

BON ACCORD JEWEL. A magnificent variety but is now rarely to be found. The large deep purple flowers are shaded with crimson on the reverse side.

BON ACCORD LAVENDER. The blooms are large with attractively waved petals and are of a lovely shade of purest lavender with a golden centre.

BON ACCORD LILAC. Of easy culture, the blooms are flat and are of a pleasing shade of lilac-mauve, the petals being marked with yellow at the base.

BON ACCORD PURITY. This must surely be one of the loveliest of all double primroses. The blooms, which are large and fully petalled, are of a lovely shade of creamy-white tinged with green, providing a most attractive appearance where growing in partial shade. The blooms, with their frilled petals, are held on sturdy footstalks so that they are held above the bright green foliage.

BON ACCORD PURPLE. The blooms are large and borne on polyanthus stems and are of a glorious shade of burgundy-purple flushed with crimson on the reverse side.

BON ACCORD ROSE. This variety of robust constitution and of semi-polyanthus habit was sent to me as 'Old Rose', yet I can find no trace of such a variety in any gardening literature, whilst 'Bon Accord Rose' is said to be extinct. My belief, however, that it is 'Bon Accord Rose' is because of the dainty, rounded bloom which is rather flat like most of the Bon Accords and has the same conspicuous orange markings at the base of each petal. The habit, too, is the same as the Bon Accords. The colour is deep old rose.

BURGUNDY. The plants in my garden bear a large well-shaped bloom of deep burgundy-red, flecked with white. This is an old variety which retains its original name but is tending to lose vigour unless well grown.

BUXTON'S BLUE. It bears bloom of pure turquoise-blue and is a plant of vigorous habit, though now rarely to be found. It is a 'sport' from the single blue primrose, found in a garden at Bettws-y-Coed at the beginning of the century.

CASTLEDERG. A chance seedling raised by Mrs Scott of Castlederg, Co. Tyrone, the large star-shaped blooms being of deep sulphur-yellow, splashed with pink and brown. The early blooms may open single, but it is quite lovely and a plant of vigorous habit.

CHEVITHORNE PINK. Found in the gardens of Chevithorne Barton in

Devon, it makes a compact plant and is of easy culture. It remains long in bloom and bears small, beautifully formed flowers of a lovely shade of orchid-pink held on short polyanthus stems.

CHEVITHORNE PURPLE. This plant now seems much more rare than its pink counterpart but is of similar habit. The blooms are of deep purple-blue, the petals being edged with white.

CLOTH OF GOLD. A variety of robust constitution and very free flowering. Thought to be a 'sport' from the 'Double Sulphur', for it possesses the same habit, having large pale green leaves. The bloom is bright yellow but is of not so deep a colour as Carter's 'Cloth of Gold' which is now extinct.

CRATHES CRIMSON. This lovely variety was said to have been found in the grounds of Crathes Castle in Scotland, though from the familiar shape of the blooms it could be 'Bon Accord Brightness', thought to be extinct. The round, neat flowers are of bright purple-crimson and possess a sweet perfume.

CRIMSON EMPEROR. It possesses a rather stronger constitution than 'Crimson King', whilst the bloom has a slightly more purple tint. It is, however, said to be the same variety which may reveal variations of colour and vigour where growing in different gardens.

CRIMSON KING. This is the same variety as the Old Scottish Double Red. It makes a plant of reasonably sturdy constitution and bears large numbers of big, fully double blooms on a short polyanthus stem, the colour being deep ruby-red.

CURIOSITY. Also known as 'Golden Pheasant', this variety was at one time grown in my garden in large numbers, but over-propagation to supply the demand has reduced its vigour. A pity, for the bloom is most interesting, the deep yellow ground colour being flecked with rose.

DOUBLE WHITE. Gerard's 'Double White' and a plant of true primrose habit, the fully double blooms being of paper white and held on long, single footstalks making them ideal for posy bunches, for which purpose the plants should be grown under cloches (see plate 51).

DOWNSHILL ENSIGN. It was raised with a number of others by Mr Murray Thomson early this century. The then recently introduced *P. juliae* seems to have had some influence in their raising for each of the dozen or more varieties bear bloom of various shades of blue and purple. The variety 'Bluebird', now lost but which received an Award of Merit in 1930, appears to have been the best for its blooms were 1½ in. (4 cm) across and were of a lovely shade of lavender-blue. 'Ensign' is the last of the double primroses to bloom, its rather shaggy blooms being of bright violet-blue and held on very long footstalks from a short polyanthus stem. The foliage is smooth and of brilliant green.

FRENCH GREY. The earlier writers named this variety 'Dingy', and the name fits the description well, the blooms being of a dirty white shade or French white or grey as it is called. The blooms are borne with freedom on long footstalks whilst the plant is of easy culture.

MADAME POMPADOUR. This lovely variety, so difficult to grow, bears a large double bloom of deep velvety crimson on single foot-stalks. The texture of its bloom is unique, hence it is always in demand at more than one pound a root. It originated in France about 150 years ago and is also known as 'Crimson Velvet'.

MARIE CROUSSE. A strong, easy grower, increasing rapidly, and bears a large, densely double bloom on short, sturdy polyanthus stems. The blooms are of a lovely shade of Parma-violet, splashed and edged with white, and they carry delicious perfume.

The description given in *The Garden* for April 1882, shortly after the plant had earned the Award of Merit from the R.H.S., describes it well; 'the blooms are 1 in. across and perfectly double, the petals forming a compact rosette . . .'

OUR PAT. Found amongst a batch of *P. juliae* at the Daisy Hill Nurseries, Newry, and was named after the owner's daughter. It is unusual in that the olive-green foliage is veined with crimson-bronze. The small sapphire-blue flowers are borne in profusion on long footstalks. The plant is of vigorous habit and is of easy culture and is one of the last of the doubles to bloom, thus extending the season.

PRINCE SILVERWINGS. A variety of polyanthus habit, the crimson-lilac blooms being flaked with white and edged with silver. The petals are tinted with orange at the base to form a bloom of great beauty, but though in no way difficult to grow it is now rare. It sometimes bears semi-double blooms which yield pollen and so may be used for hybridizing.

QUAKER'S BONNET. A very old variety probably a 'sport' from *P. rubra*. The bloom is the most beautifully formed of all the doubles, making a rosette of perfect symmetry and being of purest lilac-mauve. It blooms with as great a freedom as 'Double White' and is a plant of sturdy constitution.

RED PADDY. It is the *Sanguinea plena* and *Rubra plena* of old Irish gardens and a charming variety. It is a strong grower and bears large numbers of small, symmetrical rose-red blooms which have a salmon-pink flush and an attractive edge of silver. The blooms are flat and dainty and possess a sweet perfume.

TYRIAN PURPLE. One of the best of all primroses, indeed it is one of the best plants in the garden. The blooms are the size of a ten-penny piece and held on sturdy primrose footstalks. The colour of the bloom is bright purple, flushed with crimson, whilst the foliage is brilliant green. It makes a large plant of robust constitution. Is said to have been found in Cornwall.

WILLIAM CHALMERS. Raised at Stonehaven in Scotland, it is a magnificent variety, the large well-formed blooms being of deep midnight blue, flushed with purple.

POLYANTHUS

The polyanthus has both *Primula veris,* the cowslip and *P. vulgaris,* the primrose in its parentage which combine to form the hybrid oxlip (as distinct from the true oxlip, *P. elatior*) from which the polyanthus has been evolved. The French botanist Clusius (De l'Ecluse) in his *Rariorium Plantarum Historia* (1601) described it as *Primula veris pallida flore elatior,* the Larger Pale-flowered Cowslip and like the cowslip, it enjoys a more open situation than *P. elatior.* It was possibly from a crossing of the hybrid oxlip with John Tradescant's Red or Turkie-purple primrose obtained from the Caucasus, that the first red polyanthus or 'big oxlip' was obtained and was described by Rea in his *Complete Florilege* (1665). He wrote: 'The red cowslip or oxlip is of several sorts, all bearing many flowers on one stalk . . . some bigger, like oxlips' and which must have resembled a polyanthus of inferior quality. It was not until Miss Gertrude Jekyll in the year 1880, found in her garden a plant bearing yellow flowers, that a polyanthus of this colour appeared.

GOLD LACED POLYANTHUS

It was about the year 1750 that the polyanthus took on the now familiar striking gold lacing or edging, at the same time that the auricula took its paste-like centre. The Gold Laced Polyanthus was possibly the result of a crossing of the red polyanthus (the yellow was not then known) with the natural hybrid *Primula pubescens,* to be found growing wild in the Austrian Tyrol. Several varieties of *P. pubescens* have the same golden centre as if the centre disc had been treated with gold leaf. The Gold Laced Polyanthus has a similar golden centre and an edging of gold which appears quite brilliant against the ground colour of crimson or black (see plate 52). It has also the same sweet perfume as *P. pubescens* and of all the auriculas, so it would appear that they have a common ancestry.

The polyanthus being the result of a cross between the primrose and cowslip may be said to have taken on the characteristics of both plants. It requires plenty of moisture at its roots during the summer months and whilst it will flourish in shade it is happiest, in the words of Thomas Hogg, 'in a situation exposed to the morning rays of the sun and excluded from them for the rest of the day'. Here again, in choice of situation the polyanthus may be said to come halfway between the primrose and cowslip, the former appreciating dappled shade, the latter a position of full sun. The primrose is a flower of the hedgerow, the cowslip of the open meadow especially where low lying, to enable it to receive sufficient moisture. Provided the plants are given an abundance of humus about the roots, the polyanthus will be quite happy in full sun for in such situation are the plants set out by the cut-flower growers of Cornwall. Where a garden is exposed to the direct rays of the sun, the polyanthus may be grown to the same perfection as where grown in shade, though shelter from the midday sun will prevent the blooms from fading and will ensure an extended flowering season.

An orchard is the ideal place in which to grow on the plants during summer after they have finished flowering in the beds, when they must be divided and replanted into a humus-laden soil and kept well-watered until the new roots have formed.

A light, sandy soil, which tends to dry out in summer, will never grow such sturdy plants as a heavier loam, unless well-fortified with humus materials. With a heavy soil, humus is needed to disperse the clay particles so preventing the soil from 'panning' and improving drainage; whilst with a sandy soil, humus is required for the retention of as much summer moisture as possible. Where in partial shade, a light soil well-enriched with humus should be capable of supporting healthy plant growth even during periods of prolonged dryness. Liberal quantities of peat, used hops, shoddy, old mushroom bed compost and decayed farmyard manure should be used as liberally as possible, incorporating to a depth of at least 18 in. (45 cm).

GROWING GOLD LACED POLYANTHUS FROM SEED

Polyanthuses are best raised from seed sown in boxes or pans in a cold frame. If seed is sown in April, the seedlings will be ready for transplanting to a frame or into the open ground during June. The John Innes Sowing Compost is suitable. Careful watering at all times after sowing is essential for the seed will not germinate if the compost is dry, whilst the young plants will suffer irreparable harm if allowed to become dry at the roots.

The seed will germinate in about three weeks and the seedlings will be ready to transplant in about a month, spacing them 3 in. (7.5 cm) apart into fresh compost. The plants will come into bloom the following spring (see plates 52–7).

DIVIDING GOLD-LEAFED POLYANTHUS

Spring or early in July is the best time to divide old plants, when the ground is usually damp.

The plants should be lifted with care so that the fibrous roots, especially those near the surface, will be in no way damaged. Carefully shake away all surplus soil then, firmly holding the plant, pull apart the various crowns. In the apt words of that great gardener, Miss Frances Perry, 'tease them apart' (see plates 58 and 59). In this way each crown will have its full quota of fibrous roots, whilst there will be no open wounds as may be the case where the plants are cut into sections. Every crown, however small, will grow into a flowering plant.

Where the plants have formed large clumps which they will do if given good cultivation, they should be divided like any other herbaceous plant, by placing two border forks back to back at the centre of the plant and gently prise apart. The sections may then be divided as described into numerous offsets and these should be removed to a cool-shaded place without delay and where they remain until ready for replanting as soon as possible. Any unduly large leaves may be screwed off about 3 in. (7.5 cm) above the crown before replanting in the same way as when lifting and removing the tops of beetroot, for it is not necessary for the plants to have to re-establish these coarse outer leaves. Offsets with new and smaller leaves should be replanted without the removal of any foliage, for these young leaves will catch the dew and rains and direct the moisture to the roots thus enabling the plants to become more quickly established. The offsets should be set well into the ground and pressed firmly with the hand, or with the foot where the soil is friable. They should be kept well-watered until established.

POLYANTHUS STRAINS

BLUE STRAINS. In Blackmore and Langdon's strain, the blooms are large, with a conspicuous yellow centre, the colours ranging from dark purple-blue to palest sky blue, the paler colours being more attractive. The American 'Marine Blue' strain raised at the Barnhaven Gardens from G. F. Wilson's original 'blue' has not so sturdy a habit, neither is the bloom quite so large, but the colour is fixed somewhere between sky- and china-blue. They possess an attractive 'smoky' sheen whilst being free from the red and purple colouring which spoils so many blue strains. The 'South Sea Island Blue' strain of Tasmania is another combining the best of the blue strains.

BRILLIANCY. Excellent for cutting, the trusses are borne on long, wiry stems and whilst the individual pips are not large they are of the most vivid colourings imaginable; terracotta, tangerine, rust and orange being represented, whilst all other colours have been eliminated. One of the brightest flowers of the garden, and useful for cloche culture.

CROWN PINK. This beautiful strain was introduced by Miss Linda Eickmann shortly before her death in 1956, and was the first real pink strain of uniformity. The colour is clearest pink without any trace of rose or salmon. The blooms are large and well formed. Miss Eickmann introduced a number of named pinks from her strain such as 'Warm Laughter' and 'Radiance'.

MUNSTEAD'S STRAIN. Miss Jekyll's original Munstead strain is still marketed by Messrs Carter's Tested Seeds, who during recent years have improved it out of all recognition to that of fifty years ago. Though the colour range comprises only the original white and yellow tones, new oranges and pale yellows and much larger blooms now make this a most attractive strain, especially for planting in shaded positions where those of more sombre colouring would not be shown to advantage.

PACIFIC STRAIN. Raised in California by Messrs Vetterle and Reinelt, the colour range is wide, embracing blues, pinks, crimsons and pastel shades with a large golden centre. The plants are free flowering but as their place of introduction would suggest, they have not the same hardiness of British strains.

SUTTON'S FANCY. The blooms or pips are large with the truss flat and compact and the stems 9 in. (23 cm) in length. The stems are extremely thick to withstand adverse weather, whilst the colour range includes chiefly the art or pastel shades of peach, pink, apricot, lilac and cream.

SUTTON'S SUPERB. Whilst the 'Fancy' strain is composed chiefly of pastel shades, the 'Superb' strain is made up of more brilliant colourings – bronzes, maroons and orange tints, the habit being similar to that of the 'Fancy' strain and so making it very suitable for bedding.

TOOGOOD'S GIANT EXCELSIOR STRAIN. The large flower trusses are held on 10 in. (25 cm) stems and so are ideal for cutting or bedding though preferably for the latter purpose. The colour range includes most of the pastel shades of apricot, peach, pink and cream, the magentas having been eliminated.

NAMED POLYANTHUSES

There are a number of polyanthuses which may be the result of a primrose-polyanthus cross, but which because of their long stalk and the formation of a compact flower truss may be classed as of true polyanthus habit. The blooms of each are somewhat daintier than those of the modern polyanthus strains and the plants are of more compact habit. With their intensely rich colouring they are amongst the most attractive of all garden plants. All possess extreme hardiness, whilst they are very flowering.

BARROWBY GEM. with 'Beltany Red' this is one of the finest of all spring plants, ideal for window-box or rockery with its sturdy habit. It was raised in Scotland by Mrs McColl and is now rarely seen. It is the first polyanthus to come into bloom, the first pips opening on a mild February day, whilst its large umbrella-like heads remain colourful until June. The bloom may be described as primrose-yellow shaded green, and it carries a pleasant almond-like perfume.

BARTIMEUS. This is believed to be a polyanthus of the eighteenth century. It bears a bloom of velvety crimson-black and has no eye; in its place is a region of bronzy red. The blooms are not large nor does it form a large head in comparison with modern standards. This is the old Eyeless polyanthus, a connoisseur's plant.

BELTANY RED. Its origin is unknown, but it is one of the finest of all garden plants. It forms a stocky, compact plant and bears a large truss of tangerine-red blooms, which have an unusual green centre and an attractive wire-edge of gold. The leaves are vivid green. The plant remains ten weeks in bloom, and two or three planted together can be seen from afar.

FAIR MAID. It originated in Perthshire, so it is well-named. The small, but beautifully rounded blooms are freely produced on numerous 16 in. (40 cm) stems, their colour being burnt orange-scarlet with a most striking double centre of gold. The blooms remain fresh in water for fully two weeks.

HUNTER'S MOON. A modern polyanthus and a beauty, for like 'Barrowby Gem' it comes into bloom before all others and carries a fragrance the equal of 'Ena Harkness' rose. Of sturdy habit, the bloom is of a lovely shade of apricot with a chrome yellow centre.

INDOOR PRIMULAS

Of all greenhouse plants, the Chinese primulas are the most accommodating and after the cyclamen are the most popular of all flowering pot plants. They are also the easiest of greenhouse plants to raise from seed. Not only will the plants bloom profusely in a greenhouse which has just sufficient heat to keep out winter frosts, but they are also excellent house plants requiring a cool, northerly window where they will bloom for weeks on end. They are at their best between Christmas and early summer. All the attention they require is careful watering, for too much moisture will cause the plants to decay at the crown. They also require cool conditions. It is important that the plants do not become frosted although they will not tolerate a warm, close, atmosphere. The indoor primulas should be given a minimum temperature of 42°F. (5°C.) and a maximum temperature of 52°F. (11°C.), somewhere between the two being ideal. They are plants for a cool but not a cold greenhouse and the same may be said of their requirements where growing in the home. They will not tolerate forcing in any stage of their growth.

There are four types of greenhouse primula. Two are annuals, *P. sinensis* and *P. stellata*; whilst *P. malacoides* is usually given biennial treatment. The fourth, *P. obconica,* is of perennial habit and is the most important of the group. It makes an excellent house plant for it remains almost perpetually in bloom, until the plant appears to have exhausted its compost, and after two or three years it may have flowered itself to death.

With greenhouse primulas the plants must receive no check throughout their life, whilst they should be grown as cool as possible. Each of the primulas will be grown in the same way apart from a slight variation in the composition of the compost, and whereas *P. obconica* should be sown in spring to come into bloom at the beginning of the following year, *P. malacoides* will be sown in May and June to bloom the following year. Both *P. sinensis* and *P. stellata* are sown in March to bloom at Christmas. If a second sowing of each is made two months later in the year, the plants being treated as biennials, they will come into bloom the following spring and will continue the succession of bloom right into summer with *P. obconica* flowering continuously the whole year round.

Apart from the ability of the plants to reach perfection within ten months of sowing the seed, the cleanliness of the plants contributes to their popularity, for unlike the cineraria and certain other greenhouse plants, they carry no filth in any form. Each of these indoor primulas requires much the same culture, but it is *P. obconica* which is the most popular, not only on account of its greatly extended flowering period, but because, as far as the nurseryman is concerned, it transports so much better. The blooms do not drop their petals with the same readiness as those of *P. malacoides* when being moved.

RAISING INDOOR PRIMULAS

Little or no artificial heat will be required for the germination of the seed, for a greenhouse which has been lined with polythene will obtain sufficient warmth from the spring and early summer sunshine. The seed is sown in a box or pan and as the seed is small it should be sown as thinly as possible. Place the seed in the palm of one hand and transfer with the thumb and first finger of the other hand, scattering it as evenly as possible over the level surface of the compost.

For *P. obconica*, John Innes Seed Sowing Compost is suitable.

Whilst *P. obconica* will not tolerate leaf mould, *P. sinensis, P. malacoides* and *P. stellata* thrive on leaf mould and when preparing their compost, leaf mould should be substituted for the peat. Should it not be possible to obtain sterilised soil, the compost should be watered with Cheshunt Compound before sowing, to prevent Damping Off disease. It will act as a preventive rather than a cure for if the seedlings are attacked there is no remedy. It is advisable to water the compost and seed box (or pan) with Cheshunt Compound before and after the seed is sown whether the compost has been sterilised or not, for the disease may be introduced in many ways: by the boxes, by the hands or by water, and it may be readily present in leaf mould.

After sowing, only lightly cover the seed with dry compost, then give a sprinkling with water and cover with a sheet of clean glass and brown paper. The compost should be inspected daily for it must never be allowed to dry out. On the other hand it must not be kept in a saturated condition so never give water unless really necessary.

The seedlings will germinate in about a month from an April or early summer sowing, but will take ten days longer if sown in March. Until the seed has germinated, the greenhouse should be kept closed unless other plants requiring fresh air are growing alongside. But as soon as the seed has germinated, the coverings should be removed for it is essential that the tiny plants receive ample ventilation from then onwards. Should the sunlight be too strong, the glass on the sunny side of the greenhouse should be whitened and this may be washed off at the end of summer.

GROWING ON INDOOR PRIMULAS

Primulas must be grown as cool as possible but they enjoy a humid atmosphere so long as it is not stuffy. To reduce the need for watering to a minimum, it will be advisable to damp down the floor of the greenhouse on all sunny days. Maintain the compost in a comfortably moist condition and as soon as the seedlings are large enough to handle they should be transferred to slightly deeper boxes; spacing them just under two inches apart. The John Innes Potting Compost No. 1 should be used, substituting leaf mould for the peat for all except *P. obconica*.

When transplanting primulas it is important to remember that at no stage will the plants tolerate deep planting or the crowns will decay. Merely set the fibrous roots into the compost and water them in. In this way the plants will almost 'sit' on top of the soil.

After about four weeks in the boxes, the plants will be ready for potting, the size 60 pot being suitable and in this size pot the plants will bloom. For a compost use the John Innes No. 2, once again, substituting leaf mould for the peat for all except *P. obconica*. The pots must be clean, well crocked and deep planting must be guarded against. Make the plants quite firm but do not cover the crowns and as a further safeguard against decay, it is advisable to give water around the side of the pot. Plenty of fresh air and a moist atmosphere will be essential for the building up of a sturdy plant and with the advent of the cooler weather of late autumn, often accompanied by fog, it will be advisable to discontinue damping down and to raise the night temperature by artificial means. A temperature of between 45°–50°F. (7–10°C.) should be maintained and never at any time give excessive heat in order to bring the plants into bloom more quickly. Allow the plants to take their time, feeding occasionally with soot water to enhance the colour of the bloom and giving fresh air on all suitable occasions.

By Christmas, those plants raised from a spring or early summer sowing will be showing their first bloom when several may be taken into the home to be placed in a light, sunny position. There, with the minimum of attention, they will bloom for months, *P. obconica* remaining in bloom for two years or more. To prolong the display and to keep the plants tidy, dead blooms should be removed as they form, together with any leaves which show signs of decay. It should be said that no water must come into contact with the foliage after the end of September, otherwise spotting and damping off may occur.

SPECIES AND VARIETIES OF INDOOR PRIMULAS

Primula malacoides. It has a dainty, feathery habit, the small flat blooms being borne in great profusion on 9 in. (22 cm) stems above the attractive hairy foliage. A well-grown plant will carry more than two dozen flower stems on each of which are a dozen or more individual blooms borne in whorls or tiers in the Chinese pagoda style. Great improvement has been made with this plant in recent years and there are now some lovely varieties.

CONGRATULATIONS. A new variety of dwarf; compact habit and the earliest to bloom. The flowers, which are borne in profusion are of a lovely shade of salmon-pink.

DOUBLE VIOLET. It is of recent introduction coming early into bloom and bearing flowers of an intense shade of violet-blue which are fully double to give a most graceful effect.

JEAN RUSSELL. It makes a compact plant and bears a profusion of flower stems, the blooms being of deep carmine-rose with a striking yellow eye.

PINK QUEEN. It comes early into bloom and is most free flowering. The blooms are large and are of an attractive shade of clear rose-pink with an unusual green eye.

ROYAL PURPLE. Of compact habit it comes early into bloom. The flowers which open pale mauve, deepen with age.

ROSE BOUGUET. It has small leaves and makes a compact free-flowering plant, the blooms being of brilliant carmine-rose.

ROSITA. One of Sutton's finest introductions, the large glowing rosy-purple flowers being heavily blotched with maroon whilst they have a bright golden eye.

P. obconica. The plant is almost of polyanthus habit, having bright green leaves whilst the flowers, which are the size of a ten-penny piece, are borne in clusters on 10 in. (25 cm) stems above the foliage. A well-grown plant will bear a dozen or more flowering stems at one time and for two years or more a plant will rarely be out of bloom.

AVALANCHE. Beautiful in the purity of its white flowers which provide a marked contrast to the crimson varieties.

BERLINER BLUT. Raised by Ernst Benary at his famous Erfurt Nurseries, the blooms are of medium size and are a rich deep crimson red, freely produced.

BLUE GROTTO. An outstanding variety of German introduction, the blooms being of a lovely shade of Cambridge blue with a striking golden eye.

DURENER LACHSROT. Of vigorous habit, the flower heads are extremely large, the blooms being of a lovely shade of salmon pink.

MULLER'S ROSEA. It is extremely free flowering, the medium-sized blooms being of bright rosy-red.

PEARL OF NEIDERHEIM. The blooms are large, freely produced and are of a unique shade of cherry-red.

WYASTON WONDER. For size of bloom and freedom of flowering, it has never been surpassed. The blooms are of bright rich crimson.

P. sinensis. It is of more compact habit than the others, having hairy, serrated leaves and bearing clusters of brilliantly coloured blooms on 6 in. (15 cm) stems. Though more exacting in its requirements than the others, it is one of the finest of all indoor plants when brought to perfection on account of the clear colour of its flowers.

CARDINAL. An Ernst Benary introduction with short, compact foliage and bearing blooms of pure vermilion-red.

DAZZLER. It received an Award of Merit in 1950, its intense orange-scarlet flowers making it the brightest of all indoor flowering plants. The double form is even more brilliant.

HIS MAJESTY. The blooms are large and are of rich, deep crimson, enhanced by attractive scarlet stems.

PINK ENCHANTRESS. It makes a plant of robust habit and is most free flowering. The rose-pink flowers are suffused with salmon and have an attractive green eye.

Primula stellata. It is the least well known of the greenhouse primulas for it is of less robust habit. The tiny, star-like blooms, however, borne on 12 in. (30 cm) stems, give it a grace not to be found in the others. The bloom may be obtained in a mixture of rich colourings to be found in *P. malacoides* to which it much resembles. The plants are rarely to be found in the florist's shop for they do not travel well but are worthy of growing in the greenhouse.

PRIMULA VERIS

It is the cowslip which is so readily raised from seed sown in a cold frame, or in a seed pan early in April that it is rarely propagated by other means. If kept moist, the seed will germinate within a month, the seedlings being transplanted to boxes of John Innes Compost where they may be grown on for planting into their flowering quarters during October. Or as soon as large enough, the seedlings may be transplanted to open ground beds, spacing them 6 in. (15 cm) apart where they will bloom. The ground should be made clean of weeds and the soil brought into a friable condition, incorporating some peat and a little decayed manure. Old mushroom bed compost is excellent. The plants like a more open situation than primroses but will flourish in an orchard, planting them in beds to provide cut bloom for the home. They are valuable to have in the garden for they bloom several weeks later than primroses, bridging the gap between the spring and summer flowering plants.

PRINCE ALBERT'S SPRUCE, *see* Tsuga

PRINCE OF WALES' FEATHERS, *see* Celosia

PRIVET, *see* Ligustrum

PRUNELLA

Happy growing in ordinary soil and in partial shade, it blooms early in summer and is most attractive for the front of the border. It forms a tufted plant and should be planted 1 ft. (30 cm) apart in November. *P. grandiflora* bears short spikes of purple-mauve flowers during June and July, the form 'Pink Loveliness' bearing spikes of clearest pink.

PRUNUS

A large genus of deciduous trees, ornamental in their foliage and blossom and suitable for large and small gardens, most growing well under town conditions. The genus includes the cherry, plum and almond but under this heading they are grown for their ornamental qualities rather than for their fruit. They grow from 10–20 ft. (3–6 m) tall and bear their flowers in spring and early summer before the leaves appear. The young leaves show red or bronze tinting, and in autumn take on golden colouring. They are amongst the last leaves to fall. The trees should be given little hard pruning for they 'bleed' where cut. When pruning is to be done has always caused controversy. Some do so in midwinter when the flow of sap is lowest, whilst others wait until early spring for then the bark will callous more quickly and heal the wound. Light pruning will cause no trouble and to build up a good head to the tree, tip back any unduly long shoots. This will cause the stems to form side growth and close up any gaps.

It will also be advisable to tip back any side shoots to encourage them to form flowering spurs. The prunus grows well in chalk and limestone soils, hence its liking for the eastern side of Britain but needs reasonable protection from cold winds. It requires an open sunny situation. Propagation is by grafting in spring or by budding in summer.

The almond, *Prunus amygdalus* bears its bright pink blossom in March on the previous season's wood. The double form 'Rosea Plena' is outstanding whilst 'Pollardii' smothers itself in masses of large deep pink flowers and is the earliest almond to bloom.

Of the ornamental peaches, *Prunus persica* 'Russell's Red' with its double blooms of rich crimson is outstanding, whilst 'Clara Meyer' is also double flowering, the blooms being of soft peach-pink. 'Windle Weeping' is a distinctive form bearing double pink flowers.

The ornamental apricot, *P. mume* makes a neat tree and from late February (where sheltered) bears showers of pale pink flowers. The form 'Alphandii' bears double pink flowers and 'Albo Plena', double flowers of icy whiteness.

The Myrobalan plums should be planted for the rich colourings of their foliage. *P. cerasifera* 'Pissardii' is the Purple-leaf Plum whilst 'Nigra' has the darkest foliage of any deciduous tree. It bears its pale pink flowers before the leaves appear. The variety 'Pissardii' crossed with *P. mume* has produced *P. x blireiana* which makes a compact tree with copper coloured leaves and bears large double pink flowers.

The cherries are to be found in abundance and have considerable beauty. *P. cerasus*, one of the parents of the Morello cherry makes a small tree and bears white flowers. The form 'Rhexii' is outstanding, forming a low spreading tree and bearing large trusses of circular double flowers of purest white, like tiny carnations but they have no scent. This is one of the most useful of all ornamental trees for a small garden for it never grows too large and retains its shape without pruning.

The winter flowering cherry, *P. subhirtella,* should find a place in every garden, especially the variety *autumnalis* which bears its semi-double white flowers intermittently from October until March. The form 'Rosea' bears pale pink flowers whilst 'Stellata' bears larger flowers of rose pink which are borne at the ends of the branches.

Of the Japanese cherries, varieties of *P. serrulata,* 'Amanogawa' resembles a Lombardy Poplar in its habit. Several may be planted only 3–4 ft. (1 m) from each other and may be used to hide an unsightly wall or building. They bear fragrant semi-double blooms of shell pink. Of quite different habit is 'Cheal's Weeping', its graceful arching branches being wreathed in double flowers of deepest pink. Resembling *P. subhirtella autumnalis* in its winter flowering is 'Fudanzakura', which makes a small neat head and bears single white flowers whenever the winter weather is mild. 'Kanzan' is of quite different habit, quickly reaching a height of 30 ft. (9 m) and growing 15 ft. (4.5 m) across. It bears semi-double rose-pink flowers mid-season. In the past it has been wrongly named Hisakura. Plant with it 'Mount Fuji' ('Shirotae') which bears dazzling snow-white flowers on almost horizontal branches. These two are suitable only for a large garden. Different in that its flowers are of a lovely shade of creamy-yellow is 'Ukon' ('Luteo-virens') whilst its leaves retain their copper tinting. To extend the flowering season, 'Daikoku' bears large double flowers of rose-pink.

Prunus lusitanica is the Portugal Laurel which will grow up to 30 ft. (9 m) in height and has large ovate glossy leaves and bears, in July, white flowers in long racemes.

PSEUDOLARIX

This the Golden Larch of east China, ***P. amabilis***, resembles the Common Larch with its lance-shaped leaves borne in tufts. They are of bright green, turning to rich golden-yellow in autumn before they fall. The shoots are also bright yellow and are effective after the leaves have fallen.

Plant November to March in a lime-free soil and in a sheltered situation. It requires a deeply dug soil containing peat. Propagate from seed sown in outdoor beds in spring and where the young plants remain for two years before transplanting.

PSEUDOTSUGA

The Douglas Fir which requires a moist, lime-free soil. It is a tree of extreme hardiness, used for afforestation almost to the Arctic and from sea level to 10,000 ft. (3,000 m). Of majestic proportions, it has shining reddish bark and drooping branches with glossy leaves arranged in two rows. When crushed, the foliage is powerfully scented. The oblong cones are 4 in. (10 cm) long with rounded scales. ***P. douglasii*** (syn: ***P. menziesii***) is the fastest growing of all conifers and in the wild will attain a height of 300 ft. (90 m) where given the protection of other trees.

Plant in April and if grown as a specimen, it should be given a sheltered situation. Propagate from seed sown in spring in outdoor beds, leaving the young plants two years before transplanting.

PTEROCARYA

The Wing Nuts, fast-growing deciduous trees which have many attractions. They require a deep, moist loam and are suitable for planting by the side of ponds and streams. With their deeply furrowed bark and large pinnate leaves they are most handsome trees whilst they make rapid growth and provide valuable shade. Prune only to thin out over-crowded wood. Propagate by seed or by cuttings of the half-ripened wood.

P. fraxinifolia has furrowed, grey bark and leaves which measure up to 2 ft. (60 m) in length. It bears its pale green flowers in drooping catkins. *P. stenoptera* will grow as fast as a poplar and has large pinnate leaves like the black walnut and with winged stalks.

PULMONARIA

Valuable plants in that they bloom during early spring and will flourish in partial shade and in a moist soil. Or plant in beds edged with yellow polyanthus or yellow primroses. Plant in autumn 12 in. (30 cm) apart for with their tufty habit the plants remain compact. Propagate by root division.

SPECIES AND VARIETIES

P. angustifolia azurea. During March and April it bears its funnel-shaped flowers of rich gentian-blue in great profusion. A new variety, 'Munstead Blue', bears rich blue flowers on 10 in. (25 cm) stems during April and May.

P. saccharata rubra. The Bethlehem Sage, which bears sage-green leaves strangely spotted with white. Its salmon-coloured, funnel-shaped flowers are borne during early spring.

PUNICA

The Pomegranate, the 'Many-seeded Apple' of ancient Carthage, *P. granatum* will not fruit in Britain outdoors, but south-west of Bristol, it will grow to a height of 15 ft. (4.5 m) and is a magnificent sight with its dark green leaves and fiery scarlet flowers produced late in summer. There is a double-flowered form which is even more arresting.

The punica is deciduous and plants should be set out late in spring. Prune by removing any dead wood and cutting back straggling shoots in autumn.

PURPLE CONE FLOWER, *see* Echinacea

PURPLE LOOSESTRIFE, *see* Lythrum

PURSLANE, *see* Portulaca

PUSCHKINIA

The Lebanon Squill which blooms throughout spring enjoys a position where sunshine may reach it, also a well-drained soil, for an excess of winter moisture will cause the bulbs to decay. Work in plenty of grit and coarse sand in addition to some peat or leaf mould, and plant the bulbs 4 in. (10 cm) deep in October. They may be planted about a rockery, in the alpine house or beneath trees where the grass has died back and will not need cutting. But for the plants to be a success, it is necessary for the bulbs to be exposed to summer sunshine to become well ripened.

Native of the hilly country stretching from Afghanistan to the Lebanon, the individual blooms are of such beauty (see plate 60) that they should be enjoyed at close quarters, and so are at their best in the trough garden or window box, or in small pots indoors. A raised rockery will also show them to full advantage. Easy to grow, the bulbs cost very little and should be freely planted.

SPECIES

Puschkinia scilloides. Known as the Lebanon or Striped Squill, its blooms of palest sky-blue, are striped with porcelain-blue down the centre and at the sides of each petal. Several blooms, which measure $\frac{1}{2}$ in. (1.25 cm) in diameter when fully open, are borne on each stem of 6 in. (15 cm) above upright leaves of rich, glossy green.

PYRACANTHA

The pyracanthas, with their brilliant orange, scarlet, or yellow berries in autumn and winter, their small evergreen, glossy leaves, and the ability of the plants to cover a wall with the minimum of support, are amongst the best of all wall plants. They will grow on any wall except one facing south – for they do not like baking – but will withstand severe winds and the hardest frosts. The plants will provide colour almost the whole year round, the small white flowers, with their slight honey fragrance, appearing during June and July, followed by clusters of brilliantly coloured berries from September until almost the end of winter. And all the time the box-like foliage remains green and glossy

Though attaining a height of 15 ft. (4.5 m) but making slow growth and being of compact habit, the pyracanthas are suitable to grow on either side of a doorway. They may be grown against a trellis or on any cool wall, where they may be kept in bounds and will require the minimum of support. With its similarity in bloom and berries, the pyracantha may be classed as an evergreen hawthorn, and in fact is closely related, having the same formidable spikes on its stems.

Plant in March, when the berries have either fallen or have been removed by birds. A pot-grown plant will establish itself more readily. The pyracantha is quite happy in a sandy soil, but the incorporation of some decayed manure and a little bone meal at planting time will help for quicker growth.

The only member of the family which may be classed as being in any way tender is *Pyracantha angustifolia,* which should be planted in the angle of a wall to protect it from cold winds. It retains its yellow, flushed orange berries right through winter. *P. coccinea* 'Lalandii' is extremely hardy, but of recent years has been susceptible to black fly, which may disfigure the berries. Dusting with Lindex will keep the plant healthy, and for its abundance of vivid orange berries it should be widely planted.

More vigorous is *P. gibbsii,* which has slightly larger leaves and bears clusters of berries, the colour of which eventually turns a shade of crimson-red. *P. rogersiana* is more compact, and is ideal for a low wall. It forms brilliant orange berries, and may be clipped to form a thick flat bush. Plant near *P. flava,* of similar habit, for its bright yellow berries are a pleasing change from the red colours and they are enhanced by the plant's extremely dark foliage.

PYRETHRUM

An important plant both for garden decoration or for cutting, for it bears its bloom late in June, immediately after the spring flowering bulbs have finished, and before the border comes into its best. Like the pinks, scabious and *Chrysanthemum maximum,* the flowers do not drop their petals and are long-lasting in water. The bloom is cut with as long a stem as possible, and is made into bunches of ten. As the stems have a habit of falling over and becoming bent and twisted, it is advisable to place long twigs about the clumps as they come into bud in May, as for the Oriental Poppy. The plants like a rich loamy soil, in which plenty of decayed manure is incorporated, and they should always be planted immediately after flowering, in July. Failing this, plant in early April just as growth commences. When growing commercially, plant 18 in. (45 cm) apart in rows and the same distance apart in the border in groups. This is not a suitable plant for a shallow soil, for in such a soil the blooms will be small and are borne only on short stems. Propagation is by root division for named varieties, or the plants may easily be raised from seed. Like the *Chrysanthemum maximum,* the roots are very fibrous and the smallest offsets will quickly grow into a large plant. Named varieties may also be raised from cuttings, taken in May and rooted in a frame in a compost of peat and sand. If kept moist, the cuttings will have rooted in time to be planted out in late July. When dividing old plants, only the outsides should be used; the centre which will have become hard and woody being discarded. When planting, spread out the fibrous roots. So many pyrethrums suffer by planting with the roots bunched together, with the result that those at the centre die back. Plants should be lifted and divided every three years to retain their vigour.

VARIETIES: DOUBLES

ANDROMEDA. A new shell-pink variety, the bloom being unsurpassed for quality but the plant takes time to settle down.

APHRODITE. Almost pure white and very free. Useful in that it is early to bloom.

CAPTAIN NARES. A new rosy-red variety of fine quality. Very early and free flowering.

J. N. TWERDY. Deep crimson which will become the most outstanding double-red as it becomes better known. Stiff habit and free flowering.

KINGSTON GRANDE. A recent introduction bearing large rose-pink blooms of sturdy habit.

PRINCESS DE LAEKEN. The bright double-red blooms are freely produced whilst the plant is a vigorous grower.

PROGRESSION. The salmon-pink blooms are the largest of all the doubles and of excellent form.

QUEEN MARY. Peach-pink of fine form.

WEGA. Bears a bloom of unusual colouring, being pale pink, flushed salmon at the centre. Very attractive under artificial light.

YVONNE CAYEAUX. The tightly double pin-cushion-like blooms are of a deep ivory-cream shade and most useful for wreath making just before the first of the *Chrysanthemum maximum* reach the markets.

VARIETIES: SINGLE

AGNES KELWAY. A colour much required in single pyrethrums, the well-formed bloom being of a deep rose-pink.

AVALANCHE. The best single white. It is early into bloom, free flowering and of pure white.

COMET. The first of the reds to come into bloom, the colour being rich scarlet. Valuable on account of its earliness.

CRIMSON KING. Raised by Mr Herbert Robinson, it is one of the finest of all singles, bearing large handsome blooms with stiff well-placed petals of richest crimson, on 3–4 ft. (1 m) stems.

E. M. ROBINSON. An old favourite, in demand for its clear china-pink colour and great freedom of flowering.

EVENGLOW. Recently raised in Holland and unique in every way. The colour of the bloom is rich shrimp-pink whilst a touch of scarlet in the petals brings the bloom to life under artificial light. It is of vigorous constitution and nitrogenous manures should be used sparingly.

FIREFLY. A new variety, bearing a bloom of a glowing crimson-scarlet with a touch of orange. Greatly in demand by top class florists.

HAROLD ROBINSON. A favourite on the markets for its cherry-red flowers, flushed crimson, give it a luminous appearance. Late into bloom, and valuable for extending the season.

JAMES KELWAY. A fine variety in every way which should be in every collection. The large well-formed blooms are of a deepest blood-red.

JUBILEE GEM. Has been ably described as warm cerise which makes it a most popular variety.

KELWAY'S GLORIOUS. The glowing scarlet blooms are always in demand whilst the habit is free, and stems stiff.

MARGARET DEED. The anemone centre gives it the appearance of being almost double. The petals have great substance and are of a rich crimson colour.

MARGARET MOORE. Unusual and attractive in that its deep pink flowers have a pure white edge.

MAY QUEEN. Not the best of pyrethrums but exceedingly useful in that it is the first of all to come into bloom and so should be in every collection.

MRS J. LEAKE. Deep salmon-pink is as near a description as possible: the large blooms are of excellent form.

SALMON BEAUTY. The colour is distinctive amongst singles, being of a pure salmon-pink, with an attractive sheen. A most popular variety.

SAM ROBINSON. The deep pink colouring comes somewhere between 'Agnes Kelway' and 'E. M. Robinson'. A top class flower for cutting.

SCARLET GLOW. It really does glow, the well-formed blooms being of a bright crimson-red colour very freely produced. Early into bloom.

PYRUS

The ornamental crabs, *Pyrus malus*, are amongst the most colourful plants of the garden both in flower and in fruit. They bloom during April and May. They grow well in a chalk and limestone soil and always remain tidy in growth. They require limited pruning, sufficient only to maintain the shape of the tree. Named varieties are propagated by grafting on to crab stock in spring.

Of many lovely hybrids, *P. coronaria* 'Charlotte' bears large semi-double blooms of shell pink which are strongly violet-scented whilst the large leaves take on rich autumn tints. *P. hupehensis* is a quick growing tree bearing large, scented, white flowers in May whilst 'Profusion' is also scented, bearing large wine-red flowers in generous clusters. The hybrid *P. x eleyi* is arresting, being purple in leaf whilst it is heavily laden with large purple fruits in autumn. It is similar to *P. x purpurea* which bears its rosy-red flowers in spring amongst purple stems and foliage and a heavy crop of crimson fruits in autumn.

QUERCUS

Oaks are slow-growing evergreen or deciduous trees which flourish in a deep loam but which are tolerant of a chalk or limestone soil provided it has depth. The deeply furrowed bark and golden leaves in autumn make them worthy of planting in the larger garden. The deciduous species require no pruning but the evergreen *Q. ilex*, the Ilex or Holm Oak will tolerate hard clipping and is valuable for a hedge. Propagation of the deciduous forms is by means of the 'acorns' sown in October when ripe, whilst the evergreens are best propagated by cuttings removed with a 'heel' and rooted in a frame.

Q. bicolor, the Swamp Oak of North America is valuable for waterside planting. It has flaking bark whilst the lobed leaves are downy beneath. Suitable for a chalk soil is *Q. cerris*, the Turkey Oak which is the fastest growing of the oaks. It has long leaves, deeply indented. For rich autumnal colour *Q. borealis maxima* has leaves which turn dull red in autumn.

Best of the evergreen oaks is *Q. ilex* which makes a dense tree and grows well near the coast. The dark green glossy leaves are downy on the underside. Like the holly, its old leaves fall in early summer. In warmer parts, *Q. suber*, the Cork Oak, may be planted. From its furrowed bark, cork is obtained. Its handsome leaves are dark glossy green above, downy on the underside.

QUICKTHORN, *see* Crataegus

RANUNCULUS

It requires a well manured soil, containing a large amount of humus, but a position facing south protected from cold winds. Planted in a sandy soil containing ample supplies of humus and especially in sheltered gardens, the corms may be left in the ground undisturbed for several years; in those gardens with a heavier soil and in an exposed position, the tubers should be lifted in October and replanted mid-March each year. Those that are allowed to remain in the ground should be given a peat mulch during October.

PLANTING

The tubers have peculiar claws which should be placed downwards when planting them 3 in. (7.5 cm) deep and 4 in. (10 cm) apart. A mixture of sand and peat sprinkled round each corm will be beneficial. The cut-flower grower will plant in beds 5 ft. (1.5 m) wide, to enable picking to be done to the middle of the bed without damaging the outside rows. Plant the corms 6 in. (15 cm) apart each way and make a bed of six rows. If cloches are to be used for covering the rows early in May, make the rows only 4 in. (10 cm) wide and set out the corms in three-row beds leaving a path of 18 in. (20 cm) between each section.

Ranunculus asiaticus has claw-like roots and ternate leaves divided into deeply toothed segments. The flowers, borne on 12 in. (30 cm) stems are like small pompon dahlias and are of the same brilliant colours. The Turban form bears flowers of scarlet, spotted with gold. From this is derived the Scottish form, the blooms being edged with gold. The Turkish strain is now known as Paeony-flowered. It is semi-double and embraces scarlet, gold, orange and white. They may be raised from seed sown in spring in boxes of John Innes Compost but will take three to four years to come into bloom.

RAOULIA

These tiniest of all alpines enjoy a moist root run and as little moisture on the foliage as possible, so are happiest planted amongst crazy paving. *R. australis,* with its silver foliage and fluffy pale yellow flowers, is charming, and so is the prostrate *R. lutescens,* with its grey foliage and golden flowers produced in autumn. On the rock garden, they enjoy scree conditions.

RECHSTEINERIA

One of the best of all greenhouse plants, coming into bloom within ten months of sowing seed which will take a month to germinate. The seed is minute and should be sown in the John Innes Compost but not covered. Maintain a temperature of 65°F. (18°C.) and water carefully. Transplant to small pots when large enough to handle and again to larger pots containing the John Innes Potting Compost. The plants do well in the home, being happy in partial shade. Growing 18 in. (45 cm) tall, the large silver leaves are borne in whorls of four all the way up the stem and from the axils appear trusses of salmon-pink tubular bells. In the home, maintain a temperature of 50°–55°F. (10°–13°C.) and water sparingly.

RED CEDAR, *see* Juniperus

RED-HOT POKER, *see* Kniphofia

RED PEPPER, *see* Capsicum

REHMANNIA

The drooping, tubular flowers of *R. angulata* resemble those of the foxglove, being cerise-red with a scarlet band on the upper lip and orange spots on the lower lip. There is a variety called 'Pink Perfection' which bears tubes of dusky pink. The plants grow about 2 ft. (60 cm) tall and bloom throughout spring and summer, during which time they should be shaded from strong sunlight and frequently syringed.

Seed will germinate readily if sown in May and covered with glass and brown paper. Use the John Innes Compost which must be kept moist, removing the glass and paper upon germination. As soon as large enough, the seedlings should be transferred to small pots. There they remain during winter being given only sufficient moisture to keep the plants alive, whilst a temperature of 42°F. (5°C.) should be provided.

Early in April, the plants should be moved to larger pots containing the John Innes Potting Compost and given additional moisture at the roots will stimulate growth and bring the plants into bloom. Shade from strong sunlight and feed once each week with diluted manure water.

RESEDA

Mignonettes are annuals which are valuable for sowing in a limy soil; they are as fond of lime as the dianthus. They also like a soil containing some humus. Though sweetly scented, the mignonette, *Reseda odorata*, has never, until recently, been conspicuous for the quality of its flowers. With the introduction of the large flowered varieties, this plant has now become attractive in addition to its fragrance. The deep yellow, 'Golden Goliath'; the richly-coloured 'Crimson Giant'; and the bright, 'Red Monarch', growing to a height of 12 in. (30 cm) and making large branching plants, are a great improvement on the old mignonette. Sow early April to bloom from mid-June until mid-August; and to continue the display, make a second sowing in May.

RESURRECTION PLANT, *see* Selaginella

RHAMNUS

The Buckthorn, a genus of deciduous and evergreen trees suitable for planting in exposed coastal gardens. They also grow well in a chalky soil. They are quick-growing and long-living trees. Almost no pruning is necessary. Propagate by layering or by cuttings, taken with a 'heel' and rooted under glass.

R. alaternus 'Argenteo-variegata' grows 15 ft. (4.5 m) tall and has glossy leaves margined with white whilst *R. purshiana* has large prominently veined leaves and red berries. It is the Cascara of commerce.

RHODANTHE

An annual, *R. manglesii* grows only 12 in. (30 cm) tall and may be grown on a sunny rockery or in sandy soil in an open situation. The pretty pink flowers of papery texture may be cut and dried at the end of summer to use for winter decoration. Sow seed in the open in April where the plants are to bloom, or in pans of sandy compost under glass early in March. Plant out late in May after hardening.

RHODODENDRON

Hardy and evergreen, it is happiest in partial shade, sheltered from cold, northerly winds and shielded from the warmth of the midday sun. It requires a deep, loamy soil of an acid nature and containing plenty of peat or leaf mould which should be packed about the roots at planting time. A heavy soil should be brought into condition by incorporating ample supplies of humus forming materials. Of the many hundreds of species and varieties, the naturalised *R. ponticum* which bears its lilac-pink flowers in May, makes a valuable hedge with its large glossy leaves. In bloom shortly after is *R. camplyocarpum* which grows 6 ft. (2 m) tall and bears clear yellow funnel-shaped flowers, even on young plants. In June and July *R. discolor* is in bloom, its scented flowers of deep pink appearing in large trusses.

Of the hybrids, 'Break o' Day', raised at Exbury, bears bright orange flowers in loose trusses whilst 'Coronation Day' has flowers of bright shell-pink. 'Tally-Ho' has large funnel-shaped flowers of brilliant scarlet and 'Diane', flowers of palest primrose.

Plant October to April, preferably early in spring, taking care not to damage the soil ball. Plant firmly and keep the roots well watered until established. Give little pruning apart from the removal of dead flowers. Propagate from seed (do not cover) sown under glass or by layering in autumn which is the best method of increasing stock.

RHUS

Small deciduous trees or shrubs, closely related to cotinus and known as the Sumach. They succeed in poor soils and in coastal gardens where they make small trees and are noted for the rich autumn colouring of their foliage. Apart from the removal of dead wood, no pruning will be necessary. Propagate by sowing seeds in spring or by cuttings rooted under glass.

R. glabra, the Smooth Sumach is perhaps the best species, its pinnate leaves turning crimson in autumn whilst the plume-like clusters of seed vessels are equally striking. *R. typhina*, the Stag's Horn Sumach is an attractive tree for town gardens, its pinnate leaves often measuring 2 ft. (60 cm) long and turning orange, crimson and purple in autumn.

RIBES

The flowering currants are tolerant of every soil and of every situation; and with their attractive pale green pungent foliage and flowers like tiny bunches of grapes, are amongst the brightest plants of the garden. *Ribes sanguineum*, 'Pulborough Scarlet' bears numerous richly-coloured crimson flowers in April and May, whilst 'King Edward VII' bears flowers of deep crimson-red. As a contrast the Buffalo Currant, *R. aureum*, bears fragrant yellowish-buff-coloured flowers in April. Each grows about 4 ft. (1 m) tall and they may be clipped into whatever shape is required. The plants may be used to form a colourful hedge.

ROBINIA

The False Acacia, a genus of deciduous trees, native of North America which grow well in dry soils but as their wood is brittle, they should receive protection from prevailing winds. The pinnate leaves and drooping racemes of pea-like flowers make this genus one of the great attractions of the garden. Apart from the removal of dead wood, do not prune. Propagation is by seeds sown when ripe or by suckers.

R. pseudoacacia is a valuable tree for a town garden with its elegant foliage and bearing fragrant flowers in June. Its branches are covered in spines. The variety 'Decaiseana' bears pink flowers whilst 'Pendula' is of weeping form. Outstanding is 'Frisia' which has leaves of brilliant gold.

ROCK CRESS, *see* Arabis

ROCK JASMINE, *see* Androsace

ROCK PURSLANE, *see* Calandrinia

ROCK ROSE, *see* Helianthemum

ROMNEYA

The Californian or Tree Poppy, *R. coulteri* grows 5 ft. (1.5 m) tall and should be confined to southern gardens where it requires a sheltered, sunny situation and a well-drained sandy soil. It blooms July to October, bearing scented white poppy-like flowers 4 in. (10 cm) across and which are enhanced by their golden stamens. The large compound leaves are of bluish-grey.

Plant in April and as it makes a spreading plant, allow it room to develop. Do not prune except to remove dead wood and propagate from root cuttings as for oriental poppies.

ROSA

To provide colour through summer and autumn, no plant is more labour-saving, nor bears more bloom over so prolonged a period, than the rose. The earliest of the hybrid tea roses to come into bloom will be preceded by *Rosa hugonis*, the Yellow Rose of China, which bears its blooms during May and early June and should be planted either in a border devoted entirely to the shrub roses, or in a mixed shrubbery. A number of the shrubby species, which are of compact habit, also carry a second crop of bloom in autumn whilst others flower continuously from June until well into October, the bloom being replaced by brilliant scarlet hips which persist through winter, adding a touch of warmth to the shrubbery, which tends to be devoid of colour at this time. The old shrub roses have been used in recent years to perpetuate their long-flowering qualities in the hybrid tea rose. Twenty-five years ago bedding roses bloomed during midsummer only; today, they may be expected to remain in flower with but few rest periods, through summer and autumn, and until almost the year end. In a sheltered garden, with certain varieties, it is often possible to gather roses at Christmas.

MAKING A ROSE GARDEN

Thought should be given to the making of rose beds for not only are roses happier in beds to themselves, but the beds should also be made wherever possible in groups, where no other plants will compete. They may be surrounded by paths of crazy paving or of concrete stones; any formality in this design may be removed by planting prostrate thymes about the stones. They will in no way detract from the old-world charm of the roses. A small part of the garden devoted to roses in their numerous delightful forms will give the utmost pleasure. Construct the rose garden by degrees as one's time and finances permit; use rustic poles and clothe them in ramblers, and if possible erect a summer-house and let the climbers cover it in an abundance of summer colour.

PREPARATION OF THE SOIL

Contributing to the popularity of the rose is that once it is planted, it will require little attention apart from yearly pruning, whilst it will do well in almost all soils and in every part of Britain. It blooms as well by the sea as it does inland; in a town garden as in the country; whilst the heavier and colder the soil, the better it grows.

Roses in all soils should be given liberal dressings of humus, particularly well-decayed farmyard manure, whilst when planting into a heavy clay soil it is advisable to incorporate some grit or coarse sand to help with drainage. But farmyard manure must not be omitted and where this is not readily obtainable, wool shoddy, or straw composted with an activator to which is added pig or poultry manure, even composted garden refuse, are most valuable. Peat and leaf mould may also be added. 'They luxuriate in rich manure – coarse fare', as Walter Wright put it so ably in *Popular*

Garden Flowers. Rank manure suits them well, the more the better. But though the rose delights in a heavy soil, this should be brought to as fine a tilth as possible by preparing the beds in November and allowing the soil to become weathered by winter frosts. Roses have a fibreless tap root which is difficult to establish if the soil is excessively lumpy, and during a period of drought the plants may die back if the roots do not make full contact with the soil.

A chalky soil, often above gravel, or a light, sandy soil will grow good roses, but to do so will require humus and decayed manure which is dug in each year, preferably in late autumn. Roses in all soils will benefit from a liberal dressing of humus applied as a mulch around the plants in early summer. This will suppress annual weeds and will conserve moisture in the soil, as well as feeding the plants. Where manure cannot be obtained in quantity, plants growing in light soil should be given a 2 oz. per sq. yd. dressing of sulphate of potash. This will build up a bloom able to stand up to adverse weather conditions, and will also improve the colour and quality.

In a friable loamy soil, roses may be planted at almost any time between November and early April, provided there is no frost in the soil and it is not too wet. Heavy clay loam will rarely be suitable for working between Christmas and early March; the same may be said of November planting, though frequently December is a dry month with few hard frosts, and is possibly the best time for the operation. But the ability of the rose to overcome almost all conditions was made plain when planting my own beds.

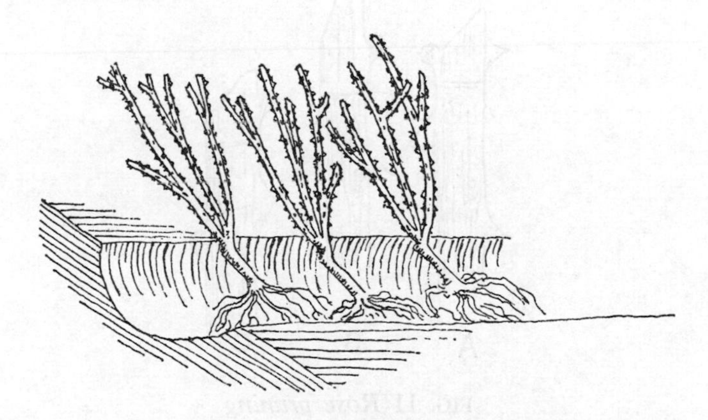

FIG. 10 *Rose trees – roots to be covered awaiting planting*

Not having the land ready before early April, and not wishing to lose a season by delaying planting until the autumn, the trees were planted on 28th April, pruned rather more severely than usual, and given a heavy manure mulch. Several hundred trees came into bloom in early July with only one per cent loss.

When planting, it is important to determine the point at which the trees were previously planted, and to plant at the same level. Too deep planting must not be done. Again, the trees must be very firmly trodden in, even in heavy soil (see plates 61–4). Loose planting will cause sucker shoots to form, and the plants may become badly damaged at the roots and above ground if there is excessive movement during windy weather. After planting, the beds are often given a mulch and are left undisturbed until the end of March, when they are pruned. Do not prune roses for at least several weeks after planting. Heavy pruning and moving the plants at the same time may cause a check from which the plants may take some time to recover. Planting distances vary with almost every variety. The dwarf polyantha roses may be planted at 20 in. (50 cm) apart, so that when they come into bloom in June there will be little or no soil to be seen – nothing but a carpet of bloom. The weaker of the hybrid teas should be planted at about 2 ft. (60 cm) apart, the more vigorous at almost 3 ft. (1 m).

PRUNING

With roses, this operation gives rise to controversy. Whether to prune

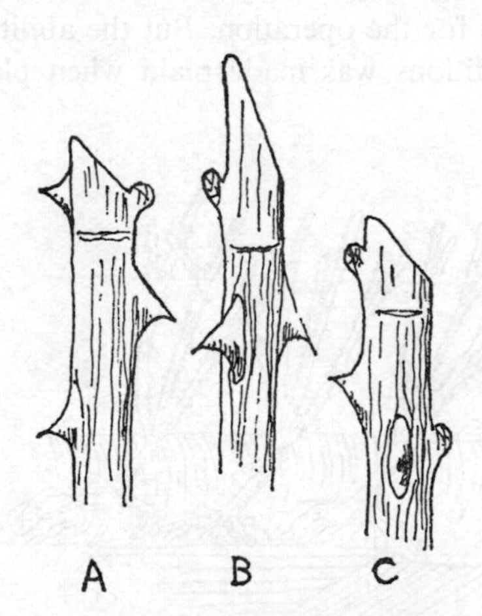

FIG. 11 *Rose pruning*
A. Too near a bud
B. Too far from a bud
C. Correct

hard or moderately hard has still to be decided definitely, and whereas one gardener suggests pruning hard to obtain few blooms of exhibition quality,

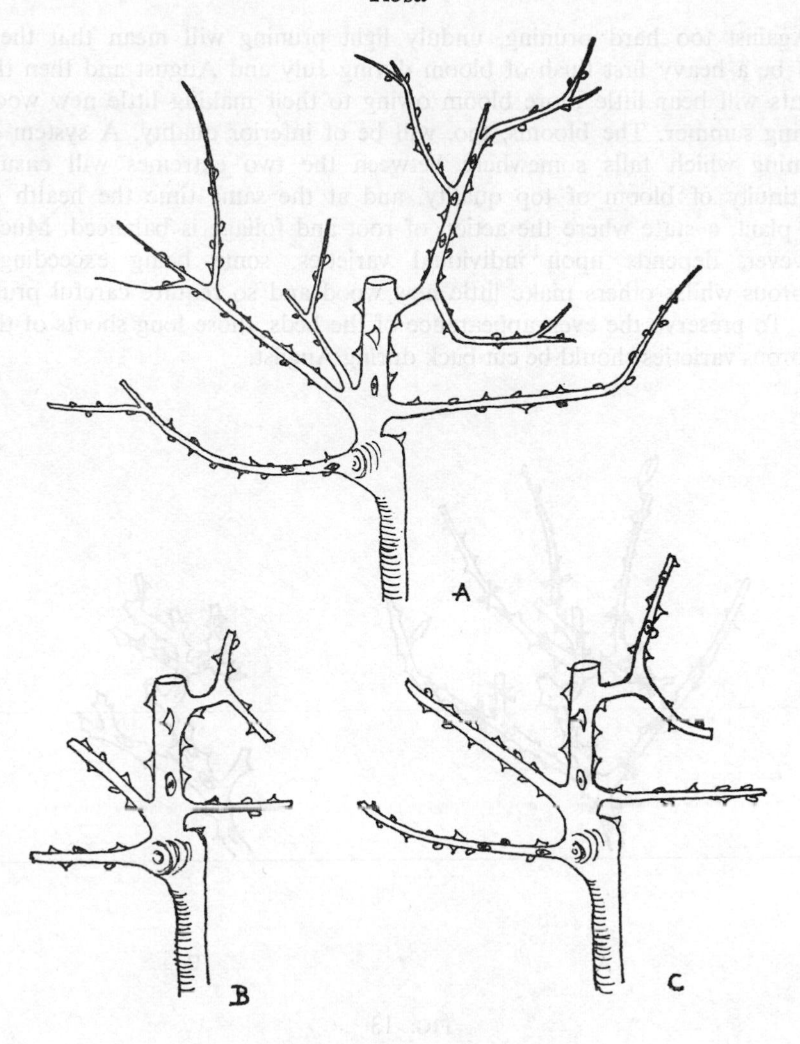

FIG. 12
A. Pruning a standard rose
B. Hard pruning
C. Moderate pruning

others are equally enthusiastic about pruning hardly at all. The most satisfactory way to maintain a healthy, vigorous plant would appear to rest somewhere between the two methods. Severe pruning not only deprives the plant of wood and foliage which is valuable in converting nutrition from the atmosphere, and so necessary in maintaining the health of the plant, but will also restrict the amount of bloom, which for a bedding plant is not desirable.

Against too hard pruning, unduly light pruning will mean that there will be a heavy first flush of bloom during July and August and then the plants will bear little more bloom owing to their making little new wood during summer. The blooms, too, will be of inferior quality. A system of pruning which falls somewhere between the two extremes will ensure continuity of bloom of top quality, and at the same time the health of the plant, a state where the action of root and foliage is balanced. Much, however, depends upon individual varieties, some being exceedingly vigorous whilst others make little new wood and so require careful pruning. To preserve the even appearance of the beds, those long shoots of the vigorous varieties should be cut back during August.

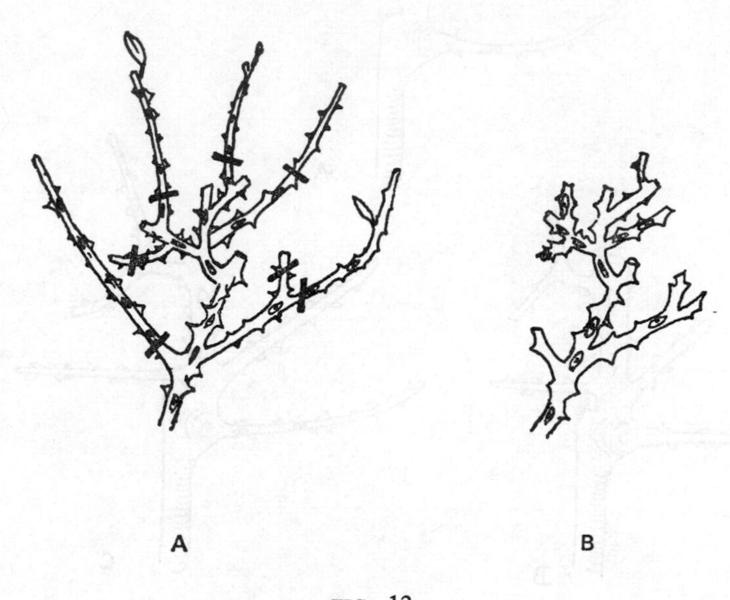

FIG. 13
A. Hybrid tea rose ready for pruning
B. After hard pruning

When to prune also gives rise to discussion. Some growers advise pruning during the dormant or midwinter period. But as all shoots should be cut back to a strong bud pointing away from the centre of the plant, this may be damaged by frost or cold winds if pruning is done at this time. To prune in early spring, about mid-March, when fear of severe frost has vanished, will make for a better plant. All dead wood should be removed, together with any damaged wood or dead blooms, towards the year end as soon as flowering has finished. Where pruning has been done lightly, the plants will be helped to bear bloom over an extended period if the first bloom is cut with as long a stem as possible, or a considerable length of stem is removed with the dead flowers. This will stimulate new growth and act in the same way as if the pruning had been moderate.

With standard roses, original pruning will consist of cutting the head hard back in early spring after the trees have been planted so as not to put too much on the constitution of the plant until the rooting system has been built up. Afterwards, pruning should be moderate; only carry out the removal of decayed wood and those unwanted shoots which may tend to crowd the centre of the head.

With weeping standards, which are so pleasing when used as a centre piece for a bed, the twiggy growth should be thinned out if it becomes overcrowded, as the weeping forms tend to do.

When pruning bush roses, the cut should be made about ½ in. (1.25 cm) above the selected bud and should slope inwards.

VARIETIES OF HYBRID TEA ROSE

BALLET (1958). One of the finest of pink roses which is tolerant of long periods of dull, wet weather. It bears blooms 6 in. (15 cm) across with a fifty-two petal count and they are of pure candy-pink.

BONSOIR (1968). An outstanding pink rose bearing long lasting high centred blooms of apple blossom pink shading to shell pink and with pronounced scent.

CHICAGO PEACE (1962). A 'sport' of the famous 'Peace' and inheriting its vigour and freedom from disease. The high centred blooms are of deep rosy-pink, shaded with yellow.

COLOUR WONDER (1964). Of compact habit, it has healthy glossy foliage and bears large globular blooms of orange-red, shaded at the centre with deep yellow. A valuable bedding rose but with very prickly stems.

DUKE OF WINDSOR (1968). An outstanding bedding rose bearing masses of medium sized blooms of glowing vermilion and of rich fragrance.

ERNEST H. MORSE (1964). The outstanding red rose of recent years, the large globular blooms being of deepest red with pronounced scent.

FRAGRANT CLOUD (1964). Won the President's Trophy for a most outstanding rose, with plenty of bronzy-green foliage and bearing flowers 6 in. across, of rich coral-red with outstanding perfume.

GRANDPA DICKSON (1966). Won the President of the Royal National Rose Society's award for the finest rose of the year, the blooms are of enormous size and of deep uniform lemon yellow.

ISABEL DE ORTIZ (1962). A bedding rose of beauty, bearing large handsome blooms of cerise pink shaded silver on the reverse.

KING'S RANSOM (1961). Bearing beautifully formed flowers of medium size and of deepest yellow, it is a fine bedding rose though is late into bloom.

PASCALI (1963). Of perfect exhibition form, the medium-sized blooms are of blush white and are better able to withstand adverse weather than any other white.

PEER GYNT (1968). Raised from 'Colour Wonder', it has the same glossy crinkled foliage and bears large globular blooms of deep golden yellow.

PICCADILLY (1959). The best bi-colour ever introduced and of short, bedding habit. The large blooms with their rolled petals being of vermilion, shaded gold on the reverse.

RED DEVIL (1967). A rose of rich perfume and great beauty with glossy disease resistant foliage and bearing blooms of brightest red with attractively rolled petals.

ROSE GAUJARD (1957). A fine bedding rose with plenty of glossy foliage and bearing on elegant stems blooms of plum-red with a silver reverse.

SHIRALEE (1965). Of ideal bedding form, it is a rose with a difference, the saffron yellow bloom being flushed with apricot and marigold. It is scented.

SHOT SILK (1914). It still remains an outstanding bedding rose being of neat upright habit, mildew resistant and coming early into bloom. Its blooms of cherry, orange, salmon and yellow are amongst the gayest of the garden.

SUPER STAR (1960). One of the great roses of all time and a President's Trophy winner. Of vigorous habit, it is without fault with an abundance of bronzy-green foliage and bearing large flowers of unfading vermilion which are sweetly scented.

TIMOTHY EATON (1968). One of the finest exhibitor's roses, bearing a high centred bloom of deepest pink above dark green glossy foliage.

VALENCIA (1969). The first of its colour of sturdy, vigorous habit, the high centred blooms being of rich apricot-orange flushed with yellow and sweetly scented.

WENDY CUSSONS (1959). Won the President's Trophy, it is a truly fine rose with a high centre and rolled petals and of a lovely shade of cherry-red with the rich damask rose perfume. Of ideal bedding habit.

FLORIBUNDAS

The crossing of the polyantha rose with the hybrid teas resulted in the introduction of the floribundas, those of compact habit being excellent plants for providing a long period of colour in the small garden. With the floribundas there will be few days during summer and autumn which will not be colourful, for, unlike the hybrid teas, they do not bloom in flushes but continuously, requiring no rest period.

The floribunda has been carried a stage further with the introduction of the grandiflora rose, which possesses the same freedom of flowering yet with a greater resistance to disease owing to the enormous vigour of the plants. The grandiflora also has more of the characteristics of the hybrid tea, that whilst the blooms are borne in clusters they possess the true hybrid tea form, though they are not so large. For a small garden where the maximum amount of colour is desirable these plants should be given serious attention, for though they do not bear so refined a bloom as the hybrid teas, they possess greater value as a bedding plant. The plants are clothed in healthy foliage, possess a compact habit, and bloom without interruption no matter what the weather.

Some striking colour combinations may be obtained by planting circles of contrasting colours, using plants of a similar height, but with the bed slightly raised at the centre so that the maximum beauty of the blooms may be enjoyed. Striking when there is a collection together in full bloom is 'Allgold', with its rich golden-yellow flowers and the orange-flowered 'Spartan'. Or plant 'Gletscher', which bears large fragrant blooms of luminous lilac-pink, with either 'Brownie', which bears chocolate-coloured blooms, or with 'Border Coral', with its clusters of deep salmon-pink flowers. The beds will remain a mass of colour from early June until the end of autumn.

In comparison with the hybrid teas, the floribunda roses require little by way of pruning, for it is required of the plants that they bear a large amount of bloom rather than that of outstanding quality. So merely shorten any unduly long shoots after flowering, and cut out all dead and over-crowded wood in spring.

VARIETIES OF FLORIBUNDAS

CIRCUS. A new rose of hybrid tea or grandiflora form, the well-formed buds, borne in clusters and in profusion, opening as chrome-yellow and changing to orange and pink. Of vigorous but dwarf bushy habit, it is a grand bedding plant.

CONCERTO. Awarded a Gold Medal and International Trophy for the Best Rose of its year, it makes a vigorous, bushy plant, and bears masses of vivid-glowing, scarlet blooms throughout the season.

DAILY SKETCH. It makes a bushy plant with bronze-green foliage and bears hybrid tea type blooms of plum and silver.

DICKSON'S FLAME. Awarded the President's Trophy for the best new rose of 1958, it remains one of the outstanding bedding roses of all time. First to bloom, it bears its trusses of brilliant orange-scarlet from June until Christmas, almost without a rest period.

ELIZABETH OF GLAMIS. A 1964 introduction bearing blooms of salmon-pink and the only rose ever to win the President's Trophy of the N.R.S. for the best rose of the year and the Clay Vase for a new rose with the most pronounced fragrance.

EVELYN FISON. With bronze foliage and bearing trusses of vivid scarlet-red through summer and autumn, it is a rose of ideal bedding form.

FASHION. Of compact habit, the colour is unique, the blooms of glowing coral, flushed with salmon and orange, being borne in clusters. The blooms carry a delicious wild rose fragrance. Liable to rust, it is not a strong grower.

FIRECRACKER. This American floribunda is of ideal bedding habit, the cherry-orange blooms with their striking golden base being borne in large clusters right through summer.

FUSILIER. Winner of the All-American award for the Best Rose of 1957, it makes a plant of vigorous upright habit, and bears its trusses of flaming orange-red throughout summer. It has become a great favourite in Britain.

GOLD CUP. A superb floribunda of the grandiflora or 'Queen Elizabeth' type, bearing beautiful tapering buds of the deepest golden-yellow on long graceful stems, making it ideal for cutting or for garden decoration.

POULSEN'S BEDDER. Of upright habit, the blooms are borne in clusters and are of a lovely shade of clear porcelain-pink, untroubled by prolonged wet weather.

SALUTE. A McGredy introduction after the style of 'Masquerade', the clusters of semi-double blooms being first yellow and pink, later changing to orange and crimson to give a most striking effect. It forms an abundance of disease-resistant foliage.

SARABANDE. During wet seasons it has proved itself to be one of the most colourful of roses, bearing its pillar-box red blooms, with their golden stamens, in generous trusses amidst dark foliage.

SCARLET O'HARA. Dwarf and compact, it covers the ground with its dark green foliage and trusses of crimson flowers.

SPARTAN. A magnificent variety, which received the National Rose Society President's Trophy for the best new rose of 1954. The blooms, of hybrid tea form, are borne in profusion and are of a brilliant salmon-orange which does not fade.

TAPIS ROSE. It makes a dwarf bushy plant and bears tight double flowers of almost hybrid tea form, which are of a glorious shade of bright cherry-pink.

ZAMBRA. Of ideal bedding habit, with plenty of dark green glossy foliage, its chrome yellow buds open to flowers of miniature hybrid tea form of glowing orange and gold.

DWARF POLYANTHUS

Most of the old dwarf polyantha roses, with their dainty button-like flowers borne in dense clusters and of dwarf bushy habit, could well be used for the smallest of beds. They require very little pruning apart from the removal of dead and over-crowded wood, and when in bloom seem little troubled by adverse weather.

Here is a selection for very small beds:

'Cameo'	'Mrs Cutbush'
'Coral Cluster'	'Pigmy Gold'
'Jean Mermoz'	'Sunshine'
'Little Dorritt'	'The Fairy'
'Margo Koster'	

MINIATURE ROSES

The Fairy Roses are now most popular and those with a very small garden, possibly only a courtyard surrounded by a narrow raised border, or even where there is no garden whatsoever, a trough or window-box providing the only means of colour, may also enjoy their roses. Requiring little attention and remaining long in bloom, the inexpensive miniature roses are hardy and long lasting and require nothing elaborate in their culture. The plants have so neat a habit that they may be used to make the tiniest rose garden it is possible to imagine. Small pieces of crazy-paving stone may be used to divide the ground into sections of the required design, the stone being used as a pathway, though it should not be more than 6 in. (15 cm) wide to keep it in scale with the miniature rose trees. Strips of grass, formed by laying turf or sowing seed, may be used as an alternative method of laying out the tiny garden and will be kept neat by clipping with shears.

The size of the rose garden will depend upon available space, but a sunny corner of a garden, however small, will prove suitable, whilst the miniature roses are most attractive, used to make a 'rose-garden' in a trough. There are miniature standard roses which greatly adds to the charm of the garden, and to obtain the best effect tiny 'hedges' of evergreen box and dwarf conifers may be used. Most attractive is the dwarf juniper, *Juniperus communis* 'Compressa', which forms a pencil-like tree only 9 in. (23 cm) high, its feathery glaucous green foliage making a pleasing foil for the roses.

Where a miniature rose garden is being made, it is important to keep everything to scale, using the most dwarf of the roses for a trough or window-box and where space in the garden is at a premium. These grow from 8 in. (20 cm) tall and separate varieties or colours may be planted to each bed, or they are most attractive planted in a small raised circular bed, surrounded by a path of crazy paving or a tiny 'hedge' or dwarf box. These delightful little roses may also be used to edge a bed of the more compact of the hybrid tea and polyantha roses where room permits the use of these more vigorous roses. They may also be used about a rockery in groups of three or four together with the dwarf junipers. At all times they require a position of full sun, but they like a soil which does not allow their roots to dry out. For this reason plants always do well when on a rockery, where pockets of soil between the stones may be prepared by the addition of humus-forming materials.

Plants used in a trough or window-box or to make up a small bed or garden in the open will require the same attention as to soil conditions, the addition of leaf mould or peat augmented by some well-decayed manure, especially cow manure, being necessary to maintain summer moisture. Only dry conditions at their roots will harm the plants, neither frost nor cold winds causing them trouble. An excess of manure, however, should not be given or the plants will form too much foliage at the expense of bloom. If a little decayed manure cannot be obtained, incorporate 2 oz. of bone meal to each sq. yd. of ground and a similar amount where a trough garden is being made. If a little Kettering loam, yellow and fibrous, can be obtained, so much the better, but apart from these simple requirements the plants need nothing more. It is, however, important to see that they do not suffer from lack of moisture during summer, whilst an occasional application of weak manure water, obtainable in concentrated form in bottles from any sundriesman, will do much to enhance the bloom and prolong the display.

The miniature roses seem to resent root disturbance more than other roses and so pot-grown plants should be used whenever possible. Plants from pots will become more quickly established and may be planted at any time except when the soil is frozen, though possibly October or March will be the most suitable months to make a miniature rose-garden. The plants should of course be knocked from their pots or they will suffer from lack of moisture.

When planting, place the ball of compost holding the roots just below soil level, for shoots which appear from below the ground will also bear bloom. These should be cut back halfway each autumn, likewise any unduly long or decayed shoots, to maintain the shape and health of the tree. Also, to prolong the display, apart from occasional feeding, all dead bloom should be removed as it forms. This will also keep the plants compact and tidy. The plants should be allowed sufficient room to develop for though they will not grow tall, they will spread out in bush-like fashion and will grow as wide as they grow tall, which will be about 8 in. (20 cm).

PROPAGATION OF MINIATURE ROSES

The miniature roses are best raised from cuttings without budding on to a rootstock. This will prevent them from making too vigorous growth. The cuttings should be taken early in autumn and inserted in a sandy compost. The new season's shoots should be used, taking cuttings about 3 in. (7.5 cm) in length and trimming to a leaf bud. They will root more readily if the base of the shoots are treated with hormone powder before inserting round the sides of pots or pans, or they may be planted in frames or under cloches. Plant firmly and keep the compost moist. Provide shade should strong sunshine be experienced before rooting has taken place. As soon as the cuttings have formed roots, they should be moved to small pots and grown-on until ready to be planted out the following spring, first pinching out the growing point to encourage bushy growth. Clean, fibrous loam containing a little sand for drainage is all that is required for growing-on the plants in pots, whilst the roots must never be allowed to lack moisture (see plates 65–7).

Miniature roses may also be quite easily raised from seed and though not coming true to name some lovely varieties may be raised in this way. A good strain of *Rosa polyantha 'nana'* may be obtained for ten pence a packet which will give about twelve plants. Seed should be sown in boxes or pans containing the John Innes Sowing Compost, the time to sow being from mid-April to mid-May. The compost should be kept moist and shaded until germination takes place, which may be slow in comparison with most seeds. If growing outdoors, cover the box with a piece of glass until the seed has germinated.

The seedlings should be transplanted to small pots containing a good loamy soil to which a little leaf mould has been incorporated and they should be kept comfortably moist. They will have formed plants suitable for setting out by the following spring or early summer.

VARIETIES OF MINIATURE ROSES

BABY FAURAX. It grows only 9 in. (23 cm) tall and bears masses of tiny button-like flowers, semi-double in form, which are of a lovely shade of pale blue.

BABY GOLD STAR. Likely to become a favourite, it grows to a height of 8 in. (20 cm) and bears tapering blooms of a deep golden colour, shaded apricot.

BABY MASQUERADE. A dwarf form of the now famous floribunda, 'Masquerade', the small golden-yellow blooms being splashed with pink and red as they age (see plate 69).

BO-PEEP. It bears sprays of double flowers of deep pink and remains in bloom throughout summer.

CINDERELLA. It grows to a height of 9 in. (23 cm) and bears attractively shaped blooms of white, edged with rose-pink.

HUMORESKE. A McGredy introduction, it makes a bushy plant and bears masses of dainty deep rose-pink flowers.

HUMPTY-DUMPTY. A *R. multiflora 'nana'* hybrid, growing only 6 in. (15 cm) tall and bearing clusters of carmine-pink flowers throughout summer.

Josephine Wheatcroft. This delightful little rose is the result of crossing the species *R. rouletti* with 'Eduardo Toda' and may be described as having the true hybrid tea form. The beautifully formed tapering blooms, about the size of a ten-penny piece, are of a rich buttercup-yellow shade and borne through summer on 12 in. (30 cm) stems. With their glossy green foliage, the blooms are ideal for table decoration or for a button-hole and retain their shape for a considerable time.

Little Sunset. Raised by Kordes, it grows 8 in. (20 cm) tall and bears fully double hybrid tea type blooms of salmon pink, shaded with orange and yellow.

Maid Marion. Syn: 'Red Imp'. One of the many hybrids raised by de Vink, the result of crossing the variety 'Tom Thumb' with the polyantha rose, 'Ellen Poulsen'. It grows only 6 in. (15 cm) tall and is one of the finest of all the miniatures, the well-formed flowers being of a rich crimson colour.

Midget. It makes a tiny plant only 4–5 in. (10–12 cm) tall and bears tiny double button-like blooms of deep rosy-red.

Mon Petit. Like a tiny rambler, the fully double blooms are of bright carmine-pink.

Perla De Alcanda. It makes a bushy plant 9 in. (23 cm) tall and the same distance in width and bears shapely blooms of bright crimson.

Perla De Monserrat. It grows 9 in. (23 cm) tall and bears globular flowers of pink, shaded with deeper pink.

Pixie. The blooms are the size of a new penny piece and are like double white rambler roses.

Pour Toi. One of the best miniatures, its beautifully shaped blooms are of a rich ivory-cream colour. It grows 10 in. (25 cm) tall.

Presumida. It grows 9 in. (23 cm) tall and bears a bloom of deep apricot, similar in shape to 'Josephine Wheatcroft'.

Rosemarin. It makes a neat rounded bush 6 in. (15 cm) tall with plenty of glossy foliage and bears flowers, like tiny cabbage roses, of softest pink flushed with lavender.

R. rouletti. It grows only 6 in. (15 cm) tall and bears masses of deep pink flowers of true hybrid tea form.

Sweet Fairy. A most charming variety, it grows only 5 in. (12 cm) tall and bears its deliciously fragrant shell-pink blooms throughout summer.

Tinkerbell. A recent introduction, it makes a bushy plant 12 in. (30 cm) tall and bears many-petalled flowers of deepest pink.

Tom Thumb. Also known as *Rosa peon,* it is an old free-flowering variety, bearing semi-double carmine-pink blooms which have a striking white centre.

EXHIBITING ROSES

The Royal National Rose Society's Show is held in June each year in London and again in autumn, and attracts thousands of visitors. There are prizes to be won for numerous classes, for beginners and for long established growers, for specimen blooms and for decorative roses. For the specimen bloom classes, the blooms are presented without foliage and they are judged entirely on their beauty as individual blooms. They may be displayed in boxes, arranged individually or singly in vases.

Decorative blooms should be exhibited in such a way as to show the natural habit of growth and foliage of the particular variety. Whether single or double, the blooms are judged on their merit and though disbudding may have been done it is not essential, for size of bloom should be a fair representation of the particular variety.

Not all rose growers can spare the time to exhibit their blooms but it does add interest to one's hobby, and the experience of presenting a winning exhibit is a never to be forgotten thrill. It is no easy task to grow and bring to a state of perfection, in the usually difficult British summer, rose blooms fit to win an award in a national contest. To exhibit roses successfully calls for skill in their cultivation from the time the plants are ordered and the ground prepared, until the very day of the show.

PRUNING FOR EXHIBITION

To grow blooms of exhibition quality, the roses must attain as near perfection as possible though one may compromise between obtaining blooms of exhibition quality and in merely allowing the plants to bloom without any degree of control. The more plants of a certain variety that can be grown, the better chance will there be of having one or more blooms of exhibition quality at peak of perfection on show day. It will also lengthen the peak flowering period if the plants are pruned over a period of about a month, some early in spring, others later, for it is otherwise difficult to time a plant to give perfect blooms on a given date. The weather can be responsible for so much variation.

With experience, it will be found that certain roses produce their best bloom late in the season, being at their best for the August shows and in this case pruning should be delayed as long as possible, as late as mid-April. 'Mme Kriloff' and 'Juno' are both at their best in the latter weeks of summer. Heed should also be given to the pruning requirements of the respective varieties, those such as 'McGredy's Yellow' and 'The Doctor', requiring only light pruning.

SHADING AND DISBUDDING FOR EXHIBITION

Disbudding will be necessary for exhibition quality bloom, so as soon as the side buds may conveniently be handled, they should be nipped out to leave the large terminal bud. When showing in the best company it will be advisable to remove the side shoots appearing on each stem so that the full energies of the plant will be directed to the selected shoot and bud. The removal of side shoots will, however, not be necessary, where showing locally or where the plants are being grown chiefly for garden display or home decoration.

During a dry season it is important to keep the plants moist at the roots, and this should be maintained throughout summer. It is not advisable to give the plants copious quantities of moisture when the ground is dry just before the buds begin to open for this will cause 'splitting'. Should the season be advanced and the first flush of bloom appear likely to mature too rapidly for the early shows, it will be better to select a side bud and to remove the terminal bud which will always open a week or ten days before a side bud.

As the time for the buds to open approaches, when the guard petals begin to unfold, the plants should be inspected as often as possible so that the blooms may be shaded from the hot sun if retarding is necessary, or if fading is likely to occur due to a prolonged period of sunshine. Again, should the weather be wet, protection will also be advisable when once the shield petals commence to open. Fix the conical shade just above the bud so that there will be no fear of the bloom knocking against the sides in strong wind. Fixing the shade too low over the bloom may cause fading of the bloom through lack of light.

About a fortnight before the buds begin to open it is advisable to place some wood wool around the plants. This will help to maintain soil moisture and suppress annual weeds, but more important, it will protect the bloom of those more dwarf growing varieties, e.g. 'Lady Belper' and 'Picture' which bear much of their bloom quite close to the ground, from splashing by heavy rains which may occur during a sudden storm.

Some exhibitors tie their blooms when half open with a piece of thick wool taken around the inner or cone petals and held in place with a double twist. This results in longer inner petals and a greater depth to the centre. One must, however, guard against the centre opening out rapidly when once the wool is removed. It is better to play safe and trust to good timing and good culture. It is not advisable to tie the blooms during a damp season for if the wool gets wet, it may cause marking of the petals. To prevent grooves forming in the petals and to allow the centre at least some room to expand, the wool should be loosened and re-tied each day.

CUTTING THE BLOOMS FOR EXHIBITION

Where showing locally, it will be possible to cut the bloom early on the day of the show, but at the large shows, judging generally takes place early and the blooms must be staged on the previous day. This will also be necessary when transporting bloom a distance. In any case it will be advisable to cut on the previous day should there be any doubt about the weather, for it would be most disappointing to have one's bloom spoilt by heavy rains at the last moment.

The blooms should be cut with as long a stem as possible, the stems being immersed in water to more than half their length and placed in a cool, shaded position indoors. Any unduly large thorns should be carefully removed or shortened, while the stems should be wiped clean of any deposits caused by spraying which should of course, have been suspended as soon as the blooms begin to open. The blooms should be given a long drink so that they will not droop their heads on the show bench and will remain fresh for as long a time as possible. It is desirable to split the very end of the stems to enable the bloom to take up moisture more readily.

TRANSPORTING THE BLOOMS TO EXHIBITIONS

The blooms should be taken to the show in small green painted metal flower buckets readily obtainable from sundriesmen. Stand the containers in a strong wooden lettuce crate and pack newspaper between to prevent movement. As the cut blooms are placed in the containers a small piece of sulphite paper should be fastened round each bloom, and to prevent them from being unduly jolted about, reeds may be placed in the containers before inserting the stems which will then be held securely. Additional reeds should be taken to the show for use in the vases when exhibiting in the decorative classes.

When transporting the cut bloom by rail, they will need to be taken in flower boxes lined with a considerable thickness of tissue paper. The end of each stem should be wrapped in paper. Strips of paper should also be placed beneath each flower head to prevent them breaking at the neck, whilst the stems will be held in place with canes which will exactly fit the box. Make sure that the blooms receive ample ventilation and always place the boxes where the sun cannot cause the blooms to expand nor in the smoke-laden atmosphere of a railway carriage. When packing the blooms make quite sure that they are dry, for moisture on the petals will cause them to turn brown, making the blooms useless for exhibition.

Take along some wires for supporting the blooms and some labels, and allow plenty of time for arranging the blooms. Especially is this necessary where exhibiting in the decorative classes, for a pleasing arrangement will decide in one's favour amongst competitors if all other points are equal.

Wires may be used to support the blooms where there is any weakness of stem, the top of the wire being shaped to fit close round the calyx. The wire is then twisted round the stem at intervals.

ARRANGING THE BLOOM AT EXHIBITIONS

When the blooms are to be displayed in vases, these are obtainable from the show secretary upon arrival. They must be quite clean, and before filling with water, place in each a bundle of reeds making sure they do not show over the top of the vase. The flower stems, which should have had the lower thorns either shortened or removed, are then carefully inserted, the reeds enabling the stems to remain in place so that the blooms may be arranged without bunching. No blooms should be used which reveal an open centre, and for this reason blooms of a variety should be taken to the show in various stages of maturity, for what appear to be one's best blooms may be too advanced when the exhibits are actually made up. After arranging the bloom, the vases are filled with water.

When arranging in open bowls, a piece of clean wire netting rolled up, should be fixed in the vase, into which the stems are inserted to make a pleasing all-round effect. The blooms should be artistically arranged without bunching and without there being any unduly large gaps. In a given class there may be as many as twenty-four blooms to arrange in a bowl, and when complete, the effect should be rounded and almost ball-like in appearance.

Blooms of the hybrid teas may require some 'dressing'. The outer petals are opened slightly, revealing the inside of the petals which is generally of brighter and richer colouring than the reverse. It will be advisable to carefully remove any outer petal which may have been damaged and which would otherwise spoil the appearance of the bloom. Pressing down with a camel hair brush the next row of petals will hide the removal of the damaged petal. When 'dressing' the bloom do not over-do it or the natural beauty of the bloom will be harmed.

POINTS OF A GOOD ROSE

In the decorative classes the blooms must be fresh and neither too old nor too young. The outer petals must be fully opened yet the centre must be closed, in spiral or cone form, which will be offset by the outer petals. If split or badly formed, or if sufficiently open to reveal the stamens, the bloom will not be considered for an award. The bloom should be of the size and form typical of the variety. Strong stems and plenty of clean, healthy looking foliage is also essential, while the colour of the bloom should be rich which will be possible where the soil is not lacking in potash.

Specimen blooms may also be exhibited in boxes. These are specially fitted with tubes which are filled with water and which are correctly spaced to accommodate a dozen or half a dozen blooms as the class demands. Like the boxes used for exhibiting violas, they are made slightly higher at the back so that the full beauty of the blooms may be revealed. Always place the largest and best blooms to the back, for in this position they will better catch the judges' eye. Only a short length of stem will be necessary and, of course, no foliage will be used. The space between the tubes should be filled in with fresh green moss. Each bloom must be provided with a name label. Where showing locally, it will be possible to make up the boxes at home, though this is not advisable when transporting a distance.

A specimen bloom should be of maximum size for the variety. It should have petals of good substance, 'regularly and gracefully arranged within a circular outline'. It should be fresh, free from blemish and must be of bright colouring. But do not be afraid of showing a bloom slightly below normal size if it is in better condition than a larger bloom. Floribunda and polyantha roses which are to be shown in the decorative classes present little difficulty, though bloom and foliage must be fresh and free from blemish and must be of rich colouring. To obtain as many blooms on a spray in as perfect a condition at one time, it will be advisable to remove the larger terminal bud about a fortnight before the sprays are to be cut, otherwise they may be left to take care of themselves.

CLIMBING AND RAMBLER ROSES

With their ability to cover a wall more quickly than most climbers and with their complete hardiness and freedom of flowering, the climbing and rambler roses are rightly more widely planted than any other wall plant. Blooming equally as well in the far north as in the south, provided the plants are not too exposed to cold winds, they thrive on a northerly wall.

Though many of the hybrid tea roses may be obtained in the climbing form, they are in most cases not so free flowering as the true climbers and ramblers, those containing *R. wichuraiana* 'blood', introduced to Britain towards the end of the nineteenth century. Those still popular ramblers 'Dorothy Perkins', 'Dr W. Van Fleet', 'American Pillar' and 'Jersey Beauty', a parent of 'Emily Gray', were all raised from *R. wichuraiana,* and appeared, with many other ramblers, at the beginning of the century. The plants are propagated from cuttings. Being susceptible to mildew, these ramblers should only be planted against a wall where trellis is used, or plant them in an open situation and train them up rustic work. They also require heavy pruning each year, for they make much dead wood and this also makes them less suitable for planting against a wall. Easlea's 'Golden Rambler' and 'Emily Gray', both of which have glossy foliage are exceptions to this for both are perfectly happy against a wall, as are 'Albertine', 'Albéric Barbier' and 'Félicité et Perpétué'.

To plant as an alternative to 'Emily Gray' for retaining its foliage until the New Year, is the old 'Félicité et Perpétué' of similar vigour and which also bears clusters of cream-coloured flowers.

Guard against planting the climbing forms of the popular hybrid teas in the expectation that they will prove as free flowering as in the bush form. They will not. Nor will they prove to be as free flowering as the true ramblers and climbers.

The climbing forms of the hybrid tea roses are 'sports', at some time having produced an unduly long stem which is without terminal buds. Eyes are then removed from this shoot for budding on to a vigorous root stock. A number of them make valuable wall plants but require very different treatment after planting than do the true climbers or ramblers, otherwise they will produce few flowers and may even revert to the bush form. They will bloom in flushes during the summer months but will bear few flowers during autumn. Perhaps the best of the climbing hybrid tea 'sports' are:

'Etoile de Hollande'	'Lady Forteviot'
'Forty-Niner'	'Lorraine Lee'
'Golden Dawn'	'Mrs Sam McGredy'
'Lady Eve Price'	'Souvenir de Mme Boullet'

Roses planted against a wall will be seen at their best when they receive plenty of sunshine and a free circulation of air. They are neither happy when in shade cast by tall overhanging trees nor in shade cast by neighbouring buildings. This, however, does not mean that it is impossible to grow a climbing rose against a north wall, for which there are a number of varieties noted for their vigour and hardiness, but they do not like overhead shade.

A rose, planted in a soil which has received some attention, will reach a height of 12 ft. (3.5 m) and a similar distance in width, within two to three years after planting and so should not be placed where there might be overcrowding from other climbing plants. Climbing roses, unlike the clematis, are not good mixers. They are happiest in a position where they remain on their own, though more than one variety may be planted within reasonable distance of each other.

Though wall roses may be trained against and be fastened directly to a wall, they will remain healthier and grow with more vigour if trained against a lath trellis. This will prevent scorching of the foliage and, by providing a greater circulation of air, will keep the troublesome mildew to a minimum. A fully-grown climbing rose will make considerable growth and so the means of support must be efficient, otherwise the weight of foliage and bloom will cause the shoots to come away from their supports with considerable loss of appearance. In addition, they are difficult to tie back when in bloom. When using trellis, it will be advisable to plug the wall with 1 in. (2.5 cm) square wooden pegs which should protrude 1 in. (2.5 cm) from the wall. This will give the plants a free movement of air and at the same time, by screwing the trellis to the pegs, greater security will be obtained.

The importance of the climbing rose is that the extended shoots may be trained exactly where they are needed to cover a wall. It may be required to cover a low wall or to train the plant beneath a bay window and possibly up either side of the window. Here any one of the less vigorous varieties may be planted and easily be kept within limits. Another method is to train shoots of a more vigorous variety in horizontal fashion beneath a window and then in an upwards direction on either side.

PLANTING CLIMBING ROSES

November is the best month in which to plant, and almost all climbing roses are sent out from open ground beds, not in pots. The exceptions are the lovely single yellow flowered 'Mermaid', and the black-crimson 'Guinée', which are slow starters and are always best when pot grown. With others, the roots should always be inspected before planting and any which are damaged should be cut away with a sharp knife. At the same time any unduly long roots should be shortened.

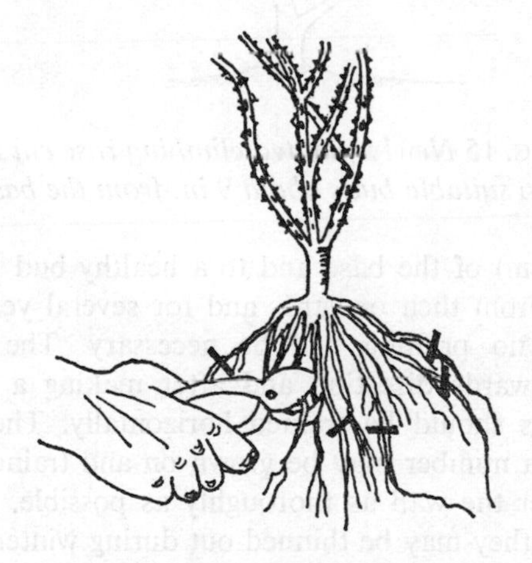

FIG. 14 *Root pruning a rose*

When planting, carefully spread out the roots, for roses will not grow well if the roots are bunched together. Make the hole large enough so that there is no need to turn up the ends and be sure to set the plant at the same level it occupied in the nursery bed. If the soil is of poor quality which is often the case near the walls of a newly built house, some friable loam should first be placed over the roots. Then should come the manure, finally filling in with the excavated soil and treading it down. The soil should be compact about the roots, yet should be sufficiently friable so as not to deprive the roots of oxygen. If the soil is dry, water the plant in and it should not then be necessary to water again until the following summer.

PRUNING CLIMBING ROSES

The climbing roses bear their next season's bloom on the old wood and require little pruning. Nor is it advisable to cut back a newly planted tree too vigorously, otherwise it may revert to the bush form in the case of the hybrid tea 'sports'. A newly planted climber should be cut back to

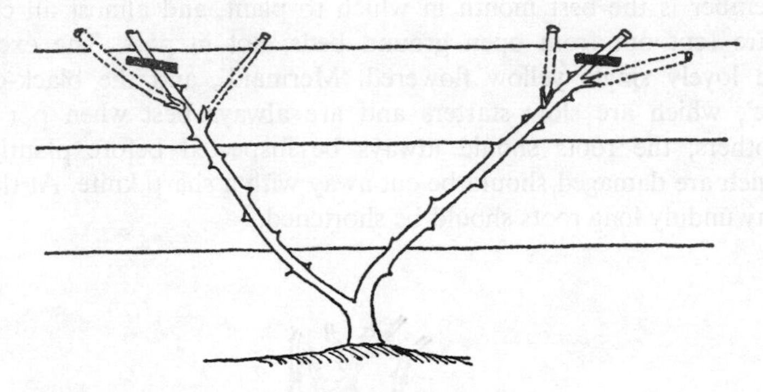

FIG. 15 *Newly-planted climbing rose cut back to suitable buds about 9 in. from the base*

about 15 in. (38 cm) of the base and to a healthy bud pointing to an outwards direction. From then onwards and for several years until the plants require thinning, no pruning will be necessary. The shoots should be trained in an outwards direction and after making a certain amount of growth, the shoots should be trained horizontally. The laterals will bear the bloom whilst a number may be grown on and trained in a similar way and so as to cover the wall as thoroughly as possible. Should the laterals be too numerous, they may be thinned out during winter.

The *R. wichuraiana* ramblers bloom only on the new wood formed during the previous summer and all old and useless wood – except the wood which comprises the framework of the tree – should be removed in autumn when the plant has finished flowering.

The newly planted tree may be cut back more vigorously than in the case of the climbing tea roses, to about 8 in. (20 cm) of the base and to suitable buds. Later, any weak growth should be pinched back and the stronger stems be allowed to grow unchecked except if they become too vigorous. The aim should be to cover the wall with bloom from the bottom upwards and only when the main stems make too much growth should they be cut back to persuade lateral buds to 'break'.

Certain varieties – such as the charming single yellow 'Mermaid' – will need almost no pruning after the newly planted tree has been cut back to vigorous buds. Some varieties are vigorous by nature, others make little growth. Each variety must be pruned on its merits.

During summer, when the plants will be making new growth at the same time as foliage and flowers, an abundance of moisture is essential. The base of a wall always suffers from lack of moisture at this time of the year. Shrubs, bedding plants, or maybe a bed of roses planted almost up to the wall will deprive the climbers of all natural moisture. Artificial watering will be necessary until the plant is thoroughly established. Give the plants a soaking at least once each week during June, July and August. Failure to provide sufficient moisture will not only prevent the formation of new growth but will bring the flowering season to an end sooner than should be so. This may happen with the ever-popular 'Paul's Scarlet Climber' or with 'Allen Chandler', which bears a profusion of bloom in June and, if not lacking moisture, will do so again in early autumn.

The plants will also obtain great benefit from a mulch of decayed manure given at the end of May and again about mid-July. This will provide plant food in addition to preventing evaporation of moisture from the soil. The plants will also benefit from an occasional application of manure water, best given immediately after they have received a heavy watering. The quality and colour of bloom and the ability of the plants to resist mildew will be enhanced by a 1 oz. application of sulphate of potash given in March each year.

As the new shoots are formed they must be fastened to the wall or trellis in whatever direction they are required to grow. If left until they become long and woody, they will be difficult to train and fasten back without their breaking. To train a climber up a rustic post, this should be done by training it round and round in clockwise fashion, whereas a rambler may be fastened to the post with raffia in the vertical position. To prevent the ramblers making too much growth, it will be advisable to pinch back the laterals as they form.

VARIETIES OF CLIMBING ROSES

Red and Crimson

ALLEN CHANDLER. It blooms early in summer and again in autumn, bearing large single blooms of bright crimson-red.

COUNTESS OF STRADBROKE. Of vigorous habit, it has glossy green foliage and bears large crimson blooms which are richly fragrant.

CRAMOISIE SUPERIEURE. An old China rose of vigorous habit, bearing its rich red blooms throughout summer and autumn.

DANSE DU FEU. This new climber will attain a height of 10 ft. (3 m) and is one of the best of all climbing roses. Its vermilion coloured blooms are borne throughout summer and autumn and withstand both hot and wet weather. It is very hardy.

CLIMBING. DUQUESA DE PENARANDA. Possibly the best of the hybrid tea 'sports' for it is free-flowering and blooms throughout summer. The blooms are an attractive combination of scarlet and gold.

GUINÉE. It does well in a sun-baked position and bears huge crimson-black flowers which possess a rich, fruity perfume. It also has attractive crimson foliage.

HAMBURGER PHOENIX. Of vigorous habit, it has glossy dark green foliage and bears its crimson flowers in floribunda clusters throughout summer.

PARADE. A new double red which is extremely free flowering and enhanced by its glossy dark green foliage which is very resistant to disease.

PAUL'S SCARLET CLIMBER. Hardy and vigorous, it covers itself with bright rosy-scarlet blooms for about a month but is not perpetual flowering.

RED FLARE. Vigorous and perpetual flowering, this American climber bears large blooms of richest red.

SOLDIER BOY. Perpetual flowering, the single blooms are large and of a brilliant scarlet-red colour with golden anthers.

SOUVENIR DE CLAUDIUS DENOYEL. Almost like its offspring 'Guinée' except that it is of more vigorous habit. The almost black flowers possess the same delicious scent, but are not so freely produced.

THOR. It blooms with great freedom in summer but not in autumn. It is an excellent wall plant, soon reaching a height of 12 ft. (3.5 m) and bears flowers similar to those of 'Ena Harkness'.

ZWEIBRUCKEN. Vigorous, free flowering and very disease-resistant, its brilliant crimson blooms have a striking white eye.

Pink

ALOHA. Raised in the U.S.A., it has 'New Dawn' for a parent. The large, fragrant fully double blooms are of an attractive shade of bright rose pink borne with freedom through summer and autumn.

CUPID. An interesting variety for a low wall. Its large single blooms with their yellow anthers are like water-lilies and are of a lovely shade of shell pink. After flowering in midsummer, the blooms turn to hips as large as tomatoes.

DREAM GIRL. The blooms are of unique colouring, being salmon-pink flushed with yellow and are borne over a long period.

FELLEMBERG. A noisette rose but extremely hardy and does well along wires to form a hedge. The deep reddish-pink blooms are produced during the latter six months of the year and are sweetly scented.

KATHLEEN HARROP. A wonderful pillar or hedge rose being a 'Zéphirine Drouhin' 'sport'. It bears large shell-pink blooms which are strongly perfumed.

LADY WATERLOW. It will attain a height of 12 ft. (3.5 m) and bears its semi-double flowers of soft-pink edged with carmine in July and again in September.

MME GREGOIRE STAECHELIN. An outstanding wall rose for it is of vigorous habit and comes very early into bloom bearing masses of strongly fragrant bright rose-pink blooms. Even in a wet season the foliage remains free from mildew.

RADMONA. A 'sport' from 'Sinica Anemone', having the same attractive foliage and requiring the same conditions. It bears huge single blooms of vivid carmine-pink with golden stamens.

RENAL. An excellent wall rose being free flowering whilst it is almost thornless. The blooms are of an exquisite shade of grey-pink.

SINICA ANEMONE. It should be grown against a warm wall in a sheltered position. It bears single flowers of rich pink amidst attractive slender foliage.

ZEPHIRINE DROUHIN. A bourbon and though raised in 1868 it remains one of the finest of all wall roses. The large, cabbage-like blooms of rose-pink are borne throughout summer in great profusion on thornless stems. The flowers are strongly scented.

Yellow and Gold

ELEGANCE. It will grow to a height of 20 ft. (6 m) and throughout summer bears its clusters of pale yellow blooms which are handsomely pointed in the bud form.

EMILY GRAY. Its dark green glossy foliage is resistant to mildew, whilst it is one of the hardiest of all climbers. The double blooms are of a rich golden buff shade.

GLOIRE DE DIJON. It will grow to a height of 10 ft. (3 m) on a north wall and bear its strongly scented yellow-orange flowers in profusion. Though more than a century old, some vigorous plants are still to be found at specialist nurseries. It blooms early and again very late.

CLIMBING GOLDILOCKS. A fine wall rose of floribunda form, its clusters of pure yellow flowers borne throughout summer and autumn.

LAWRENCE JOHNSTON. A fine but little known climber of strong, vigorous habit which will quickly cover a wall. It blooms most profusely early in summer, its flowers being of deep butter-yellow.

LEMON PILLAR. Like 'Mme Staechelin' it has the vigorous 'Frau Karl

Druschki' for a parent, hence its vigour. The beautiful blooms are palest lemon, borne during July.

MEG. Though not quite so beautiful, may be said to be an easily grown 'Mermaid'. The large yellow and pink blooms are produced with great freedom throughout the season. A new climbing rose with a great future.

MERMAID. Its glossy foliage and large single primrose-yellow blooms with their striking orange stamens make this a unique wall rose. It is, however, difficult to get started, the wood being brittle and it should never be pruned.

SILVER MOON. One of the most vigorous of all climbers, it will cover a wall to a height of 18 ft. (5.5 m) and bears large fragrant primrose-yellow flowers.

SUN GOLD. A modern climber of vigorous habit. The brilliant golden blooms enhanced by the glossy green foliage, are produced throughout summer.

W. A. RICHARDSON. It needs a warm wall where it will make only a small amount of growth but its cream and apricot flowers are so richly fragrant that it is worth growing for their scent alone.

White

MME ALFRED CARRIERE. Almost a century old, and until the introduction of 'Patricia Macoun', it was the best white climber though the blooms may be said to be blush white. It will quickly cover a large wall and bear its large fragrant blooms through summer and autumn. It grows well in almost any soil.

PATRICIA MACOUN. This is a new climber and the first to bear pure white flowers. It is a hardy variety and is extremely free flowering besides being a vigorous grower. It is the ideal contrast for such brilliantly coloured climbers as 'Danse du Feu' and 'Red Flare'.

VARIETIES OF RAMBLER ROSES

ADELAIDE D'ORLÉANS. An old rambler, bearing flesh pink buds which open out to large flat flowers with striking golden centres. The blooms are enhanced by red tinged foliage.

ALBERIC BARBIER. Extremely vigorous and free flowering. The pale cream blooms are very fragrant, the foliage glossy, being free from mildew.

ALBERTINE. One of the finest ramblers for a wall, the richly fragrant bloom is of an unusual coppery-pink colour, the foliage bright and glossy.

AMERICAN PILLAR. A grand rose of great vigour and extreme hardiness. The deep pink blooms have an attractive white base.

CARPET OF GOLD. Resembles 'Mermaid' with its large single yellow blooms with their golden stamens, and its dark shiny green foliage.

CHAPLIN'S PINK CLIMBER. Bears large, semi-double pink blooms with a striking golden centre. A strong, healthy grower.

CLIMBING ORANGE TRIUMPH. A polyantha rose with dark glossy green foliage, and orange-red flowers borne in large clusters throughout summer.

CRIMSON CONQUEST. Probably the best dark red rambler, the double flowers appearing first as attractive pointed buds then revealing striking golden stamens.

DR W. VAN FLEET. A grand vigorous rose bearing large double silvery-pink blooms over a long period.

EASLEA'S GOLDEN RAMBLER. A fine variety with its glossy foliage and clusters of warm yellow flowers.

EXCELSA. An outstanding pillar rose or weeping standard. It comes late into bloom and bears huge clusters of bright crimson-red flowers.

FELICITE ET PERPETUE. A superb wall or hedging rose raised by the gardener to Louis Philippe at Château Neuilly in 1828. It has shining dark green leaves which it retains through nine months of the year and bears great clusters of blush-white flowers.

FRANCOIS JURANVILLE. An outstanding rambler having striking red stems, shining green foliage and bearing large pale pink flowers.

GOLDEN GLOW. An outstanding rambler with glossy foliage and bearing masses of bright double yellow flowers throughout summer.

LADY GODIVA. A pale pink 'sport' from Dorothy Perkins but not in the same way troubled by mildew. It comes late into bloom.

NEW DAWN. A grand variety continuously in bloom from July until well into October and bearing large clusters of fragrant pale pink flowers.

ORANGE EVERGLOW. Its deep coppery-orange blooms really do glow and remain colourful for several months.

PURITY. It bears double pure white blooms and has glossy dark green foliage. It is a plant of immense vigour and will quickly cover a wall or pagoda.

SANDERS' WHITE RAMBLER. Late to bloom, the double pure white flowers are a pleasing contrast to its bright green foliage.

WHITE DAWN. A seedling from 'The New Dawn', and bearing fragrant pure white flowers over a similar period.

RAMBLERS FOR BEDDING

Rambler roses have yet another use and that is to cover a bank or bed, using the rambler as a bedding plant. A most attractive display may be enjoyed by planting a rambler at the centre of a large circular bed and as the shoots are sent out in octopus fashion, these are pegged down to the soil. The shoots should be arranged so that they are spaced out as evenly as possible and so that they will cover the maximum area of ground in the shortest possible time. The plant will bloom profusely and all the attention it will require will be to cut away any shoots where there is overcrowding and to remove any dead wood at the end of summer when giving the plant a top dressing. The display will be enhanced by edging the bed with a dwarf polyantha or fairy rose of contrasting colouring. 'Pour Toi', with its dainty flowers of rich ivory cream will look most attractive used to border a red rambler.

Ramblers used in this way will be equally valuable if planted on a sunny bank which would possibly grow little else during the warm summer months. No other plant would provide so prolonged or so free flowering a display. Plant ramblers of contrasting colours 6 ft. (2 m) apart and keep them separate from each other by removing the ends of the shoots as they reach a given point. Planting an exposed bank will mean that during a dry period the plants will require as much moisture as possible at their roots, so ample supplies of humus should be incorporated at planting time and additional supplies should be added when the plants are cut back after flowering. 'Purity', 'Golden Glow' and 'Excelsa' are excellent for covering a sunny bank.

PROPAGATING ROSES

Roses may be increased by each of the usual vegetative methods of plant propagation:

by striking cuttings
by budding on to a rootstock
from seed.

There is nothing mysterious about rose propagation and there is no reason at all why plants may not be propagated simply from cuttings. Most of the polyantha and floribunda roses, the ramblers and the shrub roses are readily increased by this method and will make plants in every way as vigorous and as healthy where budded on to a rootstock. The hybrid teas, especially the bi-colours and those bearing bloom of orange and apricot shades, may take considerably longer to root and are best budded on to a rootstock. Budding is more certain of success, though the use of modern hormone preparations now makes it easier to root cuttings. One great advantage of budding is that a strong flowering plant may be built up in little more than a year, about half the time taken for the plant which is being propagated from cuttings.

ROOTSTOCKS

The use of a suitable rootstock is of importance in obtaining the most vigorous plants for a particular soil. For example, briar or *R. canina* seedlings, which are possibly in greater use than any other form of rootstock, tend to be dwarfing in the type of plant they produce. Especially is this stock unsuitable for growing roses in light soils such as those of a sandy nature or over limestone. *R. canina* will make a vigorous plant where the soil is heavy, which accounts for the often disappointing results obtained from plants raised in such soils when moved to lighter soils. *R. canina* stock is almost always raised from seedlings and they make the best understock for heavy soils and for cool conditions. The vigour of the more modern introductions of the 'Peace' type tend to overcome the dwarfing habit of *R. canina* stock and if light land is fortified with humus materials plants budded on to this stock will make reasonably sturdy growth. *R. canina* will not produce as many suckers as the *rugosa* stocks.

R. rugosa strikes readily from cuttings inserted in trenches of sandy soil, so does *R. multiflora* stock which is apt to cause cankering if the cuttings are struck in ground where roses have grown previously. The *R. multiflora* stock makes a vigorous plant, especially in sandy soils, but the plants, after making an abundance of growth and flowering profusely, tend to have only a short life.

More recently, a hybrid form of *R. canina*, known as 'Laxa' stock has become widely used. It forms a plant which is long lived and which does not throw suckers. It is, however, prone to rust disease, so as that which may be described as being the perfect rootstock has still to be found. We must compromise and suit the available stocks to one's particular soil as far as possible. Under warm conditions such as may be experienced in Italy and the south of France and in parts of America, *R. odorata* 'Major' is usually selected as the rootstock, for it is able to withstand very dry soil conditions and extreme heat such as not experienced in Britain. As an all-round stock, *R. canina* is the most widely used but on light sandy or chalk land it should be given as much help as possible.

TAKING AND ROOTING CUTTINGS

Rose stocks from cuttings are taken at the end of summer from the new season's wood. They should be removed immediately below an eye or bud and tipped back to a bud. Before inserting into trenches of sandy soil, the three lower buds should be removed in the same way as when rooting gooseberries or red currants. The cuttings, which should be about 9 in. (23 cm) in length, are inserted 3 in. (7.5 cm) deep and the same distance apart and must be made quite firm or they will not root quickly. It should be said that seedling rootstocks will show greater vigour and will have a longer life than cuttings, and where seedlings can be obtained they should be used.

For the rooting of all rose cuttings, and this includes the shrub roses, trenches should be prepared in an open, sunny situation and should contain coarse sand with which is incorporated some moist peat. This will prevent the cuttings from drying out, but during a dry period the compost must be kept quite moist by daily watering.

When taking the cuttings of floribundas, polyantha, shrub and rambler roses, they should be taken about 12 in. (30 cm) in length and are prepared in the same way. To encourage rapid rooting it is advisable to treat them with a hormone, Seradix B being that to use. To ensure rapid rooting, rootstock cuttings should also be treated. The rooting end of the stem should be twisted on to a pad of wet cotton wool to make it sufficiently damp to which the hormone powder will stick. Plant without delay, using a garden line and a trowel and plant a double row to each trench. Be sure to plant the cuttings in an upright position and make quite certain that the base of the cuttings is in direct contact with the compost, otherwise rooting cannot take place. To ensure contact tread in each row of cuttings as it is completed. Cuttings of those roses which are to be grown on their own roots should be planted about 6 in. (15 cm) apart each way, staggering the rows.

Early September is the best time for taking and rooting cuttings, for they will then have formed at least a few roots before the arrival of the frosts. By the last days of March or early in April if the ground is not yet in condition following the winter rains and snow, the cuttings should be lifted and planted out in beds 12 in. (30 cm) apart, allowing 2 ft. (60 cm) between the rows. Do not bury the stem too deeply, just sufficiently to cover the roots and make it comfortably firm. Plants growing on their own roots will remain in position for the next eighteen months when they should be lifted in November and planted in their permanent or flowering quarters.

BUDDING

Budding of the rootstocks will commence towards the middle of July and will continue until the end of September though where using Laxa stock, budding will not be possible after the end of August. The right time for budding will depend upon the weather, for if a period of dry weather is experienced, successful budding will not be possible. Delay operations until there is some rain or provide the rooted plants with an abundance of water when they will be ready for budding in three or four days.

The size of stock for budding should be about ½ in. (1.25 cm) in diameter and first make it free of soil which surrounds the base. Also wipe the stem with a clean rag at the part where the budding is to be done. This will prevent any grit from entering the incision. The selected stock must be quite healthy and any which show signs of weakness must be removed. The rootstocks should be as even as possible. The rootstock should be budded as near to the roots as possible, *rugosa* standards being budded at any height on the stem.

The shoot from which the bud is to be taken must also be healthy whilst the shoot must have borne a bloom of quality. The whole length of stem is removed and the leaves and thorns stripped off. Only a small portion of stems should be left when the leaves are removed. The stem is then placed in water to maintain the buds in fresh condition. If the thorns have come away without tearing the bark this will denote that the wood is in the right condition for taking the buds.

A sharp budding knife should be at hand, also lengths of raffia about ½ in. (1.25 cm) wide and 18 in. (45 cm) in length. The raffia must be fresh so that it will not break or decay too soon. A professional grower will bud around a thousand roses a day but the amateur will be able to take his time, for at first the work may prove tedious and exacting.

A short period in water will make the buds fresh and plump and this is how they should be. The best buds will be those to the middle of the stem, being in the right condition; those at the top of the stem being too advanced and those at the base not sufficiently mature. Insert the budding knife about ½ in. (1.25 cm) above the lower of the three or four buds to be used and holding the stem firmly, make the cut to a point about ½ in. (1.25 cm) below the eye. Do not cut through the stem but place the thumb of the free hand on to the bud and pull away the bark. This will leave the shield. The strip of bark holding the bud should now be removed so as to free the bud from its wood. Using the short length of leaf stem as a handle, the bark should be trimmed so that it will fit snugly into the 1 in. (2.5 cm) long incision to be made on the rootstock. The eye of the bud should protrude from the bark and be free from damage for this is to form the new plant.

On the bark of the rootstock a horizontal cut is made and immediately below make a vertical cut, not too deeply but to enable the bark to be lifted. Open up the bark with the budding knife, insert the bud and tie firmly in with the raffia. Then replace the soil at the base of the rootstock until it is level with the scion. In three to four weeks it may be found that the bud is swelling and if so, the raffia should be untied and slightly loosened.

The plants will require no further attention until the end of February when the rootstock head is removed immediately above the point of budding. This is done before the sap commences to rise for otherwise it would cause bleeding which would greatly reduce the vitality of the plant. It is, however, advisable to delay operations until the arrival of a dry, frost free day, and it may be better to wait until early March in the north before heading back. By this time, the raffia will have decayed and come away but where this has not occurred it should be carefully removed. Care should be taken not to cut back too close to the bud or eye otherwise canker may enter and cause the eye to die back. Should the bud begin to make growth before the plant is headed back the shoot should be cut back to two eyes so that it will develop as strongly as possible.

It will still take several months for the new shoot to become quite secure on the rootstock and if the garden is in any way exposed, a short cane should be inserted near the shoot to which it is tied, to prevent it from being blown about. When the shoot has attained a length of about 4 in. (10 cm) it should be pinched back to two eyes if this has not previously been done. The shoots will come into bloom at the beginning of July and at the end of autumn the plants are removed to their permanent quarters.

Pruning a 'maiden' plant will consist of cutting back the shoots to within 6 in. (15 cm) of ground level, to a healthy eye pointing in an outwards direction. It is a mistake to cut back too severely; this may be done the following year, but first let the maiden plants become thoroughly established. There is, however, nothing to be gained by too hard pruning at any age.

GROWING FROM SEED

The species, as well as those plants intended for use as rootstocks, may be raised from seed which is contained in the hips. The seed is covered with a strong outer covering and if old and dry, germination may not take place. Always sow freshly harvested seed though this should have been dried-off fully, otherwise it may decay on account of mildew. Sow in early spring after the seed has been harvested in autumn and has been correctly dried off.

The seed may be sown broadcast in a frame or in the open in drills made 1 in. (2.5 cm) deep and lined with peat or leaf mould and dusted with superphosphate of lime. A partially shaded position will be quite suitable for here the young seedlings will not dry out too quickly.

Sow reasonably thickly as quite a few may not germinate, and after covering the seed, maintain the soil in a moist condition. M'Intosh tells us that Joseph Paxton, then gardener at Chatsworth, would cover his seeds with fresh moss which was kept damp and removed as soon as the seed had germinated. Where there is overcrowding the seedlings must be thinned to prevent mildewing and in twelve months' time they should be transplanted to beds 12 in. (30 cm) apart. Those grown for budding will be ready for working in about sixteen months from sowing time, whilst shrub roses raised from seed may be transferred to their flowering quarters after being twelve months in the rows when they will begin to bloom in another twelve months' time.

SPECIE ROSES AND VARIETIES

Rosa alba. The White Rose about which the Roman historian Pliny said, 'The Isle of Albion is so called from the white roses with which it abounds', and he was the first to give an account of the culture of roses, saying that 'they should be transplanted in February... and replanted with a foot distance one from another'. This would be too close for the White Rose which makes a plant almost as bushy as it grows tall and so requires planting about 3–4 ft. (about 1 m) apart, any time when the ground is in suitable condition, between November and March. The white roses require little attention as to their pruning, merely cutting out dead and straggling wood in spring. The flowers are followed by handsome hips. *R. alba maxima* is the Great Double White or Jacobite Rose, and was used as their symbol by the supporters of Bonnie Prince Charlie, the Young Pretender. It makes a large bush 6 ft. (2 m) tall and the same across and blooms early in June, being one of the earliest of all roses. The large flat flowers are folded at the centre. *R. alba semi-plena* is the semi-double White Rose of York. The blooms with their attractive golden stamens are borne on graceful arching stems.

BELLE AMOUR. Thought to have originated in a convent garden in Germany, it grows 4–5 ft. (1.5 m) tall and bears attractively shaped blooms of a lovely shade of palest pink with a clove-like perfume.

FÉLICITÉ PARMENTIER. A delightful rose, growing 3–4 ft. (about 1 m) tall and 3–4 ft. (about 1 m) across. Its cream-coloured buds open to flowers of rosette-shape, like those of the ranunculus and which are of palest pink with the scent of honeysuckle.

GREAT MAIDENS' BLUSH. It was grown during medieval times and the late Lady Nicolson (Miss V. Sackville-West) of Sissinghurst Castle has said that it retains its petals longer than any other rose. It has blue-green foliage and bears double flowers of blush-white with a pink centre.

R. bourboniana is said to have come about by a natural crossing of a form of *R. chinensis,* the China rose, and the autumn-flowering Damask, on the French island of Reunion (then called Bourbon) in 1817. The result is the wonderful Bourbon roses which, unlike most of the old shrub roses, are of recurrent-flowering habit. There are those who consider them the most exquisite of all roses, the blooms being quartered at the centre and filled with overlapping petals. They are also deliciously scented. Requiring little or no pruning they were during the last century planted in many cottage gardens, often as a hedge or against a wall where their 'old-fashioned' flowers enhance the white-washed walls of the cottage.

BOULE DE NEIGE. Introduced a century ago, it does best against a wall which it will cover to a height of 6–7 ft. (about 2 m). It has dark green foliage and bears its symmetrical camellia-shaped flowers of snow-white from June until October.

LA REINE VICTORIA. A delightful Victorian plant and with its compact habit is suitable for the smallest of cottage gardens. The circular cup-shaped blooms of rosy-pink are rightly packed with small petals and filled with exquisite perfume.

LOUISE ODIER. It is one of the loveliest of garden roses growing 4–5 ft. (about 1.5 m) tall with dark leathery foliage and bears circular quartered blooms of rose-pink, flushed with lilac and with a soft sweet perfume.

MME ISAAC PEREIRE. Against a wall it will reach a height of 12 ft. (3.5 m) or more and bears handsome flowers of rosy pink flushed with crimson and which appear in profusion in autumn.

MME PIERRE OGER. It is a 'sport' from 'La Reine Victoria' and like 'Zéphyrine Drouhin' is always in bloom whatever the weather. The cup-shaped blooms have almost transparent petals of creamy-pink with outstanding perfume.

ZEPHYRINE DROUHIN. It is possibly the most valuable of all garden roses and may be Queen Elizabeth I's 'Rose-without-a-thorn'. It is thornless and bears, from June until November, double Hybrid Tea-type flowers of brilliant cerise-pink with a silvery sheen and pronounced fragrance. It may be used against a wall or as a hedge or wind-break for only during three to four months of the year is it without foliage.

R. centifolia. The old cabbage rose, so called because the petals fold over like the leaves which form the 'heart' of a cabbage. It is also the rose of Provence, the most southerly of French states, situated close to the Italian border where the rose grew in abundance having reached there from the mountainous regions of Greece and Bulgaria where the plant grows naturally. It takes its botanical name from the flowers having one hundred petals. It was known to Pliny who said that the blooms were the largest of all roses. *R. centifolia* grows 4–5 ft. (about 1.5 m) tall and bears blooms of China-pink which were depicted by the Dutch flower painters of old. The blooms are borne on short lateral shoots during June and July. The plants with their thorny stems require the minimum of pruning; occasionally shortening the shoots and cutting out dead wood.

BULLATA. With its large pale green leaves deeply grooved and with their serrated edges it is known as the Lettuce-leaf rose and figured in Redouté's great work. It was known in the seventeenth century, the large globular blooms being of a lovely shade of cherry-blossom-pink and are powerfully scented.

FANTIN-LATOUR. Its arching stems and cup-like blooms of shell-pink, made it a great favourite of the French painter after whom it was named.

GROS CHOUX D'HOLLANDE. It is the Dutch Hundred-leaved rose, known to the flower painters of the seventeenth century and grown in gardens in earlier times. It makes a compact bush 3–4 ft. (about 1 m) tall with small pale green leaves and bears tightly petalled flowers of a unique shade of mushroom-pink.

R. centifolia 'Muscosa'. With its distinctive musky perfume, the Moss rose has occupied a special place in the cottage garden since early in the nineteenth century. It is a mutation of the Provençe rose, the moss-like hairs appearing on the calyx and on the stems and leaves. Its origin is undecided. Edward Bunyard in his *Old Garden Roses*, suggested that it was discovered in Holland about the year 1700 but Mme de Genlis said that the mutation appeared first in Provençe from which plants reached England early in the eighteenth century. Most, however, date it from the time of the Napoleonic Wars and by 1860 there were, as William Paul the famous rose grower of Cheshunt told us, more than fifty named varieties, the rich purple, pink and burgundy colouring of their blooms providing a striking effect against the crimson moss on the stems.

The plants should be harder pruned than most of the old roses, cutting back the old wood to about half way and the side shoots to the third bud.

BLANCHE MOREAU. One of the loveliest of the moss roses, bearing pure white cup-shaped blooms and being covered with striking crimson moss.

CAPTAIN JOHN INGRAM. More than a century old, it remains one of the loveliest roses of all with purple-tinted foliage and bearing rosette-shaped blooms of deep velvety purple.

COMMON MOSS. It grows 3–4 ft. (1 m) tall and the same wide, the chalice-shaped buds opening to large cup-shaped flowers of clearest pink.

MARÉCHAL DAVOUST. One of the loveliest roses in cultivation, it grows 3–4 ft. (about 1 m) tall and is covered in crimson moss whilst its cup-shaped blooms are of carmine-pink with a lilac edge.

WHITE BATH. It was introduced in 1817 by William Shailer, a nurseryman and was known originally as the White Moss rose, being a 'sport' from the Common Pink Moss. It makes a compact bush 3–4 ft. (about 1 m) tall, its heavily mossed pink buds opening to large cups of ivory-white.

R. damascena. The Damask Rose is known as the Holy Rose for it is usually shown surrounding the Virgin as when she appeared to St Bernadette at Lourdes. Whilst in the Wilton Diptych, which shows the oldest-known portrait of an English king, Richard II is being presented to the Virgin and Child with St Edward the Confessor, St Edmund of Bury and St John the Baptist in attendance and with the Virgin crowned with pink Damask roses whilst the blooms litter the ground about her feet.

The Damasks should be pruned early in March, merely tipping back unduly long shoots and removing any dead wood.

VARIETIES

BLUSH DAMASK. Attaining a height of 6 ft. (2 m) it makes an excellent hedge. It begins to flower in June, the flowers being of a lovely shade of clearest pale pink with a deeper flush at the centre.

CHATEAU GAILLARD. It was discovered growing on the walls of Richard the Lion Heart's castle in Normandy and may have been brought back by him from the early Crusades. It has emerald foliage and bears flowers of fuchsia-pink which turn lavender with age.

LEDA. Know as the Painted Damask, it is right for a small cottage garden for it grows only 3 ft. (90 cm) tall and bears, in profusion, blooms of purest white tipped with crimson to give a picotee appearance.

MME HARDY. It was raised by M. Hardy in the Luxembourg Gardens and G. S. Thomas in *The Old Shrub Roses* says that 'it is unsurpassed by any rose'. It bears circular cup-shaped blooms of ivory-white with a tight button eye of jade-green. The petals are attractively folded.

YORK AND LANCASTER. Syn. *R. damascena variegata*. It has a history surpassing that of almost any other rose and grows 6 ft. (2 m) tall, its blooms being a mixture of rose-pink and white, splashed and in stripes and with delicious perfume.

R. gallica. It has a history which goes back to the beginning of time and it was used by Persian warriors some 100 B.C. to adorn their shields. It is believed to have been introduced into Britain by the Romans. Shakespeare knew it well and it would be familiar to him as the Apothecary's rose for it was widely grown for its medicinal qualities, especially around the old walled town of Provins, near Paris, hence the poets reference in Hamlet to 'two Provincial roses in my razed shoes'. The petals of the rose retain their perfume when dry longer than any other and were in great demand for making pot-pourris to place in the often musty-smelling rooms of cottage and manor house and for strewing over the floors.

The gallicas flourish in all soils and are completely hardy and easy to manage. They grow especially well in a sandy soil. They should be pruned in March, shortening the previous season's wood to half their length. The plants have no real thorns but are covered in strong hairs. They make neat upright bushes about 3–4 ft. (about 1 m) tall and are midsummer-flowering only.

BELLE DE CRECY. One of the most beautiful of the old roses, it has bottle-green foliage and bears flat honey-scented blooms of violet-pink flushed with grey.

CAMAIEUX. It grows only 2 ft. (60 cm) tall and bears pretty cup-shaped flowers of crimson-maroon, striped with white; the stripes later turn to a lavender colour.

CHAMOISIE PICOTEE. It makes a small upright plant with dainty leaves and bears rosette-like flowers of light red, edged with crimson and spotted with brown.

DUCHESSE D'ANGOULEME. Growing 3–4 ft. (about 1 m) tall, it is thornless and bears dainty rosettes of soft Malmaison-pink.

ROSA MUNDI. Redouté rightly names it Fair Rosamond's rose and it is one of the oldest of all named roses, named after Rosamond Clifford, mistress of Henry II who was buried in the old Nunnery at Godstow in Oxfordshire. The best and most free-flowering of all the striped roses, the white blooms are striped and blotched with purple, pink and red and it is most free flowering.

R. moyesii. It grows 10–12 ft. (about 3 m) tall and the same in depth and makes an excellent hedge if its arching stems are trained along wires. It has blue-green foliage and bears crimson-red flowers in midsummer, followed by bottle-shaped hips of brilliant scarlet. There is a pink form, 'Rosea' and a number of interesting hybrids. Crossed with *R. rubiginosa*, Eos has graceful arching branches covered with large flat blooms of coral red with orange anthers; and 'Geranium' (raised at Wisley) bears scarlet flowers amidst bright green foliage.

R. rubiginosa. A native rose and like *R. canina*, the Dog rose, it bears flowers of clearest pink which are delicately fragrant. It sends out its long arching stems to 10 ft. (2.5 m) or more in length and which may be trained to form a canopy or bower beneath which the refreshing fruity fragrance of its leaves may be enjoyed during the warmth of summertime for which purpose it was planted in cottage gardens of old.

'The vulgar Sweet Briar,' wrote John Worlidge in his *Systema Horticulturae* (1677), 'for its excellent odour in the spring, deserves a place near your house or places or repose, yet not so much as that which bears a double blossom for which it is preferred to it, and is one of the best of odoriferous plants.'

Not least of its attractive qualities is its scarlet hips which follow the blooms in autumn and persist right through winter, providing colour when the plant has shed its leaves. Gerard wrote that 'the fruit when ripe maketh most pleasant, meats and banqueting dishes (using it as an alternative to red currant jelly), as tarts and such-like; the making whereof I commit to the cunning cooke . . .'

To form an impenetrable hedge, the plants should be set 3–4 ft. (about 1 m) apart when they will quickly close up the gaps and will eventually grow to a height of about 6–7 ft. (about 2 m) and the same distance in width. With their large hooked thorns, such a hedge will deter the most savage beast as well as providing colour almost the whole year round.

JANET'S PRIDE. It was illustrated in Miss Ellen Willmott's book, *The Genus Rosa* (1910) in which she quotes the Rev. Wolley-Dod that this variety of the Eglantine was found growing in a Cheshire lane. Of similar habit to the parent, the flowers are of a lovely shade of bright pink with a white centre but the foliage is less aromatic than that of the parent.

LA BELLE DISTINGUÉE. It is the Double Scarlet Sweet Briar, less vigorous than the parent and not nearly so thorny. The deep red flowers are followed by hips of similar colouring. The foliage however is not very fragrant.

MORNING'S BLUSH. Of compact habit, rarely exceeding 5 ft. (1.5 m) in height, the fully double blooms are quite small and of a lovely shade of blush-pink, enhanced by the neat bottle-green leaves.

In 1890, Lord Penzance began a long programme of breeding, using *R. rubiginosa* as a parent and a number of Hybrid Perpetual and Bourbon roses as the pollinating parents. The result was a number of fine hybrid varieties bearing scented flowers and with the same strongly aromatic foliage of the sweet-briar. Amongst the best are:

AMY ROBSART. The blooms are large and semi-double and of a rich shade of reddish-pink.

FLORA MCIVOR. The large single flowers are white, attractively flushed with pale pink.

JEANNIE DEANS. One of the best, the semi-double blooms being of a startling shade of crimson-scarlet.

LUCY ASHTON. One of the loveliest of all roses for a hedge, the blooms being of purest white with a distinctive rose-pink edge to the petals.

R. rugosa. The Ramanas or Japanese rose, the hardiest of all the species, troubled neither by black spot nor mildew and which does well in light sandy soil and in a wind-swept garden where it may be used as a hedge. Continuous flowering, the large magenta-red blooms are clove scented and are followed by large hips with the highest vitamin-C content of all fruits.

There are numerous hybrids including 'Agnes', raised by crossing with the yellow Persian rose and bearing pompon-shaped flowers of yellow and amber. Outstanding too, is 'Conrad F. Meyer' which bears hybrid-tea-type blooms of silvery pink and 'Roseraie de l'Hay' which grows 6 ft. (2 m) tall and is one of the best of all shrub roses, bearing double blooms of purple-crimson through summer and autumn.

R. spinosissima. The Burnet or Scottish rose which increases by suckers and is adaptable to a dry, sandy soil and exposed coastal situations. With its heavily thorned stems it makes a dense and efficient hedge. Growing 6 ft. (2 m) tall, it blooms before the end of May, bearing single creamy-white flowers followed by black hips.

The best form is 'Stanwell Perpetual', the only Burnet which is perpetual flowering, probably due to a chance crossing with the 'Autumn Damask'. Found in a London garden, it grows 6 ft. (2 m) tall and the same in depth, its graceful arching stems being covered in grey foliage and blush-white flowers of sweetest perfume. Outstanding too, is 'Claus Groth', raised by Tantau and which has reddish-purple foliage and bears flowers of apricot orange whilst 'Frühlingsgold', raised by Kordes, has grey-green foliage and bears semi-double blooms of clearest gold.

ROSE, *see* ROSA

ROSEBAY, *see* Nerium

ROSEMARY, *see* Rosmarinus

ROSMARINUS

A plant of the Mediterranean, Rosemary is at its best in the dry sandy soil and salt-laden atmosphere of the seashore from which it takes its name, 'dew of the sea'. But it loves a wall more than any plant, especially a walled east coast garden where the climate is drier than any other part of Britain and where the plants will receive the maximum amount of sunshine. And to bring out its fragrance to the full it must have some lime or lime rubble about its roots. As the young plants, until established, may become 'burnt' by cold winds, late April or early May is the best planting time, and so long as the soil is deeply dug and friable do not dig in any manure. If the soil is heavy add some coarse sand. Space the plants 2 ft. (60 cm) apart and select as sunny a position as can be found.

Rosemary may also be easily grown from seed sown in spring in shallow drills in a sunny position; also by cuttings which will strike readily in a trench of sand and peat if taken between May and August and kept moist.

The best time to gather the leafy stems is early in September when its fragrance is more pronounced. Distilled oil of rosemary is one of the ingredients of Eau-de-Cologne and the water from simmered leaves acts as an excellent hair tonic and as a freshener and purifer of the face. The ancients recommended simmering in white wine. But it is the freshly gathered foliage that is most valuable, for if placed under the pillow case it will bring about rapid sleep. Banckes' *Herbal*, the first in the English language, mentions that 'the leaves laid under the pillow deliver one from evil dreams', and Miss Sinclair Rohde has told us that in the same way the potent seeds may be gathered in September, dried and placed in muslin bags and hung about the bedroom to encourage sleep.

ROWAN, *see* Sorbus

RUBUS

Ornamental plants of the large family of brambles of which the raspberry and blackberry are members. There are several species for growing against a cold, exposed wall or fence, which are planted in autumn and winter and grown almost entirely for their leaf and stem colourings. They are deciduous, and, if cut back each winter, will form new shoots from the base up to 10 ft. (3 m) in length during the summer. Of these, *Rubus thyrsoideus* 'Alba Plena' covers its shoots in double white blossoms from July until September, whilst *R. ulmifolius* 'Bellidiflorus' bears the same button-like double flowers, but in shell-pink. Of those with coloured stems, *R. biflorus*, native of the Himalayas, has white-washed stems which should be allowed to remain without pruning until spring. They look most attractive against a dark wall or fence in winter. Another is the Japanese wine berry, *R. phoenicolasius*, the stems being covered with handsome crimson hairs, whilst in autumn the amber-coloured fruits are equally attractive and are valuable for making into preserves.

RUDBECKIA

ANNUAL

Growing to a height of 3–4 ft. (about 1 m), the annual form of the better-known perennial, *R. bicolor*, is an excellent plant for the back of the border and also for cutting. The two best varieties are 'My Joy', large golden-yellow; and 'Kelvedon Star', the yellow blooms having a striking bronze disc. Seed is sown towards the end of March, the young plants being thinned to 16 in. (40 cm) apart.

The plants may also be treated as biennial and sown in early September, but as they are naturally late to come into bloom, there is little to be gained. The rudbeckia is valuable in that it is at its best during late autumn, when most other annuals are past their prime.

PERENNIAL

Those varieties of *R. purpurea* which bear rose-coloured blooms are now classed as *Echinacea purpurea*, and are described under this heading. The golden flowered varieties of *R. speciosa*, the Cone-flowers, are outstanding plants for the back of the border. They thrive in a heavy clay soil and bloom from July until November. Propagation is by root division in November which is the best time to plant.

VARIETIES OF PERENNIAL RUDBECKIAS

AUTUMN SUN. The flowers are single and of a deep golden-yellow colour, in bloom throughout late summer and autumn and borne on 6 ft. (2 m) stems.

GOLDEN BALL. The blooms are fully double and are of a lovely shade of lemon yellow. The best of all the rudbeckias, it grows 6 ft. (2 m) tall.

GOLDEN GLOW. Similar in colour to 'Golden Ball' but not so tall of habit. The blooms are fully double.

GOLDSTURM. A continental variety, one of the best border plants introduced in recent years. The blooms are double and of a rich buttercup colour, the habit of the plant being very compact. The first of the rudbeckias to bloom in mid-July.

SAGE, *see* Salvia

ST BRUNO'S LILY, *see* Anthericum

ST JOHN'S WORT, *see* Hypericum

SAINTPAULIA

Of the same family as the Gloxinia, the Streptocarpus and the Achemene, all of which have lost much of the popularity they enjoyed during the hey-day of the conservatory, the Saintpaulia or African Violet has, of recent years, made rapid progress both in Europe and in America. In the U.S.A., the plant is more popular than any other pot plant, the first hybrid varieties of *Saintpaulia ionantha* having been sent over from Holland and Belgium where the plants at first achieved little popularity. In America the dainty habit and long flowering period of the plant, and the fact that it does well in the city apartment, quickly earned for it considerable popularity, and it was not long before many lovely new varieties were being re-introduced into Europe. Today in the U.S.A. more than 2,000 varieties are in commerce, and specialist nursery growers everywhere cater for a vast trade. The plant was discovered in East Africa in 1892 by the German botanist Baron von Saint Paul-Illaire, and it was exhibited at the Ghent Floral Exhibition the following year when it created much interest.

CULTURAL METHODS

As the plants flower between early August and the end of March, a time when flowers are expensive, this has contributed to their popularity; also the fact that the plants are perennial and evergreen, and given the minimum of attention may be expected to bloom for several years. They are also happy growing under home conditions so long as the plants receive some light, though with the Saintpaulia, sunlight is not necessary and window space may be used for other plants. But light it must have and electric light will prove quite satisfactory. One specialist grower cultivates the plants on trays fixed one above another, and in a cellar, the only light reaching the plants being fluorescent lighting.

The method is to stand the plants in their pots on trays covered with pebbles which are kept constantly moist. Each 6 ft. (2 m) long tray is lighted by an 80-watt tube fitted to a 6 ft. reflector which is suspended 12 in. (30 cm) above the actual plants, one hundred of which can be accommodated in each tray. The lights are switched on at 7.30 a.m. upon rising and are turned off at 10.30 p.m. upon going to bed, and with fifteen hours' light plant growth is sturdy whilst the plants flower profusely. The temperature of the cellar is maintained at around 58°F. (15°C.) so that the plants will receive no check when taken into the living-room, which will maintain a similar temperature during winter. Though this information is given for those who have a cellar or poorly lighted outhouse and who may wish to make it profitable as an alternative to mushroom growing or forcing rhubarb, it is chiefly suggested as a method by which Saintpaulias may be grown in the living-room, where an alcove may be fitted with glass shelves and lighted by concealed fluorescent lighting which will make a pleasing alternative to glass or china on display.

Saintpaulias require rather more humidity than most indoor plants and so pots should be stood in glazed earthenware saucers on a layer of shingle, or on trays containing shingle, which should be kept constantly moist. The plants should be given moisture at the base rather than at the top for it is important that moisture does not come into contact with the foliage. If so, Ring Spot Disease may break out when eventually the leaf will die back altogether. An excess of moisture around the crown of the plant may also cause it to decay, so wherever possible water from the base of the pot, or very carefully around the side.

Saintpaulias are acid-loving plants and require a lime-free soil. If the John Innes potting compost is to be used, extra peat should be added. A suitable compost will contain 3 parts loam, taken from pastureland; 2 parts well decayed cow manure; 2 parts peat and 1 part coarse sand. The cow manure may be obtained from a farmer when out in one's car, and the other ingredients from a nurseryman. They should be mixed well together in a deep box.

Potting of established plants should take place in April or May after flowering has finished, but do not re-pot until the roots are showing through the bottom of the pot. Saintpaulias bloom best when they have become slightly pot-bound, which will generally be after the second winter of flowering. Do not attempt to propagate by dividing the plants, for once established they will resent root disturbance, whilst to divide them will upset the balance and spoil their appearance. The plants should be made firm in the pots, transferring them from a small pot, in which they will bloom their first season, to a slightly larger pot. It is advisable to place a layer of sand, preferably silver sand, over the top of the compost and around the crown to prevent damping-off, to which all members of the group to which the Saintpaulia belongs are rather prone.

If the plants are to be grown in a window, it will be necessary to shield them from the direct rays of the sun during summer when undergoing a partial rest period, otherwise leaf scorching will occur. For this reason it is advisable to grow the plants in a window which does not have a southerly aspect, and where the plants are being grown in a greenhouse or sun room it will be advisable to shade them from mid-April after flowering, until they come into bloom again late in August.

PROPAGATION

The plant may be propagated from seed, though this will not produce plants true to the varieties from which the seed has been saved. In any case it will take longer for the plants to come into bloom than where they have been propagated vegetatively. By far the best method of increasing stock is from leaf cuttings which is an inexpensive method of forming a collection of named varieties. It is possible to obtain from specialist growers, leaves taken from named varieties. Leaves which have been rooted may also be obtained from specialists and will be ready for potting into small pots to grow-on. Those who have not the facilities for rooting will find this an inexpensive method of forming a collection.

Saintpaulia leaves are heart-shaped with slightly serrated edges. Leaves for propagating should be removed from the plants with care so as not to upset the balance. From mid-May until mid-July is the best time for rooting, though they will root at any time provided a temperature of 60°F. (16°C.) can be maintained for three weeks, in which time the leaves will have rooted. But early summer, after the plants have flowered, is the time to remove the leaves which should be healthy and mature. Small leaves and any which show signs of being too old should not be used. The leaves should be of the same size and have the petiole or leaf stalk attached. This is inserted into the rooting medium with that part of the leaf adjoining the petiole, being planted 1 in. (2.5 cm) deep and made quite firm so that the leaf will be held in an upright position.

The rooting medium should consist of equal parts of peat, first made quite moist, and silver sand, and to guard against damping-off, to which the leaves are prone, the surface of the rooting compost should be lightly covered with pure silver sand. Soil should not be used and definitely not so if it has not been sterilized. Ordinary clean kipper boxes may be used and they will accommodate four rows each of a dozen leaves, but they may also be inserted round the side of a size 48 pot where rooting only a few.

Though the compost should be kept moist, over-watering must be guarded against. It is advisable to place the boxes (or pots) away from the direct rays of the sun. If the leaves are being rooted during early summer they may be given a partially shaded position where, at this time, they may still enjoy a naturally warm temperature. The leaves, should, however, be allowed to come into direct contact with moisture as little as possible and if inserted into a moist compost they should require little additional moisture until rooted. A small sheet of polythene placed over and around the boxes or pots will greatly help to conserve moisture, but to prevent sweating and excessive humidity it will be advisable to remove the polythene every now and again.

The leaves will root in about three weeks, a little longer if the temperature falls below 60°F. (16°C.) and tiny plantlets will form along that portion of the leaf which was inserted in the soil. In about eight weeks the leaves are carefully lifted and each plantlet is removed from the remainder of the leaf, which is then discarded.

Pot up the tiny plants without delay so that the roots do not dry out, using a small pot containing a compost similar to that suggested for re-potting larger plants. Make certain that the pots are well crocked and give a sprinkling of silver sand around the crown of the plant after making it firm in the pots. The plants will remain small during their first winter when they should be watered as little as possible, giving them only sufficient moisture to keep them alive and growing. In spring they will begin to make more leaf and will require additional moisture and by June they will need to be transferred to larger pots in which they will come into bloom at the end of August and will continue until March. The plants should not be re-potted for another twelve months. At all times care must be taken with watering and to guard against draughts. It is also important that for their first year the growing temperature should not fall below 48°F. (9°C.) and in winter, if growing indoors near a window, it will be advisable to lift the plants to a warmer place in the room at nightfall. Plants in small pots may be grown in trays where the pots are surrounded with peat or moss which is always kept moist. The pots may be quite close together though extra care must be taken with the watering. Plenty of light, even if it is artificial, and a constant temperature is the secret of success with the Saintpaulia.

GROWING FROM SEED

Those who wish to raise plants from seed will find that sowings of the true species *S. ionantha* will bear bloom true to the deep blue colouring, numerous small blooms appearing from each stem and which are held above the foliage. A well-grown plant will be covered in bloom but obtain the Grandiflora strain which will produce more refined flowers. It should be said that plants raised from seed do not generally grow as large as those raised from leaf cuttings.

Specialist growers sowing in peat, prefer to do so at the turn of the year, for then the seedlings will be ready for transferring to $2\frac{1}{2}$ in. pots towards the end of March and will have made sufficiently large plants to come into bloom in autumn. Those who have to raise plants in the home should sow in May in a sunny window in a propagator or in pans which are kept covered with a glass bell-jar or cloche until the seedlings are ready for moving to small pots some time in July. They will not bloom the same year but will build up into large plants if kept growing through the winter and will bear a mass of bloom the following year.

The sowing compost should contain a mixture of moist peat and silver sand which should be made quite level as the seed is small. Do not cover the seed unless perhaps with a very light sprinkling of silver sand and on no account water from above. Whenever the surface of the compost begins to dry out, the pans or boxes should be immersed at the base into a bowl of tepid water for just so long as to allow the moisture to penetrate up to the surface. Keep the containers covered either with a sheet of clean glass or with a bell-jar which should be used after the seed has germinated. This will be in about a month.

Interesting new varieties may be raised by hand-pollinating indoor flowering plants which should take place when the plants are young. After collecting the seeds it is equally important to sow almost immediately, otherwise germination may take many months.

VARIETIES

BEATRICE. A fine variety, the large double flowers being of deep purple-blue.

BIANCA. The blooms are single and are white, splashed with purple.

BICOLOR. Most unusual, the top petals being deep purple whilst the lower petals are pale mauve.

BLUE BOY. The single blooms are of an exquisite shade of bright blue with a striking yellow centre. One of the earliest introductions and still excellent.

BLUSHING MAIDEN. The double blooms are white, suffused with cerise-pink.

DOUBLE DELIGHT. The double blooms are produced with freedom and are of a lovely shade of mid-blue.

FANTASY. Unusual in that the large blooms are of palest mauve speckled with blue.

FINLANDIA. The blooms are single and are of deep amethyst-blue, the petals being attractively fringed.

GATTON PARK. Of very compact habit, the dark amethyst-blue flowers have striking golden stamens.

GREY BLUE. A lovely variety, the single blooms being of pale blue-grey, most attractive when growing close to the pink-flowered varieties.

GYPSY ALMA GIRL. It is a beauty, bearing bright lavender-blue flowers and has attractive bronze foliage.

HERMIONE. Very free-flowering, the double white blooms have an attractive mauve centre.

INDIANOLA. The blooms are single and large and are of a lovely pale lavender colour which is accentuated by the bottle-green foliage.

IRAS. The large blooms are double and are of a lovely shade of lavender-mauve.

LADY ANNE. Lovely in that the double blooms are of a pure lilac-mauve shade.

LADY GENEVA. One of the finest varieties, the large blooms being of an attractive shade of deep mauve-pink, the petals being shaded with white.

ORCHID BEAUTY. The large single blooms are a striking shade of bright reddish-violet.

PINK IDEAL. Probably the best pink, the double blooms being of pure deep pink, a shade to be found in so few plants.

PINK PEARL. The single pink flowers are produced with freedom during fully eight months of the year.

PINK ROCKET. The blooms are double and of a deep rosy-pink colour. One of the best.

PUCK. Of compact habit, it has serrated leaves and bears masses of single mid-blue flowers.

PURPLE PRINCE. The deep purple blooms are double and are in bloom over a long period.

RAINBOW ROSE. Most attractive in that the double blooms are of white, heavily suffused with rose-pink.

RED COMET. A new introduction, it is a strong grower and quite outstanding, bearing very double blooms of a bright wine-red colour.

RUFFLED QUEEN. An unusual variety, the large double blooms are of deep purple with an attractive ruffled or waved edge.

SAILOR'S DELIGHT. A particular favourite, the blooms are double and of a soft shade of pale blue, the foliage being attractively serrated.

SILVER LINING. The first of an important new strain, producing masses of double flowers of richest blue, the petals being edged with white, like lace.

SNOW WHITE. Probably the best single white, the blooms being produced in profusion.

WHITE PRIDE. The best double white, the blooms being produced over a long period.

SALIX

Hardy deciduous trees or shrubs valuable for growing near water, whilst the weeping forms make specimen trees for a lawn or other focal point. They are amongst the first trees to come into leaf, the young leaves being grey-green whilst the shoots are bright orange or yellow and have an attractive appearance in winter. Willows like a heavy soil which does not dry out in summer and whilst they are tolerant of chalk and lime, this is so only where the soil has a high moisture content. Prune to maintain the shape of the trees but they may be cut hard back late in autumn or in spring if more colourful new growth is desired. Propagate by removing leafless shoots in November and inserting into the ground where they are to remain. Shoots up to 6 ft. (2 m) long may be used and if left moist, they root quickly.

S. acutifolia is the Violet Willow so called from the purple colour of its shoots which are overlaid with white 'bloom', like black grapes. It has long narrow leaves and bears golden yellow male catkins. *S. alba,* the White Willow, has greyish-white foliage and makes a small upright tree. The form 'Aurea' has golden leaves whilst 'Tristis' is the Weeping Willow of Minton's willow-pattern china. It forms long arching branches which reach the ground and which are bright yellow in winter. *S. babylonica,* also of weeping form has similar drooping branches. Crossed with *S. fragilis,* it produced *S.* × *blanda,* the Wisconsin Willow which is of similar habit.

S. caprea is the Sallow or Palm of the hedgerows. The female bears silver catkins; the male, golden catkins which can be seen from afar when lit by the early morning sun. There is also a weeping form, 'Pendula'. Equally handsome with its showy grey and red catkins is *S. gracilistyla,* a small Japanese willow. *S. pentandra,* the Bay Willow has glossy leaves which have the aromatic scent of the Bay Laurel when crushed, whilst the male trees bear golden catkins in April.

SALLOW, *see* Salix

SALPIGLOSSIS

No annual is more exotic-looking with its large tubular flowers of richest colouring, of velvet-like texture and veined in the manner of orchids. They grow to a height of 2 ft. (60 cm) and though when massed in a bed they hold each other up and require no staking, a bed should only be made up where sheltered from strong winds. The plants are raised in heat and planted out in early June or seed may be sown in a border early in May and the plants thinned to 8 in. (20 cm) apart. This is a late summer flowering plant, the best strain being the Mixed Large-flowered.

SALVIA (Ornamental Sage)

In association with scarlet flowers and where the soil is dry and sandy, there are no finer border plants than the ornamental sages, readily propagated by cuttings, taken in September and rooted in a closed frame. They may also be raised from seed. Several, such as the tuberous rooted *S. patens* are not hardy and to be successful, require a warm climate. They are delightful plants all the same. Flowering at various heights the hardy sages may be used throughout the border, whilst their bloom may be enjoyed from early June until the end of October.

SPECIES

Salvia azurea. Growing to a height of 3–4 ft. (about 1 m), it is a lovely plant for the back of a border. Its sky-blue spikes are borne above attractive hairy grey foliage throughout autumn.

S. glutinosa. Difficult if not given a sunny, sheltered position and a dry, sandy soil, it is an outstanding plant, bearing spires of rich yellow on 3–4 ft. (about 1 m) stems during July and August.

S. haematodes. An outstanding introduction from Greece growing to a height of 2 ft. (60 cm) and bearing silvery-lilac blooms on branched stems from July until October.

S. nemorosa. For ten weeks or more from early July, it bears spikes of violet-purple above its grey-green foliage at a height of 3–4 ft. (about 1 m). There is a dwarf form, *S. compacta,* exactly the same but growing to a height of only 20 in. (50 cm) and so may be used towards the front of the border.

S. sclarea. This is sometimes called Clary and bears large grey leaves with the pungent smell of grapefruit, and bracts of bright purple at a height of about 3 ft. (90 cm). Oil is extracted from its leaves for use in the perfumery trade, whilst the foliage is used in scent bags and pot-pourris. The flowers are much visited by bees. The Silver Clary, *S. argentea,* is also a lovely border plant, its large silvery leaves possessing a rich fragrance. The plants like a dry, sandy soil and an open, sunny position. They are propagated from cuttings taken and rooted in frames or under cloches during summer, the plants being set out in spring, for like most blue plants the sages prefer spring planting. The plants may also be raised from seed sown early in summer.

S. uliginosa. It is the tallest of the sages and a lovely back row plant for a dry soil and a sunny position. Its rich azure-blue spikes are borne from early August until October and for its beauty to be appreciated to the full it should be planted near the taller crimson michaelmas daisies. It is of branching habit and should be planted 3 ft. (90 cm) apart.

SAMBUCUS

Valuable deciduous trees and shrubs for planting in partial shade and in town gardens. Of the Common Elder there are several forms which are handsome in leaf, whilst a number of American species are suitable for the wild garden. They grow well in all soils and the only attention needed is to keep them free of suckers and to cut out any branches which become too large. Propagate by removing ripened shoots in November and inserting them in trenches of sandy soil.

S. canadensis is attractive in flower, bearing large flat white heads followed by red berries. The variety 'Maxima' has large pinnate leaves whilst its flowers measure 12 in. (30 cm) across and have contrasting purple stems.

S. nigra, the Common Elder, bears golden leaves in the form 'Aurea' whilst 'Albo-variegata' bears dark green leaves with a contrasting margin of white. The variety 'Alba-plena' has double white flowers which have the familiar honey scent of the plant.

SANDERSONIA

A genus of a single species, native of Natal and named in honour of Mr J. Sanderson, first secretary of the Natal Horticultural Society. Of trailing habit and closely related to gloriosa, it should be grown in a warm greenhouse or garden room in the British Isles where climbing by leaf tendrils, it may be trained up a pillar or trellis or the stems supported by stakes and twine inserted around the side of a pot. It has a small tuberous root and in its native land, bears its bell-shaped blooms at Christmas, hence its name, Christmas Bells. In the British Isles it blooms from July until September.

It requires protection from winter frosts and rain and where grown indoors may be left in its pot undisturbed, to come into new growth each spring; or the roots may be lifted when the plant dies back in autumn, to be stored in boxes of sand during winter and started into growth again in spring. Plant in April, one tuber to a 4 in. (10 cm) pot containing a compost made up of fibrous loam, leaf mould, decayed cow manure and coarse sand in equal parts. Plant 2 in. (5 cm) deep and water copiously as the sun increases in strength. In autumn, the stems will die back when the plants should be gradually dried off and stored in boxes of sand in a frost-free room.

In the warmest parts, the roots may be planted at the base of a warm wall in early April and the stems grown against a trellis up which they will climb to a height of 6 ft. (2 m). The roots are lifted and stored when the plants die back in autumn.

Division of the small tubers in spring, before they are planted is the most satisfactory method, treating the cut part with flowers of sulphur and planting immediately.

SPECIES

Sandersonia aurantiaca. Native of Natal, it forms a short tuberous rootstock from which arise climbing stems up to 5 ft. (1.5 m) in length furnished in alternate lance-shaped leaves. The flowers, inflated like small balloons and in shape like a cardinal's cap, are borne at the leaf axils on wire-like stems and droop down. They are of deep golden-orange and are borne several to a stem.

SANSEVIERIA

Native of Ceylon and of south and west Africa, the Snake Plants are amongst the most beaufiful and useful of all indoor foliage plants. The strikingly marked leaves grow straight up from rhizome-like roots and will often reach 3–4 ft. (about 1 m) in length. Like the aspidistra, the plants prefer a poor soil and very little moisture; they are able to exist months without watering and are tolerant of almost complete shade. As the leaves grow straight up, as many as twenty growing from a mature plant in a size 48 pot, they may be placed on a narrow ledge so long as they have room to grow upright. The plants are not quite as hardy as the aspidistra and require a winter temperature of not less than 45°F. (7°C.), during which time they will require almost no moisture.

S. trifasciata laurentii, the Bowstring Hemp, is slow-growing and bears fleshy leaves which grow quite erect and taper to a point. The dark green leaves are edged with gold and banded with grey. *S. zeylanica* has narrower leaves which are marbled with white, whilst *S. thyrsiflora*, which does not grow more than 15 in. (38 cm) tall, has wider banded leaves, margined with creamy white.

Propagation is by root division, but the plants may occupy the same pots for several years and it is never advisable to disturb a healthy plant. Wipe the leaves occasionally to free them from dust. Give very little moisture throughout the year; where in a shaded place, only once a month. This plant is happy in partial shade.

SANTOLINA

The Cotton or French lavender, *Santolina incana*, which has nothing to do with the true lavender is an interesting plant. It has beautifully serrated silvery-grey leaves which are retained through winter and makes it a valuable plant for a low hedge. It grows almost 2 ft. (60 cm) tall, but may be clipped in spring and maintained at a height of no more than 15 in. (38 cm). Its leaves are pleasantly pungent and may be used to keep moths from clothes, for which purpose it is dried and mixed with lavender flowers and the leaves of southernwood.

To make a hedge, the plants should be set 18 in. (45 cm) apart in autumn and it will be happy in ordinary soil provided it is well drained. The cuttings, removed with a heel, strike readily in a sandy soil in the open and if inserted during midsummer will have become well rooted in time for transplanting before winter. The plant bears masses of bright yellow daisy-like flowers which will be lost if the plants are clipped later than early April. In the less exposed areas the plants may be clipped in autumn.

SANVITALIA

Sow where it is to bloom or under glass in February, setting out the plants in April 8 in. (20 cm) apart. An annual, it begins to bloom early in June and continues until October, forming a dense grey-green carpet and smothering all annual weeds. Its daisy-like flowers are pale yellow with a contrasting black centre and are produced on 6 in. (15 cm) stems.

SAPONARIA

ANNUAL

It is *S. vaccaria* and its variety 'Pink Beauty' is useful as a cut flower. It grows to a height of 2 ft. (60 cm) or more and bears its long sprays of deep pink flowers through late spring and early summer, when cut flowers are most appreciated.

The seed is sown in drills 15 in. (38 cm) apart in August. Sow thinly making thinning unnecessary, nor should the seedlings be transplanted. In favourable districts they will winter unprotected but it is advisable to cover the plants with cloches in exposed areas.

The bloom is cut just as it is opening, like gypsophila, taking a handful of stems and cutting them just above ground level.

PERENNIAL

The Soapwort, *S. officinalis* 'Flore Plena' cannot be recommended for a border of choice plants, for it is of rambling habit and when established, tends to choke nearby plants. Its double pink flowers, however, are very pretty and appear during the latter weeks of summer when the border may be somewhat devoid of colour. Seed, by which method it is generally propagated, should be sown in a frame or in boxes or pans early in April, the plants being set out towards the middle of the border in autumn to bloom the following year.

SARCOCOCCA

The Christmas Box, it is a dwarf evergreen shrub, happy in partial shade and which will flourish in all soils. *S. confusa* grows 3–4 ft. (about 1 m) tall with small glossy box-like leaves and it bears its small sweetly scented white flowers throughout winter. *S. ruscifolia* is similar in habit and foliage. If several are planted together, the females will bear oval fruits; those of *S. confusa* being black and *S. ruscifolia*, crimson.

Plant in October and prune only to remove dead wood. Propagate by root division or by cuttings of the half-ripened wood removed in August.

SARRACENIA

The Pitcher Plant, an insectivorous plant of which the pitcher-shaped leaves act as traps for flies which the plant devours. *S. rubra* grows nearly 2 ft. (60 cm) tall, its leaves being veined with crimson, whilst it bears showy red flowers during spring and early summer. The plants may be grown from seed sown in a mixture of peat and chopped sphagnum moss and germinated in a temperature of 60°F. (16°C.). A similar compost should be used for potting on the plants.

During winter the plants like a temperature of 48°F. (9°C.) whilst they should be kept as cool as possible in summer, shading and damping down the greenhouse and syringing the plants on all sunny days. During summer the plants are copious drinkers, but require very little moisture during winter.

SAVIN, *see* Juniperus

SAXIFRAGA

A large genus of the temperate regions, which may be divided into three main groups:

(a) Mossy saxifrages. They are of moss-like habit and prefer a position of semi-shade and ordinary well-drained soil. They may be used for edging small beds or for planting about the rock garden or in a trough. They make small hummocks and have bright green foliage. They bloom April-June.

Saxifraga densa makes a tiny plant 3 in. (7.5 cm) tall, the flowers being white whilst the foliage turns crimson in autumn. Equally compact is 'Peter Pan' which bears crimson flowers whilst 'Dartington Pink' bears fully double blooms of deepest pink. Outstanding too, is 'Sir Douglas Haig' which bears large flowers of crimson-red.

(b) Encrusted. The foliage is silver whilst a white encrustation surrounds the leaves. They require an open, sunny situation and a well-drained soil containing plenty of grit. They bloom May-July and in this section are several of the most beautiful plants of the rock garden where they are happiest growing between the stones and cascading over the rocks.

S. cotyledon 'Southside Seedling' bears its elegant sprays of white flowers, spotted with crimson on 15 in. (38 cm) stems whilst 'Tumbling Waters' bears its spikes of pure white in tumbling sprays 2 ft. (60 cm) long.

(c) 'Kabschia 'or 'Cushion'. Ideal for pans and sinks, to plant between paving stones or on the rock garden, they flourish in sunshine or shade and in a gritty, humus-laden soil. They are amongst the earliest plants to bloom, the first blooms appearing before the end of winter.

S. burseriana 'Sulphurea' forms a mat of bright green and bears large, flat sulphur-yellow flowers on 2 in. (5 cm) stems of reddish-crimson, whilst Cranbourne forms grey-green rosettes above which are borne rose-pink flowers. *S. kellerei* bears its soft pink flowers in February whilst *S. grisebachii*, 'Wisley' variety bears spikes of deep crimson.

SAXIFRAGE, *see* Saxifraga

SCABIOUS, *see* Scabiosa

SCABIOSA

ALPINE

A delightful little plant is *S. alpina* which throughout summer bears its tiny heads of pale mauve flowers. Then there is *S. lucida* which bears its pale pink bloom right into autumn and *S. graminifolia* with its attractive silver-green foliage and deep violet flowers. Equally charming is *S. columbaria* which bears its pale blue heads during May and Jane. All appreciate a dressing of lime every autumn and they should be increased by division in March rather than in autumn just like their bigger brothers.

ANNUAL

The annual form of the scabious has never become as popular as the perennial, yet it possesses all the blooms of richer colouring than most flowers. The bloom is produced on 3 ft. (90 cm) stems. It is best sown where it is to bloom, the seed being sown late in August in drills 12 in.

(30 cm) apart, the seedlings being thinned out to 6 in. (15 cm) apart in early spring. Any well drained soil is suitable, provided it is not lacking in lime. The plants will appreciate some lime-rubble worked in before sowing.

Lovely for cutting is 'Blue Moon', of a glorious shade of sky-blue. Equally attractive is 'Cherry Red', well-named; and 'Rosette', rose-pink, shaded with salmon, a bloom similar in shape to the 'Esther Read' chrysanthemum. They bloom twelve months after sowing.

PERENNIAL

The perennial scabious, *S. caucasica* which reached this country at the turn of the nineteenth century has only during the past twenty-five years enjoyed much popularity. Today with its freedom of flowering, the improvement of the bloom, and its long lasting qualities when cut and in water, combine to make it the most widely grown of all flowers for cutting. It also travels well. The plants come into bloom towards the end of June and will be carrying at least some bloom when the border is cleaned in November. It is an excellent plant for a chalky soil and where lime is not naturally present, as much as a bucketful may be given to each plant in November each year without fear of overdoing it. It enjoys as much lime as the dianthus.

Ordinary soil, neither too heavy nor too sandy, suits the plant best, and lime rather than manure. A little decayed cow manure worked into the soil at planting time is all that is required. Of all plants, the scabious resents moving at any time of the year other than early in spring. To plant in autumn, however well-drained the soil, will be to court disaster. The plant makes a long woody rootstock which comes into action only in spring. Plant 18 in. (45 cm) apart and do not disturb the plants more than necessary. Propagation is by division of the woody rootstock in March or April.

The bloom should be cut when fully open and with as long a stem as possible, making up bunches of a dozen blooms. The ability of the plant to bloom over a period of twenty weeks makes it an indispensable plant for the border and for the commercial grower. Though plants of *S. goldingensis* will grow reasonably true from seed, those of the more recent introductions are obtainable only by root division.

VARIETIES OF PERENNIAL SCABIOSA

CLIVE GREAVES. Bearing its rich blue flowers on very long, sturdy stems this is not only one of the best of all scabious, but one of the loveliest of all cut flowers. Very free flowering, the blooms are large and semi-double and the colour does not fade even during strong sunlight (see plate 70).

CONSTANCY. Free flowering, the beautifully shaped blooms are of a distinct shade of powder-blue. A popular commercial variety.

DIAMOND. Deep, almost navy-blue, a much sought after colour, but not so freely produced as 'Moerheim Blue'.

EMILY. A new white variety, the bloom is more icy-white than that of 'Ivory Queen' or 'Miss Willmot', and is extremely free flowering. Lovely for mixing with 'Imperial Purple'.

FLORAL CHARM. A lovely shade of powdery-mauve of good form and freely produced.

GEORGE SOUTER. A new variety of deepest blue with a very long flowering season.

IDA STATHER. The colour is distinct, being of a bright lavender, flushed with rose. The habit is compact and the bloom is always in demand though it is not so freely produced as some of the others.

IMPERIAL PURPLE. A superb introduction, growing to a height of nearly 3 ft. (1 m) and bearing huge blooms of deep royal purple.

ISAAC HOUSE. A fine variety, the deep mauve flowers being borne in profusion on long stems.

IVORY QUEEN. A new scabious, coming early into bloom and bearing its ivory coloured blooms on long, sturdy stems.

MISS WILLMOTT. The large, beautifully shaped blooms are of an attractive shade of creamy-white.

MOERHEIM BLUE. A new introduction, the huge violet-blue flowers being produced over a period of six months and are borne on long, sturdy stems.

PENHILL BLUE. The colour is a shade deeper than 'Clive Greaves', quite as freely produced, although the bloom is not quite so large.

PRIDE OF EXMOUTH. Sky-blue would describe its colour, the blooms being borne on long stems though the habit of the plant is compact.

WANDA. The semi-double blooms are of an attractive shade of mid-blue, freely produced.

SCARBOROUGH LILY, *see* Vallota

SCHIZANTHUS

This, the Butterfly Flower or Poor Man's Orchid, is one of the loveliest of all greenhouse annuals, for no plant has a more dainty habit and none is more free flowering. The new strains are of dwarf habit, but to grow those compact bushy plants, attention to detail must be given. Sow in gentle heat early March, pricking off the seedlings into small pots as early as possible. As the plant makes growth stop once or twice, and stake to prevent untidiness. Keep the plant growing and water with liquid manure when the buds form. A most striking variety is the new 'Crimson Cardinal', growing to a height of 16 in. (40 cm), whilst 'Dwarf Bouquet', more dwarf, is obtainable in a wide range of colours, which includes rose, salmon-pink, amber and apple blossom. Also growing to a height of about 15 in. (38 cm) is the 'Danbury Park strain, bearing flowers in shades of pansy purple, mauve, etc.

SCHIZOSTYLIS

A genus of two species, native of South Africa and taking its name from *schizo*, to cut and *stylos*, a column, in reference to the thread-like styles. It is almost hardy in the British Isles but as it blooms late in September until Christmas, it should either be grown indoors or planted only in those gardens enjoying a favourable autumn and winter climate.

If growing indoors, plant the corms (really a swollen rhizome) in April, 2 in. (5 cm) deep and 2 in. (5 cm) apart in pots or deep boxes which may be allowed to stand outdoors at the foot of a warm wall until they begin to bloom in September. They are then lifted indoors to a sunny greenhouse or garden room where, in a temperature of 42°F. (5°C.) they will bloom until Christmas, to provide cut bloom for home decoration. The spikes remain fresh in water for several weeks. The compost should consist of fibrous loam, peat and decayed manure to which is added a liberal sprinkling of sand and at all times it must be kept moist. After flowering, protect the plants from frost and keep the compost just moist. In April, they may be placed in the open to repeat the process.

Where growing outdoors in a warm border where the plants are to bloom late in autumn, plant in April 4 in. (10 cm) deep and 3 in. (7.5 cm) apart in a sandy soil containing some peat or leaf mould to maintain summer moisture for the plant is a lover of moist conditions. Where possible, rear over the blooms in autumn, a frame light to protect them from adverse weather.

Plants that have been undisturbed for two or more years will have formed a tuft of leaves and stems at the base of which is a swollen rhizome. The plants are divided in spring and re-planted without delay.

Schizostylis coccinea is the only species in general cultivation, it has sheathing sword-like leaves and bears elegant spikes of crimson-red flowers on a 2 ft. (60 cm) stem. The flowers open flat and measure 2 in. (5 cm) across.

VARIETIES

DR BARNARD. An improved *S. coccinea*, the flowers are individually larger, of a deeper colour and are borne on slightly longer stems.

MRS HEGARTY. This is a charming rich pink form of the above, at its best throughout October.

VISCOUNTESS BYNG. A superb plant for cutting. the stems being covered with bloom of a pale apple-blossom pink colour. It commences to flower late in October and in a warm border will continue to bloom until the year ends.

SCILLA

Being inexpensive the scillas may be planted in drifts and there is no better method of planting than by removing a series of turves and planting the bulbs 3 in. (7.5 cm) deep and the same distance apart. Mix plenty of leaf mould or peat into the soil should this not be present.

The bulbs may be planted in short grass which will not be cut until the foliage has died down late in summer. Unlike the crocus and snowdrop which finishes in early May when the lawn may be cut without harming the bulbs, the scillas will only be in full bloom and so are not suitable for lawn planting. Plant on a bank or beneath trees or about the rock garden. There is little need for frequent lifting, though to increase the more expensive and choice varieties, lifting and dividing the bulbs just previous to the decaying of the foliage will quickly increase one's stock.

For home or cold greenhouse, several species are suitable. The best are *Scilla siberica* and *S. verna*, the former being the most appreciated in that it will bloom indoors in February. The bulbs should be planted in October and they are best planted in a deep seed pan or shallow bowl. At all times they must be grown cool otherwise they will become drawn and lanky and make too much leaf. Plant 2 in. (5 cm) deep into a compost made up of loam, peat and sand in equal parts. After two months in the dark the pots should be removed to a cold frame or taken to the window of a cool room early in December.

SPECIES

Scilla bifolia. It produces its bright blue sprays on 4 in. (10 cm) stems between two bronze coloured leaves. Flowering in March it increases rapidly. It should be planted with snowdrops which will enhance the beauty of its bright blue flowers. In sheltered districts of the west, it will come into bloom late in January. The variety 'Rosea' bears pale pink flowers.

S. peruviana. Though having the appearance of a hot-house plant, this scilla is as hardy as the others and should be given the same cultural treatment. It must have a light, well-drained soil. It will bear its blue lily-like flowers in June.

S. pratensis. It produces its spikes of indigo-blue during April and May and seems to do best in the scree garden where it can be left undisturbed for years.

S. siberica. From the northern shores of the Caspian Sea, it is a superb plant for massing under trees. It produces its brilliant blue flowers during March and is of dainty, dwarf habit. The white form, 'Alba', if planted with the blue will create an arresting display. It grows 4 in. (10 cm) tall and is lovely in pots.

S. tubergeniana. From Persia and a beauty for pot culture, bearing spikes of deep turquoise blue. Each bulb will throw an average of four spikes. An ideal plant for the rockery or window-box, it blooms early in March.

SCINDAPSUS

S. aureus, may be grown in a basket but it is perhaps more attractive allowed to grow up a stout cane when it will attain a height of 3–4 ft. (1 m) and send out numerous trailing stems. It makes a plant of thick growth and bears large glossy leaves of dark green variegated with gold, the leaf being heart-shaped and pointed. It is one of the finest of all indoor plants for a shady position and will be happy in a winter room temperature of no more than 45°F. (7°C.) but of all plants it hates a direct draught, so select a position away from a door or window; a corner at the back of the room suits it admirably.

Joseph's Coat, as this lovely plant is called, enjoys a soil made up of 3 parts fibrous loam, 2 parts peat and 1 part sand and, where it can be obtained, mix in a small quantity of decayed manure. Do not overwater, especially during winter, and there should be no need to re-pot for several years. Like all the glossy-leaf plants, it will benefit from the wiping of dust from the leaves at the end of each winter and this should be done with a damp cloth.

Propagation is by means of young shoots which, during summer, will root round the side of a pot in a sunny window.

SCUTELLARIA

At its best in a light soil, *S. carescens*, the Skull Cap, is similar in habit and form to the pentstemons, but is hardier. It grows to a height of 2 ft. (60 cm) and during July and August bears its handsome spikes of sky-blue tubular flowers on upright stems. It is an attractive blue-flowering plant to grow behind the front row plants.

SEA HOLLY, *see* Eryngium

SEA LAVENDER, *see* Statice

SEDUM

For a starved soil and a sun-baked situation, the sedums are without rival in diversity of form and colouring. They take their name from the Latin, 'to sit', for they are at their best when 'sitting' upon a stone wall with their roots in the mortar and baked by the sun and wind. At one time they were widely planted on the walls of cottage and manor house and were known as stonecrops of wall-peppers. The best known is *Sedum acre*, the native stonecrop or wall-pepper, so named from its acrid leaves. It bears sprays of tiny brilliant golden flowers on stems less than 3 in. (7.5 cm) long. Like all the cushion stonecrops, it is a splendid plant to use for a colourful 'lawn', planting it with the creeping thymes when they will quickly cover the soil and present a picture of brilliance throughout summer.

Like the sempervivums, the stone crops are succulents, members of the Crassulaceae family and where baked by the sun, the fleshy foliage will take on the richest shades of bronze and crimson towards the end of a dry summer. They may be planted on a sun-baked bank where little else would grow and are delightful growing between the crevices of paving stone. Many are ideal for growing in troughs and are always happy where allowed to grow around the sides. A gritty, well-drained soil is all they require but they must have sun. Only *S. pulchellum* enjoys a moist soil and partial shade where it will produce its pink flower heads above emerald-green foliage from June until September.

Sedum rhodiola or *S. roseum,* is a plant which has been grown in cottage gardens since earliest times. It was known as Rose-root for the dried roots have the unmistakable smell of a pot-pourri made of rose petals. From the roots, a rose-scented toilet water was made. It frequents mountainous slopes and sea cliffs in north England and Ireland and is a perennial plant with glaucous leaves above which it bears on 9 in. (22 cm) stems, greenish-yellow flowers in terminal cymes. It is in bloom from May until August and like most of the stonecrops may be planted between the cracks of an old wall.

Of those most suitable for a trough garden, outstanding is *S. rubrotinctum* which grows only 3 in. (7.5 cm) tall and has the most unusual salmon-pink foliage which turns brilliant scarlet during a dry, sunny summer. The tiny star-like flowers are yellow. Another beauty is *S. oreganum* which bears heads of golden flowers above orange rosettes. *S. lydium,* also grows less than 3 in. (7.5 cm) tall and forms an erect tuft of narrow bright green leaves which turn red in autumn.

S. spathulifolium, another native of Oregon, in its various forms is a beauty. 'Aureum' has bright yellow foliage whilst 'Purpureum' forms rosettes of crimson-purple covered with grey 'bloom' like a black grape. The two provide a pleasing contrast. Lovely too, is *S. dasyphyllum* 'Album', found by the late Mr Clarence Elliott at St Martin Vesubie. It bears snow-white flowers amidst beautiful grey leaves and carpets the ground. Striking in that its glistening white flowers have black anthers is *S. nevii* which makes a tiny hummock 2 in. (5 cm) high.

Of the more upright sedums, *S. rupestre* is a handsome European species and is to be found growing between limestone rocks. Its pale green leaves take on purple tinting during sunny weather whilst it sends up its yellow flowers on 6 in. (15 cm) stems. Outstanding too, is *S. aizoon* which is said to have reached Britain in 1753. It has handsome grey foliage, the stems reaching a height of 18 in. (45 cm) at the end of which it bears large heads of golden-orange. It is a useful plant to edge a dry shrub border and will die back in winter coming up fresh in spring.

S. tatarinowii is a handsome Chinese species, bearing on 6 in. (15 cm) stems, heads of pinky-white whilst *S. cauticolum* is from Japan and has grey-green glaucous foliage. It bears its heads of rich rose-red on 6 in. (15 cm) stems during autumn when the rock garden will have lost the fully beauty of spring and summer and so is especially valuable.

SELAGINELLA

An evergreen moss-like plant known as the Resurrection Plant for if hung up without planting into soil and watered freely, it will produce a mass of luxuriant fern-like growth.

SEMPERVIVUM

Natives of the alpine regions of central Europe but naturalized in England since earliest times, *S. tectorum,* the Common Houseleek, is the best known, to be found growing from ancient walls and on the roofs of houses, for it was thought to protect the home from witches.

Of the same Crassulaceae family as the sedum and cotyledon, both native plants, it lends charm to its surroundings wherever it grows, forming dense clusters of spiny rosettes of darkest green tinted with crimson and it remains colourful all the year through. It takes its name from two Latin words *semper* (always) and *vivo* (I live), for it is almost indestructible. It is usually seen on walls but is charming where used for summer bedding, planting 6 in. (15 cm) apart, with blue lobelia between. Other species may also be used and never fail to create interest. After the lobelia has finished flowering it may be replaced with dwarf early flowering tulips in red or yellow and for which the sempervivums provide a pleasing ground cover. The plants like a soil containing lime rubble and one which is well-drained in winter. Or they may be planted in the cracks of old walls and on tiles becoming anchored with the minimum of soil about their roots. To increase them, offsets may be removed without disturbing the main plant. They are charming where growing in the company of *Sedum acre* which bears tiny sprays of golden-yellow flowers. The leaves of sempervivum quickly stay bleeding and are also effective in the treatment of skin diseases.

SPECIES AND VARIETIES

Sempervivum arachnoideum 'Stansfieldii'. This distinct species has wedge-shaped leaves formed in dense rosettes of crimson-bronze and veiled with white cobweb-like hairs. In June it bears bright red flowers on 4 in. (10 cm) stems.

S. atlanticum. Native of the Atlas Mountains, it has been growing in English gardens for at least a century. It forms pale green rosettes of 2 in. (5 cm) diameter, the leaves being tipped with bronze and fringed. In June it bears pinky-red flowers on 12 in. (30 cm) stems.

S. fimbriatum. An old favourite of the European Alps forming a neat dark green rosette, tipped with purple and fringed with hairs. In July it bears brilliant red flowers on hairy stems 9 in. (22 cm) long.

S. glaucum. Native of the Alpine regions of central Europe, it forms a large rosette of glaucous mauve tipped with brown, above which it bears in July, bright red flowers on 6 in. (15 cm) stems.

S. schlehanii 'Rubrifolium'. One of the most striking of all, forming huge rosettes of plum-red and bearing crimson flowers on 12 in. (30 cm) stems.

S. tectorum. The Common Houseleek, forming pale green rosettes of 3 in. (7.5 cm) diameter, the leaves being tipped with reddish-brown. In June it bears pale red flowers on hairy stems 12 in. (30 cm) tall, the flowers being enhanced by their bright purple filaments.

SENECIO

This is one of the most valuable plants for a sea-coast hedge, for its silver leaves provide a marked contrast to the various green and ornamental shrubs. Extremely tolerant of salt winds and spray and enjoying a light, sandy soil, *Senecio greyi* will make a hedge 5 ft. (1.5 m) tall, whilst *S. eleagnifolius* is rather taller-growing. Both bear golden-yellow flowers during the latter part of summer. The plants require a sunny position, and they should be trimmed in April, clipping more closely every fourth year to prevent the formation of too much thick wood. Plant 3 ft. (1 m) apart.

SENECIO ELEGANS, *see* Jacobaea

SENSITIVE PLANT, *see* Mimosa

SEQUOIA

The Californian Redwood, a genus of a single species, it is the largest of all trees, attaining a height of nearly 400 ft. (120 m) in its native land. *S. sempervirens* requires a deep moist loam, being intolerant of dry places. If cut down, it will send up new shoots from the base. It has shaggy reddish bark and horizontal branches with drooping branchlets, covered in flat shining leaves arranged in two rows and which are bronze tinted in winter. In the form 'Adpressa', the tips of the shoots are cream coloured.

Plant in May. Propagation is by seed sown in outdoor beds; or by cuttings of young shoots removed with a 'heel' and inserted in a sandy compost under glass.

SEQUOIADENDRON

The Wellingtonia which, like the Redwood, will attain mammoth proportions. The horizontal branches have drooping branchlets covered in needle-like leaves, spirally arranged. The cones are oval and are borne solitary at the ends of the branchlets. *S. giganteum* will not flourish in a calcareous soil as will the Redwood whilst it is less hardy. Plant in May. Propagation is by seed sown in outdoor beds in spring.

SHELL FLOWER, *see* Molucella

SHOOTING STAR, *see* Dodecatheon

SHORTIA

Delightful little evergreen shrubs enjoying conditions of almost complete shade and flowering late in spring. *S. grandiflora* is perhaps the best, with foliage which turns crimson at the edges and waxy deep pink flowers. *S. galacifolia* has glossy green foliage and bears waxy white bells in profusion. Both are readily increased from cuttings.

SHRIMP PLANT, *see* Beloperone

SHRUBS

FLOWERING TIMES

Botanical name	*Popular name*	*Height*	*Colour*	*Month*
WINTER				
Azara microphylla	Azara	8 ft. (2.5 m)	pale yellow	Mar-May
Chimonanthus fragrans	Wintersweet	6 ft. (2 m)	purple/yellow	Dec-Feb
Corylopsis pauciflora	Corylopsis	4 ft. (1.2 m)	yellow	Mar-Apr
Corylopsis spicata	Corylopsis	4 ft. (1.2 m)	sulphur	Feb
Daphne mezereum	Daphne	3 ft. (90 cm)	purple/pink	Feb-Mar
Daphne 'Somerset'	Daphne	3 ft. (90 cm)	rose/pink	Apr-May
Erica carnea	Winter Heath	1 ft. (30 cm)	red/pink	Nov-Apr
Garrya elliptica	Californian Garrya	8 ft. (2.5 m)	yellow catkins	Nov-Mar
Hamamelis brevipetala	Witch Hazel	5 ft. (1.5 m)	yellow	Jan-Feb
Hamamelis mollis	Witch Hazel	5 ft. (1.5 m)	yellow	Dec-Mar
Hamamelis vernalis	Witch Hazel	5 ft. (1.5 m)	pale yellow	Dec-Mar
Mahonia aquifolium	Berberis (Oregon Grape)	2½ ft. (75 cm)	yellow	Feb-Mar
Rhododendron praecox	Rhododendron	3 ft. (90 cm)	pink	Feb-Mar
Viburnum bodnantense	Viburnum	5 ft. (1.5 m)	pink/white	Dec-Mar
Viburnum fragrans	Viburnum	5 ft. (1.5 m)	pink/white	Jan-Mar
Viburnum tinus	Viburnum	4 ft. (1.2 m)	white	Oct-Mar
SPRING				
Azalea 'Kurume'	Azalea	1½ ft. (45 cm)	orange pink	Apr-May
Azalea malvatica	Azalea	2 ft. (60 m)	various	Apr-May
Berberis darwinii	Barberry	2 ft. (60 cm)	orange	Mar-May
Cydonia japonica	Flowering Quince	3 ft. (90 cm)	crimson/pink	Mar-Apr
Cytisus albus	Portugal Broom	5 ft. (1.5 m)	white	Apr-May
Cytisus praecox	Broom	4 ft. (1.2 m)	yellow	Apr-May
Exochorda racemosa	Pearl Bush	6 ft. (2 m)	white	Apr-May
Forsythia giraldi	Golden Bell Bush	4 ft. (1.2 m)	yellow	Apr-May
Forsythia spectabilis	Golden Bell Bush	5 ft. (1.5 m)	yellow	Apr-May
Forsythia suspensa	Golden Bell Bush	5 ft. (1.5 m)	primrose	Apr-May
Fothergilla gardenii	Fothergilla	3 ft. (90 cm)	white	Apr-May
Osmanthus delavayi	Osmanthus	3 ft. (90 cm)	white	Apr-May
Pieris forrestii	Pieris	4 ft. (1.2 m)	white	May
Pieris taiwanensis	Pieris	4 ft. (1.2 m)	white	May
Ribes aureum	Buffalo Currant	4 ft. (1.2 m)	buff	Apr-May
Ribes sanguineum	Flowering Currant	4 ft. (1.2 m)	crimson	Apr-May
Viburnum burkwoodii	Viburnum	5 ft. (1.5 m)	white	Apr-May
SUMMER				
Buddleia globosa	Orange Ball Bush	4 ft. (1.2 m)	orange	May-June
Cistus 'Sunset'	Rock Rose	2 ft. (60 cm)	rose-pink	June-Sept
Cytisus hybrids	Broom	5 ft. (1.5 m)	various	June
Deutzia campanulata	Deutzia	3 ft. (90 cm)	white	June-July
Deutzia macrothyrsa	Deutzia	3 ft. (90 cm)	white	June-July
Erica vagans	Cornish Heath	3 ft. (90 cm)	pink/white	Aug
Escallonia hybrids	Chilean Gum	3–5 ft. (90 cm–1.5 m)	rose/red	June-Aug

Botanical name	Popular name	Height	Colour	Month
Hedysarum multijugum	French Honeysuckle	4 ft. (1.2 m)	purple	June-Sept
Hydrangea macrophylla	Hydrangea	3 ft. (90 cm)	blue-red	Aug-Nov
Kerria japonica	Jew's Mallow	4 ft. (1.2 m)	orange	May-June
Kolkwitzia amabilis	Beauty Bush	4 ft. (1.2 m)	pink	June
Paeonia arborea	Tree Paeony	5 ft. (1.5 m)	pink/crimson	May
Philadelphus hybrids	Mock Orange	8 ft. (2.5 m)	white	August
Rhododendron hybrids	Rhododendron	3 ft. (90 cm)	various	May-June
Syringa hybrids	Lilac	8 ft. (2.5 m)	various	May-June
Veronica hybrids	Shrubby Veronica	2–3 ft. (about 75 cm)	purple-pink	July-Nov
Weigela hybrids	Bush Honeysuckle	4 ft. (1.2 m)	pink/crimson	May-July
AUTUMN				
Buddleia alternifolia	Butterfly Bush	8 ft (2.5 m)	purple/pink	Aug-Sept
Clematis Cote d'Azur'	Clematis	3 ft. (90 cm)	blue	Aug-Oct
Colletia armata	Colletia	4 ft. (1.2 m)	white	Sept-Oct
Desfontainea spinosa	Desfontainea	6 ft. (2 m)	red/yellow	Aug-Sept
Erica vulgaris	Ling	2 ft. (60 cm)	crimson/pink	Aug-Oct
Fuchsia hybrids	Fuchsia	2–4 ft. (about 1 m)	red/purple	July-Dec
Potentilla hybrids	Potentilla	2–4 ft. (about 1 m)	yellow	July-Oct

SIDALCEA

Natives of California and with their marshmallow-like flowers in various shades of pink, *S. malvaeflora* and its varieties are delightful plants for using towards the back of a border. They will flourish in ordinary soil and may be left undisturbed for four years or more, being propagated by root division. Plant in November and provide an open, sunny situation.

VARIETIES

LOVELINESS. Very compact, it grows to a height of only 2 ft. (60 cm), and bears its attractive shell-pink flowers when the others are coming to an end.

MRS GALLOWAY. A new variety, in bloom during July and August and bearing masses of bloom of a clear shell-pink colour.

THE DUCHESS. A fine variety, bearing flowers of rich carmine-rose on 3–4 ft. (1 m) stems. It makes a bushy plant and comes early into bloom, before the end of June.

THE PRINCE. Possibly the best for a small border for it is of compact habit, with the deep rose-red flower spikes set close together.

WENSLEYDALE. Growing to a height of 3–4 ft. (about 1 m) this should be confined to the back of the border where it will bear its rosy-red flowers in profusion.

WILLIAM SMITH. A new variety and the first to bear flowers which are of a warm shade of salmon-pink.

SILENE

Known as the Catchfly owing to the sticky substance on the leaves of several varieties which gives the plant its ability to catch flies. A well-drained soil in full sun suits it best. The plants of all species are very compact, making most of them suitable to plant between paving stones. They flower throughout early summer, but one species, *S. keiskii*, which should be given cloche protection in winter, will bloom from July until October. Producing tiny cushions and wee star-like white flowers is *S. acaulis* 'Alba', whilst *S. saxatilis* covers itself with masses of pink flowers. Different in its requirements in that it prefers a shaded rockery is *S. alpestris* which bears taller stems covered with starry white blooms. Suitable for a rockery made by the seashore, where salt spray and a sandy soil suit it admirably, is *S. maritima*, also white flowered, whilst there is also a double form, very beautiful. One of the few of the silenes to bear a coloured flower is *S. schaftae*, a native of the Caucasus which bears rosy-lavender flowers, often into late autumn.

SILVER FOLIAGE

Colour of foliage is as important in the border as is colour of bloom. In the same way as it is possible to obtain the best use of foliage by planting near each other, those plants which have broad and those which have long and narrow foliage, so those plants should be introduced to the border which may possess greater beauty of foliage than of bloom. A number of the plants with silvery foliage possess an interest all their own and act as a pleasing contrast to the dark foliage plants and to those bearing brilliantly coloured flowers. There are plants suitable for all parts of the border and none should be forgotten when making out the plan. Include as many as the size of the border permits. Those of dwarf habit may also be used with begonias and geraniums for summer bedding. The following plants have silver foliage:

Achillea argenta	*Artemisia schmidtii* 'Nana'
Achillea 'Moonshine'	*Artemisia splendens*
Achillea 'Weston'	*Centaurea gymnocarpa*
Anaphalis nubigena	*Centaurea maritima*
Anthemis cupaniana	*Chrysanthemum poterifolium*
Anthemis rudolphiana	*Cineraria maritima*
Artemisia absinthium	*Helichrysum alveolatum*
Artemisia arborescens	*Helichrysum fontanesii*
Artemisia glacialis	*Helichrysum italicum*
Artemisia ludoviciana	*Helichrysum plicatum*

Salvia argentea
Santolina incana
Santolina neapolitana
Senecio greyi
Senecio leucostachys
Stachys lanata
Teucrium fruticans
Teucrium polium

SKIMMIA

Slow-growing hardy evergreen shrubs which make low-spreading plants 3–4 ft. (1 m) tall. They will flourish in sunlight or shade, in a lime-laden or acid soil. *S. japonica* has oval pale green leaves and bears male and female flowers on separate plants so to have berries, several should be planted together. In *S. japonica* 'Fragrans', the flowers are scented like lily-of-the-valley and are followed by crimson berries in autumn.

Plant in spring, the time of year when any dead wood should be removed. Propagate from seed or by layering in autumn; or take cuttings in July of the new season's wood and insert in sandy compost under glass.

SOIL

ACID

Such a soil will be of high acid reaction. It will be well-drained in winter and be retentive of moisture in summer but unless some preparation is done to counteract the acidity, it will grow only a limited range of plants. A load of heavy loam, sometimes to be obtained from a building site will do much to correct the acidity of the black peat soil. Slow-acting fertilisers such as shoddy, fish waste, seaweed and farmyard manure should be incorporated but all fertilisers of an acid nature should be omitted. A heavy dressing of lime each year will enable a wider selection of plants to be grown but this must be omitted where growing only those plants requiring acid conditions such as the following:

Calluna vulgaris
Camellia japonica
Clethra acuminata
Cyrilla racemiflora
Daphne mezereum
Desfontainea spinosa
Embothrium coccineum
Enkianthus perulatus
Erica
Fothergilla gardenii
Gaultheria fragrantissima
Hamamelis mollis
Kalmia latifolia
Lindera benzoin
Liquidambar styraciflua
Magnolia macrophylla
Pernettya mucronata
Pieris floribunda
Rhododendron (including *Azalea*)
Vaccinium corymbosum
Zenobia pulverulenta

CALCAREOUS

Calcareous soils are found south of the Thames, especially in the regions of the Chiltern Hills and South Downs. Similar soils are found in the Cotswolds whilst limestone formations cover the north of England from a line drawn from Barrow-in-Furness to Flamborough Head. These are the most difficult areas in which to garden for they are exposed to strong winds whilst the soil will usually lack depth. Also, the soil will contain seventy per cent silica and will usually be dry and hungry though on the credit side, there will be no fear of losing plants through excess moisture during winter. Green manuring will increase the depth and add to the humus content whilst decayed manure, fish or bone meal (at the rate of 4 oz. per sq. yd.) will provide nitrogen which a limestone soil will almost always lack. In addition, any humus forming materials such as peat, used hops or poplar bark fibre should be dug in when the ground is prepared. Indeed, any materials should be added which will increase the depth of soil and to ensure that it will be cool in summer.

Trees and shrubs

Acer palmatum
Aesculus hippocastanum
Aucuba japonica
Berberis darwinii
Betula pendula
Buxus sempervirens
Cedrus atlantica
Cornus mas
Cotoneaster
Crataegus oxyacantha
Cydonia japonica
Fagus sylvatica
Forsythia x intermedia
Ilex aquifolium
Juhlans regia
Juniperus communis
Ligustrum vulgare
Philadelphus coronarius
Pinus laricio
Prunus (Almond, plum, etc.)
Pyracantha coccinea
Pyrus nivalis
Sorbus aucuparia
Syringa
Taxus baccata

Perennials

Anchusa italica
Aster amellus
Campanula lactiflora
Dianthus (carnations, pinks)
Dictamnus fraxinella
Galega officinalis
Gypsophila paniculata
Lavatera olbia
Lilium candidum
Lilium chalcedonicum
Linum perenne
Lychnis coronaria
Nepata mussinii
Ostrowskia magnifica
Pentstemon heterophyllus
Salvia superba
Scabiosa caucasica

SMOKE TREE, *see* Cotinus

SNAKE PLANT, *see* Sansevieria

SNAKEROOT, *see* Liatris

SNAKE-ROOT, *see* Cimicifuga

SNAKE'S-HEAD LILY, *see* Fritillaria

SNAP-DRAGON, *see* Antirrhinum

SNEEZEWEED, *see* Helenium

SNOWBALL TREE, *see* Viburnum

SNOWDROP, *see* Galanthus

SNOWFLAKE, *see* Leucojum

SNOWY MESPILUS, *see* Amelanchier

SOAPWORT, *see* Saponaria

SOLANUM

CLIMBING

S. jasminoides is so named because its flowers resemble those of the jasmine, though they are of a pleasing shade of blue-grey. There is also a white form, the blooms having a yellow centre. On a sunny wall in a sheltered district the plant will be able to withstand an average winter and will quickly reach a height of 15 ft. (4.5 m). It is semi-evergreen and has a very long flowering season, from mid-June until winter.

More vigorous is *S. crispum*, also semi-evergreen, which bears navy-blue flowers with yellow centres from June until late in August. Both should be planted in April into a sandy loam enriched with some humus as the plants must not lack moisture during summer.

Propagate from cuttings inserted under a bell-cloche in July, and prune by cutting back unduly long or dead shoots in March. If dry conditions continue through summer, the plants will benefit from an occasional mulch.

UNDER GLASS

Solanum Capsicastrum or the Winter Cherry is one of the most satisfactory of greenhouse plants to be raised from seed. It makes a shrub-like plant 18 in. (45 cm) tall and grows to almost the same in width. During winter it is covered with scarlet berries of the size of a small cherry. As the berries will generally have ripened for Christmas, the plant is also known as the Christmas Cherry. It is a plant with few vices, but if it is to be grown well, attention to detail should be given from the time the seed is sown. The solanum is a member of the potato family, having the same familiar small white star-like flowers whilst its glossy oval leaves possess a smell similar to those of the potato. The bright green leaves are an added attraction to the beauty of the plant in winter, but if the plant does not receive the treatment it requires, it may drop its leaves as well as its fruit just when reaching its best.

It is most important to sow a reliable strain, for poor quality seed will produce plants of poor quality, with small narrow leaves and bearing small dull coloured berries. Weatherill's Hybrid strain is outstanding, the plants having strong well coloured foliage and bearing large richly coloured berries in all shades of crimson, scarlet and orange.

SOWING THE SEED UNDER GLASS

It is most important to allow the plant a long season to develop and this means sowing the seed almost twelve months before the plants are required to ripen their berries. Early February is the correct time to sow and as the seed will require a temperature of around 62°F. (17°C.) for germination, a propagator should be used and this may be dispensed with once complete germination has taken place.

The John Innes Sowing Compost is suitable, but as the solanum enjoys a gritty compost throughout its life, the sand should be of a more than usual gritty nature if the loam is on the heavy side, as it should be for solanums.

As it is important to keep the plant bushy from the beginning, so the seed should be sown as thinly as possible and as the seed is quite large in comparison with that of most greenhouse plants, it may quite easily be spaced out if the surface of the compost is made firm and level. The seed should be lightly covered with dry compost and watered in. Never at any time must the seed be allowed to lack moisture or germination will be delayed.

The seed will have germinated in about a month and by mid-March the seedlings will be ready to transplant, before they become overcrowded. They should be moved as soon as large enough to handle, transferring them into boxes containing the John Innes Potting Compost No. 1, for solanums are gross feeders from the beginning. Plant the seedlings two inches apart and make quite firm, watering in and growing on in a temperature of around 55°F. (13°C.) until the plants have become established.

From early April, artificial heat will not be necessary if the inside of the greenhouse is lined with polythene and the ventilators are closed during late afternoon to conserve warmth for the night. A moist atmosphere will encourage the plants to make headway and during the daytime, whenever the sun is shining, ample ventilation must be given so that the plants may develop sturdily. The plants should be shaded from strong sunlight by whitening the glass.

GROWING ON THE PLANTS UNDER GLASS

When the seedlings have grown about 3 in. (7.5 cm) high, the growing point should be pinched out to encourage the plant to make more bushy growth. Then as soon as new growth begins to form, the plants should be moved to small size 60 pots containing the John Innes No. 2 Compost, substituting grit for the sand. This compost will ensure that the plants receive ample food requirements for they will quickly exhaust that provided by the previous compost. At this stage, in particular, the loam should be on the heavy side to prevent the plants 'running away' during summer instead of forming stocky growth. Tall plants with long-jointed shoots are not required, yet if growth is not encouraged by constantly spraying the plants and maintaining them in a comfortably moist condition, they will not make sufficient growth to form a bushy plant of attractive appearance during the growing season.

By the end of May, the plants will be ready to move to the size 48 pot in which they will fruit. Here, the John Innes No. 3 Compost should be used to which should be added a small quantity of well-decayed cow manure if readily obtainable. The plants are given frequent waterings and the greenhouse damped down daily, whilst as much fresh air as possible should be admitted until the beginning of October.

When established in their final pots, those shoots which have made most headway should again be pinched back in order to build up a bushy plant. 'Stopping', however, should be completed by the end of June, for early in July the plants will come into bloom. This is a most exacting time for the plants for it is necessary for the blooms to set their fruit well for them to be a success. Solanums which do not berry well are of little use. The setting of the fruit should be encouraged by hand fertilisation of the blooms with a camel hair brush, this being done about midday when the blooms are quite dry and as soon as they have fully opened. If left too late, hand fertilisation will prove unsuccessful.

SETTING AND RIPENING THE FRUIT UNDER GLASS

Regular syringing of the plants after pollination will also greatly help the blooms to set fruit, whilst the plants will benefit from a weekly watering with dilute soot water and liquid manure given alternately. At this time, both the compost and the plants should be kept in a comfortably moist condition for any dryness at the roots will cause the flowers to fall before they have set fruit. The maintenance of a humid atmosphere will also prevent an outbreak of red spider which is the solanum's worst enemy. An attack may cause considerable damage where dry conditions are experienced.

It is now required that the flowers set as many berries as possible and that once set, the fruits will continue to swell. The plants must be given ample supplies of moisture to make this possible, for if the plants ever become dry at the roots the berries will shrivel and fall off. To assist with the swelling, any shoots which continue to grow should not be pinched back so that the plant can devote all its energies into the formation of the fruit.

By mid-October the plants should be given some warmth, a temperature of 50°F. (10°C.) by day, falling to around 42°F. (5°C.) by night, which will be sufficient to maintain a buoyant atmosphere. From then onwards, overhead watering should be discontinued, though the plants must never be allowed to lack moisture at the roots. During sunny days, the ventilators should be opened so that the atmosphere will not become too warm and stuffy. Excessive humidity at this time of the year should be avoided.

By the end of October, the berries will have swelled to about 1 in. (2.5 cm) diameter and will grow no larger. Watering is then reduced to a minimum, but to allow the roots to become dry will cause the plant to drop both its leaves and its fruit. The greenhouse should be kept warm to encourage the fruit to colour, for at this time it will be deep green. Gradually, however, it will turn yellow then pale pink, until towards Christmas the berries will have become a glowing orange-scarlet which colour will be enhanced by the bright green foliage.

As soon as the berries have ripened, the plants may be taken indoors where they should be given a room temperature of between 50°–55°F.

(10°–13°C.) and a position where the plants will receive plenty of light. Draughts must be guarded against or the plants may drop their leaves and they must neither be over-watered nor given too little moisture as this will have the same effect.

The plants will retain their berries for about two months, but after they have shrivelled they may be removed and the plants retained until early May for their foliage alone. They should then be cut back and the plants stood outside in their pots which, to prevent excessive attention as to their watering, will best be inserted in soil. There, in a partially shaded position, the plants will grow on again through summer to be taken into the greenhouse in early autumn for the berries to swell and ripen again.

VARIETIES GROWN UNDER GLASS

BIG BOY. It makes a compact bushy plant growing less than 12 in. (30 cm) tall whilst the scarlet fruits are cherry size, measuring 1 in. (2.5 cm) across. They contrast well with the dark green foliage.

JUBILEE. It is entirely different in habit and foliage, the leaves being narrow and pointed and grey-green, with drooping branches. The fruits are first white, then yellow before turning to brilliant orange.

SOLDANELLA

These dainty little plants should be given scree conditions and a soil containing plenty of lime. If planted in a dry, sunny area, such as in East Anglia, they should be given partial shade. In the north they will be happy anywhere, though should be covered with a cloche or sheet of glass during winter, not to protect them against the cold, but to give shelter from excessive moisture. The plants, which do not exceed 3 in. (7.5 cm) in height, come into bloom early in spring. One of the loveliest is *S. alpina,* with its tufts of round green leaves and nodding violet bells. Equally lovely is *S. pindicola* which bears bells of rich lavender, whilst *S. minima* appears as a carpet of pale lilac tubes which are veined with purple inside. *S. montana* has a slightly taller habit and bears many rich lavender-mauve bells which have attractive frilled petals.

SOLIDAGO

S. vigaurea, the golden rod native to our islands was widely used in ancient times for its healing properties. It is, however, *S. canadensis,* introduced from North America by John Tradescant three centuries ago, which is that widely planted in the border. It is valuable in that it may be planted for succession, to bloom from mid-July until the end of October, whilst the plants, so easily increased by root division, will flourish in almost any soil and in partial shade. The new dwarf flowering varieties will bring new popularity to this cut flower plant. Plant in November and allow 2 ft. (60 cm) between the plants.

SPECIES AND VARIETIES

GOLDEN FALLS. The deep golden sprays, at their best during August, are borne on 20 in. (50 cm) stems.

GOLDENMOSA. The first to bloom late in July, its deep golden sprays being borne on 3 ft. (90 cm) stems.

GOLDEN WINGS. The value of this golden-rod is in its lateness in blooming, acting as a foil, with its deep yellow sprays, to the tall michaelmas daisies during September and October. 5 ft. (1.5 m).

LEMORE. Its wide branching sprays of soft primrose, tinged with green, are borne of 2 ft. (60 cm) stems during July and August.

LENA. It bears its yellow sprays, tinged with green over a long period and well into autumn. 3 ft. (90 cm).

LINERALIS. The small, dainty spikes of pure golden-yellow are borne on 20 in. (50 cm) stems during July and August.

WENDY. Its delicate yellow sprays are borne on stems only 18 in. (45 cm) tall and remain in bloom during August and September.

S. ballardi. In a moist soil, this golden-rod will attain a height of 6 ft. (2 m) and will bear its flower heads into November. The blooms are produced from laterals all the way up the 2 ft. (60 cm) stem.

SOLLYA

The form *S. parviflora* is known as the Australian Bluebell Creeper, and through late spring and summer, bears masses of deep blue flowers. It is of similar habit to the maurandia and manettia, being an evergreen of slender growth and a most delightful plant for twining up a trellis by the side of a window where it may receive the necessary light. It requires a compost similar to that used for the manettia, and whilst it must be given ample supplies of moisture during summer, it requires little in winter. Though a native of Australia, the plant will be quite happy given a winter temperature which does not fall below 42°F. (5°C.).

SOLOMON'S SEAL, *see* Polygonatum

SOPHORA

The Pagoda tree of China and Japan, *S. japonica,* is a handsome deciduous tree with large pinnate leaves and creamy white pea-like flowers. The form 'Pendula' is of weeping habit and makes a striking tree for a lawn or other focal point. It requires an open sunny situation and a well-drained loam. No pruning is necessary other than to maintain the shape. Propagate by seed sown in a frame in spring or by shoots removed with a heel and rooted under glass.

SORBARIA

Under this name are grouped those spiraeas with pinnate leaves and which are tall growing. They require a deep, moist loam containing some peat and decayed manure. *S. arborea* grows up to 15 ft. (4.5 m) tall and during August and September bears large plumes of creamy-white flowers whilst *S. sorbifolia* 'Stellipila' has large pinnate leaves, downy on the underside and grows 5 ft. (1.5 m) tall. It bears creamy-white panicles during June and July.

Plant in November and prune hard back early March. Propagate by division or by suckers.

SORBUS

The name given to the Mountain Ash and Whitebeam, native deciduous trees and valuable for a calcareous soil and for small gardens. They are also suitable for planting in wind-swept gardens. They have attractive grey foliage and bear their flowers in corymbose cymes followed by scarlet or yellow berries in autumn. If planting in lime deficient soil, add lime rubble (mortar) or hydrated lime and some humus. Little pruning will be necessary apart from the removal of dead wood. Propagate by seed sown when ripe or by budding in July or grafting.

S. alnifolia, native of Japan makes a neat round-headed tree with foliage which turns crimson in autumn and at the same time the tree is covered in trusses of bright red berries. *S. aucuparia* is the Mountain Ash or Rowan, the best form being 'Asplenifolia' with its deeply cut fern-like leaves and crimson berries. Plant near it, the form 'Xanthocarpa' which bears contrasting yellow berries.

S. aria is the Whitebeam, its leaves being glossy above and like white velvet beneath but in the form 'Lutescens', the upper surface is also covered in tomentum. The leaves are similar in the Service Tree, giving them a soft sage-green appearance whilst the crimson-brown fruits resemble small pears.

SOUTHERN BEECH, *see* Nothofagus

SOUTHERN BUGLE LILY, *see* Watsonia

SOUTHERNWOOD, *see* Artemesia

SPANISH BROOM, *see* Spartium

SPARMANNIA

It is an excellent house plant, but though it will be happy in partial shade, it does not like too dry conditions. It must be kept away from draughts, and it must never be allowed to become dry at the roots. For this reason it prefers a compost containing 2 parts peat or leaf mould, 2 parts fibrous loam and 1 part coarse sand. Whilst the plants will require copious amounts of moisture during summer, the compost must also be kept moist in winter, when a minimum temperature of 42°F. (5°C.) will be sufficient to keep it healthy.

The plants, which grow up several feet on a single stem, have pale green leaves which are large and shaped like those of the maple and are covered in small hairs. Where conditions suit them the plants will bear small white flowers in summer.

Propagation is by cuttings, which will root easily around the side of a pot in a sunny window.

SPARTIUM

The yellow Spanish Broom, *S. junceum* will grow well anywhere, in a chalk-laden soil and in a coastal garden. It will also flourish in a sandy soil. It grows 8 ft. (2.5 m) tall, the slender green shoots being almost leafless, whilst the bright yellow flowers are borne along the whole length of the stems in July and August. They are sweetly scented.

Plant in spring, preferably from pots and do no pruning. Propagate from seed sown in small pots from which the plants are set out in their flowering quarters.

SPATHIPHYLLUM

Like the dieffenbachia, this is a member of the large aroid group of plants and may be likened to the native Arum lily, only it is evergreen.

It may be said to be one of the few house plants of an evergreen nature to bear flowers, which its does in summer. The blooms are pure white, like those of the Arum lily, and last for several weeks, whilst the foliage of the plant is dark green, ribbed and glossy. It will grow well in shade as will all the aroids, and is quite happy in a winter temperature of 50°F. (10°C.). It likes a soil containing peat or leaf mould and fibrous loam in equal parts and to which is added a small amount of silver sand. During summer give plenty of water, but keep the compost only just moist during winter.

SPECULARIA

An annual, Venus's Looking Glass, as *S. speculum* is called, is a delightful edging plant, from an early April sowing flowering from mid-July until September on 8 in. (20 cm) stems. The deep violet bell-shaped flowers sow themselves as readily as forget-me-nots and quickly come into bloom.

SPEEDWELL, *see* Veronica

SPIDER PLANT, *see* Chlorophytum

SPIDER PLANT, *see* Cleome

SPIDERWORT, *see* Tradescantia

SPINDLE TREE, *see* Euonymous

SPIRAEA

HERBACEOUS

Closely related to the astilbe, the herbaceous spiraeas are as valuable at the back of the border during midsummer months as is the Pampus Grass in autumn. Bearing their graceful plumes from mid-June until mid-August, the plants must be given a cool, moist soil. They cannot be given too much moisture during their flowering period, but during winter stagnant water should not remain about their roots. A well-drained loam in which is incorporated large quantities of humus suits them best. Plant 3–4 ft. (about 1 m) apart in November and propagate by root division.

SPECIES OF HERBACEOUS SPIRAEA

S. aruncus. The Goat's Beard, it has exquisitely toothed foliage and bears its long white plumes on 4–5 ft. (1.5 m) stems during June and July.

S. palmata. In a moist loam it reaches a height of 3–4 ft. (about 1 m) and is one of the most handsome plants of the border with large palm-shaped leaves and bearing branched plumes of carmine-red during July and August.

S. venustum 'Magnificum'. It has attractively lobed leaves and during July and August bears its feathery plumes of deep peach-pink on 4–5 ft. (1.5 m) stems.

SHRUBBERY

Possessing extreme hardiness, the dwarf, twiggy spiraeas require a moist soil and will grow well alongside water so long as their roots are not submerged. By planting various species, they may be obtained in bloom from mid-March until October, bearing their flowers in billowing panicles whilst the foliage is equally handsome. They may be divided into two groups:

Firstly, those which bloom in spring and early summer and which should have their flowering stems cut back after blooming. Included here is *S. henryi* which grows 6 ft. (2 m) tall and in May and June bears masses of foam-like creamy white flowers; and *S. thunbergii*, the first to bloom in March and which grows 3–4 ft. (about 1 m) tall, its slender stems being clothed with dainty pale green foliage whilst it bears plumes of purest white. Its hybrid, *S.* × *arguta* covers itself in a silvery foam during April and May.

Secondly those flowering in late summer and autumn and which should be pruned hard in early March. Here is included *S. japonica* which grows 3–4 ft. (1 m) tall and has narrow dark green leaves. In July and August it bears large flat heads of rose-pink whilst 'Atrosanguinea' bears flowers of crimson-red. 'Anthony Waterer' grows only 2 ft. (60 cm) tall and bears crimson flowers.

Plant in November and propagate by division or by offsets. Cuttings may also be taken of the half-ripened wood and inserted in sandy compost for rooting under glass.

SPRUCE, *see* Picea

SPURGE, *see* Euphorbia

SQUILL, *see* Scilla

STAR OF BETHLEHEM, *see* Ornithogalum

STAR OF THE VELDT, *see* Dimorphotheca

STATICE

ANNUAL

Though biennial, the well-known *S. sinuata,* so popular for drying for winter decoration, should, like the helichrysum, be treated as a half-hardy annual. The seed is sown in gentle heat in early March, the young plants being moved to the open ground mid-May. Like all everlasting flowers they prefer a light, sandy soil. The plants grow to a height of 18 in. (45 cm) and should be planted the same distance apart.

The bloom is cut at ground level as soon as the bloom is nicely showing colour. The best varieties are 'Market Rose', 'Market Grower's Blue' and 'Lavender Queen'.

PERENNIAL

Admirable plants to provide cut bloom for drying and for use during winter, the two perennial forms require different soil conditions though both prefer spring planting. The bloom should be cut during early autumn, before it is past its best when it is hung up in an airy shed to dry. Propagation is by root cuttings taken in March.

SPECIES AND VARIETIES OF PERENNIAL STATICE

S. incana 'Dumosa'. Its deep pink flowers are produced on 12 in. (30 cm) stems during the last weeks of summer. It should be given a light, sandy soil and is perhaps best planted in beds to itself.

S. lanata. Its more popular name is Donkey's Ears on account of the shape of its woolly silver leaves. Colourful from early June until early September when it bears its bloom of deep pink. It makes a plant of compact habit and with its attractive foliage could be used more often for the front of the border. For which purpose plant in November 12 in. (30 cm) apart and propagate by root division. Ordinary soil is suitable.

S. latifolia. Growing to a height of 2 ft. (60 cm) and proving most vigorous in a heavy soil retentive of moisture, it forms clouds of tiny purple flowers, most attractive in the border or when cut and dried. It will remain colourful from July until late in September. Plant 2 ft. (60 cm) apart; a new variety, 'Collyer's Pink', bears a mass of deep pink flowers.

STERNBERGIA

The Lily of the Field of Biblical times, *S. lutea* is a crocus-like plant with shining lemon yellow flowers and it loves a dry, sunny position. The drier and the hotter it is, the more lavishly will it produce its shining yellow blooms. They take a little time to become established and should rarely be lifted where growing in a dry well-drained soil for they will increase only slowly. They are almost hardy but in an exposed garden, the bulbs should be planted 4 in. (10 cm) deep and given a light peat mulch during November. As it is necessary for the continued vigour of the bulb that they become thoroughly ripened before growth commences again the following year, it will be necessary to lift the bulbs in November where they are growing in a cold, heavy soil. They should be carefully cleaned and dried in a warm room and replanted the following spring to bloom in autumn.

STOCK, *see* Matthiola

STOKES' ASTER, *see* Stokesia

STOKESIA

Bearing flowers like those of the cornflower, with a double row of petals, *S. cyanea* is a front row plant for September and October flowering. If given a light, sandy soil and a sunny position it will bear a profusion of lavender-blue flowers which appear on sturdy stems above the grey basal foliage. There is a rare white form, 'Album', which is equally lovely. Plant in March, 18 in. (45 cm) apart.

STONE CRESS, *see* Aethionema

STONECROP, *see* Sedum

STRAWBERRY TREE, *see* Arbutus

STREPTOCARPUS

A beautiful plant of a lovely family which blooms over a long period. It is rarely to be found in florists' shops for it does not travel well, the foliage being brittle whilst the tubular blooms bruise easily. The Wiesmoor hybrids, obtainable in almost twenty different self colours and bi-colours and bearing as many as two dozen blooms at the same time, are outstanding. The plants will come into bloom within nine months from the time the seed is sown, the dainty frilled tuberous blooms being held above the foliage on wiry stems to give the plant a rather lighter, more dainty effect than that of the gloxinia. With its pretty tubes of pink, blue, purple, red and white, the streptocarpus is known as the Cape Primrose and will be happy growing in lower temperatures than required by the other members of the group with the exception of the achimene. If the seed is sown towards the end of July, the plants will come into bloom early the following May and will continue to bloom until the end of October, a period of fully six months.

SOWING THE SEED

The seed is sown in a compost similar to that used for the gloxinia and as the seed is small, being almost dust-like, it should be mixed with silver sand to facilitate thin sowing. The pans or boxes should be covered with clean glass after the seed has been lightly watered in. A piece of brown paper should be placed over the glass to encourage germination which will take about a month. As soon as germination is observed, the paper must be removed but the glass should remain in position until the seedlings have become established and will be ready for transplanting.

The seedlings will have become used to a moist atmosphere and this should be maintained as far as possible after the seedlings have been transplanted 2 in. (5 cm) apart. Again, the compost should be like that suggested for the gloxinia and also for the first potting into 3 in. (7.5 cm) pots in which the plants will grow throughout winter. The young plants will be ready for their first pots by early October and as soon as established, watering must be as sparingly as possible, keeping the compost moist and the plants just growing on.

At this time the temperature of the greenhouse should not fall below 50°F. (10°C.) either by day or night and this is maintained until early May when the plants come into bloom.

Towards the end of March with the greater warmth of the sun, the plants should be moved to larger pots and be given more moisture. On all suitable days the greenhouse should be damped down each morning and the ventilators opened. The plants enjoy a more buoyant atmosphere than the gloxinia and like that plant they must always be shaded from strong sunlight. This will not only prevent scorching but will enable the blooms to remain in a fresh condition for a longer time than when exposed to the sunlight. The quality and colour of the blooms will be enhanced and the plants kept free from pests if watered with soot water once each week from the time the plants come into bud in April. The plants should be occasionally dusted with 'Lindex Dust' to prevent an attack of green fly.

If at the end of winter the plants should appear somewhat backward, the first buds to form should be pinched out to allow the plant more time to concentrate its energies upon development rather than on forming bloom. In any case, the first bloom may not be of outstanding quality and is better if not allowed to form.

STREPTOSOLEN

A delightful climber for covering a shady wall indoors, *S. jamesonii* bears clusters of orange-coloured flowers during spring and early summer – a time when there are few wall plants in bloom. Grown up a trellis or against a wall, where it remains green all year, it demands only sufficient warmth to keep out the frost and to bring it into bloom in early April. Afterwards it may be left to the natural warmth of the sun and provided it is given plenty of moisture at its roots it will bear an abundance of bloom.

The plants reach a height of 5–6 ft. (about 1.5 m) and will thrive in any good loamy compost enriched with a small quantity of decayed manure. Pruning should be done immediately after flowering, shortening back any long shoots and removing any dead wood. This is best done in August, when shoots of the new season's wood are propagated by inserting in sandy compost.

STYRAX

Deciduous trees or shrubs, native of China and Japan and requiring an acid soil. The plants are tolerant of partial shade and bear their flowers in early summer. Slow growing, they require no pruning except to remove any dead wood. Propagate by layering or by cuttings of the half-ripened wood taken in July.

S. japonica makes a small spreading tree and in June bears pendulous bell-shaped white flowers whilst *S. hemsleyana* has large toothed leaves and bears its bell-shaped white flowers in long racemes. *S. americana* blooms in July, bearing nodding white bells 1 in. (2.5 cm) across.

SUMACH, *see* Rhus

SUNFLOWER, *see* Helianthus

SWAMP CYPRESS, *see* Taxodium

SWAN RIVER DAISY, *see* Brachycome

SWEET CHESTNUT, *see* Castania

SWEET GUM, *see* Liquidambar

SWEET PEA, *see* Lathyrus

SWEET ROCKET, *see* Hesperis

SWEET VIOLET, *see* Viola

SWEET WILLIAM, *see* Dianthus

SYRINGA

Native of Persia and the far east, the lilacs are with but few exceptions, completely hardy in the British Isles and will flourish in well drained ordinary soil which has been lightly manured. The modern lilacs are descended from *S. vulgaris* and *S. persica* and are amongst the most handsome and sweetly scented of all flowers, appearing in May and June in large panicles. 'Clarke's Giant' bears its lilac-blue flowers in pyramidal clusters whilst 'Sensation' bears large flowers of purple-red, edged white. Plant near it, 'Primrose' with its clusters of pale primrose-yellow or 'Jan van Thol' with its panicles of snowy whiteness. These are single lilacs. Of the doubles, 'Charles Joly' is an old favourite with its deep crimson flowers whilst 'Mme Buchner' is late into bloom and bears flowers of carmine-rose. Other lovely species for a small garden include *S. yunnanensis* 'Rosea 'which in June, bears slender panicles of clear rose pink; and *S. microphylla* which has small leaves and bears short panicles of rosy-lilac flowers which last almost through summer.

Lilacs sucker freely from grafted plants and to maintain vigour, these should be removed when seen. Cut away the dead flowers in midsummer but no further pruning should be necessary. Plant November to March 6 ft. (2 m) apart and with at least 6 in. (15 cm) of soil over the roots. Top dress in autumn with decayed manure for as next year's flowers are borne on the new season's wood, it is essential for the plant to produce as much new wood as possible.

Propagate from suckers if on their own roots or by layering; also by cuttings about 9 in. (23 cm) long, removed with a 'heel' and inserted in trenches of sandy compost in the open or in a frame. At all times ensure that lilacs do not lack moisture in spring when in bloom.

TAGETES

This includes the annual French and African marigolds which have become so popular for bedding and have been improved more than any annuals in recent years. They may be raised in gentle heat from an early March sowing or seed is sown mid-March either in a cold greenhouse or frame. The plants may be classed as being almost hardy and require no special care in any way. Both the African and French varieties are ideal plants for a northerly garden, being planted out towards the end of May and remaining in bloom until November. The French varieties are ideal for small beds and with their vivid orange and yellow colourings no plants make a brighter display nor over so long a time.

Most of the African marigolds are tall growing, reaching a height of 2 ft. (60 cm) and more and so are suitable for cutting or for planting in the herbaceous border, or in large beds where they are not troubled by wind. Being of sturdy habit, staking is not necessary. The blooms are globular, almost as large as a cricket ball and freely produced.

Two of the most compact varieties are 'Orange Queen', which does not exceed a height of 20 in. (50 cm) and the golden-yellow, 'Crown of Gold', with its reflexed under petals. There is also a dwarf form, 'Cupid' which bears equally large blooms on 6 in. (15 cm) stems.

The French marigolds may be obtained in the most arresting colours of mahogany and gold, the markings being unique amongst flowers. In the double-flowered section, 'Gold Laced', the golden flowers having an attractive edging of carmine-red; 'Harmony', rich orange, with the outer petals deep mahogany; and 'Rusty Red', the crimson petals being edged with orange, are outstanding and rarely exceed a height of 12 in. (30 cm). In the single-flowered, 'Naughty Marietta', a tiny compact plant, the golden blooms blotched and striped maroon being unusual.

T. signata 'Pumila' is a tiny compact plant which is generally known as tagetes. The two outstanding varieties are 'Golden Ring', brilliant orange and 'Lulu', lemon-yellow. Though of branching habit they need not be planted more than 6 in. (15 cm) apart.

TAMARISK, *see* Tamarix

TAMARIX

Hardy evergreen or deciduous shrubs or small trees bearing slender branches of feathery leaves and tolerant of sandy soil and a coastal situation. It will form a thick hedge 6 ft. (2 m) tall within three years and may be clipped into shape after flowering. Propagation is by cuttings 12 in. (30 cm) long and taken in October for insertion into sandy soil where the plants are to be grown on. If used for a hedge, plant 16 in. (40 cm) apart.

T. gallica, the Common Tamarix of south France which is hardy in Britain, is almost evergreen. Its scale-like leaves turn yellow in autumn whilst it bears small pink flowers at the same time. *T. pentandra* is also autumn flowering, bearing rose-pink flowers amongst its glaucous foliage.

T. tetrandra bears bright pink flowers in May when *T. parviflora* will be in bloom, bearing deep pink flowers amongst glaucous foliage.

TARRAGON, *see* Artemesia

TASSEL FLOWER, *see* Cacalia

TAXODIUM

The Swamp Cypress, native of Florida and Central America and which will grow in damp situations. It is a deciduous, loosely branched conifer with bright green leaves arranged fern-like and with hard, round cones. *T. distichum* has drooping branchlets, its bright green leaves turning bronzy yellow before falling. The form 'Pendens' is more drooping.

Plant November to March in an acid soil containing peat or leaf mould. Propagation is by seed sown in outdoor beds in spring.

TAXUS

The Common Yew, *T. baccata,* is one of the most useful evergreens for a calcareous soil. It makes a valuable hedge for it will withstand hard clipping whilst it is readily adaptable to situation. Distributed throughout the Northern Hemisphere, it grows to a height of 30 ft., (3 m), furnished with sickle-shaped leaves of deep shining green. The bony seed is enclosed in a fleshy scarlet cup or 'aril' from which it is released by immersing in water for forty-eight hours before shaking up in dry sand before sowing. The yew makes a durable hedge but should not be planted where cattle can reach it for the leaves are poisonous. Plant in April as an inner hedge. It is tolerant of shade. Propagation is from seed sown in spring; or by layering in autumn whilst cuttings may be taken in August and inserted in a sandy compost under glass.

Of numerous varieties of *T. baccata,* 'Erecta' is of dense pyramidal habit and makes a good hedge whilst 'Elegantissima' is the Golden Yew. 'Fructolutea' is the Yellow-berried Yew and 'Jacksonii' is a tall tree with spreading branches, pendant at the tips.

TECOMA, *see* Campsis

TECOPHILAEA

T. cyanocrocus is native of Chile, and a superb plant for the alpine house or for a sheltered corner in the garden. It bears large crocus-like flowers of richest sky-blue with an attractive white centre and it carries a delicious sweet fragrance. It should be planted in pans containing a mixture of loam, sand and peat in equal quantities and these may either be placed in the alpine house or under a frame light which should be partially removed in early March to allow the maximum amount of air to enter. As soon as they begin to show colour the plants should be transferred to a position in the home where their fragrance and gorgeous colouring will be appreciated.

TEUCRIUM

The Germander, a small genus of perennial herbs or under shrubs with downy silvered leaves. *T. fruticans* will grow 3–4 ft. (1 m) in two years and has purple bracts and highly silvered leaves and stems. *T. polium* grows only 4 in. (10 cm) tall, its attractively notched leaves being covered in silvery down whilst it bears pale yellow flowers in summer. The plants require an open, sunny situation and a well-drained sandy soil. They can survive for weeks without water. Propagate by division or by cuttings, rooted in a sandy compost under glass and planted out in May.

THALICTRUM

For the back of the border there are no finer plants than the Meadow Rues with their maidenhair fern-like foliage and myriads of tiny gypsophila-like blooms produced on long wiry stems from June until August. The plants like a cool, moist soil and will be happy in partial shade. They should be planted in autumn 3 ft. (1 m) apart. Propagation is by root division, but when once established the plants should not be disturbed more than necessary.

SPECIES AND VARIETIES

T. aquilegifolium. With its attractive foliage like that of the aquilegia, it is a most valuable plant in that it provides mid-border colour during May and June. The plants grow well in any ordinary soil and in partial shade. 'Bee's Purple' bears clusters of fluffy rosy-mauve flowers held on wiry stems. There is a white form, 'Album'; and a more compact variety, 'Dwarf Purple', growing to a height of 2 ft. (60 cm) and which may be used towards the front of the border.

Thalictrum alpinum. Bears yellow tasselled flowers on 6 in. (15 cm) stems amidst tufts of fern-like foliage. Plant with it *T. tuberosum* which is similar in habit and which produces interesting milky-white tassels.

T. dipterocarpum. The masses of tiny rose-purple flowers have attractive yellow anthers. Another form, 'Album', bears pure white flowers; whilst 'Hewitt's Double' bears fully double rose-red flowers in profusion. They are excellent for cutting, lasting well in water where they are so useful for 'mixing' with other flowers.

T. flavum. It has attractive grey-blue foliage and bears clusters of bright yellow flowers on long wiry stems.

T. glaucum 'Illuminator'. Similar to *T. flavum,* its lovely lemon flowers brightening up the darkest corner of the border or garden whilst it has attractive blue-green foliage.

THRIFT, *see* Armeria

THUNBERGIA

Though perennial, it is usually treated as a half-hardy annual and is used in hanging baskets or to cover a trellis under glass or in a sheltered garden outdoors where it may also be allowed to trail over the ground, to cover a sunny bank. Sow seed early in the year and transplant to small pots, planting out towards the end of May. *T. alata,* known as Black-eyed Susan, has heart-shaped leaves and bears masses of creamy-white flowers with a black centre and it blooms from mid-July until mid-October (see plate 71). There is an attractive yellow form, 'Sulphurea'.

THUYA

Arbor-vitae are evergreen trees or shrubs, native of China and America with flattened branches and small scale-like leaves. The cones are conical and smooth with a projection below the tip of each scale. Valuable for hedges owing to their dense growth, they flourish in well-drained soils and withstand clipping. Several of the dwarf forms are suitable for the rock garden.

Plant in spring and propagate from seed sown in April or from cuttings of half-ripened wood, removed in July with a 'heel' and inserted in sandy compost under glass.

SPECIES AND VARIETIES

Thuya occidentalis. The Arbor-vitae of North America, an erect-growing tree which requires a moist soil and may be planted by the side of ponds and streams. Its small blunt leaves are thickly imbricating along the branches whilst the cones are less than 1 in. (2.5 cm) long. The form 'Fastigiata' is a pleasing upright tree for a small garden whilst Rheingold is of slow growing pyramidal habit and in autumn, turns to rich old gold. 'Filiformis' has pendent branches; and 'Robusta Nana' is an attractive crimson dwarf, the branchlets recurving at the tips.

T. orientalis. The Chinese Arbor-vitae which makes a dense pyramidal bush 20 ft. (6 m) high, its branches growing vertical. A fully grown plant is oval in outline. The foliage is similar to *T. occidentalis* but the cones are round. In the form 'Hillieri', the pale green leaves turn to dark green in winter whilst in 'Elegantissima', the foliage is yellowish-green in summer. 'Meldensis' makes a low globe-shaped plant, its blue-green foliage turning bronzy-purple in winter.

T. plicata. The Western Arbor-vitae, a fast growing tree which is tolerant of hard clipping and so makes a valuable hedge. The long slender branches are furnished with glaucous green leaves. The form 'Fastigiata' makes an erect column whilst the dense 'Hillieri' has short fern-like branchlets. In 'Semperaurescens', the leaves are yellow-tinted and change to deep bronze in winter.

THYMUS

Those thymes which are of upright, shrubby habit possess individual aromatic qualities and are most valuable for stuffings and for flavouring meats. Natives of the Mediterranean shores, the thymes prefer a sandy soil and like so many of the shrubby herbs, they grow well over limestone formations. Whilst the grey-leaved thyme may be raised true from seed, sown in shallow drills under cloches or in frames during summer, the species and varieties are best propagated from cuttings. These should be removed with a heel and inserted into sandy soil either in the open or in a frame during summer. Alternatively, plants may be increased by root division, the outer portions of the clumps being the most vigorous. Select a sunny position and a well-drained soil and allow 15 in. (38 cm) between the plants. To prevent the plants becoming untidy, cut the shoots almost to ground level early in June and again at the end of August, and where growing commercially make up into bunches of a thickness that will fit between finger and thumb joined together.

For use in the home, mixtures of the various thymes will prove interesting though *T. herba barona,* used in olden times to rub on Baron of Beef, hence its name, should always be used by itself, for it possesses a distinct caraway aroma. From Spain came two thymes having a pungent aroma, *T. micans,* which forms a little green mat and possesses a distinct pine scent and *T. membranaceus,* which has a similar perfume and bears large white flowers.

T. carnosus, which forms a compact, upright plant possesses a powerful aromatic fragrance, whilst *T. fragrantissimus* has attractive grey leaves which have a fresh orange pungency. *T. nitidus,* is also grey-leaved and has the same pungency as the Black or English thyme, *T. vulgaris,* of which there is also a golden leaf form. There is also a French variety, having narrow grey leaves. But it is the lemon-scented thymes which are the most pleasing for their perfume is most refreshing. *T. citriodorus* 'Silver Queen', is perhaps the best form, its leaves being of a lovely silver colour. There is also a gold leaf form, 'Aureus', the leaves taking on their golden colouring during early winter, for all the thymes are evergreen and may be planted about the rockery to give winter colour. The well known culinary thymes grow about 8 in. (20 cm) high and as they tend to form much old wood after four years, young plants should be grown on from cuttings to take their place or they should be divided frequently.

The thymes may be readily raised from cuttings taken with a heel from the woody stems and when about 2 in. (5 cm) in length. They will quickly root in boxes, or around the sides of a pot, or planted in frames in a sandy compost. When rooted they may be transferred to individual pots from which they may be planted out at almost any time, though they do seem to enjoy spring planting best. To plant amongst paving stones or to make a thyme 'lawn', the creeping thymes are suitable.

The form *T. serpyllum* 'Coccineus' is more robust and bears larger and more deeply-coloured flowers. Another, *T. coccineus minus,* is more dwarf than the type. It bears rich pink flowers and grows only ½ in. (1.25 cm) tall. It is ideal for a trough. Almost as prostrate is the variety 'Annie Hall', which forms mats of fleshy-pink flowers whilst 'Pink Chintz' is of similar habit and bears mats of a beautiful shade of salmon-pink. With

T. coccineus, plant 'Snowdrift', its pure white counterpart, or *T. serpyllum* 'Sandersi' of prostrate habit and which also bears white flowers. Another lovely white form, *T. serpyllum* 'Album', has leaves of a paler green colouring than the others.

Another creeping thyme is *T. serpyllum* 'Lanuginosus', which has silvery leaves and which are as if covered with wool. Also having woolly leaves is *T. doeffleri* its carmine pink flowers, which are sweetly fragrant, being accentuated by the woolly white foliage. *T. micans* is interesting in that it forms large grey-green mats, studded with attractive mauve blooms and carries a distinct aromatic pine fragrance. It is not quite as prostrate as *T. serpyllum* but may be planted with the creeping thymes, as may *T.* 'Lemon Curd', a hybrid which carries the real fragrance of lemons. It forms a tiny hummock.

Two more thymes of prostrate habit are *T. odoratus* and *T. azoricus,* both of which form bright green mats, possess a powerful fragrance and bear bright mauve flowers. Both are lovely plants for crazy paving or for planting on the top of a dry wall.

To plant between crazy paving, soil should be removed to accommodate the plant with its ball of soil as it is shaken from the pot. Where the plants have not been pot grown, the soil between the stones should be removed to be replaced by prepared compost made up of a mixture of loam, sand and a little decayed manure into which the young plants are set out. This will provide the plants with a friable soil where the original soil may be heavy or not too well-drained. Early spring is the best time to plant between crazy paving stones.

TIARELLA

Flowering during May and June, *T. cordifolia* is a delightful front of the border plant having attractive golden heart-shaped foliage and bearing plumes of creamy flowers very much like those of the astilbe. The plants are happy in ordinary soil and are readily increased by division of the roots. Plant in November, 18 in. (45 cm) apart.

A new hybrid, the result of a cross with the heuchera, has the same attractive foliage and habit, but bears a pale pink spike which is charming planted with the true species. This plant has been named 'Bridget Bloom' and will bear its spikes intermittently throughout the summer.

A charming hybrid for the alpine garden is *T. wherryi,* which bears its creamy-white spikes on 8 in. (20 cm) stems and is long in bloom.

TIGER IRIS, *see* Tigridia

TIGRIDIA

The Mexican Tiger Flower, or the Tiger Iris as is its more popular name. For late summer and early autumn flowering it should be more often planted, especially in the south-west where the bulbs may be left in the ground through winter. Elsewhere they need some protection, and if the soil is heavy, the bulbs should be lifted each October and stored in boxes of peat in a frost-proof room. The plants love full sun and such a position should be given to them. If the spring is cold, planting is better delayed until early April or later, for the soil should first have the chance of becoming warm. The bulbs should be planted 4 in. (10 cm) deep on to sand. If the pockets are filled in with a sand and peat mixture there will be no need for any further preparation of the soil, provided it is not of an unduly heavy nature. The important point is to plant in a position where the soil and the plants can obtain the maximum amount of sunshine. Where this can be provided the tigridias will be found of perfectly easy culture and will continue to produce their multi-coloured blooms throughout late summer.

The flowers only last for about two days, but each bud is really a magazine full, 'firing a long burst', of dazzlingly brilliant blooms, each roughly the shape of a three-bladed ship's propeller. As they will not open except in sun, and do not travel when cut, tigridias are only seen in gardens. They cannot appear at a flower show or on the market, so few know of their beauty. It seems to take bees a long while to learn their way about the blooms, with the stigma towering like a mast in the centre of the strange and gaudy 'propeller', but seed is usually set towards the end of the season, and with spring-sowing and one season to grow in, will make flowering-size bulbs the following year.

This is an excellent plant for a cool greenhouse. The bulbs should be planted in March and treated in much the same way as freesias. Plant three to four bulbs to a size 48 pot and stand either in a cold-house or frame. Give no water whatsoever until the foliage appears, then give only just enough to bring on the plants. More water may be given as the blooms begin to form and an occasional application of diluted liquid manure will prove beneficial.

SPECIES

Tigridia alba. It bears bloom of purest white with a carmine-spotted centre.

T. canariensis. A vigorous form, the blooms being of rich golden yellow, spotted in the centre.

T. carminea. An unusual colour, the blooms are of rich orange, flushed yellow.

T. lutea. It bears large rich yellow flowers, devoid of all markings.

T. 'Rose Giant'. A hybrid of strong constitution, which bears a lovely rich rose-pink bloom.

S. speciosa. A most striking species, the bright red blooms having a golden centre which is spotted with scarlet.

T. wheelari. The blooms of vivid scarlet are densely spotted with crimson.

TILIA

Deciduous trees, Limes are amongst the most beautiful in cultivation with their heart-shaped leaves of 'lime' green and which turn lemon-yellow in autumn, and bearing small sweetly-scented flowers of greenish-yellow which are attractive to bees. Limes are valuable for roadside planting for they withstand hard pruning and clipping. They may be cut back repeatedly to form a 'pleached' alley (mentioned by Shakespeare). Limes need a deep loamy soil and an open position. Propagation is by layering the shoots which form around the base or by seed sown in frames in autumn.

Tilia argentea, the Silver Lime, has leaves which are downy on the underside and when rustled by the wind, produce a silver-like appearance. The variety 'Pendula' (*T. petiolaris*) is perhaps the most handsome of all weeping trees. *T. dasystyla* has bright glossy green leaves and red stems and similar is *T. platyphyllos* 'Rubra', the Red-twigged lime, its crimson shoots being most striking in winter. The variety 'Aurea' is the Yellow-twigged lime. *T. cordata* is the Heart-leaf lime, the small leaves being pronouncedly heart-shaped.

TITHONIA

A brilliantly-coloured Mexican annual resembling the zinnia and requiring similar culture. Sow in gentle heat early in the year, transplanting when large enough to handle and planting out, after hardening, early June. *T. speciosa* 'Torch' is the best. It grows 3 ft. (90 cm) tall and bears glowing orange flowers with a golden central cone from late July until September.

The dwarf variety, 'Sungold', bears double ball-shaped flowers of bright golden-orange, grows only 2 ft. (60 cm) tall, and is a valuable border plant. Sow in April where it is to bloom, thinning the seedlings to 18 in. (45 cm) apart or under glass early in spring and planting out early May.

TOADFLAX, *see* Linaria

TOBACCO PLANT, *see* Nicotiana

TOLMEIA

In habit *T. menziesii* is like *Saxifraga sarmentosa,* its leaves being borne on long stems and from the point where the stem joins the leaf young plantlets arise, which take root when the leaves touch the soil, as they do where the plant is growing in the natural state. It is a pretty plant of spreading habit, its hairy, heart-shaped leaves being of brilliant green. It presents no difficulty with its culture, requiring partial shade, a sandy compost and plenty of moisture during summer. It is best propagated by pegging down the leaves into a sandy compost in small pots placed around the parent plant. They will quickly root when they may be severed from the parent plant.

TOPIARY

It takes its name from the Latin, *topiarius* or pleacher who was a person of importance in the Roman garden. It is the art of clipping trees into various designs and is mentioned by Pliny in his account of his Tuscan villa. By Elizabethan times, the clipping of box and yew was an established practice of English gardens. Whitethorn and privet were used, also rosemary which Parkinson said was 'the chiefest beauty of gardens, sette by women for their pleasure to grow in sundry proportions as in the fashion of a peacock, or such things as they fancy'.

Since earliest times topiary work has been a feature of the cottage garden, to decorate the tops of hedges, the plants being shaped like hens and peacocks whilst balls and squares are often seen resting on the hedge. The Dutch are the masters of topiary and today, clipped box and privet plants are imported into Britain.

Almost any evergreen which will withstand clipping may be used and the plants may be made into spirals and pyramids. The trees should be planted in spring and well watered until established. Careful clipping with small shears kept very sharp will maintain their shape but this must be done once each month during summer for at no time should they be allowed to lose their shape. Apart from privet, other evergreens will be slow growing and will require only the minimum of clipping if done regularly.

To make a ball or square of privet, obtain a young plant as a single stem and grow on to a height of 4–5 ft. (about 1.5 m), removing all side growths except those required to form the head. At the required height, the leader shoot is nipped out to encourage side shoots to form at the top and the head is allowed to grow quite large. It may then be clipped into the required shape and maintained by frequent clipping during summer.

TORCH LILY, *see* Kniphofia

TORREYA

The Californian Nutmeg, *T. californica,* makes a small shrub-like tree which grows well in calcareous soil. It is a strong smelling evergreen, known as the Foetid Yew for in its habit and foliage it resembles the yew. It has lance-shaped leaves, sharply pointed and of pale yellow whilst it has green damson-like fruits 2 in. (5 cm) long.

Plant in spring in a well-drained soil and in a warm sheltered situation. Propagate from seed (after removing the fleshy covering); by layering in autumn; or from cuttings inserted in sandy compost under glass.

TOUCH-ME-NOT, *see* Impatiens

TRACHELIUM

T. caeruleum makes an attractive pot plant, growing to a height of about 20 in. (50 cm) and bearing, during July and August, dense heads of small, brilliant blue flowers.

Seed should be sown in June, the pans or boxes being covered with glass and brown paper to hasten germination. If the compost is kept moist, the seed will soon germinate and by late July the seedlings will be ready to move to small pots. The compost should be composed of 3 parts sterilised fibrous loam, 1 part leaf mould and 1 part silver sand and a similar compost should also be used when the plants are moved to larger pots at the end of summer or early in autumn, potting firmly.

During winter the plants should be given little moisture whilst a temperature of 42°F. (5°C.) will be sufficient to keep them growing. In spring, the plants should be given more moisture and frequent syringing to stimulate them into growth and to bring the plants into bloom during which time they will prove to be copious drinkers. From early June the display will be enhanced if the plants are given alternate weekly feedings with dilute manure water and soot water.

Stopping will depend upon whether it is required to have one large flower head or numerous small heads produced on side shoots.

TRACHELOSPERMUM

Evergreen twining plants which may be grown in all but the coldest of gardens. They have long leathery leaves and bear jasmine-like flowers which are sweetly scented. They are best grown against a warm wall and allowed to climb over a trellis. *T. asiaticum* bears in July and August, a profusion of creamy-white flowers and is the hardiest species. *T. majus* is self-clinging and will cover a wall as quickly as ivy. Established plants will bear sweetly-scented white flowers. *T. jasminoides* 'Variegatum' has leaves edged with cream.

Plant out in April from pots, into a deeply worked soil and do no pruning apart from thinning out the dead wood. Propagate by layering or by cuttings of half-ripened wood removed with a 'heel' and inserted in a sandy compost under glass.

TRACHYCARPUS

The Chusan Palm, hardy in the British Isles in the south-west; in west Scotland and Ireland. It will reach a height of 25 ft. (7 m) and has large fan-shaped leaves up to to 3 ft. (90 cm) long and 3 ft. (90 cm) wide. In summer it bears yellow flowers in drooping panicles, followed by black grape-like fruits. *T. fortunei* requires a deep moist loam containing plenty of humus. Propagate from seed sown in spring in small pots which are kept under glass until germination has taken place.

TRADESCANTIA

Like so many border plants, *T. virginiana* the Spiderwort, was introduced from Virginia by John Tradescant, gardener to Charles I. For long it remained a plant of interest rather than of beauty, flowering in a moist soil and from early June until the end of October. From amidst its thin strap-like foliage the interesting flowers are continuously produced on 20 in. (50 cm) stems, for they remain colourful only for twenty-four hours when others take their place (see plate 72). The three-petalled flowers, like a ship's propeller, are held in clusters, and some quite charming colours have given the plant a new popularity. Plant in November, 18 in. (45 cm) apart and propagate by root division.

SPECIES AND VARIETIES

BLUESTONE. Similar to 'Leonora', the blooms are of a lovely clear mid-blue colour.

T. coerulea 'Plena'. The only double variety, the blooms being of a true Cambridge-blue.

IRIS. This variety bears the largest flowers of any and which are of a deep Oxford-blue colour.

LEONORA. The large blooms of an attractive sky-blue colour make this one of the few really true blue border plants.

OSPREY. The large white flowers have an interesting blue centre. Plant near 'Iris' or 'Purewell Giant'.

PUREWELL GIANT. The blooms are of a rich purple-red colour freely produced.

The form *T. alba-vittata* has white and pale green striped leaves whilst *T. fluminensis* 'Aurea' has leaves with golden stripes. *T. quadri-color* is interesting in that its small oval leaves are striped with pale green, pink and purple, and *T. albiflora* has tiny green and mauve leaves and bears dainty white flowers. *T. blossfeldiana* is rather different in form, being more upright with thick fleshy stems, its rich green leaves being shaded with mauve on the underside. *T. purpurea* is also very fine, its large leaves being of rich purple-green, and it is a plant of vigorous habit.

TREE OF HEAVEN, *see* Ailanthus

TREE POPPY, *see* Romneya

TREES

DECIDUOUS TREES WITH COLOURFUL FOLIAGE

GOLD

Acer drummondii
Acer negundo 'Aurea'
Catalpa bignonioides
Corylus avellana 'Aurea'
Populus serotina 'Aurea'
Robina pseudoacacia 'Frisia'
Ulmus campestris 'Louis van Houtte'

CRIMSON, PURPLE and BRONZE

Acer colchicum 'Rubrum'
Acer 'Goldsworth Purple'
Acer linearilobum
Acer ozakazuki
Acer plamatum 'Purpureum'
Fagus latifolia
Fagus purpurea
Prunus nigra
Prunus pissardii
Prunus pubescens
Pyrus eleyi
Salix purpurea

SILVER and GREY

Acer negundo 'Argentea'
Crataegus orientalis
Populus alba
Salix cinerea 'Tricolor'
Salix smithiana
Sorbus aria 'Lutescens'

AUTUMNAL TINTS

Acer rubrum
Aesculus briottii (Horse Chestnut)
Carpinus caroliniana
Crataegus prunifolia
Liquidamber styraciflua
Liriodendron tulipifera
Photinia beauverdiana
Photinia villosa
Populus tremula
Pyrus crataegifolia
Quercus coccinea (Scarlet Oak)
Sorbus discolor

DECIDUOUS TREES WITH VARIEGATED FOLIAGE

Acer negundo 'Variegatum'
Acer negundo 'Elegantissimum'
Cornus alba 'Sibirica Variegata'
Cornus mas 'Aurea'
Cornus mas 'Elegantissima'
Fraxinus pensylvanica 'Variegata'
Ilex aquifolium 'Argentea Marginata'
Ilex aquifolium 'Aurea Regina'
Quercus cerris 'Variegata'
Salix cinerea 'Tricolor'
Sambucus nigra 'Albo-marginata'
Ulmus campestris 'Variegata'

DECIDUOUS TREES WITH COLOURFUL BARK

Acer davidii (white stripes)
Acer griseum (coppery-orange)
Acer rufinerve (green)
Arbutus andrachnoides (bright brown)
Betula alba (silver)
Betula albo-sinensis (orange-brown)
Betula coerulea-grandis (silvery-white)
Betula lutea (bright yellow)
Betula mandschurica 'Japonica' (white)
Betula pendula (silver, peeling)
Carpinus caroliniana (grey, fluted)
Cornus alba (red)
Crataegus carrierei (silver)
Populus alba (grey)
Populus tremuloides (yellow)
Prunus sargentii (glossy brown)
Prunus serrula (mahogany)
Salix vitellina (bright yellow)
Sorbus moravica (golden-yellow)

DECIDUOUS TREES WITH COLOURED FRUITS AND BERRIES

Ailanthus glandulosa (orange-red)
Arbutus unedo (orange-red)
Cornus mas (red)
Cotoneaster frigida (scarlet)
Crataegus carrieri (orange)
Crataegus crus-gallii (crimson-red)
Euonymus europaeus (pinkish-red)
Fraxinus mariesii (bronze)
Hippophae rhamnoides (orange)
Ilex aquifolum (scarlet)
Pyrus 'Goldsworth Red' (crimson)
Pyrus hillieri (yellow)
Pyrus 'John Downie' (yellow)
Pyrus purpurea (purple)
Sambucus racemosa (scarlet)
Sorbus aucuparia (orange-red)
Sorbus commixtra (orange)
Sorbus pohuashanensis (Scarlet)

DECIDUOUS CATKIN-BEARING TREES

Alnus glutinosa
Alnus japonica
Alnus sitchensis
Betula nigra
Betula papyrifera
Betula verrucosa

Carpinus betulus
Carpinus japonica
Castanea sativa
Corylus avellana
Corylus colurna
Garrya elliptica
Populus tremula 'Pendula'
Pterocaya fraxinifolia
Salix acutifolia
Salix apoda
Salix bockii
Salix caprea
Salix gracilistylosa
Salix humilis
Salix pentandra
Salix spaethii

DECIDUOUS TREES OF PENDULOUS OR WEEPING HABIT

Betula pendula
Betula pendula 'Youngii'
Caragana arborescens 'Pendula'
Crataegus monogyna 'Pendula'
Fagus sylvatica 'Pendula'
Fagus sylvatica 'Purpurea Pendula'
Fraxinus excelsior 'Pendula'
Ilex argenteo-marginata 'Pendula'
Juglans regia 'Lasciniata'
Morus alba 'Pendula'
Populus tremuloides 'Pendula'
Prunus amygdalus 'Pendulus'
Prunus serrulata 'Rosea'
Prunus subhirtella 'Pendula'
Prunus yedoensis 'Perpendens'
Robinia pseudoacacia 'Pendula'
Salix babylonica
Salix alba 'Tristis'
Salix caprea 'Pendula'
Sophora japonica 'Pendula'
Tilia petiolaris 'Argentea pendula'
Ulmus montana 'Pendula'

DECIDUOUS TREES FOR AN ACID SOIL

Arbutus andrachnoides
Arbutus unedo
Halesia monticola
Magnolia grandiflora
Magnolia sargentiana
Magnolia soulangeana
Oxydendrum arboreum
Pieris formosa
Pieris taiwanensis
Styrax japonica

DECIDUOUS TREES FOR A CALCAREOUS SOIL

Acer cappadocicum
Acer carpinifolium
Acer forrestii
Acer grosseri
Acer palmatum
Acer platanoides
Aesculus glabra
Aesculus hippocastanum
Aesculus indica
Betula alba
Betula coerulea-grandis
Betula ermanii
Betula lutea
Betula nigra
Betula pendula
Buxus sempervirens
Carpinus betulus
Cornus florida
Cornus mas
Cotoneaster frigida
Crataegus arkansana
Crataegus maximowiczii
Crataegus monogyna
Crataegus oxyacantha
Ehretia dicksonii
Fagus sylvatica

DECIDUOUS TREES FOR A CALCAREOUS SOIL (*contd.*)

Juglans regia
Phellodendron amuriensis
Populus alba
Populus canescens
Populus nigra
Populus tacamahaca
Populus tremuloides
Prunus amygdalus
Prunus cerasifera
Prunus cerasus
Prunus serrula
Prunus subhirtella
Rhamnus cathartica
Ulmus carpinifolia
Ulmus glabra
Ulmus procera

DECIDUOUS TREES FOR WINDSWEPT GARDENS

Acer pseudoplatanus
Crataegus monogyna
Crataegus oxyacantha
Populus alba
Populus canescens
Populus trichocarpa
Prunus spinosa
Rhamnus cathartica
Sambucus nigra
Ulmus montana

DECIDUOUS TREES FOR COASTAL PLANTING

Acer pseudoplatanus (Sycamore)
Arbutus unedo
Cornus mas
Cornus sanguinea
Crataegus oxyacantha
Fraxinus excelsior (Ash)
Garrya elliptica
Hippophae rhamnoides
Ilex aquifolium (Holly)
Laburnum alpinum
Photinia beauverdiana
Populus alba
Populus serotina
Prunus psinosa
Rhamnus alaternus
Sambucus nigra
Tamarix pentandra
Tamarix tetrandra
Ulmus montana

DECIDUOUS TREES WITH SCENTED FLOWERS

Acacia armata
Acacia baileyana
Azara microphylla
Camellia sasanqua
Castanea sativa (male catkins)
Chimonanthus fragrans syn. *praecox*
Chionanthus virginicus
Corylopsis sinensis
Crataegus monogyna
Eriobotrya japonica
Eucryphia billardieri
Hamamelis mollis
Hamamelis virginiana
Laburnum vossii
Magnolia delavayi
Magnolia denudata
Magnolia grandiflora
Magnolia sinensis
Magnolia stellata
Malus coronaria
Paulownia fargesii
Prunus yedoensis
Tilia petiolaris

TRICUSPIDARIA

This native of South America likes warmth and shade. When planted in the south, an east wall is admirable. There the plants will reach a height of 10 ft. (3 m) in a lime-free soil containing some peat, which should be packed round the roots at planting time. The plants are evergreen with glossy, toothed leaves. Planting is best done towards the end of April, and no pruning is necessary except to cut back any dead wood.

If the two better-known species are planted, they will bloom from mid-May until late October. *T. lanceolata* bears from the axils of the leaves hanging coral-red fritillaria-like flowers on similar coloured stems. It blooms from May until early August when *T. dependens* comes into bloom and remains colourful until the end of October, bearing pure white bells. Propagation is by inserting cuttings into sand and peat under a bell cloche in early August.

TRINITY FLOWER, *see* Hepatica

TRIPTERIS

One of the easiest of half-hardy annuals, it has dark green scented foliage and forms a dense mat above which are borne large flowers of brilliant orange. The variety 'Gaiety' is especially striking. Growing 15 in. (38 cm) tall, it blooms June to September and is raised by sowing seed under glass early in the year and planting out mid-May after hardening.

TRITELIA

This almost-unknown April to October flowering bulb is suited to town gardens for it is tolerant of a barren soil and is perfectly hardy. Inexpensive, it may be planted in drifts where more fastidious flowers would achieve little. Though it will bloom in almost any well-drained soil, it will flower better where it is given some humus, either peat or spent hops. The bulbs should be planted 3 in. (7.5 cm) deep in late August in a sunny position. The great value of this plant is that it will bloom continuously during periods of sunshine from early April until well into autumn and should be planted in all the odd corners where it may be left undisturbed. During dull weather the blooms close up.

Tritlia uniflora bears its pale blue narcissus-like flowers on 6 in. (15 cm) stems. The petals are divided by a purple stripe which continues right down the small tube of the bloom. It hails from South America and favours a dry, sandy soil, but is so accommodating that it will flower well even in a clay soil.

TROCHODENDRON

A small hardy Japanese tree with aromatic bark and leathery glossy green leaves like those of the rhododendron. It bears clusters of bright green flowers early in summer. *T. aralioides* will grow well in an acid soil or one of a neutral reaction but a limestone soil should be avoided. It will grow in partial shade.

TROLLIUS

A native of this country, *T. europaeus* was grown in cottage gardens during earliest times. The Globe Flower is one of the loveliest of border plants and bears its bloom from mid-April until early June, a time when there is little colour in the border. It is a valuable flower for the commercial grower, especially where it may be marketed locally. The trollius prefers a cool, moist soil and is a useful plant for those 'difficult' soils, and where the land is low lying. A rich loam, enriched with some humus does, however, suit it best and where the soil is light and sandy, liberal quantities of humus and decayed manure should be worked in.

Plant in July, after flowering, a month which has above average rainfall. The plants will then bloom liberally the following May. Plant 18 in. (45 cm) apart, spreading out the fibrous roots. Propagation is by division of the roots, and like pyrethrums the smallest pieces may be used. When marketing the bloom, cut when the buds are just showing colour, for they then open rapidly and if allowed to become too open, will have only a short life in water before dropping their petals.

VARIETIES

ALABASTER. A German variety and a new colour break, the bloom being of rich Jersey cream colour, borne on 3 ft. (90 cm) stems.

CANARY BIRD. Early to bloom, the flowers are clear lemon-yellow. 2 ft. (60 cm).

ETNA. Taller than most and mid-season flowering. The colour is outstanding, being of a deep orange with deeper orange anthers. 3 ft. (90 cm).

FIRST LANCERS. Like 'Etna', it is tall growing, with the bloom a bright orange colour. At its best during June and early July.

GOLDEN QUEEN. The last to flower, taller growing and very free. At its best late in July and even into August in the north. The colour is bright yellow, flushed orange, the inner petals pure orange. 3 ft. (90 cm).

TROPAEOLUM

ANNUAL

All members of the nasturtium family are hardy and though the Canary Creeper, *T. peregrinum,* with its tiny yellow flowers, will thrive in a cold, sunless position, the climbing nasturtium *T. majus* prefers some sun. Both, however, thrive in a sun-baked soil, and will bloom more freely on the almost humus-devoid soil of a town garden than where the soil is of better quality. All forms may be sown where they are to bloom, sowing the seed individually at 6 in. (15 cm) apart. Where plants of *T. peregrinum* are to be grown in shade they are best sown in boxes which are covered with a sheet of clean glass and placed in full sunlight. Planting in their permanent quarters may be done when the seedlings are about 3 in. (7.5 cm) tall. All species of the climbing nasturtium may, however, be sown together to cover a screen where they are most attractive. With *T. peregrinum,* sow 'Climbing Vesuvius' with its scarlet flowers and 'Indian Chief' with its rich bronzy foliage. Excellent, too, is 'Lobb's Nasturtium', *T. lobbianum,* which has smaller foliage than the hybrids and bears vivid orange-red flowers. Other varieties in this section are 'Lucifer', with its crimson flowers and dark foliage, and the yellow, 'Golden Queen'. Seed should be sown early in April and as with most climbers a few twiggy sticks placed about the plants as soon as they appear will help them to grow away rapidly.

The 'Tom Thumb' varieties growing to a height of 10 in. (25 cm), may be used for bedding, spacing the large seeds 9 in. (22 cm) apart and sowing early in April. Outstanding is 'Ryburgh Perfection', with its vivid scarlet flowers and silver foliage.

Almost as compact are the Dwarf Double Hybrids, comprising lemon, salmon, rose, orange and crimson shades and a host of intermediate colours. Outstanding is 'Baby Salmon', bearing semi-double flowers of an unusual shade of deep salmon and with dark foliage.

PERENNIAL

T. speciosum, the Flame Flower, native of the Chilean Andes, will grow well only when given a cool, moist soil. It is seen only at its best in the cool climate of Scotland, where it grows about the walls of old houses and castles, particularly in the Highlands. It is difficult to transplant, but when once the plants take hold they will make rapid growth and, like the honeysuckle and clematis, love to grow over and down to festoon a wall, rather than grow continually up it.

Planting the tuberous roots should be done in April, setting them 6 in. (15 cm) deep into a well-drained soil containing some peat or leaf mould and a little decayed manure. Never allow the plants to suffer from lack of moisture. Although difficult to establish, those who live in cool districts should make an effort to grow this plant, for it has no rival with its abundance of brilliant scarlet flowers and pale green shamrock-like leaves during summer.

TROUGH GARDENS

Where space is limited, the joys of the alpine garden with its trees and shrubs and plants in miniature may be enjoyed by making a trough garden. Similar results may also be obtained from a window-box constructed of concrete and made permanent by the use of miniature trees, stones and dwarf perennial plants. In ideal surroundings, placed about the garden of the cottage with its roof of thatch or stone tiles where the old stone troughs blend so well, the same will look equally attractive placed on a balcony or about the courtyard of a town house. Neither need the trough be of stone. Though there is nothing comparable to weathered stone, the plants will grow well in an old sink, provided the surface is made rough, or in a trough made of concrete. A concrete trough will have one great advantage in that it will be porous. This will ensure efficient drainage, thus keeping the soil sweet and the roots healthy.

Of whatever material the trough is to be constructed, it should be provided with a drainage hole and will be more efficient and more easily managed if raised on a suitable base 3–4 ft. (about 1 m) from the ground. If the trough is to be of concrete then the base should be built of breeze blocks. These blocks may also have crushed-stone incorporated which will give them the appearance of stone and will thus also be suitable for a base for an old stone trough. A sink may be supported on rustic bricks to good effect, especially where it is to be used against the walls of a newly-built house of the same material.

The trough may be placed on a balcony or at the centre of a yard with tubs arranged round the sides of the yard. Or along the base of a house where there is no soil to make a border, several troughs may be placed, especially beneath a window where they may be observed from inside the house. It is important to ensure that the trough is not placed where rain water may drip on to the plants and so tend to wash them out of the soil. Unless the eaves of the house are wide, then the trough is best placed about 2 ft. (60 cm) from the wall. Also, the trough should not be placed beneath overhanging trees, from which melting snow and rain will drip almost continuously through the winter months.

CONSTRUCTING THE TROUGH

The most satisfactory way of constructing a concrete trough is to make two boxes of matured timber, making one about 1½ in. (4 cm) smaller in all dimensions. The smaller box will fit inside the other. For the base and walls of the trough mix up 2 parts of sand to 1 part of cement, adding sufficient water to make it into a paste so that it will pour, yet at the same time it must not be too sloppy. The cement should be poured inside the first box to a depth of 1½ in. (4 cm), two large corks being held in position to provide the drainage holes. To reinforce the base and sides, a length of wire netting should be pressed into the cement mixture just before it begins to set and this should extend almost to the top of the sides. A second piece of netting should be pressed into the mixture so that it extends up the other two sides in a similar way. Use 2 in. (5 cm) mesh netting, cutting the pieces to the exact measurements of the mould. Thus for a box 30 in. (75 cm) long x 18 in (45 cm) wide x 6 in. (15 cm) deep, the two pieces of netting will measure 42 in. (1.6 m) and 30 in. (75 cm). The smaller box does not require a base; make the four sides, and hold them together by small pieces of wood nailed across each corner and long enough to stretch across the corners of the first box to prevent it pressing into the concrete base. The cement is then poured between the two boxes. Insert a piece of stick to hold the netting away from the mould and this will prevent it showing when the concrete has finally set hard, which it will do in about twenty-four hours if not made too thin. Just before the cement has set completely, the sides of the boxes are carefully removed, the inner mould being left in position until thoroughly hardened. The wooden base and the corks are then pressed off.

A glazed sink which is to be used for a trough garden should be chipped as much as possible so that the sink will be more porous, and the roots of the plants will be able to hug the sides in the same way as do plants in earthenware pots. For the same reason plants growing in old stone and concrete troughs have a vigorous root action and as a result, top growth is equally vigorous and healthy.

MAKING UP THE TROUGH

Troughs are extremely heavy, so of whatever material they are constructed, they should be given a position of permanency before they are prepared for the plants. Make certain that they are quite firm on the single or double pedestal, using wedges where necessary. Over the drainage holes place several large crocks, or pieces of brick or stone, then over the base place a layer of small stones or crocks to a depth of 6 in. (15 cm). Over this lay old turves upside down and then fill to the top with the prepared compost, pressing it well down round the sides. This should be composed of 2 parts fresh loam, sterilised if possible; 1 part top grade peat, which is superior to leaf mould because it contains no weed spores; and 1 part coarse sand and grit. Add a sprinkling of superphosphate to encourage root action, and lime to keep it sweet; also a little bone meal, a slow acting fertiliser, and mix the whole together. If one has no garden, sterilised loam, peat and sand may readily be obtained from a local nursery and as a general guide, a barrowful of compost will be required for a sink of approximately 36 in. (90 cm) x 18 in. (45 cm) x 6 in. (15 cm). Most stone troughs will be several inches deeper so that compost requirements will be much the same. Some troughs have no drainage holes, and should be given a greater depth of drainage material, whilst some pieces of broken charcoal should be mixed with the compost to maintain its sweetness.

So many ways of making up a trough or sink are available that there is something for every taste. The troughs may be used in much the same way as window-boxes or tubs, planting them with spring flowering plants such as primroses and winter flowering pansies to be followed by multiflora begonias, ivy-leaf geraniums or suitable annuals such as verbena and *Phlox drummondii.* However, it is as a more permanent garden that the sink or trough is most suitable. The dwarf roses may be used by themselves to make a miniature rose garden, using possibly four different varieties and dividing them by tiny paths of stone. Or the trough or sink may be made into a miniature rockery, complete with weathered stone, dwarf conifers or shrubs and tiny perennial plants to provide colour all the year round. The smallest of the miniature bulbs may also be used, the whole so laid out as to give a completely natural appearance.

When planting, care must be taken in selecting the most suitable plants for certain soil conditions and situations. Plants requiring a soil containing plenty of lime, such as the fragrant little *Iris reticulata,* and the alpine pinks should be given some lime rubble in the compost, and the lime lovers should be planted together. Those which are lime haters must also be kept together. There are plants which prefer shade, others a position of full sun. Only the really hardy plants should be used, for the range is large enough to satisfy all tastes, and the less hardy plants may fail to come up to expectations if the winter is unduly wet or cold.

Where possible use plants from small pots so that root disturbance is at a minimum. Planting may be done at almost any time of the year, though autumn and spring would seem to be the most suitable periods. However, before planting allow the soil time to settle down so that there are no air pockets and when making a miniature rockery, place in several stones, grouping them as naturally as possible. Tufa stone from the hills of Derbyshire and Westmorland is the most suitable, but weathered red sandstone or Cotswold stone is good. Set the stone well into the soil so that the plants may send down their roots into the moisture, almost always to be found around those portions of the stone which are under soil level. Placing the stones well into the soil will also prevent them from being moved when the trough is receiving attention, for established plants will occasionally be replaced with those which may be considered more attractive, whilst others will need trimming to keep them in bounds. The soil will also require stirring to prevent it 'panning', whilst all weeds must be removed immediately they are noticed. The soil should never be allowed to dry out during the summer months.

The best effect will be obtained from the trough and sink garden by grouping the plants as naturally as possible. Plant one or two conifers together, with two or three primulas around a stone, with possibly a miniature shrub to give partial shade to the primulas, so building up the garden in such a way that the plants will fit into the scheme just as if nature herself had done the planting.

TROUT LILY, *see* Erythronium

TRUMPET FLOWER, *see* Campsis

TSUGA

Hemlock Spruce are stately evergreen trees, with yew-like leaves and horizontally arranged branches with pendent branchlets. They may be used for specimen planting and require a deep moist soil and a sheltered situation. The leaves are stalked and arranged in two ranks; the cones are drooping, usually borne solitary at the ends of the branches. Plant in spring and propagate from seed sown outdoors in April, allowing the young plants to remain two to three years in the seed bed before transplanting.

SPECIES AND VARIETIES

Tsuga canadensis. The Eastern Hemlock which grows well in a dry chalky soil and in an exposed situation. It is recognised by its trunk which is divided at the base and by its long feathery branchlets clothed in flat leaves. The slow-growing 'Pendula' with its widely spreading drooping branches will grow wider than it grows tall.

T. caroliniana. Distinguished from the Eastern Hemlock by its larger, more glossy leaves borne in flattened sprays whilst the cones are larger. It is lime-tolerant and makes a short pyramidal tree.

T. dumosa. The Fragrant Fir of the Himalayas which requires protection from spring frosts. It has slender drooping branches clothed in leaves which are glossy dark green above; silver on the underside.

T. heterophylla. Prince Albert's Spruce or Western Hemlock which grows as far north as Alaska attaining a height of 200 ft. (60 m). It is recognised by its red bark and drooping branches clothed in feathery sprays of dark green leaves, silver on the underside. It requires a deep moist soil.

TUBEROSE, *see* Polianthes

TUBS

As an alternative, or in addition to a window-box or trough garden, to be used on a terrace or verandah, or around the walls of a paved yard where there is no garden, tubs filled with plants will be an attraction all the year round. Their value lies in their depth and size, making it possible to grow many plants which would be unsuitable for window-box culture on account of their greater height and their need for a deeper root run. Members of the lily family and the bedding dahlias will flourish in the deep compost of a tub. The taller antirrhinums may also be grown, and the Cottage tulips. The height of the plants should be similar to the height and size of the tub; dwarf plants, unless used as an edging, will appear out of perspective. The size of the tubs will depend upon where they are to be used. On a terrace, possibly running the length of a house, the tubs should be as large as possible but need not be too deep. A tub which is too deep will take away much of its charm, the need is for a greater surface of soil to accommodate as many plants as possible. To be used on a verandah, the tubs should be smaller in diameter but deeper, so as to keep them in perspective with their surroundings.

FIG. 16 *Evergreens in tubs hiding the walls of a town courtyard*

One of the most attractive town 'gardens' imaginable was composed of a small paved courtyard where numerous species of thyme grew between the stones. Arranged round the old stone walls were about a dozen tubs filled with all manner of brilliantly coloured flowers, a truly wonderful effort in the use of bedding plants which were seen at their best against the bare walls with not an inch of soil to be seen anywhere.

PREPARATION OF THE TUBS

The most suitable tubs to use for planting are the oak casks, sawn in two, which have been previously used for storing cider. They are of seasoned wood and will not require painting though this may be done on the outside if the tubs are to be used round the whitewashed walls of a courtyard or against the walls of a house. Old cider casks are readily obtainable from most west country cider firms.

To enable the plants to take advantage of the greater depth of soil, the tubs must receive the same thorough preparation as given the window-box. In the same way, holes should be drilled at the base, over which is placed a layer of crocks or broken brick to a depth of 2 in. (5 cm), so that in the event of wet weather, ample drainage is available. Over the crocks should be placed partially rotted turves and a layer of decayed stable manure which will fill up half the tub. The balance will be of prepared compost, depending upon the requirements of the plants that are to occupy the tubs. Dahlias, for instance, prefer a soil enriched with decayed manure and peat, whereas geraniums, though having a liking for a little rotten manure, prefer a soil which is drier and contains less humus. But as a rule a soil could be made up similar to that recommended for window-boxes. Tubs possess another advantage over window-boxes in that they do not require so much attention as to their watering. Being in a more open situation, they obtain natural moisture more easily, and possessing a greater depth of soil may be left unwatered for a longer period. A top dressing with peat after planting will also help to conserve moisture in the soil.

Before filling with soil, the tubs should be treated with a reliable wood preservative, both on the inside and outside and at least a fortnight should be allowed before the compost is added. To allow the soil to consolidate before planting, several days should be allowed after the tubs have been filled. If filled to the brim, this will allow the compost to settle down to about an inch below, which will allow for a summer top dressing and still prevent the soil from being splashed over the sides by heavy rains or watering. To prevent the soil from becoming sour it should be given a dressing of lime each autumn after the summer flowering plants have been removed. This, together with the greater depth of soil will make it unnecessary to remove and replace the soil as frequently as for window-boxes. Every five years should be sufficient.

TULIPA

INDOOR

The amateur will plant tulips either in pots or bowls containing bulb fibre or into a compost of soil, sand and one-third part by bulk of peat or leaf mould. Peat is almost sterile and contains neither disease spores nor weeds. Make the peat thoroughly moist before mixing with the soil, otherwise it will be impossible to bring to the desired moist condition after the bulbs are planted. Four or five bulbs to a bowl is usual and they should be of the 4¾ in. (12 cm) size, as used by the commercial grower. When set in the bowls, the tops should be just covered with the compost. It is advisable to line the bowls with a layer of peat which will help drainage, and to the compost, to keep it sweet, should be added several small pieces of charcoal. Handle the bulbs carefully as tulips bruise easily. After planting, water lightly so that water remains in the bottom of the bowl to stagnate; then place in a darkened cupboard or in the cellar.

Bulbs should be introduced to heat and light gradually. Usually, failures are the result of the bowls being taken from darkness and a cool place and plunged immediately into the strong light and warmth of a greenhouse, or large, sunny window. The commercial grower too, is all too often guilty of this in his desire to have the bulbs in bloom in the shortest time. At first a temperature of 48°F. (9°C.) is sufficient and this can be gradually increased to 60°F. (16°C.) by raising it about 1° each day. Likewise where a cool house is being used for growing on the bulbs for cutting in boxes. These will generally follow chrysanthemums, which will have finished flowering by the year end. The Darwins will be used and they may be taken indoors early in January to bloom in March. When first introduced to the greenhouse they should be shaded either by hessian canvas nailed to the sash bars of the house, or by sheets of brown paper placed over the boxes. The shading may be removed after ten days, though in a house where no heat is used, hessian stretched over the inside of the roof will help to keep out frost and should remain in position until early March when the spring sunshine will bring the bulbs into bloom.

In the heated greenhouse, the temperature may be raised to 65°F. (18°C.) for all forcing varieties as soon as the green flower buds are observed, but at this temperature opening of the buds will be rapid and care must be taken to see that they are not too open before marketing.

As a rule, the short-stemmed varieties should be given some shade almost throughout their period in the greenhouse, while the Darwins and long-stemmed varieties should be grown on without shade under forcing conditions. In a cold house early spring cultivation may be helped by the use of hessian to keep out frost, but this should be discarded as soon as possible. For this reason, tulips grown in the semi-shade of the living-room should be confined to the short-stemmed varieties.

Care must be taken when growing on a large scale, to determine the exact forcing requirements of each variety. It will not do to select a variety for forcing merely because its colour is attractive. For instance, the striking golden-yellow, flushed red variety, 'Prince Carnival', will not stand forcing at all and if early bloom is required, select the orange 'Prince of Austria', which may be forced at a temperature of 80°F. (27°C.), whereas for most Darwins forced for cut bloom, the maximum temperature should be about 58°F. (14°C.). Fluctuations in temperatures will be fatal to bulbs being forced. A temperature of 55°F. (12°C.) kept constant is far better than one which might fluctuate between 50°–65°F. (10–18C.). Draughts in the home may cause the formation of a badly drooping bloom or of a yellowing of the foliage. The same troubles will be experienced where greenhouse temperatures fluctuate too much. If the air in a warm room or greenhouse falls suddenly, the rate at which the plant loses water by evaporation is slowed up. The roots, however, continue to work in the warm soil taking in water which cannot be transpired. The cells of the plant become waterlogged and cannot support the bloom with the result that it falls over, making it a total loss. Careful watering, correct ventilation and even temperatures make for successful tulip forcing.

The commercial grower will use mass-production methods to plant the bulbs which work out more cheaply in proportion to the number of a variety purchased. The bulbs will be set out in bulb boxes made up of wood 1 in (2.5 cm) thick and 6 in. (15 cm) deep and of a size which can be comfortably handled and which will hold about 3 dozen bulbs. A 1 in. (2.5 cm) space must be left between the bulbs so that correct air circulation may be allowed, for remember that bulbs being forced must contend with conditions of warmth and high humidity, both of which will encourage disease unless correct ventilation is provided. Bulb boxes should be drilled with holes, one to every 6 sq. in., (15 sq. cm) to make for correct drainage for the boxes are placed in the open exposed to the often wet weather of late October and early winter.

Clean virgin loam to which has been mixed some coarse sand and moist peat should be used, the soil should preferably have been taken from a depth of 10 in. (25 cm) to ensure almost complete freedom from disease and weed seeds. The bulbs should be planted with the tops just below the soil level. The boxes are then placed outdoors over a bed of boiler clinker or similar material to ensure drainage and where they may receive some protection from strong winds. Over the boxes is placed a 4 in. (10 cm) covering of weathered ash, sand or soil to exclude the light and to protect the bulbs from drying out. They will rarely need any artificial watering, rain and heavy dews providing all that is necessary. They will remain in the open until the correct times for taking into heat or the cool-house. The covering of soil or ash is then knocked completely away, taking care not to injure the growing points of the bulbs which will be about 2 in. (5 cm) tall. The gradual introduction to heat and light is essential.

OUTDOOR

The tulip enjoys a heavier soil than most bulbs; this in no way means a sticky, badly-drained soil, but one which may be described as a well-drained heavy loam. Some sand or grit and a small quantity of peat or decayed manure will give it the necessary aid to good drainage. Where growing for profit in the open, earliness will be assisted by a lightening of the soil when some peat, sand and cow manure should be worked in. Should the soil be of a sandy nature, peat or leaf mould and some decayed manure will increase the humus and moisture content of the soil.

As tulips do not grow well in soil that has continually been used for tulips, the ground should first be deeply worked, bringing up to the surface the lower soil and it is to this that is added the various humus-forming ingredients. Tulips often being grown in large unheated houses and frames, the beds will require similar attention, but must be given ample water before the bulbs are planted in November. Where it can be obtained, a quantity of wood ash is most beneficial to tulips or a 1 oz. per sq. yd. dressing of sulphate of potash will be of equal value.

PLANTING IN OPEN GROUND

Tulips are unlike narcissus in that they should be given individual attention as to times of planting and correct depths. Many varieties seem to grow better from early November planting, yet 'Inglescombe Yellow' and 'Golden Harvest' both prefer to be planted in early October. Soil depths vary too, 'William Copeland' and 'William Pitt', both Darwin tulips, are happy in a soil depth of no more than 3 in. (7.5 cm); the cherry-red 'King George V' likes 4 in. (10 cm) of soil over it, and the bulb of larger proportions, Farncombe Sanders, a grand cold-house tulip, likes a 5 in. (12.5 cm) covering, whilst some growers plant it almost 8 in. (20 cm) deep. As a general rule 4 in. (10 cm) seems to suit most varieties, rather deeper in light soil, and shallower in clay soil. In the warmer parts of Britain where Tulip Fire disease occasionally makes its presence felt, the bulbs may be planted as late as early December, for deeper planting will help to keep the trouble at a minimum. Always plant with a wide trowel so that no air pocket is left beneath the bulb. They may be spaced 6 in. (15 cm) apart. When growing for cutting, the beds should be planted 5 ft. (1.5 m) wide with a path down either side to allow for ease in cutting.

A 4¼ in. (11 cm) size bulb will give a bloom of top size for outdoor planting, though for commercial cutting outdoors a 4 in. (10 cm) bulb is used.

Tulips bruise easily and care should be taken in handling the bulbs. They should be firm and clean, with the skin bright brown in colour. They must not be exposed to the elements at planting-time more than is necessary. After planting, the top soil is raked over to leave a bed with a fine tilth. Planting late in autumn will mean that little attention will be given the beds, possibly nothing more than hoeing between the bulbs in spring when growth begins.

Cloche and cold-frame culture calls for care in planting distances. The bulb may be set only 4 in. (10 cm) apart and in rows of the same distance, a barn-type cloche taking three rows. Late November planting is preferable for cloche work, for the plant should not become too tall too soon. If it is necessary to remove the cloches whilst the weather is cold, the cloches will have lost much of their value for uncovered plants will have almost caught up with those that began under glass.

PROPAGATION

After lifting, the tiny bulblets which should be examined for any form of disease, are then planted in prepared beds in early September. Plenty of sand and peat should be worked into the soil and the bulbs must not be allowed to suffer from lack of moisture. Planted 3 in. (7.5 cm) apart and 3 in. (7.5 cm) deep, they will remain in the nursery beds for two years and should be fed regularly with liquid manure water. The flower buds are removed when they form so that the energy of the plant is conserved for the bulb. They may then be lifted in the normal way and replanted in beds where they are to bloom.

SINGLE EARLY TULIPS

Growing to a height of less than 12 in. (30 cm), this section is especially useful for bedding, though it must be said that as most varieties are suitable for forcing they are more often used for this purpose than for outside flowering. They have little value as cut flowers for the blooms lack substance and are too short in the stem. Amongst the best varieties are:

GOLDEN MASCOT (4 in. 10 cm). Pure golden yellow.

IBIS (3 in. 7.5 cm). Peach pink, shaded carmine.

KEIZERSKROON (4 in. 10 cm). Striking red and yellow striped, slightly taller growing and later flowering.

PRINCE CARNIVAL (4 in. 10 cm). Yellow, flushed red.

PRINCE OF AUSTRIA (3 in. 7.5 cm). Pure orange and possessing a sweet perfume.

VAN DER NEER (3 in. 7.5 cm). Dwarf-growing and bearing a bloom of deep purple.

WHITE HAWK (4 in. 10 cm). Purest white.

These are all bedding varieties and not very suitable for forcing, with the exception of 'Prince of Austria', though they may be cool-grown in pots and will bloom from the end of March.

DOUBLE EARLY TULIPS

These are perhaps the most valuable of all tulips. They force well and will follow on the earlier flowering varieties of the singles. They may be cool-grown in pots when they will give a colourful display and they may also be used for the cut-flower trade as they grow to a height of about 14 in. (35 cm). Outside they remain long in bloom and are ideal for bedding. In this way, they look most attractive when planted with the April-flowering mossy saxifrages, whose daintiness acts as a contrast to the fullness of the tulip bloom. Or try the early April-flowering double daisies, especially the brilliant scarlet 'Rob Roy', the salmon-pink 'Dresden China', and the white counterpart, 'Alba'. The bulbs should be planted about 8 in. (20 cm) apart, slightly further apart than the single varieties. They may also be grown under frames for cutting throughout April and even earlier in favourable districts, and even without a greenhouse a display may be enjoyed almost uninterrupted from the late February flowering of the *kaufmanniana* species, to the completion of the late flowering Parrot Tulip 'Fantasy', which is the last of all the tulips to remain in bloom in northern gardens. Some brilliantly coloured varieties, all of which should be planted 3 in. (7.5 cm) deep are:

BONANZA. Deep carmine-pink, edged yellow. Taller growing than most doubles.

BOULE DE NEIGE. Very dwarf and bearing a large bloom of purest white.

DANTE. A new variety of intense fiery scarlet.

ELECTRA. A glorious shade of dark cherry red.

MR VAN DER HOEFF. Probably the best pure yellow.

MURILLO. White, tinted rose and possessing a sweet perfume.

PEACH BLOSSOM. Rich rose-pink.

TEA ROSE. Soft yellow, flushed pink and salmon. Though an older variety, still one of the best.

WILHELM KORDES. The colour is intense yellow, flushed orange-red.

WILLEMSOORD. A variety of recent introduction, a grand bedder, bearing bloom of a rich crimson shade.

DOUBLE LATE TULIPS

Coming into bloom towards the middle of May and flowering on stems 20 in. (50 cm) tall, this is a section that should be more widely used, for the bloom is valuable for cutting, whilst the flowering period will bridge the gap between the early flowering varieties and the later May flowering Darwin and Cottage tulips. They do not force well except the variety 'Clara Carder', which is more dwarf-growing and bears a bloom of an attractive shade of Tyrian purple. This is a useful tulip for following on the early flowering doubles under glass. Those recommended for bedding are:

COXA. Scarlet, tipped white and fairly dwarf.

NISSA. Rich yellow, shaded crimson, outstanding for bedding.
SYMPHONEA. Delicate shell pink. The double form of 'Pride of Haarlem'.

TRIUMPH TULIPS

This is a section of more recent introduction with the ability to withstand adverse weather when they come into bloom, which is during the last days of April. The blooms are brilliantly coloured and borne on strong, thick stems 16 in. (40 cm) tall. They are attractive when planted with the miniature polyanthus, 'Lady Greer' which comes into bloom at the same time. The dainty yellow 'Lady Greer' looks particularly delightful with the soft mauve 'Algiba'; and red polyanthuses make a pleasing contrast with the pure white 'Kansas'. The Triumph tulips are useful for the exposed garden, as a substitute for the taller Darwins; or the Darwins may be planted in a position where they will be partially sheltered from strong May winds, the Triumphs being used for a more open situation.

Of the varieties most suited to bedding other than those mentioned, the following are selected for their sturdy habit:

DENBOLA (3 in. 7.5 cm). Striking combination of deep crimson, edged cream. A new variety of great merit.

HINDENBERG (4 in. 10 cm). Recent introduction, the large blooms being of garnet red, edged yellow.

KORNEFORUS (4 in. 10 cm). A lovely variety of bright cherry red.

PRINCESS BEATRIX (3 in. 7.5 cm). Vivid scarlet, edged orange.

WINTER GOLD (3 in. 7.5 cm). A wonderful shade of deep lemon yellow.

ZIMMERMAN (5 in. 12.5 cm). Shell pink, flushed silver. Possibly the best April tulip for growing in frames or under cloches, but it does not like forcing conditions.

PARROT TULIPS

They are a distinct break from the Darwin range from which most are 'sports'. With their unusual fringed petals and vivid stripings, the Parrots should be more widely grown. The blooms are large and whilst most attractive as cut bloom, they are at their best planted in beds of separate varieties. Or to be more original, try a harlequin display. Flowering longer than any tulips, well into June, the Parrots will provide a display of the utmost brilliance when planted against a background of cupressus trees or to the front of a sheltered border. One or two varieties grow rather tall, the lilac-rose 'Discovery' is one, growing to a height of about 30 in. (75 cm) and the large blooms tend to fall over in wind and rain. None is lovelier than the first of the Parrots, 'Fantasy', a 'sport' from the old favourite 'Clara Butt', its salmon pink splashed with green; also the vivid 'Orange Favourite', which carries a rich perfume.

Others, which are planted 4 in. (10 cm) deep, are:

BLUE PARROT. Vivid purple with the petal edges particularly broken and fringed.

FIRE BIRD. A vivid scarlet 'sport' from 'Fantasy'.

RED CHAMPION. Rosy red, a variety which forces well.

SUNSHINE. A golden yellow of great beauty only 18 in. (45 cm) tall; is a grand bedding variety.

MENDEL TULIPS

The result of a cross between the early flowering singles and the Darwins, their flowering period falls between the two. They open their bloom in the open at the end of April, while several varieties may be forced indoors as early as the Early Flowering Singles. 'Early Queen' and 'King of the Reds' may be taken indoors the first week of December, followed by 'Orange Wonder' and 'Mrs E. H. Krelage', about the last day of the year, with 'Mozart' and 'Scarlet Admiral', towards the end of January. Other varieties can be brought on in pots in the home or cold greenhouse and will bloom during March and early April until the outdoor plantings are showing colour late in the month. Some of the best are:

HER GRACE. The brightest possible scarlet.

IMPERATOR. Old rose.

ORANGE WONDER. Rich orange, shading to a lighter edge.

PINK PICTURE. Ivory-white delicately margined with rose pink.

PIQUANTE. A striking new variety of crimson-red with a gold margin.

WHITE SAIL. Rich cream shading off to full white.

BREEDER TULIPS

For those preferring the more sedate colours, the bronze, coffee, port wine and Burgundy shades, this section will satisfy their tastes. Tallest growing of all tulips, they are best planted in beds against a wall or a wattle fence. They look splendid in small beds against the house and never more charming than when mixed with Darwins of a colour that will tone with each other. For instance, the real brown-coloured 'Don Pedro', which also possesses a noticeable fragrance, combines with the rich yellow Darwin tulip, 'Golden Age', and both grow to a height of 2 ft. (60 cm). Or try the orange and brown shaded 'Dillenburg', with the vermilion-coloured Darwin, 'City of Haarlem'. Other Breeder tulips of unusual colouring are:

CHERBOURG. Rich golden bronze, flushed purple.

J. J. BOWMAN. Crushed tomato red, edged gold.

PANORAMA. Produces a large crimson-red flower on only 18 in. (45 cm) stems and is ideal for an exposed position.

TANTALUS. Dullest yellow, flushed bronze and lilac.

The Breeders will not force well, but may be grown in a cold-house frame or under cloches for cutting. Plant 4 in. (10 cm) deep.

REMBRANDT TULIPS

A small section flowering during May and June on 2 ft. (60 cm) stems. The bloom is striped with contrasting colours. They will bloom well under conditions of fairly gentle forcing and will come into bloom in early March indoors. Arresting is 'Cordell Hull' of a rich red colour, feathered and striped white.

DARWIN TULIPS

This is the largest and most important section covering hundreds of varieties, many being forced in vast quantities to supply the early spring market, others being planted in beds by the million, both for outdoor cutting and display. They will in this way give a succession of bloom from early March until the end of June. Those grown under frames or cloches in favourable districts should be covered early in February and will be in bloom from mid-March. The amateur who has no greenhouse and who wishes to enjoy cut bloom in the home, should plant them against a wall or hurdle fence in a position of full sun. Then late in February rear a frame or Dutch light over the bulbs as soon as showing above the soil. Draught can be excluded from the ends by draping a sack and holding it into position by a large stone which will prevent it from flapping about. By this simple method, bloom will be ready for cutting from mid-April and lights can then be used for hardening-off plants from early May. Some of the loveliest tulips in this section with planting depths and correct month are:

ALLARD PIERSON (4 in. 10 cm Oct.). Rich crimson-maroon.

ARISTOCRAT (3 in. 7.5 cm Oct.). A lovely new Darwin of deep rose pink, silver at the edges.

CITY OF HAARLEM (3 in. 7.5 cm Nov.). Vivid vermilion with an attractive black base.

CLARA BUTT (4 in. 10 cm Nov.). Salmon rose and grown by the million for cold-frame culture, outdoor cutting and for bedding.

DEMETER (3 in. 7.5 cm Nov.). New, of a rich violet shade.

GOLDEN AGE (4 in. 10 cm Nov.). A superb yellow, the blooms carried on sturdy stems. Ideal for cloche work.

KING GEORGE V (4in. 10 cm Nov.). Large, tall-growing, cherry red, the best of its colour.

MRS GRULLEMANS (3 in. 7.5 cm Oct.). Rich creamy white, the bloom being of beautiful form.

NIPHETOS (3 in. 7.5 cm Nov.). A superb variety, the bloom being the colour of Jersey cream, flushed deeper yellow.

SULTAN (3 in. 7.5 cm Oct.). Of darkest maroon, almost black. Most attractive when planted with a white or cream variety.

VICTOIRE D'OLIVIERA (4 in. 10 cm Nov.). Magnificent deep crimson.

WILLIAM COPELAND (3 in. 7.5 cm Nov.). Pure lavender, the earliest forcing Darwin.

YELLOW GIANT (4 in. 10 cm Nov.). Bright buttercup yellow.

COTTAGE TULIPS

Also known as single late or May-flowering tulips. They are similar to the Darwins in every way and a number of varieties will readily force. Several varieties have an oblong-shaped bloom, others with almost pointed petals, while a few are more lily-flowered. Art shades, especially yellows, predominate and look delightful when two or three varieties are planted around a raised bed, those with a dwarf habit growing to the centre. A delightful display is to be obtained from a circular bed of Cottage tulips in which half a dozen varieties are used. In the centre is the late-flowering 'Caledonia', which bears a vivid scarlet flower on 18 in. (45 cm) stems. Next is 'Orange King', and then the attractively coloured 'Princess Margaret', with its rich yellow blooms, edged and shaded orange. Then the pale yellow 'Mother's Day'; the orange red 'Dido'; finally the deep buttercup 'Yellow Emperor'. For a yellow tulip to plant with the scarlet or violet-coloured Darwins, a large selection will be found in this section, ranging from the creamy yellow 'Ellen Willmott' to the deep golden yellow of 'Mrs Moon'. Others with their correct planting depths are:

ARGO (3 in. 7.5 cm Nov.). Golden yellow, splashed red.

BELLE JAUNE (4 in. 10 cm Oct.). Clear primrose, with a long bloom.

GOLDEN HARVEST (3 in. 7.5 cm Oct.). Lemon-yellow, the bloom being of largest proportions.

INGLESCOMBE YELLOW (4 in. 10 cm Oct.). The latest yellow.

MRS JOHN T. SCHEEPERS (4 in. 10 cm Nov.). Bright golden yellow.

VLAMMENSPEL (3 in. 7.5 cm Oct.). Rich yellow, shaded red.

SPECIES AND VARIETIES

T. australis. It produces its yellow, red-flushed, star-shaped blooms on 6 in. (15 cm) stems. When in the bud stage they nod in the wind like snowdrops and are at their best in April.

T. batalinii. A species for the alpine garden, producing in April its dainty flowers of soft creamy-yellow with bright golden centres. It rarely exceeds 4 in. (10 cm) in height, and its leaves trail on the ground.

T. biflora. A dainty species from the Caucasus, having narrow leaves and bearing from the main stems two to four starry creamy-white blooms shaded green on the exterior. 5 in. (12.5 cm).

T. chrysantha. This little tulip from the Himalayas blooms in May and is rich yellow on the inside, shaded cherry-red on the outside. 8 in. (20 cm).

T. dasystemon. A gem for the rock garden, bearing several flowers on each stem. They are palest yellow, shaded green and grey on the outside. It is April flowering and grows 6 in. (15 cm) tall.

T. eichleri. A brilliant April-flowering tulip for bedding or naturalising, with large scarlet blooms having a black centre surrounded with a golden band. It grows 10 in. (25 cm) high.

T. fosteriana 'Princeps'. It grows only 8 in. (20 cm) tall and is the most dwarf member of the group. The large scarlet blooms have a black and yellow centre and are at their best during April.

T. greigii. The blooms are the largest of tulips and are of glowing orange, blotched with maroon at the centre. The foliage, too, is spotted with maroon. April flowering. 9 in. (22 cm).

Of many lovely hybrid varieties, 'Plaisir' is outstanding, growing 9 in. (22 cm) tall, the pointed petals of vermilion-red being edged with gold. 'Red Riding Hood' bears a bloom of ox-blood red and has foliage striped with purple-brown. 'Oratoria' is another beauty, its large blooms being of an unusual shade of rosy-apricot.

T. hageri. A most beautiful tulip from Greece, with small, globular blooms of dull coppery-red shaded with olive green outside. It blooms in April. 9 in. (23 cm).

T. kaufmanniana. Native of the hilly regions of central Asia, it is known as the 'Water Lily Tulip' on account of the shape of the blooms when open (see plate 73). The petals are broad and the flowers open to an enormous size. They are the first of all the tulips to bloom, showing colour by mid-March. There are a number of excellent new varieties and hybrids of which 'Gaiety' is outstanding, the deep cream blooms, striped red on the outside almost resting on the leaves.

Another gem is 'Scarlet Elegance', the vivid scarlet flowers with golden base, held on 6 in. (15 cm) stems. Growing to a similar height is 'Shakespeare', the flowers an exquisite blending of apricot, salmon and orange. Extremely showy is 'Elliott', its white blooms with red markings on the outside, and most striking of all is the new hybrid 'Robert Schumann', the latest to bloom, its flowers of chrome-yellow flushed with rose on the exterior, the centre splashed with black blotches.

T. kolpakowskiana. A splendid tulip from Asia Minor flowering in April; its deep golden-yellow flowers, unusually long, being shaded with rose on the exterior and held on 9 in. (22 cm) stems.

T. linifolia. It bears during May, its tiny blooms of glowing scarlet with a blue centre, on 6 in. (15 cm) stems. Glaucous foliage.

T. maximowiczii. Though growing under 6 in. (15 cm), it has large flowers of shining deep scarlet, the petals having a dark blotch at the base. In flower during April.

T. persica. The 'Persian Tulip', charming for a trough garden or rockery. Fragrant blooms in May of deep yellow, shaded with bronze on the outside. It grows only 3 in. (7.5 cm) high.

T. primulina. Two or more flowers are produced from each bulb and are white, flushed with green and margined with rose. The glaucous foliage adds to its beauty. It grows 9 in. (22 cm) tall and blooms during April.

T. pulchella. It comes into bloom at the end of March, its long urn-shaped blooms, of a lovely shade of soft mauve with a pale yellow centre, opening flat in the sun. 4 in. (10 cm).

T. saxatilis. It must have a dry, warm position and a soil containing plenty of grit and sand, and the bulbs should be planted deeply. The delightful silvery lavender flowers are produced throughout May, and its broad glossy green foliage, unique in the genus, adds to its charm. 9 in. (22 cm).

T. tubergeniana. This is a most striking tulip which grows 10 in. (25 cm) high, its handsome scarlet flowers, with black blotches at the base, opening as wide as the flower grows tall. It blooms during May.

T. turkestanica. Similar to *T. biflora* in that three or four blooms appear from the main stem. It is one of the earliest to bloom. The cream-coloured flowers, attractively flushed with green and rose on the exterior, are borne on 6 in. (15 cm) stems during March.

TULIP TREE, *see* Liriodendron

TWINSPUR, *see* Diascia

TICKWEED, *see* Coreopsis.

ULEX

The Gorse or Whin, valuable for coastal planting and for a wind-swept situation. *U. europaeus*, the Common Gorse, grows 5 ft. (1.5 m) tall and makes a dense prickly shrub with its spiny scale-like foliage. It bears brilliant yellow pea-like flowers intermittently throughout the year and is a blaze of colour from February until June. The double form is even more colourful. *U. nanus* is the Dwarf Gorse, which grows 18 in. (45 cm) tall and is most free flowering, being in bloom from August until Christmas. It should be planted in a dry sun-baked soil, with the autumn flowering heathers. It is valuable for coastal planting to bind the soil.

Plant in November from pots otherwise it will prove difficult to establish. Long-established and straggling plants should be cut hard back in May. Propagate by cuttings removed with a 'heel' and inserted in sandy compost under glass.

ULMUS

Deciduous trees of stature, elms grow well in any soil (including chalk) with some depth. The Wych Elm, *U. montana* (syn: *U. glabra*) is one of the most successful trees for an exposed coastal garden. Propagate from seed or by suckers which are removed from the parent plant in autumn.

Ulmus carpinifolia forms a large pyramidal tree with small glossy leaves and is valuable for coastal gardens. The form 'Pendula' has slender drooping branches whilst 'Aurea', the Golden Elm has bright yellow leaves. The English Elm, *U. procera*, has leaves which turn yellow in autumn whilst the variety 'Vanhouttei' (Louis van Houtte) has leaves of pure gold colouring.

U. glabra is the Wych Elm which has leaves like the hazel with serrated edges and winged seed vessels. The variety 'Lutescens' has leaves of creamy yellow whilst 'Pendula' is the weeping form, making a flat-topped head with graceful pendulous branches.

UMBRELLA PLANT, *see* Cyperus

URSINIA

With their orange, daisy-like flowers and the bushy habit of the plants, the annual ursinias lend distinction to any border. They may be sown at the end of April where they are to bloom, or raised in heat for they will readily transplant. *U. anethoides* and its hybrids in shades of ruby and purple, bloom on 12 in. (30 cm) stems, with the orange-flowered 'Golden Bedder' rather more dwarf. Allow 10 in. (25 cm) between the plants.

VACCINIUM

Deciduous and evergreen shrubs requiring an acid soil and tolerant of shade. They have foliage which turns brilliant red in autumn whilst their large edible fruits are an additional attraction. *V. arctostaphylos*, native of the *Caucasus*, grows 7–8 ft. (about 2 m) tall with leaves 4 in. (10 cm) long whilst in June and again in September it bears waxy-white bell-shaped flowers in large racemes which are followed by shiny black egg-shaped fruits. Of more compact habit is *V. angustifolium*, a dwarf deciduous shrub which in June bears white flowers tinted with red followed by blue-black berries in autumn. Excellent for ground cover is *V. vitis-idaea*, or Mountain Cranberry, a creeping form which bears pale pink bells in June and red edible fruits in autumn.

Plant in March, in a well-drained soil containing plenty of peat. At this time of year, remove any dead wood. Propagate by division or by layering in autumn; or from cuttings removed with a 'heel' and inserted in sandy compost under glass.

VALLOTA

Producing its spikes of brilliant scarlet late in summer, the Scarborough Lily, as it is called, is a valuable plant for a cool greenhouse. The bulbs are planted into a compost similar to that enjoyed by the nerine. After flowering, the plants should be grown on throughout winter and early spring by keeping the compost slightly moist, but it must not be allowed to dry out. The bulbs do not like disturbing when once established in the pots. All they require is to be kept free from frost during winter which may be possible by transferring the pots to a room in the home. *V. purpurea* 'Major' is the true Scarborough Lily.

VARIEGATED LAUREL, *see* Aucuba

VASES, ORNAMENTAL

Ornamental vases were used to great effect during Victorian times and now that our gardening has to be done in more confined spaces, the vogue for vases is returning.

One of the most attractive ranges of vases and urns is that produced by Minster Ltd. of Ilminster, Somerset, made in a large variety of shapes and sizes. Cast in stone and hand-finished by craftsman masons, they are obtainable in three colours – the warm buff of Ham Hill stone, the cream of Doulting stone, and the familiar grey-white of Portland stone. Several designs follow the Roman and Tudor periods, whilst others are more modern in their presentation and are more accurately described as plant containers rather than vases or urns. They may be placed on a terrace or verandah, about a courtyard or near the entrance to a house, or they make the focal point of a lawn, adding dignity and charm to wherever they are placed.

To be successful, the vase should be at least 12 in. (30 cm) deep. A shallow vase will grow little, other than the spring flowering plants such as double daisies and primroses, which are shallow rooting, and which are confined to the vases only during the less warm and dry periods of the year. But as vases, like window-boxes, are used throughout the year, and will be at their best during summer and autumn, there must be a reasonable depth of soil and this must be prepared with as much care as is given to the compost for window-box or tub.

All too often vases are planted year after year without the soil, which is often from an old town garden, being replenished. Where situated in an industrial city, the compost will soon become acid due to the constant deposits of soot and sulphur. It should therefore be replaced at least in alternate years. Care must also be taken to make up the compost so that winter and spring flowering plants will receive ample drainage, and summer flowering plants will receive sufficient moisture. Especially is this important where one may be away from home for several weeks at a time.

Drainage is provided by placing crocks at the bottom of the vase to a depth of about 2 in. (5 cm). Over this the compost is placed and this should consist of fresh loam which will be free from deposits of soot and sulphur, and to which has been mixed 1 oz. of bone meal, some gritty sand and a small quantity of peat or leaf mould. The vase should be filled to the top, but before planting, allow the compost to settle down. After a few days it will be found to have sunk about ½ in. (1.25 cm) below the rim of the vase, which will allow for summer watering without splashing the soil over the sides.

William Robinson, the Victorian authority, has suggested that trailing plants, always at their best when used in a vase or hanging basket, may be left permanently in position planted round the side, for which purpose he suggested the ivies and periwinkles. Against this is the need for providing fresh soil every two or three years, and the fact that in a town garden the foliage will tend to become covered with soot deposits which will spoil the display.

The ivies are most attractive when used either in a vase or when grown from the ends of a window-box and trained round a small window. In this way the plants should be planted permanently, and the most satisfactory method is to make two small partitions at either end of the box, allowing sufficient room to accommodate a plant from a small pot. A compost enriched with a little manure should be provided, for with the addition it is surprising how much more vigorous growth will be. The plants may be grown up the wall without support, in fact only those plants which are able to support themselves should be planted. There is thus no need for any unsightly supports or for damage to the window surrounds. The ivy being evergreen is more suitable than the Virginian Creeper, but in sheltered positions in the south *Campsis radicans*, also self-clinging and bearing vivid scarlet trumpets in summer, is a glorious sight grown around a window, particularly where the walls are washed. In the south, too, the Passion Flower, *Passiflora caerulea*, may be used to garland a window, but it must be grown up canes which may be held in position by placing at an angle and fastening the top to the wall. Another delightful plant is the ivy-leaf geranium 'L'Elegante'. More like an ivy than a geranium, it makes rapid growth though it requires supporting. It will, however, climb to a height of 4–5 ft. (about 1.5 m) in a single season. It should be grown permanently in a pot, for it will not survive the winter outdoors. However, if taken indoors it may be used to cover a trellis, or the wall of a room and may be planted outside again in June. All the ivy leaf pelargoniums are suitable for a vase or urn and are most charming when planted with the silver leafed *Cineraria maritima* or with the pink flowering fuchsia, 'Fascination', with its elegant pendulous habit. Petunias and verbena and *Phlox drummondii* and the trailing lobelias will enhance the ivy-leaf geraniums and pendant begonias. Where growing in those districts noted for an equable climate, the geraniums may be left out the whole year round. Tulips and polyanthus may be used to provide the spring display.

VENETIAN SUMACH, *see* Cotinus

VENIDIUM

Native of South Africa, it is an annual and requires a light, sandy soil and a position of full sun. At their best in a dry soil, the large daisy-like blooms with their striking black centre remain constantly colourful through the latter part of summer and early autumn. For a small bed, the Dwarf Hybrids should be planted for they grow to a height of only 12 in. (30 cm). They may be obtained in a wide variety of colours; ivory, straw, yellow and orange and intermediate shades.

Taller growing, reaching a height of 2 ft. (60 cm) is *V. fastuosum*, the large, orange flowers held above the interesting silky-grey foliage.

The plants are raised in heat in March, hardened in the usual way and should be set out 9 in. (22 cm) apart at the end of May; a week later in the north.

VENUS'S LOOKING GLASS, *see* Specularia

VERBASCUM

With their thick felt-like leaves and bearing stately spires of bloom of the most intriguing shades of biscuit, apricot, yellow and pink, the mulleins are amongst the loveliest of back row border plants. Like the anchusa, where the soil is well-drained it is perennial, where of a heavy nature the plants tend to be short-lived though they may seed themselves each year. Several of the newer varieties, however, are so beautiful that they should be included in every border. Plant in spring, allowing 3–4 ft. (1 m) between each. The taller growing varieties will require staking in an exposed garden.

VARIETIES

BOADICEA. A striking variety, its coppery-orange flowers being borne on 5 ft. (1.5 m) stems.

BRIDAL BOUQUET. It grows to a height of 4–5 ft. (about 1.5 m) and is in bloom at the same time as the delphinium with which it should be planted. The blooms are of a delicate creamy-white, the anthers being tipped with orange.

C. L. ADAMS. The tallest growing variety, reaching a height of 6 ft. (2 m) or more and it comes later into bloom, prolonging the season through August. The great spires of bloom with their branching habit are of a deep golden colour.

COTSWOLD BEAUTY. Growing to a height of 3–4 ft. (about 1 m), it is in bloom from mid-June until mid-August, its biscuit-coloured blooms having attractive purple anthers.

GAINSBOROUGH. Of more compact habit than the others, for it reaches a height of only 3–4 ft. (about 1 m). Its graceful blooms are of a delicate primrose-yellow.

MISS WILLMOTT. A delightful companion to 'Boadicea', for its graceful flower spikes are pure white.

PINK DOMINO. A dainty variety, its soft peachy-pink blooms are borne on 3–4 ft. (about 1 m) stems.

VERBENA

This bedding plant is one of those tender perennials which is given half-hardy treatment. In the north there is a tendency for the plants to come late into bloom, being at their best in August and September. In the south the plants, from a February sowing in heat, come into bloom at the end of July. Like *Phlox drummondii*, the shoots may be pegged down to fill the bed with colour, their flat heads in red, pink, purple and white composed of numerous tiny flowers, each with a striking white eye, being most effective, especially when used to underplant a bed of gladioli. The plant enjoys a position of full sun, but likes a soil containing some moisture.

'Blaze' bears bold trusses 3 in. (7.6 cm) across, of brilliant scarlet, and 'Mme du Barry', bears deep purple-red flowers, a new colour in the verbena. The Olympia dwarf strain makes a compact plant 8 in. (20 cm) tall and bears flowers in all the rich verbena colours.

VERONIA

V. vicrinita blooms so late that it should be planted only in the most sheltered borders which receive the maximum of autumn sunshine. It has dark green leaves and comes into bloom during September, its large flat heads of reddish-purple being borne on tall, erect stems. It will remain in bloom in a favourable district well into November. The plants will flourish in ordinary garden soil and in a warm climate the plants may be grown in partial shade. It grows 5 ft. (1.5 m) tall.

VERONICA

ALPINE

Several species and varieties of this shrubby plant are suitable for the rockery either in full sun, in shade or on a northerly slope. There are species to bloom from May until November and so they are indispensable members of the rock garden. The first is the May flowering *V. rupestris* and its several varieties, the two loveliest of which are the sky-blue 'Silver Queen' and the deep shell pink 'Mrs Holt' which produce their spikes on 4 in. (10 cm) stems. Flowering at the same time is *V. cinerea* which bears spikes of vivid blue from its silver-green foliage. Plant with it the tiny shrublet *V.* 'Cranleigh Gem' which with its grey foliage and silver-white flowers produces a most charming effect when planted near red sandstone. For late summer flowering, plant *V. incana* with its violet spikes and silvery foliage, and the delightful pink form 'Rosea'.

BORDER

Amongst the finest of border plants are the veronicas, being hardy and tolerant of most soils though a well-drained sandy loam suits them best. By careful selection, plants may be obtained which bloom from early April until October and with their compact habit are ideal plants for a small border. They may be planted either in November or early in spring; those flowering early being planted 18 in. (45 cm) apart in November. Propagation is by division of the roots.

SPECIES AND VARIETIES OF BORDER VERONICAS

V. alata. Happy in ordinary soil and in a sunny position, it is the tallest of the border veronicas and bears its spikes of sky-blue flowers from July until well into autumn. Plant in November 2 ft. (60 cm) apart and increase by division of the roots. 4–5 ft. (about 1.5 m).

V. amethystina. It bears its dainty spikes of brilliant mid-blue on 12 in. (30 cm) stems during June and early July.

V. gentianoides. So called because of the gentian-blue colour of its flowers and the similarity of its leaves. The tall flower spikes reach a height of 2 ft. (60 cm) and are at their best from mid-April to early June.

BARCAROLLE. Its slender spikes of rich rose-pink reach a height of 12 in. (30 cm) and like all in this section, bloom during July and August, there being few plants of similar colour for the border at this time.

CRATER LAKE BLUE. A variety of beauty, growing 12 in. (30 cm) tall and bearing in May and June, a profusion of short spikes of true cornflower blue.

MINUET. It grows to a height of 18 in. (45 cm), its shell-pink sprays growing above its grey foliage producing a charming effect.

PAVANNE. Reaching a height of 2 ft. (60 cm), its flowers are of a bright shade of rose-pink which persist into September.

SHIRLEY BLUE. Its gentian-blue flowers are borne on 12 in. (30 cm) stems throughout summer.

VERONICA, *see also* Hebe

VIRBURNUM

The Winter Viburnum, *V. tinus*, is an evergreen for a sunless garden. It has large oval glossy green leaves and bears trusses of white flowers from October until March. Also evergreen but in bloom during May, when it bears fragrant white flowers, is *V. chenaulti*, which makes a small compact plant.

One of the best deciduous forms is *V. fragrans*, a compact plant of vigorous upright growth, which bears its fragrant pink and white flowers from Christmas until the end of March. A hybrid of this species is *V. bodnantense*, which also bears pink and white flowers throughout winter. Do no pruning, apart from the removal of dead branches.

V. burkwoodii is also valuable. It is semi-evergreen and bears its large fragrant white flowers during spring, whilst *V. opulus*, the 'Guelder Rose', flowers in May, followed by red translucent berries and brilliant coloured foliage in autumn. The form 'Sterile', the Snowball tree is one of the most popular of summer-flowering shrubs.

Plant in November in a well-drained loam and keep in shape by cutting out the dead wood after flowering. Propagate by layering in autumn or by cuttings of half ripened wood, rooted under glass.

VINCA

Trailing evergreen plants which will flourish in all soils and in full sun or partial shade. *V. major* is of vigorous habit and will quickly cover an unsightly bank, bearing its bright blue flowers from June to September.

It will bear few flowers in shade and under such conditions, the variety 'Variegata' should be planted for its dark green leaves margined with cream will add brightness to a dull corner.

V. minor, the Lesser Periwinkle, is more dwarf, making a compact mound. The variety 'Azurea Flore Plena' bears double flowers of sky blue whilst 'Alba' bears pure white flowers. A number of hybrids are suitable for bedding and for a window-box, whilst they also do well in pots in a cool greenhouse. 'Little Pinkie' bears rose-red flowers and grows 6 in. (15 cm) tall whilst 'Little Bright Eye' bears white flowers with a pink eye.

Plant in November, 12 in. (30 cm) apart, into a well-drained soil containing leaf mould. Prune *V. major* by cutting back after flowering to encourage the plant to form new wood at the base. Propagate by root division or by cuttings of the half-ripened wood inserted under glass whilst the dwarf vincas are readily raised from seed.

VINE, *see* Vitis

VIOLA

In 1867, James Grieve of the Scottish firm of Dicksons of Edinburgh, after whom the famous apple was named, began to improve the habit of the pansy to make it more suitable for bedding. For his work he was to become 'the father of the bedding viola' as we know it today. James Grieve was to make use of as many species as possible in his effort to produce a pansy of more compact habit. He collected plants of *Viola lutea* on the Pentland Hills; *V. amoena* at Moffat; and *V. cornuta,* which had been introduced in 1776 from Spain, especially the variety 'Perfection' which had been raised by Mr B. S. Williams, gardener at Rotherfield Park.

To each of the numerous species, James Grieve crossed the most modern of the Show and Fancy pansies, taking pollen from the pansies and applying it to the species in turn. Pollen of a dark Show pansy applied to *V. cornuta* 'Perfection' produced violas 'Tory' and 'Vanguard', whilst applied to *V. striata* it gave that still popular yellow bedding viola 'Bullion', which was rightly hailed at that time as an introduction of enormous value, though the bloom by modern standards is not of the best. From *V. lutea* he obtained 'Golden Gem' and *V. grievii,* two fine yellows. With their compact habit and forming a mass of fibrous roots, Grieve's introductions were first known as tufted pansies, later violas.

EXHIBITION VARIETIES

ADA JACKSON. A new colour break. The large white blooms have an edge of rosy-mauve on the three lower petals, the upper petals being suffused rosy-mauve.

ANDREW JACKSON. The refined bloom is of a rich shade of purple, striped with amethyst and pale mauve.

A. S. FRATER. The large blooms are creamy-white margined with mauve.

BARBARA BENNETT. A crimson purple self of refined form.

C. S. ROBERTSON. A pure cream self of excellent form.

DOUGLAS UPTON. The rose-pink blooms are striped with crimson.

ELIZABETH WILLIAMS. The blooms are deep mauve with a cream centre.

HELENE COCHRANE. Possibly the finest pure white ever raised.

H. H. HODGE. A fine variety having a lemon ground with a wire edge of lavender.

JOHN ADAMSON. A pure golden-yellow of great size.

LADY TENNYSON. An outstanding pure white.

LOIS MILNER. A lovely yellow, edged bright blue, the two upper petals are being suffused with blue.

MARY ANDERSON. An outstanding viola, the pale sulphur-yellow blooms are edged with mauve.

MAY CHEETHAM. A variety of superb form and likely to be the most popular viola of its colour on the show bench for many years. The huge bloom is pure primrose-yellow, rayless and of great substance.

MAY JACKSON. The primrose-yellow lower petals are edged with mauve, the upper petals suffused-mauve.

MILNER'S FANCY. The bloom is most striking, being large and of deep purple-red, striped with rose.

MOSELEY PERFECTION. The beautifully formed golden-yellow blooms are entirely rayless.

MRS A. BLEARS. The pale cream blooms have a wire edge of purple.

MRS A. COCHRANE. The large blooms have a cream centre and are edged and suffused pale lavender.

MRS J. H. LITTLE. The primrose-yellow blooms are heavily banded with an unusual slate-blue colour.

MRS J. ROBERTSON. The sulphur-yellow blooms have an attractive picotee edge of lavender. The top petals are also pencilled lavender.

MRS M. WALLACE. Unusual in that the bloom is lavender, speckled or marbled with purple.

MRS T. BATES. A fine pure yellow self of lovely form.

PICKERING BLUE. One of the most popular of all violas for exhibition and bedding. The large, sweetly scented blooms are of deep sky-blue.

R. N. DENBY. An outstanding variety, the pale lemon yellow bloom being edged with pale blue.

SUE STEVENSON. The large blooms are of a rich shade of violet with a large clear yellow centre.

SUSAN. A seedling from the popular 'Ada Jackson'. The white bloom has a margin of mid-blue on the lower petals, the top petals being suffused blue.

WILLIAM JACKSON. The refined blooms are brilliant golden-yellow with a picotee edge of white.

BEDDING VARIETIES

ADMIRAL OF THE BLUES. A fine old exhibition variety, now used chiefly for bedding. The mid-blue flowers suffused with crimson and a striking yellow eye make this one of the best violas ever introduced.

AMY BARR. A fine bedding variety. The blooms are deep pink with a white centre.

ARABELLA. A beautiful viola the bloom being pale mauve deepening to violet at the edges.

BARBARA. A colourful variety, the top petals being mauve-blue, the side petals mauve-pink, the lower petals yellow, edged with mauve.

BLUESTONE. An old variety of compact habit. The clear mid-blue colour of the medium-sized blooms with their striking golden eye surrounded by a small purple blotch make it unbeatable for bedding. Very long in bloom.

BRENDA RUSHWORTH. An unusual variety, the large lemon-yellow blooms being flushed with lilac at the edges.

BULLION. Almost a hundred years old and for its freedom of flowering still widely planted. Brilliant golden-yellow and rayed. Very compact and early to bloom.

CRIMSON BEDDER. The blooms are of rich crimson-purple.

DOBBIE'S BRONZE. Probably the best bronze for bedding. The blooms are bright, with a terracotta flush, and with a bronze blotch at the centre.

MAGGIE MOTT. Long and free flowering, it bears silvery-mauve blooms of great beauty. One of the best ever introduced.

V. CORNUTA

Known as the Alpine or Horned Violet of the Pyrenees, it was introduced into England in 1776 and was used in the breeding of the garden viola and the violettas. A fibrous rooted plant of short tufted habit with heart-shaped toothed leaves, it blooms from May until August, the pale blue flowers having an awl-shaped spur. There is also a white form, 'Alba', both of which bear sweetly scented flowers.

Using the tufted *V. cornuta* as the seed-bearing parent and by crossing this with a pansy called 'Blue King', Dr Stuart of Chairnside obtained a strain of rayless violas which he called violettas. The plants have the tufty habit as *V. cornuta*, yet the blooms are larger, being longer in shape, quite rayless and possessing a delicious fragrance, like that of vanilla.

D. B. Crane continued with violettas where Dr Stuart had left off, and amongst his finest introductions were 'Diana', with its lovely clear primrose-yellow blooms; 'Eileen', pale blue with a gold eye; and the blue and white 'Winifred Phillips'. Though longer flowering than any of the pansies and violas, and being of more perennial habit it is surprising that these charming plants have never become as popular as the ordinary bedding violas. They come into bloom in April, with the primroses and continue until the frosts, the dainty flowers hovering like butterflies about the neat dark green foliage. On a year-old plant, as many as fifty-six blooms have been counted at one time. Also, with their fibrous roots and tufted habit the plants present no difficulty in their propagation, being lifted and divided in March like any other herbaceous plant whilst they are extremely long lasting. Above all is their delicious vanilla perfume.

VARIETIES OF V. CORNUTA

ADMIRATION. The larger than usual flowers are of a rich shade of purple-violet.

BABY GRANDE. The oval-shaped blooms are pale crimson-pink.

BUTTERCUP. One of the loveliest of all the violettas, the oval blooms being of a rich orange-yellow colour.

GERTRUDE JEKYLL. A lovely bi-colour with the upper petals pale primrose; the lower petals golden-yellow.

HEATHER BELL. The blooms are of rich mauve-pink.

IDEN GEM. The dark blue flowers are held on long stems.

JERSEY GEM. Also known as Blue Gem. The dainty flowers are of deep aniline-blue.

LE GRANDEUR. The blooms, which are rather larger than usual, are of a lovely shade of mid-blue.

LITTLE DAVID. The cream-coloured blooms held well above the foliage possess the rich fragrance of freesias.

LORNA. The blooms are of a lovely shade of deep lavender-blue.

LYRIC. An original variety, the pale lilac blooms being attractively marbled with lavender-mauve.

MAUVE QUEEN. The flowers are reddish-violet borne in profusion.

MRS GRIMSHAW. A recent introduction, bearing blooms of a lovely shade of rose-pink.

VIOLETTA. The original hybrid and still obtainable. The blooms are white-suffused-yellow on the lower petals, and are produced over a long period.

V. ODORATA

Included in this genus of annual or perennial plants are many of which must be numbered amongst the most delightful of all plants to be found growing in the wild and in the garden. Mostly of the temperate regions they are to be found across North America and in north Europe, extending from the British Isles to the Urals, inhabiting those countries with a cool climate, for with but one or two exceptions they are intolerant of hot, dry conditions. Even where growing under cold conditions the plants will seek the coolness and moisture of the hedgerows and woodlands and crevices between rocks, for they must have moisture at their roots. Many of the species of viola are of tufted or creeping habit, the dainty flowers being held above the foliage on thin wiry stems. In their native haunts, the plants remain inconspicuous, like the violet itself, requiring careful search and for this reason are all the more appreciated. They also possess the charming habit of blooming at the most unexpected times, generally when there is little other bloom in the countryside and garden. The flowers are cleistogamic, i.e., the first to appear do not set seed. Seed is set by the smaller blooms which appear later. They are not scented and are self-pollinating whilst the plants bury their own seed-capsules. The earlier flowers are insect pollinated. drawn by the sweet perfume in their search for nectar, which is stored in the spur formed by the front or lower petal.

Most important of the species is *V. Odorata*, the Sweet Violet, which long before the birth of Christ was in commercial cultivation, for the blooms were used by the ancient Greeks for sweetening purposes, and it was so highly regarded that it became the symbol of ancient Athens.

The blooms, dried and crystallized have been used for cake decoration and as a sweetmeat since medieval times. Today they are used to decorate chocolates usually containing a violet-flavoured cream. Candied violets are made by dipping the flower heads in a solution of gum arabic and rose water, then sprinkling with fine sugar. They are placed in a slightly warm oven to dry.

V. odorata is propagated by runners, the first of which will have rooted by the end of April. Only those which have formed a 'rosette' or cluster of leaves should be used. The removal of the runners continues throughout summer.

The plants may also be increased by division of the crowns, a method usually adopted by the gardeners of old. The double flowering varieties especially lend themselves to this method for they form tight clumps of numerous crowns.

Usually to be found in their natural state growing in the filtered shade of deciduous woodlands, or on grassy banks canopied by hedgerow plants, similar conditions should be provided for those to be grown in the garden and an old orchard is ideal. Here the plants will receive the fullness of the early spring sunshine yet will be protected from the heat of the sun in summer.

April is the most suitable time to make a planting, small raised beds being most suitable, setting out the runners 12 x 12 in. (30 x 30 cm) apart in each direction.

The plants enjoy best a soil containing ample supplies of humus and one giving a slightly acid reaction (like the strawberry). Peat is a valuable source of humus, whilst hop manure and clearings from ditches is also of value. In addition, some decayed farmyard manure should be incorporated but where this is unobtainable, give a dressing of 2 oz. per sq. yd. of bone meal, together with 1 oz. of super-phosphate of lime and 1 oz. of sulphate of potash which should be raked into the ground just prior to planting.

Plant when the soil is in a moist but friable condition, placing the crowns level with the surface of the soil. Keeping the soil stirred with the hoe and in a moist condition during summer, together with an occasional application of dilute manure water will ensure a plant capable of blooming profusely the following year.

VARIETIES OF V. ODORATA

Of many lovely garden varieties, 'Admiral Avellan' with its blooms of reddish-purple is one of the most sweetly scented; likewise 'John Raddenbury' with its blooms of brilliant china-blue. 'Rawson's White' is equally strongly scented; also 'Sulphurea', the only yellow violet, the pale yellow blooms being flushed with apricot. These are single varieties. In double flowering, outstanding is 'Countess of Shaftesbury', the lavender flowers having a rose-pink centre, whilst 'Marie Louise' also bears lavender-blue flowers but with a white centre. Exquisite too, is 'Mrs J. J. Astor', the markedly double blooms being of an unusual shade of rose-pink.

V. TRICOLOR (PANSY)

The pansy may be divided into two classes: the Show or Scottish Pansy, and the Fancy or Continental Pansy.

The show pansy was established by 1841 with the formation of the first Pansy Society whose definition of an exhibition variety was that the bloom should have 'a white or cream ground to the lower petals, with the upper petals of the same colour'. Since that time, however, the pansy has been greatly developed so that the show pansy may again be sub-divided into: Belted or Margined bi-colour pansies, and Self colours.

The blooms should be thick, smooth and circular, with no waviness in the petals, whilst they should possess a glossy, velvet-like appearance. The face of the bloom should be slightly arched or convex, with a small eye. The two centre petals should meet above the eye and reach well up the top petals. The lower petal should be deep and broad, to balance the others and each should lie evenly upon each other. To elaborate further, the top of the lower petal should be straight and horizontal, with the two centre petals arranged evenly on either side of an imaginary perpendicular line drawn through the eye. The tops of these petals should reach to the same height on the upper petals so that the whole of the bloom is evenly balanced. As to the correct marking of the Margined or Two-colour varieties, the ground colour should be the same throughout, with the margin well defined and of uniform width, and the same colour as that of the two upper petals. The blotches should be dense, solid and as near circular as possible. The eye should be well defined and circular and be of a bright golden yellow colour. It is important that there is no suffusion of colour of the margin into the ground colour.

The Self colours may again be sub-divided into light and dark self-coloured varieties, where the upper and lower petals are of the same dark colour, free of any blotch. The same remarks as to smoothness of petal, form, and texture of bloom apply equally with the selfs as with the belted bi-colours.

Here is a selection of the true Show pansies which should be grown where open competition is desired. The plants, however, have not the habit which makes them suitable for bedding and where this is required, the modern hybrid strains raised from seed will prove more satisfactory in this respect (see plates 74–6).

VARIETIES OF SHOW PANSIES

ALICE RUTHERFORD. A deep golden yellow self of great beauty.

BLUE BELL. A dark blue self colour of great beauty.

CHARLES MC'CRERIE. A belted variety, the ground colour is pale sulphur, with a margin of violet.

GOLDEN GIFT. The ground is rich yellow, the belting and upper petals being of violet.

JAMES FERGUSON. The ground is of rich Jersey cream colouring, the belting and upper petals being of violet.

JAMES GRAME. The ground is deep creamy-yellow, the belting and upper petals being deep purple.

JAMES RUTHERFORD. This is a pure yellow self.

JAMES THOM. An old favourite having a bright yellow ground, the belting and upper petals being of a chocolate-violet colour.

MAGGIE MYERS. A pure yellow self of fine form.

MRS PETER THOMSON. The ground is primrose-yellow with the margin and upper petals being rosy-purple.

As with the Show pansies, the Fancy or Belgian pansies should bear a circular bloom with smooth, velvet-like petals, lying evenly over each other. It is in this respect that the Exhibition pansies made most headway during the nineteenth century. The earliest introductions had a ragged appearance, the petals being spaced apart whilst the blooms were oval or long in shape rather than circular, which is the hallmark of the Exhibition pansy. The two centre petals should reach well up on the two upper petals, whilst the lower petal should be broad and deep to give a balanced or circular effect to the bloom. There is no definition as to colourings. The blotch of violet or chocolate colour should almost cover the whole of the three lower petals with the exception of a wide margin which may be of any colour or of more than one colour. The top petals need not be the same colour as that of the margin of the lower petals and may be rose, cream, gold, purple or intermediate shades. The eye should be bright yellow and clearly defined.

VARIETIES OF FANCY PANSIES

ADAM WHITE. The large blotches are of chocolate, the edges and top petals being of golden-yellow, suffused with violet.

ALDERLEY. A Jackson introduction, the refined blooms having large circular blotches of deep plum, the lower petals being margined rosy-pink, the upper petals are white, flushed with purple and edged white.

ALEX LISTER. The dense blotches are of chocolate, the edges rose and cream with the top petals rosy-purple.

ANNIE LISTER. The bloom has a large almost black blotch and a margin of cream. The upper petals are of violet-mauve also with a broad margin of cream.

CATHERINE. The blotches are violet, the edges ruby with the upper petals a rich shade of purple.

DR MCKINNON. The blotches are violet, the petals being edged with cream, with the upper petals cream and rosy-violet.

ENA WHITELAW. The blotches are chocolate, the petals being edged with gold with the upper petals soft purple.

ERNEST CHEETHAM. A magnificent new variety of great size with deep plum blotches and a narrow edging of white round the lower petals; the upper petals being plum-coloured.

MRS A. B. COCHRANE. A huge bloom with purple blotches and cream margins; the upper petals being purple and cream.

MRS CAMPBELL. A grand yellow self with huge circular claret-coloured blotches.

NEILL MC'COLL. The blotches are plum, the lower petals being margined with rose and white. The top petals are cream and purple.

T. B. COCHRANE. The blotches are dark blue, the lower petals being margined white with the top petals purple and white.

VIOLET, *see* Viola

VIPER'S BUGLOSS, *see* Echium

VIRGINIA COWSLIP, *see* Dodecatheon

VIRGINIAN CREEPER, *see* Ampelopsis

VIRGINIA STOCK, *see* Malcomia

VISCARIA

Any garden soil suits this annual but it does like an open situation. *V. oculata* and its hybrids reach a height of 15 in. (38 cm) and may be obtained in shades of blue, mauve and deep pink, whilst the Tom Thumb hybrids in white, rose and blue, flower on only 6 in. (15 cm) stems. Seed is sown early in May and the plants thinned to 6 in. (15 cm) apart.

VISCUM

Mistletoe is *Viscum album*, a parasitic herb which bears tiny green flowers during April and May which are followed by semi-transparent berries which become ripe by the year end. It fixes itself to the branches of certain host trees, usually as undigested seed left by birds but it may be planted artificially by pressing the seeds into cracks in the bark, preferably of apple trees when the sticky juice surrounding the seed will enable the seed to remain in position and to germinate. To prevent birds taking the seed, a mixture of soil and cow manure made moist, should be smeared over the seed when 'set'.

In spring, the seedlings will send out yellowish branches shaped like the antlers of a stag. It is mostly found in apple orchards of Somerset, Herefordshire and Worcestershire, situated in England's western side which are the chief suppliers of cut mistletoe for Christmas festivities. The plant is mostly to be found on neglected apple trees so often associated with cottage gardens.

VITIS

ORNAMENTAL

The Vitis, the ornamental vines which will twine about a trellis, rustic wall, or over a wall or fence, supported by wires. Their value lies in the rich and interesting colourings of their foliage, especially in autumn, for their flowers are insignificant. They are deciduous and like some protection from cold winds, hence their value for planting against a west wall, though in the south they are happy on a northerly wall. Whilst it will be advisable to keep the wall of a house for more exotic plants, the vines are excellent for covering a garden wall, against which they make rapid growth from a moist, deeply worked soil. They must have moisture, as they make only restricted growth in a dry, sandy soil. In a moist soil they will grow 10 ft. (3 m) or more in two years.

Plant towards the end of November, but first work some decayed compost and some peat or leaf mould into the soil – a town garden should be given a dusting with lime. Pruning will consist of shortening back the shoots to about a third of their new growth each year, and support should be given by tying the shoots to rustic poles or twining them over wires.

With its large, heart-shaped leaves, which turn a rich crimson in autumn, *Vitis coignetiae* will quickly cover a large wall. *V. henryana* is closely allied to the Virginian Creepers and is self-clinging; its leaves are of a dark velvety green veined with silver and pink, and they, too, turn vivid crimson in autumn. *V. purpurea* has a purple-red leaf and small black fruits at the end of summer. This is the Wild Vine, which figures prominently in ancient sculptures. Equally charming is 'Miller's Burgundy', the leaves of which are heavily dusted with white powder, and this, too, bears clusters of small black grapes. *V. vinifera* 'Brandt' has orange-coloured leaves in autumn and bears quite large blue-black grapes which make delicious wine. Attractive, too, is the Parsley-Leaf Vine, the large pale green leaves having the same indentations as the ordinary Grape Vine. All are extremely useful about a garden where there is a large wall or fence to cover as quickly as possible.

WALLFLOWER, *see* Cheiranthus

WALL PENNYWORT, *see* Cotyledon

WALL-PEPPER, *see* Sedum

WALL PLANTS

TIMES OF FLOWERING

JANUARY
Chimonanthus fragrans
Garrya elliptica
Jasminium nudiflorum

FEBRUARY
Azara gilliesii
Chimonanthus fragrans
Forsythia giraldi
Garrya elliptica
Jasminum nudiflorum

MARCH
Azara gilliesii
Berberis aquifolium
Clematis armandii
Forsythia intermedia
Forsythia suspensa
Fothergilla gardenii
Fothergilla major

APRIL
Berberis aquifolium
Clematis armandii
Clematis macropetala
Cydonia japonica
Ercilla nolubilis
Escallonia 'Donard Brilliance'
Forsythia intermedia
Forsythia suspensa
Fothergilla gardenii
Fothergilla major

MAY
Aegle sepiaria
Akebia quinata
Berberis pruinosa
Ceanothus dentatus
Clematis macropetala
Clematis montana
Clianthus puniceus
Convolvulus cneorum (until Sept.)
Cydonia japonica
Decumaria barbara
Enkianthus campanulatus
Fremontia californica (until Nov.)
Kolkwitzia amabilis
Lathyrus latifolius (until Sept.)
Lonicera americana (until Sept.)
Lonicera caprifolium
Lonicera periclymenum
Lonicera syringantha
Lonicera tragophylla (until Oct.)
Photinia beauverdiana
Photinia davidsoniae
Piptanthus nepalensis
Tricuspidaria lanceolata
Wistaria floribunda
Wistaria sinensis

JUNE
Abelia floribunda
Abutilon vitifolium
Actinidia chinensis
Akebia quinata
Aristolochia sipho
Carpenteria californica
Ceanothus cyaneus
Ceanothus veitchianus
Clematis hybrids
Cotoneaster species
Escallonias
Jasminum officinale (until Oct.)
Jasminum revolutum
Lonicera caprifolium
Lonicera japonica
Lonicera periclymenum
Magnolia delavayi
Phygelius capensis (until Oct.)
Piptanthus nepalensis

TIMES OF FLOWERING (*contd.*)
Pyracantha angustifolia
Rose, in variety (until Sept.)
Tricuspidaria lanceolata
Wistaria floribunda
Wistaria sinensis

JULY
Abelia grandiflora
Abutilon vitifolium
Berberidopsis corallina
Campsis grandiflora
Carpenteria californica
Ceanothus thyrsiflorus
Clematis hybrids
Clematis tangutica (until Nov.)
Cytisus battandieri
Jasminum polyanthum
Jasminum revolutum
Lonicera brownii
Lonicera japonica
Magnolia grandiflora
Myrtus communis
Pyracantha coccinea
Pyracantha gibbsii
Solanum jasminoides (until Oct.)
Tricuspidaria lanceolata

AUGUST
Abelia grandiflora
Aloysia citriodora
Berberidopsis corallina
Buddleia davidii
Campsis grandiflora
Campsis radicans
Ceanothus burkwoodii
Ceanothus 'Charles Detriche'
Ceanothus 'Gloire de Versailles'
Ceanothus thyrsiflorus
Clematis hybrids
Cytisus battandieri
Desfontainea spinosa
Jasminum revolutum
Leycesteria formosa
Lonicera brownii
Lonicera japonica
Magnolia grandiflora
Polygonum baldschuanicum
Punica granatum
Rubus thyrsoideus

SEPTEMBER
Abelia grandiflora
Berberidopsis corallina
Buddleia davidii
Campsis grandiflora
Campsis radicans
Ceanothus burkwoodii
Ceanothus 'Charles Detriche'
Ceanothus 'Gloire de Versailles'
Clematis hybrids
Colquhounia coccinea
Cytisus battandieri
Desfontainea spinosa
Leycesteria formosa
Lonicera brownii
Polygonum baldschuanicum
Punica granatum
Rubus thyrsoideus
Tricuspidaria dependens

OCTOBER
Ceanothus burkwoodii
Ceanothus 'Gloire de Versailles'
Clematis jackmanii
Clematis lanuginosa
Clematis viticella
Colquhounia coccinea
Fremontia californica
Solanum jasminoides
Tricuspidaria dependens

NOVEMBER
Clematis balearica (until March)
Clematis viticella
Fremontia californica

DECEMBER
Chimonanthus fragrans
Jasminum nudiflorum

EVERGREEN WALL PLANTS
* denotes semi-evergreen

Abelia floribunda
Abelia grandiflora
**Akebia quinata*
Ampelopsis striata
Azara gilliesii
Bellardiera longiflora
Berberidopsis corallina
Berberis darwinii
Berberis pruinosa
Camellia japonica
Carpenteria californica
Ceanothus cyaneus
Ceanothus veitchianus
Choisya ternata
Clematis armandii
Clematis balearica
Clianthus puniceus
Convolvulus cneorum
Cotoneaster exburyensis
Cotoneaster henryana
**Decumaria barbara*
Desfontainea spinosa
Ercilla volubilis
Escallonias
**Fremontia californica*
Garrya elliptica
Hedera species
**Jasminum officinale*
Lonicera americana
Magnolia delavayi
Magnolia grandiflora
Myrtus communis
Photinia beauverdiana
Phygelius capensis
**Piptanthus nepalensis*
Pyracanthas
**Rosa*, 'Felicité et Perpetue'
**Solanum crispum*
**Solanum jasminoides*
Viburnum tinus

POSSESSING PERFUME

Abelia chinensis
Aegle sepiaria
Akebia quinata
Aloysia citriodora
Azara gilliesii
Carpenteria californica
Chimonanthus fragrans
Choisya ternata
Clematis flammula
Cytisus battandieri
Elaeagnus pungens
Forsythia giraldiana
Fothergilla gardenii
Fothergilla major
Jasminum nudiflorum
Jasminum officinale
Jasminum polyanthum
Jasminum revolutum
Lonicera americana
Lonicera brownii
Lonicera caprifolium
Lonicera japonica
Lonicera periclymenum
Magnolia delavayi
Magnolia grandiflora
Myrtus communis
Rhynchospermum jasminoides
Rosa, 'Crimson Glory'
Rosa, 'Ena Harkness'
Rosa, 'Guinée'
Rosa, 'Lady Hillingdon'
Rosa, 'Mme G. Staechelin'
Rosa, 'Souvenir de Claudius Denoyel'
Rose, 'William A. Richardson'
Rose, 'Zéphirine Drouhin'
Viburnum tinus

BEARING WINTER FRUIT

Aegle sepiaria
Akebia quinata
Berberis species
Clematis macropetala
Cotoneaster species
Elaeagnus pungens
Lonicera (deciduous species)
Photinia beaverdiana
Photinia davidsoniae
Pyracantha species
Rose, climbing, 'Cupid'
Vitis 'Miller's Burgundy'
Vitis purpurea
Vitis vinifera

SELF-CLINGING

Ampelopsis
Campsis
Decumaria
Hedera
Vitis henryiana

COLOURFUL AUTUMN FOLIAGE

Ampelopsis quinquefolia
Ampelopsis veitchii
Berberis thunbergii 'Atropurpureum'
Celastrus articulatus
Clematis balearica
Cotoneaster simonsi
Enkianthus campanulatus
Fothergilla gardenii
Rubus biflorus (stems)
Rubus phoenicolasius (stems)
Vitis coignetiae
Vitis purpurea
Vitis vinifera

VARIEGATED FOLIAGE

Elaeagnus pungens
Hedera angularis 'Aurea'
Hedera dentata 'Variegata'
Hedera 'Silver Queen'
Vitis henryana

FOR LIME-FREE OR ACID SOILS

Berberidopsis
Camellia
Enkianthus
Fothergilla
Magnolia
Tricuspidaria

WANDERING JEW, *see* Tradescantia and Zebrina

WAND FLOWER, *see* Dierama

WATER GARDEN

This is one of the most satisfying of all forms of gardening and whilst few gardens have a pond that is a natural feature of the landscape, the smallest of gardens may have their pools constructed from fibreglass, concrete or plastolene whilst a delightful water garden may be made in an oak tub. The use of a tub is the most inexpensive and simple of methods. It may be sited as a centre-piece and surrounded by grass or crazy paving or it may be placed in the corner of a garden where it may be readily reached by a path. It should be away from tall trees for leaves are a nuisance. It will be necessary to obtain a tub which is water-tight and it should be re-inforced by lining it with a sheet of heavy-grade polythene.

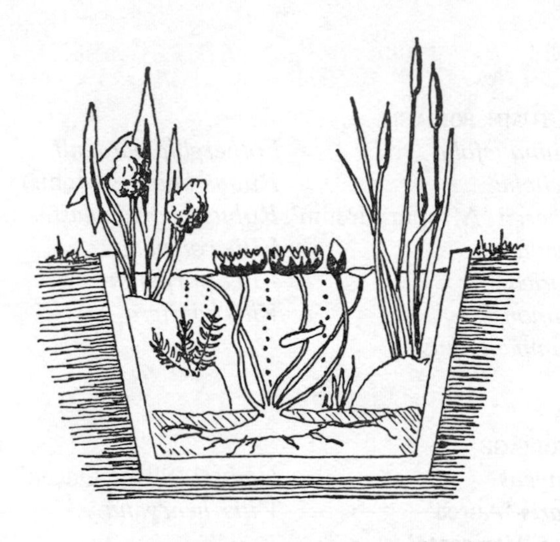

FIG. 17 *Water garden in a tub*

The tub should be inserted in the ground with its rim 2 in. (5 cm) above ground level to allow for paving stone to be laid right up to it and which will prevent soil from splashing over the side of the tub. Where using stone, set it on a layer of sand or cement to prevent the appearance of weeds. When the tub is in the ground, the dampness of the surrounding soil will cause the staves to swell, thus closing any spaces.

Before filling the tub with water, it must be planted. No pool, however small, will be complete without its water lily and the charming *Nymphaea pygmaea* 'Alba' with its dainty flower of purest white, is at its loveliest in a small pond. To plant, a 3 in. (7.5 cm) layer of soil is placed at the bottom of the tub and pressed well down. Into this the roots are planted. Or the plants may be set out in soil containers which are placed about the pond at various depths before filling with water. In this way, there will be less soil to cause muddiness and trouble with algae whilst the plants may be moved about and re-arranged as required. So that the plants may reach the surface as soon as possible, it may be necessary to stand the containers on bricks which can be removed at a later date. Around the inside of the tub should be placed stones to a depth of 8 in. (20 cm) and around them, more soil is pressed. It is here that the marginal or water-side plants are set for they do not like more than 6 in. (15 cm) of water over their roots. Between the stones set two or three oxygenating plants to maintain the purity of the water and one or two taller growing plants such as the dwarf bullrush, *Typha minima.*

Planting should be done in June and at first the water (after filling) will appear muddy. It will soon clear but will then take on a green appearance. This is due to algae, minute plants which flourish in sunlight and on the mineral salts in the water. As soon as the water lilies spread their leaves, they cut out the sunlight and when the oxygenating plants become active, the 'green' will disappear. Another form of algae is Blanketweed which appears as long green ribbons. This can be prevented by treating the water with Algimycin P.L.L., using 2 fl. oz. to a small pool and which will not require repeating for a year or more. It is harmless to plants and to fish. Once planted, the pond will require no further attention for several years apart from the removal of dead foliage and flowers but leaves should not be allowed to contaminate the pond. From mid-June when the water lilies begin to bloom, until autumn, the pond will be continuously colourful and ornamental fish may be introduced as an additional attraction. Suitable for a small pond are the goldfish, nymph, golden and silver orfe. All are hardy and will live together on friendly terms. The fish should not be introduced until the oxygenating plants have become established which will take about a month. It is recommended there be one fish of reasonable size to each sq. ft. of surface of the pond. Until the pond is well established, the fish should be fed regularly during summer and autumn but will require little food in winter.

To prevent suffocation by formation of ice during cold weather, a pool heater operating on 200–250 volts a.c. mains and used with waterproof flex will keep the pond free of ice.

Additional plants such as *Primula denticulata* and other moisture-loving plants may be set around the pond if paving stones are not laid to the edge. They will provide additional beauty during the seasons.

To make a concrete pool, it must be remembered that for every sq. ft. of surface area, it should have a corresponding depth in inches. Thus a pond made 5 ft. (1.5 m) x 4 ft. (1.2 m) should be 20 in. (50 cm) deep with 3 ft. (90 cm) the maximum depth. If the pond is made with a shelf all the way round, this should be made at half the depth and will be planted with marginal plants. It will greatly add to the area available for planting and so that the main part may be utilised for water lilies and other floating plants.

A rectangular pond is more easily constructed and after removing the soil to the required depth, take out an additional 4 in. (10 cm) at the base and fill this with crushed brick made quite firm. The sides should be made at an angle of 30°. To the concrete mixture of 5 parts aggregate, 2 parts sand, 1 part cement, add waterproofing powder which is mixed in after the water has been added. Do not make the mixture too sloppy. First cover the floor to a depth of 4 in. (10 cm) and allow it to set firm before making up more mixture. Then cover the sides with fine mesh wire netting and add the concrete to a depth of 4 in. (10 cm), holding it in place by boarding. If the pond is made with shelves, it simplifies the making of the walls.

When the walls have set, give a ¾ in. (2 cm) thick coating all over, using 3 parts sand and 1 part cement with additional waterproofing powder and leave this with a smooth finish. Allow to dry completely, then neutralise the lime content of the concrete by treating with Silglaze which will also cause the glazing of the concrete and act as a seal. The structure should be left two to three weeks before planting and filling with water, during which time paving squares of stone or concrete, may be set around the pond to enable the plants to be more easily tended and the pond enjoyed in wet weather. A concrete pond will usually be very durable and require only the minimum of attention after it has been made. The plants are set out as described.

Ponds of all sizes and shapes, made of bonding fibreglass with resins which set hard to give the container complete rigidity, may be purchased to one's requirements. After removing the soil to the required depth, take out an additional 3 in. (7.5 cm) and fill in the space with sand or gravel, made as firm as possible and which will act as a base. The pond should be sunk so that the sides are at or just below ground level but where the ground is stony and the pond can be only partially sunk, soil should be banked up to the outside lip to provide additional support and to enable plants to be set around the sides.

By banking the excavated soil at one end of the pond and making a rock garden in which a fibreglass channel is inserted, a cascading waterfall may be made. A small electric pump is placed in the pond which forces the water to the top of the cascade by means of a plastic tube hidden in the rock garden. The water tumbles down the channel and back into the pond but it should not be made too steeply as to cause the falling water to trouble either fish or plants.

WATER LILY, *see* Nymphae

WATSONIA

The Southern Bugle Lily, a native of South Africa, and requiring a dry, well-drained soil. The flowers are held on 18 in. (45 cm) stems and in appearance are similar to those of the freesia, also in their culture. The bulbs may be grown in pots in a cold greenhouse or in frames. Planting takes place in early April and until growth appears, little or no water should be given. As growth advances, the bulbs are given more water. They will come into bloom in midsummer and are supported by thin canes as the spikes tend to fall when the flowers open.

The plants should be grown as cool as possible from the beginning and should be given shade if the sun makes the greenhouse temperature excessively high.

In sheltered gardens of the south, the watsonia will bloom in the open during July, but it is advisable to lift the bulbs in October and winter them in a frost-proof room. There are a number of interesting species. Amongst the easiest are:

Watsonia aletroides. It bears a long stem covered with blooms of rich scarlet.

W. ardernei. It produces its glossy white blooms on long stems and is easy to manage.

W. moreana. A hardy species, producing an attractive spike of orange pink blooms on 3 ft. (90 cm) stems.

W. rosea. The more commonly known pink-watsonia found on Table Mountain in South Africa. There is also a most attractive white form, 'Alba'.

WEIGELA, *see* Diervilla

WELLINGTONIA, *see* Sequoiadendron

WELSH POPPY, *see* Mecanopsis

WHIN, *see* Ulex

WHITEBEAM, *see* Sorbus

WHITE SATIN FLOWER, *see* Lunaria

WILD OLIVE, *see* Elaeagnus

WILLOW, *see* Salix

WINDFLOWER, *see* Anemone

WINDOW-BOXES

If consideration was given to the wide variety of plants suitable for window-box culture and their soil and climatic requirements, a wider use could be made of window-box gardening. Many factories, especially those which are not troubled by an excess of smoke, and town offices, could be made more attractive by the use of window-boxes even if they grew nothing but the hardiest of evergreen plants. But primroses in spring, followed by the bright ornamental-leaf geraniums in summer and autumn would provide colour and interest with the minimum of attention, whilst both plants remain quite untroubled by soot deposits.

Again, it is not necessary for the window-box to be situated in full sun. There are plants for a shaded and semi-shaded position where members of the primrose family and bulbs for spring flowering, followed by pansies and violas for summer, will prove most suitable. Even where the window box is exposed to strong winds, there are many suitable plants.

PLANTING SCHEMES

The window-box may be filled with compost and filled either with plants in their season, or with permanent plants in the form of a miniature garden; or it may be used merely as a container to hold plants which are grown and flowered entirely in their pots. In this case it will not be necessary to fill the box with compost, though peat packed round the pots during summer will help to reduce artificial watering to a minimum. Nor will it be necessary to raise one's own plants to use in this way. Pots of cinerarias or hydrangeas to bloom during May and early June may be followed by zonal or ivy-leaf pelargoniums, to bloom from mid-June until September. The new dwarf pompon chrysanthemums in pots, may be used to provide autumn colour, to be followed by the hardy heathers to bloom through winter. Where it is required to raise one's own plants in pots, daffodils, double tulips and hyacinths may be raised in a dark cupboard or cellar, and may be brought into bud in a cool airy room for transferring to the window-box in April. There they will bloom until

June. The bulbs may then be replaced by geraniums which have been wintered in a frost-free room indoors and will be in bloom when transferred to the window-boxes. Given the protection of a wall, the geraniums should not be troubled by frost until early November when the plants may be taken indoors again to be replaced with the winter-flowering heaths. These may be obtained in November and will only require potting. If obtained with as much soil on the roots as possible, they will remain green and colourful until spring.

The same use may also be made of the window-box where it is required to plant directly into the made-up box. There is no need to wait until summer to enjoy colour. Late in autumn, the boxes can be planted with suitable varieties of bulbs and primroses, to bloom from the New Year. Or wallflowers, retaining much of their foliage throughout winter, may be planted in November to bloom throughout late spring. They may then be followed by annuals which will either have been raised in one's greenhouse or frame, or will be purchased from a local nurseryman where plant hybridizers have catered for the new trend in gardening.

Until more recently, the choice of annual plants of suitable habit for window-box culture was extremely limited. Such plants as the dwarf African marigold; the Tom Thumb antirrhinum; the 'Waldersee' aster; the 'Ramona Super' dwarf petunias and the 'Scarlet Pigmy' salvia were undreamed of in their compactness and freedom of flowering. Now the choice is immense, with plants for every situation and soil.

TYPE OF BOX

The type of box will be governed by several factors. That requiring the minimum of trouble will be the plant container, constructed of painted wrought iron which will give a continental look to one's property. A window-box of this type will not be suitable for filling with soil and so will be used entirely as a container for pot plants. Against the deep windows of a Georgian house, such a 'box' will be seen at its best, but would not be so happy if used with mullioned windows, or against the small windows of a country cottage. Here the box made of wood or concrete would be superior, and, the depth must be carefully considered. That to be used for a long, low window will need to be constructed less deeply than a box for a large, deep window. A shallow box provided for a large window would appear out of place and however attractively the box was filled this would not compensate for lack of proportion. Whether the container type box, or the more usual type of box is to be used, will depend upon what facilities there are for bringing on the young plants. Also upon whether one prefers the pleasures of making and tending a miniature garden rather than enjoying colour from pot plants with the minimum of trouble. But whichever method is preferred, thought should be given to the most suitable plants which will provide colour the whole year round, and to the type of property before the boxes are made and the window-box gardening commenced.

The plant rack or container in wrought iron will present no difficulties in its making, for it will have been purchased already made up to the size of the window. It will be held in place by brackets at both ends of the rack which should be painted black or white, the same as the rack. To make the container quite secure, the wall should be plugged with hardwood to a depth of not less than 2 in. (5 cm), into which the brackets are screwed. It must be remembered that on occasions a box may be almost 4 ft. (1.25 m) in length and 6 in. (15 cm) in depth, and so will contain a considerable quantity of soil. This will mean that the supports will have to carry a great weight, especially when the soil is wet. Any inefficiency in the fixing of the box will lead to trouble, and even if no harm is done, a box which will break away from the walls when in all its glory, will cause disappointment.

Where there is sufficient width to place the box in position beneath a window without the necessity of using brackets, precaution must be taken by fixing strong wire to the outer corners of the box. This is fastened to hooks fixed in the wall on either side of the window at a suitable position between the stones or bricks. A box which may be placed on the flat top of a porch or door canopy and which may be tended from a window immediately above, should also be given additional support in the same way. It must be remembered that a window-box in an exposed position will be subjected to strong winds and may also have to take a heavy weight of snow falling from the roof, so make certain that the box is thoroughly secure before it is filled.

The container of plants in pots will not be required to carry the same great weight as will say a concrete box filled with soil, and additional support for the latter should be made.

Another method, is to fix a 5 in. (12.5 cm) iron bracket along the top of the box and screw it into the window frame. It is, of course, necessary to have the box made to the exact measurements of the window, otherwise not only will it appear most unattractive, but will be difficult to fix correctly.

Take care, too, to ensure that the box is in such a position as to allow the window to open without difficulty, though this is not quite so important where the window opens up and down rather than outwards.

To support a concrete box, strong iron brackets should be securely fixed to the wall beneath the window, and on these the box will rest. If the box is to be made more than 2 ft. 6 in. (75 cm) in length, it should be given an additional support at the centre. The weight of the box filled with soil should hold it in position, but as an extra precaution, an eye screw should be fixed at the top of each side of the box as the concrete

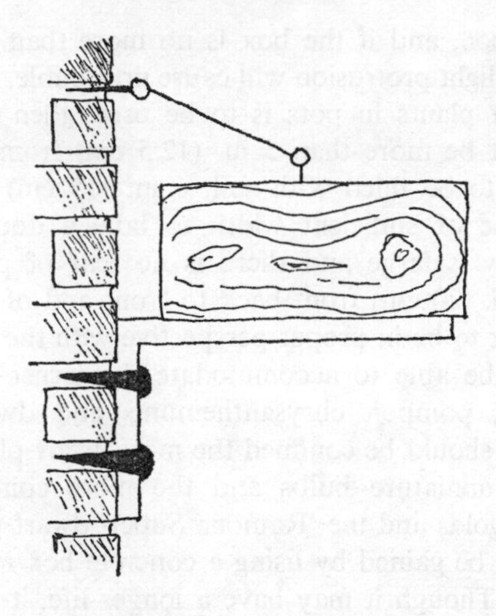

FIG. 18 *Fixing a window box*

hardens. Through the eye, strong wire is placed and taken to a similar eye or staple fixed to the wall or window frame. If the wire is made quite tight, this will prevent any movement of the box, and if required it is quite an easy matter to remove the box at any time. Where the window is more than 3 ft. 6 in. (1 m) in length, two separate boxes should be made and fixed separately alongside each other.

MEASUREMENTS OF THE BOX

In making the box, the exact measurements of the window must first be taken so that the sides of the box will coincide with the framework of the window. The depth of the box will be governed by the depth of the window. For a deep window in factory or hotel, the boxes may be 8 in. (20 cm) deep and this will allow a greater depth of soil so that the plants will require less attention than where the boxes are less deep. For the windows of most town houses and where the house is of Georgian architecture, the boxes could be 6 in. (15 cm) deep; whilst for the generally long, low window of a medieval building or for the small, possibly mullioned window of a cottage, then the box should be about 5 in. (12.5 cm) deep. The deeper the box, the less attention it will require as to watering.

Distance from back to front is important and this will be governed either by the construction of the window frame, where there may be a wide ledge; also by situation. Where there are low windows and the property is alongside a public footpath or highway, then the box must not unduly protrude. With many windows there is a ledge the width of a brick for a portion of the box to rest upon, and if the box is no more than 6 in. (15 cm) from back to front, the slight protrusion will cause no trouble.

Where a row of plants in pots is to be used, then the window-box or container need not be more than 5 in. (12.5 cm) from back to front, but where the box is to be filled with soil, 6 in. (15 cm) would prove more satisfactory and be of sufficient width to take a double row of plants. Where the window is large and there is no fear of protrusion, then the boxes may be 8 in. (20 cm) from back to front and of similar depth. This will permit the box to be in proper perspective with the size of the window. A large box will be able to accommodate the larger type of plant such as the hydrangea, pompon chrysanthemums and dwarf dahlias, whilst to the smaller box should be confined the most dwarf plants such as certain Juliae primroses, miniature bulbs and the more compact annuals such as the Cupid marigolds and the 'Ramona Super' dwarf petunias.

There is little to be gained by using a concrete box rather than one constructed of wood. Though it may have a longer life, it will be heavier and more difficult to construct and secure. A wooden box made of 1 in. (2.5 cm) timber, treated with preservative and painted on the outside will prove long lasting and will present no difficulty in its construction and fixing. A suggestion is to use two boxes to maintain the display, one to fit inside the other. This will permit for greater elasticity in the year-round display. For example, whilst the original box will continue with the summer display of annuals or geraniums throughout autumn, these plants may be replaced by an inner box made up of bulbs and other winter flowering plants which will come into bloom as soon as placed in the permanent box, possibly in November. Or again, whilst the fixed box may be planted with bulbs to bloom from Christmas until March, the second box will be planted to provide colour from April until early June. Another method is to allow the original spring display to continue until the end of May, and then to replace it with an inner box made up of annuals sown directly in the box in early April, possibly in a frame. There will thus be few periods when the box will not be colourful. The inner box, made of wood, may be fitted inside a fixed box constructed either of wood or of concrete. It is placed in by 'handles' of cord at either end and which are removed when the box is in position. A second box may be used in much the same way as where plants are grown in pots and brought on either in the home or in the greenhouse to maintain a succession of bloom. It is preferable that the inner or replacement box be of wood rather than of metal, for plants are never happy when their roots are in a non-porous container, however well-drained it may be. For the same reason, a window box should be constructed either of concrete or of wood. These are porous materials and also possess a rough inner surface which is conducive to vigorous root action.

MAKING THE BOX

As a window box has to carry a considerable weight of soil, it should be constructed of 1 in. (2.5 cm) wood, cut to the correct lengths and planed. The front and back of the box should be cut to the length required, the ends fitting inside (see plates 77–78). When cutting the ends, allow for the thickness of wood so as to keep to the correct overall measurements required. If any attempt is made to dovetail the corners it should be remembered that the strength of a dovetail lies in the perfection of its construction and a water-resistant glue should be used. Always bear in mind that a good, simple job is better than a bad complicated one. When securing the two ends, no advantage will be gained by using screws instead of 2 in. (5 cm) nails, for when driven in they only tend to part the grain, thus splitting the wood. For additional strength at the corners an angle bracket should be screwed either on the inside or outside of the box.

Adequate drainage holes should be made in the base, preferably making a dozen or so holes of ½ in. (1.25 cm) diameter rather than half the number of 1 in. (2.5 cm) diameter, to ensure that there will be little of the compost escaping. Where possible always use hard wood, such as seasoned oak, in the construction of the box, or failing that, American Red Cedar, both of which will remain almost impervious to moisture through the years and neither of which require painting as a preservative.

After the box has been made up it should be treated on the inside with a wood preservative, Cuprinol being most efficient. This treatment is especially necessary if the box has been constructed of deal or other soft wood. Thoroughly soak the inside of the box and allow it to remain in the open after treating for at least ten days until it has become thoroughly weathered and any fumes, which might be injurious to plant life, will have escaped. The box may then be painted on the outside only, to conform to one's tastes, or to the colour scheme of the house. The window-box should be of the same colour as the window frames; turquoise, grey, or pale blue being most attractive, also cream or white. These shades seem to bring out the rich colours of the blooms to the utmost advantage. Against stone or mullioned windows then a box made of oak presents a better appearance if it is not painted. Filled with crimson geraniums the effect is extremely rich. Where the boxes are to be used against the white or cream washed walls of a cottage, a pleasing effect will be to paint the boxes and window frames pale blue or pink. Always use one of the more delicate colours, deep greens and brown never looking right for window-boxes.

Owing to the weight of soil, the boxes are always fixed before they are filled, though it is preferable to add the drainage materials first. And always remember to place the boxes where they can be easily attended. To place them in some inaccessible position which necessitates the use of a pair of steps to give them attention, or where the watering-can has to be held at arm's length above one's head, will be to make window-box gardening a toil rather than a pleasure.

It should be said that a box which is to be used inside the fixed box should be constructed of ¾ in. (2 cm) wood to make its manipulation as easy as possible. An inner box, in which plants are already coming into bloom must of course be moved with the compost already in, and for this reason will be more easily used for those windows which may easily be reached from outside.

The concrete box will be heavier and should not be made more than 6 in. (15 cm) deep, the base and sides being 1 in. (2.5 cm), the same as for a wooden box. The concrete box, however, should be reinforced with wire netting.

The box is made from a wood mould, the sides being lightly held together so that they may easily be removed when the concrete has set. The base should not be nailed to the sides. The method is to mix the concrete, using 2 parts of sand to 1 part of cement, so that it is of such a texture as to pour without it being too thin or sloppy. The base is first made to a depth of 1 in. (5 cm), corks at regular intervals being pressed into the concrete and removed when it has set. These are for the drainage holes. Then 1 in. (5 cm) mesh wire netting is placed across the base before the concrete has set and this will also be used for reinforcing the sides.

When the base has partly hardened, an inner 'box' or mould containing no base is then placed inside, its dimensions being such that there will be a 1 in. (5 cm) space between the sides of the two boxes. Into this the concrete is poured, the wire netting reinforcement being tacked to the top of the inner mould so that it will not show on the outside of the box when the cement has dried. This it will do in about twenty-four hours when the moulds are carefully removed and the wire netting reinforcement is trimmed level with the top of the sides. The box should be moved and the base prised off only when it is thoroughly dry. It should then be weathered for at least a fortnight by allowing it to stand exposed to the elements, but not to frost which would cause it to disintegrate before it has set. Do not forget to insert the eye screws at the top of the sides, somewhere near the centre, before the concrete has set, so that they may be used for the wire supports. The boxes may be painted on the outside to match the paintwork of the window frames and to take away their unattractive appearance.

FILLING THE BOX

Before placing any compost in the boxes it will be necessary to ensure drainage. First, the holes in the base must be made so that the soil does not fall through, and this is best done by placing a piece of fine mesh wire netting over the base. Then add a layer of crocks to a depth of about $\frac{1}{2}$ in. (1.25 cm) to ensure efficient drainage during winter. Over the crocks, a layer of turves placed grass downwards will occupy another $1\frac{1}{2}$ in. (4 cm) of the box. The remaining space, depending upon the depth of box, is filled with prepared compost.

The soil should preferably be taken from pasture, or be a good quality loam from a country garden and where the soil is not troubled by deposits of soot and sulphur. Soil taken from a town garden will generally be of an acid nature and will also contain a large number of weed seeds. Therefore pasture loam is to be preferred. This should be stacked under cover when it is mixed with some peat and grit. A satisfactory compost will be made up by mixing 3 parts loam, 1 part peat, 1 part grit or coarse sand (by weight). Allow 2 lb. of ground limestone or lime rubble to a box 3 ft. (90 cm) long and 6 in. (15 cm) deep; 4 oz. of bone meal, a slow-acting fertiliser and a sprinkling of superphosphate which encourages a vigorous root action.

Whilst the same compost will not suit all plants, a compromise may be made. For instance, geraniums prefer decayed manure to peat, whilst with begonias the reverse is the case. Most annuals, bulbs and spring-flowering plants will be happy where either peat or decayed manure is used, or half and half of each, which mixture would also prove acceptable both to begonias and geraniums.

Make up a well-drained, friable compost but it must also be able to retain moisture during dry periods in summer. For this reason the peat or decayed manure, or a little of both, should not be omitted, and where the loam is of a light, sandy nature, than a large proportion of humus materials should be added. It is important to keep the compost sweet as long as possible without the continual changing of the soil. Therefore lime in some form should not be omitted. A few pieces of charcoal in the soil will also help to maintain sweetness. Correctly prepared, the compost will not only make for healthy and vigorous plant growth, but will require changing only once every three years.

The window-boxes, after they have been made secure, should be filled several days before they are to be planted so that the compost is allowed time to settle down. The compost is best taken to the boxes in a small bucket, and as it is placed in the box it should be pressed around the sides so that all air pockets will be eliminated. The box should be filled to the top and be allowed three or four days in which to settle down before any planting is done. The compost will then sink to about $\frac{1}{2}$ in. (1.25 cm) below the top of the box which will allow for watering without the soil splashing over the side.

CARE OF THE BOXES

Planting will take place late in autumn for the spring flowering plants and bulbs, and again in early summer when the plants are replaced by those which bloom through summer and autumn. Where planting is to be done directly into the soil, pot grown plants should be knocked from the pots and taken to upstairs boxes in a clean wooden box, so that there will be the minimum of mess inside the house. Lower window-boxes will be planted from outside. Plant firmly and allow the plants room to develop.

Where window-boxes are being used as containers, the pots should be placed on a layer of peat which should also be pressed round the pots and kept comfortably moist to prevent undue evaporation of moisture in the pots.

The boxes should be given a dressing with lime each autumn as the plants are changed, and a small quantity of bone meal should also be worked into the soil which should be thoroughly stirred up. After three years the compost is best replaced and wooden boxes given another dressing on the inside with preservative. First remove the old compost, then take out the crocks which are replaced after the boxes have been treated. After ten days refill with freshly-prepared compost when, after it has had time to settle down, planting may be done.

Mention should be made here about watering the boxes. Where it is thought necessary to water the boxes give a thorough soaking to enable the moisture to reach right down to the roots so that they do not turn upwards in search of it. However, on no account give so much water that it will soak through the soil and drip from the bottom of the box.

WING NUTS, *see* Pterocarya

WINTER CHERRY, *see* Solanum

WINTERSWEET, *see* Chimonanthus

WISTARIA

W. sinensis is a climbing plant which will grow to large proportions. An enormous plant completely covers two walls 50 ft. (15 m) long in the old gardens of the George Hotel, Stamford. But it is a plant which needs understanding, otherwise it may prove disappointing. If not kept well watered and syringed throughout its first summer after planting, it may have some difficulty in breaking into growth. Again, it is usual to allow the new shoots to run on entirely without stopping. This is a mistake for unless the wistaria is encouraged to make spurs it will bear no blossom. This is done by pinching back the shoots during summer to the second or third leaf, and it is at the axils of these leaves that the flower buds are formed. It will take time for a wistaria to cover a wall and bear the quantity of bloom of which it is capable, but if the shoots are allowed to grow unchecked they will quickly cover the wall with foliage though it will bear little bloom.

The wistaria likes a rich, deeply-worked soil, one containing plenty of humus to ensure that the roots are kept moist in summer. The plant does not like a dry, arid soil, and like all plants with similar requirements, it will appreciate a mulch of decayed straw manure given in June. The plant is deciduous, and so should be planted any time between November and March. Use pot-grown plants. Besides the pinching back of the new wood during summer, winter pruning should consist of cutting back all shoots to within 3 in. (7.5 cm) of the old wood in February each year. The wistaria may take several years to establish and come into bloom; afterwards it will grow rapidly, and if pruned as suggested, it will cover itself with racemes of rich mauve from May until July. Both *W. sinensis* and *W. floribunda* bloom at the same time, and may be obtained in shades of mauve, pink and white – the latter possibly being the loveliest of them all. The plants may be trained to a wall by means of strong wires to which the shoots are entwined or fastened, or against a trellis or lath frame. A west wall will prevent any burning of the rather succulent wood and foliage.

WISTERIA, *see* Wistaria

WOLFSBANE, *see* Aconitum

WOOD HYACINTH, *see* Endymion

WORMWOOD, *see* Artemesia

XERANTHEMUM

An annual and native of South Africa, *X annuum* is a pretty everlasting flower, sending up its daisy-like flowers, borne singly on 2 ft. (60 cm) stems, during July and August. The flowers have an outer ring of petals and with numerous petaloids at the centre. They appear in all shades of crimson, rose and pink, also blush-white. They require a sunny position and a dry, sandy soil. Sow in early April where the plants are to bloom.

YARROW, *see* Achillea ptartmica

YELLOW SCABIOUS, *see* Kephalaria

YEW, *see* Taxus

YUCCA

Though not usually considered a border plant, this compact form of the exotic desert plant is in no way out of place at the back of a large border, especially the double border. Two or three plants at regular intervals will, when bearing their huge branched buff panicles in autumn, bring an exotic touch to the border or it may be used as a focal point on a lawn. *Y. filamentosa* is free flowering and its long leaves are narrow and compact so that the plant occupies only about a sq. yd. of ground. It likes an ordinary soil which is well drained and has been enriched with some humus. Planting is best done in early spring for the leaves are evergreen and this is the most suitable time to move all evergreens.

ZANTEDESCHIA

Zantedescia aethiopica, in the warmer parts of the British Isles, may be grown outdoors to bloom in summer, to be lifted indoors in autumn like chrysanthemums. Known as the Calla or Arum Lily, it is not a lily but an aroid, growing from a corm. It may be given cold house culture throughout when it will bloom early in summer, though with a temperature of 55°F. (13°C.), it will bloom for Easter when it is much in demand for church decoration.

The Arum Lily requires a moist situation where growing outdoors. It may be planted by the side of a pond or stream and is quite happy where submerged in water for long periods.

Indoors, plant one corm to a 6 in. (15 cm) pot; two to a 10 in. (25 cm) pot, using a compost made up of 6 parts turf loam, 1 part peat, and 1 part decayed cow manure to which is added a liberal amount of coarse sand or gravel. Plant 3 in. (7.5 cm) deep with the top of the compost 1 in. (2.5 cm) below the rim of the pot to allow for watering. Use no lime in the compost. Arums like a slightly acid soil whilst lime will encourage soft rot of the corms (*Bacterium carotovorum*).

The corms are potted in August and are stood out of doors until mid-September when they are moved to a greenhouse heated to 50°F. (10°C.) at night, 55°F. (13°C.) by day. As the plants will make growth quickly water frequently and from Christmas, give a fortnightly application of dilute manure water. The plants will begin to bloom by mid-March but if they appear retarded, increase the day temperature to 60°F. (16°C.) from early March. One plant will produce five or six blooms with a 9 in. (22 cm) long spathe of purest white on a stem 2 ft. (60 cm) in length.

After flowering, place the pots on their side outdoors in a sunny position for the bulbs to ripen and take indoors again in September after topping up with fresh compost. Every two years, divide the plants and re-pot into fresh compost. The best for outdoors and a cold greenhouse is 'Crowborough'.

ZAUSCHNERIA

The species *Z. californica* is a superb plant for a dry rockery and the most colourful of autumn alpines. Propagated by division, it bears flaming orange-red blooms held above its grey-green foliage, a single plant providing a brilliant effect when the rockery is beginning to look bare. It requires a well-drained humus-laden soil.

ZEBRA-STRIPED RUSH, *see* Eulalia

ZEBRINA

With its small, fleshy leaves striped with gold or silver, it is one of the best of all indoor trailing plants and is quick to grow. Though of trailing habit it makes a compact, bushy plant and requires more light and moisture than most of the trailers. It is able to withstand a smoky atmosphere provided it does not lack moisture, but it is also able to tolerate cooler temperatures than most indoor plants and will come to no harm so long as a temperature just above freezing is maintained.

Propagation is by means of short stem cuttings placed in a sandy compost around the side of a pot. Water sparingly until rooting takes place, then pot separately into small pots and grow-on in a light position, watering more frequently as the plants make growth.

ZENOBIA

Closely allied to vaccinum and requiring the same acid soil conditions, *Z. speciosa* 'Pulverulenta' is a handsome shrub growing 3–4 ft. (1 m) tall, its leaves and stems covered in glaucous 'bloom'. It blooms in midsummer, bearing its scented bell-shaped flowers of waxy-white all along the branches.

Plant in March, in full sun or partial shade and prune only to remove dead wood. Propagate by division or offsets. This is a valuable plant for waterside or bog garden for it enjoys moist conditions.

ZEPHYRANTHES

Known as the Zephyr Flower, the blooms are as lovely as the name of the plant. Those who know it, often refrain from planting it on account of its reputation of being difficult. In a sheltered position, and in a well-drained soil which is not of too heavy a nature, it will prove almost completely hardy in all districts, though where exposed it could be given a little protection over winter. A peat mulch in early November and a light covering of straw or bracken will be suitable protection or if the soil is heavy, the bulbs may be lifted in November, dried and wintered in a frost-proof room until planting time in April. There is a species that blooms during late spring which should be planted in October. Plant the bulbs 4 in. (10 cm) deep and around each bulb place a mixture of peat and sand and some shingle if the soil is heavy. Being natives of the tropical areas of South America, the zephyranthes need all the sun possible, so plant with this in view and they will do well either on a rockery, in the border, or in grass. In the south they will increase quite rapidly by sowing their own seed. In the north they may be lifted every three years for dividing and replanting in April.

SPECIES

Zephyranthes andersonii. It is a magnificent rock garden plant, bearing its shiny coppery orange bloom at a height of only 5 in. (12.5 cm). At its best in August.

Z. atamasco. Flowering in late spring and where some shading can be given, the later blooms may join up with those of *Z. carinata*, thus making possible a display from May until the end of October. It bears large white flowers tinged with lilac.

Z. candida. A most charming species especially when planted in grass or under trees where it will receive some sunshine. The flowers are of purest white with rich orange stamens and are produced from a bed of grass-like leaves. In bloom throughout September and October.

ZEPHYR FLOWER, *see* Zephranthes

ZINNIA

This half-hardy annual, native of Mexico, likes a sunny position, a light sandy soil, and an absence from rain. In the north it is happy only in a summer where sunshine is above average and then to see it at its best, it should be given a seaside or country garden where the air is in no way polluted. Where it flourishes, the modern zinnia far surpasses all other summer annuals in the brilliance of its display with the rubbery texture and vivid colourings of the blooms.

Though growing to a height of 2 ft. (60 cm), the flower stems are so sturdy and erect that staking is not necessary. As the zinnia resents transplanting, seed is germinated in a heated greenhouse in March and as soon as large enough to handle, the seedlings are transplanted into small pots and kept growing. Later they are moved to larger pots and hardened off in May. As for stocks and asters it is advisable to water sparingly and then with Cheshunt Compound.

The plants should not be set out until early June, being planted from the pots containing a ball of moist soil. Plant 12 in. (30 cm) apart. They will come into bloom mid-July.

For a sunny window-box and for bedding, the 'Thumbelina' strain is delightful. Growing only 6 in. (15 cm) high, it bears flowers, like tiny pompon dahlias in brilliant clear colours including rose-pink, scarlet, orange, yellow and white. Also for bedding is the 'Early Wonder' strain, bearing flowers 3 in. (7.5 cm) across, like those of the small decorative dahlia.

For cutting, and the zinnia lasts fully a week in water, the Giant Chrysanthemum-flowered strain bearing huge blooms 6 in. (15 cm) across with twisted petals and in all the brilliant zinnia colours, is outstanding, whilst for sheer brilliance for cutting and for bedding. 'Sombrero', bearing flowers of gold and red, and on 20 in. (50 cm) stems is most striking.

ZYGOCACTUS

Known as the Christmas Cactus, *Z. truncatus* is one of the most interesting of indoor plants, being easily grown and propagated, and its attractive cerise-pink flowers, which are borne in winter, make it one of the best of all indoor plants; and preferring semi-shade it could be more widely grown. It is a cactus, being one of the Epiphyllum group, the plants being almost spineless and having flat, fleshy stems. In their natural state they are to be found growing on trees in the forests of South America, where they are able to survive long periods without moisture. The zygocactus forms its hose-in-hose flowers at the end of the stems. These are composed of numerous flat segments and are produced in weeping fashion which makes this an excellent plant for an indoor hanging basket. It is also attractive if grown in a size 48 pot placed on an inverted pot where its drooping habit can be fully appreciated.

More than for any other form of cactus this plant requires a soil containing plenty of humus. The compost should be composed of 3 parts fibrous loam, 2 parts peat and 1 part sand, and never at any time allow the compost to dry out. Unlike the desert cacti, the plants do not require lime. Throughout summer and until the plant has finished flowering towards the end of winter, the compost must always be kept moist and during summer the plant will require plenty of moisture. After flowering, it must be given a rest period of several weeks, but this does not mean that all moisture be withheld. Give sufficient to keep the plant alive, then, as the weather becomes warmer, the amount of water is increased. Take care not to water the plant itself for moisture remaining on the segments in winter will cause decay, whilst during sunny weather scorching may occur. A minimum temperature of 45°F. (7°C.) should be provided but one nearer 50°F. (10°C.) will make for a better flowering display and humidity may be encouraged by standing the plants in saucers containing damp moss.

Propagation is by the removal of the segments at the joints and inserting into a compost of peat and sand. They will root quickly during early summer, the rooted segments being planted into small pots for growing-on when they will come into bloom by the year end.